Ducks, Geese, and Swans of North America

REVISED AND UPDATED EDITION

Ducks, Geese, and Swans of North America

REVISED AND UPDATED EDITION

Volume One

Guy Baldassarre

With Assistance from Susan Sheaffer

A WILDLIFE MANAGEMENT INSTITUTE BOOK

JOHNS HOPKINS UNIVERSITY PRESS / BALTIMORE

© 2014 Wildlife Management Institute
All rights reserved. Published 2014
Printed in China on acid-free paper
9 8 7 6 5 4 3 2 1

Johns Hopkins University Press
2715 North Charles Street
Baltimore, Maryland 21218-4363
www.press.jhu.edu

Library of Congress Cataloging-in-Publication Data
Baldassarre, Guy A.
Ducks, geese, and swans of North America /
Guy Baldassarre. — Revised and updated edition.
pages cm
Includes bibliographical references and index.
ISBN 978-1-4214-0751-7 (hardbound : acid-free paper) —
ISBN 978-1-4214-0808-8 (electronic) —
ISBN 1-4214-0751-5 (hardbound : acid-free paper) —
ISBN 1-4214-0808-2 (electronic)
1. Anatidae—North America.
2. Waterfowl—North America.
I. Title.
QL696.A52B3348 2013
598.4′1—dc23   2012027052

A catalog record for this book is available from the
British Library.

Maps of North America (p. 2), Alaska, Canada, the
continental United States, and Mexico (see CD-ROM)
by Robert Cronan, Lucidity Information Design

Distribution maps by Josh Stiller

*Special discounts are available for bulk purchases of this
book. For more information, please contact Special Sales at
410-516-6936 or specialsales@press.jhu.edu.*

Johns Hopkins University Press uses environmentally
friendly book materials, including recycled text
paper that is composed of at least 30 percent post-
consumer waste, whenever possible.

For Eileen, Dan, and Adam
Family is everything

Contents

Foreword

Stephen P. Havera
ILLINOIS NATURAL HISTORY SURVEY (EMERITUS)
HAVANA, ILLINOIS

As the youngest member of a remarkable research staff at the Illinois Natural History Survey in the mid-1970s, I was particularly interested in all the discussion and excitement associated with the activities of Frank Bellrose's revision of Francis H. Kortright's 1942 classic volume, *The Ducks, Geese, and Swans of North America.* Glen Sanderson was head of the Section of Wildlife Research and he had granted Frank permission to assume the daunting task of revising and updating Kortright's benchmark, a masterwork that underwent 14 printings, indicating the importance and popularity of the book. Glen was exceptionally fond of Frank and waterfowl, especially after he first met the already formidable Bellrose in the summer of 1947 when Glen was a graduate student at the Delta Waterfowl Research Station in Manitoba and Frank was conducting some summer research. Over the ensuing years, Glen joined the Natural History Survey and eventually became the leader of the Wildlife Section, and, as such, Frank's boss. Glen offered his support to Frank as well as that of our incomparable wildlife editor, Helen Schultz. Consequently, there were a lot of communications and materials exchanged between Glen, located at our main office on the University of Illinois campus, and Frank, who was housed at our biological station on the Chautauqua National Wildlife Refuge along the Illinois River, 100 miles to the west. Frank had been at this site since 1938, when he was hired by Art Hawkins, who joined the Survey only a month before.

Of course, the decade of the 1970s was well before the days of word processing, fax machines, email, and even copy machines. Frank's original longhand drafts submitted to Glen and their numerous edited and typed versions are still housed in the files of the station, now known as the Frank C. Bellrose Waterfowl Research Center. The amount of effort that went into Bellrose's 1976 edition of *Ducks, Geese, and Swans* was incredible. I was particularly interested in this undertaking since my home was in Peoria, on the Illinois River about 40 miles north of the Chautauqua Refuge headquarters in Havana, and I had been following Frank's career for many years. Several of my neighbors were ardent duck hunters, and that only added fuel to my desire to work with Frank at Havana. As a result, I attended the University of Illinois for graduate school and soon became a student under the direction of Glen Sanderson. My intent to study Wood Ducks with Frank for my graduate degrees did not materialize because of the time constraints on Frank and Glen necessitated by the workload of *Ducks, Geese, and Swans.* Glen was also the editor of the *Journal of Wildlife Management* at that time. Eventually I became a Survey wildlife research scientist and, as such, witnessed first-hand the activities and responsibilities involved with the 1976 version of the book. *Ducks, Geese, and Swans of North America* became, once again, a huge success, to the extent that the 1976 second edition was followed by a revised version and second printing in 1978, and a third edition in 1980, which updated some species accounts. The 1980 revision also replaced the original color plates of the various waterfowl species drawn by T. M. Shortt for Kortright's original edition with those of Robert W. Hines, one of America's best and most popular artists. Ironically, for the 1978 and 1980 versions I was then working in Havana with Frank and was fortunate to experience the other end of the working corridor for the book, at the biological station.

In the following decade, Frank focused on writing his and coauthor Dan Holm's book, *Ecology and Management of the Wood Duck*, published in 1994, but *Ducks, Geese, and Swans* was never far from his heart. Although Frank semiretired in August 1982, he continued to work full time on partial salary until February 1991, when he formally retired after 53 years with the Survey. Nevertheless, he remained involved with his passion—the welfare of waterfowl—and with revising the 1980 edition of *Ducks, Geese, and Swans* until his passing in February 2005 at the age of 88, a sad day indeed for waterfowl.

Subsequently, discussions ensued about whether the revision should proceed, and, if so, by whom. Guy Baldassarre stepped forward to accept such an invitation from Richard McCabe of the Wildlife Management Institute and assumed the momentous task. The proliferation of scientific waterfowl literature since 1980 had been enormous, and the revision would be intimidating. But Guy had decades of waterfowl knowledge, a thirst for their well being, and exceptional writing and publishing experience, including two editions of the highly acclaimed *Waterfowl Ecology and Management*, coauthored with waterfowl expert Eric Bolen. Guy was an obviously perfect fit to accept the challenge.

I first met Guy at the "Waterfowl in Winter" symposium held in Galveston, Texas, back in January of 1985. He made a remarkable impression on me and his peers, as Guy's presentations were not only informative, but entertaining as well. His energy and positive attitude were infectious. In one session, I was sitting with Dennis Thornburg, who was the head waterfowl biologist for the Illinois Department of Conservation and well aware of Guy and his work. Dennis chuckled and commented to me that Guy was the next speaker, adding, "He'll really get things rolling!" How true! After the symposium, I had the fortuitous opportunity to sit next to Guy on our flight home. During a stimulating and enjoyable conversation, we imprinted upon each other and subsequently became friends for life.

In the early years of his career, from 1978 to 1979, Guy taught a course in wildlife management at the University of Wisconsin–Stevens Point. He received his master's degree there in 1978, studying waterfowl ecology on man-made impoundments in the central part of the state. While a doctoral student of Eric Bolen's at Texas Tech University, Guy strengthened his foundation for a career in waterfowl research and education with studies on the field-feeding ecology of waterfowl wintering in the Playa Lakes Region of the Texas Panhandle. His teaching and research skills in waterfowl and wildlife ecology were further honed at Auburn University, where he was employed after completing his doctoral program at Tech in 1982. In 1987, Guy matriculated to the State University of New York, College of Environmental Science and Forestry (SUNY-ESF), where he rose to the rank of Distinguished Teaching Professor while continuing as a noted publisher of scientific investigations on various aspects of waterfowl and waterbird ecology and as the mentor of some 40 graduate students, along with teaching courses in waterfowl and wetland ecology, ornithology, and wildlife management. Guy also expanded his research program to include studies on flamingos, waterfowl in Mexico, and waterbirds in Venezuela.

Revising *Ducks, Geese, and Swans* was not an easy task. As Guy said, "It's been a beast to do, but loads of fun." First of all, preceding editions of the book were highly successful and popular because they appealed to both professional biologists and sportsmen, but it is difficult to write for such a diverse audience. Secondly, the amount of information regarding waterfowl published in scientific outlets and popular venues over the past three decades has been overwhelming. Lastly, finding sufficient quality time to dedicate to such an endeavor is nearly impossible in this rapidly accelerating world.

To complete a successful revision, Guy made some adjustments in his version. He focused on expanding the species accounts of waterfowl in order to include the staggering amount of new information that had appeared over the last 30 or so years, but concurrently chose to omit the introductory chapters present in earlier editions to satisfy space constraints. I concur with such an approach, since the species accounts were really the central elements of interest for the readership in previous editions, and the information in these introductory chapters can be found in his and Bolen's book, *Waterfowl Ecology and Management*, and in others. Consequently, this edition moves the all-important species accounts to immediately after the details given in the section Basis for the Book.

With his primary focus on species accounts, Guy's results are stunning. He presents an in-depth discussion of the biology of each species, embellished with knockout color photos and color range maps. Additionally, the noted species drawings by Bob Hines were retained, rescanned at high resolution, and moved forward to the head of each account rather than being imbedded in the center. Guy generally used the organization of the accounts provided by Bellrose but added a species synopsis at the beginning, as well as specific sections on habitat, courtship displays, survival and recruitment, brood parasitism/amalgamation, and molts and plumages. To allow for these additions, the descriptive material was reduced somewhat for each species because it is more prevalent today in alternative sources. He also put less emphasis on Frank's exhaustive distribution descriptions, but Guy enhanced the biology and added more quantifying information, such as sample sizes and years of study for the references cited.

The text of each species account is thus presented with an initial species synopsis, followed by sections on Identification (with subsections labeled At a Glance, Adult Males, Adult Females, Juveniles, Ducklings/Cygnets/Goslings, Voice, Similar Species, Weights and Measurements), Distribution (Breeding, Winter, Migration),

Migration Behavior (Molt Migration), Habitat, Population Status (Harvest), Breeding Biology (Behavior, Mating Systems, Sex Ratio, Site Fidelity and Territory, Courtship Displays, Nesting, Nest Sites, Clutch Size and Eggs, Incubation and Energetic Costs, Nest Chronology, Nest Success, Brood Parasitism, Renesting), Rearing of Young (Brood Habitat and Care, Brood Amalgamation, Development), Recruitment and Survival, Food Habits and Feeding Ecology (Breeding, Migration and Winter), Molts and Plumages, and, lastly, Conservation and Management. The literature cited section (provided on a CD-ROM) is exhaustive. All in all, this revision is a thorough, in-depth presentation of the biology of the various species and a benchmark of information with which to address the many challenges facing the ecology, management, research, and habitats of North American waterfowl in the twenty-first century.

As the consummate professional and an exceptionally qualified author, Guy undertook the revision like the proverbial duck takes to water. He has performed marvelously. Generations of biologists, managers, sportsmen, students, lay people, avian enthusiasts, and naturalists will reap the benefits from Guy's tireless efforts and persistent dedication to this remarkable endeavor. As usual, this edition of *Ducks, Geese, and Swans of North America* is a job well done by Guy Baldassarre.

Regretfully, in August 2012, at the age of 59, Guy lost a valiant effort fighting complications resulting from chronic lymphocytic leukemia. Fortunately, this book, which was the focus of his professional career in his last years, was in its final stages and, with the assistance of others, was completed. Guy's passing was a monumental loss not only to his family, colleagues, peers, and all whose lives he touched, but to the natural world, especially waterbirds and wetlands, which he loved so dearly. This book will remain a testimony to Guy's legacy.

Introduction

Richard E. McCabe

VICE PRESIDENT (RETIRED)
WILDLIFE MANAGEMENT INSTITUTE

Few books deserve the title classic, but in the years since Canadian outdoorsman Francis H. Kortright crafted the first edition of *Ducks, Geese, and Swans of North America*, the word has been appropriately applied to the various editions of the book you are now reading. Kortright's version was released in 1942 by the American Wildlife Institute, predecessor of the Wildlife Management Institute. In the first printing of that first edition, Kortright admitted that he wasn't an ornithologist. For that reason, he also expressed trepidation about compiling "the type of book that the majority of gunners and many naturalists would find useful."

In those days, as now, a book like this one was a team effort, and Kortright was the grand synthesizer. His finished work conveyed that fact: "Nothing original is contributed by this book, as everything it contains has been gleaned from the researches and writings of others." Kortright dedicated the book to the "Sportsmen Gunners and the Naturalists" of Canada and the United States and added his hope "that it prove interesting and if, through the increased knowledge it imparts, it assists even in a small measure in the great work of conservation of our wildlife it will accomplish a useful purpose." The publication of Kortright's book, illustrated by T. M. Shortt of the Royal Ontario Museum of Zoology, was immediately popular and ultimately required 14 reprintings.

The genesis of a second edition of the book occurred when Frank C. Bellrose, senior biologist with the Illinois Natural History Survey (INHS), fortuitously encountered Wildlife Management Institute (WMI) Vice President C. R. "Pink" Gutermuth as both were waiting to depart the Shamrock Hotel in Houston, Texas, after attending the 33rd North American Wildlife and Natural Resources Conference in March 1968. In the course of conversation, Frank recommended that the WMI update *Ducks, Geese, and Swans of North America*, inasmuch as it needed to reflect the results of ornithological investigations since 1942. Indeed, Frank called the two-plus decades since 1942 a period that "probably represented the 'golden age' of fact finding." (The computer age has since preempted the title, but Frank's assessment at the time was reasonable, considering advancements in science "figuratively exploded" following World War II.)

Pink, an astute but tight-fisted former banker, wanted a simple revision. But Frank argued for a thorough, new treatment. Dr. Laurence R. Jahn, then WMI's Director of Conservation, supported Frank's position, and Pink, somewhat uncharacteristically, acquiesced with one condition: that Frank take on the revised edition.

As Frank later wrote, "many times since C. R. 'Pink' Gutermuth requested that I write this book, I have regretted my acceptance."

As Kortright had been before him, Frank Bellrose was the right person at the right time for the monumental assignment. Frank was widely regarded as North America's foremost waterfowl authority. But the undertaking was outside the scope and purview of his responsibilities for the INHS. He admitted that "I had made so little a dent in the task the first year of working nights and weekends that I sought permission from Dr. George Sprugel Jr., Chief of the Illinois Natural History Survey, to make the work a Survey project with the stipulation that there would be no personal remuneration from its publication." Dr. Sprugel agreed.

Eight years of work at the INHS and during Frank's personal time (and a number of missed deadlines) were necessary to bring the new waterfowl picture to fruition. Two INHS stalwarts, Dr. Glen Sanderson and Helen Schultz, provided editorial support. Frank leaned especially on Art Hawkins, the U.S. Fish and Wildlife Service's Mississippi Flyway representative and "one of the most knowledgeable waterfowl biologists in North America," for his fact checking. Art also wrote a chapter on "The Role of Hunting Regulations." Dr. Sanderson penned an informative chapter on "Conservation of Waterfowl." Dr. Milton Weller of the University of Minnesota prepared a chapter on "Molts and Plumages of Waterfowl." The additional, vast amount of new data that Frank mustered for the individual species accounts (and otherwise) arose from the contributions by more than 100 waterfowl biologists. The second edition again used T. M. Shortt's color and line art, and Frank Bellrose Jr. provided some art as well.

Despite its professional acclaim and commercial success (including a second, revised printing in 1978), there was demand for improved color plates of the species and a new color dust jacket. By then, artists in various media, photographers, bird watchers, and even philatelists were significant audiences for the book, so making it more attractive and, in some instances, more accurate were reasonable requests. Frank used the opportunity of a third edition to provide various nominal data updates, but that third edition mainly presented some new artwork, including silhouettes, and a more durable publication by virtue of better paper stock and an improved binding. Bob Hines, famed artist with the U.S. Fish and Wildlife Service, was enlisted by the WMI to provide a new dust jacket, end sheets, and color plates. The third edition was released in 1981 and received the Wildlife Society's annual book award in the year of its publication.

Twenty-four years later, "on 19 February 2005, a great era of waterfowl conservation came to an end with the passing of Frank C. Bellrose." He was 88 at the time of his death and had been in the process of preparing a new edition, being unselfishly assisted by two friends, Dan and Angella Moorehouse.

In 2008, I met with Vince Burke, Executive Editor at Johns Hopkins University (JHU) Press, with whom I had worked previously on several other very successful WMI books. Also in attendance were two leading waterfowl biologists—Bob Blohm and Jerry Serie—in Washington, D.C., to talk about the new edition, and whether or not there might be someone who could undertake it. Half a dozen names of prospective candidates to assume the task emerged, but one name was the ideal: Dr. Guy A. Baldassarre of the State University of New York College of Environment Science and Forestry.

Guy was widely known for his waterfowl expertise, research, and writing abilities. After some thought, he agreed and signed on to become successor of the Kortright-Bellrose legacy, with one important proviso—that he be given control of the book's content and also have a say in decisions regarding illustrations. This was agreeable to the WMI and the JHU Press. Guy's plan was to take time off from teaching and complete the job in a year. It didn't happen.

The work took much longer than Guy expected, despite constant effort, but also in part because Guy had a vision for what the new edition might be: big and bold, everything every reader would want. He soon charted the course for new maps, produced by his student Josh Stiller, and for photos he said he would somehow get from great photographers around the continent. He wanted to see Bob Hines's art moved into the individual species accounts. Four years later, in early 2012, Guy submitted the final manuscript and all its various pieces. It was ready for copyediting. But all was not well.

Guy was stricken with the effects of chronic lymphocytic leukemia. His enthusiasm for the new edition never wavered, even when his stamina did. That he was able to submit a complete manuscript at all was extraordinary.

Guy Andrew Baldassarre, scientist, scholar, author, friend, father, and husband, died peacefully at home on August 20, 2012, at the age of 59, well before the new edition was published. Guy knew the importance of the book and, just before he died, he asked that Dr. Sue Sheaffer work with the copyeditor and then check the page proofs. He surely knew the waterfowl research community would rally to help Sue, and he was right.

I trust that Guy is looking on, satisfied that his magnum opus was completed the way he wished it to be. And I am confident that he shares the sentiment of Francis Kortright, who expressed hope 70 years previously that the book might assist "even in a small measure in the great work of conservation of our wildlife."

The various editions of *Ducks, Geese, and Swans of North America* are tributes to Francis H. Kortright, Frank C. Bellrose, Guy A. Baldassarre, and the legion of waterfowl biologists over time who have dedicated their careers to the betterment of waterfowl habitats and populations. May we all use this magnificent edition in the finest spirit of recreation, education, and conservation—and salute the memories of Francis, Frank, and Guy.

A Remembrance

Richard M. Kaminski

JAMES C. KENNEDY ENDOWED CHAIR IN WATERFOWL AND WETLANDS CONSERVATION
MISSISSIPPI STATE UNIVERSITY

One of my best friends and colleagues in life, Guy A. Baldassarre, authored this volume of *Ducks, Geese, and Swans of North America*. It's been said frequently that "the good, they die young," and Guy was and did on 20 August 2012, only several months after submitting the manuscript for this book. Guy knew the significance of filling the shoes of Francis H. Kortright and Frank C. Bellrose, authors of this book's previous editions. These three scholars, Francis, Frank, and Guy, each possessed the unique ability to absorb voluminous information and synthesize it into what can only be called classic literature on waterfowl. With the publication of this book, Guy will be known for two classic publications, his other being the textbook *Waterfowl Ecology and Management*, coauthored with Eric G. Bolen. No one besides Guy has ever produced two comprehensive treatises on waterfowl and their habitats.

I think Guy also would have liked to be remembered for at least two other keystone traits. He was a devoted family man and husband to Eileen, his best friend for over 30 years. His son Daniel eulogized his father by saying: "Dad often said, the first priority in life is family; the second priority—there is none." Running third, then, was Guy's tireless passion for teaching and mentoring students using his worldly wisdom of waterfowl and ornithology. When he lectured, he neither needed notes nor a microphone. With his booming voice and Boston accent, he always found ways to weave his witty humor into his communiques. Those of us who knew Guy Baldassarre firsthand would agree he was a peerless professional and person. As I said to Guy in spring 2012, "Thank you for rising to the challenge and need to write this book; it is a classic, timeless contribution to waterfowl science and conservation." Cherish what you hold, for it is Guy's gift to all who care about our continent's magnificent ducks, geese, and swans.

Preface

I was flattered when, in 2009, Dick McCabe from the Wildlife Management Institute and Vince Burke, senior editor at Johns Hopkins University Press, asked me to undertake the daunting task of authoring the revised and updated edition of *Ducks, Geese, and Swans of North America*, especially because it was begun by one of my personal friends and professional heroes, Frank Bellrose. I quickly agreed, however, because I believed a current book on the life histories of all of North America's waterfowl was in demand and thus should exist, replete with artists' drawings of each species, accompanied by photographs.

Like all my colleagues with an interest in waterfowl—professional and nonprofessional alike—I was thoroughly familiar with earlier editions of *Ducks, Geese, and Swans of North America*. As an undergraduate at the University of Maine in the early 1970s, I had purchased the first edition, which was authored by Francis Kortright in 1942. For nearly 35 years and through 14 printings, that first edition served as the "bible" for everyone interested in waterfowl until Frank Bellrose's landmark revision was published in 1976. Frank's revision was an extraordinary synthesis, condensing the voluminous new information on waterfowl ecology and management that had accumulated since Kortright's first edition. Frank updated his edition in 1980, most notably by replacing the original color plates by T. M. Shortt with those of Bob Hines, adding some new drawings, and adjusting some species accounts, but the major revision occurred with the second edition. Frank then began work on a new edition in 1994 and continued that work until his passing on 19 February 2005. Despite his efforts, his further revision of the volume was far from completion. So I dove in, and after nearly 3 years of continuous effort, I sent an updated treatise off to the publisher.

My goal was a technical and in-depth treatment of each species, but not to the point of being overly complex and cumbersome for users, especially the many nonprofessionals with whom the earlier editions were very popular. I wanted to simplify the range maps and supplement the artwork of Bob Hines with multiple photographs of each species, so as to display their habits and beauty in a way that differs from drawings. In sum, I wanted a book that would equally appeal to and be used by professional waterfowl biologists, waterfowl hunters, and waterfowl watchers, an attractive book that one could comfortably sit down with and rapidly find information on all of North America's breeding waterfowl.

Readers will find this edition replete with an abundance of new studies that are moving our understanding of waterfowl ecology from descriptive to explanatory. Starting from about 900 literature citations in the 1980 edition, this version now approaches 3,500. Major research tools have also emerged since the appearance of the third edition, and these, too, have begun to revolutionize our understanding of North American waterfowl. Significant among them have been advances in genetics and molecular biology, the use of satellite radiotransmitters, population modeling, and large-scale studies involving many birds over a broad geographic area. In 2004, for example, evidence from genetic studies on the various subspecies of the Canada Goose led the American Ornithologists' Union to split the then Canada Goose into 2 separate species: the Canada Goose and the Cackling Goose, with the former including the 7 larger to medium-sized subspecies and the latter including the 4 smaller ones. Satellite transmitters now allow biologists to track waterfowl movements over vast, often inaccessible areas. In eastern Canada, for example, satellite transmitters fitted into Harlequin Ducks have revealed that those breeding in northern Québec and Labrador overwinter in southwestern Greenland, whereas those breeding farther south winter along the Atlantic coast. Satellite transmitters have also revealed more of the remarkable migratory feats of waterfowl. For instance, a female Lesser Scaup located on Devils Lake in North Dakota was in Cuba 3 days later, having traveled 3,328 km. Large-scale studies and associated modeling of breeding Mallards in the Prairie Pothole Region are revolutionizing our understanding of habitat use and recruitment for that species.

All in all, this edition very much reflects the tremendous surge in waterfowl research and conservation efforts that have occurred since the mid-1970s, and my hope is that this version, like its predecessors, will serve those who hunt, study, enjoy, and otherwise work to conserve waterfowl and their habitats. These birds and the areas where they live and breed are constantly under siege from old threats; they will also face new threats as we advance into the twenty-first century, such as climate change and increasing pressures on their wetland and upland habitats that stem from an expanding human population. But these challenges are no cause for pessimism. Indeed, they reflect a great paradox in conservation, in that the challenges have never been greater, but our ability to solve them has likewise never been stronger. Hence I am optimistic, *very optimistic*, in large part because each generation of waterfowl enthusiasts has always answered the call to protect waterfowl and waterfowl habitat in North America, from the tundra to the tropics and everywhere in between.

We are winning this conservation and management battle everywhere the waterfowl community engages, especially when one looks back and considers the long-term and collective effect of such efforts. A National Wildlife Refuge System that began in 1903 with tiny Pelican Island (1.2 ha) off the eastern coast of Florida now encompasses >60 million ha in 555 units that occur in all 50 states and U.S. territories. The Duck Stamp Act, enacted in 1934, had spent >750 million dollars and protected 2.1 million ha of wetland habitat by 2011. Formed in 1937, by 2011 the efforts of Ducks Unlimited had conserved a stunning 5.3 million ha of waterfowl habitat in North America and influenced conservation on 24.8 million ha in Canada. The North American Waterfowl Management Plan, begun in 1985, had expended 4.5 billion dollars by 2010 and protected, restored, or enhanced 6.4 million ha of waterfowl habitat. The Plan is a cooperative venture that involves Canada, the United States, and Mexico and, thus, is a truly hemispheric approach to water-fowl conservation that addresses conservation issues affecting virtually all breeding, migrating, and wintering habitat used by North America's waterfowl. Since its inception in 1992, >11,000 private landowners have enrolled nearly 1 million ha of wetland habitat into the Wetlands Reserve Program of the Natural Resources Conservation Service in the United States. This list could go on to include waterfowl habitats protected by other organizations as well as by state and provincial governments, but the collective result has been expansive protection of habitat across the range of North American waterfowl, such that those who come after us will inherit the same legacy as we inherited from those before us. Such habitat protection on behalf of the waterfowl resource is unrivaled anywhere else in the world.

In understanding the cumulative impact of these conservation efforts, one might consider a fundamental tenet of waterfowl management: it is an endeavor of abundance focused on ecosystem conservation. Every nature enthusiast is versed in the plight of endangered and threatened species and well aware of recently extinct species, including waterfowl. But what about the species of abundance? The Mallard, for example, had a breeding population of about 11 million in North America in 2011, and a much larger fall population. Such sizeable numbers exist in the twenty-first century, despite a burgeoning human population and the resultant loss and degradation of waterfowl habitat, in large part because of the collective efforts of the waterfowl conservation community. A huge habitat base is required to support these birds, yet this base is an umbrella under which numerous other species of plants and animals—many of which are threatened and endangered—also exist. These protected wetlands and uplands for waterfowl also have a positive and direct effect on humans, via flood and erosion control, water purification, opportunities for nature study and outdoor recreation, and open space. Therein lies the cumulative impact of the North American Waterfowl Management Plan and other waterfowl conservation and management endeavors: ecosystem conservation at its finest, with multiple benefits to waterfowl, other wildlife, and people.

Plans and practices of conservation, however, are not sustained without the participation of people who are highly knowledgeable and passionate about waterfowl conservation. Universities must continue to train waterfowl professionals, and I hope that this book will become part of their curriculum. Waterfowl hunters and other waterfowl enthusiasts must also continue to be informed and committed to waterfowl conservation as they pass this heritage on to the next generation, just as it was handed down to them. And let us not forget what individuals can do for waterfowl and their habitats, as all conservation efforts can trace their roots to a single person's idea that is promoted and then nurtured until it attracts others and leads to significant action and change. Our best computer models, especially those with frequent dire predictions, are often wrong because they cannot factor in the power of the dedicated conservationists who, again and again, have come forward for the benefit of waterfowl resources. If this book accompanies them along the way, then my goal in writing it will have been accomplished.

Acknowledgments

My efforts in producing the revised and updated edition of *Ducks, Geese, and Swans of North America* were assisted by an incredible number of people to whom my expression of appreciation cannot improve on the words used by Frank Bellrose for the same purpose, so let me repeat them here: "The greatest compensation to me for the blood, sweat, and toil that went into this book has been the wholehearted enthusiasm of every biologist who has been asked to help. Not one refused." Such an incredible outpouring of assistance led me to realize that the present edition was, in reality, a group effort reflective of the achievements by the waterfowl research and management community since the mid-1970s. I was just the privileged synthesizer.

Among all those who provided assistance, none did more than my two recent graduate students, Josh Stiller and Ian Gereg, without whose efforts this book would not have been completed. Josh produced all of the range maps through countless hours of work, and he also was the one analyzing databases and band-recovery data. Ian exerted tremendous effort in acquiring hundreds of photographs and selecting from them those that ultimately appeared in the species accounts, which features the work of over 35 different photographers. These individuals are acknowledged with their photographs, all of which were contributed free of charge and added immensely to the appearance of the text. My sincere appreciation for everyone's skill and generosity, with a special thanks to Gary Kramer. I also recognize here the earlier work of Frank Bellrose. Frank was assisted by Dan and Angella Moorehouse, who were supported with funding from the Wildlife Management Institute. On occasion, some of their work appears in this volume, and I am grateful for that contribution.

For writing the foreword, I thank a longtime friend, Steve Havera, now an emeritus biologist with the Illinois Natural History Survey, where he has had a long and productive career. His book, *Waterfowl of Illinois*, is among the classic waterfowl books in North America. Steve was intimately familiar with the genesis of the second and third editions of *Ducks, Geese, and Swans of North America*, and that history is expressed in the foreword. There was no better person anywhere to produce a foreword for the revised and updated edition.

Databases and reports from the flyways were provided by biologists from the U.S. Fish and Wildlife Service (USFWS), Division of Migratory Bird Management (MBM): Dan Collins (Pacific Flyway), Kammie Kruse (Central Flyway), Dave Fronczak (Mississippi Flyway), and Jon Klimstra (Atlantic Flyway). All were incredibly helpful and prompt in providing databases and reports, answering questions, and putting me in touch with other experts. For reviewing species accounts in their areas of expertise, as well as directing me to pertinent studies and answering many, many questions, I am grateful to Ray Alisauskas, Environment Canada Prairie and Northern Wildlife Research Centre; Jane Austin, U.S. Geological Survey (USGS) Northern Prairie Wildlife Research Center; John Cornely, Trumpeter Swan Society; Brian Davis, Mississippi State University; Bruce Dugger, Oregon State University; Craig Ely, USGS Alaska Science Center; Paul Flint, USGS Alaska Science Center; Larry Hindman, Maryland Department of Natural Resources; Bill Hohman, U.S. Department of Agriculture (USDA) Natural Resources Conservation Service; Rich Malecki, Livingston Ripley Waterfowl Conservancy; Carl Mitchell, USFWS; and Jonathan Thompson, Ducks Unlimited. For proofreading several of the accounts, I thank undergraduates from the SUNY College of Environmental Science and Forestry (SUNY-ESF), especially Kayla Miloy but also Colin Hoffman and John Vanek. Colin also summarized the harvest data from the Canadian Wildlife Service (CWS) website. Also from SUNY-ESF, Jim Williamson at Moon Library was incredible with his skill and timeliness in procuring my several hundred library loan requests. Eileen Baldassarre, who assisted me when I was editor-in-chief of the *Journal of Wildlife Management*, once again came to my rescue in checking my entire literature cited section and its agreement with the text.

For major contributions in providing numerous databases, reports, and publications, as well as answering my many inquiries, I express appreciation to Ken Abraham, Ontario Ministry of Natural Resources; Al Afton, USGS Louisiana Cooperative Fish & Wildlife Research Unit; Brad Allen, Maine Department of Inland Fisheries and Wildlife; Mike Anderson, Ducks Unlimited Canada; Bart Ballard, Caesar Kleberg Wildlife Research Institute; Joe Benedict, Florida Fish and Wildlife Conservation Commission; Ron Bielefeld, Florida Fish and Wildlife Conservation Commission; Tim Bowman, USFWS; André Breault, CWS; Myke Chutter, British Columbia Ministry of Forestry, Lands and Natural Resource Operations; Jorge Coppen, USFWS Patuxent Wildlife Research Center; Dirk Derksen, USGS Alaska Science Center; Kathy Dickson, CWS; Julian Fischer, USFWS Division of Migratory Bird Management, Alaska; Kathy Fleming, USFWS MBM; Mike Fournier, CWS; Tony Fox, Denmark Ministry of Environment and Energy; Michel Gendron, CWS; Mary Gustafson, American Bird Conservancy; Mike Haramis, USGS Patuxent Wildlife Research Center; Dean Harrigal, South Carolina Department of Natural Resources; Gary Hepp, Auburn University; John Hodges, USFWS MBM, Alaska; Dale Humburg, Ducks Unlimited; Joe Johnson, Kellogg Biological Station of Michigan State University; Jón Einar Jónsson, University of Iceland Snæfellsnes Research Centre; Rick Kaminski, Mississippi State University; Walt Koenig, Cornell Lab of Ornithology; Mark Koneff, USFWS MBM; Kevin Kraai, Texas Parks and Wildlife Department; Ken Kriese, USFWS Division of Bird Habitat Conservation; Bill Larned, USFWS MBM, Alaska; Christine Lepage, CWS; Ed Mallek, USFWS MBM, Alaska; Mark Mallory, CWS; Scott Melvin, Massachusetts Division of Fisheries and Wildlife; Tommy Michot, University of Louisiana at Lafayette; Tom Moorman, Ducks Unlimited; Tim Moser, USFWS; Paul Padding, USFWS Atlantic Flyway

Representative; Margaret Petersen, USGS Alaska Science Center; Scott Petrie, Long Point Waterfowl; Bob Platte, USFWS MBM, Alaska; Bob Raftovich, USFWS MBM; Eric Reed, CWS; Fred Roetker, USFWS MBM; Jean-Pierre Savard, CWS; Joel Schmutz, USGS Alaska Science Center; Mike Schummer, Long Point Waterfowl; Sue Sheaffer, Livingston Ripley Waterfowl Conservancy; Emily Silverman, USFWS MBM; Bonnie Swarbrick, USFWS Buenos Aires National Wildlife Refuge; Bryan Swift, New York State Department of Environmental Conservation; John Takekawa, USGS Western Ecological Research Center; Phil Thorpe, USFWS MBM; Francisco Vilella, USGS Mississippi Cooperative Fish and Wildlife Research Unit; Sandy Williams, Museum of Southwestern Biology; Heather Wilson, USFWS MBM, Alaska; Rick Wishart, Ducks Unlimited; Dan Yparraguirre, California Department of Fish and Game; and Guthrie Zimmerman, USFWS MBM.

Other people making an array of contributions to the book were Martin Acosta, Tom Aldrich, Dave Ankney, Barbara Avers, Shannon Badzinski, Dan Baldassarre, Greg Balkcom, Dick Banks, Pedro Blanco, Eric Bolen, Caroline Brady, David Brandt, Mike Brasher, David Brown, Dawn Browne, Bob Brua, Mike Conover, Steve Cordts, Chris Dau, Susan De La Cruz, Patrick Devers, Deanna Dixon, Jamie Dozier, Jim Dubovsky, Susan Earnst, Dan Esler, Gilles Falardeau, Ron Giegerich, Barry Grand, Kevin Hartke, Jim Heffelfinger, H. Heusmann, Bill Hoffman, Liz Huggins, Jerry Hupp, Matt Kaminski, Don Kraege, Bill Krohn, Karen Laing, Bryce Lake, Tom Langschied, Jim Leafloor, Jo Anna Lutmerding, David Luukkonen, Ray Marshalla, Ramesh Maruthalingam, Guy McCaskie, Bob McLandress, Don McNicol, Katherine Mehl, Lourdes Mugica, Wayne Norling, Shaun Oldenburger, John Pearce, Aaron Pearse, Bruce Peterjohn, Adam Phelps, Pam Pietz, Bruce Pollard, Bill Pranty, Hal Prince, Mike Rabe, Kelly Rathburn, Barnett Rattner, Brandon Reishus, Ken Richkus, Neil Ringler, Frank Rohwer, Nora Rojek, Jon Runge, Doug Ryan, Dave Santillo, John Sauer, Jason Schamber, Dave Scott, Hélène Sénéchal, Dave Sherman, Bill Shields, Charlie Smith, Don Stewart, Kelsey Sullivan, Brian Tefft, John Thompson, Josh Traylor, Bob Trost, Scott Turner, Josh Vest, John Vradenburg, Richard Webster, Maria Wieloch, Khristi Wilkins, Barry Wilson, and Ben Zuckerberg.

At Johns Hopkins University Press, Vince Burke, my editor, and his editorial assistant, Jennifer Malat, were wonderful to work with from start to finish, as was Julie McCarthy, the managing editor. As for copyediting, the skilled and tireless efforts of Kathleen Capels greatly improved the quality of the text. I could not possibly have worked with a better wordsmith.

Final acknowledgments go to an extraordinary group of people, the doctors and nurses whose skills and compassion and caring saw me through my medical journey, much of which occurred while I was writing this book: Drs. Philippe Armand, Dan Bingham, Jennifer Brown, Dennis Daly, Jonathan Friedberg, Shahrukh Hashmi, Steve Ladenheim, Tony Lombardo, Tom Maher, and Paul Richardson, and the most spectacular nurses in the world, especially Anna Morrison, Lynn Rich, Toni Dubeau, Katey Stephens, Kathy Close, Peter Sullivan, and the rest of the Brigham and Women's Hospital 6A staff.

I wish to thank the following organizations for their financial support of this book:

California Waterfowl Association
Central New York Wildfowlers
Connecticut Waterfowlers Association
Delta Waterfowl Association
Ducks Unlimited
Ducks Unlimited Canada
Illinois Natural History Survey, Frank C. Bellrose Waterfowl Research
    Laboratory of the Forbes Biological Station
Livingston Ripley Waterfowl Conservancy
Long Point Waterfowl
State University of New York–College of Environmental Science and Forestry
Tudor Farms, Cambridge, Maryland
U.S. Fish and Wildlife Service, Division of Migratory Bird Management
Wildlife Management Institute

I am also grateful to the following individuals for their financial contributions:

Frank C. Bellrose
Wesley M. Dixon Jr.
Charles C. Haffner III
Rosemary Ripley

Basis for the Book

A vast amount of new information on waterfowl biology and management had accrued since the appearance of the third edition of *Ducks, Geese, and Swans of North America*, yet the accessibility of that information was dramatically enhanced via the Internet. In particular, online literature search engines and library databases put virtually all published papers and reports at my fingertips. All federal, state, provincial, and nongovernmental organizations focused on waterfowl and wetlands research and management now have websites replete with an abundance of reliable information and databases, with many of the databases set up for easy extraction of summary information.

Nonetheless, like the now overabundant population of Snow Geese in North America, this tremendous amount and accessibility of new information created somewhat of an "embarrassment of riches." Although the nature of the information sources did not need to be explained in detail in the current edition, as they were in previous versions, choices were necessary on what time frames to review, what information to use, and when and where to include it. I also added several sections under the species accounts, some of which required background information. Hence the purpose of this brief introduction to the revised and updated edition is to provide a synopsis of the overall plan of the book and the information contained therein.

MAKING CHOICES

Revising a book of this magnitude necessitated choices consistent with my desire to produce a thorough treatment of each species while not generating a version that was so large, cumbersome, and expensive as to discourage potential readers. I also wanted room for the many photographs that enhance this edition, so there are no introductory chapters on general aspects of waterfowl ecology and management. These topics were thoroughly treated in *Waterfowl Ecology and Management*, which I coauthored along with Eric Bolen in 2006. General topics in waterfowl ecology and management also appeared in the first volume of a two-volume set, *Ducks, Geese, and Swans* (2005), edited by Janet Kear, although the treatment there has a decided bent toward the European waterfowl literature. I have retained an identical layout for each species, with the same general headings and subheadings, so as to facilitate finding information on a given species and to increase the speed with which readers may become familiar with the book.

All measurements in this edition are metric, which reflects the system used in most scientific publications, including those in the United States since about 1970. Equivalents in U.S. measurements are readily available online, but readers unfamiliar with the metric system might do better in equating metric units to more familiar objects. An acre, for example, is about the size of an American football field from goal line to goal line, and there are 0.4 hectares (ha) in an acre; hence a hectare is roughly the area of 2.5 football fields. The interior area surrounded by a standard athletic track is about 1 ha. A kilogram (kg), or 1,000 grams (g), is approximately the weight of an adult female Mallard. An inch = 2.54 centimeters (cm); thus 25 cm is about the wing chord length of a Northern Shoveler. A mile = 1.6 kilometers (km); hence there are 2.6 km² in a square mile or section of land. A meter (m) = 3.3 feet, with 2 m being equivalent to the wingspan of a Trumpeter Swan from wing tip to wing tip.

TAXONOMY

I only developed accounts for species conclusively known to currently breed in North America, except for those in the Caribbean, which thus excluded the West Indian Whistling-Duck (*Dendrocygna arborea*) and White-cheeked Pintail (*Anas bahamensis*). Common and scientific names of waterfowl (including capitalization) follow the seventh edition of the American Ornithologists' Union (AOU) *Check-list of North American Birds* and subsequent supplements as published in the journal *Auk*; this information is also available on the AOU website. The only exception was my use of American Green-winged Teal for Green-winged Teal, as genetic and morphological evidence indicate that the American Green-winged Teal is a separate species from the European Green-winged Teal, and it is treated as such by several taxonomic authorities. I presented the species accounts in the order used by the AOU, except that swans are featured first, which followed the convention of most field guides. Also, because the common names of North American waterfowl are standardized by the AOU, the accompanying scientific names appear just once, at the head of each species account, rather than being repeated over and over again in other accounts, although I did use scientific names for other waterfowl species and to differentiate subspecies where necessary. For all other animals and plants, the scientific names are as given in the original publications from which this information was cited; occasionally an alternative, more familiar name is added, but I did not insert common names where only scientific names were provided. For consistency with current botanical practice, some words in common names have been combined (such as "widgeon grass" to "widgeongrass").

The family Anatidae, within the order Anseriformes, contains nearly all of the world's current and recently extinct species of ducks, geese, and swans. I consider the work of Livezey (1997) to be the most thorough and authoritative source on global waterfowl taxonomy, and he recognized 171 species in the family Anatidae, including 2 recently extinct species: the Labrador Duck (*Camptorhynchus labradorius*) of North America (declared extinct in 1875) and the Auckland Islands Merganser (*Mergus australis*) of New Zealand (in 1902). Livezey grouped the Anatidae into 5 subfamilies, 3 of which occur in North America: the Dendrocygninae (whistling-ducks), the Anserinae (swans and true geese), and the Anatinae (typical ducks);

the latter subfamily includes the majority of the Anatidae (112 species). Below the subfamily level, waterfowl are grouped into 13 tribes, which are related groups of genera. The 7 tribes that occur in North America are briefly described below, to set the occurrence of North American species into a global perspective.

The tribe Dendrocygnini (whistling-ducks, formerly called tree ducks) contains 8 species worldwide, 2 of which occur in North America: the Black-bellied Whistling-Duck and the Fulvous Whistling-Duck. All species of whistling-duck are in the same genus, *Dendrocygna*, which literally means "tree swan" and reflects some of the similarities shared between swans and whistling-ducks. Whistling-ducks are primarily tropical in distribution and thus are at the northern edge of their range in North America. They are moderate-sized ducks, with 7 of the 8 species averaging less than 850 g. The sexes are similar in appearance (monomorphic); males and females participate in incubation, a feature virtually unique to the group; and pair bonds appear to be for life. Unlike swans and geese, however, the downy young are distinctly patterned. Both sexes give a clear, usually multisyllabic whistle; hence the name whistling-ducks. Most species never occur in trees (the Black-bellied Whistling-Duck is an exception), so whistling-duck is a more appropriate name than "tree duck," although the latter is still in use.

Under the subfamily Anserinae are the swans (Cygnini), with 8 species worldwide, 3 of which occur in North America, including the introduced Mute Swan; and also the true geese (Anserini), with 17 species worldwide (including the newly recognized Cackling Goose by the AOU), 7 in North America. The swans are the largest of the waterfowl, among which the Trumpeter Swan is the heaviest, with large males reaching 14 kg. Most swans and geese are found in the Northern Hemisphere, where they tend to occur in large marshes. The sexes are alike in appearance and voice and have an irregular reticulate pattern of scales on the tarsus. They mate for life, an adaptation not only for breeding during the characteristically short summers in arctic and subarctic regions, but also because adults and their offspring remain together as a family group through their first winter and upon their return to the breeding grounds the following spring. Swans, geese, and whistling-ducks undergo only 1 molt of their plumage each year. In contrast, most ducks undergo 2 body molts per year but only a single wing molt. Swans and geese do not breed until they are at least 2 years old, with most not breeding until they are 3 or older; Trumpeter Swans generally do not breed until age 4–7.

The Anatinae are grouped into 5 tribes, 4 of which occur in North America: the Anatini (puddle ducks, or surface-feeding ducks), the Aythyini (pochards, or diving ducks), the Mergini (sea ducks), and the Oxyurini (stiff-tailed ducks). The puddle ducks contain 62 species, of which 12 occur in North America (13 if the Mexican Duck is included); many of the puddle ducks are in the genus *Anas*. Males and females of most species have different plumages (dimorphic), with the males more brightly colored and distinguished by shiny coloration on the secondary flight feathers, an area known as the speculum. The pattern of scales on the tarsus is arranged in a linear (scutellate) pattern. Puddle ducks are denizens of shallow, freshwater marshes and small wetlands, where they characteristically tip up to feed. They form seasonally monogamous pair bonds that are usually broken at some time during incubation. Most species breed at 1 year of age. The puddle ducks are ecologically well diversified and considered the most successful waterfowl, because the group contains more species and greater numbers of individuals than any other tribe, and they occur on all continents except Antarctica. Major groupings within the Anatini include the Mallard-like ducks, wigeons, pintails, and blue-winged ducks.

The Aythyini are correctly referred to as pochards, but in North America they are more commonly known as diving ducks. There are 16 species worldwide (12 in the genus *Aythya*); 5 occur in North America. The Aythyini dive for food and do so with large feet that are placed farther back on the body than in puddle ducks. They often feed extensively on vegetative matter. The sexes are dimorphic, with the males more brightly colored, but not as bright as most puddle ducks. They form seasonally monogamous pair bonds and generally breed at 1 year of age. Diving ducks are heavy in relation to the surface area of their wings, which reduces lift. Hence they characteristically must run along the water to gain enough lift before takeoff. In contrast, the lower wing load of puddle ducks allows them to immediately spring from the water. In flight, the high wing load of diving ducks also requires them to beat their wings faster than puddle ducks, which increases lift.

The Mergini, or sea ducks, contain the eiders, scoters, goldeneyes, mergansers, and a few other species. There are 22 species worldwide, of which 15 occur in North America (16 counting the extinct Labrador Duck). Sea ducks are the largest of the ducks, with male Common Eiders exceeding 2,500 g. Sea ducks largely breed on freshwater wetlands in arctic and subarctic regions or on offshore islands, but they primarily winter on the ocean, although goldeneyes also winter on large rivers, and Long-tailed Ducks are common during winter on the Great Lakes. They are accomplished divers, with the Long-tailed Duck recovered in fishing nets and lobster traps set 70 m underwater. Males and females are dimorphic, with some of the males being among the most colorful of all ducks (e.g., Harlequin Ducks). Sea ducks do not begin to breed until they are 2–3 years old. They form annual pair bonds that terminate during the breeding season but, in some species, are known to then reform during winter. Sea ducks primarily feed on animal matter, with the mergansers having strongly serrated bill lamellae as an adaptation for a diet of fish.

The Oxyurini, or stiff-tailed ducks, contain 9 species worldwide, of which 2 occur in North America, although only the Ruddy Duck is common and widespread. The Masked Duck is rare, with only a few reported breeding records from southern coastal Texas and Florida. Stiff-tailed ducks breed on freshwater wetlands, where they construct overwater nests in dense vegetation. They have short, rigid retrices (tail feathers) and large feet, both of which are adaptations for diving. The sexes are dimorphic, with all species characterized by short thick necks. Most males having bright blue bills during the breeding season that are used during a unique courtship display, where the bill is rapidly slapped against the breast to create a tapping sound audible at close range. The stiff-tailed ducks are well-known nest parasitizers that readily lay eggs in each others' nests and in the nests of other species.

WATERFOWL POPULATION SURVEYS

Various published and online sources were consulted to report the population size, distribution, and harvest of waterfowl; so, too, were several databases provided by agency personnel. Methodologies associated with these surveys are presented in their associated reports or elsewhere and need not be repeated here, although a brief overview is given below so that readers can understand the time frame and extent of the various surveys. All of these referenced reports and databases are readily available online.

For breeding populations, I used the Waterfowl Breeding Population and Habitat Survey (WBPHS) which is an aerial survey conducted annually in May and June by the U.S. Fish and Wildlife Service and the Canadian Wildlife Service. The WBPHS consists of two parts: the Traditional Survey and the Eastern Survey. The Traditional Survey has been conducted annually since 1955 and covers about 3.7 million km² of prairies, parklands, taiga, open boreal forest, and tundra, largely from the Dakotas and Montana in the south, then northward into Canada through most of Manitoba, Saskatchewan, Alberta, the Northwest Territories, a small part of northeastern British Columbia and northwestern Yukon, and on into Alaska to the Bering Sea. The eastern boundary covers most of western Ontario. The Eastern Survey was begun in 1990 and covers about 1.8 million km² in parts of eastern Ontario, Québec, and the Maritime Provinces, as well as Maine and a small part of northern New York State. The results of these surveys are published annually by the U.S. Fish and Wildlife Service under the general title *Waterfowl Population Status* and *Trends in Duck Breeding Populations*, with an appropriately identified year or years (e.g., *Waterfowl Population Status, 2011*; *Trends in Duck Breeding Populations, 1955–2011*). The Canadian Wildlife Service annually publishes a similar report, titled *Population Status of Migratory Game Birds in Canada*. These reports are easily accessible via the Internet.

Another helpful data source was the Atlantic Flyway Breeding Waterfowl Plot Survey, which is a ground survey of 1 km² plots that became operational in 1993 and covers 11 states from New Hampshire to Virginia. These data, and a wealth of additional information, are available in the *Atlantic Flyway Waterfowl Harvest and Population Survey Data* annual reports. I also regularly consulted the Flyway Data Books for the Mississippi, Central, and Pacific Flyways, which are published annually by the U.S. Fish and Wildlife Service. Additional numbers were generated by surveys conducted in states and provinces outside the WBPHS or the Atlantic Flyway Breeding Waterfowl Plot Survey, with such data available in Flyway Data Books, reports from individual states, or Joint Venture reports associated with the North American Waterfowl Management Plan. I occasionally used data from the North American Breeding Bird Survey, which began in 1966 and is available online. The Breeding Bird Survey is ground based, with over 4,000 survey routes in the United States and Canada, of which about 3,000 are canvassed each spring. This survey is primarily used to assess songbirds, but it was helpful for some species (e.g., whistling-ducks) that occurred well south of traditional breeding waterfowl surveys.

Estimates of winter populations and their distribution were derived from data in the annual Midwinter Waterfowl Survey, which has been conducted by the U.S. Fish and Wildlife Service since 1955, and the Christmas Bird Count, which has taken place under the auspices of the National Audubon Society since 1900. The Midwinter Survey is very useful for delineating the regional distribution of ducks and geese at that time of the year, and it provides data on some populations of geese and swans that are otherwise difficult to survey on remote breeding areas in arctic and subarctic regions. The Midwinter Survey consistently underestimates Snow and Ross's Goose populations, however, and it has other methodological problems (Eggeman and Johnson 1989, Heusmann 1999), but nonetheless provides valuable long-term data. I generally reported summary data from 2000 to 2010. The Christmas Bird Count data were especially helpful in determining numbers and distributions for species that are combined during the Midwinter Survey. For example, the Midwinter Survey lumps the mergansers and scaup, but the associated species are separated in the Christmas Bird Count. About 1,500 Christmas Bird Counts are conducted annually in the United States and Canada. Midwinter Survey data were provided by each flyway, whereas Christmas Bird Count data were from an online database. The Midwinter Survey is also conducted in Mexico, although only at about 3-year intervals since the mid-1980s. I reported results on that country's east coast, which were taken from a database provided by the U.S. Fish and Wildlife Service, and from survey reports for the interior highlands and mainland west coast of Mexico.

RANGE MAPS

My intent was to present fairly simple range maps, so that readers could instantly determine the general breeding and wintering ranges of a species, along with high-density areas within the breeding range and concentration areas within the wintering range. Distributions of both breeding and wintering areas are further detailed in the text, along with estimates of population numbers within areas of both breeding and wintering range. Distribution maps always leave room for questions, especially for such mobile species as waterfowl, but I relied on virtually every possible source of information to produce the maps in this revised and updated edition.

For breeding distributions, I examined data from the Traditional Survey as far back as the 1960s for the species covered by that survey. Data were also included from the Eastern Survey, the Christmas Bird Counts, and the Breeding Bird Survey, as well as information from Breeding Bird Atlases, other published range maps, and professional opinion. Winter distribution data were primarily taken from the Midwinter Survey (since 1990), but the Christmas Bird Counts were also helpful, along with published range maps and professional opinion.

A map of the entire North American continent appears just before the species accounts in both volumes. This map and four others (Alaska, Canada, the continental United States, and Mexico) appear on the CD-ROM at the back of volume 2.

WATERFOWL HARVEST

Harvest information came from a database provided by the U.S. Fish and Wildlife Service to determine harvest in the United States, and from an online database by the Canadian Wildlife Service. The U.S. Fish and Wildlife Service also publishes an annual comprehensive report on waterfowl harvest that is easily accessible online. In an effort to provide more accurate and reliable information on the harvest of all migratory birds, the U.S. Fish and Wildlife Service replaced the long-running (since 1952) Waterfowl Harvest Survey, or Mail Survey, with a new survey known as the Harvest Information Program (HIP). The HIP Survey was introduced in 1992 and adopted by all states, beginning with the 1999 hunting season. The HIP Survey retained the Cooperative Parts Collection Survey of the original survey approach, which contacts a small percentage of randomly selected hunters and asks them to mail in a wing from each duck and a tail from each goose harvested during the season. These parts are then examined at annual "wing bees" within each flyway and used to estimate the overall species, sex, and age composition of the harvested waterfowl. The Canadian Wildlife Service also conducts a Parts Collection Survey that is similarly used to determine the composition of the harvest. Lastly, I used band-recovery data (usually from 1950 to 2010) from the Gamebirds Program of the U.S. Geological Survey Bird Banding Laboratory, available on the Internet. These data were especially useful in determining migration routes and destinations.

MOLTS AND PLUMAGES

This edition includes a new Molts and Plumages section for each species, as replacement of plumage is an essential component in the annual cycle of waterfowl. Waterfowl acquire different plumages and associated coloration to meet their needs at various stages of this cycle, from hatching until they attain the definitive plumage by which we recognize adults. Waterfowl, along with some other waterbirds, also share the unusual trait of molting all their wing feathers simultaneously, which renders them flightless during the wing molt. This simultaneous loss of flight feathers, which completes the wing molt faster than would an asynchronous molt, is advantageous, because the high wing load of waterfowl would render them flightless with loss of only a few flight feathers. Regrowth of new flight feathers requires 20–49 days, depending on the species involved (Hohman et al. 1992a). In geese, the flightless period associated with the wing molt occurs when adults are rearing their broods. In contrast, most male ducks undergo wing molt after they desert their nesting females, whereas most females undergo wing molt after the brood-rearing period. Many species also exhibit a northward molt migration; hence, there is a subsection on that topic under the Migration Behavior section.

The Molts and Plumages section provides only brief summaries of the actual plumages and the timing of their acquisition, although it emphasizes information on the ecological strategies associated with molting (if available). Thorough descriptions of molts and plumages of each species are available in two waterfowl volumes of Ralph Palmer's *Handbook of North American Birds* (1976a, 1976b), as well as in individual species accounts in the *Birds of North America* series. Regardless, some of the terminology used to describe molts and plumages may be unfamiliar to most readers, so I provide a brief overview here.

This terminology follows that put forth by Humphrey and Parkes (1959), which was subsequently used to describe waterfowl plumages by Palmer (1976a, 1976b) and Hohman et al. (1992a). The Humphrey and Parkes (H-P) system of describing molts is elegantly simple and avoids confusing terms like nuptial, eclipse, winter, and transitional plumages. For example, the breeding plumage of a male Mallard is worn throughout most of the fall and winter, in addition to the breeding season in spring and early summer; hence the term "breeding plumage" is misleading. Geese and swans wear the same plumage all year, so their plumage cannot be unambiguously labeled breeding or nonbreeding. The H-P system thus recognizes only 5 major plumages—(1) natal, (2) juvenal, (3) basic, (4) alternate, and (5) supplemental—for which there are synonymously named molts (e.g., the prealternate molt leads to the alternate plumage). Basic describes the plumage of birds that, as adults, have 1 plumage per cycle (whistling-ducks, swans, and geese) that is replaced by a complete molt (prebasic molt) involving all the body and wing feathers; hence there is a flightless period. For species that, as adults, have 2 plumages per year (ducks), the prebasic molt results in basic plumage and the prealternate molt results in alternate plumage, which replaces only part of the plumage and not the wing feathers; thus individuals are not flightless during this molt. The timing of acquisition of these plumages varies between males and females, with males acquiring basic plumage in mid- to late summer; females often begin the prebasic molt of their body feathers in late winter and early spring, but they do not replace the wing feathers until during or after brood rearing. Some species, such as the Long-tailed Duck, have 3 plumages per cycle, with the third known as the supplemental plumage.

Plumages such as natal and juvenal are worn just once in a lifetime. Natal down is replaced with the more rigid contour feathers of the juvenal plumage and typically first develops on the ventral area exposed to the water. The feathers of the juvenal plumage originate in the same feather follicles as the natal down: the down feathers are pushed out of their bases and, in some cases, remain attached to the shafts (rachises) of the new juvenal feathers. This phenomenon is most conspicuous on tail feathers, because of their larger shafts, which often results in a severing of the two shafts to produce a notch at the tip of each tail feather that, in autumn, serves to distinguish young-of-the-year from older birds. The replacement of natal down by the juvenal plumage and attainment of flight naturally varies by species from about 6 weeks in Blue-winged Teal to about 9 weeks in Canvasbacks, 8–10 weeks in Canada Geese, and 13–15 weeks in Trumpeter Swans.

Basic plumage follows juvenal plumage. For first-year swans, geese, and (to a lesser extent) most species of ducks, the initial basic and sometimes the initial alternate plumages are recognizably different from the basic and alternate plumages of adults, and also worn only once. Because these one-time plumages can be differentiated, the H-P system refers to them as first basic (sometimes seen as Basic I), and first alternate (sometimes seen as Alternate I), after which plumages do not change further with age. Adult birds then repeatedly replace the same plumage via the same molt; such repetitive plumages are identified with the term definitive. When subsequent basic or alternate plumages are still different from definitive plumages, the H-P system then names them second basic (sometimes seen as Basic

II), second alternate (sometimes seen as Alternate II), and so forth. First basic is easily identifiable in swans, but such plumages can be difficult for inexperienced observers to identify in puddle ducks and geese. First alternate is readily identifiable among male sea ducks.

Molts and plumages should be seen as occurring in a continuum, although there is substantial variation among individuals in the timing of plumage acquisitions, especially in the sea ducks. Canada Geese, for example, molt body feathers during all parts of the year, except the nesting season. Also, individuals can wear several different plumages at the same time. Thus a young Mallard in late summer may wear a mix of juvenal and first basic plumages.

Howell et al. (2003) noted a few inconsistencies in the H-P system that occur because of the highly variable molt that replaces juvenal plumage with first basic plumage. The latter can even be absent (or nearly so) in some species of ducks (Pyle 2005). Hence these authors suggest that the juvenal plumage should be considered the first plumage in the plumage cycle, synonymous with first basic in the H-P system. Howell et al. (2003) also proposed the term formative plumage for any plumage present in the first cycle that is not a basic plumage and does not occur in subsequent cycles. Thus juvenal plumage in puddle ducks and pochards is succeeded by a formative plumage that is very limited in many species but is a more-or-less continuous transition to the next molt, which produces the bright first alternate of the H-P system. However, because these proposed changes are not in widespread use, and because the H-P system has long been used to describe molts and plumages of waterfowl without much ambiguity, I followed the latter system in presenting molt and plumage information for each species.

A BRIEF GLOSSARY

As an aid to readers who may be unfamiliar with certain terms used in this book, I've provided an explanation of some that appear regularly in the species accounts: brood amalgamation (a mixing of two or more broods), brood parasitism (2 or more females laying eggs in the same nest, commonly called dump nesting), brood success (the percentage of broods fledging at least 1 bird), comfort movements (behaviors such as preening, wing stretching, etc.), conspecific (the same species), crèche (a group of flightless young from more than one brood), culmen (bill), dimorphism (occurs in 2 forms, such as different sizes or colors), extirpation (local elimination of a species' population), feral (escaped from captivity but breeding in the wild), fledging (time when flight is first attained), hatching success (percentage of eggs that hatch), incubation constancy (time spent on the nest), nest success (percentage of nests that hatch at least 1 egg), philopatric (return to the natal area), remiges (flight feathers), retrices (tail feathers), speculum (iridescent coloration on the secondary flight feathers), and tarsus (leg).

OTHER ISSUES

Different species of waterfowl have different life spans, with swans and geese generally living much longer than ducks. The oldest known individual of a given species is not that valuable in the study of population structure and dynamics, but such ages are of deserved interest as part of each species' life history. Thus the Recruitment and Survival section of the species accounts reports longevity records for each species as listed in the online database of the U.S. Geological Survey Bird Banding Laboratory (Lutmerding and Love 2011). This database includes longevity records for waterfowl published in Clapp et al. (1982) and Klimkiewicz and Futcher (1989).

There is a section on nest success for each species, as nest success is an especially important component of population demographics. Biologists have used different techniques for quantifying nest success; I provide a brief overview here so readers can understand the derivation of the results and judge comparisons among studies. Early studies simply reported what is now termed "apparent nest success," which is the proportion of all nests discovered that eventually hatch. Not all nests will be located at the same stage of incubation, however, which can bias the overall estimate of nest success for the sample. For example, a nest located near the end of incubation has a much greater probability of hatching than one located during laying, because the former is very near "success," while the latter has many days to go. To account for this difference in days of exposure to potential predators, almost all studies since the mid-1970s report nest success that is estimated via the Mayfield Method, which takes into consideration the length of time each nest is exposed to predation (Mayfield 1961, 1975), or a closely related alternative (Jehle et al. 2004). In general, apparent nest-success estimates are higher than Mayfield estimates.

The plumage-development classification system created by Gollop and Marshall (1954) is often reported in my species accounts to categorize duckling ages that are associated with growth and survival or with some other aspect of duckling age. The system relies on recognizable changes in down and plumage as ducklings mature from hatching to being near flight capable (fledging); it was developed for 7 species of puddle ducks and 4 species of diving ducks. The system involves 3 age classes and several subclasses that vary in age somewhat among species but principally are as follows: Class Ia ducklings are covered in brightly colored natal down (about 1–6 days old); Class Ib are still down covered but the down color is fading (about 7–12 days old); Class Ic are still primarily covered with faded down but the body is elongated (about 13–18 days old); Class IIa show their first feathers, which appear on the sides and the tail (about 19–26 days old); Class IIb have over half the body covered with feathers (about 27–35 days old); Class IIc are primarily fully feathered, except for some down on the back (36–44 days old); and Class III are fully feathered but not capable of flight (45–55 days old). In general, diving ducks take longer to develop than dabbling ducks.

Lastly, the extensive literature citations reflect the amount of new information that has come forth since the third edition of *Ducks, Geese, and Swans of North America*, and of course they give proper credit to those producing that information. More than at any time in history, such literature also is at virtually everyone's fingertips, due to the Internet. Hence individuals desiring more details (even those without access to a subscription-style database) can enter a title into a general Internet search engine and almost always obtain a given paper's abstract, if not the entire paper. The majority of federal, state, and private entity reports are readily available in their entirety, so those readers seeking more detailed information on a given topic can do so more easily than ever before. Furthermore, the CD-ROM of the literature cited, found at the back of volume 2, allows readers to search the citations by topic and author(s). Text citations with a full name but not a date refer to personal communications, and the individuals who supplied this helpful information are listed in the acknowledgments.

Ducks, Geese, and Swans of North America

REVISED AND UPDATED EDITION

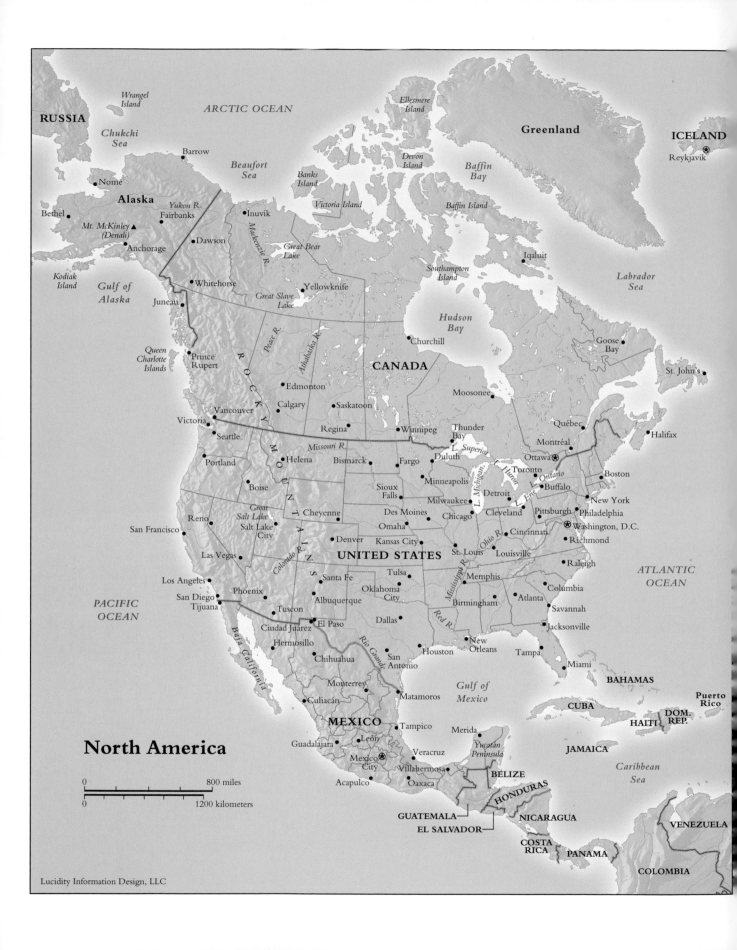

North America

RUSSIA

Chukchi Sea

ARCTIC OCEAN

Wrangel Island

Barrow

Nome

Alaska

Yukon R.
Fairbanks

Bethel

Mt. McKinley (Denali) ▲

Anchorage

Dawson

Kodiak Island

Gulf of Alaska

Juneau

Queen Charlotte Islands

Prince Rupert

Beaufort Sea

Inuvik

Mackenzie R.

Great Bear Lake

Whitehorse

Yellowknife

Great Slave Lake

Banks Island

Victoria Island

Devon Island

Ellesmere Island

Greenland

ICELAND

Reykjavik

Baffin Bay

Baffin Island

Iqaluit

Southampton Island

Labrador Sea

Hudson Bay

Churchill

Goose Bay

St. John's

CANADA

Peace R.

Athabasca R.

Edmonton

Calgary

Saskatoon

Regina

Winnipeg

Moosonee

Québec

Montréal

Halifax

Vancouver

Victoria

Seattle

ROCKY MOUNTAINS

Missouri R.

Helena

Bismarck

Fargo

Duluth

Thunder Bay

L. Superior

L. Huron

Ottawa ⊛

Toronto

L. Ontario

L. Erie

Buffalo

Boston

New York

Portland

Boise

Great Salt Lake

Salt Lake City

Cheyenne

Sioux Falls

Minneapolis

Milwaukee

Detroit

Cleveland

Pittsburgh

Philadelphia

L. Michigan

Reno

San Francisco

Las Vegas

Colorado R.

Denver

Des Moines

Omaha

Chicago

Cincinnati

Washington, D.C. ⊛

Richmond

Kansas City

St. Louis

Louisville

UNITED STATES

Raleigh

Los Angeles

Phoenix

San Diego

Tijuana

Tuscon

Santa Fe

Albuquerque

Oklahoma City

Tulsa

Mississippi R.

Ohio R.

Memphis

Columbia

ATLANTIC OCEAN

PACIFIC OCEAN

El Paso

Ciudad Juarez

Hermosillo

Rio Grande

Dallas

Red R.

Birmingham

Atlanta

Savannah

Jacksonville

Chihuahua

San Antonio

Houston

New Orleans

Tampa

Miami

BAHAMAS

Monterrey

Culiacán

Matamoros

Gulf of Mexico

CUBA

Puerto Rico

DOM. REP.

HAITI

MEXICO

Tampico

Merida

Yucatán Peninsula

JAMAICA

Caribbean Sea

VENEZUELA

Guadalajara

León

Mexico City ⊛

Veracruz

Villahermosa

BELIZE

Acapulco

Oaxaca

HONDURAS

GUATEMALA

EL SALVADOR

NICARAGUA

COSTA RICA

PANAMA

COLOMBIA

0 ———— 800 miles

0 ———— 1200 kilometers

Lucidity Information Design, LLC

Black-bellied Whistling-Duck

Dendrocygna autumnalis (Linnaeus 1758)

The Black-bellied Whistling-Duck is a medium-sized, tropical species of waterfowl endemic to the Western Hemisphere, where its northern range just reaches the United States, principally in Texas. As in all whistling-ducks, both sexes are virtually identical in size and plumage, and they undergo only 1 molt/year. They pair for life and both sexes share incubation duties; among waterfowl, this is a near-unique feature of whistling-ducks. Black-bellied Whistling-Ducks are gregarious and vociferous—it is uncommon to encounter a flock that is silent. They nest in tree cavities or nest boxes, and also on the ground, but brood parasitism is prevalent, especially where they use nest boxes. Black-bellied Whistling-Ducks are largely vegetarians and will readily eat cereal grains,

3

especially corn and rice. More than any other species of North American waterfowl, they have dramatically expanded their numbers and range in the United States since reports by Bellrose (1980). Wild populations are now established in Arizona, Louisiana, and Florida; breeding records are documented from several other states. Postbreeding birds wander widely, with sight records documented in the United States and in several Canadian provinces. They are not an important game species in the United States because of their relatively small numbers and restricted distribution. In Mexico, they are important to local people, who harvest both adults and eggs.

IDENTIFICATION

At a Glance. Black-bellied Whistling Ducks are unmistakable, and the sexes are similar. On the ground, note the erect goose-like posture, the black belly, the pink bill and feet, and the white eye ring. At rest, extensive white on the wings contrasts sharply with the black belly. In flight, the wing beat is slow, and the white on the wings is highly visible against the dark underside. The feet trail beyond the tail during flight, and flight is frequently accompanied by vocalizations. Black-bellied Whistling-Ducks are gregarious and thus are usually encountered in flocks.

Adults. A medium-sized duck, males and females are virtually identical in appearance and size. The erect posture and dark black belly and sides are especially diagnostic, and contrast with the chestnut breast and lower neck. The head and the upper neck are a light brownish gray, with a white eye ring. A black streak extends up the back of the neck to the brownish crown. Most of the coverts and the base of the outer primaries are bright white; the remaining parts of the primaries are dark black. This contrasting black and white on the wings is very noticeable in flight, as are the long legs that extend beyond the tail. The tarsus is reticulate. The legs and the feet are pink. The bill is red. Their posture on the ground is very erect, almost heron-like. The abdominal plumage of females is somewhat more dully colored than that of males, but it is not a reliable characteristic. The southern subspecies is distinctively gray breasted (see Distribution).

Juveniles. Juveniles are readily distinguishable from adults by their overall dull brown body and gray head. The breast is buffy brown; the belly is grayish white, with some cross-barring. Juveniles have the distinctive contrasting black, white, and olive upperwing coloration of adults, although the white component is less extensive. The legs, the feet, and the bill are gray.

Ducklings. The ducklings are bright yellow, with striking black markings about the face, the back, and the body, including a black eye line (Nelson 1993). As in all *Dendrocygna*, a distinct dark T-stripe occurs on the back of the head and the neck, separated from the crown by a light stripe around the head.

Voice. Black-bellied Whistling-Ducks are vociferous in flight, repeatedly uttering an unmistakable whistling *pee-chee-chee*, which provides them with 1 of their local names. They also commonly call when on the ground.

Similar Species. Adults are unmistakable. Juveniles somewhat resemble other whistling-ducks, including the Fulvous Whistling-Duck, but the latter does not contain any white on the wings. The overall grayish appearance of the juvenile Black-bellied Whistling-Duck is also diagnostic.

Right:
Black-bellied Whistling-
Duck duckling.
Ian Gereg

Below:
Black-bellied Whistling-
Duck pair (*foreground*)
and juveniles.
GaryKramer.net

5

Weights and Measurements. Black-bellied Whistling Ducks are medium-sized. In southern Texas (northern subspecies), males ($n = 35$) averaged 816 g (range = 680–907 g); females ($n = 37$) averaged 839 g (range = 652–1,021 g; Bolen 1964). During the breeding season in southern Texas, males averaged 785 g ($n = 43$), and females, 792 g ($n = 38$; Chronister 1985). In Guyana (southern subspecies), males ($n = 78$) averaged 741 g (range = 530–890 g) and females ($n = 82$), 725 g (range = 530–890 g; Bourne 1979). James and Thompson (2001) also reported weights from Texas as averaging 796 g for males ($n = 61$) and 760 g for females ($n = 59$). Wing chord length averaged 24.8 cm for males and 24.4 cm for females; tarsus, 60.6 mm for males and 59.5 mm for females; culmen, 51.5 mm for males and 50.8 mm for females. Bourne (1979) reported total length as 47.1 cm for males and 46.5 cm for females, with significant individual overlap between the sexes.

DISTRIBUTION

The Black-bellied Whistling-Duck is restricted to the Western Hemisphere, but it is at the very northern edge of its range in the United States. Two subspecies are recognized: the northern (*Dendrocygna autumnalis fulgens*) and the southern (*D. a. autumnalis*), although interbreeding may occur where their ranges overlap in Panama (Wetmore 1965, Banks 1978).

Breeding. The eastern population of the northern subspecies breeds from the Gulf Coast of southwestern Louisiana through much of eastern and southern Texas. There is also a small breeding population recently established in central Florida, but the largest breeding population in the United States is in southern coastal Texas. South of Texas, this subspecies extends south along the east coast of Mexico through to the coastal regions of Costa Rica. The western population of this subspecies breeds from southern Arizona south along the west coast of Mexico through to the western coast of Panama. The southern subspecies breeds in coastal and lowland areas from Panama to central

Argentina. They are very rare wanderers throughout the West Indies, but breeding has been recorded (Raffaele et al. 2003).

Black-bellied Whistling-Ducks have dramatically expanded their breeding range in the United States since the early 1900s. The initial range was largely the lower Rio Grande valley (Bolen and Rylander 1983), with the map in Kortright (1942) fixing the Rio Grande River as the northern edge of this species' range. By the mid-1960s, however, the greatest density of breeding birds had shifted north to Texas, near Corpus Christi, where the presence of Lake Corpus Christi and the construction of stock ponds and irrigation impoundments may have facilitated their northward expansion in the state (Bolen et al. 1964). Still, Bellrose (1980) only showed the U.S. range as "southern coastal Texas." Since then, a population was established in the southwestern corner of Louisiana, largely as a result of releases from refuges (Wiedenfeld and Swan 2000). Breeding is also recorded in adjacent Arkansas, where sightings are increasing (Purrington 1996).

Since 1968, Black-bellied Whistling-Ducks have been frequently encountered in central and southern Florida and appear to be expanding their range in that state. Escapees from zoos established the initial Florida population, but most are now considered to be wild birds from Mexico (James and Thompson 2001) that have established a population near Sarasota, although breeding has been reported in northern Florida as well (Bergstrom 1999). In Arizona, breeding was first documented in 1949 near Tucson, and a population became established in the Santa Cruz Valley; population expansion occurred after 1960. They are currently considered common from March through November in southeastern and central Arizona (Brown 1985). Along the East Coast, Black-bellied Whistling-Ducks expanded their range into South Carolina in 1994 (Harrigal et al. 1995), where breeding was documented during 2003 and 2004 (Harrigal and Cely 2004). On 2 July 2011, a Black-bellied Whistling-Duck with 12 ducklings was

Breeding Distribution ■
Year-round Distribution ■

north as California, Minnesota, New Jersey, and Delaware (American Ornithologists' Union 1998), and in Canada have reached Québec and Nova Scotia (Bergstrom 1999). In Mexico, Black-bellied Whistling-Ducks are a common year-round resident along both coasts at elevations up to 750 m (Howell and Webb 1995).

Winter. Black-bellied Whistling-Ducks winter throughout their range in Central and South America. Midwinter Surveys have been conducted in Mexico by the U.S. Fish and Wildlife Service, in cooperation with appropriate Mexican agencies, within 3 major areas since the 1930s: the east coast, west coast, and interior highlands (Saunders and Saunders 1981). Surveys along the east coast did not separate species of whistling-ducks until 2000, when 64,304 Black-bellied Whistling-Ducks were recorded, mostly in the Tabasco Lagoons (54%) and the Campeche-Yucatán Lagoons (31%). On the next (2006) east coast survey 190,057 were recorded, of which 66% were in the Tabasco Lagoons and 24% were in the Campeche-Yucatán Lagoons. The interior highlands recorded an average of only 144 from 1948 to 2006, and the lower west coast averaged 5,808 (1947–2006); Thorpe et al. (2006) noted 9,061 in 2006. On the west coast of Mexico, surveys averaged only 1,257 Black-bellied Whistling-Ducks from 1997 to 2006 (Conant and Volzer 2000, Conant and King 2006), well below the 1981–94 average of 14,000. Although tallied during winter, most of these birds are breeding residents.

Migration. In the United States, some Black-bellied Whistling-Ducks regularly winter in southern Louisiana, but most of them winter along the Gulf Coast of Texas and in the lower Rio Grande valley, where the peak Christmas Bird Count was 20,054 birds in 2008/9. The Florida birds apparently are resident year round, with the highest Christmas Bird Count (4,169) in 2009/10.

photographed in the Santee delta / Winyah Bay region at the Tom Yawkey Wildlife Center (on a pond at Cat Island), which is the northernmost observation of nesting along the East Coast (Dozier 2012). Sporadic reports of nesting Black-bellied Whistling-Ducks have also come from southwestern Arkansas and probably represent a northward expansion of wild birds from Texas (James and Thompson 2001). The first breeding in Oklahoma occurred in 1999 (Kamp and Loyd 2001). Unpublished reports from Tennessee recorded breeding in 1988 and 2008.

Cain (1973) proposed Dallas as the potential northern breeding limit for Black-bellied Whistling-Ducks in Texas, due to low spring temperatures that restricted energy available for reproduction. More recent work on the distribution of breeding Black-bellied Whistling-Ducks in Texas has documented significant range expansion, although the primary range is still largely south of Dallas (Schneider et al. 1993). Black-bellied Whistling-Ducks wander widely, especially during the postbreeding period. The influence of climate change, however, cannot be discounted as another factor facilitating the northward expansion of their range. They have been recorded as far

At least some Black-bellied Whistling-Ducks breeding in the United States are migratory. Bolen (1967a) reported that 8 out of 9 recoveries from

Black-bellied Whistling-Ducks inspecting a nest box. *GaryKramer.net*

birds banded near Corpus Christi, Texas, occurred in the Mexican state of Tamaulipas, and 1 was recovered in San Luis Potosí; travel distances were 257–684 km. Given the number of wintering birds in Texas, however, many must be year-round residents. Among Black-bellied Whistling-Ducks banded from 1980 to mid-2001 in Texas (n = 1,807), 20 out of 25 out-of-state recoveries (direct and indirect) occurred in Louisiana and 5 were in Mexico. Of birds banded in Louisiana over that time period (n = 554), only 2 out of 24 recoveries occurred outside the state. Black-bellied Whistling-Ducks are year-round residents in Mexico and elsewhere in Central and South America.

In Texas, Black-bellied Whistling-Ducks arrive on breeding areas from early March to April. Over a 10-year period, arrival at the Santa Anna National Wildlife Refuge on the Rio Grande River occurred between 1 March and 12 April (average = 25 Mar; Bolen 1967a). Departure from breeding sites in Texas typically occurs from August to October. Leopold (1959) considered Black-bellied Whistling-Ducks in Mexico to be nonmigratory, but with a tendency to wander.

MIGRATION BEHAVIOR

There are no specific studies of migration behavior; molt migrations, if they occur, have not been documented.

HABITAT

Black-bellied Whistling-Ducks prefer tropical lowlands characterized by shallow lagoons, floating vegetation, and mudflats (Palmer 1976a). In Venezuela, they are reported to occur from sea level to 600 m (de Schauensee and Phelps 1978). Bolen (2005) remarked that they can be found on lakes, reservoirs, and various types of wetlands at low elevations above sea level, but only rarely in deep forest or mountains. They are also common in tropical savannas, such as the Llanos of Venezuela. James and Thompson (2001) reported nesting Black-bellied Whistling Ducks on shallow freshwater ponds and lakes that may be devoid of

vegetation but often contain water hyacinth (*Eichhornia crassipes*), waterlilies (*Castalia* spp.), and cattails (*Typha* spp.). Howell and Webb (1995) reported their presence on wooded marshes, swampy forests, lagoons, mangrove swamps, and flooded fields.

POPULATION STATUS

Black-bellied Whistling-Ducks have been increasing in numbers within the United States. Their current range, however, is outside that of the population surveys conducted by the U.S. Fish and Wildlife Service, except for the Midwinter Survey; hence one needs to rely on other indices to assess population status. In the United States, most Black-bellied Whistling-Ducks occur in Texas, followed by Florida. For example, the Christmas Bird Count data for 2009/10 tallied 20,396 Black-bellied Whistling-Ducks in the United States, of which 73% were in Texas and 20% in Florida; 6% were in Louisiana. In 2007 the Midwinter Survey along the Texas Gulf Coast recorded 52,607 Black-bellied Whistling-Ducks; a high of 81,102 was recorded in 2008 from southern Texas.

The rate of increase in the numbers of Black-bellied Whistling-Ducks is readily seen from the Christmas Bird Count data in Texas, where the average number of birds recorded rose from 133/year in the 1970s, to 1,822/year in the 1980s, 4,568/year in the 1990s, and 12,227 from 2000 to 2009/10. The Breeding Bird Survey estimated that the Black-bellied Whistling-Duck population increased in the United States at an annual rate of 8.2% from 1966 to 2009. In Texas, that rise was 6.8%/year. As for range expansion into Texas, Christmas Bird Counts began in the state in 1903/4, but no Black-bellied Whistling-Ducks were recorded until 1959/60 (143 birds). In Florida, the first appearance of Black-bellied Whistling-Ducks on Christmas Bird Counts occurred in 1970/71 (6 birds) and then averaged only 9/year for the 1970s and 1980s. The average was 224/year in the 1990s but increased sharply to 1,932/year from 2000 to 2009/10. An overall population number has been assessed as 100,000–1,000,000 for the northern subspecies and >1,000,000 for the southern subspecies (Rose and Scott 1997).

In Mexico, the Midwinter Survey recorded an average of 58,000 Black-bellied Whistling-Ducks from 1978 to 1994; 44,000 on the east coast, 14,000 on the west coast, and <100 in the interior highlands. On the east coast, the largest concentrations (of almost 50,000) occurred between Veracruz and the Tamiahua Lagoon.

Harvest

Black-bellied Whistling-Ducks were legally protected from hunting within the United States until 1984. From 2000 to 2007, the average annual harvest was 1,345 in Florida (range = 252–3,740 birds), 1,974 in Louisiana (range = 568–6,225 birds), and 5,745 in Texas (range = 2,492–11,872 birds). The harvest from other states during this time period was 370 in Arizona (in 2000), 450 in South Carolina (in 2004), and 637 in Arkansas (in 2004). The 2010/11 U.S. harvest was 17,550, with virtually all of it from three states: Texas (39.8%), Florida (37.0%), and Louisiana (22.8%). The waterfowl harvest in Mexico is not well documented, but it is very low (<1%) in comparison with the United States, with an estimated harvest of 4,956 whistling-ducks (Black-bellied and Fulvous) generated from 1987–93 surveys (Kramer et al. 1995). Migoya and Baldassarre (1993) reported a harvest of 3,336 Black-bellied Whistling-Ducks at a hunting club in Sinaloa during the 1987/88 hunting season, when the daily bag limit in Mexico was 15 birds (Guy Baldassarre).

BREEDING BIOLOGY

Behavior

Mating System. Black-bellied Whistling-Ducks breed during their first year of life, and the pair bond is generally lifelong. Pairing evidently occurs during winter, because migrants were paired upon arrival in Texas in late March (Bolen 1967a). Bolen (1971) also presented evidence from banded birds

that pair bonds remain intact from year to year. He captured the same 7 pairs in the same nest boxes for 2 years in a row, and in subsequent studies (e.g., Delnicki and Bolen 1976), Bolen documented that 15 pairs remained as mates at least through a second year, and 2 pairs stayed together for 4 years. Black-bellied Whistling-Ducks will form new pairs if a member of a pair dies; there are 2 reports, however, where new pairs formed even though both mates were still alive (Delnicki 1983). The longest documented pair-bond duration is 4 years (Bolen 1971).

Sex Ratio. The sex ratio in Black-bellied Whistling-Ducks is more balanced than in other ducks. At hatching, the sex ratio for 337 ducklings was 51% males, 49% females. The adult sex ratio, determined from 631 birds trapped during the spring in Texas, was 51.8% males and 48.2% females (Bolen 1970).

Site Fidelity and Territory. Males and females exhibit equal fidelity to their breeding areas, which is expected in a species with lifelong pair bonds. This conclusion initially stems from a sample of 31 banded adults that returned a second year to the area of banding, of which 42% were males and 58% females (Bolen 1971). More recently, 22 out of 34 (65%) marked adult males and 19 out of 33 (58%) marked adult females returned to their previous breeding area in southern Texas (James and Thompson 2001). Black-bellied Whistling-Ducks are not territorial per se, but they will defend the nest cavity during egg laying and incubation (Chronister 1985).

Courtship Displays. Displays prior to pair formation are mutual and include *neck-stretching*, *head-dipping*, and *diving* (Bolen et al. 1964). The usual threat display is a *head-low-and-forward* posture (Johnsgard 1965). Copulation occurs in shallow water or on the shore and is preceded by precopulatory movements closely resembling normal drinking behavior. Postcopulatory displays are mutual and consist of the pair standing side-by-side, mutually calling. Meanley and Meanley (1958) described the postcopulatory display, as did Bolen et al. (1964:80–81): "The birds, standing side-by-side in shallow water, rapidly tread and splash while puffing out their breasts and sharply curving their necks in a deep S."

Nesting

Nest Sites. Both the male and the female are involved in nest site selection, which is typically a tree cavity but can occur on the ground; nest boxes are also readily used. Black-bellied Whistling-Ducks have several adaptations to an arboreal existence, notably scales on their foot webs and a comparatively small foot (Rylander and Bolen 1970). Such adaptations, and their resultant dexterity, help explain observations of Black-bellied Whistling-Ducks perched on strands of wire fence, loops of Spanish moss (*Tillandsia usneoides*), and even on a telephone line.

In southern Texas, 45% of 83 occupied cavity nests occurred in live oaks (*Quercus virginiana*), 19% in ebony (Ebenaceae), 16% in willows (*Salix* spp.), 13% in elms (*Ulmus* spp.), 6% in mesquite (*Prosopis glandulosa*), and 1% in hackberries (*Celtis* sp.; Delnicki and Bolen 1975). Cavity entrances used by Black-bellied Whistling-Ducks were large, averaging 17 × 31 cm; the smallest was 10 × 12 cm. Cavity depth averaged 58 cm, and the cavity floor was large, averaging 25.9 × 25.6 cm (663 cm²); total cavity space averaged 749 cm³. Trees containing these big cavities were obviously large themselves, averaging 64 cm in diameter at breast height and 52 cm at the cavity. The use of cavities was not influenced by distance to water, as >30% of all nests were 500 m or more from the nearest water body (Delnicki and Bolen 1975). Eggs were deposited directly on the bottoms of the cavities, often without benefit of nest material and always without down feathers.

Black-bellied Whistling-Ducks also lay eggs

Black-bellied Whistling-Duck pair in copulation posture; note the penis of the male. *Ursula Dubrick*

in nest boxes. The most extensive report of nest-box use by Black-bellied Whistling-Ducks was a 12-year study (1964–75) in Texas, where box occupancy over the period was 81% (McCamant and Bolen 1979). The authors evaluated 778 box nests, of which 279 were incubated; 75% of those hatched.

Black-bellied Whistling-Ducks will readily nest on the ground, as recorded in Texas (Bolen 1967a) and Mexico (Markum and Baldassarre 1989a, Feekes et al. 1992). The largest evaluation of ground nesting comes from the Laguna la Nacha in northeastern Tamaulipas, Mexico, where 496 ground nests were reported from 3 small islands (Markum and Baldassarre 1989a). Natural cavities did not exist on the islands, but there were 168 nest boxes specifically designed to attract Muscovy Ducks that were also used by Black-bellied Whistling-Ducks. Most ground nests were located in thickets of pricklypear cactus (*Opuntia* spp.), where density reached a high of 42.6 nests/ha. Overall nest density was 15.7/ha, and nest success was 28.3%–41.7%. All nests were not followed to completion, but 41.8% of the monitored nests were parasitized and 58.4% were lost to desertion. Only 59 nests were located in the island's nest boxes (35% occupancy); hence Black-bellied Whistling-Ducks appear unique among waterfowl in nesting frequently both on the ground and in nest cavities. Extensive ground nesting was perhaps unnoticed in other studies because research objectives were focused on nest-box use.

In Tamaulipas, ground nests in pricklypear cactus consisted of shallow unlined depressions in the soil, whereas nests in bunchgrass sacaton (*Sporobolus* spp.) were constructed from grass stems and leaves located at the base of the plant. Ground nests in Texas were composed of dead grasses woven into shallow bowls about 19 cm in diameter (Bolen et al. 1964, 1967a). Most nests were in grazed brush pastures and were usually well hidden under low shrubs. A few nests occurred in such odd sites as a chimney, a cotton-gin exhaust pipe, and a pigeon loft.

Clutch Size and Eggs. Initial studies of Black-bellied Whistling-Ducks in southern Texas reported that the size of first clutches by single females ranged from 9 to 18 eggs and averaged 13.4 ($n = 58$; Bolen 1967a). Early studies may have underestimated the incidence of brood parasitism, as subsequent work in southern Texas reported an average clutch size of 6–7 ($n = 134$ nests; James 2000).

Black-bellied Whistling Duck eggs are ovate, white, and average 5.2 × 3.8 cm ($n = 2,982$; James and Thompson 2001). Eggs were laid at the rate of 1/day, usually between 17:00 and 19:00 (Bolen 1967a).

Incubation and Energetic Costs. Both sexes incubate the eggs, which occurs in all 8 species of whistling-ducks and is virtually unique among all waterfowl, except for the closely related White-backed Duck (*Thalassornis leuconotus*) of Africa and the Black Swan (*Cygnus atratus*) of Australia and New Zealand (Johnsgard 1978), and occasionally among the white swans (Kear 1972). Black-bellied Whistling-Ducks (and most likely all other whistling-ducks) do not have a defeathered incubation patch, but the lower abdominal region in both sexes becomes highly vascularized during the incubation period (Rylander et al. 1980). No down is added to the nest, however, which is a unique feature among waterfowl and probably reflects the continuous incubation provided by both sexes. Incubation in the wild was 25–30 days in Texas, with an average of 27.5 days, but removal of either of the incubating birds resulted in abandonment of the nest (Bolen 1967a). Eggs in incubators hatched in 31–32 days. Radiotelemetry monitoring of an incubating pair of Black-bellied Whistling-Ducks in Texas found that nest attentiveness over 8 days of observation, beginning at week 2 of incubation, averaged 9.9 hr/day for the male and 10.1 hr/day for the female; the nest was unattended for 3.9 hr/day (Bolen and Smith 1979). Males suffered a small but significant loss of body mass during incubation, whereas female body mass did not decline similarly (Chronister 1985).

At 5 camera-monitored nests in southern Texas, males and females typically alternated incubation duties that averaged a total of 24.9 hours, with half of all nest exchanges occurring within 2 hours of sunset (Chronister 1985). Daytime recesses averaged 1.0/day for males and 0.14/day for females. Recess duration averaged 1.2 hours and did not differ between the sexes, although recesses occasionally lasted up to 3 hours. On average, nests were unattended for 0.8 hr/day.

Nesting Chronology. Black-bellied Whistling-Ducks commenced nesting 33–80 days (average = 56 days) after arriving on sites in lower coastal Texas (Bolen 1967a). First nests were started between 25 April and 26 May (average = 5 May); the last nests were initiated in late August. Thus Black-bellied Whistling-Ducks initiated nesting over a 5-month period, but such an extended breeding season is not unusual for tropical species of waterfowl. Attempts to nest were most frequent between mid-May and mid-June, but drought conditions can result in significant delays to nest initiation (James and Thompson 2001).

Nest Success. Data are plentiful on the nest success of Black-bellied Whistling-Ducks using nest boxes. In contrast, there are scant data on those choosing natural cavities, but this information nonetheless reveals much more about the nesting biology of this species, because only a miniscule percentage of the population can use the few nest boxes available. In Texas, success of 32 nests located in natural cavities was 44%; 41% were lost to predation, 9% were abandoned, and 6% were lost to other causes (Bolen 1967b).

In contrast to natural nests, nest boxes in Texas that were equipped with predator guards exhibited a 77% success rate, and none were lost to predation. The nest-success rate in unprotected nest boxes was 46%, virtually identical to the rate in natural cavities (Bolen 1967a). In the larger, 12-year study (1964–75) of nest-box use, the success rate was 75% for incubated nests and 28% of all nests discovered

(McCamant and Bolen 1979). In a 1998–99 study in southern Texas, James (2000) reported that nest success of clutches in nest boxes protected by predator shields ranged annually from 41% to 56%. All unsuccessful nests were abandoned, due to excessive brood parasitism rather than predation.

The principal predators of Black-bellied Whistling-Duck nests were black rat snakes (*Elaphe obsoleta obsoleta*) and raccoons (*Procyon lotor*; Bolen 1967b).

Brood Parasitism. Natural cavities are usually in short supply relative to the number of females seeking to use them, and they can be difficult to locate; hence females have evolved a strategy of brood parasitism. Females will lay eggs in the nest of another female if they cannot find a cavity of their own, yet they often do so when they are also laying eggs in their own nest. Nest boxes exacerbate this behavior, because they usually are installed in significant numbers and often are placed in the open; females thus have a plethora of easily available opportunities to engage in brood parasitism, and they do so extensively. Extensive rates of parasitism interfere with normal incubation practices, however, and limit the hatchability of both nests and eggs.

During a 12-year study (1964–75) of nest-box use by Black-bellied Whistling-Ducks in Texas, at least 70% of 778 nests were parasitized, with the parasitized nests containing >15 eggs: 1 nest held 101 eggs, and 45% of the 778 nests contained more than 31 eggs (McCamant and Bolen 1979). In addition, only 40% of the 778 nests were incubated, and only 75% of the incubated nests were successful (hatched at least one egg). Furthermore, of the incubated nests, only 48% of the eggs hatched. Among the successful nests (210), only 63% of the eggs hatched; overall, only 20% of the 21,982 eggs laid during the 12 years hatched. During a 1998–99 nest-box study in southern Texas, 57%–58% of all nests were lost to abandonment or poor incubation efficiency, again due to brood parasitism; only 42% of 2,983 eggs hatched (James 2000).

The reported high rates of brood parasitism in Black-bellied Whistling-Ducks may be underestimated when clutch size alone is used as the criterion to identify parasitized nests. When clutches with 14 or more eggs were considered to be parasitized nests, the percentage of such nests in southern Texas was 70% ($n = 778$ nests; McCamant and Bolen 1979). Brood parasitism may be even more prevalent than reported in some populations, however, because the use of "normal clutch size" as a criterion can dramatically underestimate brood parasitism of cavity-nesting waterfowl, as Semel and Sherman (1992) documented for Wood Ducks that used nest boxes in Illinois. James (2000) used egg measurements (length and width) to identify individual Black-bellied Whistling-Duck females depositing eggs in a given nest and found the parasitism rate to be 100% (134 nests), with an average of 6–8 females contributing eggs to each nest. The resultant clutch size of host females was only 6–7. Brood parasitism also extends to nests on the ground. On islands in Tamaulipas, 41.8% of 484 nests were parasitized and only 28.3%–41.7% of all nests hatched successfully (Markum and Baldassarre 1989a). All observed nest loss was due to desertion.

In southern Texas, camera monitors revealed that 18 out of 20 Black-bellied Whistling-Duck nests contained eggs laid by more than 1 female, with an average of 2.3 females depositing eggs in each parasitized nest; the average clutch size of incubated nests was 29.2 eggs (Chronister 1985). Marked pairs visited an average of 3.8 nest sites during the egg-laying period; only 10% of the visits were by single birds, usually the egg-laying female. Females were in next boxes an average of 12.3 minutes when eggs could potentially be laid, and in 4 instances deposited eggs in only 3 minutes.

Interspecific brood parasitism between Black-bellied Whistling-Ducks and other species is uncommon but has been reported. In Texas, 2 instances of mixed clutches between Wood Ducks and Black-bellied Whistling-Ducks in nest boxes have been documented, both of which were incubated by the Wood Duck (Bolen and Cain 1968, Labuda 1969). In Tamaulipas, mixed clutches in nest boxes have been reported with Muscovy Ducks; at least 1 such clutch was successfully incubated by a Muscovy Duck (Woodyard and Bolen 1984, Markum and Baldassarre 1989a). Perhaps the most unusual case of interspecific parasitism is that with a Laughing Gull (*Larus atricilla*) on the periphery of a 56-nest colony in the upper Laguna Madre of Texas (Ballard 2001).

Renesting. Black-bellied Whistling-Ducks will renest when their first nests are destroyed. In Texas, at least 19% of 57 pairs that lost initial clutches renested; 2 pairs renested a third time when the clutch from their first renest was removed (Delnicki and Bolen 1976). There were also 3 instances where renesting occurred following successful hatching of the first clutch, with 1 pair renesting 6 weeks after their first nest hatched, and another pair waiting 8 weeks. These renests probably did not represent second broods, because ducklings remain with their parents for at least 6 months after hatching. James et al. (2012), however, documented the first incidence of double brooding by a pair of Black-bellied Whistling-Ducks marked on the Welder Wildlife Refuge in southern Texas. They noted that this behavior is not a common reproductive strategy in this species, which is somewhat surprising, given the long breeding season that characterizes their subtropical and tropical habitats.

REARING OF YOUNG

Brood Habitat and Care. As with many species of ducks, brood-rearing sites consist of wetlands with dense emergent vegetation and open-water areas. In a study comparing habitat preference by Black-bellied Whistling-Duck broods on high-versus low-use ponds in southern Texas, high-use ponds were smaller (22.5 ha vs. 54.2 ha), shallower (54.7 cm vs. 64.2 cm), and had higher habitat interspersion (Heins 1984).

Ducklings remain in their nest cavity 18–24 hours after hatching and then jump from the cavity entrance to the ground or water (Bolen 1967a). Both parents escort the brood and remain with the young for at least 6 months, 4 months after flight is reached (Delnicki and Bolen 1976).

Brood Amalgamation (Crèches). There has only been a single documented case of brood amalgamation in Black-bellied Whistling-Ducks (Bergman 1994). The observation occurred in southern Texas, where a brood of 20 captive-reared ducklings near flight stage was released on a nearby 4 ha pond with 4 naturally occurring broods of 3, 4, 5, and 7 ducklings, each with 2 adults. The 4 natural broods and the released brood then amalgamated with 2 of the wild adults (assumed to be male and female).

Development. Cain (1968, 1970) conducted a detailed study of the growth and development of Black-bellied Whistling-Ducks. The body mass of ducklings averaged 30.8 g at hatching, but the ducklings' weight increased rapidly from days 5 to 30. Flight was attained at 53–63 days, although flight was possible for pen-reared birds when the first primary reached 6.9 cm (56 days for larger ducklings, 63 days for smaller ones); final primary length (12.5 cm) was not achieved until the ducklings were about 100 days old. Faster-growing ducklings flew several days in advance of slower-growing birds.

There were 2 pauses in duckling growth: at 31–35 days, when remiges first emerged; and 56–63 days, when flight was initiated. The emergence of contour feathers during week 4 necessitated large amounts of energy; a pause in growth probably occurred then as a result of competing nutritional requirements for contour feather development.

RECRUITMENT AND SURVIVAL

In southern Texas, Bolen (1967a) reported that broods of Black-bellied Whistling-Ducks tended by both parents averaged 10.7 ducklings ($n = 15$) the first week, 8.8 from weeks 2 through 4, 10.3 ($n = 21$) from weeks 5 through 7, and 9.7 ($n = 45$) for flying broods. Nine broods accompanied by only 1 parent averaged only 7.7 ducklings. Also in southern Texas, Heins-Loy (1986a) reported an average brood size of 9.5 ($n = 10$) that ranged from 10.5 for Class I ducklings to 8.7 in Class III.

The age ratio in the fall population of Black-bellied Whistling-Ducks in 1963 was 1 adult to 3.9 young (Bolen 1967a), a much higher ratio than those found for most species of ducks. A later study in southern Texas observed flocks of Black-bellied Whistling-Ducks totaling 1,665 birds and reported an age ratio of 3.1 young/adult (Heins-Loy 1986b). Harvest data in the United States from 1998 through 2007 ($n = 76{,}131$) yielded an age ratio in the harvest of 1 adult to 1.53 young.

Bolen and McCamant (1977) used band recoveries from hunters and recaptures of marked birds in nest boxes to estimate annual mortality rates. Both methods yielded similar results: an annual mortality rate of 48%–54%. Sample sizes were very low, however ($n = 48$), which could bias these estimates (James and Thompson 2001). The longevity record for a wild Black-bellied Whistling-Duck is 8 years and 2 months.

FOOD HABITS AND FEEDING ECOLOGY

Adults are primarily herbivorous and often take advantage of available grain at planting time, as well as grain left in the fields after harvest (Johnsgard 1975). They forage both during the day and at night. Kramer and Euliss (1986) reported results from 29 birds taken during the winter in the state of Sinaloa in Mexico, where 75% of the foods eaten were the cereal grains rice and corn; plant foods formed 97% of their diet. Bolen and Forsyth (1967) reported that plant material composed 92% of the diet of Black-bellied Whistling-Ducks collected in southern Texas, with a preponderance of sorghum and the seeds of bermudagrass (*Cynodon* spp.).

Black-bellied Whistling-Ducks are especially fond of rice, and they can cause damage when consuming seeds before the rice germinates. In Guyana, Black-bellied Whistling-Ducks ate 54 g of pregerminated rice/day, which was potentially 2% of the planted crop (Bourne and Osborne 1978). Plant food formed 97% of their diet, but most (86%) of that was pregerminated rice (Bourne 1981).

Pregerminated rice is not available for long, but harvested rice is readily utilized thereafter. In a rice-growing region of Venezuela, nonbreeding Black-bellied Whistling-Ducks consumed 95% rice by volume, along with distinctly lesser volumes of 13 identified plants (Bruzual and Bruzual 1983). Although they prefer rice, other plant foods, and some invertebrates, are also readily available in rice fields and will be used by Black-bellied Whistling-Ducks. In Guyana rice fields, plants other than rice made up 11% of the diet, and aquatic insects and snails totaled 3%; young apple snails (*Pomacea* sp.) were the most important food for juveniles (Bourne 1981). Bourne (1981) noted that female Black-bellied Whistling-Ducks increased their consumption of animal foods during the breeding season, but the total intake remained <10% of their diet.

As with all ducks, young ducklings prefer animal matter, but that diet shifts toward plant material as ducklings age. In Texas, Bolen and Beecham (1970) reported that 21-day-old ducklings fed primarily on the seeds of wild millet (*Echinochloa muricata*) and yerba-de-tajo (*Eclipta prostrata*). Older (35 days) ducklings consumed arrowhead (*Sagittaria* sp.) tubers and water stargrass (*Heteranthera liebmannii*) seeds. Animal food items included insects, spiders, pond snails (Gastropoda), and fingernail clams (Sphaeriidae). The older young also ate oligochaete worms (Oligochaeta) and freshwater shrimp (Palaemonidae).

Black-bellied Whistling-Ducks are primarily terrestrial grazers as opposed to aquatic sievers, and they have several adaptations to do so: a more attenuated maxilla, a less pronounced bill nail, and more highly developed bone and integument at the distal end of the maxilla (Rylander and Bolen 1974b). The Black-bellied Whistling-Duck is also the most nocturnal of the 8 species of whistling-ducks, and again has adaptations to do so: they have a duplex retina with a preponderance of rods, which indicates visual acuity at very low light intensities (Hersloff et al. 1974). They also have relatively large optic and vestibulocochlear lobes, which may provide greater optical and auditory capabilities that are useful during nocturnal feeding (Rylander and Bolen 1974b).

MOLTS AND PLUMAGES

As adults, Black-bellied Whistling-Ducks, like swans and geese, have only 1 plumage (basic) that is molted each year via the prebasic molt. Hence there is no alternate (nonbreeding) plumage, such as is observed in puddle ducks. The descriptions of molts and plumages are primarily from summaries and data in Palmer (1976a) and James and Thompson (2001), which should be consulted for further details.

Hatchlings are down covered and uniquely patterned, unlike the single-color down of hatchling swans and geese. Juvenal feathers first appear in 10–13 days and develop rapidly from week 3 to week 9. Ducklings are completely feathered in Class III but still flightless at 7–8 weeks; the plumage is completed in about 10–13 weeks (Cain 1968, 1970). Remiges begin to emerge at about day 30. Juvenal plumage is briefly described under Identification, with juveniles characterized by an overall dull brown body and gray head.

In pen-reared birds, the first prebasic molt begins on the head and tail during weeks 13–14 and is completed by weeks 34–35 with first basic (Cain 1970). The juvenal remiges are retained. Birds appear spotted during the peak of the first prebasic molt, but final first basic is similar to definitive basic and worn for about 1 year. First basic is then molted prior to postbreeding movements or fall migration to yield definitive basic, which will be the undistinguishable plumage worn for the remainder of a bird's life. Individuals molt all their juvenal remiges at this time (at about 19 months of age), which renders birds flightless for about 20 days (Cain 1968).

CONSERVATION AND MANAGEMENT

In the United States, management associated with Black-bellied Whistling-Ducks has focused on population monitoring via the various avian surveys conducted within the birds' range, and on the erection of nest boxes to enhance nesting opportunities. In Texas, Bolen (1967b) constructed boxes from marine plywood that were 28 cm² on the bottom, with a 56 cm front and 51 cm back; a 12.7 cm circular entrance was near the top. Several centimeters of sawdust were placed on the bottom (Bolen 1967b). The boxes were mounted on a 5 cm metal well pipe equipped with a 91 cm conical predator guard. James (2000) evaluated the use of nest-box types and observed greater occupancy in large (floor area = 778 cm²) versus small (floor area = 643 cm²) boxes (58%–86% vs. 32%–46%). Cavity depth in the boxes was also different (55.9 cm vs. 49.5 cm) and may have been a significant factor affecting use, because the depth in large boxes approached that of natural cavities used by Black-bellied Whistling-Ducks (Delnicki and Bolen 1975).

Another issue affecting nest-box occupancy is that of placement. Boxes situated in the open and in close proximity to each other are so readily visible as to promote extremely high rates of brood parasitism (reaching 100% in some instances). Such parasitism interferes with incubation patterns and the hatchability of eggs, and it causes nest abandonment.

The relationship between nest-box placement and the incidence of brood parasitism has been well studied in Wood Ducks, and those results offer guidance for Black-bellied Whistling-Ducks as well as other cavity-nesting waterfowl. Specifically, the rate of brood parasitism for Wood Ducks using nest boxes in Illinois was 49.5% when boxes—whether singly or in groups—were all placed in the open, where they were readily visible to female Wood Ducks (Semel et al. 1988). In contrast, parasitism was only 29.8% where boxes were placed singly in visually occluded habitat.

The recommendation of this study was that nest boxes should be placed in habitats and at densities that resemble the natural conditions under which Wood Ducks evolved. James (2000) found, however, that isolated nest boxes used by Black-bellied Whistling-Ducks still were highly parasitized, perhaps because there is a naturally high rate of competition for the limited number of cavities. Delnicki (1973) noted that brood parasitism contributed 94% more Black-bellied Whistling-Duck ducklings to the population he studied in southern Texas versus an equal number of unparasitized nests, but he ultimately concluded that parasitism was detrimental, due to the decreased hatchability of eggs. Further, it is unknown if there are differences in brood survival between parasitized and unparasitized nests. (See the Wood Duck account for more information on the topic of brood parasitism in cavity-nesting waterfowl.)

A future management issue involves the potential competition for nest sites between Black-bellied Whistling-Ducks and Wood Ducks. Wood Ducks have recently expanded their breeding range southward along the Gulf Coast, and Black-bellied Whistling-Ducks have expanded their breeding range northward, thus overlapping more significantly with the breeding range of Wood Ducks (Schneider et al. 1993, James and Thompson 2001).

In Mexico, nesting habitat is being lost through large-scale deforestation (Feekes 1991), and the reduction of forested areas to allow for cultivation in the lower Rio Grande valley has long had a negative impact on local populations of Black-bellied Whistling-Ducks (Palmer 1976a). Due to a scarcity of lumber, attempts to start a nest-box program have instead used readily available natural materials, including hollow logs of palm trees and coco trees (*Cocos nucifera*; Feekes et al. 1992). Initial results indicated that Black-bellied Whistling-Ducks readily accepted these nest boxes, but nest-success rates were only 7%–11%. Recommendations to increase success included providing predator shields and placing nest boxes at sites where the harvest of eggs by local peoples would be reduced. Organo-chlorine pesticides were documented in resident Black-bellied Whistling-Ducks collected in the state of Sinaloa in Mexico, in 1981–82, but the levels were not high enough to suggest adverse effects on survival or reproduction (Mora et al. 1987).

Additional research should further assess the biology of brood parasitism in association with nest boxes, similar to what has been accomplished for Wood Ducks. There is reason to expect different behaviors and overall effects on reproduction in a tropical species, such as the Black-bellied Whistling-Duck, versus the largely temperate-nesting Wood Duck. The extent and contribution of ground nesting to population dynamics has also not been assessed in this species, but it may prove to be significant. Radiomarking pre-breeding females in areas where no nest boxes are available would yield fruitful data on the extent of ground nesting. Information on brood survival is also sparse, as are studies during the nonbreeding season, particularly winter. In Mexico, research on the interactions between Black-bellied Whistling-Ducks and Muscovy Ducks would be of interest, given the interspecific brood parasitism between these species (Woodyard and Bolen 1984, Markum and Baldassarre 1989a). Delacour (1959) reported wild Muscovy Duck males were dangerous to other birds, attempting to mate with several waterfowl species and even killing weaker birds.

Fulvous Whistling-Duck

Dendrocygna bicolor (Vieillot 1816)

The Fulvous Whistling-Duck is a medium-sized, tropical to subtropical species that occurs on 4 continents but is at the northern edge of its range in the United States, where it is principally found in Texas, Louisiana, and Florida. These birds are migratory in the northern portions of their range but year-round residents elsewhere. Both sexes are virtually identical in size and plumage. They most likely pair for life, and both sexes share incubation duties, which, among waterfowl, is a near-unique feature of whistling-ducks. They undergo only 1 molt/year. This species, formerly called the Fulvous Tree Duck, nests on the ground and rarely, if ever, perches in trees. They are, however, gregarious and quite vociferous, with a characteristic high-pitched, whistled *pee-chee*, accented on the second syllable. More than any other species of waterfowl, Fulvous Whistling-Ducks are linked with human agricultural activity, specifically rice fields, where they readily nest and feed. They are primarily vegetarians. They have expanded their range in the United States, especially in Florida, but former populations in California are virtually non-existent. They are an excellent pioneering species, with postbreeding individuals and flocks commonly wandering well north of breeding sites, with sight records in virtually every state and province. Fulvous Whistling-Ducks are not an important game species in the United States because of their relatively small numbers, restricted distribution, and early fall departure from Louisiana and Texas. In Mexico and South America they are important to local peoples, who harvest both adults and eggs.

IDENTIFICATION

At a Glance. This species is unmistakable, and the sexes are similar. A long-legged duck with an erect goose-like posture, Fulvous Whistling-Ducks have a rich tawny color overall, with a blackish-brown back scalloped with the same tawny color. A line of white feathers extends along the flank and contrasts with the overall tawny color. Birds in flight show a very conspicuous white rump, with the legs extending beyond the tail. Flight is slow, lumbering, and typically at low altitude.

Adults. A medium-sized duck, the males and the females are virtually identical in appearance and size, except that the black stripe along the back of the neck is continuous in females but usually inter-rupted at the rear of the head in males (McCartney 1963). When both sexes are together, the marginally larger size of the males is apparent, and the males are also somewhat more vividly colored (Palmer 1976a). The feathering along the neck is furrowed, as in some species of geese. The head, the chest, the breast, and the belly are tawny brown; the back is blackish brown. White-edged side and flank feathers form a striking border between the sides and the back. The white rump band is conspicuous in flight, at takeoff, and at landing. In flight, the long blue-gray legs extend beyond the tail. The bill color is a dark slate blue overall.

Fulvous Whistling-Ducks have an erect goose-like posture, but stand less erect than Black-bellied Whistling-Ducks (Rylander and Bolen 1974a). Fulvous Whistling-Ducks also walk without the waddle so characteristic of other ducks. Unlike Black-bellied Whistling-Ducks, Fulvous Whistling-Ducks are not arboreal; their feet are comparatively larger, which is a poor adaptation for perching. Rylander and Bolen (1970:87) noted, "We have never seen *bicolor* [Fulvous Whistling-Ducks] in trees or even perched on the ground."

Juveniles. In the field, juvenal plumage appears identical to that of adults (Clark 1976). In the hand, however, juveniles are notably duller and grayer. The uppertail coverts are gray rather than white. The white flank stripe on juveniles is more subtle, and the belly color is more uniformly buff. The colors of the legs, the feet, the bill, and the irises are identical to those of adults.

Ducklings. Most have a medium-gray and white plumage, but some individuals are silver gray and white, and still others are light buffy yellow (Lynch 1943, Nelson 1993). They usually lack conspicuous dorsal spots and wing patches. The dark coloration on the head and the nape extends to the base of the neck to form a very distinctive T stripe that is characteristic of all *Dendrocygna*. The cheeks and the throat are a light yellow that extends below the crown and around to the back of head. The feet are grayish.

Right:
Fulvous
Whistling-Duck
pair.
GaryKramer.net

Below:
Fulvous
Whistling-Ducks
in flight.
Larry Wan

Voice. Fulvous Whistling-Ducks call incessantly in flight, a 2-note squealing whistle, *pit-tu, kit-tee,* or *pee-chee,* accented on the second syllable (Palmer 1976a). The male utters a louder and lower-pitched whistle, in contrast to the soft, higher-pitched call of the female.

Similar Species. None.

Weights and Measurements. Males are somewhat larger than females, but body measurements overlap too extensively to be conclusive. In the United States, weights for Fulvous Whistling-Ducks from southwestern Louisiana averaged 771 g (range = 545−958 g) for males (*n* = 138) and 743 g (range = 595−964 g) for females (*n* = 148; Hohman and Lee 2001). Wing chord length averaged 21.2 cm for males and 20.7 cm for females; tarsus, 62.2 mm for males and 60.3 mm for females; culmen, 46.6 mm for males and 45.5 mm for females. Seasonal fluctuations in body mass were estimated to be 16% for adult females and 17% for adult males. In Cuba, body mass averaged 714 g for adult males (*n* ≥ 280) and 720 g for adult females (*n* ≥ 229; Acosta Cruz et al. 1989).

DISTRIBUTION

Fulvous Whistling-Ducks have a unique distribution that encompasses 4 continents. In Asia they occur in southern and eastern India, Bangladesh, and Myanmar (Burma). In Africa they occupy a broad band extending across the central part of the continent south of the Sahara Desert and then reaching down through eastern Africa, including Madagascar. They are resident in South America from northern Peru eastward across Columbia, Venezuela, Guyana, Suriname, and French Guiana to Brazil (largely outside the Amazon basin), and then south to northern Argentina. In North America they occur in the West Indies, where they are common in Cuba; along both coasts of Mexico; and then sporadically south to Costa Rica. In the United States they are found in the Gulf Coast states and parts of California (Hohman and Lee 2001).

This broad and disjunct distribution has not produced any recognized subspecies, or even a significant variation in plumage or size. Such variations almost invariably develop with isolation, yet Fulvous Whistling-Ducks somehow have remained virtually the same in appearance over their vast range.

Breeding. In the United States, Fulvous Whistling-Ducks breed along the Gulf Coast from Brownsville, Texas, to central Louisiana, largely west of the Atchafalaya River. There is also a well-established population in Florida. Earlier breeding colonies in California are all but nonexistent, due to tremendous habitat alteration, although a few Fulvous Whistling-Ducks still occur locally in the Salton Sea area in far southern California. There were reports of 2 broods with an adult in the southern San Joaquin Valley in 1983 (Gerstenberg and Rey 2004), but Fulvous Whistling-Ducks have not otherwise been recorded outside the San Joaquin Valley since 1976. As many as 5 pairs potentially nested in the Imperial Valley during the 1990s, but the last confirmed breeding record was a female with 10 young observed at Finney Lake on 27 June 1999 (Patten et al. 2003). Fulvous Whistling-Ducks in California are now considered to be close to extirpated as a breeding bird by Hamilton (2008), who detailed the historical and current status of this species in that state, where it is listed as a Species of Special Concern (breeding).

The largest breeding numbers of Fulvous Whistling-Ducks in the United States occur in the Rice Belt of eastern Texas and southwestern Louisiana. In Texas the rice lands extend 80−128 km inland from the coastal marshes; in Louisiana they lie in a 24−112 km zone north of the coastal marshes. Bolen and Rylander (1983) noted the importance of rice agriculture to the dispersal of Fulvous Whistling-Ducks and the subsequent expansion of their breeding range. In Louisiana, early sightings in the state occurred shortly after the Civil War, as the original coastal prairies were initially converted to rice agriculture; the first breed-

Breeding Distribution ■
Year-round Distribution ■

In the West Indies, they were vagrants until breeding was recorded in the Greater Antilles in 1943 (Bond 1993), followed by a breeding expansion in Cuba during the 1960s that coincided with a significant increase in rice agriculture (Peris et al. 1998). They were reportedly most abundant in Cuba during the 1970s, with observations of flocks of "many thousands," but control and hunting decreased their numbers (Acosta and Mugica 2006a). Elsewhere in the West Indies they are locally common in Hispanola and uncommon in Puerto Rico, but rare and nonbreeding on other islands (Raffaele et al. 2003).

Winter. Fulvous Whistling-Ducks from Louisiana and Texas primarily winter in Mexico, although some winter in southern Texas. Those from Florida mainly appear to winter in Cuba and on other Caribbean islands, while others are year-round residents in the state.

In Mexico, the U.S. Fish and Wildlife Service, in cooperation with appropriate Mexican agencies, has conducted Midwinter Waterfowl Surveys since the 1930s within 3 major areas: the east coast, the west coast, and the interior highlands (Saunders and Saunders 1981). Recent surveys (since 1982) have been conducted at 3-year intervals. Observations along the east coast did not separate species of whistling-ducks until the 2000 survey, during which 17,691 Fulvous Whistling-Ducks were recorded, nearly all (98%) in the Tabasco Lagoons survey area. On the 2006 east coast survey, 9,124 were counted, of which 91% were again in the Tabasco Lagoons. Although tallied during winter, most of these birds are breeding residents.

Very few Fulvous Whistling-Ducks occur in the interior highlands or along the lower west coast of Mexico (Guerrero, Oaxaca, Chiapas). An average of only 597 birds occurred in the interior highlands (1948–2006), and merely 995 (1947–2006) were seen along the lower west coast; the 2006 total for the interior highlands was 898, and the 2003 total, 1,964. On the west coast of Mexico, the number of Fulvous Whistling-Ducks averaged only 210 from

ing record was not established, however, until 1939 (Lynch 1943). By the 1950s they were reported as a locally common breeding duck throughout the rice-growing region in the southwestern part of the state (Meanley and Meanley 1959).

In Florida, Fulvous Whistling-Ducks first appeared in the mid-1950s (Jones 1966), with the first breeding record noted near Lake Okeechobee in 1965 (Ogden and Stevenson 1965). By the mid-1980s, a breeding population was well established south of Lake Okeechobee in the Everglades Agricultural Area, where they were closely associated with rice agriculture (Turnbull et al. 1989a). They have now expanded to wetlands associated with the St. Johns and Kissimmee Rivers and other parts of southern and eastcentral Florida (Stevenson and Anderson 1994, Hohman and Lee 2001).

In Mexico, Fulvous Whistling-Ducks are a common year-round resident along both coasts, at elevations up to 2,300 m. Along the Pacific coast their range extends from southern Sonora to Chiapas (just south of the Gulf of Tehuantepec), and along the Gulf Coast from the Texas border south to Campeche (Howell and Webb 1995). In Central America they occur in disjunct populations from Guatemala to Costa Rica (Howell and Webb 1995).

1997 to 2006 (Conant and Volzer 2000, Conant and King 2006), well below the 1981–84 average of 9,500. Most Fulvous Whistling-Ducks on the west coast surveys were tallied near Los Mochis, Sinaloa.

Island-wide ground surveys in Cuba estimated 10,000–25,000 Fulvous Whistling-Ducks in 2006 (Acosta and Mugica 2006b), but some are probably resident birds. They also occur as residents, and probably wintering birds, on virtually all the other Caribbean islands (Raffaele et al. 2003).

Migration. In Louisiana, 16 birds fitted with radio-transmitters all moved in a southwesterly direction and left the state during winter (Hohman and Richard 1994). Among Fulvous Whistling-Ducks banded from 1980 to 2009 in Texas ($n = 1,546$), Florida ($n = 1,281$), and Louisiana ($n = 198$), nearly all out-of-state recoveries (direct and indirect) of Texas birds occurred in Louisiana (22 out of 25), and all out-of-state recoveries ($n = 39$) of Florida birds occurred in Cuba. There were only 4 recoveries of Fulvous Whistling-Ducks banded in Louisiana: 2 in Louisiana, 1 in Texas, and 1 in Wisconsin. They also are known to migrate between Texas and the Yucatán Peninsula of Mexico (Flickinger et al. 1973).

Fulvous Whistling-Ducks are well known for their northward postbreeding movements, which began in the mid-1950s (Bolen and Rylander 1983). Prior to 1949, the only Atlantic coastal records of this duck were several from Florida and a single bird from North Carolina (McCartney 1963). Starting in 1955 and extending for over a decade, a succession of invasions occurred along the Atlantic coast, with Fulvous Whistling-Ducks recorded from the tip of Florida to New Brunswick and Nova Scotia, as well as a few inland reports (McCartney 1963, Jones 1966, Munro 1967, Tufts 1986). Most were observed during the fall and winter, but a few appeared during spring and summer. These records were followed by sightings along the Pacific coast and the Great Lakes basin. Historical records from 1905 showed Fulvous Whistling-Ducks wandering along the Pacific coast as far north as Washington

and British Columbia (Roberson 1980). There are now certified sight records in virtually all states and provinces. Sightings often involve small flocks as well as individual birds. Bolen and Rylander (1983) speculated that these vagrants were juvenile birds.

In the fall, Fulvous Whistling-Ducks depart the rice fields of Louisiana from September through early October, heading for the coastal marshes, and reach peak numbers at the Lacassine National Wildlife Refuge (NWR) in the third week in October (McCartney 1963). The Lacassine NWR appears to be their most important area of concentration in Louisiana during the fall and spring. In spring, Fulvous Whistling-Ducks begin to return to the Lacassine NWR in late March or early April and increase in number through the month.

MIGRATION BEHAVIOR

There are no specific studies of migration behavior; molt migrations, if they occur, have never been documented.

HABITAT

Fulvous Whistling-Ducks are found in open habitats characterized by hot summers and mild winters, preferring shallow fresh or brackish waters (Palmer 1976a). They especially prefer extensive prairie and savanna-like habitats, such as rice fields, where Fulvous Whistling-Ducks will be abundant within their range. Hohman and Lee (2001) reported that quality habitats in Louisiana are freshwater marshes with depths <0.5 m, impoundments managed for rice production, and seasonally flooded pastures and grasslands. Jarrett (2005) remarked that Fulvous Whistling-Ducks are most commonly encountered on shallow fresh or brackish wetlands characterized by extensive grass habitat or rice cultivation. In Mexico, Howell and Webb (1995) observed these ducks in open marshes, flooded fields, and lagoons.

POPULATION STATUS

Like Black-bellied Whistling-Ducks, the range of Fulvous Whistling-Ducks is outside the areas covered by the population surveys conducted by the U.S. Fish and Wildlife Service, except for the Midwinter Survey; hence one needs to rely on other data to assess their population status. In Louisiana and Texas the population has fluctuated, due to changes in rice agriculture, especially the use of pesticides. By the 1950s the peak postbreeding population was estimated at 3,000 birds in Louisiana (Meanley and Meanley 1959) and 3,500–5,000 in Texas (Singleton 1953). These numbers declined rapidly, however, during the 1960s, due to the extensive use of aldrin-treated rice to control the rice water weevil (*Lissorhoptrus oryzophilus*). Aldrin was highly toxic to wildlife, and Fulvous Whistling-Ducks were no exception (Flickinger and King 1972). A 1960 eyewitness account (Buckley and Springer 1964:459) following application of aldrin on a rice field in Texas reveals the lethality of this insecticide: "Fulvous Tree Ducks . . . soon started to die, quickly and violently. Some flew a short way, then flopped around. Others fell from the air." By 1968, at the height of the use of aldrin-treated rice, the Louisiana/Texas population had reached a low of only 1,123 birds (Flickinger et al. 1977). Numbers began to increase in 1970, however, with a voluntary ban on the use of aldrin, and late summer numbers reached 7,000 in Texas and about 10,000 in Louisiana (Flickinger et al. 1977). The U.S. Environmental Protection Agency formally banned all use of aldrin in 1974. Thereafter, the number of Fulvous Whistling-Ducks continued to rebound, with 7,300 tallied in southwestern Louisiana during April 1985 (Zwank et al. 1988) and at least 18,000 in Texas during March 1993 (Anderson et al. 1998). The Midwinter Survey in southern Texas recorded a high of 19,505 Fulvous Whistling-Ducks in 2005, and the Gulf Coast survey area had a high of 3,507 in 2008. Christmas Bird Counts have not detected many Fulvous Whistling-Ducks; the total U.S. count averaged only 770 from 2005/6 to 2009/10, 92% of which were from Florida and 7% from Texas.

After widespread dispersion along the Atlantic coast in the mid-1950s and early 1960s, small numbers of Fulvous Whistling-Ducks began to remain in Florida throughout the year. By the 1980s, Turnbull et al. (1989a) reported a minimum population of over 6,000 residents in Florida. At the Loxahatchee NWR just south of Lake Okeechobee, the number of Fulvous Whistling-Ducks has increased from less than 300 in the 1960s to 10,700 in 1990 (Hohman and Lee 2001). As in Louisiana and Texas, the increase in Florida corresponded to the expansion in the amount of rice cultivation.

In California, Fulvous Whistling-Ducks were common residents in the southern and central parts of the state from the late 1890s into the 1920s (Barnhart 1901, Hoffmann 1927); by the mid-1900s, they usually bred from the Imperial and San Joaquin Valleys northward to San Francisco Bay (Grinnell and Miller 1944, Cogswell 1977). By the early 1960s, however, only 2 small breeding colonies were reported: 1 near Mendota (Fresno County), and the other near the Salton Sea (McCartney 1963). Based on data from 1950 to 2000, Patten et al. (2004) demonstrated the dramatic decline of Fulvous Whistling-Ducks in the Salton Sea area and predicted its extirpation as a breeding bird in California, a finding supported in the review by Hamilton (2008). Bent (1925) reported breeding in Nevada (Washoe Lake) and southern Arizona (Fort Whipple), but there are no recent records from these states.

Harvest

From 1961 to 1998, the estimated annual sport harvest of Fulvous Whistling-Ducks ranged from 200 to 4,600 birds within the United States (average = 1,412 birds). More than 75% of this harvest occurred in Florida, Louisiana, Texas, and California. No birds have been reported as being harvested in California since 1976 (Hohman and Lee 2001). From 2000 to 2007, the average harvest/year was 190 in Texas (range = 0–718 birds), 405 in Florida (range = 0–797 birds), and 651 in Louisiana (range = 0–1,926 birds). The 2010/11 U.S. harvest was 1,740 (73% from Florida and 27% from Texas).

During the 1987/88 hunting season in the state of Sinaloa in Mexico, Fulvous Whistling-Ducks formed 9% (3,003) of the total waterfowl harvest at a duck-hunting club at Pabellón Bay (Migoya and Baldassarre 1993). During 1987–93 surveys in Mexico, Kramer et al. (1995) estimated a harvest of 4,956 whistling-ducks (Black-bellied and Fulvous).

BREEDING BIOLOGY

Behavior

Mating System. Fulvous Whistling-Ducks can breed at 1 year of age, but a detailed study of age-related breeding activity has yet to be conducted. The timing of pair formation is not well known, but it does occur rapidly once birds arrive on breeding areas. In Louisiana, Fulvous Whistling-Ducks arrive from their wintering areas in small flocks of <50 birds, but pairs become evident by mid-April (Hohman and Lee 2001). Some Fulvous Whistling-Ducks probably remain paired for life (as is known for Black-bellied Whistling-Ducks), but others form pair bonds after the birds arrive on breeding areas (Meanley and Meanley 1959). Biologists do not know whether these are reunited pairs from prior breeding seasons or individuals forming pairs for the first time.

Sex Ratio. The sex ratio (males:females) of wild-trapped birds in Florida from 1983 to 1987 was 347:340 (50.5% male) for prefledged young (Turnbull et al. 1989a). In Cuba, the sex ratio of adults collected from 1983 to 1985 was 280:229 (55.0% male; Acosta Cruz et al. 1989).

Site Fidelity and Territory. No information is available on site fidelity. Studies have shown no evidence of territoriality when birds arrive in groups to breeding areas. Pairs, however, do defend a small area around feeding sites, and males defend the nest site (Hohman and Lee 2001). Home-range size is probably quite variable, depending on the distances between nest sites and feeding areas and the stage of reproduction, but it is not well studied.

Courtship Displays. The courtship displays of Fulvous Whistling-Ducks are not extensive, which is characteristic of all species of *Dendrocygna* (Palmer 1976a). On the water, precopulatory behavior consists of mutual *head-dipping* by the pair, followed shortly thereafter by mounting (Johnsgard 1965). Copulation is followed by an elaborate, spectacular *step-dance* that involves both the male and the female rising out of the water side by side, with their chests and necks extended into an S shape (Meanley and Meanley 1958). Clark (1978) described a *chin-back* display where the head is laid back with the bill resting on the base of the neck. The display was observed in groups of 2–6 birds swimming together and was performed before aggressive movements, copulation, and the reunion of pair members.

Nesting

Nest Sites. Fulvous Whistling-Ducks nest on the ground, preferably in dense flooded vegetation. Earlier reports of cavity nesting probably represented misidentification—only the eggs were used as the basis of identification (Rylander and Bolen 1970). Over most of their range in Texas and Louisiana, Fulvous Whistling-Ducks usually nest in rice fields, where nest densities can be substantial. In Louisiana rice fields, nest densities of Fulvous Whistling-Ducks averaged 15.1 nests/km² (Hohman et al. 1994) and ranged from 3.0 to 13.7 nests/km² (Pierluissi et al. 2010). The highest nest densities were in agricultural landscapes with an abundance of rice fields, fallow fields, and canals, but no trees and minimal residential areas. Within the rice fields, nests are placed on low contour levees, as well as over water among the rice plants and weeds growing between levees (Meanley and Meanley 1959). In coastal marshes, Fulvous Whistling-Ducks nest in dense vegetation, usually over water. In southwestern Louisiana, nests are also established in pastures, old fields, and areas planted in small grains adjacent to fields being prepared for rice cultivation (Hohman and Lee 2001). In marshes at the Welder Wildlife Foundation Refuge, near Sinton, Texas, they nested in dense stands of beadgrass (*Paspalum* spp.), cutgrass (*Leersia* spp.), and cattails (*Typha* spp.), almost always over water (Cottam and Glazener 1959).

Nests are usually constructed of the surrounding vegetation and are 0–0.5 m above the water (Clark 1976); nest height is often increased if the water rises. Some nests have canopies of vegetation over them and ramps of vegetation leading to the nest bowls. The nests are constructed as egg laying proceeds, beginning with a weak platform and then a nest cup, which appears in about 3–4 days; the final nest is bowl shaped (Hohman and Lee 2001). Early-laid eggs sometimes fall through the flimsy floors into the water below (Cottam and Glazener 1959). Unlike other species of waterfowl, Fulvous Whistling-Ducks (like other whistling-ducks) do not add down to their nests.

Clutch Size and Eggs. Clutches in Louisiana averaged 14.1 eggs (*n* = 193) for ground nests and 13.4 (*n* = 296) for overwater nests, but the range was 2–44 eggs (Hohman and Lee 2001). In Texas, clutch size averaged 9.6 (*n* = 17) and ranged from 6 to 16 eggs (Cottam and Glazener 1959).

Fulvous Whistling-Duck eggs are bluntly ovate and white to buffy white. In Louisiana they averaged 46.1 g (*n* = 28); hence an average clutch of 13 eggs was equivalent to 84.4% of the mass of the incubating female (Hohman and Lee 2001). Eggs are laid at a rate of 1/day.

Incubation and Energetic Costs. Both sexes incubate the eggs, which is virtually unique among all waterfowl, except for the closely related White-backed Duck (*Thalassornis leuconotus*) of Africa and the Black Swan (*Cygnus atratus*) of Australia and New Zealand—where the male also definitely incubates the clutch (Johnsgard 1978)—and occasionally among the white swans (Kear 1972). The minimum incubation period is 24–25 days, but it is often extended for larger clutches that result from brood parasitism (Hohman and Lee 2001). In El Salvador, the mean incubation period was 29.4 days (range = 19–44 days): 25 days for clutches of

8 eggs and 44 days for 22 eggs (Gómez Ventura and de Mendoza 1982). The mean incubation period for the average clutch size (12.1 eggs) was 28.6 days. In Louisiana, incubation constancy was >96% (Hohman and Lee 2001). Eggs in Texas were left uncovered when not being incubated (Cottam and Glazener 1959).

Nesting Chronology. In the United States, the breeding season may last up to 120 days, with nest initiation ranging from early April in Louisiana through mid-September in Texas. In southwestern Louisiana (1991–93), nest initiation occurred from 5 April to 10 May for ground nests (median = 24 Apr; *n* = 261 nests) and from 10 May to 20 August for nests over water in rice fields (median = 9 Jun; *n* = 426 nests; Hohman and Lee 2001). Initiation of early nests may await rice plants growing to an adequate concealing height. In Florida, nests were not established until the rice plants reached a height of >60 cm (Wyss 1996). In southwestern Louisiana, the median date of nest initiation was 1 June in 2004 (*n* = 37 nests) and 7 June in 2005 (*n* = 139 nests; Pierluissi et al. 2010).

Nest Success. By far the most extensive study evaluating nest success for Fulvous Whistling-Ducks comes from work in southwestern Louisiana (1991–93). Apparent nest success varied: 6.3% for ground nests (*n* = 353), 14.5% for overwater nests (*n* = 733), 8.8% for nests in dry-seeded rice fields (*n* = 56), and 15.9% for nests in water-seeded rice fields (*n* = 266; Hohman et al. 1994, Hohman and Lee 2001). More recent estimates of nest success in Louisiana rice fields (Mayfield Method) were 38.8% in 2004 (*n* = 26) and 42.6% in 2005 (*n* = 74; Pierluissi et al. 2010). In 2005, 24 nests were found underwater, but it was not known if they were abandoned before they sank. An additional 22 were depredated, all by mammals. In Florida, nest success in rice fields averaged 5.6% (*n* = 116; Wyss 1996). Nest predators in Louisiana include Yellow-crowned Night-Herons (*Nycticorax violacea*) and other wading birds, snakes, mink (*Mustela vison*), raccoons (*Procyon lotor*), Virginia opossums (*Didelphis virginiana*), striped skunks (*Mephitis mephitis*), and dogs (Hohman and Lee 2001). Nest predators reported in Florida were raccoons and bobcats (*Felis rufus*; Wyss 1996).

Brood Parasitism. Intraspecific parasitism is common among Fulvous Whistling-Ducks. Nests containing more than 13 eggs were considered to be parasitized, with parasitism rates reported at 34% in uplands (*n* = 193) and 42% in rice fields (*n* = 489; Hohman and Lee 2001). Interspecifically, other duck species responsible for parasitizing Fulvous Whistling-Duck nests include Redheads, Ruddy Ducks, and Northern Pintails (Dickey and Van Rossem 1923). There are no studies on the age and reproductive status of parasitizing females or the effects of parasitism on host females.

Renesting. Despite such a long nesting season, renesting is not documented for Fulvous Whistling-Ducks, although it probably does occur (Hohman and Lee 2001).

REARING OF YOUNG

Brood Habitat and Care. In Louisiana broods may remain in rice fields used for nesting, but the time interval from seed germination to harvest is often insufficient to allow birds to nest and rear their young to fledging in the same field. Rice matures in about 75–90 days, depending on the variety, and several more weeks are needed after maturity for the grain to ripen before harvest (Dunand and Saichuk 2009, Louisiana State University Agricultural Center 2009). Rice fields are drawn down 2–3 weeks before harvest, however, so adults must often move their young to nearby flooded fields. Young birds unable to fly at this time probably suffer heavy mortality because of heat exposure and increased vulnerability to predators.

Like geese and swans, both parents assist in brood rearing and remain with their brood until fledging. Ducklings forage primarily by surface feeding, but they are capable of diving for food at 2 days old (Hohman and Lee 2001).

Brood Amalgamation (Crèches). There is no evidence of brood amalgamation in Fulvous Whistling-Ducks.

Development. Incubator-hatched ducklings (*n* = 24) averaged 27.8 g at hatching (Smart 1965a). Growth of a captive female duckling was 28.4 g at 4 days, 34.7 g at 8 days, 223.8 g at 33 days, and 523.0 at 60 days; fledging occurred at about 63 days (Meanley and Meanley 1959). The body mass of prefledged young with fully grown flight feathers (about 60 days old) averaged 531 g for males (*n* = 9) and 485 g for females (*n* = 5; Hohman and Lee 2001).

RECRUITMENT AND SURVIVAL

There are no studies of duckling or brood survival for Fulvous Whistling-Ducks. Similarly, mortality rates have not been estimated for this species, because so few birds are harvested in the United States. U.S. harvest data from 1998 through 2007 (*n* = 10,953 birds) yielded an age ratio of 1 adult to 1.45 young. The longevity record for a wild Fulvous Whistling-Duck is 11 years and 2 months for a male banded in Florida and shot in Cuba.

FOOD HABITS AND FEEDING ECOLOGY

Adult Fulvous Whistling-Ducks are almost exclusively vegetarian, even during the breeding season, which is when most female ducks increase their intake of invertebrate foods to obtain the protein needed for egg production. Fulvous Whistling-Ducks forage primarily by dabbling below or at the surface, tipping up, and diving, in contrast to Black-bellied Whistling-Ducks, which are more adapted to grazing (Rylander and Bolen 1974a). These different foraging niches are reflected in structural differences that include a larger foot in Fulvous Whistling-Ducks (a swimming advantage), as well as a more expansive maxillary cavity, more lamellae along the inner edge of the upper mandible, and other adaptations that facilitate sieving/filter-feeding for food (Rylander and Bolen 1974b). The diving capability of Fulvous Whistling-Ducks was observed in deepwater habitats (such as lagoons and floodplains) in Zambia in Africa, where diving was the primary (70%) feeding method, observed at water depths of 45–90 cm and as deep as 170 cm (Douthwaite 1977).

The major foods eaten by Fulvous Whistling-Ducks are seeds from moist-soil plants and cultivated rice. In Zambia their diet ($n = 33$) was 97% seeds—principally from *Nymphoides indica*, *Echinochloa stagnina*, and *Nymphaea capensis*—and 3% other vegetative matter (Douthwaite 1977). In Louisiana, the major plant materials that they consumed were seeds of *Panicum* spp., *Cyperus* spp., *Brachiaria* sp., *Brasenia schreberi*, and *Paspalum* spp. (Hohman and Lee 2001). In southern Florida, food items included cultivated rice, smartweeds (*Polygonum* spp.), Walter's millet (*Echinochloa walteri*), beggarticks (*Bidens* sp.), midges (Chironomidae), water beetles (*Hydrocanthus* sp.), and mosquitofish (*Gambusia affinis*; Turnbull et al. 1989a).

Fulvous Whistling-Ducks heavily consume cultivated rice, and they are most concentrated and visible in rice fields during planting and harvesting periods; they are regarded as pests, due to perceived crop depredation. The most detailed study of this issue in the United States was conducted in the Rice Belt region of southwestern Louisiana, where Fulvous Whistling-Ducks feeding on planted rice are considered to be a nuisance by farmers (Hohman et al. 1996). This study noted that feeding sites in rice fields that were tilled and flooded in preparation for planting contained a diverse and abundant food source that averaged 109 g/m^2, 101 g of which were seeds from at least 28 taxa. Junglerice barnyardgrass (*Echinochloa colonum*) was the preferred plant food of 81 Fulvous Whistling-Ducks that were collected; earthworms were the favorite animal matter but made up only 1% of the food available. Seeds from signalgrass (*Brachiaria extensa*), flatsedge (*Cyperus iria*), and beakrush (*Rhynchospora* sp.) also formed an important part of their diet. Plant food consumption was similar between males and females.

Females did eat slightly more animal material during the egg-production period, but plant use never decreased below 96%. Rice was eaten in proportion to its availability; <4% of the diet of birds collected before and during the period of egg development consisted of rice, although consumption was 25% during the incubation period. Nonetheless, overall crop depredation by Fulvous Whistling-Ducks was deemed miniscule (≤0.1% crop loss), compared with other factors affecting rice production.

Nonetheless, localized depredation can occur when large flocks of whistling-ducks use rice fields, especially at night, when feeding is undisturbed, but such events most likely are not common (Hohman et al. 1996). In the Llanos region of Venezuela, Fulvous Whistling-Ducks consumed both planted and mature rice (Dallmeier 1991). Subsequent interviews with farmers yielded reports of 32%–45% crop loss due to whistling-ducks, but these were probably exaggerated. In the huge rice-growing region of Calabozo, Bruzual and Bruzual (1983) noted the high diversity of plant seeds eaten by 3 species of whistling-ducks, including the Fulvous Whistling-Duck, and concluded that the aquatic environment was more attractive to the birds than rice per se. They also speculated that whistling-ducks could potentially be used to control unwanted weed seeds before the rice is planted.

MOLTS AND PLUMAGES

As adults, Fulvous Whistling-Ducks have only 1 plumage (basic) that is molted each year via the prebasic molt. Hence there is no alternate plumage, such as is observed in puddle ducks. The descriptions of molts and plumages are primarily from summaries and data in Palmer (1976a) and Hohman and Lee (2001), which should be consulted for further details.

Juvenal plumage is fully developed in 9–10 weeks, with remiges fully grown by 63 days, when the birds are fully flight capable. Juvenal plumage resembles definitive basic but is somewhat paler overall. First basic involves all feathers except those on the wings and is initiated after departure from the breeding grounds. This plumage is then worn until the following summer, when a full body and wing molt yields definitive basic. The subsequent molting of definitive basic is thought to occur continuously, except on the wings. The flightless period during the wing molt is about 20 days. In Louisiana, 8 adults captured with broods of various ages exhibited a light body molt (<10% of feathers) but no molt of the remiges (Hohman and Richard 1994). Subsequent data on 16 radiomarked birds, none of which underwent a flightless period in Louisiana before departure by 28 September, supported indications that Fulvous Whistling-Ducks are unique among North American waterfowl in undergoing the wing molt on wintering areas.

CONSERVATION AND MANAGEMENT

More than any other species of waterfowl, Fulvous Whistling-Ducks are closely associated with rice during all stages of their life cycle; hence these birds can incur significant exposure to pesticides. The use of aldrin-treated rice caused a serious decline of Fulvous Whistling-Ducks in Louisiana and Texas during the late 1960s (Flickinger and King 1972). White et al. (1983) reported a 1982 incident where the organophosphate pesticide azodrin was deliberately used to poison waterfowl and other waterbirds on a rice field in Texas. More recently, 18 different contaminants were detected in the livers and breast muscles of Fulvous Whistling-Ducks collected in the Everglades Agricultural Area in 1984–85 (Turnbull et al. 1989b). Significantly, the U.S. Environmental Protection Agency had previously banned several of these compounds from agricultural use as insecticides: DDT in 1972, aldrin in 1974, mirex in 1977, and heptachlor in 1983. The source of the contaminants, however, is difficult to identify, because at least some Fulvous Whistling-Ducks in Florida migrate to and from Cuba and possibly other islands in the West Indies, as well as to Mexico or even South America (Turnbull et al. 1989b). These regions continued to use pesticides banned in the United States (White

et al. 1981, Mora et al. 1987), so pesticide burdens in adult birds could reflect exposure while the birds were outside the United States. Aldrin is rapidly (2–3 weeks) metabolized into dieldrin; hence its detection in Florida most likely signaled illegal use. Contaminants found in recently fledged and local birds probably were also due to local exposure. Lastly, low levels of some contaminants may indicate bioaccumulation through sediment residues, since DDT, its breakdown products dichlorodiphenyldichloroethylene (DDE) and dichlorodiphenyldichloroethane (DDD), and dieldrin were all detected in sediment samples collected in and around the Everglades Agricultural Area (Pfeuffer 1985). The levels of pesticides found in Fulvous Whistling-Ducks in Florida were below those known to cause a direct threat to birds or humans, but the strong association of this species with agricultural habitats warrants continued monitoring and assessment.

Rice fields provide essential habitat for Fulvous Whistling-Ducks, as well as significant breeding habitat for a plethora of other waterfowl, shorebirds, rails, and wading birds (Taft and Elphick 2007, Pierluissi and King 2008, Pierluissi et al. 2010). In California at least 118 bird species, representing 38 families, have been recorded in rice fields during winter (Eadie et al. 2007). Among the world's agricultural crops, rice is the food most consumed by waterbirds, especially waterfowl. In the United States, roughly 1.2 million ha of rice were planted annually from 2010 to 2012, largely in Arkansas (47%), California (19%), and Louisiana (16%; U.S Department of Agriculture 2012), and an average of 344–491 kg/ha of residual rice remains after harvest (Manley et al. 2004). Rice fields also contain substantial densities of moist-soil seeds, as well as populations of aquatic macroinvertebrates. In Arkansas, Reinecke et al. (1989) documented 12–37 kg/ha of moist-soil plant seeds in harvested rice fields. In Louisiana, Hohman et al. (1996) reported seed densities of 101 g/m² at feeding sites in rice fields used by Fulvous Whistling-Ducks. Invertebrate densities in rice fields averaged 6.3 kg/ha (with a maximum of 21.1–31.7 kg/ha) in Mississippi (Manley et al. 2004) and 22.0 kg/ha in Louisiana (Hohman et al. 1996). The seed densities in Louisiana were comparable to the range (90–134 g/m²) found in impoundments specifically managed to produce moist-soil plants (Reid et al. 1989).

Waterfowl management in rice-growing states has focused on the significant resources that rice fields provide for ducks and geese. Along the Gulf Coast in the United States, the U.S. Fish and Wildlife Service began leasing rice fields in 1988 through the Gulf Coast Joint Venture of the North American Waterfowl Management Plan (Hohman et al. 1994). Subsequent management practices called for flooding rice fields after harvest, which makes both waste rice and moist-soil food sources available to waterfowl. Flooding may also benefit rice farmers, because feeding waterfowl remove undesirable moist-soil seeds, which can reduce the need for costly herbicide treatments (Hohman et al. 1996). Hohman et al. (1996) also suggested actions by farmers that can potentially reduce damage to seeded rice. The use of pregerminated seed, for example, significantly shortens the time period over which fields need to be flooded after seeding. The removal of water within 48 hours of planting was highly recommended—especially early in the growing season (before 1 Apr), when whistling-ducks occur in large flocks—as it greatly reduces use by these birds and thus prevents both the ingestion and trampling of seed. Hohman et al. (1996) never observed feeding by Fulvous Whistling-Ducks in dewatered rice fields.

In southwestern Louisiana, most rice fields are planted by aerially dispersing seeds over flooded fields that are then drained within 24 hours of planting but reflooded within 7–14 days of germination, remaining that way until about 2–3 weeks before harvest. Elsewhere, fields are "dry seeded"; this approach involves planting before the field is actually flooded, which is done after germination. In Louisiana, Pierluissi (2006) recommended no-tillage for rice fields, because leaving stubble undisturbed probably keeps more of the invertebrate and seed communities intact. No-till fields also averaged a higher Fulvous Whistling-Duck nest density (8.3/km²) than tilled fields (0.9/km²), but no-tillage was not commonly practiced. Waterbird nest densities in rice fields in Louisiana tended to be greater in dense versus less dense stands, probably because dense stands provided a better nest substrate (Hohman et al. 1994).

Mute Swan

Cygnus olor (Gmelin 1789)

Left, adult; *right,* juvenile

The Mute Swan, native to Eurasia, is an invasive species in North America. These swans were first recorded breeding in the wild along the lower Hudson River in 1910. Significant populations now occur in the mid-Atlantic and Chesapeake Bay regions of the Atlantic Flyway, and along the lower Great Lakes in Michigan; additional populations are scattered on the West Coast and elsewhere. The Atlantic Flyway population numbered 10,541 in 2008 and about 9,700 in 2011, but Michigan has the most Mute Swans of any state or province, with an average spring population of 12,167 from 2007 to 2011, numbering 15,420 in 2011 (Barbara Avers). The continental population is probably around 22,000–25,000. Large all-white birds, Mute Swans are monogamous and mate for life, although separation is not uncommon, especially in young pairs. Females return to their natal areas to nest, and most pairs are highly territorial and aggressive toward other swans and waterfowl. Mute Swans are adaptable, using habitats ranging from freshwater ponds to saline coastal ponds and estuaries. Territories are usually occupied year-round, as long as open water is available. They build large nests, within which 5–6 eggs are incubated for about 36 days. Mute Swans are vegetarians, which, coupled with their large size, translates into the consumption of significant aquatic vegetation, about 4 kg/bird each day. Hence

32

they negatively affect submerged aquatic vegetation (SAV), to the detriment of native waterfowl and other fauna. Territorial swans are also aggressive toward humans, which has led to negative encounters. Mute Swan populations continue to increase in some areas and expand their distribution, which—together with their impact on SAV, native fauna, and humans—has led several states to control swans by egg addling and the removal of adults.

IDENTIFICATION

At a Glance. Mute Swans are large and unmistakable. Both sexes, like all species of swans in the Northern Hemisphere, have all-white plumage and long necks. The deeply curved neck and prominent black knob at the base of an orange bill readily identify this species. Juveniles are commonly grayish brown, with gray feet and a gray bill; some juveniles have white plumage, with light brown feet and a light brown bill.

Adults. Males are larger than females, but otherwise the sexes are similar in their plumage. Adults have an all-white plumage. Like other swan species, the head may be stained reddish from feeding in iron-rich waters. The neck is held in an S shape, with the bill pointed down. The bill is orange, except for the black base that extends to a fleshy black knob on the forehead. The knob, or berry (in England), is less prominent in females than in males, and the orange coloration of the bill is brighter in males during the spring. The legs and the feet range from dark gray to black to pinkish gray. The irises are brown. Mute Swans often swim with their wings arched over the back. In flight, the motion of the wings produces a unique whistling sound, sometimes audible from 1.6 km away, and the pointed white tail extends noticeably beyond the dark feet. The male Mute Swan is referred to as the cob, and the female as the pen.

Juveniles. Young Mute Swans have 2 distinct plumage colors. Some have a dingy pale gray-brown plumage that becomes browner before the first winter molt, while others have a white plumage

as juveniles and adults. White largely replaces the gray-brown plumage by mid-November. The feet and the bill are gray. The bill of juveniles turns pinkish during the first winter, but the frontal knob is relatively small. The adult-sized frontal knob and adult bill coloration are not attained until the second winter. The irises are brown.

Cygnets. Mute Swan cygnets occur in 2 color morphs: a gray (Royal) morph and a white, or leucistic (Polish), morph. The gray morph is more common, but both color morphs can appear in the same brood. The white morph was dubbed "Polish" by London poultry dealers, who imported captive-reared swans from the Polish coast of the Baltic Sea. Selective breeding of the white morph had been practiced for centuries in Europe, because they were more attractive to collectors (Birkhead and Perrins 1986).

Voice. Generally silent but not "mute," Mute Swans differ from the 4 northern species of white swans in not being as vocal, which is reflected in a trachea that is relatively simple and not convoluted (Johnsgard 1961a). Nonetheless, adults have 8–10 different calls that are variously given during courtship and greeting, as well as in group interactions. They also have a resonant loud cry (similar to that of cranes), a solicitation call (*glock, glock*) given by the female to her mate, a lost call uttered when separated from other swans, and a brood call made by the female to her young (Ciaranca et al. 1997).

Similar Species. The orange bill and curved neck readily distinguish Mute Swans from the other species of swans in North America.

Weights and Measurements. The males are larger than the females. In Rhode Island, weights of Mute Swans captured before their annul molt ($n = 68$) averaged 10.9 kg for gray-phase males, 10.7 kg for white-phase males, and 7.9 kg for females (Willey 1968). The wingspan averaged 241 cm for males (range = 235–249 cm) and 223 cm for females (range = 216–239 cm). Molting Mute Swans in Massachusetts averaged 10.2 kg (range = 7.7–14.1 kg)

Mute Swan in a territorial display posture. *Ian Gereg*

for males ($n = 42$) versus 8.4 kg (range = 6.1–11.3 kg) for females ($n = 52$; Ciaranca et al. 1997). In Sweden, adult males averaged 10.5–10.8 kg ($n = 137$), and adult females, 8.1–8.6 kg ($n = 93$; Mathiasson 1981). From the Massachusetts data, bill length averaged 103.5 mm for males and 97.7 mm for females; tarsal length, 97.0 mm for males and 90.8 mm for females; knob length, 23.3 mm for males and 19.7 mm for females; and knob width, 14.8 mm for males and 12.3 mm for females. Heavier breeding birds have larger knobs than nonbreeding birds. Among adult males in Sweden, breeders ($n = 22$–25) averaged 12.3 kg and 17.4 mm in knob size, compared with 11.0 kg and 11.5 mm for nonbreeders ($n = 23$–24; Mathiasson 1981). Some nonbreeding males ($n = 8$) had extremely small knobs that averaged only 7.8 mm, and their body weights averaged only

9.8 kg. A similar relationship was observed among females.

DISTRIBUTION

Native. There are no subspecies of Mute Swans. These birds are native to northern and central Eurasia, where they are found in a discontinuous range from Great Britain east to central Asia. Wild populations breed in southern Sweden, Denmark, the Netherlands, northern Germany, and Poland, as well as locally in the Soviet Union and Siberia. They are less abundant in Belgium, northern France, Switzerland, Austria, and Turkey, where populations have been influenced by a long history of domestication and introduction (Ogilvie 1970, Johnsgard 1978). This species also occurs in Asia Minor and Iran, reaching east through Afghani-

Year-round Distribution ■

stan to Inner Mongolia. Feral and semiferal populations breed locally, especially in Great Britain, France, the Netherlands, and central Europe. Mute Swans are introduced in South Africa, New Zealand, Australia, and Japan. Most populations are nonmigratory. Mute Swans breeding in Siberia and Mongolia (east of Lake Baikal), however, probably winter 1,600 km away, along the Pacific coast from Korea south to the coastal marshes of the Yellow Sea (where it borders Korea) and the Yangtze River (Palmer 1976a, Johnsgard 1978). They also winter along the eastern Mediterranean, the Persian (Arabian) Gulf, and the northern Arabian Sea.

Mute Swans are easily tamed and generally flourish in association with civilization, in marked contrast to Trumpeter and Tundra Swans. For example, the Mute Swans in the moat surrounding the Bishop's Palace in Wells, England, ring a bell set in the wall when they want to be fed, a practice they have been trained to do since 1850 (Kear 1990). Elsewhere in the United Kingdom, wild Mute Swans had disappeared prior to the thirteenth century, because they were trapped, pinioned, and captive-bred in such numbers that the population became semidomesticated, which is largely the state of the population today. During this period swans were a delicacy in England, jealously reserved for important royal banquets and other medieval feasts, giving further credence to the phrase "food fit for a king." There were 125 young swans at a Christmas dinner for Henry III's court in 1251, and 400 were consumed when the Archbishop of York was installed in 1466 (Kear 1990).

Ownership of Mute Swans in medieval England thus became strictly monitored and enforced, because it was a very lucrative business. By the twelfth century the Mute Swan was given royal status, and swans escaping from a landowner became the property of the Crown; hence to own swans on one's property was a status symbol of the times. By an act of King Edward IV, however, individuals could not own swans, unless their land was valued at a minimum of 5 marks (Dawnay 1972). By 1378 an officer known as the Keeper of the King's Swans was appointed, and laws were set in place assigning ownership of swans in association with swan farming. Young birds were chosen for the table; hence cygnets were taken from their parents during an annual event (known as swan-upping), pinioned, and their bills given the same unique mark as that of their parents, for identification and ownership purposes. As many as 630 marks were in use between 1450 and 1600 (Ticehurst 1957). Cygnets were then placed in pens to be fattened on barley for feasts and celebrations.

A Swan Master and numerous deputies were appointed by the King to supervise the swan-upping and, thus, the division of cygnets, before the birds could fly. Swan-upping declined throughout the eighteenth century, because luxury foods such as goose and turkey became more available and it was expensive to keep cygnets. Late in the fifteenth century, however, 2 trade guilds were granted a royal charter for owning swans (the Worshipful Dyers and the Worshipful Vinters Companies of London). These companies continue that tradition today. The current Queen also retains ownership rights to swans along certain stretches of the River Thames and its tributaries. Swan-upping along the Thames today is a colorful reenactment of an ancient custom, replete with costumes and traditional Thames rowing skiffs that fly relevant flags and pennants. The Queen's Swan Marker and Swan Uppers are accompanied by the Vinters and Dyers, and the swans are rounded up and marked for ownership. The Dyers make a nick on one side of the bill, while the Vinters make two; the beaks of royal swans are left unmarked. Lastly, an annual report is produced that details the number of swans, broods, and other information on the cygnets.

The famous Abbotsbury Swannery in Dorset, on the southern coast of England, also retains the right to own swans, with records dating back 600 years. The first written record was in 1393, but the Swannery may have been in existence since 1320. The Benedictine monks in Abbotsbury raised swans for meat and used their quills for pens. The monastery was dissolved by King Henry VIII in 1539, but it came into private ownership by the Strangways family in 1543 and is still owned by them today. The Swannery is now managed as a tourist attraction and visited by over 100,000 people annually (Fair and Moxom 1993).

North America. Mute Swan populations in North America are the result of multiple introductions, and they now breed naturally in the wild. They are nonmigratory, but will move to larger bodies of water when smaller ponds and marshes freeze over. They were first brought to the United States in the late 1800s, and by the early 1900s these swans were common in city parks, zoos, and country estates. More than 500 Mute Swans were imported into the United States between 1910 and 1912, which eventually led to escapes and their subsequent breeding in the wild (Phillips 1928). The earliest records of Mute Swans in the wild appear to be those reported in New York: along the lower Hudson River in 1910 and on Long Island in 1912 (Bull 1964). Elsewhere in the Atlantic Flyway, Mute Swans first appeared in New Jersey in 1919, Massachusetts (Martha's Vineyard) in 1922, Rhode Island (Block Island in 1923, the mainland in 1930), Maryland in 1954,

Virginia in 1955, and Pennsylvania in the 1930s, but they were not seen farther south along the Atlantic coast (the Carolinas) until 1989–93 (Stewart and Robbins 1958, Allin 1981, Atlantic Flyway Council 2003). The Chesapeake Bay population, which peaked at 3,955 in 1999, originated from 5 pinioned Mute Swans that escaped in 1962 from a private estate along the Miles River in Talbot County, Maryland, and then successfully bred in the wild (Reese 1969). A pair of Mute Swans first bred in Ontario in 1958, but large-scale colonization within the lower Great Lakes did not occur until the mid-1960s and 1970s; they are concentrated in the coastal wetlands associated with Lakes St. Clair, Erie, and Ontario (Petrie 2004).

In the Atlantic Flyway, Mute Swans breed from New Hampshire south to the Carolinas and in Florida. Since 1986, breeding populations (adults and cygnets) in this flyway have been surveyed in summer, at 3-year intervals, in most states and in the province of Ontario. During the 2008 survey, the highest numbers of adult Mute Swans were found in New York (2,624, mostly on Long Island), followed by Ontario (2,357, mostly along the northern shores of Lakes Ontario and Erie), New Jersey (1,253), Massachusetts (1,046; estimated), and Connecticut (1,012); there were 954 in Chesapeake Bay (Maryland and Virginia). In 2011, survey figures were about 2,200 in New York, 3,062 in Ontario, 1,059 in New Jersey, 452 in Connecticut, and 317 in Chesapeake Bay. Massachusetts did not conduct a survey in 2011, but numbers there appear to be declining along the coast while increasing inland (H. Heusmann). Not many Mute Swans occur south of Virginia: about 100 along the coast of the Carolinas (mostly North Carolina); none in Georgia (Greg Balkcom); and an estimated 22–99 pairs in Florida, mostly in the southcentral part (in and around Polk County), but with scattered birds elsewhere in the state (Bill Pranty). The total Atlantic Flyway population was 10,541 (90% adults) in 2008, compared with 13,649 in 2005, and about 9,700 in 2011.

There are no Mute Swans in Minnesota (Steve

Cordts). Along the lower Great Lakes in the United States, Mute Swans occur in comparatively small numbers from Wisconsin to Ohio (Nelson 1997). Those in Wisconsin are largely in the southeast, with a few in the northwestern part of the state. Mute Swans are very common in Michigan, especially near Traverse City, on lakes in the lower third of the Lower Peninsula, and around Lake St. Clair. The first pair was introduced there in 1919, near Charlevoix (along the northeastern shore of Lake Michigan), and increased to a flock of 47 by the mid-1940s (Wood and Gelston 1972). This population spread over the northern portion of lower Michigan, with the flock near Traverse City numbering 450–500 birds by 1972. They then continued to expand into much of southern Michigan. The statewide population averaged 5,917 birds from 1992 to 2003 (10,200 from 2002 to 2008) and had reached 15,400 by 2011. They now occur in every county on both the Upper and Lower Peninsulas (Barbara Avers). Mute Swans in Illinois are primarily found in the northwest (around Chicago), but a few are on strip-mine lakes in the center of the state. Those in Indiana (700–1,000) are in the northern part of the state (Adam Phelps). In Ohio about 200 are mostly along Lake Erie, but also around some of the major cities (Dave Sherman).

In the Pacific Flyway, Mute Swans have established viable populations in Washington (<100), with most located around Puget Sound (Don Kraege). In British Columbia (2002), about 200 primarily occurred around southern Vancouver Island, the nearby Gulf Islands, and the lower southwestern mainland, although there are a few records of nonbreeding birds from the southcentral interior (along the border with Washington), and about 100 more in the possession of avicultural permit holders (Campbell et al. 1990, Myke Chutter). About 200–400 occur in California, mostly in the San Francisco Bay area (Petaluma and Suisun marshes), but they are expanding in the state (Shaun Oldenburger); there are no feral breeding populations to the north in Oregon (Brandon Reishus). Scattered pairs (<10) can be found across

southcentral Idaho (Carl Mitchell), but a feral population on the Yellowstone River in Montana was removed by the mid-1990s.

The Midwinter Survey averaged 12,214 Mute Swans from 2000 to 2010, with virtually all birds located in the Atlantic Flyway (63.0%) and the Mississippi Flyway (36.9%). Most of the Atlantic Flyway birds were recorded along the upper mid-Atlantic coastal states between Massachusetts and Maryland (94.8%); most from the Mississippi Flyway were observed in Michigan (92.4%). Ontario averaged 4,502 over this time period.

MIGRATION BEHAVIOR

Mute Swans in the United States are not truly migratory. Pairs generally remain in breeding areas as long as open water and food are available, although failed breeders will join flocks of nonbreeding birds during the molt. Interior populations move to nearby ice-free locations, such as rivers and river deltas. East Coast swans fly short distances to coastal bays during the winter. Many Mute Swans breeding in Michigan winter in Traverse Bay, which requires a flight of up to 160 km from some breeding areas.

Molt Migration. Mute Swans do not undergo a molt migration per se, but they can concentrate at traditional molting sites. In Chesapeake Bay, Mute Swans will move up to 50 km to a molting site (Larry Hindman). These sites are usually shallow estuarine bays and marshes where SAV is abundant and disturbance from boat traffic is minimal.

HABITAT

Mute Swans use an array of wetland habitats, all having the common feature of abundant aquatic vegetation and shallow water. In coastal areas, Mute Swans will breed on ponds, estuaries, backwaters, and tributaries that range in salinity from freshwater to salt. During winter, Mute Swans often remain on breeding areas but will move to open water during periods of ice cover. Mute Swans are cold tolerant, with a lower critical temperature in winter of about 5.5°C (Bech 1980). In Scotland, breeding and nonbreeding Mute Swans mainly were found on saline lochs, with pairs using 86 different lochs during spring (Jenkins et al. 1976).

Sousa et al. (2008) used the Global Positioning System (GPS) to track 5 male Mute Swans and identify habitat use in a 217,500 ha area of Chesapeake Bay in 2002 and 2003. Habitats were categorized as SAV, open water, shoreline, and upland. Four swans remained within 8 km of their capture site, while the fifth moved nearly 24 km. Habitat use was 47% in SAV, 32% in open water, 13% on shorelines, and 8% in upland areas (probably on ponds). There was no difference between diurnal and nocturnal habitat use. The authors believed the GPS observations were representative of other Mute Swans in the Bay, because marked individuals were always found within flocks of 30–400 other Mute Swans.

POPULATION STATUS

Mute Swan populations have increased dramatically in both numbers and distribution since the early introductions of this species. In Chesapeake Bay, their numbers rose from an initial 5 birds that escaped from captivity in 1962 to 3,995 by 1999, an annual growth rate of 23% from 1986 to 1992 and 10% between 1993 and 1999 (Hindman and Harvey 2004). An analysis of Midwinter Survey data from the Atlantic Flyway revealed an annual growth rate of 5.0% from 1954 to 1963, followed by an explosive rate of 43.3% from 1963 to 1974, and then 9.4% from 1974 to 1983 (Allin et al. 1987). A Midsummer Survey of Mute Swans has also been conducted in the Atlantic Flyway at 3-year intervals since 1986. These surveys indicated an annual growth rate of about 6% from 1986 to 2002, which coincided with a population increase of 147% during that period (2.6 times the 1986 population). A 6% growth rate could potentially double the population every 12 years (Atlantic Flyway Council 2003). Recent Midsummer Surveys in the Atlantic Flyway, however, recorded 10,541 Mute Swans in 2008, down 27% from the 13,649 tallied in 2005, and only 3% above

the long-term (1986–2008) average; the 2011 figure was about 9,700. In Canada, along the lower Great Lakes, estimated growth rates from 3 long-term survey datasets projected that the Mute Swan population was increasing 10%–18% annually (Petrie and Francis 2003). The Mute Swan population in Traverse Bay in Michigan, rose 17%/year from 1969 to 1979 (Gelston and Wood 1982). At a minimum rate of 10%, the population would double every 7–8 years. Subsequent projections from the 2002 Christmas Bird Count totals indicated that the Canadian population alone would reach 30,000 in 30 years. The Michigan population increased 9.0% annually from 1949 to 2010. Growth was slower toward the end of the period (1991–2010), at 4.5%/year, but was 15,400 in 2011 (David Luukkonen). Overall, North America would appear to have a population of perhaps 22,000–25,000 Mute Swans.

Globally, the Mute Swan is the most abundant swan species. Winter population estimates are 250,000 swans in central Europe, 31,700 in the United Kingdom, 10,000 in Ireland, 45,000 along the Black Sea, 250,000 from western and central Asia to the Caspian Sea, and 1,000–3,000 in central Asia (Wetlands International 2006). Most of these populations appear to be stable or are increasing.

Harvest

There is no sport-hunting season for Mute Swans.

BREEDING BIOLOGY

Behavior

Mating System. Mute Swans are monogamous and assumed to mate for life, but there are exceptions to the latter. Pair formation occurs from January through March. In England, mate change due only to pair separation was 38% for males and 36% for females, whereas mate change because of death was 17% for males and 11% for females (Coleman et al. 2001). Mate change often does not alter breeding status, with 59% of the males and 63% of the females continuing as breeding or nonbreeding pairs with a new mate; nonetheless, 19% of the males and 37% of the females did change from nonbreeding to breeding status following a mate change. The remaining 22% of the males changed their status from breeding to nonbreeding.

Like other swans, Mute Swans form pairs at least 1 season prior to laying, and some will first nest when only 2 years old. The majority, however, do not breed until they are 2–3 years old, with females apt to breed earlier than males. During a 39-year study (1961–99) of a marked Mute Swan population in central England, males ($n = 136$) paired in an average of 3.54 years and females ($n = 152$) paired in 3.23 years; about 60% of the males paired at age 2–3, versus 68% of the females (Coleman et al. 2001). The duration of paired life, however, was only 1–2 years for 50% of the males and 53% of the females. Some of the males (21%) and the females (19%) did remain paired for >5 years, but 65% of those males and 64% of those females changed mates at least once over that time period. Only 7% of the males and 5% of the females remained with the same partner for more than 10 years, but breeding did not occur each season. The maximum paired life was 17 years for a male and 16 years for a female, during which time the male changed mates twice and the female changed mates 3 times. During a 7-year study (1961–67) in England, separation occurred in <3% of successful pairs and in 9% of either failed pairs or pairs that had not yet bred (Minton 1968).

In central England, only 8% of the males were first recorded breeding as 1-year-olds, with 74% joining the breeding population in almost equal proportions each year from ages 3 to 5 (Coleman et al. 2001). The percentage of females breeding was 8% at age 2, with 78% joining the breeding population between ages 3 and 5. Elsewhere in England, the known age at first breeding by 24 females was 2 swans at 2 years old, 13 at 3 years old, and 9 between 4 and 6 years old (Perrins and Reynolds 1967). In contrast, the known breeding age for 18 males was 9 swans at 3 years old, 5 at 4 years old, and 4 between 5 and 6 years old. At the Abbotsbury Swannery in Dorset, England, age at first breeding averaged 4.6 years old for males ($n = 332$) and 4.3 years old for females ($n = 357$; McCleery et al. 2002). In a rapidly expanding population of Mute Swans studied in Chesapeake Bay, 4% of the females paired when <1 year old, 40% at 1 year old, 76% at 2 years old, 90% at 3 years old, and 97% by age 4. Females first nested as 1-year-olds (14%), which increased to 53% at 2 years old, 79% at 3 years old, and 92% by age 4 (Reese 1980). In contrast, no males paired when <1 year old, 15% paired at 1 year old, 42% at 2 years old, 59% at 3 years old, 70% at 4 years old, 79% at 5 years old, and 91% between the ages of 6 and 20. Few males nested as 1-year-olds (2%), but the percentage thereafter increased to 30% at 2 years old, 45% at 3 years old, 66% at 4 years old, 79% at 5 years old, and 89% between the ages of 6 and 20. In England, 7 records of incestuous pairs were recorded from a marked population of Mute Swans: mother/son (3), father/daughter (1), and brother/sister from the same brood (3); most of their breeding attempts (70%, $n = 26$) produced no young (Coleman et al. 1994, Coleman et al. 2001). The parent/sibling pairs occurred on the natal site and were formed when 1 member of the parental pair was recorded as being dead or missing the previous winter.

Based on their long-term study in England, Coleman et al. (2001) classified the breeding life cycle of Mute Swans into 3 categories ($n = 86$ males, 91 females). The first was of individuals that bred continuously throughout their paired life (31% of the males, 38% of the females). The longest period of continuous breeding was 17 years for a male and 11 years for a female. The second category was of individuals with nonbreeding seasons interspersed among breeding seasons (22% of the males, 16% of the females). Most nonbreeding was associated with mate change. The third category was of individuals with nonbreeding seasons at the beginning or end (or both) of the breeding span (47% for males, 46% for females). Most nonbreeding occurred prior to the breeding span (55% for males, 65% for females). However, 28% of the males and 5% of the females had nonbreeding seasons at the end of the breeding span.

Sex Ratio. The sex ratio (male:female) is male biased, which is first expressed at 6 months of age and increases thereafter. In Rhode Island the sex ratio was 53.6:46.4 for cygnets, 65.2:34.8 for subadults, and 61.2:38.8 for adults. The percentage of males for all age groups combined was 58.1% (Willey and Halla 1972). In Chesapeake Bay, the sex ratio of cygnets ≤1 year old was 54% male, and 57% male at age 2 or older (Reese 1980).

Site Fidelity and Territory. Mute Swans are philopatric, and they aggressively maintain territories for courtship, nesting, and foraging. They may occupy those territories year-round, depending on winter weather and the availability of food (Perrins and Reynolds 1967, Scott 1984, Conover and Kania 1994). In central England, 33% of the males and 50% of the females established their own nesting territories within 5 km of their natal site; of these, 40% of the males and 42% of the females kept their same mate and territory throughout their reproductive lives, but most (50% for males, 58% for females) only bred for 1 or 2 seasons. The remaining individuals changed either their mate, their territory, or both (either simultaneously or on separate occasions) at least once during their breeding lives. The 50% that did change territory with their mates did so after a season as a nonbreeding pair, and a further 27% moved elsewhere after failing to breed successfully. Territory shifts not associated with mate change resulted in an improvement in breeding status from paired nonbreeders to paired breeders in 29% of the males and 31% of the females.

In Connecticut, Mute Swans occupied territories on small ponds (2–30 ha) throughout the year, except when the ponds froze (Conover and Kania 1994). In Rhode Island, territories on small ponds, backwaters, and tributaries ranged in size from 0.2 to 4.8 ha (Willey and Halla 1972). Territory size varies with habitat quality (Bacon 1980), but nests can be crowded together in some situations. In Rhode Island and Massachusetts, 2–8

nests were found on small coastal ponds, with a minimum distance of 91–137 m between the nests (Ciaranca et al. 1997). The average distance between nests on rivers near Oxford in the United Kingdom was 2.4–3.2 km, with a minimum separation of 90 m (Perrins and Reynolds 1967). Mute Swans have nested colonially at coastal sites in Denmark, Poland, and the United Kingdom (Kear 1972). In Scotland, 53 of the 86 lochs used by Mute Swans in spring contained only 1 pair (Jenkins et al. 1976). The home range size of 5 males tracked with GPS transmitters in Chesapeake Bay averaged 582 ha (range = 181–1,071 ha; Sousa et al. 2008).

Courtship Displays. Mute Swans are quite graceful in their courtship activities, which are among the attributes that have endeared them to aviculturalists and collectors for centuries. The most frequent courtship display by the pair involves a slow, mutual *head-turning* done while facing each other, with the neck and secondary feathers raised and their bills pointed down, almost touching the breast. The pair may vocalize during the *head-turning* display. Pairs may also exhibit courtship displays upon greeting. Precopulatory behavior consists of mutual *head-dipping* alternated with a variety of comfort movements, and it becomes more synchronized as the display progresses. The postcopulatory display is especially stunning. The pair rises out of the water, breast to breast, with necks extended and bills up. With their bills lowered, they then mutually turn their heads from side to side and slowly subside into the water (Huxley 1947, Boase 1959, Johnsgard 1965, Birkhead and Perrins 1986).

Mute Swans aggressively defend their territories from conspecifics. In Chesapeake Bay, 99.5% of 854 observed agonistic behaviors were directed toward other Mute Swans (Tatu et al. 2007a). The major threat display of the pair is called *busking*, where the head and neck feathers are erected, the head is projected backward, and the intruder is approached with an abrupt/jerky swimming action. *Busking* can escalate into *wing-flapping, chasing, slapping-feet-on-the-water*, and even a quite intimidating full-scale attack that will involve the bill and carpal joints (Delany 2005). *Wing-flapping* is also a threat display. The *triumph-ceremony* is given by both birds in the pair after an intruder is repulsed; it is similar to the male's *threat* posture but is also accompanied by *chin-lifting* and *calling* (Johnsgard 1965).

Some Mute Swans are aggressive toward other species of waterfowl, and this behavior has resulted in direct attacks, injury, and even death (Stone and Marsters 1970, Willey and Halla 1972, Ciaranca 1990, Conover and Kania 1994). In Maryland, Mute Swans have been recorded killing Mallard ducklings and Canada Goose goslings (Hindman and Harvey 2004). Conover and Kania (1994) noted 870 aggressive interactions during 804 hours of observing Mute Swans on 15 freshwater ponds (2–30 ha) in Connecticut, of which 410 (47%) were directed toward other Mute Swans and 460 (53%) were aimed at other species. Of the interspecific interactions, 249 (54%) were directed toward Mallards, American Black Ducks, and Canada Geese. Aggressive interactions with geese were the most intense and frequent (0.38/hr) and resulted in the geese swimming (>50 m) or flying, being chased into shallow water (<5 cm), and even being driven onto dry land. In contrast, aggressive interactions with ducks usually ended with the ducks moving a few meters away. There was individual variation, however, in the tolerance of Mute Swans toward other waterfowl: aggressive swans readily chased other species, while more tolerant swans did not. Still, the high rates of interspecific aggression observed by Conover and Kania (1994) are uncommon in birds (most aggression is usually intraspecific). Lastly, 38% of 174 interspecific interactions observed in Connecticut were directed toward humans and usually involved a threat display during attempts to feed the swans.

Nesting

Nest Sites. Mute Swan pairs are highly faithful to their nesting sites, returning to the natal area of the female. In England (Staffordshire), Mute Swans first recorded as pairs often came back to their natal areas: 36 out of 98 (37%) males and 60 out of 98 (61%) females returned to within 8 km, and 13 (36%) of those males and 32 (53%) of those females were within 1.6 km (Coleman and Minton 1979). Of 54 males, 20 (37%) first nested within 8 km of their natal area, 7 (13%) of which were within <1.6 km. In contrast, of 52 females, 32 (62%) first nested within 8 km of their natal area, 19 (37%) of which were within <1.6 km.

The male chooses the nest site, which may not be accepted by his mate, but, once they agree on a selection, the site is often reused in subsequent years (Reese 1975, Birkhead and Perrins 1986). Preferred nest sites have good access to water and are not susceptible to flooding. Characteristic sites include peninsulas; shorelines with dense emergent vegetation; and the edges of rivers, ponds, and brackish tributaries above the high-tide line (Ciaranca et al. 1997). In England, nests were most commonly located in or near standing water (51%) or near running water (46%; Campbell 1960). Both sexes construct the nest, but the male initiates the process by building the nest platform. The entire process takes about 10 days, and the resultant nest is large: the outside diameter can average 1.5 m. Outside height averaged 0.46–0.66 m; bowl depth, 7.7–10.3 cm; and bowl diameter, 38.5 cm ($n = 60$; Willey and Halla 1972, Gelston and Wood 1982, Ciaranca et al.1997).

Clutch Size and Eggs. Clutch size averages 5 or 6 eggs, but ranges from 4 to 8 (Johnsgard 1978). Kear (1970) reported an average of 5.7–6.2 eggs across a wide variety of sites in both Europe and North America. In North America, clutch size in an expanding Mute Swan population on Chesapeake Bay averaged 6.2 eggs from 1969 to 1979 (range = 4.8–8.0 eggs; $n = 151$), with the largest clutch containing 10 eggs (Reese 1980). Clutch size averaged 6.6 eggs (range = 6.2–6.8 eggs) in Connecticut (Conover and Kania 1999), 5.2 (range = 4.7–6.0) in Massachusetts (Ciaranca et al. 1997), and 4.3

(range = 1–8) in Michigan (Gelston and Wood 1982). Along the River Thames in England, Mute Swans laid smaller clutches after cold winters. The mean clutch size was greater for experienced pairs (those that had bred at least once) compared with inexperienced pairs (those that had not bred): 7.0 eggs (*n* = 41) versus 5.7 (*n* = 7; Birkhead et al. 1983). Elsewhere in England, adult female Mute Swans around Oxford laid earlier clutches on territories with abundant aquatic vegetation versus those with little vegetation. Clutch sizes were also larger on sites where humans provided an ample bread supply (Scott and Birkhead 1983). Clutch size ranged from 5.5 to 6.2 eggs over a 3-year period (1964–66) for Mute Swans at Oxford (Perrins and Reynolds 1967). Generally, average clutch sizes are higher in North America than in Europe (Conover and Kania 1999).

Mute Swan eggs are subelliptical (close to equally rounded at each end) and greenish blue when laid, but they then change to white (Kear 1972). In Rhode Island, egg dimensions averaged 103 × 75 mm; mass, 295 g (range = 258–365 g; Willey and Halla 1972). Johnsgard (1978) summarized information from several studies and reported egg size as 115 × 75 mm and average weight as 340 g; Kear (1972) noted size as 112.5 × 73.5 mm and weight as 340 g (3.8% of female body weight). Mute Swan eggs collected from 10 nests along the River Thames in England averaged 364 g (range = 294–397 g). The yolk averaged 135.7 g (37.2% of the wet weight of a fresh egg), and the lipid content of the dry yolk was 69.8% (Birkhead 1984). Egg laying usually begins 3 days after the nest is complete, and eggs are laid at 48-hour intervals.

Incubation and Energetic Costs. Only the female incubates the eggs. The male may sit on the nest when the female is at recess, but the male does not have a brood patch and thus does no real incubation (Willey and Halla 1972, Birkhead and Perrins 1986). The incubation period averages 36 days and has ranged from 34 to 41 days across several studies (Boase 1959, Reese 1975, Gelston and Wood 1982, Birkhead and Perrins 1986); Johnsgard (1978) reported 35–36 days. Not much is known about incubation constancy. Halla and Willey (1972:18) reported that the female was absent "only long enough to permit her to exercise and feed." Kear (1972) speculated that nest recesses in the northern species of white swans were only for about 30 minutes, during the warmest part of the day. There are no studies of reproductive energetics.

Nesting Chronology. Mute Swans in the Northern Hemisphere nest from March through June (Johnsgard 1978). In Chesapeake Bay (1968–74), the extremes for egg-laying dates were 13 March and 23 June, but 61% of all nests contained eggs from 9 to 16 April; the earliest young were seen on 4 May (Reese 1975). In Rhode Island, nesting extended from mid-April to mid-June, with a peak in mid-May (Willey and Halla 1972). Over 3 nesting seasons at the Wildfowl and Wetlands Trust in England, the first eggs were seen on 26 March and the median date of egg laying was 4 April (Kear 1972).

Nest Success. Nest success of Mute Swans is high, because their large size makes them formidable defenders against most predators. In a 9-year study (1982–90) in Connecticut, nest success averaged 81%–89% (*n* = 377), and all nests hatched in 4 of the 9 years (Conover and Kania 1999). Nest loss (determined for 33 nests) was due to flooding (46%), abandonment (27%), disturbance by humans (15%), predation (6%), and failure after the loss of a parent (6%). Mammalian predators destroyed Canada Goose and Mallard nests within 10 m of Mute Swan nests, but no swan eggs were lost. In a Chesapeake Bay study from 1969 to 1979, eggs were found in 94% of all nests (*n* = 210) and young (prefledglings) in 65% (Reese 1980). The percentage of eggs that hatched varied from 27% to 75% annually (average = 49%). The percentage of eggs that ultimately yielded fledglings ranged from 23% to 68% annually (average = 40%). Egg losses from nests were due to disappearance of the eggs between observer visits (44%), flooding by high tides (15%), eggs found as fragments (10%), and eggs found outside the nest (9%). Direct predation on eggs was rare, because nest predators in the area were much smaller than the swans. Overall, flooding ranks as the most common cause of complete nest loss (Perrins and Reynolds 1967, Reese 1980), followed by vandalism. During a study of Mute Swans in England (Staffordshire) from 1961 to 1978, vandalism caused 68.7% of 548 clutch losses (Coleman and Minton 1980).

In Rhode Island, Willey and Halla (1972) examined 142 eggs in 24 intensively monitored nests, of which 90.1% were fertile and 87.4% hatched successfully, compared with 49% of 1,101 eggs monitored in Chesapeake Bay (Reese 1980). In Connecticut, mean hatching success was 69%. In successful nests, 71% of the unhatched eggs had been depredated and 29% were abandoned (Conover and Kania 1999). In England, 94% of 103 eggs hatched and 48% produced fledglings (Reynolds 1965). Also in England, Minton (1968) found young in 58% of 456 nests monitored from 1961 to 1967.

Brood Parasitism. Brood parasitism has not been reported in the literature and is unlikely, given the territoriality and aggressiveness of breeding pairs. In Chesapeake Bay, however, a clutch of 8 eggs first discovered on 25 April was being incubated by the female and under incubation again when checked on 2 May, but 2 eggs were added on 6 May. The additional eggs were smaller than the original 8 (117.8–120.2 mm × 72.0–74.9 mm vs. 107.1–107.5 mm × 68.9–70.1 mm), of noticeably different color, and freshly laid (Larry Hindman). This observation occurred in 2007, when there were >200 Mute Swan nests.

Renesting. Mute Swans will re-lay if the first clutch is destroyed, probably because they have a longer breeding season in comparison with other northern species of swans, which primarily nest in the Arctic. As with other waterfowl, clutch size is usually smaller for the second nest. In Chesapeake Bay, a second clutch was laid where a clutch was lost before 10 May (*n* = 13 nests), with re-laying occurring 11–30 days after the loss of the first clutch

(average = 18 days; Reese 1980). Clutch size averaged 6.6 eggs for first nests and 5.2 for second nests (Reese 1980). Renests often contain smaller eggs.

REARING OF YOUNG

Brood Habitat and Care. Brood habitats are similar to those used by adults: relatively shallow water with abundant emergent and submerged vegetation. Mute Swans hatch with a yolk reserve that can be as much as 25% of their body weight, sufficient to sustain them for a week or more (Kear 1972). Cygnets are not fed directly, but they initially eat items stirred up by their parents through treading and upending. Cygnets can first dip their heads underwater at 7 days old and begin upending at 10 days (Dewar 1942, Owen and Kear 1972).

Adults brood cygnets during the hatch, which can take 2 days; they will continue brooding at night during cool, wet weather for 2–7 days posthatch, and sometimes longer (Willey and Halla 1972, Gelston and Wood 1982). The parents often carry the young cygnets on their backs, between the wings, which can provide both a dry resting place and protection from predators. Such endearing parental care is unknown among the other northern white swan species, but it does occur, albeit less commonly, in the Black-necked Swan (*Cygnus melancoryphus*) of southern South America (Delacour 1954).

Brood Amalgamation (Crèches). Brood amalgamation does not occur in Mute Swans.

Development. At the Wildfowl and Wetlands Trust in England, 1-day-old cygnets (*n* = 17) averaged 220 g (range = 180–248 g; Scott 1972). The sexes are similar in size at hatching, but males become larger at about 2 weeks of age (Kear 1972, Mathiasson 1981). Cygnets grow rapidly, with their weight increasing at a rate of 40%–50%/week for at least the first 8 weeks, and their length by 10.2 cm/week (Gelston and Wood 1982). The young fledge in 120–150 days (Kear 1972, Willey and Halla 1972, Reese 1975), at about 70%–75% of adult weight, with males averaging about 28% heavier than females (Mathiasson 1981). Juvenile subadults continue to put on weight for up to 4–6 years, at which time they reach sexual maturity. Around Staffordshire in England, males averaged 9.5 kg at fledging, their weight increased an average of 0.5 kg/year for the next 2 years (Bacon and Coleman 1986). Females averaged 8.0 kg at fledging and grew heavier by 0.25 kg/year for the next 2 years. The bill knob also increases in size and changes color from gray to orange red during this same time period (Mathiasson 1981).

RECRUITMENT AND SURVIVAL

Important information on survival, recruitment, and associated life-history attributes comes from a 39-year study (1961–99) of a Mute Swan population on a 1,440 km² area in central England (the Midlands), where much of the region is still rural (Coleman et al. 2001). Since 1961, 9,370 swans have been ringed (banded) on the study area, 7,027 (75%) of which were ringed either as pulli (nestlings) or fledged birds in their first or second year; 95% of the population is banded. The Mute Swan population in this locale ranged from a low of 213 birds in 1985 to a high of 551 in 1997. The paired population similarly ranged from a low of 53 pairs in 1985 to a high of 162 pairs in 1997—a tripling in 12 years. The number of breeding pairs changed more markedly: 53–78 during the first 15 years of the study, declining to only 34 in 1985, before rising to 120 in 1997.

This study also examined the life histories of 1,647 marked individuals with known dates of hatching and death. Annual mortality rates of these individuals were fairly constant over the first 9 years of life, ranging from 27% to 32%, except for birds in year 6 (24%). Collisions (usually with overhead wires) caused an average of 29% of all deaths each year (range = 21%–39%). The minimal survival rate for the first year of life (from hatching through Aug of the next year) averaged 41.4% from 1961 to 1977 (Coleman and Minton 1980). During the birds' first year, 43.7% of all mortality occurred from September through December (with a peak in Oct), and then again in March (about 15%); 71.4% of all deaths occurred within 3 km of the natal site.

Of the 1,647 marked individuals, 18% were ultimately observed as being paired (136 males, 152 females) and 11% (86 males, 91 females) achieved breeding status, but only 37% of the paired males and 40% of the paired females were ever seen breeding. The duration of paired life was only 1–2 years for 50% of the males and 53% of the females. Of the 288 paired swans, 60% of the males and 58% of the females survived beyond 5 years, and 16% of the males and 11% of the females lasted beyond 10. Six males and 7 females reached 15 years old; the maximum age was 19 for a male and 18 for a female. There was no statistical difference in life spans between males and females.

Cygnet production, as calculated from 177 breeding individuals, provided insightful data on the individual lifetime reproductive success of these long-lived birds (Coleman and Minton 1980). There were no statistically significant differences between the sexes; hence data were combined and revealed that 22% of the individuals produced no cygnets, and 26.7% produced only 1–5 cygnets over their breeding lives. In sharp contrast, 5.6% of the breeding birds produced 21–25 cygnets during their breeding lives, and 6.8% produced >26. The maximum number of cygnets reared over a breeding lifetime was 41 for a male and 50 for a female, yet only 20% of the breeding adults produced 59% of the fledged cygnets during the 39-year study period. The productivity of the population increased from 1.8 cygnets/breeding pair in the early 1960s to 2.7 in the late 1990s. Cygnet mortality before fledging averaged 24% from 1966 to 1968. Of 434 broods monitored over this time period, 215 (49.5%) were reared without loss, 168 (38.7%) had at least 1 cygnet die prior to fledging, and 44 (10.0%) were completely lost. Brood size at fledging averaged 3.7, with a maximum of 4.3 fledglings recorded during 2 years of the study.

Mute Swans at the Abbotsbury Swannery in Dorset, England, exhibited an average annual survival rate of 85% from 1976 to 1998 (McCleery et al. 2002). The rate was not different between the sexes, but it was greater in birds aged 8–11 years (89%), compared with birds >14 years old (60%). Very few birds, however, managed to live beyond 14 years. Survival was about 60% for swans aged 0–2 years. The mean number of breeding attempts averaged 5.3 for females and 4.4 for males, with females generally breeding for a longer time than males (females accounted for 23 of the 30 breeding attempts in swans >10 years old); 2%–28% of the birds that had bred failed to breed in subsequent years.

Near Oxford, England, Perrins and Reynolds (1967) reported first-year survivorship was 50% from July through September, 87% from October through December, 89% from January through March, and 88% from April through June. Second-year survivorship over these same 4-month periods ranged from 86% to 92%, third-year survival from 88% to 96%, and fourth-year survival from 87% to 94%. These estimates were based on ringed (banded) cygnets, except for the July–September period during the first year, which was estimated from observations of broods. On islands in the Outer Hebrides of Scotland, a 1971–74 study reported an average of 86 breeding pairs of Mute Swans in March, of which 39 (45%) had young in September–October, with an average of 1.48 young/productive adult. Cygnet loss averaged 27.8% from June/July to September/October, and 74.1% from June/July to December (Jenkins et al. 1976).

In Chesapeake Bay, an 11-year study (1969–79) of an expanding Mute Swan population reported that 82% of 535 young fledged (Reese 1980). Mean productivity was 2.2 fledglings/nest with eggs (range = 1.2–4.1 fledglings), and the mean brood size was 3.9 (n = 137). Reproductive success also increased with swan age and over the span of first- to fourth-time nest attempts (i.e., experience). Fledglings averaged only 0.77/nest for 1-year-old parents (n = 13 nests) versus 2.6–2.7/nest for 4- and 5-year-olds (n = 37 nests). Relative to experience, first-time nesters averaged 4.9 eggs, 1.2 young, and 0.84 fledglings/nest versus 6.7 eggs, 3.8 young, and 3.0 fledglings/nest for fourth-time nesters. Swans aged 5–15 years averaged 6.7 eggs (n = 40 nests), 3.1 young (n = 42 nests), and 2.4 fledglings/nest (n = 42 nests). Most of the mortality for young birds (84%) was recorded as "disappeared between visits" and occurred when the cygnets were <40 days old. Hatchlings with gray-brown plumage had statistically higher (87%) survivorship than white-plumage hatchlings (73%). Little direct predation was observed, except for some instances of cygnet loss due to snapping turtles (*Chelydra serpentina*) and diamondback terrapins (*Malaclemys terrapin*). Postfledging survival (at least 100 days old but <1 year) averaged 90% over 9 years for males (range = 83%–100%) and 89% for females (range = 84%–100%). The most common form of postfledging mortality was collision with power lines (28%). In the United Kingdom, overhead wires were the single most important (44%) cause of mortality for Mute Swans (Ogilvie 1967).

During a 9-year study (1982–90) in Connecticut, brood size averaged 4.5 at hatching but was only 3.2 at fledging; brood survival averaged 94% and cygnet survival, 74% (Conover and Kania 1999). The percentage of eggs laid that ultimately produced cygnets averaged 41% (range = 26%–50%). Mean production (2.7 young/pair) was lower than mean brood size, because the latter does not account for the complete loss of clutches or broods. There were no differences in reproductive rates between the 2 major habitats used by Mute Swans: lakes/ponds and estuaries/tidal rivers. Nonetheless, a growth rate of 2.7 young/pair was among the highest reported for Mute Swans and thus was indicative of an expanding population that was not yet close to carrying capacity.

In addition to collisions with power lines, adult and juvenile Mute Swans are also victims of lead poisoning. In Michigan, Gelston and Wood (1982) reported that 13% of fledged Mute Swans died from this cause. In England, lead poisoning of Mute Swans has long occurred on the River Thames, where the birds ingest lead sinkers used on fishing lines (Simpson et al. 1979, Sears 1989). This form of poisoning caused 24% of all Mute Swan mortality on the Thames in 1987; the 1983–88 mortality rate was reduced by 70%, however, after a ban on larger lead sinkers was instituted in 1987 (Sears 1989). Elevated levels of lead in the blood of Mute Swans were detected in 84%–89% of the swans sampled from 1983 to 1986, which declined to 44% in 1987, and 24% in 1988, although higher-than-normal amounts are still found in some swans (Perrins et al. 2003). Lead-poisoning deaths in the United States have occurred in the Atlantic Flyway from the ingestion of lead shot and fishing sinkers (Ciaranca et al. 1997). Mute Swans examined on Lakes Erie and St. Clair ($n = 50$) during 2001–4 contained an array of trace elements, including mercury, copper, and selenium, but these levels were not deemed sufficient to affect survival or reproduction (Schummer et al. 2011a). Adult Mute Swans have few natural predators except humans; most deaths (aside from disease) are caused by accidental encounters with man-made objects, such as power lines.

After their introduction to and subsequent expansion in the United States, Mute Swans demonstrated nearly exponential population growth in some areas, as might be expected for an invasive species first arriving in unoccupied habitat. Hence Mute Swans can saturate a breeding habitat, at which point density-dependent factors can act to regulate population growth. On the 16 km² island of Aasla and surrounding areas in the southeastern archipelago of Finland (40 km² total), an increasing population of Mute Swan was studied from their establishment in 1976 until 1998 (Nummi and Saari 2003). The 1998 population density in the area is the highest reported in Europe for a non-colonial population: 135/100 km², of which about 60 pairs were nesting. All production parameters correlated negatively (Spearman rank correlations) with the number of pairs: the average number of clutches/pair ($r = -0.45$), broods/clutch ($r = -0.84$), fledged cygnets/brood ($r = -0.71$), and fledged young/pair ($r = -0.88$). In a multiple regression analysis, the number of pairs and the number of clutches/pair explained 71% of the variation in the average number of fledglings/pair. In 2 of the last years of the study (1996 and 1998), 32 and 54 pairs, respectively, produced about the same number (3–4) of fledglings as did the initial 1–2 pairs in the first 2 years of the study.

In North America, the longevity record for a wild Mute Swan is 26 years and 9 months for a male banded and later found dead in Rhode Island. A female banded in Massachusetts lived for 25 years. And 20-year-old wild birds are not uncommon (Reese 1980). In a marked population of paired Mute Swans ($n = 288$) in England, 4% of the males and 5% of the females survived >15 years (Coleman et al. 2001).

FOOD HABITS AND FEEDING ECOLOGY

Mute Swans are almost entirely vegetarian, largely consuming the leaves, stems, and tubers of submerged aquatic plants (Owen and Kear 1972). They will eat small amounts of animal matter, perhaps incidentally to their consumption of plant matter. They prefer to feed in shallow water (<0.5 m), where they can reach bottom, but their long necks allow them to find foods at depths of up to 1.5 m when upending (Berglund et al. 1963, O'Brien and Askins 1985, Allin and Husband 2003). As in all swans, the neck of a Mute Swan contains a large number (25) of cervical vertebrae, which help make the neck muscular and mobile. Such flexibility, along with a long neck, allows swans to consume submerged vegetation in places that are too deep for surface-feeding ducks and yet too shallow to be exploited by diving ducks (Owen and Kear 1972). When feeding, Mute Swans also often use their feet to paddle and disturb the substrate, which facilitates the uprooting of rhizomes and roots and brings food to the surface for cygnets (Owen and Kear 1972, Birkhead and Perrins 1986).

Like other swans, Mute Swans exhibit 4 general feeding behaviors: (1) dabbling at the surface, (2) submerging the head, (3) submerging the neck, and (4) upending (Owen and Kear 1972). Dabbling Mute Swans skim items from the water's surface, whereas dipping birds submerge the head and neck, usually in water 41–69 cm deep (Dewar 1942), but also at depths of up to 70 cm (Owen and Cadbury 1975, Sears 1989). In flocks of Mute Swans observed in a rural area along the upper River Thames in England, surface feeding was the most (51%) common form of foraging, followed by neck submersion (28%), and upending (21%; Sears 1989). In contrast, swans on an urban river, where they were often fed by humans, primarily used neck submersion (43%), then surface feeding (36%), head submersion (18%), and upending (3%).

In Chesapeake Bay, 81.8% ($n = 80$) of all Mute Swan food consisted of SAV, determined from fecal analysis. Algae formed 8.4% of their diet, followed by other plants (mostly corn and miscellaneous grasses) at 8.3%, and animal matter at 0.3% (Fenwick 1983). Of the various SAV components, 47.1% was widgeongrass (*Ruppia maritima*), 18.3% sago pondweed (*Potamogeton pectinatus*), 11.2% myriophyllum (*Myriophyllum spicatum*), and 10.0% pondweed (claspingleaf pondweed; *Potamogeton perfoliatus*). Subsequent food-preference studies of 5 captive Mute Swans found that common waterweed (*Elodea canadensis*) was selected over all other SAV in 10 out of 12 tests. Other preferred foods were wild celery (*Vallisneria americana*) and widgeongrass. In contrast, myriophyllum was never chosen over other species of plants in 13 trials, and on 3 occasions it was left untouched. Food-habits analyses (gullets of collected birds) of 29 Mute Swans in Chesapeake Bay similarly found an overall dependence on SAV—widgeongrass (76%) and eelgrass (*Zostera marina*; 9%)—with some corn (<2%) and a small amount of invertebrates that were believed to have been consumed incidentally with vegetation (Perry et al. 2004). Sea lettuce (*Ulva lactuca*) was the primary dietary item eaten by Mute Swans in Connecticut (O'Brien and Askins 1985). In Rhode Island, 95%–100% of the stomach contents of 51 Mute Swans consisted of plant material (Willey and Halla 1972). In all, 21 plant foods were utilized, including pondweeds,

widgeongrass, and eelgrass, as well as sea lettuce, red algae (*Bangia fuscopurpurea*), and diatoms.

Mute Swans foraging in Chesapeake Bay during spring and summer spent 38.4% of their time feeding, 21.8% nonforaging (swimming), 18.6% in self-maintenance (such as preening), and 18.4% resting (Tatu et al. 2007a). Feeding intensity did not differ between spring and summer, but swans foraged more intensely during the morning than at midday. Mute Swans in flocks of more than 3 foraged more than pairs (45.4% vs. 34.2%), and large flocks (average size = 75) foraged more than small flocks (average size = 7; 60.1% vs. 42.2%). In England, Mute Swans feeding in agricultural fields (e.g., winter wheat) spent 67% of their time foraging, versus only 22% for swans in urban habitats, where humans often gave them bread (Sears 1989). Similarly, Mute Swans fed by humans in Krakow, Poland, spent only 4.6% of their time eating, versus 48.1% for swans outside the city that did not have similar food supplements (Jozkowicz and Gorska-Klek 1996). On the Ouse Washes in England, Mute Swan foraging time peaked at 70%–80% (Owen and Cadbury 1975).

In marshes along the lower Great Lakes, Mute Swans were collected throughout the year from 2001 through 2004 (Bailey et al. 2008). Their diets largely consisted of the aboveground biomass of SAV: pondweeds, muskgrass (*Chara* spp.), coontail (*Ceratophyllum demersum*), slender naiad (*Najas flexilis*), common waterweed, and wild celery. Wild rice (*Zizania palustris*) was also eaten, but belowground parts of SAV were infrequently used. SAV formed 66.3% of the diet in adult males ($n = 57$), 72.4% in adult females ($n = 51$), and 71.1% in cygnets ($n = 24$); >98% of the diets consisted of plant matter (invertebrates made up <1.2%). Their diets did not differ by year, season, season within a given year, or collection site, but the percentage of 10 major food items varied between adult males and females: the females ate a larger amount of pondweeds but less slender naiad and common waterweed than did the males. Adults and cygnets had similar diets during summer and autumn. All of the major items consumed by Mute Swans were also common and important foods for native waterfowl that use the lower Great Lakes marshes as a nutritional source during migration and winter.

In marshes associated with estuaries in South Devon in England, Mute Swans extensively ate emergent vegetation (Gillham 1956). Their preference was for succulent emergents, such as sea arrowgrass (*Triglochin maritimum*), sea plantain (*Plantago maritima*), sea aster (*Aster tripolium*), spurry (*Spergularia* spp.), scurvygrass (*Cochlearia anglica*), and black saltwort (*Glaux maritima*). Sea arrowgrass and sea plantain were preferred early in the season and eaten until these plants were nearly depleted, after which saltmarsh grass (seaside alkaligrass; *Puccinellia maritima*), a semisucculent, became the major food source.

The large size of Mute Swans and their preference for plant foods means that they eat significant amounts of aquatic vegetation. In studies of 5 penned Mute Swans, the males consumed 34.6% of their body weight daily and the females, 43.4%, which translated into 4.2 kg/day for the males and 4.0 kg/day for the females (Fenwick 1983). Willey and Halla (1972) presented an average intake of 3.8 kg/day for adult/subadult Mute Swans in Rhode Island, which Perry et al. (2004) used in their estimate that 4,000 Mute Swans in Chesapeake Bay would eat 5.5 million kg of SAV annually (about 13% of the estimated SAV in the upper Bay). Mute Swans in marshes along the lower Great Lakes potentially consume 8,000 kg of plant biomass daily (Bailey et al. 2008). In Sweden, Mathiasson (1973) calculated that about 40 Mute Swans ate 8,635 kg of sea lettuce during a 45-day period at a single location, and roughly 65 birds consumed 14,197 kg in another. Captive Mute Swans during the molt each ate an average of 3.66 kg of eelgrass and 4.03 kg of sea lettuce/day (7.69 kg/day total). In the Netherlands, Mute Swans removed 87% of the eelgrass beds at foraging sites (Nienhuis and Van Ierland 1978).

Mute Swans will also uproot a significant amount of vegetation that is not eaten, which is detrimental to the growth of SAV beds (Gillham 1956, Naylor 2004). Individuals can remove about 9 kg of plant matter (wet weight) but typically consume <50% of it (Mathiasson 1973, Fenwick 1983). Lastly, native species of SAV evolved concurrently with native waterfowl, where the timing of most feeding does not overlap with the reproductive cycle of SAV. In contrast, Mute Swans negatively affect SAV, because they feed on this vegetation year-round and often remove whole plants before the plants can reproduce (Naylor 2004).

In Chesapeake Bay, researchers used exclosures at 18 sites to investigate the effects of Mute Swan herbivory on SAV (Tatu et al. 2007b). At the end of their 2-year study (2003–4), the mean percentage of cover was reduced by 79%, shoot density by 76%, and canopy height by 40%. Paired swans predominantly occupied moderate-depth (0.76–0.99 m) sites, and these areas experienced less (32%–75%) SAV loss in comparison with shallower (0.50–0.75 m) sites that were occupied by flocks (75%–100% loss). Flocks also inhabited some deep-water sites and a moderate-depth locale; SAV loss was considerable (77%–93%) in these areas, too. The study demonstrated that Mute Swans severely affected SAV growth in Chesapeake Bay; it also showed that flocks were more detrimental to SAV than pairs and thus should be the focus of control efforts.

On an 85 ha coastal pond in Rhode Island, Allin and Husband (2003) used a series of exclosures to determine the effect of Mute Swans on SAV and macroinvertebrates. The SAV taxa on the pond (in order of decreasing abundance) were widgeongrass, sago pondweed, muskgrass, horned pondweed (*Zannichellia palustris*), and claspingleaf pondweed. Mute Swans used the pond throughout the year, with peak use (>300) occurring during the molt in August. Swans fed at all water depths on the pond but seemed to prefer shallow water (<0.5 m), where they reduced SAV biomass by as much as 95%. Other studies also suggest that herbivory by Mute Swans is greatest in shallow water, although they can reach food at much deeper depths (Berglund et al. 1963, Mathiasson 1973, Gelston and

Wood 1982). A direct effect on invertebrates could not be demonstrated.

In contrast to the Rhode Island study, Conover and Kania (1994) did not find that Mute Swan herbivory affected SAV on coastal ponds in Connecticut. Their study was on small (2–30 ha) ponds, however, where swans were resident year-round and highly aggressive toward conspecifics; hence only the mated pair or small numbers of swans probably used each pond during the year, which would lessen their foraging impact on SAV. Reichholf (1984) reported that Mute Swans on breeding territories only removed about 20% of the available vegetation.

MOLTS AND PLUMAGES

As adults, Mute Swans have only 1 plumage (basic) that is molted each year via the prebasic molt. The descriptions of molts and plumages are primarily from summaries and data in Palmer (1976a) and Ciaranca et al. (1997), which should be consulted for further details.

Down-covered hatchlings appear as 2 color morphs, with each capable of occurring in the same nest: gray (Royal) or white (Polish), although there can be shading of colors on either morph (Nelson 1976). The color of the cygnets is controlled by a single sex-linked, recessive gene: gray is dominant over white (Munro et al. 1968). The white morph can be common in some areas (e.g., 20% in parts of Eastern Europe), however, because selective breeding favored white swans, which were preferred by early buyers and collectors. The gene for color is also linked to the sex chromosomes; females with the Polish gene are always white, because there is no second gray gene to overrule it, while male cygnets are only white if both genes are white (Birkhead and Perrins 1986).

Replacement of natal down begins in about 42 days and leads to juvenal plumage. There is not much feather growth before 7 weeks. All feathering is white for the white morph. The bill is light brown and without a knob; the feet are also light brown. In contrast, juvenal plumage of the gray morph is a drab blend of various grays and browns. The head and the neck are grayish brown, with the remaining upperparts and the flanks margined with brown, which presents a mottled appearance; the underparts are a pale gray. The bill is dark gray and without a knob; the legs and the feet are black; and the lores are nearly bare and black, as in the white morph. The juvenile bill of both morphs becomes pinkish during the first winter.

The prebasic molt begins at about 4.5 months and yields first basic approximately 2–3 weeks thereafter. This molt replaces all or most of the body feathers and retrices, but not the remiges. The resultant feathering is all white in white-morph birds, but gray morphs have a brownish-gray or neutral-gray head and neck. Most upperparts and underparts are shades of gray and grayish brown, but the tail is pale to whitish. The second basic molt usually begins in late spring and involves a molt of body feathers and remiges, which results in a flightless period as early as June. This period lasts about 4–7 weeks (Birkhead and Perrins 1986), with flight capabilities usually restored by August, but sometimes not until early September. Mathiasson (1973) reported remigial growth rates of 5.5–7.6 mm/day for the fifth primary and 3.2–5.8 mm/day for the fifth secondary. Breeding Mute Swans molt while their cygnets are unable to fly; the female starts to molt soon after the young hatch, while the male begins to molt after his mate's flight feathers are well grown. Mute Swans often use their wings in fights and in defense of the brood; hence a staggered molt means that a member of the pair is always capable of using its wings for defense, without damage to the developing quills (Birkhead and Perrins 1986).

The differences in the subadult plumages between the 2 color morphs yield costs and benefits for individual swans. In Chesapeake Bay, gray morphs had higher survival rates from hatching to fledging (87%) than did white morphs (73%; Conover et al. 2000). Furthermore, by fall the parents often drove white morphs out of territories while allowing their gray-morph brood mates to remain for several more months. Parents may chase away their white-morph offspring because they confuse them with competitors. Alternatively, the coloration of white morphs may signal that they are ready to breed, so the parents force them from the breeding territory. Either way, about half of the white morphs driven away by parents were dead within a few months. Gray-morph males therefore had higher survival rates than white morphs during the first 2 years, but white morphs bred at an earlier age. The explanatory hypothesis for the evolution of a subadult plumage is that gray morphs advertise their true subadult status to other swans. This advertisement increases their chance of survival, because they receive a longer period of parental care and are subject to less aggression from older birds, but the cost may be to forgo breeding at an early age. In contrast, white morphs convey incorrect age information to unintended recipients such as parents and other adults, so they are prematurely excluded from their parents' territory and less tolerated by other adults: the price for appearing older than they really are is a lower survival rate.

Definitive basic is acquired by adults from June to September. Females normally begin molting well before males, usually when the cygnets are small. The associated basic molt involves both body feathers and remiges, which results in a flightless period of about 8 weeks. The knob on the bill also continues to enlarge, reaching a height of 17–25 mm in males and 14–20 mm in females. Successful breeders molt later than unsuccessful breeders and nonbreeders.

CONSERVATION AND MANAGEMENT

Mute Swans are invasive exotics in North America, where their expanding population negatively affects native waterfowl and other waterbirds because of their reliance on submerged aquatic vegetation for food and their aggressive nature toward other waterbirds, especially during the breeding season. Mute Swans also can be a threat to humans,

as a blow from the carpal wing joint of these strong birds can severely injure a person. Mute Swan attacks by breeding adults have caused alarmed recipients to upend canoes, kayaks, and even small fishing boats (Atlantic Flyway Council 2003). In Indiana, a 13-year-old girl swimming in the Lake James chain was pulled underwater for several seconds (*Chicago Sun-Times* 2006). In Norway, a Mute Swan attacked an elderly woman, dragging her 5 m into the water, resulting in an overnight stay in an area hospital for the victim (Mountford 2004). Mute Swans have killed dogs on numerous occasions, and, alarmingly, even humans: a child in Massachusetts (circa 1930), and an Indiana man in 1982, who was drowned after an attacking swan knocked him from a small boat and continued the attack thereafter (Williams 1997). In April 2012, a 37-year-old man in Illinois was knocked from a kayak and subsequently drowned by an attacking Mute Swan (*Chicago Sun-Times* 2012).

The effect of Mute Swans on SAV is well documented, as their large size and foraging habits mean that they eat large amounts of plant material, as well as pull up vegetation that is not consumed (see Food Habits and Feeding Ecology). This use and destruction of SAV by Mute Swans is significant, because SAV is an important food source for native swans (e.g., Tundra Swans and Trumpeter Swans), as well as native ducks (e.g., Redheads, Canvasbacks, American Wigeon, and others). Tatu et al. (2007b) recorded 13 species of waterbirds sharing sites used by Mute Swans in Chesapeake Bay. The loss of SAV also affects myriad other organisms, such as blue crabs (*Callinectus sapidus*), which are of significant commercial importance in Chesapeake Bay (Hurley 1991, Hindman and Harvey 2004).

Mute Swans are also strongly territorial and aggressively defend their territories, not only from other swans during the breeding season, but also from additional species of waterfowl, such as Mallards and Canada Geese. In Maryland, a molting flock of >600 Mute Swans using shell bars and beaches caused the abandonment of those sites by breeding Least Terns (*Sterna antillarum*) and Black Skimmers (*Rynchops niger*), both of which are state listed as threatened species (Therres and Brinker 2004).

The factors above add justification for control of the Mute Swan population in North America, which has been practiced in some states since the mid-1970s. In the Atlantic Flyway, the current management goal is "to reduce Mute Swan populations in the Atlantic Flyway to levels that will minimize negative ecological impacts to wetland habitats and native migratory waterfowl and to prevent further range expansion into unoccupied areas" (Atlantic Flyway Council 2003:ii). In the Mississippi Flyway, the goal is to reduce the population to 4,000 or fewer birds by 2030. Egg addling has been used in many states and is less controversial for the public than killing adult swans, but the technique is ineffective in reducing population size, because adult swans are long lived. In Rhode Island, a review of that state's efforts to control Mute Swans from 1979 to 2000 reported 9,474 eggs addled at 1,636 nests, although the population continued to grow at about 5.6% annually and to expand in its distribution (Allin and Husband 2004). Clutch reduction as a management tool is also inefficient: it is labor intensive, it must be persistent (if not continuous) over several years, and the effects may vary with immigration rates. In a Mute Swan population in England, models predicted that a reduction in clutch size to just 2 eggs/clutch would only lower breeding numbers by 30% after 10 years. The total destruction of clutches would stabilize the nonbreeding population but not eliminate it, because clutch destruction was offset by immigration and high survival rates (Watola et al. 2003).

Control measures, however, must be implemented over broad areas if effective population reductions are to occur, given the reproductive potential of Mute Swans. Reese (1980) had conservatively predicted a population of 1,832 Mute Swans in Chesapeake Bay by 1992, yet 2,245 were recorded during the Midsummer Survey of 1993,

and a high of 4,443 was tallied in 1999. For the entire Atlantic Flyway, Allin et al. (1987) used what was then a 5.6% annual growth rate to surmise that the extant Mute Swan population in 1985 could possibly double by the year 2000. The Midsummer Survey reported 12,643 Mute Swans in 1999, about 2.2 times the 1985 number, and the population still appears to have available habitat within which to expand. In Maryland, Hindman and Harvey (2004) estimated that wetland habitat in the upper Chesapeake Bay and elsewhere in the coastal zone could potentially support 18,134 pairs of Mute Swans, which, together with nonbreeders, could lead to a population of 100,000 Mute Swans at some time in the future. Population modeling suggested that an 80% reduction in recruitment was necessary to stabilize the population; a 20% reduction in annual survival rates of adults would cause the population to slowly decline over time. A model developed by Ellis and Elphick (2007) indicated that survival rates have to be reduced by >17% to achieve a 90% chance of population decline; reproductive rates would need to diminish by 72%. The model suggested that population control via a period of intensively culling adults was the most biologically effective and economically efficient option to reduce the population, and ultimately it would minimize the number of swans killed.

A coordinated flyway-wide approach is needed for effective Mute Swan management. In the Atlantic Flyway, the Chesapeake 2000 Agreement was adopted by Maryland, Pennsylvania, Virginia, the District of Columbia, the Chesapeake Bay Commission, and the U.S. Environmental Protection Agency. This agreement requires the jurisdictions to develop and implement management plans for nonnative invasive species, and the Mute Swan was identified as a priority. In 2003, the *Atlantic Flyway Mute Swan Management Plan* outlined a flyway-wide approach to control Mute Swan populations (Atlantic Flyway Council 2003). In 2004 a compatible plan, *Mute Swan (*Cygnus olor*) in the Chesapeake Bay: A Bay-Wide Management Plan*

(Chesapeake Bay Mute Swan Working Group 2004), was developed for the Chesapeake Bay area. In 2005, with the adoption of a statewide Mute Swan management plan, Maryland began using a combination of adult removal and egg oiling to reduce hatching, which has whittled the state's population from a high of 3,955 in 1999, down to 581 in 2008, and to only 76 birds in 2011. Hence the results in Maryland demonstrate that population control can be effectuated. Michigan's most recent (2012) plan calls for the removal of 13,000 Mute Swans, with a long-term goal of a spring population below 2,000 by 2030 (Michigan Department of Natural Resources 2012). Many states have active egg-addling and/or egg-oiling programs; they have also removed Mute Swans from wildlife management areas and, in some instances, private lands. Washington, Oregon, and Montana have removed Mute Swans in association with management for Trumpeter Swans. Most states also prohibit the importation of Mute Swans, as a means to prevent their escape into the wild. In 1998 the U.S. Fish and Wildlife Service issued a policy statement that directed refuge managers to control Mute Swans.

Control of Mute Swan numbers in North America is biologically defendable and should be supported, although there are issues of acceptability by the general public. Mute Swans are not federally protected in the United States, but such a policy was controversial and required an amendment, titled the Migratory Bird Treaty Reform Act of 2004, to the Migratory Bird Treaty Act of 1918. Mute Swans are a protected species throughout Canada, however, via the Migratory Birds Convention Act of 1916, which makes direct control in that nation difficult, but there is increasing discussion to remove protection for Mute Swans as their population increases in numbers and distribution, and as it potentially competes with the expansion of the native Trumpeter Swan population.

Trumpeter Swan

Cygnus buccinator (Richardson 1831)

Left, adult; *right*, juvenile

The Trumpeter Swan, endemic to North America, is the world's largest species of waterfowl. Their recovery is among the great conservation success stories of the twentieth century, as this species was on the verge of extinction in the early 1900s. By 1932, only 69 Trumpeter Swans were known to exist in the lower 48 states, at remote locations in and around Yellowstone National Park. By 2010, however, the population was 46,225 (an increase of >6%/year from 1968 to 2010), the result of decades of dedication by waterfowl biologists. Three populations are recognized: the Pacific Coast, the Rocky Mountain, and the Interior. The Pacific Coast Population breeds primarily in Alaska and is the largest of the 3 groups, containing about 58% of the continental population. The Interior Population is the result of highly successful reintroductions into its former breeding range. These large white birds can weigh over 12 kg (males), mate for life, and are long lived; the record age for a wild bird is 32 years and 6 months. They are primarily vegetarians, with sago pondweed (*Potamogeton pectinatus*) a highly preferred food year round. Protection of habitat throughout the birds' annual cycle is paramount to the welfare of Trumpeter Swans, but it is especially acute on wintering areas along the Pacific coast, where the loss of natural wetland habitat has been extensive, and for the Tristate Flock of the Rocky Mountain Population where winter overcrowding in Idaho, Montana, and Wyoming is a significant problem. Lead poisoning is a serious mortality issue, especially on the coast in western Washington and nearby British Columbia, as are collisions with power lines. The Trumpeter Swan is closely related to the Whooper Swan (*Cygnus cygnus*) of Europe and Asia, with some authorities considering them to be conspecific. The early monograph by Banko (1960) is still perhaps the best single source of information on the life history of the Trumpeter Swan.

IDENTIFICATION

At a Glance. Adults are extremely large, with snowy all-white plumage; immatures are a dull grayish brown. They almost always lack yellow on the lores. Trumpeter Swans have remarkably long necks, almost as long as their bodies. They can be confused with Tundra Swans, from which Trumpeter Swans can best be distinguished by their bill characteristics and voice.

Adults. Trumpeter Swans are large birds, with an all-white plumage that is identical in the males and the females. The head and the upper neck are often stained a rusty color from contact with ferrous soil minerals encountered while feeding. The legs and the feet can be orange, yellow, pink, gray,

gray pink, or grayish yellow (McEneaney 2005). The black bill is long and straight, not unlike the profile seen in Canvasbacks. A prominent salmon-red line is often visible along the upper edge of the lower mandible. Only rarely is there any yellow on the lores.

Juveniles. Young Trumpeter Swans have grayish-brown plumage that is well developed in about 10 weeks and retained until late spring or early summer. The legs and the feet are grayish pink, changing to olive gray and then black. The bill is fleshy pink, with black borders (Delacour 1954). Second-year Trumpeter Swans are mainly white but often have some pale grayish feathers on the head, the neck, and the body (Palmer 1976a).

Cygnets. Cygnets are usually a "mouse gray" at hatching and darker dorsally, but they are occasionally white (Banko 1960). Nelson (1993) refers to these color morphs as "gray" and "leucistic" (white). The white morph formerly occurred only in the Tristate Flock of the Rocky Mountain Population, where a 4-year study (1937–40) found that 13.0% of the cygnets were white (Banko 1960). White-morph birds now are found in the Interior Population, most likely due to translocations associated with reestablishing populations. The feet and the tarsi are grayish pink. The bill of newly hatched cygnets is pink at the base, with a grayish-black tip.

Voice. Trumpeter Swans have a deep resonant call, *ko-hoo*, emphasized on the second syllable. The syrinx is considerably larger in Trumpeter Swans than in Tundra Swans, which is probably responsible for the Trumpeters' deeper and more resonant tones (Bellrose 1980). Tundra Swans lack an upward loop of the large trachea inside the sternum, which may also cause the difference in voices (Delacour 1954).

Similar Species. The Trumpeter Swan is most similar to the Tundra Swan, from which it is separated by a longer and straighter bill, a lack of yellow on the lores, its voice, and the shape of the black

Above:
Trumpeter Swan
pair in winter.
RyanAskren.com

Right:
Trumpeter Swan
captured for marking
on the Minto Flats
wetlands in Alaska.
*Heather Wilson, U.S. Fish
and Wildlife Service*

facial skin at the base of the bill. Immatures are very similar but can be separated by their plumage and foot color (see Similar Species in the Tundra Swan account). Trumpeter Swans are closely related to the Whooper Swan of Europe and Asia, and some authorities have considered the 2 taxa to be conspecific.

Weights and Measurements. The Trumpeter Swan is the largest species of waterfowl in the world, although an exceptionally large Mute Swan can be heavier. The males are bigger than the females. The most extensive dataset on weights and measurements of Trumpeter Swans was reported from birds captured during the winter in Idaho and Montana (Drewien and Bouffard 1994). Weights averaged 11.9 kg (range = 9.1–14.5 kg) for adult males ($n = 152$); 10.3 kg for male cygnets ($n = 167$); 10.3 kg (range = 7.0–12.5 kg) for adult females ($n = 12$); and 8.7 kg for female cygnets ($n = 147$). Annual body mass varied in relation to mean air temperature during trapping periods. Weights during the summer at the Red Rock Lakes National Wildlife Refuge (NWR) in Montana averaged 11.4 kg for males ($n = 27$) versus 10.3 kg for females ($n = 47$; Barrett and Vyse 1982). Bill length (tip of the bill to the posterior edge of the nares) averaged 69.2 mm for adult males ($n = 94$), 68.3 mm for adult females ($n = 71$), 68.6 mm for male cygnets ($n = 118$), and 67.0 mm for female cygnets ($n = 98$; Drewien and Bouffard 1994). Tarsi averaged 132.2 mm for adult males ($n = 84$), 127.0 mm for adult females ($n = 64$), 130.1 mm for male cygnets ($n = 108$), and 124.7 mm for female cygnets ($n = 86$). Wing chord length averaged 61.9 cm for adult males ($n = 5$) and 62.3 cm for adult females ($n = 3$; Banko 1960); wingspans can easily reach 2.35 m (Powell and Engelhardt 2000). Mitchell and Eichholz (2010) present data from other studies reporting weights and measurements of Trumpeter Swans.

DISTRIBUTION

Historical. The present range of Trumpeter Swans is a vestige of a once-larger distribution in North America. Hansen (1973) provided a hypothetical map of the ancestral breeding range that extended from central Alaska east through the central Yukon, southwestern part of the Northwest Territories, most of British Columbia, Alberta, Saskatchewan, Manitoba, Ontario, and eastern Québec. The continental U.S. range was from eastern Washington east to Michigan, Illinois, and Indiana, and south to Idaho, Wyoming, Nebraska, northeastern Kansas, and northern Missouri. In eastern Canada, Lumsden (1984) gave the species' original range as the Hudson Bay Lowlands of Manitoba, Ontario, and Québec. There are also published reports of Trumpeter Swans as far east as New Brunswick, Nova Scotia, and Newfoundland in Canada, and as far south as Arkansas, northwestern Mississippi, and Tennessee, although the southern populations were probably more localized (Banko 1960). Trumpeter Swans once wintered on Chesapeake Bay and Currituck Sound (North Carolina), the lower Mississippi River valley, the Gulf Coast and the Rio Grande valley, the Sacramento Valley and Los Angeles basin of California, the lower Columbia River, and the Pacific coast from Puget Sound to Alaska (Mitchell and Eichholz 1994).

Present. The present breeding range of Trumpeter Swans extends from central Alaska south of the Brooks Range and east of the Yukon-Kuskokwim Delta to Canada, and locally through the southern Yukon and western edge of the Northwest Territories south to British Columbia and Alberta. There are also locally breeding flocks in Saskatchewan, Manitoba, along southern Ontario and in western Quebec. In the United States, breeding birds are found in Washington, Oregon, southern Montana, eastern Idaho, western Wyoming, western South Dakota, northern Nebraska, Minnesota, Wisconsin, Iowa, Michigan, Ohio, and New York. Their winter range extends from southern Alaska along the coast to Oregon. Inland wintering concentrations are in southcentral British Columbia, eastern Nevada, southcentral Oregon, southern Montana, eastern Idaho, northwestern Wyoming, the

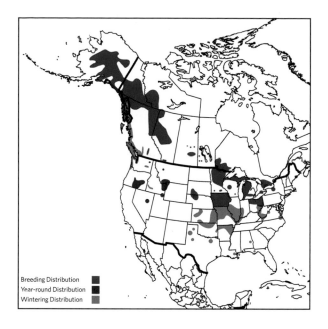

Breeding Distribution ■
Year-round Distribution ■
Wintering Distribution ■

Nebraska Sandhills, Kansas, Oklahoma, southern Minnesota, Iowa, Missouri, Arkansas, southern Illinois, and southern Indiana, with a few birds observed farther south each winter (Mitchell and Eichholz 2010).

Three populations are recognized: the Pacific Coast, the Rocky Mountain, and the Interior. The Pacific Coast Population is naturally occurring and the Rocky Mountain Population is mostly naturally occurring, while the Interior Population is entirely the result of restoration efforts that used stock from both of the natural populations. Genetic studies revealed significant differentiation between the Pacific Coast and Rocky Mountain Populations, although both had similar (and somewhat limited) levels of diversity, which indicated that they had passed through a population "bottleneck" in the past (Oyler-McCance et al. 2007). Trumpeter Swans thus have much lower mitochondrial DNA variability (and are less genetically able to adapt to changing environmental pressures) than other waterfowl studied thus far.

Pacific Coast Population

Breeding. The Pacific Coast Population (PCP) of Trumpeter Swans breeds primarily in the for-

ested wetlands of interior Alaska, south of the Brooks Range, and along the coastal plain from Cook Inlet south to southeastern Alaska. Within that range, Alaska recognizes 11 Trumpeter Swan nesting units, based on significant geographic features such as rivers and mountain ranges (Conant et al. 2002). PCP swans are much less common in the western Yukon and northwestern British Columbia, where the exact dividing line with the Rocky Mountain Population is not known (Hawkings et al. 2002). Genetic samples collected from the area where the PCP and the Rocky Mountain Population nesting areas are closest suggest that there is some genetic interchange between the two groups (Oyler-McCance et al. 2007).

In late summer 2010, the PCP (adults and cygnets) reached a record high of 26,790 birds (22% cygnets), of which 23,347 (87%) were in Alaska (Groves 2012). The Yukon and northwestern British Columbia contained 1,443 birds, which—together with the Alaska population—included 58% of all the Trumpeter Swans in North America, compared to 72% in 2005 (Moser 2006). The estimated growth rate of the PCP was 7.0% from 2000 to 2005, 1.5% from 2005 to 2010, and 5.5% from 1968 to 2010; mathematical (Bayesian) models using these same 1968–2005 data calculated an annual growth rate of 5.9% (range = 5.2%– 6.6%), and an annual growth rate of cygnets of 5.3% (range = 2.2%–8.0%; Schmidt et al. 2009a). In Alaska, censuses of both adults and cygnets on the breeding grounds showed a dramatic increase: from 1,924 in 1968 to 13,934 by 2000 (Conant et al. 2002). Important nesting units (numbers of pairs in 2005) were Lower Tanana (3,054), Gulkana (2,440), Cook Inlet (1,470), Upper Tanana (1,164), Koyukuk (950), and Gulf Coast (800; Pacific Flyway Council 2006a).

Winter. The PCP winters along the Pacific coast from the Alaska Peninsula south to the Columbia River delta. Victoria Island in British Columbia and the Skagit Valley and Puget Sound in Washington are significant wintering sites (Bailey et al.

1990, Anderson 1993, Mitchell and Eichholz 2010). They also are found inland, in central and south-central British Columbia, as well as a few winter records from California, Arizona, and New Mexico (John Cornely). The average number of Trumpeter Swans recorded in the Midwinter Survey in the Pacific Flyway states has ranged from 481 (1955–1959) to 9,156 (2004–8), with a peak of 13,068 in 2007; the 1955–2008 average was 2,125 (Trost and Sanders 2008). In 2011 the survey estimate was 11,373 (88% in Washington).

The coastal wetlands and agricultural fields of Vancouver Island and the Fraser River valley are winter habitats for over 40% of the PCP, which constitutes the largest population of wintering Trumpeter Swans in North America. Aerial surveys of these areas are usually conducted during the winter every 3 years. The 2006 survey tallied 7,570 Trumpeter Swans, an 11.7% increase from 2000/2001 (Canadian Wildlife Service Waterfowl Committee 2011). Estuaries, coastal marshes, farmland, and freshwater lakes were the most important wintering habitats on Vancouver Island; swans in the Fraser River valley were distributed about equally between tidal marshes and upland habitat. Christmas Bird Count data in British Columbia averaged 5,962 Trumpeter Swans from 2000/2001 to 2009/10. Hawkings et al. (2002) also noted increasing numbers of Trumpeter Swans wintering near Vancouver Island and in the Fraser River delta area: from about 1,000 in 1970 to at least 7,100 in 2000. PCP Trumpeter Swans also winter along rivers and at other ice-free sites in central British Columbia, where a winter 2001 survey tallied 1,293 birds (17% cygnets), largely along the Stuart, Middle, Tachie, and Nechako Rivers (Corbould 2001).

Migration. Trumpeter Swans from the PCP use several migratory routes from their breeding grounds in Alaska. The dominant route goes northeast to the Tanana River, then southwest to the Yukon River, and then east of the coastal mountains to Vancouver Island and other areas along the Pacific coast. Another route extends from the Kenai Peninsula through Prince William Sound and then south along the coast to southeastern Alaska, British Columbia, Washington, and, occasionally, to Oregon. A third path originates in eastern and central Alaska and goes southeast through the Copper River delta and then along the coast as above (McKelvey and Burton 1983, Mitchell and Eichholz 2010). Trumpeter Swans leave wintering sites along the Pacific coast in late February and early March (Jordan and Caniff 1981).

Rocky Mountain Population

Breeding. The Rocky Mountain Population (RMP) of Trumpeter Swans breeds within the Rocky Mountains, from the Northwest Territories south to Idaho, Montana, and Wyoming. The RMP consists of 3 groups: the Canadian Flock, the Tristate Flock (Idaho, Montana, Wyoming), and the Restoration Flock. Recent genetic studies found that swans from the Tristate Flock are not genetically different from those in the Canadian Flock, and need not be treated as a separate group (Oyler-McCance 2007). The Canadian Flock breeds within the Rocky Mountain region of Canada, from Alberta and British Columbia north to the Yukon and the Northwest Territories. Breeding surveys of the Trumpeter Swans in western Canada have recorded an increase: from 100 swans in 1970 to more than 3,700 by 2000 (Hawkings et al. 2002). The Canadian portion of the RMP was estimated at 10,550 birds in 2010 (Canadian Wildlife Service Waterfowl Committee 2011).

The Tristate Flock is centered in the Yellowstone / Centennial Valley (Red Rock Lakes) region of eastern Idaho, southwestern Montana, and northwestern Wyoming. The Restoration Flock includes Trumpeter Swans reintroduced from the Tristate area to the Malheur NWR in Oregon from 1939 to 1958 (Cornely et al. 1985a), the Turnbull NWR in Washington, and the Ruby Lakes NWR in Nevada in the 1950s. Beginning in 1996,

Trumpeter Swans were reintroduced to the Flathead Indian Reservation in Montana, with the first wild-nesting birds observed in 2004 (Becker and Lichtenberg 2007). Another restoration was initiated in the Blackfoot River valley in 2005, and the first successful nesting there occurred in 2011. The 2010 North American Trumpeter Swan Survey tallied 189 swans in other flocks within the RMP, largely from the Restoration Flock.

The RMP totaled 9,626 Trumpeter Swans (34% cygnets) in 2010, the highest estimate on record. The total for the Canadian Flock was 8,950 swans (93% of the RMP), versus 487 for the Tristate Flock. Breeding surveys in northeastern British Columbia estimated 1,126 RMP Trumpeter Swans in 2005, up from just 122 in 1985 (Breault et al. 2007).

The average annual growth rate for the RMP was 6.3% from 1968 to 2010, and 13.0% from 2005 to 2010 (Groves 2012). The average for the Tristate Flock (493) was below the 1968–2010 average for the RMP as a whole, but 5-year survey totals for the Tristate Flock have increased each period from 1995 through 2010. The highest number recorded for the Tristate Flock was 642 in 1954 (Banko 1960). The number of breeding Trumpeter Swans in Yellowstone National Park is declining, from 59 individuals in 1968 to only 10 in 2007, a growth rate of −36% (Proffitt et al. 2009). By 2010 there were only 5 resident Trumpeter Swans in the park, including 2 pairs, and they produced no cygnets (Smith and Chambers 2011); there was no reproduction in the park in 2011. Probable factors for the decline of Trumpeter Swans in Yellowstone National Park are human disturbance, swan management outside the park, habitat changes, and predation, but no single element seems responsible and all could be interacting (Smith and Chambers 2011).

Translocations of Trumpeter Swans or their eggs have been made from national wildlife refuges in Alaska and the Red Rock Lakes NWR in Montana to several other refuges: the Turnbull NWR in Washington, the Malheur NWR in Oregon, the Ruby Lake NWR in Nevada, the National Elk Refuge in Wyoming, and the Lacreek NWR in South Dakota. Small numbers of breeding swans occur on all these refuges. In 1987, efforts were initiated to expand the range of Trumpeter Swans in Alberta by relocating birds from Grande Prairie to Elk Island National Park (Beyersbergen and Kaye 2000).

Winter. In the Greater Yellowstone Ecosystem, wintering RMP Trumpeter Swans concentrated on wetlands in the Red Rock Lakes NWR, the Island Park country of southeastern Idaho, Yellowstone National Park, the National Elk Refuge, and the Snake River drainage of southeastern Idaho and northwestern Wyoming, where warm springs keep waters open even in the coldest winters, providing habitat in a region where winter temperatures averaged −9.3°C (Squires and Anderson 1997). During winter in the Greater Yellowstone Ecosystem, Trumpeter Swans slept (42% of the time), fed (30%), swam (12%), and preened (7%). They spent more time sleeping as temperatures decreased, with feeding largely stopped when temperatures were below −17°C (Squires and Anderson 1997). Feeding time rose to 45% in spring (17% sleeping), perhaps to increase the energy reserves used for breeding. Trumpeter Swans at the Malheur NWR largely winter locally (Ivey 1990), as do birds at the Ruby Lake NWR.

Migration. Trumpeter Swans from the Canadian Flock migrate south, largely east of the Rocky Mountains, to winter in the Tristate region (Gale et al. 1987, Slater 2006). These migrants join the relatively nonmigratory Tristate Flock in these areas. Many migrants arrive at Yellowstone Lake from late October to mid-November and remain there until the lake freezes, after which they move to ice-free areas (Shea 1979, Gale et al. 1987). The resident Tristate swans make short local movements to ice-free waters, such as thermal springs. In the spring, Canadian migrants depart the Tristate region during March. The Restoration

Flocks in Washington, Oregon, Nevada, and Montana are mostly nonmigratory.

Interior Population

Breeding. Historically, the Interior Population (IP) of Trumpeter Swans probably contained the greatest number of birds, perhaps exceeding 100,000 (Gillette and Shea 1995). Today, the IP breeds from the Great Plains eastward and was created by transplants from established populations or from artificially incubated eggs. These restoration efforts rank among the most successful, exciting, and continuing legacies of modern wildlife conservation. A 2005 summary of restoration efforts for the IP reports 19 restored populations in 8 states and 3 Canadian provinces that collectively produced over 9,600 cygnets from 1962 to 2005 (Johnson 2007a). The 2005 fall flight was estimated at 4,750 swans and has been doubling every 5 years since 1985, exceeding all stated population goals. The range and status of IP swans is discussed below on the basis of nesting flocks that occur in the High Plains, the Mississippi River valley, and the eastern Great Lakes.

Restoration efforts began with the High Plains Flock, which was the first restoration effort east of the Rocky Mountains. This group of Trumpeter Swans resulted from 57 cygnets captured at the Red Rock Lakes NWR in Montana and released at the Lacreek NWR in southern South Dakota from 1960 to 1962 (Monnie 1966). This successful transplant has steadily increased from only 15 swans in 1964 to 639 recorded during winter 2008, and Trumpeter Swans now occupy much of the available wetland habitat in the Sandhills region of nearby Nebraska as well as in the Lacreek area (Vrtiska and Comeau 2009). This population has exhibited an annual growth rate of 4.2% from 1990 to 2004, with 90% of the swans located in the Nebraska Sandhills (Comeau-Kingfisher and Koerner 2007). A second group in the High Plains occurs in eastern Saskatchewan and is probably the result of pioneering birds from Lacreek. In 2010, the High Plains Flock numbered 573 swans (Groves 2012).

In 2005, 78 Trumpeter Swans were counted in Saskatchewan, but a survey was not completed in that province in 2010.

Captive-reared birds in Minnesota, Wisconsin, and Iowa created the IP that is associated with the Mississippi River valley. The 2005 status of the Wisconsin and Iowa flocks underscores the success of these restoration efforts. The Wisconsin effort began in 1987, with a goal of attaining 20 breeding pairs by 2000, using, in part, an innovative approach that involved imprinting hatchling cygnets to life-sized adult decoys. After a slow start that yielded 11 nesting pairs in 1995 and 18 in 1998, the population expanded rapidly from 1999 on to reach 92 pairs in 16 counties (Matteson et al. 2007). Iowa initiated Trumpeter Swan restoration in 1995, with a goal of 15 nesting pairs by 2003, revised to 25 pairs by 2006. The first wild pair was reported nesting in 1998, with a second in 2000. Nesting effort subsequently increased to 26 attempts in 2005 that produced 87 cygnets, 67 of which reached flight stage (Andrews and Hoffman 2007).

Trumpeter Swans released in Ontario, Michigan, and Ohio are also included in the IP, and those restoration efforts have also been very successful. In Michigan, a restoration effort that began in 1986 established a goal of 200 birds in 2 flocks by 2000. By 2005, Trumpeter Swans nested in 23 out of 83 Michigan counties and yielded a fall flight of 728 birds (Johnson 2007). Since the initial release of Trumpeter Swans at the Seney NWR in Michigan's Upper Peninsula, the population increased at an annual rate of 24% through 2004 (Corace et al. 2006). In Ontario, which previously recorded its last wild Trumpeter Swan in 1886 (Lumsden 1984), restoration efforts had similarly notable results, with the number of swans increasing from 12 in 1990 to 303 by 2002 (Lumsden and Drever 2002). The 2005 estimate totaled 644. According to Lumsden, there were 523 Trumpeter Swans in Ontario in 2005, with breeding attempts by 48 pairs, of which 38 (79%) were successful, with an average of 3.6 cygnets/pair. By 2010 there were an estimated 839 Trumpeter Swans in southern Ontario, 274 in northwestern Ontario (west and north of Thunder Bay), and at least 54 in eastern Ontario (Canadian Wildlife Service Waterfowl Committee 2011).

Winter. Trumpeter Swans in the High Plains Flock largely winter in the Nebraska Sandhills, along spring-fed rivers like the Snake. Small flocks have been reported during the winter in Kansas, stemming from reintroductions in Nebraska and Minnesota (Thompson and Ely 1989). Small numbers of wintering Trumpeter Swans have also occurred in Texas and New Mexico (Burgess and Burgess 1997). Large numbers of IP Trumpeter Swans in Minnesota and Wisconsin winter on the Mississippi River (near Monticello, Minnesota) and on old coal-mine pits in Illinois (Varner 2008), with smaller groups scattered throughout the states to the south. Since 1976, a few (20–30) Trumpeter Swans from the High Plains Flock have been observed wintering as far south as Arkansas, Oklahoma, and Texas (Burgess 2002), which perhaps indicates establishment of a migratory tradition.

Trumpeter Swans are found year round in Minnesota, Wisconsin, Michigan, and Iowa (John Cornely). Trumpeter Swans in the Mississippi River valley and eastern Great Lakes often winter close to their breeding sites, largely because restored populations have yet to develop strong migratory traditions. Nonetheless, more and more Trumpeter Swans are wintering farther south, particularly in Oklahoma, Missouri, and Arkansas. In Ontario, a regular migration has developed between Wye Marsh and the northeastern shore of Lake Ontario, about 120 km to the south, where algae is available and the birds are fed by the public (Lumsden 2002). Other Ontario Trumpeter Swans have been observed in Ohio, Pennsylvania, New York, and several other states. The development of migration patterns in restored populations is desired by waterfowl managers, but it may take many years before the birds themselves establish such routes and traditions.

Migration. The IP is the result of restoration programs; hence these populations do not have well-developed migration traditions. Adult Trumpeter Swans lead their young to wintering grounds, but in restoration programs, young birds do not have migratory parents to lead them and thus pass on routes and traditions (Lumsden 2002). Most migrate only short distances, to find open water. For example, IP Trumpeter Swans from Wyoming, South Dakota, and Nebraska usually winter in the Nebraska Sandhills, a short distance from where they nest, although they will leave during severe winter weather for ice-free habitats as far south as Arkansas, appearing there for the first time in 1990 (Kraft 1991, Linck et al. 2007). Restored populations in the Midwest similarly exhibit short migrations, again probably because traditions are not yet well developed. Minnesota Trumpeter Swans migrate <40 km, largely to the Mississippi River northwest of Minneapolis (Hines 1991).

MIGRATION BEHAVIOR

Trumpeter Swans migrate in family groups rather than in large flocks, and they migrate either during the day or at night. The mean flock size of Trumpeter Swans on spring migration in Alaska was 6.6 birds (King and Ritchie 1992), compared with 10.6 in the fall (Cooper and Ritchie 1990). During migration, Trumpeter Swans fly in V formations in small flocks averaging 11 birds (range = 2–25 birds; King and Ritchie 1992). They fly at lower altitudes and lower speeds than Tundra Swans (King 1985), although they occasionally migrate in mixed flocks (Cooper and Ritchie 1990). Average flight speed is about 71 kph (range = 40–96 kph; King and Ritchie 1992). Their foraging time at spring staging areas was much greater than that reported for them on wintering areas (49% vs. 30%; Squires and Anderson 1997, LaMontagne et al. 2001), which could indicate that spring stopover sites are used to build up energy reserves that are accessed later for breeding.

Molt migration. Some nonbreeding birds and failed breeders at the Red Rock Lakes NWR in Montana and the Grays Lake NWR in Idaho have undertaken a short local molt migration to large reservoirs. Otherwise, molt migration has not been reported for Trumpeter Swans (Mitchell and Eichholz 2010).

HABITAT

Unlike Tundra Swans, which breed in coastal tundra habitats, fur-trade records indicate that the greatest numbers of breeding Trumpeter Swans formerly occurred in open boreal forests of Canada and Alaska (Banko 1960). The present breeding range of Trumpeter Swans also encompasses other major habitat types, such as aspen parkland, grasslands, Pacific rainforests, and arctic-alpine, montane, and eastern deciduous forests. Within these habitats, Trumpeter Swans use freshwater marshes, ponds, and lakes, and they occasionally breed on slow-moving rivers (Banko 1960, Hansen et al. 1971, Gale et al. 1987). Basic habitat features include space from which to take off (about 100 m); accessible forage; shallow, stable levels of unpolluted water; emergent vegetation; a structure for the nest site, such as a muskrat (*Ondatra zibethicus*) house or an island; and infrequent human disturbance (Mitchell and Eichholz 2010).

Across 5 major breeding areas in Alaska, Trumpeter Swans with broods occupy large closed-basin wetlands (e.g., lakes and ponds) at much greater rates than other habitat types, such as shrubby, forested, or riverine wetlands (Schmidt et al. 2009b). The probability of occupancy is greater on wetlands at higher, rather than lower, elevations. On average, pond size in these breeding areas shrank by 0.32 ha from 1982 to 1996, which may reflect the influence of climate change. Schmidt et al. (2009b) found that smaller pond size did not affect occupancy by Trumpeter Swans, although Johnson et al. (2005) had suggested that wetland shrinkage due to climate change would negatively affect most waterfowl. Sojda et al. (2002) developed a model to assess breeding habitat for Trumpeter Swans in the northern Rocky Mountains; it included wetland size and depth, length of the annual ice-free period, prelaying food resources (such as sago pondweed), nest-site availability and its vulnerability to flooding, and brood habitat. In Yellowstone National Park (1987–2007), the probability of at least 1 cygnet from a nest surviving until September ranged from 6.2% for a nest located in a wetland occupied only once historically to 27.5% in a wetland occupied 38 times, which underscored the need to maintain high-quality nesting areas (Proffitt et al. 2010).

Trumpeter Swans use an array of habitats during winter, depending on ice cover, forage availability, and the amount of human disturbance. On the Pacific coast, wintering Trumpeter Swans primarily roost in ice-free estuarine and freshwater habitats and forage in nearby agricultural fields (McKelvey 1981, Anderson 1993). Field-feeding by wintering Trumpeter Swans is becoming more common in the Tristate area, as well as for wintering birds in the IP (Varner 2008). Elsewhere, when forage is available, Trumpeter Swans have used freshwater springs, streams, rivers, lakes, ponds, and reservoirs (Gale et al. 1987, Mitchell and Eichholz 2010). Habitat use by the RMP in the Tristate area is strongly influenced by the availability of ice-free water and the number of swans present, especially on Henrys Fork of the Snake River in eastern Idaho (Snyder 1991). During migration in the spring and fall, Trumpeter Swans stage on marshes and lakes, moving to progressively larger water bodies to avoid ice formation, which restricts access to food (Gale et al. 1987, King and Ritchie 1992).

POPULATION STATUS

Historical. From 1772 to 1903, Trumpeter Swans were exploited for their skins over most of their range (Banko 1960; Houston et al. 1997, 2003). Waterfowl played a predominant part in the plume trade then, with the skins of 17,671 swans, mostly Trumpeters, sold for their plumage between 1853 and 1877, which drastically reduced their populations; only 57 skins were sold between 1888 and 1897 as the number of swans diminished throughout their former range (Banko 1960). Approximately 108,000 swan skins had been marketed by the Hudson's Bay Company between 1823 and

1880 before overhunting brought the sales to a halt (Banko and Mackay 1964). Trumpeter Swans were especially vulnerable to this trade, because they nested throughout the Canadian fur country southward into Minnesota and Iowa and were additional prey for trappers in search of pelts.

The trade in swan and goose quills to supply the quill pen industry was equally substantial. Houston et al. (2003) reported that swan and goose quills sold from the Hudson's Bay Company peaked at 1.2 million in 1834, the year when a grand total of 18.7 million quills were sold in London. The durability of swan quills made them far superior to goose quills. Hudson's Bay Company quills were the most preferred in the world at that time, with sales during the peak years of the 1830s commanding as much as 63 shillings/100 swan quills, compared with 15 shillings for domestic goose quills. Only the outer 5 primaries were used, of which the second and third were considered to be the highest-quality feathers. Quills from the left wing were the most desirable, because they curved up and away from right-handed writers (Kear 1990). John James Audubon preferred Trumpeter Swan quills for his detailed artwork, commenting that they were "so hard, and yet so elastic, that the best steel pen of the present day might have blushed, if it could, to be compared with them" (Audubon 1838:538). By 1822 there were at least 27 quill pen manufacturers and dealers in London; the last did not close until 1954, although firms had ceased using Trumpeter Swan quills before that time.

Trumpeter Swans were on the verge of extinction by the early 1900s. The eminent ornithologist E. H. Forbush (1912:476) wrote that "the trumpetings that were once heard over the breadth of a great continent . . . will soon be heard no more." By 1925, Bent (1925:293) stated, "This magnificent bird, the largest of all North American wild fowl, belongs to a vanishing race"; they only persisted in remote areas near what is now the Greater Yellowstone Ecosystem and in parts of Canada and Alaska. By 1932, only 69 Trumpeter Swans were known to exist in the lower 48 states, at remote locations in and around Yellowstone National Park (Banko 1960). These swans led to the establishment of the Red Rock Lakes NWR in 1935, which is located in the Centennial Valley of Montana. About half of the Trumpeter Swans known to exist in the lower 48 states during the 1930s inhabited this area, where hot springs and pools provided year-round open-water habitat. Shea et al. (2002) believed that about 1,000–2,000 Trumpeter Swans remained in North America, with most in Alaska and perhaps no more than 200 in Canada and the Yellowstone area.

Reports of Trumpeter Swans had always occurred in Alaska, but substantial breeding was not discovered until 1955, when 69 adults and at least 15 cygnets were found during aerial surveys for salmon in the lower Copper River basin (Monson 1956), followed by about 200 discovered in 1957 on the Kenai Peninsula (Banko and Mackay 1964). By 1959, 1,124 were located in southcentral Alaska, mostly on the lower Copper River and at Cook Inlet; by 1968, surveys tallied 2,848 Trumpeter Swans in the state (Hansen et al. 1971). In Canada, about 100 Trumpeter Swans were discovered in 1946 within the Grand Prairie region along the British Columbia / Alberta border (Mackay 1978).

Present. Trumpeter Swans are among the most well-monitored wildlife species, because their large size, white plumage, and use of open habitats make them readily countable by observers. Cooperating conservation groups and agencies survey the Trumpeter Swan population in late summer throughout its range. The first survey was conducted in 1968, followed by another in 1975, and then ones at 5-year intervals.

The number of Trumpeter Swans in North America is increasing rapidly and represents a particularly significant achievement in wildlife conservation. In 2010, a record 46,335 Trumpeter Swans were tallied in North America, an increase of 33% since the 2005 survey; Bellrose (1980) reported the 1980 continental Trumpeter Swan population at only 3,600–4,400. All 3 populations also reached record highs: 26,790 for the PCP (7% more than in 2005), 9,626 for the RMP (84% more than in 2000), and 9,809 for the IP (111% more than in 2005). In addition, 26% of all observations were cygnets, which equaled the average percentage obtained from the 1968–2005 surveys (Groves 2012).

The annual growth rate was 6.2% from 1968 to 2010: 5.5% for the PCP, 6.3% for the RMP, and 13.0% for the IP (Groves 2010). The most rapid (24.7%) annual rate of growth has occurred in the Mississippi and Atlantic Flyway flocks of the IP. During the same period, only the Tristate (−0.6%) and Restoration Flocks / other U.S. flocks (−0.7%) in the RMP decreased from 1968 to 2010, although they increased from 2005 to 2010 (1.5% and 27.1%, respectively). The 2005 survey showed a range expansion of Alaskan Trumpeter Swans: to the Mackenzie River in the Northwest Territories, as well as to small areas adjacent to their ranges in Alberta, eastern Saskatchewan, western Manitoba, western Wyoming, and southeastern Michigan.

These population levels represent a significant conservation achievement. For example, the management objectives for the PCP include maintaining a population of not less than 25,000 swans (reached in 2005), and those for the RMP aim for a 5% annual growth rate (reached in 2005). The population objective for the IP is to maintain 2,000 birds and 180 breeding pairs, which was exceeded in the 2000 survey (2,430 birds; Caithamer 2001). Flyway management plans for each population have other goals and objectives, however, that have not yet been achieved, particularly in areas of migration and winter distribution.

Harvest

The Migratory Bird Treaty Act of 1918 prohibited hunting of all swans in North America, although the harvest of Tundra Swans was reinstated in 1962. Trumpeter Swans occur in areas where Tundra Swans are legally hunted, however, which raised the issue of accidental take. Drewien et al. (1999) stated that of the 890 swans reported harvested in Montana and Utah (1992–96), only 19 (2.1%) were Trumpeter Swans. In an environmental assessment of proposed Tundra Swan hunting in the Pacific Flyway for the 1995–99 seasons, the U.S. Fish and Wildlife Service found "no significant impact" on Trumpeter Swan populations, which resulted in a 5-year season with a fixed quota of Trumpeter Swans (20/year) that could be accidentally harvested (Bartonek et al. 1995).

The growing Trumpeter Swan population, however, has now yielded opportunities for hunters to accidentally harvest them in association with hunting seasons on Tundra Swans in several western states. From 1995 to 1999, the U.S. Fish and Wildlife Service authorized an experimental permit-only Tundra Swan hunt in Utah and Nevada that had a limited allowable take of Trumpeter Swans (15 in Utah and 5 in Nevada). If the allowable take of Trumpeter Swans was reached in either state, the swan-hunting season would have been closed immediately. The Utah quota was then reduced to 10 in 2000. A final environmental assessment in 2003 continued the limited quotas in Utah and Nevada, adjusted season lengths in those states, added a limited quota in Montana, reduced the areas where swans could be hunted, and continued to require careful monitoring of the swan harvest. The intent of these actions was to reduce the incidental harvest of Trumpeter Swans while continuing the hunting of Tundra Swans in these states.

BREEDING BIOLOGY
Behavior

Mating System. Trumpeter Swans are monogamous, usually pair for life, and remain together year round. Initial pair bonds form from late March to mid-May (Lockman et al. 1987), with most pair bonds occurring 1 year before breeding (Kear 1972). In his review of case histories, Banko (1960) concluded that Trumpeter Swans may begin nesting as early as their fourth year or as late as their sixth year. At the Lacreek NWR, 2 pairs of newly introduced Trumpeter Swans established pair bonds and territories when they were 20 months old and nested the following year (Monnie 1966). Hansen (1973), however, reported that few Trumpeters in Alaska bred at 2 or 3 years of age. Gale et al. (1987) found that most first breeding occurred between 4 and 7 years old. The density of territorial pairs is thought to account for some of the variation in breeding age (Hansen et al. 1971). These authors also reported that a marked female remated the next year after the loss of her mate.

Sex Ratio. Limited data exist on sex ratios, but males outnumbered females in Wyoming (Lockman et al. 1987).

Site Fidelity and Territory. Females show fidelity to their natal sites, while males tend to disperse (Lockman et al. 1987). Territorial behavior is strikingly evident, with pairs vigorously defending mating, nesting, and cygnet-feeding grounds from conspecifics (Banko 1960). Pairs spend 0.1%–2.3% of their day defending their territory (Henson and Cooper 1992). In Yellowstone National Park and the Red Rock Lakes NWR, pairs select and defend sites in the winter, sometimes before open water is available, and some pairs guard their territories until late summer, when the cygnets are half grown (Banko 1960). Trumpeter Swans often tolerate other waterfowl within their territories and commonly allow ducks to loaf on nests, but at other times they are aggressive toward other nesting birds as well as toward mammals (Hansen et al. 1971, Gillette 1990).

Territories range in size from 1.5 to >100 ha, with their size potentially influenced by shoreline complexity and food availability (Mitchell and Eichholz 2010). Trumpeter Swan territories at the Red Rock Lakes NWR in Montana averaged 28–61 ha (Banko 1960). In Alaska, territory size was governed by the amount of wetland, with only 1 pair of territorial Trumpeter Swans on water areas ranging in size from 2.4 to 52 ha; large lakes could support >1 pair (Hansen et al. 1971). As breeding numbers increased at the Red Rock Lakes NWR, new territories were established in previously unoccupied, less suitable habitat, but existing territories were not made smaller to accommodate more breeders (Banko 1960). Banko speculated that the new, more marginal territories may have accounted for the lower reproductive success that occurred as breeding swans became more numerous.

Territories and nest sites are often reused annually. During a 4-year period at the Red Rock Lakes NWR (1954–57), 82 out of 109 (75%) nest sites were reused in subsequent years (Banko 1960). In Alaska, nest-site occupancy at 2 study areas ranged from 63% to 88% (Hansen et al. 1971). In Yellowstone National Park, on average, a territory was used for 14.7 years (range = 1–38 years; Proffitt et al. 2010). Historical occupancy of territories was correlated with higher elevation ($r = 0.55$), larger individual wetland area ($r = 0.16$), and larger total area of the wetland complex ($r = 0.47$).

Courtship Displays. Trumpeter Swans probably all begin pair formation activities by their second or third winter (Palmer 1976a). The precopulatory display consists entirely of both adults rapidly *head-dipping*, which resembles normal *bathing* movements (Johnsgard 1965). This display only lasts for about 10–15 seconds before the male mounts the female. Precopulation also involves slow, synchronized swimming by the pair, followed by *head-dipping* and *blowing* into the water (de Vos 1964). During postcopulation, the male spreads his wings and the female *calls*, then both birds rise out of the water, *calling* together and turning in an incomplete circle before settling back on the water and *bathing*. A *triumph-ceremony* is performed by the pair after repelling an intruder. Threat displays take several forms, such as *neck-stretching* and raising the wings either half or fully open, but the wings are usually fully spread before an outright attack (Johnsgard 1965).

Nesting

Nest Sites. Pairs may select potential nesting territories or ponds several years before actually nesting (Lockman et al. 1987). Nests are then typically built in extensive beds of marsh vegetation: sedges

(*Carex* spp.), bulrushes (*Scirpus* spp.), cattails (*Typha* spp.), and rushes (*Juncus* spp.) in Montana; and horsetails (*Equisetum* spp.) and sedges in Alaska (Banko 1960, Hansen et al. 1971). The first nests are built in late April to early May, before the ice has completely melted (Gale et al. 1987). In Alaska, on wetlands where there are no muskrats, nearly all nests were built in emergent vegetation surrounded by water 0.3–0.9 m deep (Hansen et al. 1971). In contrast, most nests at the Red Rock Lakes NWR in Montana were placed on muskrat houses (Banko 1960). Nest densities have ranged from a low of about 0.01/km² (Wilk 1993) to >1.0/km² (Banko 1960, Gale et al. 1987). In Yellowstone National Park (1987–2000), Trumpeter Swans nested in 44 wetlands located within 15 different wetland complexes (Proffitt et al. 2010). The wetland complexes ranged in size from 0.05 to 1.94 km², while individual wetlands varied from 0.05 to 0.07 km². Where recorded, the nesting substrate (*n* = 98) was a floating vegetation mat (36.7%), a muskrat house (28.6%), a beaver (*Castor canadensis*) lodge (12.2%), shoreline vegetation (12.2%), islands (6.1%), or emergent vegetation (4.1%).

Both sexes construct the nest, which takes 11–35 days to complete (Hansen et al. 1971, Cooper 1979). The male usually brings clumps of uprooted marsh plants to the nest site, which the female arranges into a mound and a nest bowl (Cooper 1979). The nest is largely constructed by uprooting marsh plants immediately around the nest site, which leaves a large area of open water that averaged 7.6 m around 19 nests in Alaska (Hansen et al. 1971). Pairs may construct more than 1 platform, but only 1 is used. The nest mound is large, reaching a diameter of 1.8–3.6 m and an average height of 1.5 m; the nest cup varies from 25 cm to 41 cm in diameter and 10 to 20 cm in depth (Hansen et al. 1971).

Clutch Size and Eggs. Trumpeter Swan clutches in the wild have ranged from 1 to 9 eggs, but they typically contain 4–6 (Gale et al. 1987). Clutch size averaged 5.2 eggs (*n* = 213) in Alaska and 4.7 (*n* = 457)

at the Red Rock Lakes NWR in Montana. Clutch size in Alaska varied from an average of 4.4 eggs in 1964 to 5.7 in 1965, perhaps as a result of early versus late springs (Hansen et al. 1971). During 17 years at the Red Rock Lakes NWR, clutch size ranged from an average of 3.4 eggs in 1967 to 6.0 in 1947 (Page 1976). An early spring in 1972, with consequent early nesting, may have caused the large mean clutch size of 5.4 eggs (*n* = 33) that season. Banko (1960) reported average clutch size as 5.1 eggs for 74 clutches at the Red Rock Lakes NWR. The clutch size of 18 nests in Yellowstone National Park averaged 4.2 eggs (Proffitt et al. 2010). Eggs are laid at 48-hour intervals (Hansen et al. 1971, Alisauskas and Ankney 1992a). Females take about 2 minutes to lay an egg, from first view in the cloaca to actual emergence (Lumsden 2002).

Trumpeter Swan eggs are off-white, with a shape described as subelliptical to long elliptical, but they soon take on a brownish stain from the nest material; the shells have a granular texture (Palmer 1976a). Eggs of Trumpeter Swans in Alaska measured 117.4 mm × 75.0 mm (*n* = 144; Hansen et al. 1971), while those from the Red Rock Lakes NWR were 110.9 mm × 72.4 mm (*n* = 109; Banko 1960). Egg mass averaged 363 g in Alaska (*n* = 104; Hansen 1971) and 336 g at the Red Rock Lakes NWR (*n* = 72; Mitchell and Eichholz 2010). Egg laying begins 6 days after nest building, and the eggs are laid at intervals of 39–48 hours (Cooper 1979).

Incubation and Energetic Costs. The period of incubation varies from 33 to 37 days (Banko 1960, Hansen et al. 1971). Both sexes may appear to incubate, but the male probably just sits at the nest for protection and does not actually incubate the eggs (de Vos 1964). Banko (1960) never observed incubation by male Trumpeter Swans. In the captive pair observed by de Vos (1964), the male spent <5% of his time at the nest, sitting on the nest only once. Hence most, if not all, incubation is done by the female (Mitchell and Eichholz 2010). Trumpeter Swans do not develop a brood patch, but observation of 3 captive pairs found that they incubate with their feet on top of the eggs. Trumpeter Swans have large feet that can theoretically cover 5 or 6 eggs, so regulation of blood flow to the feet may effectively provide enough warmth for the eggs (Lumsden 2002).

Diurnal incubation constancy for females averaged 88% ($n = 10$) on the Copper River delta in Alaska and 80% ($n = 6$) in the Tristate area. Females took 4–5 recesses/day that lasted 23–42 minutes each, during which they spent much (46%–48%) of their time feeding; total recess time/day averaged 117–179 minutes (Henson and Cooper 1993). Incubation constancy averaged 91% on Minto Flats in interior Alaska (Bollinger and King 2002). Females in productive Alaskan habitats took shorter recesses than females at less productive sites in the Tristate area. These differences in recess time were influenced by food quality: Alaskan birds fed on abundant emergent macrophytes, whereas Tristate swans fed on submerged vegetation that required search and handling time. These longer periods away from the nest may negatively affect the productivity of the Tristate Flock. The incubation constancy of 2 pairs of captive Trumpeter Swans observed by Cooper (1979) averaged 95%–96%, with nest recesses averaging only 21 min/day (Cooper 1979).

At Minto Flats in Alaska, females spent little (<5.3%) time eating during both incubation and hatching, but they increased their feeding to 33.9% during brood rearing (Bollinger and King 2002). The feeding rate during incubation was similar to the 8% reported from other studies of Trumpeter Swans (Henson and Cooper 1992, Grant et al. 1997). The increase in the feeding rate from incubation to brood rearing was similar to that of Trumpeter Swans on the Copper River delta (from 8% to 30%; Henson and Cooper 1992). Alert behavior was highest (51.1%) at hatch, in comparison with incubation (24.3%) and brood rearing (24.3%).

In contrast, males at Minto Flats fed during incubation (33.1%) and brood rearing (36.1%), but they were also most watchful at hatching (32.3%), compared with alertness levels of 18.4% during incubation and 23.7% during brood rearing (Bollinger and King 2002). The feeding rates of males on the Copper River delta were 36% during incubation and 31% during brood rearing (Henson and Cooper 1992).

Nesting Chronology. In southeastern Alaska (1957–59), the initiation dates of first nests ranged from 22 April to 5 May, with the first egg laid between 28 April and 5 May (Hansen et al. 1971). Banko (1960) found a similar nesting chronology at the Red Rock Lakes NWR in Montana. Although these 2 breeding areas are far apart in latitude, this difference is offset by the 2,012 m altitude of the Red Rock Lakes NWR, as compared with the near-sea-level elevation of the lower Copper River delta, leading to a similar nesting chronology. At the Red Rock Lakes NWR, nest initiation spanned 11–16 days. Nest initiation in southeastern Alaska apparently extends over a longer period (28–34 days), as indicated by the range of hatching dates from first to last nest.

Nest Success. Trumpeter Swans are large and thus can defend their nests against mammalian predators. In addition, a high rate of incubation constancy means that their eggs are rarely left exposed to avian predators. Only 4 nests were lost to predation over 7 years at the Red Rock Lakes NWR, but 34%–49% of the eggs did not hatch (Banko 1960). In a later study at the Red Rock Lakes NWR, nest success was 76% for 101 nests, but all failures were due to abandonment (Page 1976). Of all eggs laid, 55% hatched, 24% were infertile, 17% were fertile but did not hatch, and 5% were lost to predators or to an unknown fate.

Nest success on the Kenai Peninsula averaged 79% from 1965 to 1967, with 82% of the eggs hatching from successful nests (Hansen et al. 1971). In contrast, these authors found that nest success on the lower Copper River delta was 76% ($n = 38$) in 1959, with 76% of all eggs hatching. Nest success on the Copper River delta and in surrounding areas of the Chugach National Forest averaged 49% from 1968 to 2007 (Groves et al. 2008). Bart et al. (1991a) reported an average nest-success rate of 46% from breeding sites in Montana and Alaska.

Large mammals are common predators of Trumpeter Swan nests, among them black bears (*Ursus americanus*), grizzly bears (*Ursus horribilis*), coyotes (*Canis latrans*), gray wolves (*Canis lupus*), and wolverines (*Gulo luscus*). Common Ravens (*Corvus corax*) and raccoons (*Procyon lotor*) are also nest predators (Banko 1960, Hansen et al. 1971, Lockman 1987, Mitchell and Eichholz 2010). Of 85 incidents of egg failure in Yellowstone National Park, 53% were due to flooding, 9% to Common Ravens, 9% to coyotes, 4% to grizzly bears, and 4% to human disturbance, while 19% were due to unknown predators and 2% to unknown causes (Proffitt et al. 2010).

Brood Parasitism. The extent of intraspecific brood parasitism is unknown, but it is probably not extensive because of territorial defense by the swans and high incubation constancy. Canada Geese, however, have occasionally laid eggs in the nests of Trumpeter Swans (Carl Mitchell).

Renesting. There are no specific studies on renesting in Trumpeter Swans, but it probably does not occur, especially if a full clutch is lost, because of the physiological demands of producing such large eggs, the long incubation period, and the lengthy time it takes for cygnets to reach flight stage.

REARING OF YOUNG

Brood Habitat and Care. Cygnets eat alongside their parents, initially feeding in shallow water at the margins of emergent vegetation, where they consume aquatic insects, crustaceans, and some aquatic plants. The adults often tread the substrate to stir up food for the cygnets (Lockman et al. 1987). At 2 weeks of age, the cygnets' diet is largely aquatic vegetation, and it becomes very similar to that of their parents in 2–3 months (Banko 1960).

The female broods hatched cygnets at the nest for 24 hours (longer during colder weather), and at brief intervals at night or during inclement weather, up until they are a few weeks old (Hansen et al. 1971, Mitchell and Eichholz 2010).

Brood amalgamation (Crèches). Banko (1960) and Page (1976) reported that brood amalgamation rarely occurred during their studies at the Red Rock Lakes NWR in Montana. Later studies at the Red Rock Lakes NWR, however, recorded brood amalgamation among at least 11.5% of 114 broods observed between 1987 and 1990 (0%–18.8%/year; Mitchell and Rotella 1997). Recipient broods averaged 8.6 cygnets (range = 3–11 cygnets), in contrast to 3.5 (range = 1–8 cygnets) for broods not involved in amalgamation. Broods gained a minimum of 1–4 cygnets during amalgamation, which tended to occur where broods were at higher rather than lower densities (1.1 broods/100 ha and 6.5 broods/wetland vs. 0.4 broods/100 ha and 3.0 broods/wetland), or when broods moved from natal to brood-rearing wetlands (65% of the amalgamations).

Development. At hatch after drying, cygnets in Alberta averaged 210 g ($n = 51$; Ripley 1984, 1985), after which they grew very rapidly. Captive cygnets gained an average of 52 g/day up to day 26 (Ripley 1984, 1985), and 61–99 g/day between days 90 and 120 (Cochran 1970). In Alaska, cygnets grew from between 198 g and 228 g to 8.6 kg in 8–10 weeks (Hansen et al. 1971). Jobes (1990) found that growth rates of captive cygnets did not vary by sex, and there was no correlation between growth rate and hatching weight, which suggests that growth rate is controlled by environmental factors. In Alaska, cygnets were fully feathered in 63–70 days but unable to fly until they were 91–105 days old (Hansen et al. 1971).

Cygnets on Minto Flats in Alaska fed more than their parents during the brood-rearing period (52.9% vs. 33.9%–36.1%; Bollinger and King 2002), which was nearly identical to cygnet feeding time in southern Alaska (Grant et al. 1997). Cygnets remain with their parents through the fall and their first winter and return to the breeding grounds with them, but are chased away when their parents initiate territorial defense.

RECRUITMENT AND SURVIVAL

From 1968 to 2000, 3,031 broods surveyed in Alaska averaged 3.3 cygnets/brood, and 31% of 9,461 pairs had broods (Conant et al. 1991, Conant et al. 2002). Brood size averaged 3.3 cygnets ($n = 1,825$) from 1968 to 2007 on the Copper River delta in Alaska and in the nearby Chugach National Forest (Groves et al. 2008). In southeastern Alaska, Hansen (1971) reported an average brood size of 3.6 in 1968 ($n = 251$). Brood size at the Red Rock Lakes NWR in Montana averaged 3.2 (range = 1.9–4.3 cygnets) over a 12-year period (Banko 1960). Cygnet mortality from hatching to fledging ranged from 48% to 78% at breeding areas in Alaska and Montana (Bart et al. 1991a). In Alaska, cygnet mortality was 15%–20%, almost all of which occurred in the first 8 weeks (Hansen et al. 1971). On the nearby Kenai Peninsula, broods suffered losses of 20%, 29%, and 23% from 1965 to 1967. Potential predators were Bald Eagles (*Haliaetus leucocephalus*), Peregrine Falcons (*Falco peregrinus*), and Glaucous Gulls (*Larus hyperboreus*); the latter were deemed most likely to predate stray cygnets while they were still small. Page (1976) recorded very high mortality among Trumpeter Swan broods at the Red Rock Lakes NWR from 1971 to 1973. Of the 264 cygnets that hatched, only 72 survived to fledging, a mortality rate of 73%. The high number of cygnet deaths were attributed to early summer storms and a bacterial infection. In the declining population of Trumpeter Swans at Yellowstone National Park, fledging success (number of young fledged/breeding female) averaged only 0.35/year (1987–2000), with a 15% probability of a nest fledging at least 1 cygnet; an average of only 2.8 cygnets was fledged in the park each year (Proffitt et al. 2010).

Away from breeding areas, juvenile swans are easily recognized in winter because they retain their juvenal plumage; hence the ratios of juveniles to adult pairs are a good indicator of sur-

vival from first migration well into the first winter, as well as of recruitment. Over a 32-year period (1974–2005) immature Trumpeter Swans formed 19% of the RMP (Dubovsky 2005). Alaskan populations of Trumpeter Swans were first surveyed in 1968, and then surveyed at 5-year intervals since 1975. From 1975 to 2005, immatures made up 25% of the population and 31% of the pairs had broods, with an average of 3.1 young (Pacific Flyway Council 2006a). The 2010 North American survey of Trumpeter Swans recorded 2,108 broods in the PCP, with an average of 2.9 cygnets/brood (Groves 2012). There are no studies of lifetime reproductive success of Trumpeter Swans.

Annual survival of adult birds (>2 years old) is high; Mitchell and Eichholz (2010) noted that it ranged from 77% to 100%. At the Red Rock Lakes NWR, annual survival of Trumpeter Swans banded in July and August between 1949 and 1982 was 80%–88% (Anderson et al. 1986). Of Canadian Trumpeter Swans wintering in the Tristate area, the survival of juveniles from fledging until June was 43%, while subadult survival was 71%, and adult survival was 82% (Turner and Mackay 1982). Survival of local nonmigratory Trumpeter Swans in Wyoming was 60% for cygnets (from Jun to fledging), 66% for subadults, and 93% for adults (Lockman et al. 1987); minimal survival estimates of translocated Trumpeter Swans in the Greater Yellowstone region were 71% for adults and 50% for cygnets (Drewien et al. 2002). Survival of Trumpeter Swans used for restoration efforts in Ontario (1993–2000) did not differ between wild-hatched and captive-reared birds (Lumsden and Drever 2002). Minimum survival estimates ranged from 88%–98% for Ontario swans <1 year old, to 78%–89% for 1- to 2-year-olds, and to 77%–86% for swans >2 years old.

Varner and Eichholz (2012) compared the annual and seasonal survival of short-distance and long-distance migrant Trumpeter Swans marked on their breeding grounds in Wisconsin and then observed at wintering sites (2000–2008). The short-distance migrants wintered near their breeding areas in Wisconsin or below the Monticello Nuclear Generating Plan on the Mississippi River near Monticello, Minnesota, and the long-distance migrants wintered in central/southern Illinois. Annual survival of adults was similar (81%) between the two groups, but survival was higher for the long-distance subadult migrants (86%) than the short-distance ones (70%). There was little seasonal variation in survival between the groups, with annual survival >97%, which suggests that the migration period does not contribute significantly to annual mortality.

The longevity record for a wild Trumpeter Swan is 23 years and 10 months for a male banded as a second-year bird in Montana and found dead in Idaho. A female banded as a second-year bird in Montana lived for 18 years and 2 months; a captive bird was reported to have lived for 32 years and 6 months.

FOOD HABITS AND FEEDING ECOLOGY

Trumpeter Swans are primarily herbivorous, feeding in a diverse array of freshwater marshes, springs, lakes, ponds, rivers, and estuaries, as well as in agricultural fields and pastures. The bulk of their food comes from the leaves, stems, roots, and tubers of a very broad variety of marsh and aquatic plants. Mitchell and Eichholz (2010) list about 80 aquatic plants eaten by Trumpeter Swans. Important food items include the tubers, stems, and leaves of sago pondweed and other pondweeds (*Potamogeton* spp.), as well as burreeds (*Sparganium* spp.) and the tubers of duck potato (*Sagittaria* spp.). In feeding experiments at the Red Rock Lakes NWR, confined adult swans preferred waterweed (*Elodea canadensis*) over watermilfoil (*Myriophyllum exalbescens*), muskgrass (*Chara vulgaris*), and various pondweeds, consuming about 9 kg/bird/day (Page 1976). McKelvey (1985) reported the ingestion of 4.5–5.5 kg/day (wet weight) for wild Trumpeter Swans wintering in British Columbia. Trumpeter Swans that were fed supplemental wheat consumed about 227 g/day (Mitchell 1990). LaMontagne et al. (2004) reported the basal metabolic rate of Trumpeter Swans as 1,748 kJ/day; they then used time budgets to determine the daily caloric requirement for swans at a spring stopover site near Calgary, Alberta, as 2,578 kJ/day.

Trumpeter Swans generally feed in shallow water, usually by submersing the head and neck or, less frequently, by picking food items off the surface. They also obtain food by paddling their feet to loosen rhizomes (McKelvey and Verbeek 1988). In deeper water, swans feed by tipping up. During the winter in southern Illinois, they ate aquatic plants growing at depths of up to 1.5 m (Babineau 2003). In the Greater Yellowstone Ecosystem, Trumpeter Swans decreased their feeding time when temperatures fell below −17°C, and ceased eating altogether at temperatures below −23°C (Squires and Anderson 1997). Trumpeter Swans also avoid feeding in high winds and where there are strong currents (Squires 1991). Cygnets reportedly dislike eating during heavy rains (Grant 1991).

Breeding. On the Copper River delta in Alaska, breeding adult Trumpeter Swans consumed 99.9% plant matter; the inclusion of animal matter was probably incidental. Time-budget analysis revealed that prior to egg production, adults spent most of their feeding time (90.5% for males, 88.6% for females) foraging on the stems and leaves of submerged plants, primarily pondweeds, watermilfoil, and waterweed (Grant 1991, Grant et al. 1994). During incubation, foraging was dominated by consuming emergent plants (56.9% for males, 76.1% for females), with horsetails comprising the majority (43.4% for males, 82.6% for females), along with smaller amounts of sedges. During brood rearing, the time spent foraging on submerged vegetation was 19.9% for males and 12.5% for females, with that spent foraging on horsetails being 74.4% for males and 82.6% for females. Adults may have preferred emergent vegetation because it was easier to locate and handle. Incubation constancy was also greater for Trumpeter Swans feeding on horsetails in Alaska (88%) than Tristate swans feeding on submerged vegetation (80%; Hensen and Cooper 1993). In Wyoming, breeding Trumpeter Swans ate pondweed vegetation (48.2%), muskgrass (14.9%), and waterweed (8.5%), as well as small quantities of sedges, pondweed tubers, and horsetails (Squires and Anderson 1995).

Cygnets primarily spent 89.9% of their time feeding on the fruiting bodies, aerial shoots, and whorls of horsetails, as well as the seed heads of sedges (Squires and Anderson 1995). Horsetails are high in crude protein, which averaged 16.4% for fruiting bodies and 13.6% for aerial shoots. Less than 2% of the cygnets' time was spent foraging on aquatic macroinvertebrates, usually during the first few days after hatch and in association with treading by the adults. The cygnets' chief invertebrate foods included scuds (*Gammarus* spp.), arthropods (Arthropoda), midges (Chironomidae), diving beetles (Dytiscidae), mayflies (Ephemeroptera), mollusks (Mollusca), oligochaete worms (Oligochaeta), and caddisflies (Trichoptera; Grant 1991). Banko (1960) also observed that cygnets fed primarily on aquatic macroinvertebrates during the first few weeks, and Hansen et al. (1971) reported the consumption of macroinvertebrates by Trumpeter Swan cygnets in Alaska.

Migration and Winter. Only a few studies have examined the food habits of Trumpeter Swans on migration or during the winter. In the Greater Yellowstone Ecosystem, fecal analysis showed that principal food items during the spring were pondweed tubers (38.5%), muskgrass (27.4%), pondweed vegetation (16.3%), and smaller (<6%) amounts of waterweed, plant seeds, mustards (Cruciferae), and sedges (Squires and Anderson 1995). Wintering Trumpeter Swans consumed muskgrass (36.8%), waterweed (29.6%), and sago pondweed tubers (23.4%). Sago pondweed tubers were highly preferred in both seasons. The nutritional value of

sago pondweed is substantial: the metabolizable caloric content is 13.3 kJ/g for tubers and 11.8 kJ/g for rhizomes, and the crude protein content is 10.9% for tubers and 19.1% for rhizomes (LaMontagne et al. 2004).

At a stopover site near Calgary, Alberta, the use of 13 ponds by migrating Trumpeter Swans in the spring was most consistent on those ponds with the greatest biomass of sago pondweed tubers and rhizomes (LaMontagne et al. 2003a), although ice cover also influenced pond use. Swans removed 24% of the tubers and rhizomes of sago pondweeds from 2 study ponds in the area (LaMontagne et al. 2003b). Adults spent 48% of their time feeding, 26% resting, 14% in locomotion, and 12% preening. Yearlings spent 49% of their time foraging, 19% resting, 18% in locomotion, and 15% preening (LaMontagne et al. 2001). Foraging diminished at temperatures below −4°C, and sleeping became the dominant activity. Based on the nutritional values of foods and time budgets, the total daily caloric requirement for a 10.8 kg swan was estimated as 2,578 kJ (LaMontagne et al. 2004).

On coastal estuaries in British Columbia, the diet of wintering Trumpeter Swans, determined from fecal analysis, was 33.8% eelgrass (*Zostera marina*), 30.8% bulrushes, 18.5% grasses, and 13.8% sedges (McKelvey 1981). Trumpeter Swans in British Columbian estuaries also used dairy pastures, where they foraged on grasses by grazing and grubbing. The pasture grasses were high in proteins but difficult for swans to metabolize (McKelvey and Verbeek 1988). In Montana, Trumpeter Swans did not feed on the supplemental grain that was provided until all the aquatic plants within the natural wetlands had been consumed (Mitchell and Eichholz 2010). Within the Strait of Georgia in British Columbia, sedges, bulrushes, and eelgrass were the preferred winter foods (Hutchinson et al. 1989). Swans in western Washington fed heavily on corn, potatoes, and carrots (Anderson 1993). During the winter in southern Illinois, Babineau (2003) observed diurnal foraging by Trumpeter

Swans in wheat fields, and sporadically in soybean fields.

MOLTS AND PLUMAGES

Adult Trumpeter Swans have only 1 plumage (basic) that is molted annually via the prebasic molt. The descriptions of molts and plumages are primarily from summaries and data in Hansen et al. (1971), Palmer (1976a), and Mitchell and Eichholz (2010), which should be consulted for further details.

Cygnets are usually a "mouse gray" at hatching (Banko 1960), but they are occasionally white (leucistic; see Identification). In Alaska, cygnets acquired juvenal plumage between 28 and 70 days posthatching. Cygnets were fully feathered in 63–70 days, except for some down on the rump and the underwing coverts, but the cygnets did not fly until 90–105 days posthatch. The resultant plumage is variable but usually grayish brown, with white primaries. First basic is acquired via a partial molt from November to March/April that involves the replacement of some body feathers. This plumage is retained into summer. Definitive basic is not acquired until the swans are 12–13 months old; the associated molt involves the replacement of both remiges and retrices, as well as some contour feathers, which yields a flightless period of about 30 days. Second-year birds may have a delayed or prolonged molt.

Breeding pairs usually molt asynchronously, so both parents are not flightless at the same time. On the Kenai Peninsula in Alaska, the males usually started to molt early in the incubation period, prior to the females (Hansen et al. 1971). In contrast, females molted first in Montana (Banko 1960) and in Grande Prairie, Alberta (Mackay 1988). The reason for this sexual differentiation in the molt has not been determined, but it could be related to age or to reproductive success. Most nonbreeding Trumpeter Swans in Alaska gathered on large lakes to begin their molt in late June or early July. At the Red Rock Lakes NWR in Montana, the molt may be completed as early as June, but it does not peak until July.

CONSERVATION AND MANAGEMENT

Although the breeding habitat of Trumpeter Swans seems secure, continued expansion in the human population, and the inevitable development issues associated with it, may change that situation, especially for swan populations in the lower 48 states. In addition, the quality and quantity of their wintering habitat is a major issue for all 3 populations. The PCP once wintered in coastal estuaries and associated wetlands as far south as California, many of which were drained for agriculture in the early 1900s. This population now depends on agricultural croplands, some of which have been lost to industrial and urban development and conversion to crops not eaten by Trumpeter Swans (Gillette and Shea 1995, King 2000, Shea et al. 2002, Anderson 2004). The protection of coastal habitats in southern British Columbia, Washington, and Oregon would thus appear to be of paramount importance for PCP Trumpeter Swans.

Growth of the RMP is hampered by a lack of wintering habitat in the Tristate region, which can lead to overcrowding when swans from the Canadian Flock arrive and mix with year-round residents. The resultant risk of starvation is especially high during periods of extensive ice formation (Shea et al. 2002). Accordingly, since the mid-1980s managers have implemented programs to expand the winter distribution of Trumpeter Swans in the Tristate region. Major actions have included the cessation of winter feeding at the Red Rock Lakes NWR (starting in 1993); winter and summer translocations of swans to alternate wintering sites; and water-level management, both to discourage swan use at high-risk sites (such as at Harriman State Park on Henrys Fork of the Snake River in Idaho), and to address winter emergencies caused by ice buildup (Pacific Flyway Study Committee 2002). These actions, however, have resulted in only some Trumpeter Swans wintering at lower elevations or in other areas outside of the Tristate region. By 2000, most (90%) still wintered in the Greater Yellowstone area (Shea et al. 2002).

From 1990 to 1996, of the 1,127 neck-banded Trumpeter Swans relocated to 8 alternative sites from high-risk locales in the Tristate area, 683 (61%) survived at least 1 year (71% of the adults, 50% of the cygnets), and 62% of them wintered away from monitored sites (Drewien et al. 2002). Juveniles were significantly more likely to use the new sites than were adults. By 1997, however, recurrent use by >50 swans was occurring on only 2 of the 8 release sites. The study concluded that greater use of new wintering sites would be accomplished by primarily relocating juveniles to locales that had adequate food, were ice free, and contained minimal obvious mortality factors, such as power lines. These findings were confirmed by Kilpatrick (2007) who also concluded that winter habitat conservation would probably be more important than translocation. Trumpeter Swans in the Tristate area have expanded somewhat into new wintering habitats naturally, and use open water habitat on the Snake River and its tributaries north and east of American Falls Reservoir.

Restoration efforts for the IP are clearly successful, but Trumpeter Swans from this population have only just begun to develop migration traditions. Historical records suggest that most Trumpeter Swans from the IP wintered south of 40° N latitude, but only about 10% of the IP swans exhibit migratory behavior (Slater 2006). Supplemental feeding has been an effective technique in encouraging them to return to sites with reliable food resources, but this approach is controversial (Gillette and Linck 2004). A study with an ultralight aircraft showed some promise in training Trumpeter Swans to develop migration routes, as has ground transport of the birds (Sladen et al. 2002).

Lead poisoning is an important mortality factor for Trumpeter Swans, due to their feeding methods of stirring up large amounts of sediment and their reportedly high susceptibility to lead toxicosis (Blus et al. 1989). The problem is especially acute for PCP Trumpeter Swans wintering in northwestern Washington (northern Whatcom County) and the adjacent Sumas Valley in British Columbia, where 2,300 swans have been documented as killed from 1999 to 2009, due to the ingestion of spent lead shot (Trumpeter Swan Society 2009). Lead poisoning also caused 16% of the known mortality of Trumpeter Swans in the RMP from 2000 to 2003 (Whitman and Mitchell 2004). Lead toxicosis from spent shot or fishing sinkers produced nearly 50% of the deaths among Trumpeter Swans examined in western Washington from 1976 to 1987, and 20% for swans in the Tristate area (Blus et al. 1989). A Trumpeter Swan in British Columbia had 451 lead pellets in its gizzard (Munro 1925). For the IP, lead poisoning caused 38% of the deaths among 84 Trumpeter Swans necropsied from 1982 to 2000 (Lumsden and Drever 2002). Trumpeter Swans can be successfully treated for lead poisoning, however. Of the 63 Trumpeter Swans from restored populations in Wisconsin and Minnesota that received treatment for lead poisoning, 29 (46%) were subsequently released (Degernes et al. 2002), and 6 of these were later known to reproduce successfully.

Trumpeter Swans in the IP also died from disease, such as aspergillosis (13%), as well as from broken wings (12%). Collisions with power lines, accidents / highway fatalities, and predators were each responsible for 11% of IP deaths (Lumsden and Drever 2002). Power lines caused 62% of the Trumpeter Swan deaths in Wyoming from 1980 to 1986 (Lockman 1990), and they were the leading cause of mortality for Trumpeter Swans reintroduced to the Flathead Indian Reservation in Montana (Becker and Lichtenberg 2007). Of 116 known deaths of Trumpeter Swans in the Tristate region, 32% were the result of a single incident created by the intestinal protozoan *Histomonas*; another 24% were due to power line and fence collisions, 23% to shooting, and 10% to predation (Drewien et al. 2002).

Trumpeter Swans are vulnerable to human disturbances, such as birdwatching, photography, boating, aircraft passages, and other activities in nesting areas. On the Copper River delta near Cordova, Alaska, aircraft overflights and passing road traffic made Trumpeter Swans more watchful but did not cause incubating females to leave their nests (Henson and Grant 1991). Stopped vehicles, pedestrians, and researchers, however, led to changes in behavior: while undisturbed swans always covered their eggs with nesting material before recessing from the nest, 26 out of 28 (93%) of the disturbed swans did not, and disturbed females took longer recesses. Such behavioral alterations could potentially affect productivity, as they could lead to increased nest predation, embryo mortality, retarded embryo development, and changes in female energy budgets. In Alaska, the presence of transportation infrastructure had a negative effect on pond occupancy by Trumpeter Swans at important breeding areas such as the Minto Flats State Game Refuge and the Kenai and Tetlin NWRs (Schmidt et al. 2009b).

As is the case for Tundra Swans, global climate change will probably have a negative effect on breeding Trumpeter Swans, especially in north-ern areas. Predictions foresee a reduction in both pond sizes and numbers, a trend already observed over a broad area of Alaska (Klein et al. 2005, Riordan et al. 2006), but it has not yet affected Trumpeter Swans (Schmidt et al. 2009b). Climate change, however, may also make more suitable wintering habitat available in some areas. The current harvest of Trumpeter Swans is small, but there will probably be more accidental swan shootings during waterfowl hunting seasons as the number of Trumpeter Swans increases. Nonetheless, the accidental harvest of Trumpeter Swans during the hunting season for Tundra Swans is not deemed to be significant by the U.S. Fish and Wildlife Service.

Trumpeter Swans are a stunning success story of a wildlife species brought back from the brink of extinction. They stand as a living testament to a decades-long effort from dedicated waterfowl biologists determined not to let this magnificent bird disappear from the skies of North America.

Tundra Swan

Cygnus columbianus (Ord 1815)

Left, three adults showing variable amount of yellow on the lores; *right,* juvenile

The Tundra Swan, also commonly known as the Whistling Swan, is the most numerous and widespread of the 3 swan species in North America. They are large all-white birds breeding in arctic and sub-arctic tundra habitats from the Aleutians east to Baffin Island. Two populations are recognized, the Eastern Population (EP) and the Western Population (WP), with the latter breeding solely in Alaska. Tundra Swans mate for life, and families remain together into the first spring. Adults have few natural predators and thus are long lived: the record age for a wild Tundra Swan is 23 years and 7 months. They are largely vegetarians that graze on a variety of emergent and submergent aquatics, but during migration and winter these swans are also heavy consumers of agricultural grains, such as corn and rice. Their primary wintering areas are in California (for the WP) and North Carolina (for the EP); the Prairie Pothole Region, upper Mississippi River, Great Lakes area, and large river deltas in Canada are important stopover and staging sites during migration. Protection of their habitats throughout the birds' annual cycle is paramount to maintaining Tundra Swan populations, which in 2009 were 100,300 birds for the EP and 105,200 for the WP. Harvests of both populations are small and strictly controlled by a permit process. Tundra Swans are closely related to the

Bewick's Swan (*Cygnus bewickii*) of Europe and Asia, and some authorities consider them to be conspecific.

IDENTIFICATION

At a Glance. The large size, long upright neck, and all-white plumage instantly distinguish this species from all but the other 2 swans that occur in North America. Most Tundra Swans have yellow on the lores (the area between the eye ring and the bill), although the amount is highly variable among individuals. In migratory flight, Tundra Swans form Vs or oblique lines.

Adults. Tundra Swans are very large, with the males slightly larger than the females. Their plumage is all white and is identical between the sexes. A small but variable amount of yellow is visible on the lores of most (but not all) individuals. The black facial skin tapers to a fine point in front of the eye. The sides of the lower mandible are often reddish salmon. The legs, the feet, and the bill are black. The irises are brown or (rarely) pale gray.

Juveniles. First-winter birds have a brownish-gray plumage. By spring (mid-Mar), most body feathers are white, except for parts of the head and the neck. The legs and the feet are pinkish gray. The bill is also pinkish gray and black tipped, but it becomes increasingly darker through the winter and is mostly black by spring.

Cygnets. Hatchlings have pale bluish-gray and white plumage, with 2 large white shoulder spots. The bill and the feet are pink (Nelson 1993).

Voice. The voice is similar between the sexes. Although formerly called Whistling Swans, Tundra Swans do not produce a whistling call. Instead, the call is a high-pitched, often quavering *oo-ou-oo*, accentuated in the middle; the voice of first-year birds is higher pitched (Limpert and Earnst 1994). The call can be confused with that of Snow Geese, but it is more melodious.

Similar Species. Adults are most likely to be confused with Trumpeter Swans, which are much larger and virtually always lack a yellow spot in front of the eye. Drewien et al. (1999) found that only 2 out of 698 (0.3%) Trumpeter Swans captured in Idaho had yellow lores, versus 97% in Tundra Swans harvested in Utah; yellow lores were absent from about 10% of the Tundra Swans banded in Maryland (Larry Hindman). The voices are very different between Tundra and Trumpeter Swans. A major characteristic distinguishing the two species is the black facial skin, which tapers to a broad point at the eye in Trumpeter Swans, so that the eye seems to be contained in a black mask. This line comes to a narrow point in Tundra Swans. The profile of the bill is also more smoothly sloping in Trumpeter Swans (not unlike that of a Canvasback) versus the concave shape in Tundra Swans. There is almost no overlap in bill length (from the tip to the posterior edge of the nares) between the 2 species: >99% of Trumpeter Swan adults and cygnets ($n = 672$) averaged ≥ 61 mm, versus ≤ 59 mm in Tundra Swans ($n = 1,414$; Drewien et al. 1999). The Mute Swan has an orange bill with a black base and knob, and a sharply curved neck. Among other large white birds in North America, the Snow Goose is much smaller and has black wing tips.

Juvenile Tundra and Trumpeter Swans are very similar, but the overall body coloration is a bright silvery gray in Tundra Swans versus a darker sooty gray in Trumpeter Swans, especially about the head and the neck. Juvenile Tundra Swans also molt earlier and thus appear much whiter by late winter (nearly all white by mid-Mar) than do immature Trumpeter Swans. The bill of immature Tundra Swans is usually a mottled pink with a black tip, while that of Trumpeter Swans is black at the base and at the tip, with pink in the middle. Tundra Swans are closely related to the Bewick's Swan of Europe and Asia, and some authorities have considered the 2 taxa to be conspecific. Interbreeding of the two species has occurred in the wild (Evans and Sladen 1980).

Weights and Measurements. Limpert et al. (1987) published an extensive dataset on weights and

Above:
Tundra Swans in flight.
Dave Menke, U.S. Fish and Wildlife Service

Right:
Tundra Swans.
GaryKramer.net

Tundra Swan pair (*foreground*) in an aggressive interaction while feeding. *GaryKramer.net*

measurements of wintering Tundra Swans in Maryland and North Carolina. They found that males were larger than females, averaging 7.2 kg (*n* = 1,447) versus 6.3 kg (*n* = 1,290). Juveniles averaged 6.1 kg for males (*n* = 299) and 5.6 kg for females (*n* = 403). Bill length averaged 103.7 mm for adult males (*n* = 305), 101.1 mm for adult females (*n* = 164), 88.4 mm for juvenile males (*n* = 34), and 89.1 mm for juvenile females (*n* = 38). Tarsal length averaged 115.7 mm for adult males (*n* = 290), 110.3 mm for adult females (*n* = 160), 113.7 mm for juvenile males (*n* = 33), and 111.3 mm for juvenile females (*n* = 37). Average wing length, as measured by Miller et al. (1988) for Tundra Swans wintering in northern California, was 54.1 cm in adult males (*n* = 23) and 51.1 cm in adult females (*n* = 27). They also reported data on 36 internal and external morphometric variables, noting that while sexual dimorphism is slight in immatures, it is pronounced in adults (males are larger), except for bill-size measurements. Bill length averaged 68.8 mm in

adults (*n* = 354) and 67.6 in cygnets (*n* = 318; Drewien et al. 1999). The wingspan in Tundra Swans commonly reaches 1.8–2.1 m.

DISTRIBUTION

Tundra Swans usually nest in lowland habitats along much of the arctic and subarctic coasts of North America, although some breeding occurs in interior Alaska (Wilk 1993). Their extensive arctic/subarctic breeding grounds extend from the Aleutian Islands and Seward Peninsula of western Alaska east to northeastern Hudson Bay and Baffin Island. Two populations are recognized: the WP and the EP.

Western Population

Breeding. Tundra Swans in the WP breed solely in Alaska. The majority (76%) nest on the Yukon-Kuskokwim (Y-K) delta (Pacific Flyway Council 2001), where up to 100,000 swans can be found in summer. Elsewhere in Alaska, WP birds breed on

Breeding Distribution
Breeding Concentration Area
Wintering Distribution
Wintering Concentration Area

the Seward Peninsula and in the Kotzebue Sound region, the Colville River delta, Unimak Island in the eastern Aleutians, Kodiak Island, and the coast and islands of the eastern Bering Sea. Tundra Swans on the North Slope of Alaska belong to the EP (Limpert et al. 1991).

Winter. WP Tundra Swans have been reported wintering in all 12 Pacific Flyway states and in British Columbia, but they rarely travel south into Mexico (only 121 records from 1947 to 1995; Drewien and Benning 1997). The Midwinter Waterfowl Survey averaged 83,936 Tundra Swans in the Pacific Flyway from 2000 to 2010. Most (79.9%) were in California, primarily from San Francisco Bay northward through the Sacramento Valley, where they especially favor rice fields (Pacific Flyway Council 2001). Elsewhere, 8.2% of the WP occurred in Utah, 4.5% in Oregon, and 4.5% in Washington. Important wintering areas in these states are the marshes around the Great Salt Lake in Utah (especially the Bear River Migratory Bird Refuge), along the Columbia River in Oregon, and the Skagit River delta in Washington. Tundra Swans also are found locally in southcentral British Columbia, southern Idaho, and western Nevada, with smaller numbers

at scattered locations where open water and food prevail throughout winter. Winter distribution can be substantially affected by weather variations and water availability (Pacific Flyway Council 2001). A unique small group that breeds at the southern end of the Alaska Peninsula (Izembek Lagoon) usually winters nearby, at Petersen Lagoon on Unimak Island (Dau and Sarvis 2002).

Migration. WP Tundra Swans most commonly use a trans–Rocky Mountain route during both fall and spring migrations, with some variations along specific segments of it (Sherwood 1960, Bellrose 1980, Paullin and Kridler 1988). One pathway was nicely elucidated by Ely et al. (1997), who placed satellite transmitters on Tundra Swans in the Y-K delta. These birds departed the Y-K delta in late September and stopped in wetlands on the western side of the Alaska Range during early October, after which they stopped briefly in Cook Inlet. They then moved east into the Yukon in Canada, southward on a route paralleling the Wrangell Mountains and going into the Tanana River valley, and through the Yukon to a staging area in northeastern British Columbia (Fort Nelson / Liard River). From there, migration was gradual through central Alberta and southwestern Saskatchewan and across Montana to a second staging area in southeastern Idaho (Snake River), where they remained from mid-November to early December. Subsequent movement was across Nevada to the Sacramento / San Joaquin delta in California. The spring migration route was similar. Like the Tundra Swans on the Y-K delta, those breeding in northwestern Alaska depart their breeding areas in late September and move southeast up the Tanana River valley (Moermond and Spindler 1997). A more westerly route extends from southern Alberta across Idaho to Malheur Lake in Oregon, and then through the Willamette Valley and the Klamath basin to the Sacramento / San Joaquin delta (Paullin and Kridler 1988).

Coastal migrants are probably from breeding populations at Bristol Bay and the Alaska Peninsula. These birds travel through southcentral Alaska, with some stopping briefly in Cook Inlet and the Copper River delta. They then fly into southeastern Alaska, where a portion splits off to follow the interior route. Those continuing down the coast winter from British Columbia south to California (Pacific Flyway Council 2001). Fewer WP Tundra Swans are believed to follow the coastal route in spring, compared with fall. During spring migration through the Rocky Mountains, some WP birds mix with those of the EP (Moermond and Spindler 1997).

Important staging areas along the trans–Rocky Mountain route are large lakes in the Okanagan Valley and west and east of the Kootenay Valley in British Columbia, the Peace River agricultural belt on the British Columbia / Alberta border, Freezeout Lake in Montana, Ruby Lake and Carson Sink in northwestern Nevada, and the Bear River marshes north of the Great Salt Lake in Utah (Sladen 1973, Campbell et al. 1990, Pacific Flyway Council 2001). About 75% of the WP migrate through the Great Salt Lake region before their arrival on wintering areas in the Central Valley of California (Aldrich and Paul 2002). Along the coastal route, important staging areas are Puget Sound in Washington, the Malheur National Wildlife Refuge (NWR) in southwestern Oregon, and the Klamath basin on the Oregon/California border.

Eastern Population

Breeding. The breeding range of the EP extends from the North Slope of Alaska across northern Canada to Baffin Island. Small numbers have also been documented breeding on Bylot Island (Lepage et al. 1998). About 8% of the EP breeds on the Arctic Coastal Plain of Alaska (Hodges et al. 1996). Aerial surveys there averaged 9,971 Tundra Swans from 1986 to 2006 and 14,458 from 2007 to 2010 (Larned et al. 2011). In Alaska, data from band recoveries and sightings of neck-collared swans indicate that the separation of the EP and the WP occurs in the vicinity of Point Hope (Limpert et al. 1991).

Winter. The EP once wintered in large numbers in the Chesapeake Bay area along the mid-Atlantic coast, but the percentage of the population wintering there has declined steadily, decreasing from 65% in the 1960s to 20% by 2002. The EP now largely winters farther south in the Albemarle Sound / Mattamuskeet NWR area of North Carolina, where the percentage of the wintering population has increased from 25% to 70% over the same time period (Serie et al. 2002). The Midwinter Survey in the Atlantic Flyway averaged 92,299 Tundra Swans from 2000 to 2010, of which 68.7% were tallied in North Carolina and 22.6% from Chesapeake Bay (Maryland and Virginia). Other important wintering areas in the mid-Atlantic region are the lower Susquehanna River in Pennsylvania, various New Jersey bays, and Delaware Bay. The Mississippi Flyway averaged 2,514 Tundra Swans from 2000 to 2010, with the largest (45%) percentage at Lake St. Clair in Michigan, but about 25,000 were tailed in the flyway in 2012, no doubt due to an exceptionally mild winter.

Christmas Bird Counts and the Midwinter Survey have recorded smaller numbers of Tundra Swans (usually <500–1,000) wintering inland at various sites, such as on the Mississippi River near La Crosse, Wisconsin; on the Detroit River; in the Toronto, Ontario, region; at Long Point in Ontario; along Lake Erie in Ohio; and in western New York (mostly in the Finger Lakes area). Tundra Swans are considered rare winter visitors in areas north to Halifax, Nova Scotia, occurring in groups of up to 12 birds (Tufts 1986).

Migration. The EP migrates about 5,000 km from their breeding grounds to their principal wintering grounds along the mid-Atlantic coast. Although the basic route and important staging areas have been identified through resightings of neck-collared birds (Sladen 1973), the use of satellite telemetry has revolutionized our understanding of this migration. Petrie and Wilcox (2003) used satellite transmitters to track spring and fall migratory movements of 12 Tundra Swans captured at

Long Point in Ontario, with their results corroborating those identified in previous studies that used neck collars (Sladen 1973). Satellite tracking provided new information, however, on the duration of migration, the length of stay at staging areas, and migration speeds. The swans traveled between the Atlantic coast and the northern prairies along a narrow migration corridor through parts of the southern Great Lakes. In the spring the swans followed 3 migration corridors from the Great Lakes to breeding areas on the western coast of Hudson Bay, the central High Arctic, and the Mackenzie River delta. They spent 20% of their annual cycle on wintering areas, 28% on spring staging areas, 29% on breeding areas, and 23% on fall staging areas. During spring migration they spent 27% of their time on the Great Lakes and 40% on the northern prairies, but passed 48% of the fall migration period in the northern boreal forest.

The largest study to date fitted satellite transmitters on 43 EP Tundra Swans captured during winter on the East Coast and followed their movements from November 2000 to March 2002 (Wilkins et al. 2010a, 2010b). The swans in this study spent 3.5 months on breeding areas, 3.5 months on wintering areas, and 5 months on staging areas. In spring (n = 56 migrations), satellite-tracked swans left their wintering areas during the first half of March; they arrived at spring staging sites within the Susquehanna River valley in southeastern Pennsylvania from mid- to late March, when 10,000–30,000 have been reported in a single period (Limpert and Earnst 1994, McWilliams and Brauning 2000). The swans then moved northwest to the Great Lakes region of Ontario and Michigan, where they remained for 15–30 days before continuing on to prairie habitats in western Minnesota, North Dakota (Devils Lake area), and the Prairie Provinces of Canada. The birds stayed in the prairies for 30–40 days, usually until mid-April. At that point the migration routes diverged: some swans went northwest toward the Mackenzie River valley in the Northwest Territories, the western Canadian Arctic Archipelago, and the North Slope of Alaska; and others moved north or northeast to Nunavut or Hudson Bay. Swans traveling toward the Mackenzie River valley first stopped in boreal forest habitats of Saskatchewan and Manitoba and usually remained in the Athabasca River delta for 14–21 days. From the Athabasca delta, the migration paths again diverged: some swans continued to the North Slope of Alaska and the Mackenzie River valley, while others went northeast to the western Canadian Arctic Archipelago. The swans settled on breeding locations from 4 May to 18 June (median arrival date = 28 May).

Elsewhere in the Arctic, EP swans initially arrived at breeding grounds in the Beaufort Sea area from late May to early June (Johnson and Herter 1989). They first reached Churchill, on Hudson Bay in Manitoba, on 24 May (Littlefield and Pakulak 1969). At Cambridge Bay, Victoria Island, they arrived on 1 June 1960 (Parmelee et al. 1967). Farther east on the Perry River, they first arrived on 31 May and 9 June (Ryder 1967).

In the fall (n = 28 migrations), satellite-tracked EP swans left their breeding areas around mid-September and migrated south through the boreal forest of the Northwest Territories and Nunavut to northern Saskatchewan and Manitoba, where they stayed for 2–3 weeks. These swans then continued to southern Saskatchewan and Manitoba, where they again remained for several weeks. Their next major stop was in the prairies of Montana, the Dakotas, and western Minnesota, where they remained for 20–30 days before moving to the upper Mississippi River and Great Lakes regions. Most swans then migrated nonstop to the mid-Atlantic area, arriving from 27 October to 5 January (median date = 3 Dec). The most important fall stopover sites were the upper Mississippi River, the Souris River, and the Athabasca River delta. All major stopover sites were visited during the spring and fall, but use often varied with the season. In the valleys of British Columbia, Campbell et al. (1990) noted that EP Tundra Swans occurred in larger numbers during spring migration, compared with fall, which suggested a seasonal change from the coastal route by some populations breeding in interior Alaska.

The upper Mississippi River is probably used by more Tundra Swans in the fall than any other EP stopover site. A 1998–99 study revealed the dramatic importance of such key stopover areas, as peak numbers occurred during November in both years: 21,155 in 1998 and 28,115 in 1999, with substantial numbers (>5,000) present for nearly 2 months (Thorson et al. 2002). Total use days (the number of swans counted per day on an area, summed over the number of days they were there) were 788,746 in 1998 and 990,329 in 1999. Subsequent calculations suggested that 25% of the EP stopped in the upper Mississippi River area during fall migration. The turnover rate (ratio of swans using the area:peak number) was low: 1.29 in 1998, and 0.94 in 1999. Data from radiomarked swans ($n = 43$) revealed an average length of stay of 33.6 days. EP swans are probably attracted to the upper Mississippi River because of its abundant aquatic vegetation, areas with limited human disturbance, and position along the migration corridor. Emergent and submergent aquatic vegetation is especially abundant on Pools 4–9 (navigational pools in the river).

Other specific staging areas that are important to the EP, especially in spring, are Lake St. Clair, the Detroit River, and the Lake Erie marshes in Michigan and Ohio (Bellrose 1980). The wetlands of Long Point Bay on Lake Erie were used by an average of 7,177 Tundra Swans during fall migration in the 1990s, which represented 7.9% of the EP; spring use was substantially less (Petrie et al. 2002). Thousands of EP swans also concentrate on the marshes of the upper Mississippi River, as well as in Green Bay and at Horicon Marsh in Wisconsin (Thorson et al. 2002). Peaks of greater than 5,000 birds have been observed between 20 March and 10 April at Green Bay. Within Saginaw Bay, Michigan, more than 4,000 Tundra Swans are routinely seen during spring migration (Granlund et al. 1994). In western New York, I have observed Tundra Swans using the Montezuma NWR complex in increasing numbers, especially following the recovery of native submerged aquatic vegetation on Cayuga Lake.

MIGRATION BEHAVIOR

Tundra Swans migrate in moderate-sized flocks made up of family groups that remain together during the winter, with the yearlings separating from the adults near or on breeding areas (Limpert and Earnst 1994). The estimated flight speed during migration for satellite-tracked Tundra Swans captured at Long Point in Ontario ranged from 37 to 70 km/hr in spring and from 48 to 59 km/hr in fall (Petrie and Wilcox 2003). These speeds yielded estimates of 114 hours of total flight time in spring and 101 hours in the fall. Nolet (2006) used published accounts of Tundra Swans tracked by satellite and calculated an average speed of 52 km/day during spring migration, which was about the speed of melting/retreating ice. The maximum rate of movement for Tundra Swans migrating from the Y-K delta in Alaska ranged from 60 to 90 kph (Ely et al. 1997).

Molt Migration. Molt migration is not well studied in Tundra Swans. On the Y-K delta in Alaska, flocks of 100–1,000 failed breeding birds gather to molt on large lakes near their nesting areas, but large molting flocks also occur a certain distance away (near Bethel), which indicates that some birds do indeed travel from breeding areas to molting sites (Craig Ely). In northern Alaska, flocks of molting Tundra Swans gather on large lakes in the Colville River delta, with the flocks increasing in size during the molting period; hence birds probably come to molt from areas outside the delta (Susan Earnst). Resident pairs (breeding and nonbreeding) also molt on large lakes; the nonbreeders join larger flocks, but successful pairs remain with their cygnets and molt on their breeding territories. Molting flocks clearly respond to disturbances by moving to the middle of lakes, especially when they are flightless.

HABITAT

Tundra Swans primarily breed in coastal delta areas, but they also occur inland to the tree line, albeit less frequently. On tundra lowlands in northwestern Alaska, territorial Tundra Swans ($n = 24$) preferred areas with water and wetlands (41% by area), as did flocked nonbreeders (47%), although water accounted for only 19% of the available habitat. Habitats described as wet-moist tundra, tall shrub–deciduous forest, and tussock lichen tundra were used less than expected, based on availability (Spindler and Hall 1991). Shorelines along river deltas were of major regional importance as fall habitat, because areas with brackish water and river currents were the last to freeze, and submerged mudflats contained an abundant growth of pondweeds (*Potamogeton* spp.). On Bristol Bay in Alaska, breeding Tundra Swans preferred wet meadows with shallow lakes, emergent vegetation, and little elevational relief (Wilk 1988). Swans here also frequently chose shorelines and mudflats where pondweeds were plentiful. On islands and peninsulas in the central and eastern Canadian Arctic, breeding Tundra Swans preferred low-lying coastal areas (79% were <60 m above sea level), where they used shallow ponds and lakes with lush shoreline vegetation (Stewart and Bernier 1989).

Habitat use during fall migration was monitored on a 2,200 km² study area in the Prairie Pothole Region of North Dakota, where Tundra Swans encounter an array of wetland types. Foraging Tundra Swans were 4 times more common on wetlands with sago pondweed (*Potamogeton pectinatus*) than wetlands without pondweed, while nonforaging swans preferred large wetlands with contiguous open water (Earnst 1994).

At Long Point in Ontario, satellite-tracked Tundra Swans more frequently used agricultural fields in the spring (74%) than in the fall (9%). During spring, 65% of the swans observed in agricultural habitats were in cornfields, followed by 24% in fields of winter wheat. In the fall, 67% used winter wheat areas, and only 7% were in cornfields (Petrie et al. 2002). In spring, swans concentrated on wetlands that were closest to agricultural fields. In contrast, fall birds fed mostly on aquatic vegetation and thus flew up to 30 km from agricultural fields to reach habitats near the tip of Long Point.

During winter, Tundra Swans extensively use agricultural fields, but they will also occupy tidal estuarine habitats and freshwater lakes, ponds, and rivers. When the EP wintered primarily in Chesapeake Bay (prior to the 1970s), 76% of the swans were in brackish estuarine bays, 9% in salt estuarine bays, 8% in fresh estuarine bays, and 6% in slightly brackish estuarine bays; 1% used other habitats (Stewart 1962).

POPULATION STATUS

The Midwinter Survey is the primary means of assessing the Tundra Swan population. Serie et al. (2002) reviewed the status of the EP, noting that the population had more than doubled since 1955, averaging 90,625 from 1984 to 2000. Numbers of EP swans totaled 97,700 in 2010 and 97,300 in 2011 (U.S. Fish and Wildlife Service 2010, 2011). The management goal is to sustain the EP above 80,000 birds.

The WP was estimated at 49,300 birds in 2011, 36% below the 2010 tally of 76,700, but much of California was not covered during those 2 surveys, which probably accounts for the low numbers. The 2009 figure was 105,200, and the population had increased by 1%/year from 1999 to 2008. The long-term (1949–2000) average was 80,600 (Pacific Flyway Council 2001). In general, WP winter indices doubled in the 1950s, rose 50% in the 1970s and 1980s, and then began a sustained increase in the 1990s.

Harvest

The Migratory Bird Treaty Act of 1918 closed the hunting season on swans in North America. In 1962 Utah became the first state where Tundra

Swans could again be legally harvested, with 1,000 permits issued. The U.S. Fish and Wildlife Service first authorized hunting of the EP in 1983, with a small number of permits allowed in Montana and the Dakotas; only Montana selected a season and harvested 34 birds. North Carolina was authorized to initiate an experimental harvest in 1984, with other states allowed to do so soon thereafter. In the United States, Tundra Swans are harvested via a permit system (allowing 1 bird/hunter/season) in selected states. For the EP, harvests averaged 3,903 from 1983 to 2008, and 3,382 from 2000 to 2008, of which 69% were shot in North Carolina.

The harvest of WP swans averaged 1,057 birds from 1994 to 2010 (1,025 in 2010), of which 67% were from Utah, 23% from Montana, and 11% from Nevada (Collins et al. 2011). Additional losses occur from WP birds that are shot but unretrieved, which amounted to 15% from 1994 to 2003 (Trost and Drut 2004). Bartonek et al. (1991) estimated the WP subsistence harvest at 6,000–10,000 annually. There is much less subsistence harvest of EP swans, as they are more geographically dispersed than those of the WP. Bromley (1996) reported that 105–288 EP swans were harvested by native peoples during spring hunting in the Northwest Territories, but the swan harvest accounted for only 2% of the total waterfowl harvest. In contrast, WP Tundra Swans breed in areas with larger Inuit populations and thus are more accessible to a greater number of native hunters.

The harvest of WP swans is somewhat complicated by the problem of an accidental harvest of Trumpeter Swans, but the U.S. Fish and Wildlife Service reported a finding of "no significant impact" in an environmental assessment of proposed Tundra Swan hunting in the Pacific Flyway for the 1995–1999 seasons (Bartonek et al. 1995). A 5-year season was subsequently implemented, with a fixed quota of Trumpeter Swans (20 birds/year) that could be accidentally harvested; this quota was reviewed in 2000, again with a finding of "no significant impact."

BREEDING BIOLOGY
Behavior
Mating System. Tundra Swans are monogamous, with initial pair bonds thought to occur in early spring. Like other swans, Tundra Swans remain together year round and probably retain pair bonds for life (Palmer 1976a). In the closely related Bewick's Swan, 96.5% of individuals observed at Wildfowl and Wetlands Trust centers in the United Kingdom during winter retained their mates during their known lifetimes (Rees et al. 1996). Extra-pair copulations have not been observed. Age at first breeding for Tundra Swans is poorly known, but it is probably similar to that for the Bewick's Swans of Europe, which form pair bonds when they are 2–3 years old and breed at 3–5 years old; separation of pairs is very uncommon (2%; Scott 1978). Captive Tundra Swans have not bred before 4 years of age (Johnsgard 1978). Palmer (1976a) stated that the likely minimum breeding age was 2 or 3 years old.

Sex Ratio. Little information is available, but males probably slightly outnumber females.

Site Fidelity and Territory. On the North Slope of Alaska, neck-collared Tundra Swans used the same breeding territory in successive years (Sladen 1973). Breeding pairs also defend their territory throughout the breeding season (Dau 1981). At the Colville River delta in Alaska, territory size ranged from 50 to 100 ha (Earnst 1992a). On the lower Alaska Peninsula (Pavlof), 71% of neck-collared cygnets were philopatric, compared with 35% for subadults (Dau and Sarvis 2002).

Tundra Swans are aggressively territorial and will chase both conspecifics and other waterfowl from their territories. Territories are large (0.5–1.0 km²) and usually include portions of a substantial body of water (Sladen 1973, Dau 1981, Monda 1991, Limpert and Earnst 1994). The large size and conspicuous white plumage of Tundra Swans (and other northern white swan species) undoubtedly increases their visibility on open tundra habitats; these features probably help pairs advertise their presence and maintain their breeding territory (Kear 1970). Reported breeding densities of Tundra Swans are 0.3–0.9/km² on the Alaska Peninsula, 0.5–0.6/km² on the Arctic NWR, 1.4–2.5/km² in arctic Canada, and 3.7/km² on the coast of the Y-K delta (Wilk 1988). McLaren and McLaren (1984) reported 0.9/km² on the Rasmussen Lowlands in Nunavut, although certain areas had densities >2/km².

Courtship Displays. Pairs probably form in winter, and displays are used year round to maintain pair bonds. Tundra Swans are thought to arrive paired on breeding grounds, but courtship among un-paired birds seems to be most common during early spring (Scott 1977). Hawkins (1986a, 1986b) and Limpert and Earnst (1994) described court-ship and reproductive displays. What is variously termed a *quivering-wings, triumph,* or *greeting* dis-play involves *calling* while a pair or family members are facing each other with raised head and neck and their wings partially extended and rotated at the wrists. This display is given when the pair re-unites or during aggressive encounters with rivals or other threats. In the *forward-call,* the bird calls while extending the head and neck horizontally. In *head-bobbing,* the head is raised and lowered vertically, with each movement accompanied by a single call, which is given prior to long-distance moves and probably serves to alert the mate and family members (Black 1988). The pair performs a precopulatory display of *ritualized-bathing,* fol-lowed by mutual *head-dipping* as they move side by side. During copulation the male mounts the partially submerged female, grasping the back of the neck or head. Copulation takes less than 20 seconds from mounting to dismounting. After the dismount, the pair face each other in a postcopula-tory display, partially raising their wings, bowing their heads, and *calling.* Tundra swans also engage in highly ritualized *nest-building.* At times this be-havior occurs away from the actual nest and out-side the breeding period, and it often happens after nest exchange.

Nesting

Nest Sites. On the Arctic NWR in Alaska, most nests were located in marshes <1 km from coastal lagoons or large coastal lakes, versus upland or partially vegetated habitats (Monda et al. 1994). In northcentral Alaska, Tundra Swans built their nests in wet-meadow, dwarf shrub, and tussock peatland habitats (Gill et al. 1981, Wilk 1993). On the Y-K delta, nests were located in upland and wet-meadow tundra and on islands (Dau 1981). Swans frequently choose elevated hummocks, possibly because they provide a better view of the surrounding terrain, but also because hummocks and other high spots are the first areas that are free of snow. Nest densities on the Y-K delta averaged 0.7/km² over a 13-year study (1988–2000; Babcock et al. 2002). Data from other research on the Y-K delta noted an average of 0.4 nests/km² in low tun-dra and 1.5/km² in wet-meadow tundra (Dau 1981). At the Kanuti NWR in northcentral Alaska, Tun-dra Swans nested at much lower (0.01/km²) densi-ties in tussock peatland areas (Wilk 1993).

Both sexes build the nest from material gath-ered within about 3 m (Hawkins 1986). Monda et al. (1994) noted that Tundra Swans frequently used old nest mounds created in previous years, and that such nests were more successful than nests on new mounds (83% vs. 70%). Use of the previous year's nest site reduces construction time, because the mounds are very large (10–61 cm high and 40–200 cm in diameter); it probably also di-minishes the time required for nest-site selection (Owen 1980, Monda et al. 1994). The actual nests are tall—the mean height of 16 nests was 20.5 cm from the base and 38 cm from adjacent tundra—and the base width was about 25–50 cm (Limpert and Earnst 1994). Nest building commences 4–9 days before egg laying (Hawkins 1986).

Clutch Size and Eggs. Tundra Swans nesting on the Y-K delta from 1963 to 1971 laid an average of 4.3 eggs/clutch (*n* = 354; Lensink 1973). Over a subsequent 13-year study (1988–2000) on the Y-K delta, clutch size averaged 4.5 eggs (range = 3.4–

4.8 eggs; Babcock et al. 2002). From 1988 to 1990 on the Arctic NWR, clutch size on 2 study areas ($n = 110$) averaged 3.6 eggs (range = 3.1- 4.5 eggs); 15 were 5-egg clutches and 1 was a 6-egg clutch (Monda et al. 1994). Clutch sizes were larger during years with an earlier snowmelt and, therefore, earlier nest initiations. About 50% of the variation in clutch size was explained by the nest-initiation date, and clutch size was negatively correlated ($r = -0.71$) with the egg-laying date. Lensink (1973) also observed that the clutch size of Tundra Swans on the Y-K delta varied with the timing of snowmelt. Babcock et al. (2002) reported that clutch size on the Y-K delta was not correlated with the date of river ice-breakup, the date of peak arrival, or the mean nest-initiation date.

Tundra Swan eggs are large, elliptically ovate, creamy or dull white, and nonglossy. They often acquire a brownish stain from the damp plant material at the nest (Bent 1925, Monda 1991). In Alaska, fresh egg mass ($n = 320$) averaged 273 g (range = 210–340 g); egg dimensions were 10.6 × 6.8 cm (Monda et al. 1994). Tundra Swan eggs are noticeably smaller than those of Trumpeter Swans. Alisauskas and Ankney (1992a) reported an egg-laying rate of 1 every 2 days, and Limpert and Earnst (1994) noted a rate of 1 every 1.5–2 days; hence most clutches are completed within 5 to 10 days.

Incubation and Energetic Costs. Both sexes incubate the clutch, although it is uncertain if the male has an incubation patch. The female pulls a small amount of down from her breast, which creates a ruffled appearance but not a readily noticeable brood patch (Hawkins 1986b). At 5 nests monitored by video cameras on the Colville River delta in Alaska, females averaged 71% of their time on the nest (range = 60%–79%), compared with 27% for males (range = 20%–38%); nests were unattended 2% of the time (Hawkins 1986b). By the second or third day, eggs were incubated 98% of the time, but the female incubated longer than the male as the clutch approached hatching. Male in-

cubation probably prevents egg cooling and deters predation (especially avian), as well as allows the female to replenish nutrient reserves used to produce the clutch. The incubation period ($n = 63$) on the Y-K delta ranged from 26 to 33 days (Babcock et al. 2002); Johnsgard (1978) reported 30–32 days.

Nesting Chronology. The timing of snow- and ice-melt are key factors affecting arrival dates and subsequent nest initiations by Tundra Swans. During a 13-year study (1988–2000) of nesting Tundra Swans on the Y-K delta, nest initiation ranged from 1 to 27 May and was highly correlated with the date of 90% snow-free uplands ($r = 0.96$; Babcock et al. 2002). Measures of spring ice-breakup phenology varied by 15–20 days, as did peak arrival (by 17 days). Generally, nest initiation was synchronized and occurred an average of 12 days after peak arrival. Nest initiation was also synchronized on the Arctic NWR, taking place within 5–15 days of arrival (Monda et al. 1994).

Latitude is also a key factor affecting nest initiation. On the Rasmussen Lowlands in Nunavut, nest initiation was observed from mid-June to early July (McLaren and McLaren 1984). In contrast, mean nest-initiation dates over 3 years (1988–90) and 2 study areas on the Arctic NWR ranged from 20 May to 1 June; the earliest nests were initiated on 10 May and the latest on 11 June (Monda et al. 1994). Much farther south, however, at Bristol Bay on the Alaska Peninsula (1983–87), Tundra Swans exhibit the earliest breeding phenology of any major population, arriving 2–4 weeks before those in the subarctic or arctic (mid- to late Mar; Wilk 1988). A late spring delayed nest initiation by 10 days.

Nest Success. Nest success is exceptionally high in Tundra Swans. Their large size is a defense against even large mammalian nest predators, and their near-constant nest attentiveness during incubation virtually eliminates avian predation. On the Y-K delta (1988–2000), 89% of nesting pairs hatched at least 1 egg (Babcock et al. 2002). On the Arctic NWR, nest success averaged 76% (range = 58%–

84%) during a 3-year study (1988–90; Monda et al. 1994). At Izembek Refuge in Alaska (1980–90), average nest success was 80% (range = 65%–91%; Limpert and Earnst 1994).

Nest destruction captured on videotape at six nests on the Colville River delta in Alaska documented 21 encounters with predators: 14 with arctic foxes (*Alopex lagopus*), 2 with red foxes (*Vulpes vulpes*), 2 with Golden Eagles (*Aquila chrysaetos*), 2 with Parasitic Jaegers (*Stercorarius parasiticus*), and 1 with Glaucous Gulls (*Larus hyperboreus*; Hawkins 1986a). On the Arctic NWR, suspected nest predators were Glaucous Gulls, Common Ravens (*Corvus corax*), jaegers, and brown bears (*Ursus arctos*; Monda et al. 1994). Brown bears are probably significant nest predators on the Alaska Peninsula (Wilk 1988).

Brood Parasitism. There is no documentation of brood parasitism in Tundra Swans.

Renesting. Tundra Swans do not renest, most likely due to energy constraints associated with the short breeding season characteristic of arctic and subarctic summers.

REARING OF YOUNG

Brood Habitat and Care. In the Arctic NWR, preferred brood-foraging sites include aquatic marsh, graminoid marsh, and saline graminoid-shrub habitats near coastal areas. Younger broods feed within the uplands more frequently than older broods (Monda et al. 1995). Parents often use their feet to stir up submerged vegetation for the cygnets, a behavior most common during the first 2 weeks after hatching (Earnst 1992a). In northwestern Alaska, an average brood-rearing habitat covered 139 ha (Spindler and Hall 1991).

Cygnets generally leave the nest within 24–26 hours after hatching (Hawkins 1986b). Adults lead broods to lakes with emergent vegetation, and both parents guard their young. Throughout the brood-rearing period, the cygnets feed and loaf within a few body lengths of 1 or both parents (Earnst 1992a).

Brood Amalgamation (Crèches). Crèches are not documented for Tundra Swans.

Development. There is surprisingly little information on the growth and development of Tundra Swans. The weight of 1-day-old cygnets in Alaska averaged 179 g (range = 171–190 g; Smart 1965a). Growth curves developed by Lensink (*in* Bellrose 1980) showed that male cygnets increased in weight from about 170 g at hatching to 5.6 kg in 70 days; females reached about 5.0 kg in the same period. Most cygnets probably attain flight stage in 60–70 days, although some may require up to 75 days. Cygnets remain with their parents through the fall and their first winter. Cygnets have not reached adult size when they depart from the breeding grounds, so having their parents nearby protects them from foraging competition with other swans and kleptoparasitism by gulls. Families initiate spring migration, but the parents arrive alone on breeding areas; 1-year-old swans join nonbreeding flocks (Limpert and Earnst 1991).

RECRUITMENT AND SURVIVAL

On the breeding grounds, the 2 major indicators of recruitment are the percentage of Tundra Swan pairs with broods and late-summer brood size. Juvenile swans are easily recognized in winter because they retain their juvenal plumage; hence the juveniles:pairs ratio is a good indicator of both recruitment and survival from first migration well into the first winter.

On the Y-K delta in Alaska (1963–71), the percentage of pairs with broods in August averaged 31.4% (Lensink 1973). On the Alaska Peninsula (1983–87), 31%–40% of the pairs had nests or young (Wilk 1988). Average brood sizes were 3.0 cygnets on the Y-K delta, 2.7–3.3 on the Alaska Peninsula, and 2.5 on the North Slope (Bart et al. 1991b). Within the Inuvialuit Settlement Region in the Northwest Territories, an average of 21% of the breeding pairs successfully produced a brood, with an average brood size of 2.5 by late summer (Hines and Wiebe 1998). On islands (e.g., Victoria and King William) and peninsulas (e.g., Boothia) in the central and eastern Canadian Arctic (1980–85), 17%–33% of the pairs produced cygnets, with an average brood size of 1.3–1.6, which was lower than that of other populations (Stewart and Bernier 1989). On the Y-K delta, Lensink (*in* Bellrose 1980) reported that broods suffer only slight losses after the first month; broods averaged 3.15 cygnets in July, 3.01 in August, 3.01 in September, and 2.70 in October (1963–71).

At Izembek Lagoon on the lower Alaska Peninsula, the percentage of cygnets surviving to fledgling was 54%–58% during a 20-year study from 1977 to 1996 (Dau and Sarvis 2002). Of the cygnets that died ($n = 236$), 52.6% perished in the first 10 days after hatching, 32.6% between days 11 and 30, and 14.8% from day 31 to fledging. Limpert and Earnst (1994) reported that 86% of the cygnet loss at Izembek Lagoon occurred within the first 30 days. This population of Tundra Swans is unique, in that they usually (17 out of 20 years) winter at nearby Unimak Island.

From 2000 to 2009, productivity surveys in the Atlantic Flyway of EP Tundra Swans averaged 12.2% juveniles (1.5/family), varying from a low of 5.2% in 2003 to a high of 22.6% in 2006; the average long-term (1976–2009) percentage of juveniles was 13.6%, with a high of 30.2% in 1981 (Klimstra and Padding 2010). Combined productivity surveys of the WP swans averaged 14.8% juveniles from 2000 to 2009 (Huggins 2009). Area surveys over the same time period averaged 16.9% juveniles (2.1/family) in Utah, 14.2% (1.7/family) at Summer Lake in Oregon, and 12.6% (1.8/family) in the Sacramento Valley of California.

Annual survival rates (1966–90) for the EP were calculated from resightings of 5,963 neck-banded swans captured in Alaska, Maryland, and North Carolina. Survival rates were very high; they were similar (92%) for adult males and females, 81% for juvenile males, and 52% for juvenile females (Nichols et al. 1992). Bart et al. (1991b) estimated juvenile survivorship of Tundra Swans in the EP at 52%

during their first migration and 76% during their first winter.

Avian cholera has fatally affected Tundra Swans in certain years; it caused the death of 1,100 swans in California during winter 1987/88. Of 392 Tundra Swans examined by the U.S. Fish and Wildlife Service's National Wildlife Health Center (now under the auspices of the U.S. Geological Survey) from 1981 to 1988, 29% died from lead poisoning, 15% from avian cholera, 8% from emaciation, and 8% from kidney dysfunction (Bartonek et al. 1991).

The longevity record for a wild Tundra Swan is 23 years and 7 months for an adult female banded in the Northwest Territories and shot in North Carolina. A second female survived at least 23 years and 4 months, having been banded in Ohio and later recaptured and released.

FOOD HABITS AND FEEDING ECOLOGY

Tundra Swans are vegetarians, mostly eating the leaves, stems, and tubers of emergent and submergent plants, but they readily consume agricultural crops (primarily corn and rice) during the winter and on migration. They usually feed in water so shallow that immersing the head and the neck is sufficient for them to obtain desired food items. They will occasionally tip up to reach food in deeper water, and they will use their feet to dig up plant material and mollusks. Badzinski (2005) noted that Tundra Swans feeding at Long Point in Ontario exhibited 4 feeding methods: dabbling, tipping, submerging the head, and treading; young swans dabbled more than adults. Both parents and juveniles benefited from the higher social standing associated with family status: families won nearly all conflicts with nonfamily groups (Badzinski 2003). Juveniles without parents formed the lowest-ranking social group.

Breeding. During the breeding season, Tundra Swans feed in family groups or in small flocks. On the Colville River delta in Alaska, an aquatic grass (*Arctophila fulva*) and a sedge (*Carex aquatilis*) were the plants most heavily grazed by waterfowl, including Tundra Swans (Bart and Earnst 1991). Grazing was rare in wetlands with <10 cm of water or in wetlands >150 m from a large body of water. Fecal analyses from swans on the Colville River delta revealed that their principal foods were sedges (38%), *Nostoc* algae (23%), sheathed pondweed (*Potamogeton vaginatus*; 22%), smaller quantities of miscellaneous plants (16%), and arthropods (1%; Earnst 1992a).

Migration and Winter. In North Dakota, migrating Tundra Swans strongly preferred wetlands with sago pondweed (Earnst 1994). Arrowhead (*Sagittaria* spp.) tubers are especially important on the Mississippi River, along with wild celery (*Vallisneria americana*) and sago pondweed (Thorson et al. 2002). Early studies on the Great Salt Lake marshes in Utah reported that the stomach contents of 12 Tundra Swans consisted entirely of the tubers and seeds of sago pondweed (Sherwood 1960), but Nagel (1965) also observed the use of harvested cornfields when the marshes were covered with ice. Near Stockton, California, Tundra Swans fed on waste corn in both dry and flooded fields, and on unharvested potatoes (Tate and Tate 1966). Corn is eaten by Tundra Swans wherever it is available.

Historically, Tundra Swans wintering in Chesapeake Bay fed on species such as widgeongrass (*Ruppia maritima*) and sago pondweed in the brackish estuarine bays, and wild celery and pondweeds in the freshwater estuarine bays (Stewart and Manning 1958). Of 49 birds collected for food-habits analyses, submerged vegetation composed 100% of the diet in fresh water, 60% in brackish water, and 41% in estuarine marsh ponds. In more marine habitats, however, the invasive Baltic clam (*Macoma balthica*) and other clam species formed 31% of their diet.

In the late 1960s, however, Tundra Swans at Chesapeake Bay began to depart from their long-standing tradition of feeding exclusively on aquatics and began supplementary feeding on waste corn in fields on Maryland's Eastern Shore (Munro 1981). Since then, Tundra Swans field-feed almost as extensively as Canada Geese. They commonly fly well inland to glean waste corn and soybeans and browse on shoots of winter wheat.

Castelli and Applegate (1989) reported that the roots of redroot (*Lachnanthes tinctoria*) are a principal food of Tundra Swans feeding in cranberry bogs in New Jersey. Redroot is a common weed in cranberry bogs, but the swans are a nuisance because they uproot adjacent cranberry vines and cause craters in the substrate that subsequently interfere with machinery. Their use of winter wheat is probably the only instance when Tundra Swans potentially impact agricultural crops. In North Carolina, Tundra Swans feeding on winter wheat reduced aboveground biomass by 12% and seed-head mass by 11% (Crawley and Bolen 2002).

MOLTS AND PLUMAGES

As adults, Tundra Swans have only 1 plumage (basic) that is molted each year via the prebasic molt. The descriptions of molts and plumages are primarily from summaries in Palmer (1976a) and Limpert and Earnst (1994), which should be consulted for further details.

Juvenal plumage is acquired at about 9–10 weeks of age and is brownish gray, darkest on the head and the neck. This plumage is completed and flight capability achieved in about 70–75 days. First basic is acquired from October through March and replaces most of the body plumage (but not the remiges). The resultant plumage is mostly white, but there is usually a gray or grayish-brown cast to the head, the neck, and parts of the back. The molt into definitive basic occurs from July to December; young birds may begin and finish this molt about 1 month earlier, and they may retain a few juvenal or first basic feathers on the head. The wing molt begins in late July and produces a flightless period of 33–34 days, but the swans could engage in short escape flights at 23 days (Earnst 1992b). The resultant plumage is all white and retained until the following summer.

On the Colville River delta in Alaska, breeding females initiated the wing molt an average of 21 days after their broods hatched, significantly later (3.1 days) than their mates and nonbreeding birds (Earnst 1992b). Growth rate of the tenth primary was 8.96 mm/day. Females apparently delayed the wing molt so as to replenish reserves spent during reproduction, as female body condition increased through day 18 of the molt. Asynchronous molting also allows a member of the pair to retain primaries, which facilitates territorial and brood defense. The timing of the molt did not differ between breeding and nonbreeding males.

Craigie and Petrie (2003) provide 1 of the few ecological studies of molting in Tundra Swans during spring and fall migration at Long Point in Ontario. Adults, subadults, and juveniles all molted contour feathers at low intensity, and the patterns generally were similar between males and females. The authors speculated that perennial monogamy (a life-long pair bond), lack of an alternate plumage, and shared incubation and brood-rearing duties were some of the selection pressures leading to intersexual similarities in feather replacement during migration. Adults, subadults, and juveniles generally molt more intensively in the fall than in the spring, perhaps because fall migrants carried larger lipid reserves and were thus nutritionally capable of a more thorough molt. Swans probably have better access to food in the fall and thus have less need to store energy, because their wintering areas in the mid-Atlantic region are only 1,000 km away, versus their 2,400–3,500 km migration to breeding grounds in the spring.

CONSERVATION AND MANAGEMENT

The U.S. Fish and Wildlife Service and the Canadian Wildlife Service manage the harvest of Tundra Swans under a strict permit system; hence overharvest is unlikely to threaten either the EP or the WP. Habitat loss or degradation is by far the more serious threat, particularly from oil, gas, and mineral extraction activities in breeding habitats in the Arctic. Protection of key staging areas is of critical importance, given the distances Tundra Swans travel from their breeding to their wintering areas (up to 5,000 km for the EP). All winter habitats for Tundra Swans are important, but those in Chesapeake Bay and North Carolina are especially vital for the EP, as are those in California for the WP. Habitat recovery efforts in Chesapeake Bay, particularly regarding submerged aquatic vegetation, are of great consequence, because reliance on agricultural foods always subjects wintering waterfowl to the vagaries of agricultural commodity practices, and a shift in cropping patterns can disrupt their use of traditional migration and wintering areas.

Tundra Swans are victims of lead poisoning from ingesting spent lead shot and lead fishing sinkers, but such events usually have not involved large numbers of birds (Bartonek et al. 1991). Globally, a 1994 review found that nearly 10,000 swans of 6 species from 14 countries had died from lead poisoning, due either to the ingestion of fishing weights and shotgun pellets, or to contamination from mining and smelting wastes. Mortality rates from lead poisoning were higher for Tundra Swans than for any other swan species, largely because an estimated 7,200 succumbed over a 5-year period from consuming lead shot pellets on wintering areas in North Carolina (Blus 1994). Bowen and Petrie (2007) examined Tundra Swans ($n = 77$) during migration on the lower Great Lakes in Canada and reported that 4% contained nontoxic shot and 2.6% contained lead shot, but no birds had ingested lead sinkers from fishing tackle. Lead from mine wastes that had accumulated in sediments poisoned Tundra Swans in Idaho (Blus et al. 1991). Even a small number of swan deaths is significant, however, because these birds are normally long lived and thus are reproductively active for many years.

The United States banned the use of lead shot for waterfowl hunting in 1991, followed by Canada in 1999; these actions dramatically reduced the availability of spent lead shot that could be eaten by ducks, and presumably swans as well (Anderson et al. 2000). Lead fishing weights, however, are a source of mortality for a variety of large waterbirds (Franson et al. 2003, Rattner et al. 2008), and such weights are common in the environment. Twiss and Thomas (1998) estimated that 125–187 million lead sinkers are deposited *annually* in Canadian waters. Lead sinkers are an especially significant factor in the deaths of Common Loons (*Gavia immer*) in both the United States (Sidor et al. 2003) and Canada (Twiss and Thomas 1998), which has led to calls for nontoxic substitutes in both countries (Thomas 1995). Reliable nontoxic alternatives to lead sinkers are available (e.g., steel, brass, ceramics, tin) and their use is required in some areas (Scheuhammer and Norris 1995, 1996). In Canada, lead sinkers and jigs weighing <50 g are banned in national parks and wildlife refuges. The U.S. Fish and Wildlife Service has prohibited the use of lead sinkers at several national wildlife refuges, especially those frequented by swans and loons (Goddard et al. 2008). The National Park Service has banned the use of lead sinkers in Yellowstone National Park. Several New England states and New York have imposed restriction on lead sinkers, largely to protect Common Loons. In 1994, the U.S. Environmental Protection Agency

tried to institute a nationwide ban on the sale and manufacture of some lead sinkers, but the proposal was not approved (U.S. Environmental Protection Agency 1994). It appears, however, that the United States and Canada are at the beginning stages of restricting/prohibiting the use of lead fishing sinkers as a means of reducing waterbird mortality (Goddard et al. 2008). In general, sinkers <2.5 cm and weighing <50 g seem most problematic to waterbirds (Scheuhammer and Norris 1995), but a future ban on all lead sinkers seems worthwhile, given the overall toxicity of lead to flora and fauna in the environment.

Tundra Swans, and other wildlife in the Arctic, are also becoming increasingly linked with issues associated with global climate change. The Arctic is unquestionably warming, which is producing an earlier spring greenup (Myneni et al. 1997) that may already be affecting Tundra Swans. The median nest-initiation date of 17 May reported from a 13-year study (1988–2000) on the Y-K delta (Babcock et al. 2002) was much earlier than the median date of 24 May observed in an 8-year study conducted over 1972–79 (Dau 1981). The long-term effects of climate change on flora and fauna are yet to be determined, but Tundra Swans and other species of arctic-nesting waterfowl will no doubt be affected by policy decisions in this arena.

Greater White-fronted Goose

Anser albifrons (Scopoli 1769)

Left, Juvenile; *center*, adult; *right*, Tule Goose subspecies

The Greater White-fronted Goose, usually referred to simply as the White-fronted Goose or "White-front," is virtually circumpolar in its breeding distribution, which is the broadest breeding range of all species in the genus *Anser*. Hunters commonly refer to them as "Specklebellies" or "Specks." Their taxonomy is confusing and has been debated over the years, but 4 subspecies are currently recognized, 2 of which occur in North America: the Pacific White-fronted Goose (*Anser albifrons frontalis*) and the Tule White-fronted Goose (*Anser albifrons elgasi*), generally known just as the Tule Goose. There are 2 recognized populations of Pacific White-fronted Geese in North America. The Pacific Population largely nests on the Yukon-Kuskokwim (Y-K) delta in Alaska and winters in the Central Valley of California. The Midcontinent Population breeds from Hudson Bay in Canada, west to the Arctic Coastal Plain of Alaska, and then south into the boreal forest and taiga of interior and northwestern Alaska. The Midcontinent Population traditionally wintered along the Gulf Coast, but it now occurs farther inland, due to changes in agricultural practices. This population also winters in Mexico. Tule Geese have the smallest population of any subspecies of goose in the world. They nest exclusively in tundra habitats along Cook Inlet in Alaska and winter in the Central Valley. White-fronted Geese are monogamous and, like other geese, tend to mate for life. Family members have the most extended associations known among geese: 38% of 3-year-olds were observed with their parents during winter, and some associations persisted as long as 8 years. The Pacific Population declined dramatically from the mid-1960s to the mid-1980s, but it has recovered, numbering 649,785 geese in 2010; the Midcontinent Population had 709,800 birds in 2010. White-fronted Geese depend heavily on agricultural foods during migration and in the winter, but changing agricultural practices are a cause for management concern.

Greater White-fronted Goose juvenile

Greater White-fronted Goose adult
Tule Goose is similar

IDENTIFICATION

At a Glance. This medium-sized goose is easily recognized. Adults are grayish brown with a conspicuous white band posterior to the base of the bill; hence the name "White-fronted." Irregular black barring or mottling on the underparts led to the alternative common name "Specklebelly." The legs and the feet are orange; the bill is pinkish, with a white tip.

Adults. The sexes are similar in appearance, although males average 4.6%–5.7% larger than females and are 10% heavier (Ely et al. 2005). The general appearance of the body is brownish gray, with the belly and the breast streaked or mottled with black. The barring on the belly varies from light to solid black and is generally darker in males than in females, due to the higher proportion of males with near-complete dark barring (Ely and Dzubin 1994). The chest and the breast are ash gray, lighter colored than the back. The neck is generally darker, and the neck feathers are deeply furrowed. The white on the face is most apparent when the birds are facing forward. At rest, distinctive white striping is visible below and on the edge of the folded wing. A prominent horseshoe-shaped white patch on the lower rump contrasts with the brown tail; the ventral region is also white. The legs and the feet of adults are orange. The bill is pinkish. The irises are dark brown.

The Tule White-fronted Goose subspecies (usually referred to as the Tule Goose) is distinctly larger and darker than the Pacific White-fronted Goose. In Alaska, Tule Geese from the Cook Inlet lowlands were 10.8% larger overall than Pacific White-fronted Geese on the Y-K delta or the Bristol Bay lowlands, especially in body weight (24.2%), bill height (15.4%), bill width (13.4%), and culmen length (11.3%; Orthmeyer et al. 1995). Two bill measurements alone correctly differentiated 92% of the male and 96% of the female Tule Geese from the 2 other White-fronted Goose populations. In addition to these size differences, Tule Geese are noticeably browner (the head and the neck are dark chocolate brown), have less barring on the belly, and have more white on the face than Pacific White-fronted Geese (Delacour and Ripley 1975, Krogman 1975). Most specimens of Tule Geese examined by Wilbur (1966) showed a blackish-brown cap above the eye.

Juveniles. In the fall, juveniles lack the distinctive white on the face, as well as the black barring or

Flock of Greater White-fronted Geese; note the individual variations in the speckling on the belly. *GaryKramer.net*

Greater White-fronted Geese preparing to land. *GaryKramer.net*

blotches on the belly, but they acquire some barring over the winter. White flecking appears on the face in 10–12 weeks (early Sep or later) and the white band is usually completed by March. The legs, the feet, and the bill are dull yellow orange. The yellowish bill becomes pinkish in 13–16 weeks. Heavy black speckling, however, is not visible until as late as June, and it is not complete until the following fall (Palmer 1976a).

Goslings. Goslings are a variable yellowish green, with a grayish eye strip and loreal spots that sometimes join to form a mask (Nelson 1993). The bill has a slight but distinct downward droop.

Voice. White-fronted Geese are among the most vociferous of all geese. Their most distinguishing characteristic is their high-pitched melodious *klow-yo*, *leq leq*, or *klew yo-yo*, also described as *leek-leek* (Johnsgard 1978), or *kow-yow* or *kow-kik-kik* (Bellrose 1980). The Yupik Eskimo name for these geese, *leg leg*, is derived from the sound of their call. This vocalization, which led to their alternative common name of "Laughing Goose" (Ely and Dzubin 1994), is often heard during flight.

Similar Species. Adults are unlikely to be confused with other species, but juveniles can appear similar to the juvenile blue phase of the Lesser Snow Goose (*Chen caerulescens caerulescens*). The immature Lesser Snow Goose, however, has an overall dark grayish-brown body, with an even darker head and neck; lacks white on the face; and has a dark bill and dark legs, the latter attributes in contrast to the light bill and light legs of White-fronted Geese.

Weights and Measurements. Males are larger than females, for which Orthmeyer et al. (1995) provided data from 2 populations of Pacific White-fronted Geese captured during the breeding season in Alaska. On the Y-K delta, body weights averaged 2,255 g for males ($n = 217$) and 2,000 g for females ($n = 247$). For males, culmen length averaged 51.4 mm ($n = 201$); bill width, 24.4 mm ($n = 152$); and total tarsus length, 88.4 mm ($n = 152$). Female culmen length averaged 48.4 mm ($n = 228$); bill

width, 23.2 mm ($n = 183$); and total tarsus length, 83.7 mm ($n = 187$). On the Bristol Bay lowlands, body weights averaged 2,408 g for males ($n = 141$) and 2,114 g for females ($n = 204$). For males, culmen length averaged 53.0 mm ($n = 143$); bill width, 24.7 mm ($n = 136$); and total tarsus length, 89.4 mm ($n = 143$). Female culmen length averaged 49.1 mm ($n = 206$); bill width, 23.4 mm ($n = 171$); and total tarsus length, 85. 3 mm ($n = 206$). Bellrose (1980) noted average wing lengths of 43.9 m for adult males ($n = 57$) and 41.9 cm for adult females ($n = 51$). Ely and Dzubin (1994) reported substantial data on weights and measurements of White-fronted Geese sampled from multiple breeding and wintering locations.

Orthmeyer et al. (1995) also provided body measurements for Tule Geese captured during the breeding season on the Cook Inlet lowlands. Body weights averaged 3,010 g for males ($n = 105$) and 2,776 g for females ($n = 83$). For males, culmen length averaged 58.5 mm ($n = 105$); bill width, 27.5 mm ($n = 71$); and total tarsus length, 91.7 mm ($n = 74$). Female culmen length averaged 55.3 mm ($n = 78$); bill width, 26.3 mm ($n = 26.3$); and total tarsus length, 87.6 mm ($n = 44$). Other studies reported body weights of Tule Geese in Alaska as 2,735 g for males ($n = 37$) and 2,285 g for females ($n = 39$), and weights during winter in California as 3,000 g for males ($n = 53$) and 2,700 g for females ($n = 54$; Ely and Dzubin 1994).

DISTRIBUTION

Greater White-fronted Geese and Brant are the only two geese with a circumpolar distribution, with the range of Greater White-fronted Geese more widespread than that of any other taxon within *Anser*. Four subspecies (formerly 5) are recognized. Outside North America, the nominate European or Russian White-fronted Goose (*Anser albifrons albifrons*) breeds throughout the Siberian tundra, extending from the Kanin Peninsula on the Barents Sea east to the Kolyma River delta, and is slightly smaller than *A. a. frontalis* (Fox and Owen 2005). These birds winter in northwestern Europe,

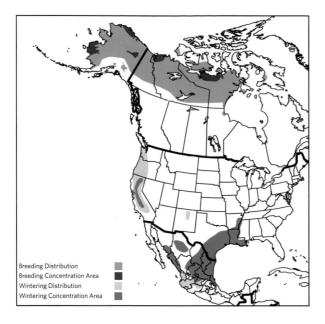

Breeding Distribution
Breeding Concentration Area
Wintering Distribution
Wintering Concentration Area

the Mediterranean region, and near the Caspian Sea and the Persian (Arabian) Gulf (Owen 1980). The Greenland White-fronted Goose (*A. a. flavirostris*) breeds in western Greenland, between 64° and 72° N latitude, and migrates through Iceland to its wintering grounds in the northwestern part of the United Kingdom and on the Wexford Slobs in southeastern Ireland (Warren et al. 1992, Fox et al. 2003). The Greenland White-fronted Goose is slightly larger and darker than *frontalis* and has a bright orange bill (Ely and Dzubin 1994). The closely related Lesser White-fronted Goose (*Anser erythropus*) is a subarctic species with a fragmented breeding range that extends from Fennoscandia to easternmost Siberia. The current world population of this species is estimated at only 25,000; hence it is the most threatened goose taxon in the Palearctic (Ruokonen et al. 2004).

Aside from widespread acceptance of the Greenland White-fronted Goose, however, the taxonomy and distribution of the group is widely debated. Some authorities believe that the nominate subspecies in Eurasia (*albifrons*) breeds across the continent from the Kanin Peninsula in Siberia east to the Bering Sea (Owen 1980, Portenko 1989). Others argue that the range of *albifrons*

ends at the Kolyma River, after which the breeding distribution of *frontalis* begins and continues uninterrupted eastward to Alaska and across the Canadian Arctic to Hudson Bay (Delacour 1954, Cramp and Simmons 1977). In further contrast, Mooij and Zockler (2000) proposed subspecific status (as *A. a. albicans*) for those Greater White-fronted Geese breeding in eastern Siberia from the Kolyma River to the Bering Sea.

Ely et al. (2005) analyzed morphological measurements of Greater White-fronted Geese from 16 populations across this taxon's range and found that the Palearctic (Eurasian) forms varied clinally; progressing geographically eastward, they increased in size from the smallest forms (on the Kanin and Taimyr [Taymyr] Peninsulas in western Eurasia) to the largest (on the Anadyr lowlands of the Chukotka [Chukchi] Peninsula). There was little clinal variation, however, in the Nearctic (North America), as both the smallest (from among the Greater White-fronted Geese) and the largest forms (Tule Geese) were from separate breeding areas in Alaska. Ely et al. (2005) take the position that the breeding distribution of Greater White-fronted Geese was altered substantially with the glacial maximum (15,000–21,000 years before the present), since much of their current breeding range was covered with glaciers or polar desert, and that changes in morphology were accelerated as the geese emerged from glacial refugia. Overall, however, there is little consensus on the taxonomy and distribution of the subspecies.

In North America, Hartlaub (1852) first separated White-fronted Geese from those in Europe, using *gambelli* to describe 3 specimens he examined in the United States. Subsequently, the name *Anser albifrons gambelli* or *gambeli* (note the difference in spellings) was applied to all North American White-fronted Geese well into the 1950s (*gambeli* is the correct spelling under the International Code of Zoological Nomenclature). Todd (1950) then reviewed the taxonomic status of the White-fronted Goose, based on an examination of 53 specimens in the Carnegie Museum; he was the first to argue that the smaller White-fronted Goose in North America differed from the European subspecies *albifrons* and therefore should be given the subspecies designation *frontalis*, which was accepted then and continues in use today (American Ornithologists' Union 1998).

In contrast, the taxonomy of the Tule Goose has been debated for nearly a century; this particular goose has been variously referred to in the literature as *Anser albifrons gambeli*, *A. a. gambelli*, and *A. a. elgasi*. The first scientific description of the Tule Goose was from specimens harvested by hunters in the Central Valley of California (Swarth and Bryant 1917). They used the name *gambeli*, but the specimens were considered synonymous with the large White-fronted Geese Hartlaub (1852) had described much earlier in Texas (under the epithet *gambelli*). Currently, the American Ornithologists' Union (1998) uses *elgasi*, which was first coined by Delacour and Ripley (1975) to formally describe the Tule Goose. Gibson and Kessel (1997), Banks (1983), and Dunn (2005) provide reviews of the taxonomic history of the White-fronted Goose, and the scientific nomenclature for this bird will probably see further refinement as more information becomes available.

In North America, Pacific White-fronted Geese breed across the tundra and taiga of Alaska and western and central Canada. Two major populations are recognized: the Pacific Population and the Midcontinent Population (Pacific Flyway Council 2003).

Pacific Population

Breeding. The Pacific Population (PP) of White-fronted Geese breeds only in Alaska, mainly from the Alaska Peninsula (Bristol Bay lowlands) north to the Yukon River; those breeding north of the Alaska Range and eastward through northern Canada belong to the Midcontinent Population. Despite this widespread distribution, >95% of the population breed on the Y-K delta (Timm and Dau 1979, Conant and Groves 2002, Eldridge and Dau 2002), with about 75% breeding along the coast

and 22% dispersing inland (Pacific Flyway Council 2003). From 2000 to 2011 the population index for White-fronted Geese on the Y-K delta averaged 133,519, ranging from a low of 90,407 in 2002 to an all-time high of 174,556 in 2010. The all-time low (since 1985) was 13,400 in 1986 (Collins et al. 2011). The indicated number of pairs averaged 62,376 from 2000 to 2011, with an all-time high of 84,551 in 2011. Except for 2002, the total population has been >100,000 since 2001, and the number of breeding pairs, >70,000 since 2006 (except for 66,759 in 2009).

Outside North America, Pacific White-fronted Geese breed in northeastern Russia (100,000–150,000), from the Kolyma River to the Anadyr River lowlands. These birds largely winter in Japan (50,000 in 1995/96 and 1997/98), but they are also found in South Korea (35,000), North Korea (3,600), and China (50,000), where Poyang Lake is an important site (Fox and Owen 2005).

Winter. California provides the principal wintering areas for the PP: specifically, the Sacramento Valley and, to a lesser extent, the Sacramento/San Joaquin delta and the northern San Joaquin delta. The Midwinter Survey averaged 287,564 White-fronted Geese from 2000 to 2010, with only a very small number occurring outside of California. During the Midwinter Survey, 283,916 were recorded, almost all (99.9%) in California: 90.2% in the Sacramento Valley, 7.4% in the San Joaquin Valley; and 2.5% (7,000) to the north, primarily in the Klamath basin on the Oregon/California border. Only about 100 winter in Washington and Oregon.

Midwinter Surveys conducted at 3-year intervals in the interior highlands in Mexico averaged 11,311 White-fronted Geese from 1985 to 2006 (Thorpe et al. 2006). Band-recovery data indicate that those populations at the southern end of their breeding range are most likely to winter in Mexico, and that winter distribution in Mexico is largely segregated, based on breeding areas (Pacific Flyway Council 2003). Of the population from the Bristol Bay lowlands, 90% winter in the interior highlands in the state of Chihuahua, primarily at Laguna Babícora, versus only 20% of the Y-K delta population (Ely and Takekawa 1996). Laguna Babícora is among the most significant natural wetlands in Mexico (Sparrowe et al. 1989). Of the Y-K delta population, band recoveries in Mexico varied from none among birds banded in the northern portion of the delta, to 2.5% in the middle, and 23% in the south. A small percentage (9%–10%) of the Y-K delta birds winter in the western coastal states of Mexico, but 20%–25% were recovered in the interior highland states of Chihuahua and Durango (Pacific Flyway Council 2003). Geese from the PP may segregate from the Midcontinent Population during winter in the interior highlands (Ely and Dzubin 1994).

Migration. Ely and Takekawa (1996) provided substantive insight into the migration of the PP, based on their 1988–92 monitoring of 262 radiotagged geese: 183 (19 males, 164 females) marked during the breeding season on the Bristol Bay lowlands and 61 (13 males, 48 females) from the Y-K delta; as well as 18 marked in the Klamath basin in early fall. Fall migration begins in early to mid-August, and the Klamath basin is the major autumn staging area. Most of the Y-K delta birds fly nonstop to the mouth of the Columbia River before moving through central Oregon and arriving at the Klamath basin. Several thousand, though, stop at the Copper River delta during fall migration (Pacific Flyway Council 2003), which is the last major use area in Alaska before their arrival at the Klamath basin. Moreover, the 2 groups differed dramatically in their temporal use of the area. In autumn 90% of the Bristol Bay birds used the Klamath basin, arriving in early September (nearly 30 days before birds from the Y-K delta) and departing before most of the Y-K delta geese had arrived. In contrast, only 70% of the Y-K delta birds stopped at the Klamath basin, while 30% bypassed it for a direct flight to the Central Valley of California. Most (85%–93%) of the Bristol Bay geese migrated directly from the Klamath basin to wintering areas in Mexico, where they arrived by late September.

In spring, Bristol Bay White-fronted Geese left Mexico and arrived in the Central Valley by the last week of January, with 55% of this group then stopping at the Klamath basin. Others staged in eastern Oregon and western Idaho. In contrast, most (>80%) of the Y-K delta group initially staged at the Klamath basin before moving on to other staging areas along the northeastern coast of the Gulf of Alaska, although up to 20% remained in the Central Valley until March, where they mixed with Bristol Bay birds arriving from Mexico. The Bristol Bay geese probably flew across the Gulf of Alaska to arrive on breeding areas by early to mid-April, while Y-K delta birds did not arrive until mid-May. Ely and Raveling (1989) noted that most White-fronted Geese had left the Central Valley by mid-April. During a nesting study from 1977 to 1979, White-fronted Geese were the first geese to reach the Y-K delta in spring, arriving in singles, pairs, and small groups, and they were commonly observed by 4–8 May (Ely and Raveling 1984).

The use of the Klamath basin as a fall staging area has declined greatly in recent years, however, with many birds bypassing the area and flying directly to the Sacramento Valley (Pacific Flyway Council 2003). Since the early 1990s, peak counts at the Klamath basin during the fall were about 100,000 birds, down from 500,000 in the late 1960s, although spring population peaks have exceeded 200,000 (Pacific Flyway Council 2003).

Midcontinent Population
Breeding. The breeding grounds for the Midcontinent Population (MCP) extend from Alaska, eastward through the Canadian Arctic, to Hudson Bay. In Alaska, these geese breed from the Arctic Coastal Plain south to the Y-K delta, as well as in the taiga of the interior, where important breeding areas are Yukon Flats; Kotzebue basin; and the Innoko, Kanuti, Koyukuk, Selawik, Tanana, and Yukon Rivers. In the Canadian Arctic, MCP geese breed on tundra and taiga habitats in the Yukon (Old Crow Flats), the western Northwest Territories (Mackenzie and Anderson Rivers, Liverpool

Bay), and then east along the coast to Coronation Gulf, the Kent Peninsula, and Victoria Island in western Nunavut. They also use the lowlands of Queen Maud Gulf and the Rasmussen Lowlands, including the Inglis River drainage, in central and eastern Nunavut; and nearly all of Keewatin, to the Geillini River on Hudson Bay (King and Derksen 1986, Ely and Dzubin 1994, Ely and Schmutz 1999, Ely et al. 2012). Ely et al. recognized 6 distinct subpopulations of the MCP: (1) interior Alaska (Innoko, Kanuti, Koyukuk, Selawik, Tanana, and Yukon River drainages, including Old Crow Flats in the Yukon); (2) the North Slope of Alaska; (3) the western Northwest Territories (Mackenzie and Anderson Rivers and the Liverpool Bay region); (4) western Nunavut (Coppermine, Victoria Island, and Kent Peninsula); (5) central Nunavut (Queen Maud Gulf); and (6) eastern Nunavut (Rasmussen Basin, including the Inglis River drainage).

MCP geese on the Arctic Coastal Plain of Alaska averaged 123,963 birds from 1986 to 2006, but, in a redesigned survey, averaged 220,202 from 2007 to 2010 (Larned et al. 2011). In interior Alaska and on the Seward Peninsula, aerial surveys from 2001 to 2009 averaged 30,343 MCP geese: in Kotzebue (10,244), Yukon Flats (8,733), Koyukuk (4,089), the Seward Peninsula (3,222), Innoko (2,611), and Tanana-Kuskokwim (1,444). In the Selawik region of northwestern Alaska, surveys in 1996, 1997, and 2005–8 averaged 9,410 geese, of which 2,541 were estimated to be breeding (Fischer 2009).

In the Canadian Arctic, MCP geese had been surveyed in early spring, but these surveys were problematic, as they were conducted at a time when the geese were too widely spread along migration routes to allow for accurate counts. Hence the spring surveys were discontinued in 1992 in favor of fall surveys at major staging areas in Alberta and Saskatchewan (Canadian Wildlife Service Waterfowl Committee 2011). In 2011, 685,700 birds were tallied during the survey, which was 4% higher than the 2009–11 average of 659,600. Aerial surveys conducted on the Inuvialuit Settlement Region of the western Canadian Arctic from 1989 to 1993

estimated an average adult population of 55,600 MCP geese in that area (Hines et al. 2006). In other surveys on the Northwest Territories mainland part of the Settlement Region, 19,864 birds were tallied at South Liverpool Bay, 15,784 on the Tuktoyaktuk Peninsula, and 9,400 on the Mackenzie River delta. There were 8,466 breeding pairs in the Inuvialuit Settlement Region, which—in years of average reproductive success—would have formed 11% of the MCP. The authors recommended that surveys of the Settlement Region be conducted at 10-year intervals. From 2006 to 2010, various surveys in the Canadian Arctic tallied 177,136 MCP geese in the Queen Maud Gulf and Rasmussen Lowland areas of the mainland (for the majority) and on nearby Banks, King William, and Victoria Islands, as well as 32,957 on the Tuktoyaktuk Peninsula (Conant et al. 2007; Groves et al. 2009a, 2009b; Groves and Mallek 2011). Farther east in the Canadian Arctic, population densities of MCP geese were >5 geese/km² between the Perry and Ellice Rivers in the Queen Maud Gulf region (Alisaukas and Lindberg 2002).

Winter. The MCP winters in the Central and Mississippi Flyways, from eastern Arkansas in the Mississippi Alluvial Valley to the coastal and inland areas of Louisiana and Texas, with smaller numbers at a few locations elsewhere. MCP geese are also common farther south into Mexico: the state of Tamaulipas along the east coast, and the states of Durango and Zacatecas in the interior highlands. The MCP is very segregated from the PP, as evidenced by band returns; <0.5% of the MCP birds banded on breeding or molting areas ($n = 10,298$) have been recovered in the Pacific Flyway (Ely and Dzubin 1994).

In the United States, the Midwinter Survey averaged 339,913 MCP geese from 2000 to 2010, distributed almost equally between the Central Flyway (50.3%) and the Mississippi Flyway (49.7%). In the Central Flyway, MCP geese wintered primarily in Texas (63.1%) and Kansas (34.4%); in the Mississippi Flyway, the birds mainly wintered in

Louisiana (50.8%), but they were also dispersed within other states in that flyway, such as Arkansas (23.9%), Illinois (7.8%), and Missouri (6.6%). The geese in Texas and Louisiana are largely found in their traditional winter habitats: the coastal marshes. In contrast, the expansion of the MCP from core coastal marshes to more northern sites was accomplished because of adaptability in their food choices, which broadened to include waste grain, especially rice, in harvested fields.

In Mexico, the number of MCP geese (as assessed by Midwinter Surveys conducted by the U.S. Fish and Wildlife Service) averaged 44,159 for the entire country from 1961 to 2000 (Pérez-Arteaga and Gaston 2003). Surveys were conducted at irregular intervals until 1985, after which they generally were repeated every 3 years. Most MCP geese in Mexico are tallied along the east coast, where 27,527 were recorded in 2005, compared with only 6,018 in 2006. In 2000, 94% of the MCP geese on Mexico's east coast were recorded from the Tamesi and Panuco River deltas, but many birds are probably missed because they are feeding in grain fields (primarily on milo, or sorghum) away from water areas covered by the surveys. As many as 180,000 MCP geese have been estimated to winter on the east coast of Mexico, primarily in Tamaulipas, where the expansion of grain sorghum agriculture, which covers about 30% of that state, has attracted large numbers of both wintering White-fronted Geese and Snow Geese (Yépez Rincón 2003). Almost no MCP geese winter south of the city of Tampico. Populations in the interior highlands of Mexico averaged 8,932 from 1991 to 2006 (in surveys conducted every 3 years), with important wintering areas in the states of Chihuahua (Laguna de Bustillos) and Durango (Laguna de Santiaguillo and the wetland complex of Durango).

Migration. Nearly all of the MCP migrates through the interior of North America. Ely et al. (2012) provide the most extensive information on the migration chronology and subsequent winter distribution of MCP geese, based on the authors' analysis of band recoveries from the 1940s through 2008, as well as resightings of nearly 19,000 neck-collared birds captured on 6 major breeding areas from 1990 to 1994 across the range of the MCP. These geese were generally spatially and temporally segregated during migration and in the winter, with the eastern breeding populations occurring farther east and migrating later than birds breeding farther west.

In the autumn (Sep–Nov), birds from all populations move to staging areas in Alberta and Saskatchewan before continuing south through the Northern Plains and midwestern states to wintering areas in Louisiana, Texas, and Mexico; the breeding populations vary markedly, however, in their use of the 2 major staging areas. There were more band recoveries of birds from interior Alaska in Alberta and Saskatchewan (60.1%) than of birds from the North Slope (38.7%), the western Northwest Territories (39.9%), western Nunavut (26.2%), central Nunavut (16.9%), and eastern Nunavut (2.0%). Similarly, observations of neck-collared birds the first autumn after marking revealed 24% of interior Alaska geese using Alberta, compared with only 10.9% of the birds from the western Northwest Territories, 9.8% of the North Slope birds, and <6% of the birds from areas in Nunavut. The timing of their arrival at staging areas also differs among populations, as resighting data of neck-collared birds revealed the mean date of arrival as late September for interior Alaska birds, compared with early October for the North Slope, the western Northwest Territories, and the Nunavut stocks.

The geese then continue on to wintering areas in Texas, Louisiana, and Arkansas, with interior Alaska birds arriving a week earlier than the other MCP geese, at a mean date of about mid-November. Mexico is a more important wintering area for MCP White-fronted Geese from interior Alaska, with 13.5% of all band recoveries reported there, versus 2.9% for all of the other stocks. An analysis of band recoveries since 1949 found that the state of Tamaulipas was the major wintering site for all MCP geese except interior Alaska birds, which wintered in the states of Durango and Zacatecas in the interior highlands (72% of all recoveries).

In the spring (Feb–Apr), Nebraska (Rainwater Basin) is the most important staging area, especially for the interior Alaska geese, which formed 85% of the observations, compared with 60%–67% for the other populations. The mean arrival date is about mid-March. Farther north, more (91.0%) interior Alaska birds use staging areas in Alberta than those from the other stocks (56.8%). In contrast, birds from eastern Nunavut are more likely to stage in Saskatchewan during their spring migration. Anderson and Haukos (2003) used observations of neck-collared White-fronted Geese to determine that the Winchester Lakes region in northwestern Texas was a fall staging area for the western portion of MCP birds prior to their traveling south to winter in the rice prairies of interior Mexico and coastal Texas.

Tule Greater White-fronted Goose Population
Breeding. The breeding grounds of the Tule Goose, near Redoubt Bay and Susitna Flats on the western side of Cook Inlet in Alaska, was first discovered in 1979 (Timm et al. 1982). By the early 1990s, however, the population declined dramatically (90%), following a major eruption of the Redoubt Volcano in 1989, which substantially altered habitats used by Tule Geese in the vicinity of Redoubt Bay. Since the eruption, breeding has shifted away from Redoubt Bay to the upper Cook Inlet basin, largely between the Susitna and Yentna Rivers (Ely et al. 2006). Earlier reports of Tule Geese from the Old Crow Flats region of the Yukon (Elgas 1970) have not been corroborated; these were probably Pacific White-fronted Geese from interior Alaska (Ely et al. 2005).

Winter. The majority of Tule Geese winter at the Sacramento, the Delevan, and the Colusa National Wildlife Refuges (NWRs) within the Sacramento Valley of California. Small wintering numbers are also reported at Suisan Marsh, the Napa Marshes, Butte Sink, the Gray Lodge Wildlife Area, the Sutter NWR, and nearby duck clubs and rice fields (Deuel and Takekawa 2008). Tule Geese share the same wintering grounds with the PP of Pacific White-fronted Geese, but they largely roost and feed within wetlands, while the PP tends to forage in agricultural fields (Bauer 1979, Ely 1992).

Migration. Radiomarked Tule Geese depart Cook Inlet from 18 to 25 August (Ely et al. 2006), after which they move south along the coast to Puget Sound in Washington, and then inland to Summer Lake and the Malheur NWR in southwestern Oregon, with a few continuing south to the Klamath basin in northern California, as well as adjacent private lands (Timm et al. 1982, Wege 1984). In the spring they return to the Klamath basin in February and March, although increasing numbers have been using areas in southeastern Oregon (Deuel and Takekawa 2008). Tule Geese arrive at coastal and interior marshes on the outer portion of the Cook Inlet basin from mid-April to early May, prior to moving inland to nest within the upper reaches of the basin (Ely et al. 2006).

MIGRATION BEHAVIOR
White-fronted Geese migrate both by day and by night, with estimated flight speeds of 70–90 kph (Ely and Dzubin 1994). They undertake extensive migrations from their arctic and subarctic breeding grounds to wintering grounds extending from central California eastward to the Gulf Coast of Texas and Louisiana; some travel as far south as Mexico. Many of those from the Y-K delta in the Pacific Flyway travel nonstop about 3,200 km across the Gulf of Alaska to the Columbia River, before then heading farther south to the Central Valley in California (Dzubin et al. 1964). Others in the Pacific Flyway travel along the coast to reach the Fraser River delta in British Columbia (Butler and Campbell 1987). On staging areas, they often are found mixed with smaller subspecies of Canada Geese and Cackling Geese.

Molt Migration. In Alaska, White-fronted Geese from on the Arctic Coastal Plain (primarily non-breeders and failed breeders) undergo a short migration to the 2,000 km² Teshekpuk Lake region, where they molt alongside other species of geese (Derksen et al. 1979, 1982). At the Teshekpuk Lake area, molting geese use tundra lakes located farther inland than those favored by molting Canada Geese and Brant; most (57%) occupy shoreline habitats by deep open-water lakes (Derksen et al. 1979). Alexander et al. (1991) reported that preferred molting areas were flat islands in deltas, meandering stream channels, meadows, and marshes dominated by sedges and grasses. The numbers of molting White-fronted Geese have risen in this area since 1990, concurrently with their overall population increase. In 2007 a molting survey of the Teshekpuk Lake region tallied 45,747 adult and 2,563 juvenile White-fronted Geese, which was 49% of all geese observed in the area (Mallek 2007).

Up to 50% of radiomarked Tule Geese breeding in the Cook Inlet basin underwent a molt migration to wetlands across the Alaska Range, around 400–600 km west of their breeding area (Ely et al. 2006). Important molting areas were the Innoko River region and near Pilot Station on the Yukon River. These geese returned to Cook Inlet in the fall, prior to their southward migration.

HABITAT

On the Y-K delta, the prenesting PP uses meltwater areas, slough banks, and river edges within 30 km of the coast (Ely and Dzubin 1994). Nesting occurs in 3 tundra habitats that are inland from those used by Brant: lowland, intermediate, and upland. Lowland tundra averages 0.5 m above mean high tide and is characterized by tidal sloughs and small shallow lakes, with lowland meadows dominated by various tundra sedges and grasses (Ely and Raveling 1984). Upland tundra averages 1.7 m above mean high tide, has steep-sided tidal sloughs and lakes, and is dominated by sphagnum moss (*Sphagnum* spp.) and lichens. Intermediate tundra is more dispersed and includes pingos, raised slough banks, and lake edges. The MCP chooses a broader mix of landscapes and vegetation types as its breeding habitat, ranging from alluvial lowlands to taiga forests and bogs. These geese will even move into very dry sites, such as hill slopes, rock fields, and heath tundra (Ely and Dzubin 1994).

During the winter, White-fronted Geese use open-water habitats for daytime and nighttime roosts. The PP consumes a variety of agricultural crops during migration, but wintering birds in the Central Valley of California rely heavily on rice and corn. MCP geese once fed almost exclusively in the coastal marshes in Texas and Louisiana, but since the 1960s they have shifted farther north to forage on agricultural foods. In Mexico, White-fronted Geese inhabit large lakes and natural wetlands but largely feed in nearby corn, sorghum, and rice fields.

POPULATION STATUS

In 2010 the total population of White-fronted Geese in North America was 1,359,585 birds: 709,800 in the MCP and 649,785 in the PP, based on a fall population index for the PP and a fall survey of the MCP on staging areas in Alberta and Saskatchewan (U.S. Fish and Wildlife Service 2011). From 2000 to 2009, the MCP averaged 708,900 geese, and the PP, 479,700. The PP has increased 6% annually since 2002; the MCP increased 1% annually from 2001 to 2010. The population of Tule Geese is the smallest of all goose subspecies and is probably not more than 10,000 birds (Deuel and Takekawa 2008); it has been considered "at risk" by the International Waterfowl Research Bureau (Callaghan and Green 1993).

Between 1965 and 1978, the size of the PP was indexed via peak fall counts in the Klamath basin, after which a coordinated survey was conducted throughout the flyway, in part because fewer birds staged at the Klamath basin in the fall. These earlier surveys documented a dramatic decrease in population size: from a high of 492,900 in 1966 to a low of 80,100 in 1983, a decline of 84%. The Alaska-Yukon Breeding Waterfowl Survey also confirmed a significant decrease in the PP (Pacific Flyway Council 2003). This decline was attributed to the overharvest of geese by native peoples on the Y-K delta, whose own numbers increased by 42% between 1960 and 1980 (Raveling 1984). The Midwinter Survey alone showed that in the Pacific Flyway, the White-fronted Goose population dropped nearly 70% from 1955 to 1973. These declines led to harvest management initiatives on the Y-K delta (Mitchell 1986, Pamplin 1986), which appear to have had a positive effect on these geese. More widespread surveys showed their population rising from an average of 122,600 in the 1980s, to 307,520 in the 1990s, and to 466,980 from 2000 to 2009; the highest count was 627,000 in 2008 (U.S. Fish and Wildlife Service 2009). On the Y-K delta, nest plot surveys showed a concomitant steady increase in nest density as populations recovered, from 1.4 nests/km² in 1985 to 12.0/km² by 1995 (Pacific Flyway Council 2003). The estimated number of White-fronted Goose nests on the Y-K delta averaged 135,008 from 2000 to 2009 (Fischer et al. 2009).

The MCP did not suffer the declines experienced by the PP. Fall surveys averaged 796,986 birds from 1992/93 to 1998/99 and 734,280 from 1999/2000 to 2008/9; peak counts were more than 1 million geese in 1995/96, 1998/99, and 2000/2001 (U.S. Fish and Wildlife Service 2009).

Harvest

The total harvest of White-fronted Geese in the United States has paralleled their population growth: from an estimated 134,000 harvested annually in the 1960s to 225,000 in the 1990s, an increase of nearly 70%. From 1999 to 2008, their total harvest in the United States averaged 262,740, with the most birds taken in the Mississippi Flyway (44.8%), followed by the Central Flyway (37.2%) and the Pacific Flyway (18.0%). White-fronted Geese are rarely harvested in the Atlantic Flyway, averaging only 9/year. The 2009/10 U.S. harvest was 268,759. The PP is largely harvested in California (92.4%), with 110,523 birds taken in 2008. The MCP is harvested in the Central and Mississippi Flyways, where the harvest in 2008 was 61,247 and 138,097 birds, respectively. In Canada, the harvest averaged 73,834 from 1999 to 2008, with a high of 196,738 in 2000. Nearly all were harvested in Saskatchewan (71.1%) and Alberta (27.7%).

The age ratio (immatures:adults) in the harvest averaged 1.29 in the PP from 1962 to 2001 (determined from the Waterfowl Parts Collection Survey) and ranged from a low of 0.8–0.9 in 1983, 1984, and 2000 to a high of 2.68 in 1977 (Pacific Flyway Council 2003). In the PP, Tim and Dau (1979) reported that immatures were 2.1–2.5 times more vulnerable to hunters than were adults, while Miller et al. (1968) noted a figure of 2.3 for the MCP.

In Alaska, the subsistence harvest of White-fronted Geese on the Y-K delta averaged 14,431 birds from 2001 to 2005 (Wentworth 2007a); at Bristol Bay it averaged 2,113 from 1999 to 2005 (Wentworth 2007b). In Canada, the subsistence harvest of White-fronted Geese in the Inuvialuit Settlement Region of the western Canadian Arctic averaged 1,761 from 1988 to 1997 (Inuvialuit Harvest Study 2003). The harvest in Mexico is not known, but it is probably very small in comparison to North America, as is generally the case for waterfowl harvests in Mexico (Kramer et al. 1995).

BREEDING BIOLOGY

Behavior

Mating System. White-fronted Geese are monogamous and generally pair for life, but they will repair if a mate is lost (Ely and Dzubin 1994). There is little information on pair-bond formation, but it probably occurs during the spring and summer (Ely and Scribner 1994). There are few data on the breeding age for White-fronted Geese in Alaska, although Barry (1967) reported that they first bred as 3-year-olds on the Anderson River delta, with some breeding as 2-year-olds during years with favorable weather conditions.

In a study of 253 Tule Geese banded with neckcollars during the molt at Cook Inlet in Alaska, birds were observed forming pair bonds as 1-year-olds, which supported an assumption that they began breeding as 3-year-olds (Campbell and Goodwin 1985). Of 66 Greenland White-fronted Geese banded as juveniles on their wintering grounds in Ireland and subsequently observed when paired, the mean age of pairing was 2.5 years, and there was no difference between males and females in their age at first pairing (Warren et al. 1992). Overall, 28 out of the 66 (42%) birds banded as juveniles bred successfully, with the mean age for successful breeding being 3.2 years. No birds bred as 1-year-olds, but 28.6% bred as 2-year-olds, 35.7% as 3-year-olds, 28.6% as 4-year-olds, and 7.1% as 5-year-olds.

Sex Ratio. The sexual composition of White-fronted Geese captured in Alaska and on their wintering grounds was 49.8% female among adults ($n = 923$) and 48.9% female among juveniles ($n = 687$: Ely 1993). For birds captured on fall staging areas in Saskatchewan, the percentage was 48.0% female among adults ($n = 5,331$) and 48.7% female among juveniles ($n = 2,869$; Ely and Dzubin 1994).

Site Fidelity and Territory. White-fronted Geese have high rates of site fidelity on their breeding grounds. At the Queen Maud Gulf Migratory Bird Sanctuary in the central Canadian Arctic, the site fidelity of White-fronted Geese that were marked from 1991 to 1997 ranged from 61% to 86% (Alisauskas and Lindberg 2002). In westcentral Alaska, the return rates of females radiocollared from 1994 to 2004 were 48%–50%. Breeding fidelity rates were also high on the Y-K delta, with most females generally nesting within 500 m of the previous year's nest (Ely and Dzubin 1994).

Unlike other arctic-nesting geese, White-fronted Geese are solitary breeders. Barry (1967) reported that most of the territorial defense against predators and human intruders was provided by nonbreeding yearlings. Breeding males, however, are known to defend a space around their mate, as well as protect their offspring. At brood-rearing areas on the Y-K delta, the average home-range size was 106 ha (Ely and Dzubin 1994).

In a 5-year study (1994–97, 2004), radiomarked Tule Geese on Cook Inlet in Alaska averaged 116 days on their breeding areas, during which time they used home ranges averaging >273,000 ha (Ely et al. 2005). The home-range sizes reported from this study are the largest ever recorded for any species of waterfowl, perhaps due to these birds' substantial size and their use of mostly unproductive bog and fen habitats that are widely dispersed within the boreal forest.

Courtship Displays. White-fronted Geese exhibit an array of *threat* displays and postures during aggressive encounters between pairs and pairs with young. Common *threat* displays are the *forward* and *diagonal-neck*; like other species of gray geese (*Anser*), they vibrate their furrowed neck plumage as a lateral threat, given most frequently when grazing (Johnsgard 1965). *Forward-head-stretching* is often accompanied by waving the head (Boyd 1953). Both sexes, but especially the males, move the head laterally toward an intruder, which can be followed by short rushes and pecking. The lateral head movements are also accompanied by a high-pitched *squee, squee, squee* (Ely and Dzubin 1994). The *triumph-ceremony* involves low head and neck movements and a series of loud calls. Small flocks (50–300, mostly adults) exhibit collective "mobbing" behavior (Ely and Dzubin 1994). Aerial *3-bird-chases* occur on nesting areas. Copulation always occurs on water and is preceded by *head-dipping* by the male, followed by a *triumph-like ceremony* (Fox and Owen 2005). Ely (1989) recorded 1 extra-pair copulation (and 9 pair copulations) during 356 hours of observations of breeding pairs on the Y-K delta.

An extensive study (1949–52) of aggressive interactions among wintering White-fronted Geese was conducted on the River Severn in the United Kingdom, during which >2,000 encounters were observed and the outcomes recorded (Boyd 1953). Initiators "won" 93% of 1,655 aggressive acts, of which about 95% did not elicit a response. Physical contact was rare, recorded in only 4.4% of 2,129 encounters. Of 2,171 conflicts where the status of all participants could be determined, parents were involved in 1,886 (as attackers or victims), nonfamily adults in 959, juveniles within families in 1,331, and juveniles without families in 166. Families were by far the most successful in winning aggressive encounters, with success ratios (number of wins/ number of defeats) increasing with family size: from 1.31 in a family of 4 to 2.90 in a family of 8. In contrast, success ratios were very low for other categories: 0.15 for single juveniles, 0.21 for single adults, and 0.65 for paired adults. These valuable data reveal not only the outcomes resulting from aggressive encounters and displays, but also the benefits of belonging to a family group during the winter, particularly for juveniles, which are subordinate to all other groups.

Nesting

Nest Sites. The female selects the nest site and constructs the nest as the eggs are laid (Barry 1967). Nest sites occur in a wide array of breeding habitats, from coastal tundra and nearby uplands to inland taiga forests. Nests are also dispersed rather than being clustered in colonies, which is the norm for other species of arctic-nesting geese, such as Snow Geese and Brant.

On the Y-K delta, Ely and Raveling (1984) noted that White-fronted Geese nested in 3 main types of tundra habitat: lowland (68%), intermediate (23%), and upland (10%). Specific nest sites were slough banks (55%), lakeshores (23%), and grass-sedge meadows (11%). Higher elevations generally were preferred for nesting sites, as nests located in lowland areas are susceptible to destruction from flood tides. Flood tides capable of large-scale nest destruction on the Y-K delta are possible, but they may only occur once within a 14-year period (Ely and Raveling 1984). Vegetative nesting cover included sedges (*Carex* spp.), ryegrass (*Elymus* spp.), bluegrasses (*Poa* spp.), bluejoint grass (*Calamagrostis canadensis*), willows (*Salix* spp.), birches (*Betula* spp.), raspberries (*Rubus* spp.), and lichens (Ely and Raveling 1984). On the Anderson River

delta and the Liverpool Bay region in the Northwest Territories, breeding White-fronted Geese spread out over the coastal plains to nest beside rivers, tributary streams, and countless lakes, nesting farther inland than Black Brant or Snow Geese (Barry 1967).

Tule Geese monitored on Cook Inlet in Alaska nested in boreal forest wetlands; they used seasonally flooded or saturated wetlands in these forests in proportion to their availability. Most (93%) nests were located >75 m from open-water ponds or lakes, and in wetlands with little or no open water (Densmore et al. 2006).

Clutch Size and Eggs. Geographically, average clutch sizes are 4.1 eggs ($n = 180$ nests) on the North Slope, 4.2 ($n = 34$) in the western Arctic, 3.8 ($n = 185$) in the central Canadian Arctic, and 4.6 ($n = 721$) on the Y-K delta (Ely and Dzubin 1994). Clutches with >8 eggs probably are the product of more than 1 female. Long-term (1985–2009) estimates of the clutch size of White-fronted Geese on the Y-K delta ranged from a low of 3.86 in 2001 to 5.01 in 2009 and averaged 4.37. Clutch size in 2009 ($n = 868$) averaged 4.30 (Fischer et al. 2009).

In a detailed 3-year nesting study (1977–79) on the Y-K delta, White-fronted Geese were the first geese to nest there each year. Average clutch size ranged from 3.7 eggs ($n = 26$) during a year with a late spring thaw to 5.2 ($n = 49$) and 5.7 ($n = 24$) in years with earlier snowmelts (Ely and Raveling 1984). Clutch size was also greater in nests initiated earlier in the season, regardless of the timing of snowmelt, with clutch size reduced by 0.20 eggs for each day a female delayed nesting after the first nest had been started. The best predictor of clutch size, however, was the time interval between when geese were first commonly observed (4–8 May) and the laying date ($r^2 = 0.57$).

Greater White-fronted Goose eggs are almost white (creamy or very pale buff) when clean, but they vary in shape between subelliptical and long elliptical (Palmer 1976a). As with other goose species, the eggs usually acquire a brown stain from the nest material during the incubation period. On the Y-K delta, eggs averaged 53.5 × 80.1 mm ($n = 369$), and their weight averaged 128 g ($n = 11$; Ely and Raveling 1984). In the Northwest Territories, egg size ($n = 546$) averaged 53.0 × 80.6 mm (Ely and Dzubin 1994). Eggs are laid at an average rate of 1 every 30–32 hours.

Incubation and Energetic Costs. As in all species of geese, only the female incubates the nest. On the Y-K delta, Ely and Raveling (1984) reported an incubation period of 26 days, while Barry (1967) noted 23 days on the Anderson River delta. The eggs in a clutch all hatch in about 28 hours ($n = 33$; Ely and Dzubin 1994). Females increase the time they spend at the nest as the egg-laying period progresses, with nest attentiveness rates ranging from 97% on the Y-K delta to 99% on the North Slope and central Arctic. Females take 0–3 recesses/day that average 10–40 minutes each (Ely and Dzubin 1994).

In northern Taimyr [Taymyr] in Russia, detailed behavioral observations and continual data on body weight (obtained via a balance inserted under a single nest) provided information on the costs of incubation for a White-fronted Goose (Spaans et al. 1999). During incubation, the female averaged 2.1 recesses/day for 10.9 minutes each and lost 25.4 g of body mass/day. The female did not appear to feed during these recesses, so stored reserves were needed to complete the incubation process.

On the Y-K delta, female White-fronted Geese, collected upon arrival and again before incubation ($n = 11$–12) in 1986 and 1987, increased in body mass by 19.6%–21.0%, lipids by 29.8%–88.8%, and protein by 10.8%–12.3% (Budeau et al. 1991). Females probably acquired more energy reserves, because ruptured follicles in the collected birds indicated that they had already laid 1–3 eggs. The total calculated cost of incubation was 4,888 kcal. Hence these data demonstrate the importance of the prenesting period as a time when White-fronted Geese acquire the nutritional reserves es-

sential for reproductive success. In contrast to the females, the body mass of males did not change, but their total lipids decreased by 29.4%.

Prenesting female White-fronted Geese on the Y-K delta spent more time (60.3%) feeding than did males (30.5%), averaging 10.8 hours during an 18-hour observation period versus only 5.5 hours for males (Budeau et al. 1991). Males, however, spent dramatically more time being alert than did females (34.1% vs. 5.0%), generally standing watchfully next to the female while she was feeding. On the Kent Peninsula in Nunavut, prenesting females of lone pairs spent 75%–81% of their time feeding, compared with only 42%–47% for males, which spent 46%–50% of their time being alert (Fox et al. 1995).

In Greenland, female Greenland White-fronted Geese fed for up to 10 days between the period of peak arrival and the onset of nesting. They spent 68% of their diurnal activity feeding on highly nutritious parts of plants, roosting only during periods of subzero temperatures (Fox and Madsen 1981).

Nesting Chronology. Pacific White-fronted Geese breeding in Alaska arrive on the breeding grounds in May, and they are the first species of goose each year to nest on the Y-K delta (Ely and Raveling 1984). Ely and Dzubin (1994) summarized the peak nest-initiation periods in Alaska as 9–25 May on the Y-K delta, 8–10 June at the Colville River delta on the North Slope, 12–28 May on Cook Inlet, and 10–14 June on the Kent Peninsula in Nunavut. In general, birds in the Bristol Bay lowlands nest up to 2 weeks before those on the Y-K delta (Ely and Takekawa 1996). Tule Geese nesting in the Susitna River valley on Cook Inlet ($n = 31$) initiated nests from 8 to 16 May (Ely et al. 2005).

White-fronted Geese are synchronous nesters, as documented during a 3-year study (1977–79) on the Y-K delta. The annual spring thaw was among the latest recorded in 1977, and among the earliest in 1978 and 1979, although White-fronted Geese were commonly observed on the study area between 4 and 8 May each year (Ely and Raveling 1984). Nest initiation was 11 days earlier in 1978 and 14 days earlier in 1979, however, than it was in 1977. Regardless, most nests were initiated within a 10-day period: 84% in 1977, 77% in 1978, and 78% in 1979. Peak initiation in 1977 occurred 24 days after the birds became common on the study area, versus 13 days in 1978 and 12 in 1979. Nest densities varied between 2.7 and 4.6/km².

These data indicated that the arrival of White-fronted Geese to the Y-K delta was independent of weather. However, the time period between when birds were commonly observed and the midpoint of nest initiation was 13–14 days in years of early snowmelt, which is about the time needed for follicle development of a clutch of eggs. Hence females apparently use the interval from arrival to nest initiation to accumulate energy for egg production. During the late-snowmelt year (1977), females probably reabsorbed egg follicles, as Barry (1962) had observed for Brant on Southampton Island.

In the High Arctic of the Kent Peninsula in Nunavut, White-fronted Geese initiated their first nests on 5 June in 1993 and 28 May in 1994 (Carrière et al. 1999). Median laying dates were 11 June in 1993 ($n = 38$) and 2 June in 1994 ($n = 26$).

Nest Success. As with other arctic-nesting geese, nest success of White-fronted Geese can fluctuate widely, due to variations in weather conditions and predation. Nest success ranged from 35.9% to 90.9% over 11 years on the Y-K delta, and from 2% to 95.7% in the central Arctic (Ely and Dzubin 1994). Mickelson (1973) found 77 White-fronted Goose nests on the Onumtuk area of the Y-K delta, of which 65 (84.4%) hatched successfully. Successful nests averaged 4.08 eggs, 0.53 fewer than the incubated clutch. In a 3-year study (1977–79) on the Y-K delta, Ely and Raveling (1984) found that nest success ($n = 90$) averaged 48% but varied from 26% to 82%. The major (28%) cause of nest failure was either direct or indirect flooding. The primary avian egg predators are jaegers (*Stercorarius* spp.),

large gulls (*Larus* spp.), and Common Ravens (*Corvus corax*); the most prevalent mammalian nest predators are arctic foxes (*Alopex lagopus*), red foxes (*Vulpes vulpes*), mink (*Mustela vison*), and other large mammals (Sargeant and Raveling 1992).

Brood Parasitism. Brood parasitism is rare or incidental for White-fronted Geese, although Emperor Geese may infrequently parasitize nests. The size and appearance of the eggs of both species are so similar, however, as to make it difficult to determine the amount of parasitism (Ely and Dzubin 1994).

Renesting. White-fronted Geese do not renest, due to the energetic constraints of reproducing during the brief arctic/subarctic summer.

REARING OF YOUNG

Brood Habitat and Care. On the Y-K delta, home ranges in brood-rearing areas averaged 106 ha and were located 1–2 km from nesting areas (Ely and Dzubin 1994). Brood-rearing areas on the Y-K delta were pond-edge communities dominated by grasses and sedges, such as *Carex mackenziei, Carex subspathacea, Triglochin palustre,* and *Puccinellia phryganodes* (Babcock and Ely 1994). The brood-rearing areas of White-fronted Geese were often located farther from water than those used by Cackling Geese (*Branta hutchinsii minima*) or Emperor Geese and consisted of tall coarse vegetation.

The female broods her goslings at the nest for 1–2 days, after which both parents lead them to brood-rearing areas. The male assumes the dominant role in brood life, becoming the more alert and watchful parent after the family's arrival at the brood-rearing area. On the Anderson River delta, broods remained well hidden in cover during the first week, before later becoming more mobile. Barry (1967) believed that broods suffer their heaviest losses during the first week or so of life, when they are most easily separated from their parents.

The offspring of White-fronted Geese remain

associated with their parents for longer periods than any other species of geese. Observations of neck-collared birds in California and southern Oregon from 1979 to 1989 revealed that 69% of yearlings, 39% of 2-year-olds, and 38% of 3-year-olds remained with their parents throughout the winter (Ely 1993). The percentage of time parents spent with their offspring declined as their young grew older, however: 76% for juveniles, 32% for yearlings, and 15% for those 2 years old and older. Bonds among siblings also persisted, with some degree of contact maintained at 1 year (74%), 2 years (50%), and 3 years (39%) of age. Furthermore, although formal observations ended when the oldest young reached 34 months, incidental observations revealed that some offspring associated with their siblings and/or parents until up to 8 years of age. The fitness benefits of these bonds were not readily apparent, as there was no difference in either the reproductive success between adults that were accompanied or not accompanied by offspring, or in survival among offspring that associated or did not associate with their parents. Individuals in family groups, however, may benefit, because such groups are more dominant at food and roost sites during the winter, and parents may gain an advantage on breeding areas if their prior offspring are involved in current nest defense. Family contacts with adults and older siblings also probably help subadults, who can learn foraging and predator-avoidance strategies from them.

Mutual benefits of associations between adults and yearlings were documented in White-fronted Geese breeding on the Kent Peninsula in Nunavut (Fox et al. 1995). Breeding pairs with 1 or more yearlings present spent significantly more time (64%–97%) feeding and less time remaining alert (in males, 29%–34% vs. 48%). Similarly, yearlings devoted less time feeding when they were alone (59%) than when with their parents (71%–76%), although such associations occurred before the adult females began searching for nest sites. During the egg-laying period, groups of geese (mostly yearlings) engaged in distraction flights over ter-

restrial predators and humans that were approaching nests. Hence the presence of groups of geese at some nest sites suggests that the continued parent/offspring relationship clearly provided an advantage for both adults and yearlings, especially in nest defense.

Development. Goslings averaged 90.3 g at hatching ($n = 129$) and grew rapidly thereafter, fledging in 38–45 days (Ely and Dzubin 1994). Barry (1967) reported that young White-fronted Geese on the Anderson River delta grew as rapidly as young Lesser Snow Geese and were able to fly at the same age: about 45 days. However, Mickelson (1973) noted that it took White-fronted Goose goslings on the Y-K delta 55–65 days to reach flight stage, which was longer than the time required for Black Brant (*Branta bernicla nigricans*), Cackling Geese (*Branta hutchinsii minima*), or Emperor Geese.

Brood Amalgamation (Crèches): White-fronted Geese infrequently mix with broods of other geese, but otherwise there is little information on brood amalgamation for these geese (Pacific Flyway Council 2003b).

RECRUITMENT AND SURVIVAL

From 1969 to 1972 on the Y-K delta, successful breeding pairs of White-fronted Geese ($n = 65$) hatched an average of 4.5 young/nest, with an average Class III fledging size of 4.0 ($n = 8$; Mickelson 1975). Ely and Raveling (1984) reported the average size of Class Ia and Ib ($\leq$15 days old) broods on the Y-K delta as 2.8 ($n = 18$) in 1977, which was a 13% reduction from the average clutch size at hatching. In another Y-K delta study, Timm and Dau (1979) noted that Class I brood size averaged 3.7 from 1969 to 1978, 15% less than the average clutch size of 4.3.

Schmutz and Ely's (1999) examination of the survival of the PP from 1979 to 1982 was significant, because it linked survival with various factors: year, season, sex, and body condition. Their study stemmed from an analysis of 1,224 geese captured and marked with neck collars in the fall, winter, and spring in Oregon and California, and during the summer on the Y-K delta. Adult survival averaged 74.9%, which was higher than the 67.9% reported from 1967 to 1969 (Timm and Dau 1979). Further modeling also suggested that the survival rate from 1985 to 1996 was as much as 10% higher than the 1979–82 estimate and corresponded to an increasing population. Monthly survival of adult females varied seasonally, being higher (98.6%) during a winter period without hunting or migration, compared with monthly survival at other times of the year (96.4%). Survival of adult males varied among years but was generally higher than that of females during the fall harvest season and the migration/breeding season, and lower during the nonharvest season. An index of body condition was positively correlated with the survival of adult females in the fall and spring, but this factor did not affect the survival of adult males or immatures. Monthly survival of immatures was lower (88.6%) during their first hunting season as compared with all other seasons (96.3%). Annual survival of immatures from 1 October (just before the hunting season) was 47.1%.

Miller et al. (1968) reported first-year mortality rates for White-fronted Geese banded near Kindersley, Saskatchewan, as 29% for adult males, 34% for adult females, 45% for immature males, and 44% for immature females. The annual mortality rate for all sex and age groups combined was 34%.

The longevity record from banding data is 23 years and 6 months for an adult bird of unknown sex banded in Nunavut in 1975 and shot in Louisiana. A hatching-year male banded in Saskatchewan and shot in Alberta survived 22 years and 4 months, and a female of unknown age when banded in California was shot in Washington 20 years and 4 months later.

FOOD HABITS AND FEEDING ECOLOGY

White-fronted Geese are herbivores, consuming an array of grasses, sedges, berries, and the underground parts of plants in summer, but relying heavily on agricultural grains while on staging and wintering areas. Their feeding behaviors are grazing; probing for underground roots and tubers; gleaning cultivated grains, such as waste corn, barley, wheat, and rice; and tipping up, dabbling, or pecking on the water's surface. Like all geese, they are social feeders outside of the breeding season; they are commonly found in association with large groups of ducks, other geese, and Sandhill Cranes (*Grus canadensis*), the latter especially during spring stopovers along the Platte River in Nebraska. White-fronted Geese are primarily diurnal foragers. Their preferred roost sites include a diverse array of aquatic habitats, freshwater impoundments, saltwater lagoons, estuarine bays, intertidal flats, river deltas, and floodplains.

Breeding. During the breeding season, the roots and underground portions of slender arrowgrass (*Triglochin palustre*) serve as an important nutritional source for White-fronted Geese (Ely and Dzubin 1994). On the Y-K delta, prebreeding foods include pendantgrass (*Arctophila fulva*) shoots, slender arrowgrass bulbs, and crowberries (*Empetrum nigrum*; Budeau et al. 1991). By August they principally eat salmonberries (*Rubus chamaemorus*) and crowberries (Pacific Flyway Council 2003). On breeding areas in the Northwest Territories, Barry (1967) reported sedges, horsetails (*Equisetum* spp.), and pondweeds (*Potamogeton* spp.) as the major food items. The buds, shoots, and catkins of willows were ingested by White-fronted Geese on the upper Mackenzie River (Ely and Dzubin 1994).

Carrière et al. (1999) compared the feeding habits of White-fronted Geese with Canada Geese nesting in the High Arctic of the Kent Peninsula in Nunavut, where both species fed intensively before nest initiation. White-fronted Geese were more destructive, however, spending 56.9% of their foraging efforts grubbing in all types of habitats and continuing this behavior throughout the breeding season. Before snowmelt, the root tillers of alkaki grass (*Puccinellia phryganodes*) were a major food item (52%), after which sedges and the stems of Fisher's dupontia (*Dupontia fisheri*) constituted their preferred forage (>50%). Live plant biomass in pond margins (30–60 g/m²) was 4–15 times greater than in habitats that became available to geese earlier in the season (mudflats and hummocks).

Migration and Winter. During migration and winter, White-fronted Geese rely on virtually all forms of cultivated grains, although Tule Geese also feed on marsh plants, especially alkali bulrush (*Scirpus maritimus*) in the spring. On staging areas in Saskatchewan, most foraging occurred within 12 km of roost sites, although some flocks fed up to 38 km away (Ely and Dzubin 1994).

During spring, a major portion of the MCP stage at the Rainwater Basin in southcentral Nebraska, where their foraging ecology was well studied in 1979 and 1980 (Krapu et al. 1995). Geese were observed in harvested cornfields (76%) and fields of emerging winter wheat (23%); corn composed 90% of the diet, and wheat 9%, for 42 geese collected during the study. The paramount importance of this staging area to the MCP was underscored by the documented changes in body composition: adult geese collected during the spring in 1979 and 1980 deposited 12.2 g of lipids/day, with about 582 g deposited between 22 February and 3 April.

These authors stated that the long-term availability of corn and other cereal grains for migrating birds on the Great Plains was uncertain, but it was probably adequate for the foreseeable future. Subsequent work on the same study area, however, concluded that the amount of waste corn available for staging White-fronted Geese, as well as other migratory birds (such as Sandhill Cranes and Snow Geese), has changed dramatically in the central Platte River valley on the Great Plains (Krapu et al. 2004a). After accounting for a 20% increase in yield, the biomass of waste corn in 1997 and 1998 was reduced by 24% and 47%, respectively—lower than that recorded in 1978—as corn-harvesting efficiency by humans increased from 96% to 98%.

In response to these extensive landscape-level changes, fat storage by White-fronted Geese decreased to 0 g/day in 1998–99 (Pearse et al. 2011a). Adults collected over a period of 3 weeks on their next staging area (in southern Saskatchewan), however, increased lipids at a rate of 11.4 g/day and protein at 1.6 g/day. The lipid levels of females ranged from 252 to 278 g in Nebraska ($n = 30$), as compared with 654–688 g in Saskatchewan ($n = 45$); the lipid levels of males were 273–282 g in Nebraska ($n = 92$), versus 627–718 g in Saskatchewan ($n = 56$). Hence White-fronted Geese have exhibited considerable flexibility in their acquisition of nutrient reserves during the spring in response to these landscape-level changes. Moreover, lipid storage had declined 46% in Sandhill Cranes, which also use the Rainwater Basin as a major staging area (Krapu et al. 1984, 1985).

In southeastern Louisiana, White-fronted Geese wintering in a 3,270 km² area of coastal prairie and marshes used 8 of the 14 habitat types identified during a study conducted in 1981–82 and 1982–83 (Leslie and Chabreck 1984). Agricultural land made up 76.6% of the study area, the major components of which were soybean and rice fields, and winter pasture. Although habitat types varied over the study period as crops were harvested and replanted, harvested wet rice fields were the major feeding site, containing 47.3% of the flocks observed in the early season, 30.3% in midseason, and 16.9% in late season. Harvested wet rice fields were preferred more often than any other habitat type during 4 of the 6 survey times covered during both wintering periods. In the early and midseasons, more (92%) flocks were found in harvested rice fields that were wet than in those that were dry. Harvested soybean fields were preferred during the early season in 1981–82, when they made up only 4.3% of the available habitat but contained 34.0% of the flocks. Winter pasture was used in proportion to its availability, while harvested dry rice fields and fallow fields were generally avoided. Soybeans were consumed in proportion to their availability during other seasons over both years. Hobaugh (1982) also reported that soybean fields were preferred by White-fronted Geese in south-eastern Texas, with the greatest use occurring between late November and December. Krapu et al. (2004a), however, did not find any soybeans in the stomachs of White-fronted Geese staging on the central Platte River valley in Nebraska and concluded that they were a poor food choice in comparison with waste corn, which made up the majority of the stomach contents.

In southeastern Texas, Hobaugh (1982) found that harvested rice fields were preferred by winter-ing White-fronted Geese, as nearly 54% of all use was in rice stubble that covered only 14% of the study area. The birds also used rice fields almost exclusively from early fall until late November, and these were the only areas where large numbers of White-fronted Geese were consistently observed between October and March.

During early autumn for the years 1979–82, over 80% of the PP observed at the Klamath basin for-aged in stubble fields of barley, wheat, or oats, but they largely switched to fields of potatoes by late autumn, although the stubble fields remained im-portant (Ely and Raveling 2009). Most geese then left the Klamath basin for the Central Valley of California, where they consumed crops rich in sol-uble carbohydrates (Ely and Raveling 2009). After arriving in the Sacramento Valley, White-fronted Geese ate rice almost exclusively (42% of the avail-able crops), while those in the Sacramento / San Joaquin delta consumed corn (22% of the available crops). Upon returning to the Klamath basin in the spring, 39% of the flocks were located in fields con-taining new growths of wild and cultivated grasses. Cereal grains and potatoes were high (17%–47%) in soluble nutrients but low (7%–14%) in protein; grasses contained 47%–49% soluble nutrients and were also high (26%–42%) in protein.

The PP winters in California, where the geese fed extensively on the seeds of rice, watergrass (millet; *Echinochloa crus-galli*), milo (sorghum), and barley (McFarland and George 1966). The consumption of rice seed in the Sacramento Valley was 7 times greater than that of millet; of millet,

1.5 times greater than milo; and milo, 10 times greater than barley. Feeding tests of 12 penned geese belonging to several species (2 were White-fronted Geese) showed a 2:1 preference for rice over millet; 5–6:1 for millet over milo, alkali bul-rush (*Scirpus paludosus*), and safflower; and 9:1 for milo over barley. Food consumption by the geese over the 49-day trial period averaged 182 g.

White-fronted Geese studied on wintering areas in southern Oregon and California from Sep-tember to May in 1980–82 generally spent 3.2–6.2 hr/day at foraging sites, but they actually engaged in feeding activity for only 1.8–5.0 hours, which was lowest in late winter and highest in late spring (Ely 1992). Foraging time was shortest when the minimum temperatures fell below freezing, but the available lipid reserves (250 g) of the geese were probably sufficient to sustain them for up to 6 days. Foraging efforts also varied among sites, ranging from 27.5% to 39.7% in the Sacramento / San Joa-quin delta to a high of 70.4% at the Klamath basin in late winter. Alert time was substantial at all sites, ranging from 21.6% to as much as 47.7%. Foraging time of White-fronted Geese in Oregon and Cali-fornia, where they fed on nutritionally rich pota-toes and cereal grains, was substantially less than that reported in the United Kingdom (10 hr/day), where they fed on poorly digestible cellulose-laden grasses (Owen 1972). At roost sites, the geese spent most of their time sleeping (24%–46%), being alert (17%–40%), walking or swimming (6%–24%), and performing comfort behaviors (3%–25%; Ely 1992). The birds left their roosts in the morning and eve-ning, commonly using the same fields and return-ing to the same roosts. They generally departed from their roosts within 30 minutes of sunrise and returned in 2–3 hours.

Coupled with these descriptions of the birds' flexible foraging strategy, Ely and Raveling (1989) documented changes in body weight and compo-sition for this same population of White-fronted Geese. In general, both sexes maintained or in-creased their body weight in autumn, lost weight from autumn through winter, and then rapidly increased in weight before the spring migration in late April, although the timing and magnitude of body fluctuations varied annually for both sexes. For females, most changes in body weight reflected changes in lipids, which were lowest in mid-March (12%–13% of wet weight) and highest in late April (24%); lipids of the males probably also increased, although they were not collected in late spring. In general, females left staging areas in the spring at comparatively lower weights than most other species of geese, but they probably then benefited from foraging after their arrival on their breeding areas (primarily the Y-K delta).

Ely (1992) concluded that White-fronted Geese were generalists in their food choices and foraging strategies, highly suited to exploit the wide array of agricultural food available to them during winter in North America. Such flexibility in food use, and the maintenance of a lipid reserve during winter, probably contributes to flexible time budgets; these factors, taken together, are a major reason why this species has increased in abundance despite high hunting mortality and both the loss and alteration of their habitat over the last century.

MOLTS AND PLUMAGES

Adult White-fronted Geese have only 1 plumage (basic), which is molted annually via the prebasic molt. The descriptions of molts and plumages are primarily from summaries in Palmer (1976a) and Ely and Dzubin (1994), which should be consulted for further details.

Juvenal plumage begins at about 20 days of age (Class II goslings) with the emergence of wing and tail feathers. The head, neck, and back feathers appear last, with flight attained in 38–45 days (usually mid-Aug). Juvenal plumage is usually complete by late August, before fall migration, although some down can remain on the back and the lower neck. Birds in juvenal plumage are easily recognized, because they lack the adults' white forehead, furrowed neck feathers, and barring on the belly. In autumn some young birds have breast feathers marked with a light or dark brown flecking, which

disappears by midwinter. The upperparts are olive green to pale brown, the sides are olive brownish, and the belly and the breast are a clear whitish gray. Juvenile birds also do not yet have a white line along the flank, immediately below the folded wings. The bill color is a dull yellow orange, with variable black streaking. The legs and the feet are a dull yellow.

First basic is gradually acquired from early autumn through winter and into spring migration, and this plumage includes all feathers except the remiges. Most of the first prebasic molt is completed by midwinter. The first few white feathers appear on the top of the forehead by mid-October and spread to both sides of the bill by mid-November, increasing in their extent through January. Most (>80%) birds have acquired some barring on the belly by March and April. First basic is similar to definitive basic, except that it is not as bright. In general the upperparts are lighter than they are in adults, and the retrices are not white tipped. The bill color changes toward fleshy pink when young birds are about 13–16 weeks old.

Definitive basic is acquired via the prebasic molt, which occurs from July through mid-August and involves all feathers, including the remiges. Successfully breeding adults molt their remiges 2–4 weeks after their goslings hatch, and these adults regain flight in 20–30 days, which is about the time the goslings fledge. The body molt, however, continues through the autumn. This plumage is then retained until the following summer. In definitive basic, the head is a medium brown in White-fronted Geese, as compared with a dark chocolate brown in Tule Geese. Tule Geese also have less barring on the belly.

On their breeding area at the Anderson River, Barry (1967) reported that, like other geese, the first birds to molt were yearlings and older subadults; these had gathered in numbers of up to 20,000 on traditional areas. Wiebe and Hines (1998) noted high (>90%) site-fidelity rates to molting areas by 391 banded White-fronted Geese. Ely and Dzubin (1994) observed molting flocks of 1,000–15,000

White-fronted Geese, often separated into sub-flocks. Flightless birds can be found in mixed flocks that also contain small subspecies of Canada and Cackling Geese, but the White-fronted Geese tend to segregate and feed in different microhabitats (Derksen et al. 1979, 1982).

CONSERVATION AND MANAGEMENT

The Pacific Flyway Council (2003) has identified winter habitat protection and management, population inventories, harvest management, research, and education as management priorities for the PP. The rapid increase in competition among sympatrically nesting geese on the Y-K delta is a potential issue that can affect habitat use, gosling growth rates, and the subsequent survival of White-fronted Geese. Habitat loss and water shortages have and will continue to affect foraging and roosting areas for White-fronted Geese (and other waterfowl) in the Sacramento Valley in winter and the Klamath basin during the spring and fall. In particular, the decline in the use of the Klamath basin is recognized as a major cause for concern, and further deterioration of this significant wetland will place increasing pressure on other wetlands in the Pacific Flyway as birds are forced to find food and shelter elsewhere. Increased human activity on the Y-K delta breeding grounds can disturb birds during critical nesting and brood-rearing periods.

Management issues for the MCP call for the development of a breeding population survey. Currently, an annual census of birds staging in prairie Canada (Alberta and Saskatchewan) in late September is the primary method for assessing the annual status of the MCP. A breeding-ground survey would provide more consistent population estimates, as well as document population trends on a regional basis (Arctic Goose Joint Venture Technical Committee 2008). More research is also needed on population biology and ecology. Continued harvest assessment is essential, especially in Mexico, where the extent of harvest is basically unknown. In terms of habitat protection, the loss of wetlands in the Rainwater Basin is a significant concern, as is the continued deterioration of wetlands along the Gulf Coast.

Lastly, although the high energy content, ready availability, and vast expanse of areas planted in agricultural grains cannot be ignored by waterfowl managers, such resources are not stable. In the agricultural heartland of the Midwest and the Great Plains region of the United States, the cropland area planted in corn has remained relatively constant for the past half century, while that for soybeans has increased by 600%, and wheat fields have declined by 43% (Krapu et al. 2004a). The greater amount of land devoted to soybeans is especially significant, because it is a poor food for waterfowl. In the central Platte River valley of Nebraska, where soybeans were widely available in 1998 and 1999 (11% of the study areas), they did not appear in the esophageal contents of White-fronted Geese ($n = 198$), Lesser Snow Geese ($n = 208$), or Northern Pintails ($n = 139$). Nor were soybeans found in the esophageal contents of Sandhill Cranes ($n = 174$), which also use this area as their principal spring staging site (Krapu et al. 1984). From February through April in 1998 through 2001, <2.5% of the habitat use by arctic-nesting geese was in areas where >90% of the landscape was planted in soybeans, while 38%–83% usage occurred where >90% was in corn (Krapu et al. 2004a, 2005). Moreover, a significant concern relative to waste grain is increasing efficiency, as modern farm machinery can harvest close to 99% of the corn crop. Such efficiency reduced the amount of waste remaining in the central Platte River valley, which subsequently affected the ability of White-fronted Geese and Sandhill Cranes to accumulate lipids during the spring in this region. This decline in the availability of waste corn has also led to competition between Sandhill Cranes and the arctic-nesting goose species using the central Platte River valley during spring (Krapu et al. 2005).

In the Central Valley of California, implementation of the Central Valley Joint Venture has positively affected wintering PP geese, which traveled shorter distances between roost sites and feeding areas from 1998 to 2000—24.2 km, as compared with 32.5 km a decade earlier (1987–90)—despite a 2.2-fold increase in the size of the wintering population (Ackerman et al. 2006). The use of rice habitats for roosting and feeding also rose between these decades, probably because farmers were subjected to burning restrictions in 1991 that led to them flooding harvested fields as a way to decompose rice-straw residues. These authors cautioned, however, that dependence on agricultural crops as a feeding and roosting habitat might yield a false sense of security for managers. The current quantity of rice fields that are flooded after harvest may not be sustainable, given the ever-increasing de-

mands for water in California, as well as new uses for rice-straw residues, mosquito-control initiatives, and changing agricultural practices. Hence managers should continue to plan and implement programs that provide the natural wetland habitats that are necessary to maintain the numbers and distribution of PP geese in the Central Valley. Overall, changes in agricultural food availability, such as those seen in both the central Platte River valley and the Central Valley of California, call for landscape-level research to evaluate the possible consequences of a decline in such energy-rich resources on wildlife populations like migrating geese (Krapu et al. 2004a).

Emperor Goose

Chen canagica (Sewastianow 1802)

Left, adult; *right*, juvenile

The Emperor Goose is a maritime species restricted to coastal areas in western Alaska, the Aleutian Islands, and extreme eastern Russia; 90% of the world's population breeds on the Yukon-Kuskokwim (Y-K) delta in Alaska. The population of Emperor Geese was never large, but it has declined from about 139,000 in 1964 to 91,900 in spring 2009. Subsistence harvest on the breeding grounds, substantial gull (*Larus* spp.) predation of goslings, and competition for food from Cackling Geese (*Branta hutchinsii minima*) during the brood-rearing period are all significant factors contributing to low population levels for Emperor Geese. This species is a medium-sized, colorful goose. They mate for life and rarely stray from coastal habitats during all phases of their life cycle. Clutch size averages about 5 eggs, which are incubated for approximately 24 days. Young goslings reach flight stage in roughly 50–60 days. Their major wintering area is the Aleutian Islands, with important spring and fall staging areas being the lagoons along the Alaska Peninsula. Failed breeders and nonbreeders undergo a molt migration to the Chukotka [Chukchi] Peninsula in Russia. Unlike other geese, Emperor Geese consume a large amount of animal matter, such as bivalves (clams and mussels), especially during the winter and on staging areas.

Emperor Goose adult

Emperor Goose juvenile

IDENTIFICATION

At a Glance. Emperor Geese are easily identified. This species is a medium-sized goose with a stocky body, and a short thick neck. The head and the back of the neck are white, which contrasts sharply with the black chin and throat and the dark bluish-gray body. The legs and the feet of adults are bright yellow orange. The small bill is pink.

Adults. The sexes are similar. The white edgings on the bluish-gray body feathers form wavy bands across the back, the sides, and the flanks, which present a barred effect above and a scaled appearance below. The head and the back of the neck are white; the chin and the throat are black. During summer, the white head is often stained a reddish brown from feeding in tidal pools where concentrations of iron oxide occur. The legs and the feet are bright yellow orange; the bill is pink. The irises are brown or hazel. The short wings of Emperor Geese require rapid strokes to maintain flight, which is comparatively slow.

Juveniles. The plumage is a uniform dark gray, except for the white of the exposed retrices. The bill color is grayish blue to blackish. The legs and the feet are dull yellowish. The irises are grayish brown. The head and the upper neck of juveniles become largely white by late October, but they are still flecked with a scattering of dark feathers. Full adult plumage is attained by the first winter, but an experienced observer can identify most 10- to 12-month-old juveniles based on the residual black flecking on the head feathers.

Goslings. Hatchlings are varying shades of grayish white (Nelson 1993). A white area remains around the bill for about 3 weeks, which is then replaced by gray feathers. The feet are variably grayish / olive brown to blackish; the bill is black, with a white nail tip.

Voice. There are 2 basic calls given by Emperor Geese. A hoarse, piercing *kla-ha, kla-ha, kla-ha* is given in flight, and a deep coarse *u-lugh, u-lugh* is uttered as an alarm call.

Similar Species. The blue morph of the Lesser Snow Goose (*Chen caerulescens caerulescens*) has an all-white head and neck. In flight, the underwings of Emperor Geese are entirely gray, in contrast to the white and black underwings of Snow Geese.

Weights and Measurements. Petersen et al. (1994) recorded measurements of adult females trapped on nests at the end of incubation in 1993 near the Manokinak River on the Y-K delta in Alaska: body weight averaged 1,638 g ($n = 126$), culmen 36.0 mm ($n = 32$), and tarsus 80.2 ($n = 127$). Body weight 35−53 days after the hatch, when the adults were molting, averaged 2,370 g for adult males ($n = 13$) and 1,926 g for adult females ($n = 101$). Juvenile males ($n = 53$) averaged 1,165 g and juvenile females ($n = 56$), 1,107 g. Body weights of 136 adult females captured during the flightless period in late July and August on the Y-K delta averaged 1,900 g (range = 1,580−2,630 g; Hupp et al. 2008a). Schmutz (1993) captured Emperor Goose goslings in flocks with flightless adults at 9 sites on the Y-K

Above:
Emperor Goose flock.
GaryKramer.net

Right:
Emperor Goose
defending her nest.
RyanAskren.com

117

Emperor Goose; the head color is the result of staining from iron in the marine sediments where they forage.
RyanAskren.com

delta and fitted them with individually numbered neck collars for resightings on major staging areas. Body weights of juvenile males when they were banded averaged 1,499 g ($n = 139$) for those later seen on staging areas, versus 1,458 g ($n = 163$) for those not seen; for juvenile females, body weights at banding averaged 1,379 g ($n = 116$) for those seen again versus 1,348 g ($n = 156$) for those not seen.

DISTRIBUTION

Breeding. Emperor Geese, with minor exceptions, are restricted to the region surrounding the Bering Sea. In North America, the breeding range for this species extends from the northern side of Kotzebue Sound south to Bristol Bay, but 80%–90% of the world's population breeds on the Y-K delta (Eisenhauer and Kirkpatrick 1977). Emperor Geese are uncommon as breeders on the Seward Peninsula, and they no longer breed south of the Y-K delta (Kessel 1989).

In Russia, Emperor Geese are among the rarest and least known species of waterfowl (Kistchinski 1971). They breed principally along the northern coast of the Chukotka [Chukchi] Peninsula (between Kolyuchin Bay west to the Cape Shmidt area) and in the Anadyr lowlands along the coast of Anadyr Bay (Schmutz and Kondratyev 1995). Surveys have recorded a population of 3,000–5,000 Emperor Geese breeding in Russia, but only 127 broods were observed by Hodges and Eldridge (2001). Kistchinski (1976) suggests that up to 80% of the Emperor Geese found in Russia during the summer are nonbreeding molting birds.

Winter. Most of the Emperor Geese from both Alaska and Russia winter throughout the Aleutian Islands, but some are also found along the southern coast of the Alaska Peninsula and on Kodiak and Afognak Islands. Nelson (1961) reported 500–700 Emperor Geese around Amchitka Island in the Aleutians in November 1957, noting that the number of geese declined steadily through March and April, with the last flock seen on 21 April. A few Emperor Geese winter farther south in British Columbia, Washington, Oregon, and occasionally in northern California, but these birds probably are the result of parasitic eggs laid in the nests of other goose species nesting on the Y-K delta (Lensink 1969). Some Emperor Geese will winter in estuaries on the Alaska Peninsula when mild winters

Breeding Distribution ▓
Wintering Distribution ▓

keep habitats ice free. In Russia, some Emperor Geese winter in the Commander Islands and along the southeastern coast of Kamchatka (Eisenhauer and Kirkpatrick 1977, Petersen et al. 1994).

Migration. Virtually all Emperor Geese stage on the Alaska Peninsula during both spring and fall migrations, where important habitats are the protected bays on the northern side of the peninsula, particularly Izembek Lagoon, Nelson Lagoon, Port Heiden Bay, Cinder Lagoon, and Seal Island (Eisenhauer and Kirkpatrick 1977, Peterson and Gill 1982, Petersen et al. 1994). Three estuaries located on the southern coast of the Alaska Peninsula also serve as staging areas, but for smaller numbers of birds: Ivanof Bay, Chignik Lagoon, and Wide Bay. A few Emperor Geese use islands south of the peninsula, as well as Kodiak Island (Pacific Flyway Council 2006b). Schmutz and Kondratyev (1995) verified that Emperor Geese breeding in Russia used staging areas in western Alaska. Schmutz et al. (1997) also reported a site fidelity of 55%–80% to the previous year's staging area in Alaska.

Emperor Geese migrate only 600–750 km between their breeding and staging areas, but the distance from their staging to their wintering areas can reach 2,200 km (Schmutz 1993). Wintering Emperor Geese have been recorded from Attu (the last island in the Aleutian Archipelago), arriving in late December (Wilson 1948).

MIGRATION BEHAVIOR

Like all arctic geese, the timing of migration in Emperor Geese most likely is dictated by weather, especially ice and snow on the northerly breeding and molting sites. In spring, Emperor Geese leave the Aleutians as early as March and arrive by mid-April on the Alaska Peninsula, where some birds may remain in selected lagoons for 2–3 weeks. From there, most of these geese travel nonstop northward, over the open water of Bristol Bay to Hagmeister Strait and the lagoons around Cape Newenham, arriving on the Y-K delta by mid-May (Petersen et al. 1994).

Autumn migration follows the spring route in reverse, with birds arriving at Cape Newenham in mid-August. Most of the geese have reached the lagoons on the northern side of the Alaska Peninsula by late September, and many will remain there until late October or November (Gill et al. 1981). Headley (1967) noted that small numbers of Emperor Geese arrived at Izembek Lagoon during late August, with the largest influx between 24 and 29 September. Most birds have moved into the Aleutians by late November (Petersen et al. 1994). On Adak Island in the central Aleutians, Emperor Geese arrived from late September to early October and departed by mid-May (Byrd et al. 1974).

In 1999, 2002, and 2003, Hupp et al. (2008b) tagged 53 adult female Emperor Geese on the Y-K delta with satellite radiotransmitters to examine the annual cycle of migration, which can extend >3,000 km from Kodiak Island in Alaska to the Commander Islands in Russia. Females wintering only 650–1,010 km from the Y-K delta in lagoons on the southern side of the Alaska Peninsula either bypassed fall staging areas altogether or used them an average of 57 days. In contrast, females with longer (1,600–2,640 km) migrations to the western

Aleutians used staging areas an average of 97 days. Further, females wintering on the Alaska Peninsula averaged more days at winter sites than females in the western Aleutians (172 vs. 91 days). Females wintering in the eastern Aleutians, an intermediate distance of 930–1,610 km from the Y-K delta, averaged 77 days on staging areas and 108 days on wintering areas. Return dates to the Y-K delta, however, did not vary among birds from different wintering areas. The study concluded that coastal staging areas on the Alaska Peninsula were especially critical to migrating Emperor Geese, as they allowed fall migrants to prepare for long-distance migration to winter sites (such as the western Aleutians), and spring migrants coming from different distances to reach comparable levels of body condition before nesting.

Molt Migration. A molt migration of subadults and failed breeders occurs in mid-June, when non-breeders or breeders that fail very early in their nesting attempts migrate to the northern Chukotka [Chukchi] Peninsula in Russia for the summer wing molt, a distance of 965–1,600 km. The flight path takes the birds over St. Lawrence Island, where small numbers stop to rest, but only a few have molted there since the mid-1980s. Hupp et al. (2007) tagged 32 adult females with satellite radio-transmitters on the Y-K delta from 2000 to 2004, of which 16 unsuccessful breeders departed either for the Chukotka [Chukchi] Peninsula (15) or St. Lawrence Island (1). While molting in Chukotka, most marked geese used the coastal lagoons west of Kolyuchin Bay, particularly Tenkyrgynpilkhen Lagoon (nearly 9,000 birds). A 2002 survey in Chukotka tallied 21,150 adult-plumaged Emperor Geese, which led to the conclusion that about 20,000 adult Emperor Geese from the Y-K delta (28% of the adult population) leave Alaska annually to molt elsewhere, most likely in the Chukotka [Chukchi] Peninsula. Emperor Geese depart Chukotka from late August to early September for fall staging areas on the Alaska Peninsula. Of the 16 females that nested successfully, only 1 migrated to Russia, while 15 remained and molted on the Y-K delta.

HABITAT

Emperor Geese are a maritime species occupying habitats throughout their annual cycle that are never far from the ocean. On their principal breeding area in the Y-K delta, the vast majority are found within 15 km of the coast (Petersen et al. 1994). These coastal habitats are flat and laced with tidal rivers and sloughs, as well as a plethora of brackish and freshwater ponds. Eisenhauer and Kirkpatrick (1977) estimated that 50% of the outer Y-K delta was covered by lakes, ponds, and puddles ranging in size from <1 m² to 10,000 ha. Lensink (1966) estimated that there were 62,700 lakes and ponds totaling 380,413 ha (50%) of the 757,183 ha that he surveyed on the Y-K delta.

At Kokechik Bay on the Y-K delta, Eisenhauer and Kirkpatrick (1977) described 5 major topographic-floristic habitats: tidal sedge flats, grass flats (35% covered by small ponds), lowland pingo tundra, tall-sedge marshes, and upland tundra. The plant communities of the tidal flats were dominated by salt-tolerant species.

During their spring and fall migrations, Emperor Geese make extensive use of the lagoon system on the Alaska Peninsula. The lagoons are characterized by sandy or muddy bottoms and are subject to tidal fluctuations of up to 4 m. Low tide exposes beds of blue mussels (*Mytilus edulis*) and macoma clams (*Macoma* spp.) that are eaten by the geese (Schmutz 1994), which roost on the barrier islands and sand spits associated with the lagoons. In winter Emperor Geese are almost exclusively found on the intertidal habitats associated with the Aleutians.

POPULATION STATUS

Emperor Geese inhabit remote, inaccessible habitats; historical data, while sparse, nonetheless reveal a population that has declined. The first extensive aerial survey was conducted on spring staging areas in 1964 and tallied 139,000 Emperor Geese, although not all areas were surveyed (King 1965). A later survey of fall staging areas estimated about 150,000 Emperor Geese in Alaska (King and Lensink 1971). Eisenhauer and Kirkpatrick (1977) used these and other survey data available in the 1970s and estimated a world population of 175,000–200,000 during the fall and 140,000–160,000 in the spring. An estimated 12,000–15,000 (10%) were assumed to be from Russia (Kistchinski 1973, 1976). During 2002 aerial surveys of coastal wetlands on the Chukotka Peninsula, the population of Emperor Geese was estimated to be 12,000–30,000 (Pacific Flyway Council 2006a).

Since 1981 the U.S. Fish and Wildlife Service has surveyed Emperor Geese on both their spring and fall staging areas in Alaska. The Spring Survey is used as a management index, because the geese are concentrated for a shorter time interval in spring versus fall, so the spring count probably better reflects the actual population. The 3-year spring average has ranged from a high of 90,355 (1981–83) to a low of 49,213 (1986–88; Pacific Flyway Council 2006b). The 2009 spring estimate was 91,900, which was 42% higher than 2008 and reflected a 2% annual increase from 2000 to 2009 (U.S. Fish and Wildlife Service 2009).

Harvest

In 1985, the sport-hunting bag limit on Emperor Geese was reduced from 6 to 2/day, which was followed by a complete closure in 1986. The estimated sport harvest averaged 2,100 from 1970 to 1980 and seldom exceeded 2% of the estimated population.

The subsistence harvest, a traditional use of this resource by Native Americans in Alaska, includes both the birds and their eggs. It is especially important to the Yupik peoples on the Y-K delta, where migratory birds provide the first source of fresh meat after the long winter (Mitchell 1986). These harvests occur in both spring and fall, but the spring harvest is more extensive. The importance of migratory birds to native peoples has long been recognized; early laws, such as the first Alaska Game Act (1902), prohibited spring and

summer hunting but exempted native subsistence hunters. Native Americans remained exempt in the 1908 Alaska Game Act, and again in the 1925 Act, although the latter stipulated that they had to be in "absolute need." The Migratory Bird Treaty Act (MBTA) of 1918, however, outlawed subsistence waterfowl harvests in the spring and summer. The taking of migratory birds in Alaska was subject to the rules of the 1925 Alaska Game Act, not the MBTA, but these conflicting pieces of legislation led to a long regulatory and enforcement conflict (Mitchell 1986). Furthermore, populations of Emperor Geese and other goose species began to decline, to the concern of both native and nonnative hunters. This concern led to the Hooper Bay Agreement (HBA) in January of 1984, which was the culmination of discussions and negotiations among the U.S. Fish and Wildlife Service, the Alaska and California Departments of Fish and Game, and the Association of Village Council Presidents (AVCP). The HBA led to the voluntary curtailment of native harvests of geese and a commitment to seek reductions in recreational harvests in the Pacific Flyway, especially in California. The HBA was adopted by the Pacific Flyway Council in July 1984, but it did not include Emperor Geese until a modification was made in 1985, which led to the formation of the Yukon-Kuskokwim Delta Goose Management Plan, which develops the U.S. Fish and Wildlife Service's harvest policy (Pamplin 1986). More refinement, however, was needed in federal legislation governing the harvest of migratory birds (the MBTA).

The Migratory Bird Treaty Act Protocol Amendment of 1995 (enacted in 1997), and the resultant formation of the Alaska Migratory Bird Co-Management Council (AMBCC) in 2000, led to the formal legalization of spring/summer subsistence hunting in 2003. The AMBCC consists of the U.S. Fish and Wildlife Service, the Alaska Department of Fish and Game, and 11 native regional partner organizations. The Association of Village Council Presidents' Waterfowl Conservation Committee, which has existed on the Y-K delta since the mid 1980s, became the AMBCC's regional partner representing the Y-K delta. The Protocol Amendment specifically provides for the customary and traditional use of migratory birds and their eggs by the indigenous peoples of Alaska. The Amendment, however, also states its intent not to increase the take of migratory bird species relative to their continental population sizes. The AMBCC closed subsistence hunting of Emperor Geese in 1987, although some illegal take continues.

The first subsistence survey on the Y-K delta was conducted in 1964 and estimated a harvest of 8,200 Emperor Geese, which was about 6% of the population (Klein 1966). From 1985 to 2000, harvest on the Y-K delta averaged 2,119 birds shot and 290 eggs collected, but these data were probably underestimates, because not all villages where harvests occurred participated in the associated surveys (Wentworth and Wong 2001). A more detailed study on the Y-K delta (38 villages) estimated an average harvest of 1,162 Emperor Geese and 191 eggs from 2001 to 2005, which was lower than the average annual harvests from 1995 to 2000 (1,659) and 1990 to 1994 (2,520). The 2001–5 data were considered more accurate, because they included surveys of a greater number of villages near prime Emperor Goose habitats, with the resultant harvest yielding 2,440 kg of meat for native households (Wentworth 2007a). Most (63%) of the birds that are harvested on the Y-K delta are taken during the spring (1 Apr–30 Jun). Wolfe and Page (2002) estimated a subsistence harvest of 4,500 Emperor Geese/year in Alaska during the 1990s, which represented nearly 8% of the spring population index. The extent of the geographic coverage and sample intensity, however, suggest that this estimate is low.

BREEDING BIOLOGY
Behavior
Mating System. Emperor Geese are monogamous and pair for life (Petersen et al. 1994). Initial pair bonds probably form among subadults either during spring migration or on the breeding grounds (Headley 1967). The exact timing and the process of pair formation are unknown, although both aggressive and courtship behaviors of Emperor Geese were observed during April and May in the area of Izembek Lagoon (McKinney 1959). Emperor Geese arrive as pairs on their nesting grounds on the Y-K delta in May (Eisenhauer and Kirkpatrick 1977), but some new pairs may be formed after arrival (Petersen et al. 1994).

None of the 480 female Emperor Goose goslings marked by Schmutz (2000a) on the Y-K delta was ever observed on nests as 1- or 2-year-old geese. Subsequently, 3 were recorded on nests as 3-year-olds, 7 as 4-year-olds, and 3 as 5-year-olds. Schmutz (2000a) concluded that few (if any) Emperor Geese bred as 2-year-olds and that 3-year-olds breed at a lower frequency than older geese. Even after breeding begins, not all females will nest every year. During a 5-year study (1982–86) on the Y-K delta, the percentage of adult females that were breeding ranged from 38.5% to 52.0% (Petersen 1992a). The nesting frequency of adults ranged from 42.6% to 69.8%. Additionally, 52.5% of the females that nested in a given year also nested the next year, while 72.0% of those that failed to initiate a nest in a particular year also failed the following year. The low nesting frequency perhaps occurred because a substantial degree of annual mortality may result in a high percentage of newly paired birds attempting to nest.

Sex Ratio. Few data are available on sex ratios (males:females).

Site Fidelity and Territory. Eisenhauer and Kirkpatrick (1977) reported that up to one-third of the nest sites at Kokechik Bay were reused in subsequent years, with some sites occupied in 2–3 successive seasons. Old nest sites generally are used when nesting densities are high, but most pairs prefer to construct a new nest when pair densities are lower (Mickelson 1975). Emperor Geese did not reuse nest sites during a 5-year study (1982–86) at Kokechik Bay (Petersen 1990).

At Kokechik Bay, Emperor Goose pairs defended a 14 m² area around the nest site in the period before egg laying, with the male being more aggressive (Eisenhauer and Kirkpatrick 1977). Petersen et al. (1994) reported that pairs defend the nest area through the egg-laying and incubation periods. Nest densities at Kokechik Bay were highest in lowland pingo tundra, where distances between nests averaged 58.2 m ($n = 260$), compared with 75.6 m in other habitat types ($n = 120$; Eisenhauer and Kirkpatrick 1977). At lower nest densities, the distance between nests averaged 105.9 m ($n = 197$; Petersen 1990).

Courtship Displays. Few descriptions exist of the courtship displays of Emperor Geese. Johnsgard (1965) noted a *forward* posture, where the head is held low; the white head and neck form a conspicuous pattern against the darker body. McKinney (1959) reported precopulatory *head-dipping* on a spring staging area in Alaska (Nelson Lagoon). Eisenhauer and Kirkpatrick (1977) did not observe any sexual displays at Izembek Lagoon through 16 April, nor any copulations on the breeding grounds, which led to a conclusion that copulation occurs just before or during spring migration.

Nesting

Nest Sites. Females select the nest site. At the Kashunuk River on the Y-K delta, Emperor Goose nests ($n = 81$) were on shorelines (53.1%), peninsulas (28.4%), and islands (18.5%; Mickelson 1975). In comparison to random sites elsewhere, Emperor Goose nest sites at Kokechik Bay on the Y-K delta were farther from open water, contained more shrubs, were near ponds with fewer islands, were higher above the pond's water level, and were positioned lower along the sides of pingos (Petersen 1990). Nests were also located where abundant dead vegetation could be used to conceal the nest from avian predators during the laying period, when the female is infrequently at the nest. Nests initiated early were at higher elevations, particularly during years with heavy spring snow cover. At Kokechik Bay, 94% of 380 Emperor Goose nests were located on raised areas somewhat higher than their surroundings, which probably gave the female a better chance of spotting predators (Eisenhauer and Kirkpatrick 1977).

Females build the nest, which is a scrape lined with grasses, sedges, or other adjacent vegetation and a small amount of down. Eisenhauer (1976) reported average dimensions from 237 nests: outside diameter, 376.5 mm; inside diameter, 200.0 mm; and depth of cup, 81.6 mm.

Clutch Size and Eggs. Eisenhauer and Kirkpatrick (1977) summarized U.S. Fish and Wildlife Service reports on clutch size data for Emperor Geese on the Y-K delta (exclusive of Kokechik Bay) from 1963 to 1972. Mean clutch size was 4.8 eggs ($n = 293$) but ranged from 3.8 in 1965 to 5.3 in 1966. Clutch size varied from 1 to 10, but 71% contained 4–6 eggs. On their Kokechik Bay study area, clutch size averaged 5.2 ($n = 426$) from 1971 to 1973 and was greater during early-nesting seasons (5.4 vs. 4.3). During her 5-year study (1982–86) at Kokechik Bay, Petersen (1992a) observed an average clutch size of 4.9 ($n = 472$), ranging from 4.8 in 1984 to 5.0 in 1986; mean clutch sizes did not vary between early and late seasons. Mickelson (1975) reported average clutch size on the Y-K delta as 4.2 ($n = 72$).

Emperor Goose eggs are elliptical ovate and pure to creamy white when laid, but they gradually become stained a dull brown from nest material (Bent 1925). Eggs from nests at Kokechik Bay (1971–73) were 79.9 mm × 52.1 mm ($n = 1,399$ eggs; Eisenhauer and Kirkpatrick 1977). Petersen (1992b) reported egg measurements from Kokechik Bay (1982–86) as 86.7 mm × 57.0 mm ($n = 1,743$ from 301 nests). Individual females laid similar-shaped eggs in successive years, and eggs within the clutch of a particular female appeared alike. Fresh egg mass averaged 122.3 g. Egg volume and width were negatively correlated with spring population size, which may explain the differences in egg measurements between the 1971–1973 and 1982–1986 studies (Petersen et al. 1994). Eisenhauer and Kirkpatrick (1977) estimated that the laying interval averaged 1.2 days/egg ($n = 60$ clutches), which equates to a laying rate of 0.9 eggs/day.

Incubation and Energetic Costs. Only the female incubates the eggs. Female Emperor Geese have the highest nest attentiveness of all geese, averaging 99.5% ($n = 11$). Eisenhauer and Kirkpatrick (1977) noted an average incubation period of 24.3 days for 52 nests monitored at Kokechik Bay. Recess frequency averaged only 0.54 recesses/day, and they were short in duration (13.3 minutes), during which time the females traveled an average of 53 m from their nests (Thompson and Raveling 1987). Males were present during 56% of the recesses and were alert for 49% of that time.

High incubation constancy for Emperor Geese appears to be a defense against avian predation. These geese can repel their principal mammalian nest predator, the arctic fox (*Alopex lagopus*), but they are clumsy flyers, inefficient in pursuing aerial predators. Hence adults are not able to effectively keep aerial predators away from nest sites unless the parents are in close proximity to the nest. In contrast, other small-bodied geese, such as Black Brant (*Branta bernicla nigricans*) and Cackling Geese (*B. h. minima*), are much more agile aerially and have lower (89.6% and 93.6%, respectively) incubating constancies.

Due to such high nest attentiveness and a large clutch size, females lost 20.7% of their body mass during the nesting season. Hence they probably arrive on breeding grounds with a very substantial amount of endogenous (stored) reserves in comparison to other geese (Thompson and Raveling 1987).

Nesting Chronology. At Kokechik Bay, Eisenhauer and Kirkpatrick (1977) reported that severe weather and snow conditions delayed nest initiation by 2 weeks in 1971 and 1972, while nesting began 7 days after arrival in 1973. In all years, however, peak clutch initiation was within 9 days of peak arrival, with the total clutch-initiation period ranging from 22 to 32 days. The first eggs were laid from 20 to 31 May, and the total nesting season was 46 days in 1971, 44 days in 1972, and (except for 2 nests) 46 days in 1973. Most (80%) of the females began laying their clutches within the first 14 days of the initiation period, which is a high degree of synchrony. Cumulatively, 2.5% of nests were started before 1 June in 1971, 16.6% in 1972, and 72.6% in 1973.

Marked Emperor Geese arrived at Kokechik Bay on the same relative date each year (12–19 May) during a 5-year study (1982–86) and initiated nests 5 days after arrival (range = 0–10 days), with median dates of nest initiation varying from 20 May to 3 June (Petersen 1992a). Weather did influence nesting chronology, however, as Emperor Geese arrived and began nests later during years when freezing temperatures and snow and meltwater covered the nesting area. Based on snowmelt, the first nests of marked geese occurred on 15 and 19 May during 2 years considered to be early-nesting years, compared with 23 May and 27 May during late-nesting years.

During a subsequent 5-year study (1999–2003) on the Y-K delta, prelaying intervals were also influenced by spring snowmelt. In general, the prelaying interval declined by about 0.4 days and the nest initiation date increased by roughly 0.5 days for each day that arrival was delayed. Hence females that were the first to arrive had longer prelaying intervals (up to 4 days), but they nested up to 5 days earlier in comparison to females that arrived last. The median prelaying interval of radiomarked females was 15 days (range = 12–19 days) during a year of late snowmelt compared with 11 days (range = 4–16 days) during 2 years with an earlier snowmelt (Hupp et al. 2006). A prelaying interval of <12 days for 11 out of 15 females during warm years suggested that follicle development occurred on the spring staging areas. Follicle development in Emperor Geese requires 12 days (Alisauskas and Ankney 1992a); hence, during warm springs, some females in the Y-K delta study probably initiated follicle growth as long as a week before arrival, because their prelaying interval was only 4 days. In contrast, follicle development during cold springs probably occurred after the birds arrived on the Y-K delta.

Hupp et al. (2006) concluded that Emperor Geese could exhibit plasticity in the timing of follicle development because a distance of only 600–700 km separates their breeding grounds on the Y-K delta from their spring staging areas on the Alaska Peninsula. Weather conditions at these staging areas often reflect conditions on the Y-K delta (Petersen 1992a), and Emperor Geese appear to fly nonstop between the 2 locales. Flexibility in the timing of follicle development, perhaps cued by spring weather, is of obvious advantage in an environment where snow-cover conditions can vary by 2–3 weeks. Perhaps this flexibility is why Petersen (1992a) observed that a delayed nesting season did not change the relative arrival pattern at the Y-K delta: individuals arriving early in mild years also arrived early in cold years, and birds arriving later in mild years arrived later in cold years. Such consistency in arrival patterns indicates that individual Emperor Geese respond similarly to the environmental cues that trigger the timing of spring migration, again probably because spring staging areas on the Alaska Peninsula are close to their breeding grounds on the Y-K delta.

Nest Success. Nest success in Emperor Geese is extremely variable from year to year. Eisenhauer and Kirkpatrick (1977) reported nest success ranging from 81.6% to 94.4% (*n* = 332) during a 3-year study (1971–73) at Kokechik Bay on the Y-K delta. During a subsequent 5-year study (1982–86) on the same research area, nest success (*n* = 591) ranged from a high of 90.6% to a low of 0.1% (Petersen 1992a). In the 2 years of substantial nest predation, the percentages of nests lost to avian (53.8%–62.5%) and mammalian (37.5%–46.2%) predators were similar. In contrast, during the 3 years of diminished nest predation, the percentage destroyed by mammals (79.2%–89.5%) was high in comparison to that by avian predators (8.8%–15.1%). Most nest destruction occurred during the egg-laying period, when females were usually not at their nests.

Eisenhauer and Kirkpatrick (1977) noted that foxes caused <1% of nest losses at Kokechik Bay, but only 1 arctic fox was seen during their study. Petersen's (1992a) research in the same area found that arctic foxes were the principal mammalian predator of eggs in Emperor Goose nests. Parasitic Jaegers (*Stercorarius parasiticus*) and Glaucous Gulls (*Larus hyperboreus*) were the chief avian predators (Mickelson 1975, Eisenhauer and Kirkpatrick 1977).

Brood Parasitism. Brood parasitism is common in Emperor Geese, having occurred in 56.4%–69.2% of the nests studied on the Y-K delta from 1982 to 1986 (Petersen et al. 1994). Most (54.1%) parasitic eggs were deposited during the egg-laying interval of the host, with the remainder deposited during incubation. An average of 1.5 females laid an average of 2.2 parasitic eggs in each parasitized nest. Hatchability of the host eggs ranged from 89% to 93%, but 15% of all hatched goslings resulted from parasitically laid eggs (Petersen 1991). The host female responded by threatening and chasing potential parasitic females. The presence of a parasitic egg or eggs reduced the probability of the host eggs hatching by 4.5% (from 93.2% to 88.7%), but nests containing parasitic eggs hatch at the same level of success as nonparasitized nests. Peterson et al. (1994) noted that 49.2% of the parasitic eggs hatched (*n* = 362).

Watkins (2006) noted that some nests contained eggs that were entirely parasitic. Over half of all Emperor Goose eggs were incubated by a female other than the goose who laid them. Species of other waterfowl responsible for brood parasitism of Emperor Goose nests include Cackling Geese (*B. h. minima*), Greater White-fronted Geese, Black Brant, and Spectacled Eiders.

Renesting. There is no evidence of renesting by Emperor Geese.

REARING OF YOUNG

Brood Habitat and Care. On the Y-K delta, a 3-year study (1994–96) of radiomarked Emperor Goose families ($n = 56$) found a strong habitat preference for saline ponds, mudflats, and meadows dominated by *Carex ramenskii* (Schmutz 2001). These were the most saline locales, which covered one-third of a 70 km² study area; 43% of all locations were in *C. ramenskii* meadows, which made up only 21% of the available habitat. Broods frequently moved into a saline pond habitat in response to predators. Broods also used adjoining saline habitats dominated by *Carex ramenskii* and *C. subspathacea*, the latter species being a preferred food of Emperor Goose goslings. Emperor Goose broods selected more saline habitats than broods of Cackling Geese (*B. h. minima*) and Greater White-fronted Geese, but there was considerable overlap in habitat use among the 3 goose species (Schmutz 2000b).

Within 1 week of hatching on the Y-K delta, Laing and Raveling (1993) found that Emperor Goose goslings moved to tidally influenced salt-marshes within 500 m of Kokechik Bay, where they remained for the brood-rearing period. The goslings preferred mudflats vegetated with *Puccinellia phryganodes* and *Carex subspathacea*, although such mudflats covered <5% of these authors' study area on the Y-K delta. Goslings fed 80%–82% of the time they spent in mudflat habitats

Goslings remain in the nest and are brooded by the female for up to 48 hours after hatching, but they generally leave the nest in 12–24 hours and begin foraging (Eisenhauer and Kirkpatrick 1977). Just the female broods the goslings, but by the time they are 2–3 weeks old she only broods them during inclement weather. The male leads the brood and defends them from intruders coming within 1–2 m of the goslings. Adults use an *upward threat* display to ward off Glaucous Gulls when a gull's flight path is below 25 m (Frazer and Kirkpatrick 1979). Preferred brood-rearing habitat included sedge meadows bordered by a slough or river.

Brood Amalgamation (Crèches): Frazer and Kirkpatrick (1979) observed a high degree of integrity among Emperor Goose broods, but they noted 2 instances of adoptions of young from other broods, a behavior also seen by Eisenhauer and Kirkpatrick (1977).

Development. Body weights of 5 incubator-hatched goslings averaged 81. 8 g (range = 73.5–86.7 g; Smart 1965a). Goslings captured 43 days after hatching grew at an average rate of 26.0 g/day for males and 24.4 g/day for females (Petersen et al. 1994). Goslings fledge in 50–60 days (Mickelson 1975). During a 1993–96 study of Emperor Goose goslings near the Manokinak River on the Y-K delta, goslings fed 63%–73% of the time (Schmutz and Laing 2002). They spent 81% of that period foraging in stands of *Carex subspathacea*, their preferred food, versus 47% in stands of *Carex ramenskii*.

Gosling growth was negatively correlated with densities of other Emperor Goose broods, and especially with the density of Cackling Geese (*B. h. minima*; Schmutz and Laing 2002). In his 1990–96 study, Schmutz (2001) noted that Cackling Geese (*minima*) preferred different brood habitats, but, due to their greater abundance, they frequently occurred in habitats favored by Emperor Goose goslings. The effect of competition was readily apparent in 1994–96 measurements of the mean body mass of 6-week-old Emperor Goose goslings, which were 23.8% less in males and 23.4% less in females than in 1990, when densities of Cackling Geese (*minima*) and other geese were much lower. In 1990, 89 Emperor Goose nests, 98 Cackling Goose (*minima*) nests, and 56 Greater White-fronted Goose nests were located on the study area, compared with 1997 figures of 46 Emperor Goose nests, 339 Cackling Goose (*minima*) nests, and 93 Greater White-fronted Goose nests. The time goslings spent feeding increased during 1993–96, in comparison to their feeding time in 1985–86 on another study area on the Y-K delta, when Cackling Goose (*minima*) populations were 2–3 times greater, while the Emperor Goose population remained relatively stable (Schmutz 1993).

These effects of interspecific competition are significant, because gosling body mass at fledging is a crucial predictor of their subsequent survival.

A broader study at 3 locations across the Y-K delta from 1990 to 2004 supports the hypothesis that large-scale variations in the body mass of Emperor Goose goslings is correlated with interspecific competition for food (Lake et al. 2008). Densities of Cackling Geese (*minima*) over this period were 2–5 times those of Emperor Goose populations, which were relatively stable. Body mass of prefledging Emperor Geese was negatively correlated with the densities of both species combined and positively correlated with food availability, although grazing by geese removed ≥90% of the aboveground primary productivity that occurred on grazing areas during the brood-rearing period. Cackling Geese (*minima*) probably depress the growth of Emperor Goose goslings more from their sheer higher numbers than from any other competitive advantage, and this effect most likely occurs at a population level across the Y-K delta. Hence efforts to increase Emperor Goose populations should consider interspecific interactions from other goose species and subspecies, particularly the *minima* subspecies of Cackling Geese.

RECRUITMENT AND SURVIVAL

Class I broods of Emperor Geese on the Y-K delta averaged 3.9 goslings ($n = 73$); Class II broods, 3.9 ($n = 43$); and Class III, 3.3 ($n = 139$); the loss of young between hatching and flying was 21.9% (Mickelson 1975). From 1993 to 1996 along the Manokinak River on the Y-K delta, gosling survival to 30 days old varied from 33.2% in 1994 to 70.8% in 1995; daily survival was lowest among 0- to 5-day-old goslings (Schmutz et al. 2001). Large amounts of rainfall at the beginning of the brood-rearing period also caused increased mortality of goslings that were ≤5 days old. The number of juveniles in families during fall staging ($n = 23$ years) was negatively correlated with rainfall during the early brood-rearing period (0.085 fewer young/cm of rain). Fledgling survival until they moved to their first fall staging areas was directly correlated with prefledging body mass (Schmutz 1993).

Low gosling survival also occurred during years when Glaucous Gulls, a major predator, frequently disturbed Emperor Goose broods. Bowman et al. (2004) examined the stomach contents of 434 Glaucous Gulls collected over a 544 km² area on the Y-K delta during summer 1994 and extrapolated that this gull species ate 21,147–42,294 Emperor Goose goslings in the central Y-K delta, or a minimum of 47% of all hatchlings. Goslings are most vulnerable to predation by gulls when they become separated from their parents or when they are on land. On the water, goslings attempt to escape gull attacks by diving, while their parents ward off the aerial dives and swoops of the attackers. If the goslings remain near their parents and are old enough to dive completely underwater, the assaults by Glaucous Gulls are usually repulsed (Headley 1967).

Since 1985, the annual productivity of Emperor Geese has been assessed by photographing birds at fall staging areas on the Alaska Peninsula (Dau et al. 2006). Juveniles averaged 19.1% (range = 9.2%–35.2%) of the population from 1985 to 2006, and annual survival (Oct–Oct) for all age classes over that time period was 83%. The percentage of juveniles tallied in 2006 (38.0%) was exceeded in only 2 of the years since 1966, when age-ratio data were first collected on Emperor Geese: 1969 (41.8%) and 1977 (40.7%; Groves 2006).

Schmutz et al. (1994) determined survival of 1,115 adult and 1,578 juvenile neck-collared Emperor Geese marked on the Y-K delta from 1988 to 1990. Annual survival of adults was 63.1%, with no differences between adult males and adult females or among years (1988–91). Monthly survival of adults over winter (1 Oct–30 Apr) did not differ among years (94.0%), in contrast to monthly rates over summer (1 May–30 Sep), which were subsequently calculated at an overall rate of 98.0%. Monthly survival of juveniles also did not vary among years during their first overwinter period (71.0%), with their subsequent monthly survival similar to that of adults (94.3%). Survival rates were low in comparison to other goose species.

Annual survival of adult females trapped at or within 14 days of the hatch on the Y-K delta was also low, averaging 58.7% (range = 43.9%–67.5%) over 4 years (1983–86; Petersen 1992c). Average survival was lower (23.0%) for geese that nested the previous summer, in comparison to those that did not nest (50.0%), which suggested that harvest on the breeding grounds was the most likely cause of the disparity. Similarly, Hupp et al. (2008a) determined that the annual survival of 133 radio-marked adult female Emperor Geese on the Y-K delta was 79%–85%. Monthly survival was lowest (95%–98%) in May and August, when 44%–47% of annual mortality occurred. Monthly survival was higher (98%–100%) from September through March, when the birds were on staging or wintering areas, and in June and July, when they were nesting, brood-rearing, or molting. The high mortality in May and August corresponded to periods when subsistence harvest was probably highest. Schmutz et al. (1994) had also found that annual survival rates of adult Emperor Geese on the Y-K delta during 1988–92 (63.1%) were not different from those observed by Petersen (1992c), which suggested that continued subsistence hunting contributed to persistent low populations of Emperor Geese, despite legal agreements to end subsistence hunting in 1985.

The mean estimated life span for adult Emperor Geese (from banding data) is minimally 2.2 years (Petersen et al. 1994). The longevity record is 12 years, for an Emperor Goose banded in Alaska as an adult and later found dead. A neck-collared Emperor Goose in Alaska, however, has been observed for at least 13 years (J. A. Schmutz, U.S. Geological Survey).

FOOD HABITS AND FEEDING ECOLOGY

The Emperor Goose is a maritime species. These geese spend most of their annual cycle in close proximity to the ocean, where foraging activities, behaviors, and food choices are influenced by tidal cycles. Spring- and fall-staging Emperor Geese at Nelson Lagoon on the Alaska Peninsula fed on intertidal invertebrates from 3 hours before to 3 hours after low tide and then roosted on the adjacent beach. Most (90%–100%) of these geese fed at or near low tide (Peterson 1983). Foraging behavior in the intertidal zone involves dipping the head into water <30 cm deep or walking in shallow water and feeding with only the bill in the water. Emperor Geese also feed by grazing, grubbing for roots, and puddling, where they use their feet to create pools in soft sediments and then eat the exposed bivalves (Petersen et al. 1994, Schmutz 1994).

Breeding. A study by Cottam and Knappen (1939) of the foods present in the gizzards of 33 Emperor Geese collected largely during the summer found that algae (primarily sea lettuce) were the most important, then sedges and grasses, followed closely by eelgrass. A variety of animal foods, chiefly mollusks and crustaceans, formed 8.4% of the gizzard contents. Both adults and goslings on the Y-K delta grazed extensively on the aboveground portions of *Carex subspathacea*, but they also used *C. ramenskii* (Schmutz and Laing 2002). Laing and Raveling (1993) reported that food consumed by adults and their young during the brood-rearing period included *Carex subspathacea*, *C. ramenskii*, *Puccinellia phryganodes*, and *Elymus arenarius*. Adults with broods abandoned stands of *Triglochin palustris* within 1 week after hatching, despite a preference for *Triglochin* in feeding trials, and moved to saltmarsh habitats for the remainder of the brood-rearing period. Captive goslings selected *Carex subspathacea* (35% availability) over *Puccinellia phryganodes* (65% availability) in 1 feeding trial (76%–81% vs. 19%–24%) but not another (1%–48% vs. 52%–99%), when the availability of *Puccinellia* (84%) was much higher than that of *Carex* (16%). Molting Emperor Geese fed primarily on grasses (*Dupontia* spp.) and sedges (Kistchinski 1971).

The caloric value of blue mussels (whole wet weight) at Nelson Lagoon averaged 0.054 kcal/g. Protein averaged 5.27%; fat, 7.13%; and carbohydrates, 8.77% (Petersen et al. 1994). Macoma clams averaged 1.52 kcal/g, with protein averaging 11.72%; fat, 1.93%; and carbohydrates, 0.36%. Laing and Raveling (1993) listed the nitrogen content for various plant foods used by Emperor Geese on the Y-K delta: *Triglochin palustris* (4.7%), *Puccinellia phryganodes* (2.5%), *Elymus arenarius* (1.7%), *Carex ramenskii* (2.4%), and *C. subspathacea* (3.1%).

Migration and Winter. There are no major studies of Emperor Goose food habits on staging areas or wintering grounds, although they are comparatively unique among geese in their heavy reliance on animal foods to meet their winter energetic requirements (Schmutz 1994). On the Alaska Peninsula, Emperor Geese at Nelson Lagoon during both spring and fall probably fed on blue mussels and macoma clams, which were the dominant invertebrates within the intertidal zone (Petersen 1983). In contrast, eelgrass (*Zostera marina*) formed the entire contents of 17 crops from Emperor Geese collected in October and November 1966 at nearby Izembek Lagoon (Headley 1967). Crops and gizzards from Emperor Geese collected in 1966 on Adak Island almost exclusively contained sea lettuce (*Ulva* sp.). At Amchitka Island, Kenyon (1961) observed Emperor Geese feeding among rocks and kelp beds at low tide, as well as among kelp heaps washed ashore.

In 1991, Schmutz (1994) studied the fall feeding ecology of Emperor Geese at 3 of the most heavily used staging areas on the Alaska Peninsula: Cinder Lagoon, Port Heiden Bay, and Nelson Lagoon. Most birds fed at low tide and roosted during high tide; feeding intensity was highest when the geese were in beds of blue mussels, compared with mud/sand or vegetated habitats. In all habitats, juveniles spent 6.4% more time feeding than adults; in mussel beds, juveniles fed 77% of the time, compared with 64% for adults. Blue mussels appear to be a significant food source for Emperor Geese, which probably select them because of their high protein content. Some small flocks, especially families with juveniles, also fed in upland habitats on beach pea (*Lathyrus maritimus*) and seabeach sandwort (*Honckenya maritima*) until these plants senesced in October; the geese then switched to blue mussels, which were still available during low tides (Petersen 1981).

MOLTS AND PLUMAGES

Adult Emperor Geese have only 1 plumage (basic) that is molted annually via the prebasic molt. The descriptions of molts and plumages are primarily from summaries in Palmer (1976a) and Petersen et al. (1994), which should be consulted for further details.

Gray feathers begin to replace natal down at about 3 weeks of age, with the complete juvenal plumage acquired in approximately 50 days. Birds in juvenal plumage have a medium gray head and neck; otherwise juvenile plumage is like basic. The molt associated with first basic involves all feathering except the juvenal wing; it begins soon after the juvenal plumage is complete (in early Sep) but is not completed until October–mid-December. In September the head and the back of the neck are medium gray flecked with white; the remaining body feathers are also medium gray, but with black subterminal bands and white margins. The exposed retrices are white; the bill is dark. The chin and the front neck are medium gray with a brownish cast, but they have some small white margins.

Juveniles, nonbreeders, and failed breeders begin the definitive basic molt in mid- to late June or early July. The wing feathers are replaced by late July or early August, but the body molt continues through late August and into September. Successful breeders molt on the breeding grounds; they begin to molt when their goslings are 15–20 days old. Some males will molt earlier, right after their goslings hatch. Nonbreeders and failed breeders molt 2–4 weeks earlier than breeders; some molt on the breeding grounds, but most will undergo a molt migration. The flightless period lasts about

28–35 days (Mickelson 1975, Palmer 1976a, Eisenhauer and Kirkpatrick 1977) and is completed by mid- to late August.

CONSERVATION AND MANAGEMENT

Emperor Goose populations have declined by >50% from their historical levels (Pacific Flyway Council 2006b). Illegal subsistence harvest has continued to contribute to low population numbers in Emperor Geese (Schmutz et al. 1994, Hupp et al. 2008a). This harvest, along with a low survival of goslings because of predation by Glaucous Gulls (Bowman et al. 2004), clearly are major factors that limit population recovery. The greater abundance of Cackling Geese (*B. h. minima*) on the Y-K delta breeding grounds affects the growth of Emperor Goose goslings, which can subsequently affect their survival (Schmutz 1993, Schmutz and Laing 2002). Declining numbers of Emperor Geese in the early 1980s, however, resulted in the institution of agreements in 1986 to reduce the subsistence harvest (Schmutz et al. 1994). Emperor Geese are especially vulnerable to hunting in the spring, when they fly low to the ground; they are more wary in the fall (Petersen et al. 1994). Also, because wintering Emperor Geese forage within the intertidal zones of the Aleutians, they may be more vulnerable to chronic oil pollution than waterfowl feeding farther out at sea or within interior wetlands (Byrd et al. 1992).

Management recommendations call for continued harvest management through the Yukon-Kuskokwim Delta Goose Management Plan, which requires the cessation of all hunting when the 3-year average from the spring population survey falls below 60,000 geese, and continued closure until the population exceeds 80,000. Other measures advocate persistent monitoring of subsistence harvests, along with outreach programs designed to increase awareness of Emperor Goose management issues (Pacific Flyway Council 2006b). Additional management recommendations include maintaining the spring and fall population surveys, developing a population model, and continuing the study of life-history parameters, especially during the winter.

Snow Goose (Lesser and Greater)

Chen caerulescens (Linnaeus 1758)

Left, Lesser Snow Goose adult, blue phase
Right, Lesser Snow Goose juvenile, blue phase

The Snow Goose is among the most well studied birds in North America, as well as one of the world's most abundant waterfowl species. Snow Geese breed in colonies on open tundra from Wrangel Island (off the northeastern coast of Russia), along the coast and on islands across arctic and subarctic Alaska and Canada, to the northwestern coast of Greenland. Two subspecies are recognized: the widely distributed Lesser Snow Goose (*Chen caerulescens caerulescens*) and the Greater Snow Goose (*C. c. atlantica*), whose breeding range is restricted to a few islands in the eastern High Arctic. Lesser Snow Geese primarily winter in the Central Valley in California and the Gulf Coast of Texas and Louisiana, but they also have increasingly moved inland, in response to agricultural changes. The Wrangel Island Population largely has shifted into the Fraser and Skagit River deltas in British Columbia. Greater Snow Geese winter along the mid-Atlantic coast from New Jersey to North Carolina. Lesser Snow Geese have 2 color phases (morphs)—white and blue—whereas Greater Snow Geese

Left, Greater Snow Goose adult
Right, Greater Snow Goose juvenile

Left, Lesser Snow Goose adult
Right, Lesser Snow Goose juvenile

are only white. Snow Geese are monogamous and mate for life, with family groups remaining together through the first fall and into the following summer before the yearlings leave the adults. Snow Geese are also voracious foragers, with a bill that is especially adapted for grubbing and extracting belowground plant tubers and rhizomes. Traditional methods of population estimation revealed that both subspecies increased dramatically since 1970 in response to expanding agricultural production in the United States, to totals of nearly 7 million Lesser Snow Geese and 1 million Greater Snow Geese. A newer approach to population size estimation (the Lincoln Estimator) revealed that the population of midcontinent Lesser Snow Geese may have exceeded 20 million adults and 5 million goslings from 1999 to 2006. Overabundance has caused extensive damage to arctic breeding habitats, which has prompted special harvest regulations designed to lower the population. *The Snow Geese of La Pérouse Bay*, by Fred Cooke and colleagues, is an excellent treatise on Lesser Snow Geese, reflective of long-term studies conducted on that population since 1968

Greater Snow Goose adult

Greater Snow Goose juvenile

Lesser Snow Goose adult, white phase

Lesser Snow Goose adult, blue phase

Lesser Snow Goose juvenile, white phase

Lesser Snow Goose juvenile, blue phase

IDENTIFICATION

There are 2 subspecies of Snow Geese in North America: the broadly distributed Lesser Snow Goose (*Chen caerulescens caerulescens*), and the Greater Snow Goose (*C. c. atlantica*), which is much more restricted (see Distribution). The 2 subspecies are differentiated by size and color

133

Left:
Lesser Snow Goose flock; note the blue color morph (*center*, with head up) and the grayish immature preening. *Dave Menke, U.S. Fish and Wildlife Service*

Below:
Blue-phase Lesser Snow Goose pair with their goslings; note the variation in the adults. *Tim J. Moser*

Right:
Closeup of the "grinning patch" on a Greater Snow Goose, which aids in extracting underground plant rhizomes and tubers.
Ian Gereg

Below:
Flightless Lesser Snow Geese being rounded up for banding on the Great Plain of the Koukdjuak on southwestern Baffin Island.
Tim J. Moser

phases, although there is sufficient overlap in size to potentially warrant dropping the subspecific designations (Cooke et al. 1995).

Lesser Snow Geese have 2 color phases, or morphs: a dark-plumaged "Blue Goose" and a white-plumaged "Snow Goose." The color morphs are controlled by a single gene locus (a position within a chromosome), with the allele (the genetic coding for a particular trait) for blue incompletely dominant to the white allele (Cooke and Cooch 1968). Greater Snow Geese are generally larger and almost all are white morphs; the blue phase is extremely rare. The location of the breeding grounds for blue-phase Lesser Snow Geese was a long-running mystery in North American ornithology until nesting colonies were first documented on the western coast of Baffin Island, south of the Koukdjuak River, through extensive searches by Dewey Soper in 1929 (Soper 1930). Evidence from the early 1600s (and perhaps as far back as the Pleistocene) suggested that the 2 color phases of Lesser Snow Geese were primarily separated on breeding and wintering areas, but they began to mix around 1930 (Cooke et al. 1995). Cooch (1961) presented unquestionable evidence that the blue and white morphs of Lesser Snow Geese were simply color phases of the same species. Nonetheless, the 2 color phases were considered separate species by the American Ornithologists' Union until 1983. Many authorities (e.g., Palmer 1976a, Johnsgard 1978, Livezey 1997), use the genus name *Anser* in preference to *Chen*.

At a Glance. White-phase Snow Geese are almost entirely white, except for their gray primary coverts and black primaries. Blue-phase Snow Geese have a nearly all-white head and neck, but the body is grayish brown. The upperwing coverts are grayish, with contrasting black primaries. In flight, the white bodies and inner wings of white-phase birds contrast sharply with the black primaries. Snow Geese fly in a peculiar but characteristic undulating fashion, with individual flock members flying at staggered heights, rising and descending slightly,

which gives rise to their alternative common name, "Wavie." In flight, the flocks form Us, oblique lines, Vs, and irregular groupings, but they are seldom arranged in the well-formed Vs of Canada Geese and Cackling Geese.

White-Phase Adults. White-phase birds are virtually white, except for the black primaries and grayish primary coverts. Their heads are often stained with orange rust that results from grubbing in substrates rich in ferrous (iron) oxides. The legs and the feet are rose red. The bill is rose pink, with a pale pink or white nail; the bill also has a distinctive blackish grinning patch, or smile, formed by a broad black area on both sides of the mandibles. The irises are dark brown. In the hand, the larger size, the longer bill, and the more wedge-shaped head of the Greater Snow Goose can distinguish it from its smaller counterpart.

Blue-Phase Adults. The soft parts (bill, feet, etc.) are similar to those of white-phase birds. The head and the upper neck are white, but otherwise most of the body is a rich grayish brown, with a dark brown mantle (the back and the top of the wings) that is tinged with blue. The rump is pale gray, with gray or white uppertail coverts that contrast with the dark gray tail. The rear part of the abdomen is usually paler gray and white, compared with the overall brownish-gray body plumage. The upperwings and underwings are gray, with contrasting black primaries. Intermediate white and blue morphs can occur.

White-Phase Immatures. Juvenile white-phase birds are sooty gray on the top of the head, the back of the neck, the back, and the wing coverts, but otherwise they are mostly white below; the wing tips are black, as in adults. The legs, the feet, and the bill are grayish brown, lighter than those of the blue morph.

Blue-Phase Immatures. Blue-morph juveniles are almost entirely brownish gray, but slightly lighter below, with medium gray wing coverts. They also have a white chin spot. The legs, the feet, and the bill are grayish brown to almost black. The irises are brown.

Goslings. White-morph goslings are greenish yellow with a gray tint; blue-morph goslings are generally dusky brown to sooty black, but they have a yellow chin patch (Nelson 1993).

Voice. Snow Geese are probably the most vociferous of all waterfowl, with the incessant clamor of large flocks capable of being heard from several kilometers away. The main calls are a single or 2-syllable *kowk* or *kow-luk* that resemble the shrill barking of a small dog when heard separately, but the sound is much more musical when heard from a flock (Ogilvie 1978). The alarm call is described as a penetrating *kaah-ahh.* Greater Snow Geese reportedly produce lower-pitched and more resonant calls than Lesser Snow Geese, but most listeners do not consider them distinguishable from the calls of Lesser Snow Geese.

Similar Species. White-phase Snow Geese differ from Ross's Geese in the smaller size and bill of the latter species, which also lacks the grinning patch. Immature Ross's Geese are almost as white as adults, versus the gray or gray-brownish coloration in immature Snow Geese (see the Ross's Goose account). Immature blue-phase birds are occasionally confused with Greater White-fronted Geese, but the former have dark feet and bills, in contrast to the pinkish bills and feet of the White-fronted Geese.

Weights and Measurements. On average, Greater Snow Geese are larger than Lesser Snow Geese, although body weights vary throughout the year (Ankney 1982) and geographically (Mowbray et al. 2000). Bellrose (1980) presented average body weights for Lesser Snow Geese: 2,749 g for adult males ($n = 534$), 2,495 g for adult females ($n = 483$), 2,182 g for immature males ($n = 888$), and 2,014 g for immature females ($n = 687$). Weights for Greater Snow Geese averaged 3,375 g for males ($n = 19$) and 2,776 g for females ($n = 12$). Lesage and Gauthier (1997) noted that body weights of adult

male Greater Snow Geese averaged 12% heavier than adult females. The wing length of Lesser Snow Geese averaged 737 mm for adult males ($n = 32$) and 711 mm for adult females ($n = 36$), whereas the wing length of Greater Snow Geese averaged 787 mm for adult males ($n = 20$) and 762 mm for adult females ($n = 20$). Bill lengths of Greater Snow Geese were 65.5 mm for adult males ($n = 23$) and 64.5 mm for adult females ($n = 23$; Heyland *in* Bellrose). For Lesser Snow Geese, bill lengths averaged 56.1 mm for adult males ($n = 32$) and 53.3 mm for adult females ($n = 36$; Trauger et al. 1971).

DISTRIBUTION

The breeding range of Snow Geese is quite extensive, reaching across the Arctic from Wrangel Island (off the coast of Russia) eastward to northwestern Greenland. During the mid-1800s, hundreds of thousands of Lesser Snow Geese bred along the arctic coast of mainland Russia, from the Lena River east to the northeastern coast of Siberia; they wintered in Japan (Takekawa et al. 1994). This population was thought to be extirpated by humans; however, up to 100–300 geese may still breed in low densities along the northern coastal mainland of Russia (Baranyuk 1999, Kerbes et al. 1999a). Snow Geese undertake extensive migrations from the Arctic to the mid-Atlantic coastal states, the lower Mississippi River valley, the Gulf Coast, the Central Valley of California, and northern Mexico. With the exception of extreme southern British Columbia, most winter south of Canada.

Lesser Snow Geese have been grouped into population units that are based on shared breeding and/or wintering grounds, which has led to some confusion of terms, especially where populations overlap in distribution with Ross's Geese. When these 2 taxa occur together, they are referred to as "white geese" or "light geese," with the latter designation more commonly used by biologists. For Lesser Snow Geese, 3 populations have been identified, based on their breeding ranges: the Wrangel Island Population, the Western Arctic Population,

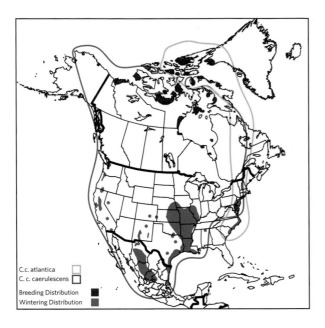

C.c. atlantica ☐
C. c. caerulescens ☐
Breeding Distribution ■
Wintering Distribution ■

and the Midcontinent Population. The Western Central Flyway Population is a fourth designation based solely on the winter distribution of Snow Geese in the western portion of the Central Flyway. Together, wintering Lesser Snow Geese and Ross's Geese from the Midcontinent and Western Central Flyway Populations are often referred to as Midcontinent light geese or Central/Mississippi Flyway light geese, although the Western Central Flyway population consisted of only about 6% of all Midcontinent Population light geese from 2001 through 2010. Generally, most North American goose biologists favor management of goose populations based on their breeding aggregations.

Greater Snow Geese, which breed on islands in the eastern Canadian Arctic and largely winter in the mid-Atlantic states of the Atlantic Flyway, are presented separately from the Lesser Snow Goose populations in this account. The breeding distribution of both Lesser Snow Geese and Greater Snow Geese has changed dramatically over the past half century, with new colonies becoming established and dynamic growth occurring in several older ones. Wintering areas and migration routes have also shifted in some populations as they respond to changing agricultural practices

and increasing numbers. Lesser Snow Geese of both color phases regularly appear together on Midcontinent Population breeding, migration, and wintering areas, but the white phase is most common elsewhere.

Greater Snow Goose
(*Chen caerulescens atlantica*)

Breeding. Greater Snow Geese breed in the High Arctic, from about 70° N latitude in the Foxe Basin northward to northern Baffin Island, then to Bylot, Somerset, Prince of Wales, Bathurst, Devon, Ellesmere, and Axel Heiberg Islands, and to the Thule district of northeastern Greenland; however, the bulk of the population occurs on Baffin and Bylot Islands (Reed et al. 1998a). During the mid-1990s, the breeding population was estimated at 157,000 birds within the core breeding colony on Bylot Island in Nunavut, which represented 15% of the total breeding population at that time (Reed et al. 1998a). Maximum densities of Greater Snow Geese nesting on Bylot Island have been reported at 1,400/km² (Lepage et al. 1996, Mowbray et al. 2000). This breeding density is estimated to be 46% of the short-term carrying capacity for the island (Masse et al. 2001). The Greater Snow Goose population has increased dramatically, from about 180,000 birds in 1980 to 800,000–1,000,000 by 2007. Recent expansion of their breeding range has also brought Greater Snow Geese in closer contact with some populations of eastern Lesser Snow Geese (Arctic Goose Joint Venture Technical Committee 2008).

Winter. Greater Snow Geese winter along the mid-Atlantic coast from New York to Georgia, but they are primarily found from the lower coast of New Jersey through Pamlico Sound in North Carolina, with a few reaching South Carolina. Most (about 90%) winter on bays, estuaries, and national wildlife refuges from southern New Jersey through the Chesapeake Bay area, where they exploit abundant waste corn in nearby agricultural fields. Especially important refuges are the Bombay Hook National

Wildlife Refuge (NWR) and the Prime Hook NWR in Delaware, the Edwin B. Forsythe NWR in New Jersey, and the Chincoteague NWR in Virginia. Up to 100,000 of these geese, however, have been reported within the Piedmont region of southeastern Pennsylvania (McWilliams and Brauning 2000), which reflects more birds wintering in the central and northern portions of their wintering range.

The Midwinter Survey in the Atlantic Flyway from 2000 to 2010 averaged 398,237 Snow Geese, most of which are presumed to be Greater Snow Geese: 37.9% in Delaware, 26.5% in New Jersey, 20.0% in Maryland, 8.3% in North Carolina, and 4.8% in Virginia.

Migration. Fall migration begins in late August and leads to 2 major staging areas. The first involves a flight of >1,000 km to the Ungava Peninsula, where they feed on berries (*Empetrum, Arctostaphylos, Vaccinium*) located on rocky heaths, and the basal portions of sedges growing in small wetlands (Reed et al. 1998a). From the Ungava Peninsula, the geese then fly another 1,000 km over the boreal forest to arrive on staging areas on the upper St. Lawrence River, as well as portions of the Richelieu River and Lake Champlain. Their principal staging area is along the St. Lawrence River, in the marshes near Cape Tourmente, about 65 km below Québec City. During the late 1980s, however, in both the spring and fall, Greater Snow Geese began to use new staging areas about 145 km upstream (around Lake Saint-Pierre) to take advantage of expanding agricultural areas in southeastern and north-central Ontario (Maisonneuve and Bédard 1993). The geese spend up to 5–7 weeks within the St. Lawrence estuary (Gauthier et al. 1984a) before heading southward across Vermont, particularly the Lake Champlain valley, to then arrive on mid-Atlantic coastal wintering areas.

During fall (1985–87) along the St. Lawrence River (between Québec City and Saint-Roch-des-Aulnaies), the first sightings of 2,150 neck-collared Greater Snow Geese occurred between 13 and 23 September; 50% had arrived by 3–13 October (Maisonneuve and Bédard 1992). Birds arrived earlier in 1986, a year of near-total breeding failure. Arrival dates did not differ between the sexes or among age categories. In 1985, however, solitary birds and individuals in established families arrived earlier (median date = 7–11 Oct) than pairs (15 Oct) or individuals in new families (12 Oct). On average, marked geese spent 15.5–19.1 days on staging areas along the St. Lawrence River; 11%–20% of the marked geese were not located on the St. Lawrence staging grounds, but were seen on the mid-Atlantic coast.

Habitat use along the St. Lawrence River has changed over the years as geese make more use of waste corn. Into the 1970s, Greater Snow Geese staged along the St. Lawrence between Île d'Orléans and Saint-Roch-des-Aulnaies, where they fed almost exclusively in bulrush (*Scirpus*) marshes, and then flew nonstop to Delaware Bay (Lemieux 1959a, Blokpoel et al. 1975). More recently, however, they have begun to disperse from this traditional stopover area in October and move southwest along the St. Lawrence River to Lake Saint-Pierre or northern Lake Champlain, where they feed on waste grain in nearby cornfields. Some birds remain in the area until well into November or even December before continuing south to the Delaware Bay area (Reed et al. 2008).

The spring migration route is essentially the reverse of the fall route. Historically, spring-migrant Greater Snow Geese flew nonstop from Delaware Bay to arrive at a traditional staging area of tidal marshes along the St. Lawrence River that extended about 80 km from Québec City to Saint-Roch-des-Aulnaies, where they fed almost exclusively on the rhizomes of common three-square bulrush (American bulrush; *Scirpus pungens*; Lemieux 1959a, Reed et al. 2008). Increasing agricultural activity is changing that pattern, however. The spring staging area now extends for >400 km along the St. Lawrence River, as well as along at least 3 of its main tributaries (especially the Richelieu River coming from Lake Champlain), and field-feeding on waste corn is much more common. Concentrations of up to 500,000 have been observed at the Baie-du-Febvre area along the southern shore of Lake Saint-Pierre, where the geese feed in nearby cornfields (Reed et al. 2008). Since the 1960s, the geese have also used *Spartina*-dominated marshes of the lower estuary. Departure from their wintering areas begins in February, with the geese then typically arriving on spring staging areas by early April and remaining there for about 6 weeks, during which time they accumulate fat reserves (Gauthier et al. 1984b). Fat reserves of birds departing for arctic breeding grounds around 18–20 May averaged 19%–20% of their body weight. Males from *Scirpus*-dominated marshes carried 23% more fat than those from *Spartina*-dominated marshes, while females carried 9% more.

Lesser Snow Goose
(*Chen caerulescens caerulescens*)

WRANGEL ISLAND POPULATION

Breeding. The Wrangel Island Population (WIP) of Lesser Snow Geese breeds on Wrangel Island, an 800,000 ha island in the Arctic Ocean off the coast of northeastern Siberia. Since 1976 the entire island has been protected as a state nature preserve administered by Russia's Ministry of Natural Resources. The principal nesting colony covers about 2,600 ha and is located on the westcentral part of the island, along the middle reaches of the Tundra River, where the Snow Geese often nest in close association with Snowy Owls (*Bubo scandiacus*) during years of high lemming abundance (Baranyuk 1999, 2000). The population fluctuates in response to weather, with the first systematic spring surveys tallying 150,000 in 1970, but only 50,000 in the mid-1970s, due to several consecutive years of breeding failures (Pacific Flyway Council 2006c). The WIP increased to about 100,000 birds in the late 1980s, declined to approximately 65,000 during the early 1990s, then increased to almost 120,000 by 2004–5, and to 140,000–150,000 in 2008.

Winter. The WIP breeds in a single colony; they then separate into 2 subpopulations that migrate to 2 major wintering areas: the Fraser and Skagit River deltas on the British Columbia/Washington border, and the Central Valley in California, although a small number (about 1,600) winter along the lower Columbia River in Washington and Oregon. In the 1960s, most (78%–90%) of the population wintered in California (Hines et al. 1999), but the percentage wintering farther north has increased steadily since then, to about 60% in the Fraser River/Skagit Bay area and 40% in the Central Valley in 2000 (Jeffrey and Kaiser 1979, Boyd and Cooke 2000). Studies of neck-banded geese indicated strong fidelity to these 2 main wintering areas, with <3% changing wintering areas from 1 year to the next (Williams et al. 2008). Historically, wintering Snow Geese in the Fraser/Skagit River deltas exhibited red facial staining, due to the high iron content of intertidal areas used for foraging, while birds in California did not (Baranyuk et al. 1999). This situation has changed in recent years, however, as Fraser/Skagit geese feed more in upland habitats (Boyd and Cooke 2000). Most recently, Williams et al. (2008) noted that wintering birds with significant levels of face staining had a high (>88%) likelihood of wintering in British Columbia and Washington, and geese with lower levels of face staining wintered in California (≥90%), but geese with intermediate staining were not consistently associated with either area.

Migration. Lesser Snow Geese on Wrangel Island undertake some of the longest migrations of any goose population in North America. Fall departure begins in late August, with most (>90%) birds traveling along coastal routes (Armstrong et al. 1999). The first major stopovers are on the Chukotka [Chukchi] Peninsula on the Siberian mainland, where they gather at Cape Billings and Koluthin Bay (Baranyuk and Takekawa 1998). From there, they cross the Bering Sea, St. Lawrence Island, and the Seward Peninsula to stage within the Yukon-Kuskokwim delta and northern portions of the Alaska Panhandle. They then continue along the coastline of southeastern Alaska and British Co-

Lesser Snow Geese in the Central Valley of California. *Clair de Beauvoir*

lumbia to the Fraser / Skagit Bay area on the border of British Columbia and Washington, which they first reach in late September, building in numbers through October and early November (Armstrong et al. 1999, Hines et al. 1999). About 80% of the WIP wintering in California will stage in the Fraser/Skagit area before proceeding down the coast to the Summer Lake / Klamath basin region of Oregon and California, and on into the Central Valley in California (Armstrong et al. 1999). A second minor migration route used by <10% of the WIP extends eastward to join the Western Arctic Population, where the geese stop on the Arctic Coastal Plain before continuing on to a major staging area in Alberta/Saskatchewan, and then south to California.

Spring migration from the Fraser/Skagit area begins in February and March, with the WIP generally following the fall migration routes in reverse, except for the California population. Major staging areas are in Alaska and include the Stikine River delta in southeastern Alaska and then upper Cook Inlet in southcentral Alaska, where the geese arrive in late April (Pacific Flyway Council 2006c). They also stage on the lower Yukon River in early May, from which they travel to mainland Siberia and then on to Wrangel Island, beginning to arrive in late May; most reach the area by early June. In mid-April, >74% of the WIP wintering in the Central Valley in California follow an inland route, staging on Freezeout Lake in Montana and in the Canadian prairies.

WESTERN ARCTIC POPULATION

Breeding. Most (>95%) of the Western Arctic Population (WAP) of Lesser Snow Geese breeds on southwestern Banks Island in the Northwest Territories, where a combination of 396,184 Snow

Geese and Ross's Geese were surveyed during 2010 (Groves and Mallek 2011). WAP geese also occur in 4 smaller mainland colonies: 2 located on the North Slope of Alaska and 2 farther eastward in the Anderson and Mackenzie River deltas of the Northwest Territories. Snow Geese from the WAP breeding population are primarily (99%) composed of white morphs.

A large nesting colony is located inland along the Egg River on Banks Island. This colony has increased nearly 2.5-fold since 1973, expanding from 200,000 breeding birds to 479,000 in 1995 (Kerbes 1983, Kerbes et al. 1999a). The number nesting at the Anderson River colony in the Northwest Territories declined from a peak of 8,360 in 1981 to only 1,200 in 2000–2001, possibly due to substantial egg predation from grizzly bears (*Ursus horribilis*). About 240 km west, approximately 2,500 nesting birds are found on Kendall Island, which is part of the Mackenzie delta (Wiebe Robertson and Hines 2006). Small numbers nest on the Arctic Coastal Plain and adjacent barrier islands of Alaska, where aerial surveys estimated an average of 897 birds from 1999 to 2009 (Dau and Bollinger 2009). About 1,000 occur on Howe Island, which is located just offshore from the Sagavanirktok River delta, east of the Prudhoe Bay oil fields (Johnson 1996).

Winter. WAP geese are widely distributed on their wintering grounds. Historically, they mostly wintered in the Central Valley in California, but a gradual shift eastward has occurred over the past several decades (Arctic Goose Joint Venture Technical Committee 2008). About 75% of the WAP geese now winter in northern and central California, with smaller numbers in southern California, New Mexico, northwestern Texas, the west coast of northern Mexico, and Mexico's interior highlands (Kerbes et al. 1999). Due to the fact that the vast majority of western Arctic breeders winter with Wrangel Island geese in northern California, they are grouped together in the same winter population.

Drewien et al. (2003) made a concerted effort to inventory the light geese (Snow Geese and Ross's Geese) wintering within the interior highlands of Mexico. During 1998 and 1999 they recorded 229,288 and 310,204 birds, respectively, of which 20% were Ross's Geese. The majority of those wintering in Mexico were found within the northern highlands (the states of Durango and Chihuahua) and along the east coast south to Veracruz. Fewer than 300 Lesser Snow Geese winter along the west coast of Mexico, but 1,000–1,500 occur on Lago de Chapala near Guadalajara, Mexico. Occasional vagrants reach Belize and Honduras (Howell and Webb 1995).

Migration. Prior to initiating their fall migration southward, WAP geese gather along the Beaufort Sea coastal plain, where the Arctic NWR provides an important staging area (Robertson et al. 1997). From 13,000 to 310,000 geese stage off the coast of northeastern Alaska and northwestern Yukon, where they remain for 2–4 weeks from late August to mid-September (Oates et al. 1987, Johnson 1996, Hupp et al. 2002). A small part of the WIP and birds from the North Slope of Alaska join other breeders from Banks Island, Liverpool Bay, and the Mackenzie River delta on a nonstop flight of about 1,900 km to stage in southeastern Alberta and southwestern Saskatchewan, where goose numbers reach 800,000 (Johnson and Herter 1989, Johnson 1996). Neck-banding studies (1987–92) revealed that 80% then depart the Alberta/Saskatchewan staging area and travel south-southwest to winter in the Central Valley in California (Armstrong et al. 1999). Large numbers stop at various staging areas, including Freezeout Lake in Montana, Summer Lake in Oregon, and the Klamath basin on the Oregon/California border.

Smaller numbers travel south to the marshes of the Great Salt Lake / Bear Lake region of Utah and then beyond, heading south-southwest to the Imperial Valley in California. The remainder travel due south through Colorado to Bosque del

Apache NWR in New Mexico, the Playa Lakes Region of western Texas, and the interior highlands of Mexico. A study of neck-collared and leg-banded WAP geese by Hines et al. (1999) provides a detailed account of their migration from the western Arctic through the Rocky Mountain states of Montana, Wyoming, Utah, and Colorado to winter either in California or farther south in New Mexico, southwestern Texas, or the interior highlands of Mexico. The WAP largely reverses these routes during spring migration.

Neck-banding studies (1987–92) found that WAP geese arrived on their main fall staging areas in Alberta and Saskatchewan by about mid-September and remained there for more than 6 weeks (Armstrong et al. 1999). From these areas, some geese moved to eastern Oregon and western Montana by late October, staying there until mid-November. They then migrated to wintering areas in the Klamath basin and the Central Valley, where they remained an average of 144 days. Most (83%) of the Snow Geese wintering farther south in the Imperial Valley migrated directly from Alberta, Saskatchewan and Montana. Bellrose (1980) noted that a few Snow Geese reach the Klamath basin by early October, but peak numbers do not occur until mid-November, followed by a gradual exodus through December to wintering grounds in the Central Valley.

MIDCONTINENT POPULATION

Breeding. The Midcontinent Population (MP) is the largest goose population in North America. It basically consists of those Lesser Snow Geese from central and eastern Arctic breeding areas, mainly east of 110° W longitude (Alisauskas et al. 2012). Most (90%) of the MP nest north of 60° N latitude in the Canadian Arctic of Nunavut, with major colonies in the Queen Maud Gulf region and on islands surrounding the Foxe Basin (Southampton and Baffin Islands); other colonies occur on Jenny Lind Island and in the Rasmussen Basin (Kerbes et al. 2006). The remainder (10%) of the MP nests south of 60° N latitude, mainly at colonies along Hudson Bay, both on the southern coast (Cape Henrietta Maria in Ontario) and the western coast (McConnell River in Nunavut and La Pérouse Bay in Manitoba). Some older designations had divided the MP into a Central Arctic Population and an Eastern Arctic, or Hudson Bay, Population, but these groupings have now been abandoned, based on the shared wintering grounds of birds from those 2 regions and an exchange of birds among the colonies. North American waterfowl managers traditionally delineated the MP according to the distribution of geese during winter in the Central and Mississippi Flyways, but this group now mixes with birds from the WAP that are shifting eastward during the winter. Hence population delineation based on breeding areas is more biologically meaningful than that determined by wintering areas (Alisauskas et al. 2012).

The largest colonies within the MP are found in the Queen Maud Gulf region of Nunavut and primarily (80.6%) consist of white morphs, based on 1998 photographic counts (Kerbes et al. 2006). Within that region, most colonies occur along the coast, where surveys in 1976 estimated nearly 56,000 nesting Lesser Snow Geese spread across 30 colonies, which increased to 280,000 geese in 57 colonies in 1988. By 1998, a photographic survey estimated a population of >800,000 spread over 80 colonies (Kerbes et al. 2006, Canadian Wildlife Service Waterfowl Committee 2010). The large colony at Karrak Lake contained 452,000 nesting Lesser Snow Geese in 2010, up from 346,000 in 2008 (U.S. Fish and Wildlife Service 2011). A rare inland colony was discovered at the Lower Garry and Pelly Lakes area, >80 km southeast of the nearest colony in the Queen Maude Gulf Bird Sanctuary (McCormick and Arner 1987).

To the east of Queen Maude Gulf, MP Lesser Snow Geese breed in sizeable colonies on Baffin and Southampton Islands, western Hudson Bay, southern Hudson Bay, and Akimiski Island in James Bay. Kerbes et al. (2006) noted the average percentage of blue-phase morphs at these colonies in 1997: 76% on Baffin Island, 32% on Southampton Island, 25% on western Hudson Bay, 53% on southern Hudson Bay, and 75% on Akimiski Island. Alisauskas et al. (2012) reported results from aerial photographs of these colonies, with estimates of 652,546 geese nesting on Southampton Island (2004), 1,318,560 on Southampton Island and Foxe Basin (2005), 261,080 on western Hudson Bay (2003), and 307,100 on southern Hudson Bay (2005).

Kerbes (1975) conducted the first surveys of the eastern Arctic in 1973 and noted MP geese nesting at only 5 widely separated areas, with the largest colonies located on the Great Plain of the Koukdjuak on southwestern Baffin Island (446,600 nesting birds), and on Southampton Island (155,800), primarily in 3 colonies along the Boas River. Since that time, however, this population has expanded both spatially and numerically, with 1.7 million nesting birds on Baffin Island and 0.7 million in Southampton Island by 1997, with similar numbers in 2004–5, although the Baffin Island population may have declined slightly (Canadian Wildlife Service Waterfowl Committee 2010). Both Lesser Snow Geese and Greater Snow Geese nest on Baffin Island, but Lesser Snow Geese principally nest south of 68° N latitude (Gaston et al. 1986).

On the western coast of Hudson Bay, a dynamic colony is centered on the McConnell River delta. The local Inuit gave 1910 as the probable date for the establishment of this colony (Kerbes 1982), which numbered 14,000 geese when first observed by Angus Gavin in 1941 (Cooch 1963), 35,000 by 1961, and at least 100,000 by 1969 (Hanson et al. 1972). When photographed from the air in June 1973, the population had reached a high of 326,000 birds and extended between the Tha-Anne and Maguse Rivers (Kerbes 1975). In 1985, the colony was down to 260,000 birds (MacInnes and Kerbes 1988), and only 154,000 were surveyed by helicopter in 1997. This decrease most likely was caused by habitat destruction by large numbers of geese feeding on the limited amount of tundra vegetation. The nesting population increased slightly between 1997 and 2003, mostly north of

the McConnell River area, and especially to the north of Arviat, Nunavut (Canadian Wildlife Service Waterfowl Committee 2010).

Along the western coast of Hudson Bay, the MP colony located at La Pérouse Bay near Cape Churchill, Manitoba, is the best known and most well-studied colony, with long-term research conducted continuously there since 1968. This site was first occupied by nesting Lesser Snow Geese in the early 1950s (about 2,000 pairs), with consistent use since 1963 (Cooke et al. 1995). This colony expanded dramatically (8%/year) from the mid-1960s through the early 1980s, and numbered nearly 50,000 birds (22,500 pairs) by 1990 (Cooke et al. 1995). There were 41,700 pairs in 1997 and 41,800 pairs in 2006; 2 small colonies near Thompson Point contained 7,100 pairs (Canadian Wildlife Service Waterfowl Committee 2010). Nesting-pair surveys of colonies on nearby West Pen Island and Shell Brook in 1997 yielded 8,500 pairs on West Pen and 2,700 on Shell Brook.

In the Hudson Bay Lowlands, MP geese nest at Cape Henrietta Maria, which juts out into Hudson Bay where it meets the western edge of James Bay. This colony was first discovered in 1947, when a few birds were photographed from the air (Hanson et al. 1972). The number of geese in the colony was subsequently estimated at 17,300 in 1957, 40,000 in 1968, and 79,000 (61,500 nesting) in 1973 (Bellrose 1980). Helicopter surveys in 1997 revealed a breeding population of 430,000 birds nesting along southern Hudson Bay, 320,000 of which were from Cape Henrietta Maria. This colony subsequently declined to an estimated 129,000 nesting pairs in 2001 and 128,000 in 2003, with similar numbers in 2005 (Canadian Wildlife Service Waterfowl Committee 2010).

Intermittent breeding developed into a permanent small colony on Akimiski Island in western James Bay in 1958. The colony averaged around 900 pairs between 1998 and 2000, 1,500 in 2001, and about the same number in 2003 (Abraham et al. 1999b, Canadian Wildlife Service Waterfowl Committee 2010). Other small colonies occur at scattered locations that were long distances from the main colonies; McLaren and McLaren (1982b) counted a high of 543 Lesser Snow Goose nests within the Rasmussen Lowlands of northern Keewatin in Nunavut.

Winter. The MP predominantly winters in Missouri, Arkansas, eastern Texas, and coastal Louisiana, with lesser numbers at scattered locations along the western Mississippi Flyway, eastern Central Flyway, and northeastern Mexico. The number of light geese (Snow Geese and Ross's Geese) indexed in the Midwinter Survey from 2000 to 2010 averaged 1,692,000 birds in the Mississippi Flyway and 963,800 in the Central Flyway, for a total of 2,655,800. Of that total, 22.5% occurred in Texas, 22.4% in Louisiana, 18.3% in Missouri, 14.6% in Arkansas, 8.2% in Kansas, and 4.4% in Mississippi. Smaller numbers of light geese winter at scattered locations in other states of the Mississippi and Central Flyways, extending as far north as South Dakota (3,300) and Illinois (50,000).

Along the Gulf Coast of Texas and Louisiana, most wintering MP Lesser Snow Geese are found close to coastal estuaries and within the rice prairies. The percentage of blue-phase versus white-phase geese on the wintering grounds differs greatly, with a much larger percentage of blue-phase geese found on the eastern portion. The majority of those wintering in southeastern Texas and into northeastern Mexico are white-phase geese (Cooke et al. 1995). Blue-phase geese are also more common within inland areas, as opposed to coastal sites.

About 10,000 MP Lesser Snow Geese from breeding areas around Hudson Bay and the Foxe Basin winter primarily along the Atlantic coast, from New Jersey to North Carolina, with smaller numbers wintering farther south in South Carolina, Georgia, and Florida (Cooke et al. 1995). In places where they overlap with Greater Snow Geese, it is virtually impossible to differentiate the 2 subspecies. Greater Snow Geese, however, often prefer to feed in marine habitats, natural marshes, and estuaries, as opposed to the agricultural habitats favored by Lesser Snow Geese.

Migration. Historically, at least some of the MP probably flew nonstop between breeding areas on Hudson and James Bay to the coastal marshes of Texas and Louisiana (Jefferies et al. 2003 has an excellent review). Changes in agricultural patterns and increasing crop yields since the 1970s, however, now allow Snow Geese (and other North American species of geese) to migrate in more of a stepping-stone pattern both northward and southward. The MP once wintered almost exclusively in the Gulf Coast marshes, with birds rarely seen >13 km from the coast. Once rice agriculture took hold along the Gulf Coast, Snow Geese began to feed in rice fields as early as the 1920s in Texas and the 1940s in Louisiana, but they did not make extensive use of these habitats until the late 1950s and early 1960s. By the 1970s, Snow Geese began to winter in the rice-growing areas of Arkansas, with wintering numbers increasingly rapidly during the 1980s. To the west, the MP began to use waste grain, especially corn, in the Missouri River valley (southwestern Iowa, southeastern Nebraska, northeastern Kansas, and northwestern Missouri) since the 1940s, with their numbers rapidly expanding since the early 1950s.

Today, MP geese from the Queen Maud Gulf region initiate their fall migration by fanning out southwest and south to agricultural lands stretching across southern Canada from eastern Alberta to western Manitoba. The majority of geese from the central Arctic travel south-southwest to staging areas in southern Saskatchewan and Manitoba, and then follow the eastern portion of the Central Flyway. These birds join with eastern Arctic breeders in southwestern Manitoba and geese at the Des Lacs NWR and Devils Lake in North Dakota; continue south to various national wildlife refuges (Sand Lake NWR in South Dakota, DeSoto NWR in Iowa, and Squaw Creek NWR in Missouri); and finally head south to coastal eastern Texas and western Louisiana (Bellrose 1980). Smaller numbers travel down the western part of the Central Flyway, where they join with some of the WAP on a southerly route through Montana, Wyoming, Colorado, New Mexico, western Texas, and into the interior highlands of Mexico. Alisauskas et al. (2012) provide maps of band-recovery distributions (1989–2002) for MP geese banded on various breeding colonies: the Rasmussen Lowlands (10 recoveries), Queen Maude Gulf (1,561), Baffin Island (650), Southampton Island (118), La Pérouse Bay (3,365), western Hudson Bay (515), Cape Henrietta Maria (406), and Akimiski Island (403). The maps basically show the MP staging in southern Saskatchewan and southern Manitoba before moving southward through the central United States and on to wintering areas.

MP geese from breeding areas on southern and southwestern Baffin Island, the south coast of Hudson Bay, and James Bay stage in large numbers during the fall at the southern end of James Bay (Bellrose 1980). From James Bay, they migrate south-southwest through Ontario, Michigan, Wisconsin, and Minnesota and join birds from the central and western Arctic at major staging areas in southern Manitoba, Saskatchewan, and the Dakotas. A smaller percentage crosses western Wisconsin in a southerly to southwesterly direction through eastern Iowa to the Swan Lake NWR and nearby areas in northcentral Missouri. As winter weather dictates, many of these birds move south into southern Missouri and Arkansas. Large numbers are also found at other major stopovers during migration: the southern end of Lake Manitoba, Arrowwood NWR in North Dakota, Sand Lake NWR in South Dakota, and DeSoto NWR in Nebraska and Iowa.

Substantial numbers depart La Pérouse Bay by mid- to late August, but most remain along the coast of Hudson and James Bays during September; some have moved on to southern Manitoba and North Dakota (Cooke et al. 1995). By October, most are concentrated in southern Manitoba and the eastern Dakotas, after which they continue southward through November and on into the Missouri River valley in Iowa, Nebraska, Kansas, and Missouri. A major flight then occurs directly to the Gulf Coast (about 800 km), arriving by December, although increasing numbers now also winter in Oklahoma and Arkansas.

There is an apparent difference between fall and spring migration routes within the Mississippi Flyway, where an increasing number of MP geese migrate north from their Mississippi and eastern Arkansas wintering grounds through southern and central Illinois. Here they change their direction to west-northwest, to join those migrating up the Missouri River valley in northwestern Missouri and southwestern Iowa. Spring migrants make fewer stops on their journey northward, with most flying directly from the Dakotas and southern Manitoba to their nesting grounds (Cooke et al. 1995). MP geese wintering on the Gulf Coast begin their spring migration in late February and early March, with peak arrival in Arkansas and Tennessee by mid- to late March, Missouri and Iowa by late March and early April, and the Dakotas and southern Manitoba (their final staging area) by 1–15 April. They depart about a month later to arrive at breeding areas on Hudson and James Bays by early to mid-May (Cooke et al. 1995, Mowbray et al. 2000).

MP geese have altered their migration routes in response to changing agricultural patterns. The spring migration route taken from central Illinois to Hudson and James Bays (about 2,400 km) is far from the most direct course, which is roughly 1,300 km. The longer route, taken by 300,000–400,000 Snow Geese (1995 to 2003), is thought to have developed to take advantage of corn and grain fields along the Missouri, Red, and James Rivers, which provide greater food resources than are available in Wisconsin, Michigan, and Ontario. Since spring comes earlier to the plains than to the east, over the decades this segment of the MP has evolved a migration pattern that potentially provides the greatest nutrition prior to the breeding season (Frank Bellrose).

From James Bay, millions of MP geese use the

wetlands along the coast of Hudson Bay during the spring for resting and foraging before dispersal to inland habitats, which are initially covered by ice and snow (Mellor and Rockwell 2006). In May 1994, Abraham and Miyaski (1994) estimated 179,300 Snow Geese on the western side of James Bay and 98,200 on the southwestern coast of Hudson Bay. These data suggest that MP geese headed for breeding grounds on Southampton and Baffin Islands use the western side of James and Hudson Bays. Here numerous rivers drain greater areas of land, bringing more nutrients to the tidal deltas, which provide needed food resources as this tremendous population progresses northward to their various breeding grounds.

WESTERN CENTRAL FLYWAY POPULATION

The Arctic Goose Joint Venture also recognizes a Western Central Flyway Population (WCFP) of Lesser Snow Geese and Ross's Geese, which, unlike the other population units, is defined solely on its winter distribution. This population mainly consists of Lesser Snow Geese from the WAP (75%), with smaller numbers from the MP (15%), and Ross's Geese from the central Canadian Arctic (10%). More of the WAP now winter in the western Central Flyway than they did historically, concomitant with a decrease in the percentage of the population wintering in the Pacific Flyway. In general, the WCFP nests from Banks Island in the western Canadian Arctic to the Queen Maud Gulf region, where they overlap with the MP.

MIGRATION BEHAVIOR

Snow Geese migrate both during the day and at night in large flocks that often number in the thousands. Flocks include individuals and family units, with the latter remaining together during both fall and spring migration the first year after successful nesting. Traditional staging areas are used along migration routes, interspersed between long-distance flights. Concentrations at night roosts often reach 100,000 birds or more. Migration usually begins in the early daylight hours or from dusk into the early nocturnal hours, but it then continues both night and day until the desired destination is reached. The flight speed of 23 different flocks was 64–80 kph (Bellrose 1980). Radar data often recorded migrating flocks at altitudes of 1,500–3,000 m, but their usual flying height is 600–900 m. Pilots occasionally observe Snow Geese at 3,350–3,660 m, with an exceptional report at 6,100 m.

Departures from wintering areas have been closely associated with maximum daily temperatures. In Texas (1972–77), Lesser Snow Geese initiated migration 3–5 days after maximum daily temperatures exceeded 28°C and thereafter remained at or above 18°C for at least 4 days (Flickinger 1981). Departures were not strongly correlated with minimum temperatures, surface wind, atmospheric pressure, relative humidity, sky cover, or precipitation. From spring staging areas in North Dakota and southern Manitoba, major departures took place following the passage of frontal systems that brought southerly and southwesterly winds, after which the geese traveled day and night over the boreal forest to reach breeding areas in the Arctic (Blokpoel 1974, Blokpoel and Gauthier 1975). Cooke et al. (1995) found that higher mean April temperatures (1972–87) in Winnipeg, Manitoba, were correlated with an earlier first arrival of Lesser Snow Geese at La Pérouse Bay. Presumably the higher temperatures caused an earlier snowmelt and thus increased the availability of forage, which allowed the geese to acquire the reserves needed for an earlier departure.

Snow Geese undergo long migrations, especially those breeding on Wrangel Island. In the fall a neck-collared adult male from Wrangel Island traveled through British Columbia to winter in Durango, Mexico (6,741 km) and then returned in the spring through Saskatchewan (6,922 km), for a total migratory distance (shortest route) of nearly 14,000 km (Armstrong et al. 1999). More typically, those wintering in California's Central Valley undergo a round-trip migration of about 11,000 km. Such migratory feats represent the longest migration distances of any North American species of goose, with the exception of some Black Brant (*Branta bernicla nigricans*).

Alisauskas et al. (2011) analyzed recovery data for Lesser Snow Geese banded at several colonies in the central and eastern Arctic from 1989 to 2006, which showed that during fall migration, these geese moved through Canada at a rate of 1° of latitude every 4.8 days, or about 25 km/day. Movement through the United States was 1° of latitude every 4.3 days, or about 22 km/day. The average date of harvest for geese breeding north of 60° N latitude occurred 15 days later in Canada, and 8 days later in the United States, than harvest dates for geese from breeding areas farther south, which reflects large differences in the timing of migration relative to breeding location.

In general, nonbreeding Snow Geese begin to depart their breeding areas in mid- to late August, with adults and their young following in early September (Bellrose 1980). Snow Geese from La Pérouse Bay depart en masse by mid- to late August to staging sites along Hudson and James Bays. They arrive in southern James Bay around mid-September, with peak populations gathered by mid-October. Large numbers remain until late October or early November, depending on the timing of freezeup. Small numbers of Lesser Snow Geese arrive on their major staging areas in southeastern Alberta and southwestern Saskatchewan in early September, with their numbers increasing rapidly until mid-October; most are gone by mid-November. Peak numbers reach Freezeout Lake in Montana by mid-November, with almost all departing by early December. The chronology of passage into the Bear River marshes of Utah parallels that at Freezeout Lake (Bellrose 1980).

During spring migration on staging areas in southcentral Alaska, Snow Geese selected areas with 10%–50% snow cover and avoided places with no snow cover (Hupp et al. 2001). The snow-free locales may have been less preferred by geese because available forage is quickly exploited once it is uncovered; moreover, the exposed soils become drier, making the extraction of belowground forage more difficult.

Molt Migration. Nonbreeding Snow Geese and some failed breeders undergo a molt migration during mid- to late June. From Southampton Island, Lesser Snow Geese make a molt migration to La Pérouse Bay (Abraham 1980). Failed and nonbreeding geese from the La Pérouse colony traveled northward, coinciding with the arrival of Canada Goose molt migrants from southern Manitoba (Cooke et al. 1995). Nonbreeders at Wrangel Island are thought to travel to the Chukotka [Chukchi] Peninsula in mainland Russia or to Alaska to molt. McLaren and McLaren (1982a) observed up to 4,800 molting Lesser Snow Geese within the Rasmussen Lowlands of Nunavut during the mid-1970s. Successful breeders remain with their broods, but they may molt up to 75 km from the breeding site (Kerbes et al. 1990).

On Bylot Island, a 5-year study (1997–2001) of 121 radiocollared female Greater Snow Geese revealed that 90% ($n = 51$) of the nonbreeders and 97% ($n = 29$) of the failed breeders left the island to undergo molting (Reed et al. 2003a). In contrast, only 2% of the successful nesters ($n = 41$) left the island to molt, and they probably had lost their goslings early in the brood-rearing period. The locale used by molt migrants, however, remains unknown. Hughes et al. (1994a) reported that on Bylot Island, molting Greater Snow Geese without goslings often formed large flocks and used pond and lake habitats almost exclusively, which probably reduced predation. These nonparents spent less time being alert than parents with goslings (3% vs. 20%), and they devoted more time to foraging (38%–71% vs. 27%–37%).

HABITAT

Breeding Snow Geese nest colonially on the flat terrain associated with subarctic and arctic tundra near the coast, although those nesting in the High Arctic can be found well inland, on more rolling terrain (Mowbray et al. 2000). Breeding colonies in the Queen Maude Gulf region often are located on islands. The timing of snow- and icemelt is an especially important factor in determining the availability of nesting habitat. Deltas of large rivers often provide the first available nesting locales, since these areas become snow free from mid- to late June. On Bylot Island, nesting Greater Snow Geese breed on moist upland and lowland tundra characterized by concave tundra polygons, with their associated wet meadows and ponds (Tremblay et al. 1997). Characteristic plant species are Arctic willow (prostrate willow; *Salix arctica*), bell-heather (*Cassiope tetragona*), northern woodrush (*Luzula confusa*), and various sedges and grasses.

Migrating Snow Geese use freshwater and brackish marshes, lakes, large impoundments, and agricultural fields. During spring migration at staging areas in southeastern South Dakota and southern Manitoba (1983–84), Lesser Snow Geese from the wintering MP roosted at night on water areas, from which they flew at dawn, primarily to nearby corn stubble but also to pasture, after which they returned to the water areas to rest, bathe, and drink (Alisauskas and Ankney 1992b). They went back to the agricultural fields in the afternoon, from which they departed after dusk and returned to the water areas, where they spent the night. Geese spent 44%–49% of their time on water areas, 15%–35% in corn stubble, and up to 8% in pasture (in southern Manitoba). Large numbers of Greater Snow Geese stage along the St. Lawrence estuary in both fall and spring, where they consume the rhizomes of American bulrush and forage on nearby waste grain in cornfields (Maisonneuve and Bédard 1992). Foraging Greater Snow Geese make more use of marine habitats, natural marshes, and estuaries than Lesser Snow Geese, which feed heavily on agricultural crops, such as corn and rice.

During winter (Oct–Apr 1978–80) in the rice prairie region of southeastern Texas, aerial observations of 1.6 million wintering Lesser Snow Geese found that a majority (59.6%) of the birds were in rice fields, which occupied only 14% of the study area (Hobaugh 1984). The geese exclusively made use of rice fields as long as rice was readily available (Oct–Nov), began to shift to soybean fields by December as the amount of rice diminished, and then moved to sprouting green vegetation in the rice stubble and in plowed fields once the rice was depleted (mid-Jan). To the north, in the middle Missouri River valley of Iowa, Kansas, and Missouri, wintering Lesser Snow Geese spent 75% of their early-winter feeding time in corn stubble and 25% in soybean fields, but increased their use of winter wheat fields later in the season (Feb–Mar; Davis et al. 1989).

On the West Coast, wintering Snow Geese in the Central Valley in California make extensive use of rice fields, but they also eat green vegetation, particularly during late winter and spring (Heitmeyer et al. 1989). To the north, those wintering in the Fraser and Skagit River deltas in British Columbia forage in tidal marshes and nearby farmland.

POPULATION STATUS

The Snow Goose population has increased tremendously over the past 5 decades, especially the MP, for which Midwinter Survey estimates rose from 0.8 million in 1969 to about 2.7 million by 1994, representing about a 3.4-fold increase in 24 years (Abraham and Jefferies 1997). Early Midwinter Surveys may have underestimated the population by at least 100% (Kerbes 1975), and Snow Geese and Ross's Geese were surveyed together as light geese. Using photographic surveys supplemented by ground counts, Kerbes et al. (2006) estimated a minimum population of 3,815,600 nesting adults in the MP in 1997–98, which was a 3-fold increase above 1979–80 estimates for colonies in the eastern Arctic plus 1982 numbers from the central Arctic. Abraham and Jefferies (1997) estimated a total of 4.5–6 million for the MP by the mid-1990s. Rockwell et al. (1997) concluded that the Midwinter Survey counts of Snow Geese increased by about 5% each year (1969–94). Estimates of MP light geese peaked in 1998 at about 3 million birds, then decreased during the period of special harvest initiatives (see Special Harvest Initiatives in the Harvest section) from 1998 to 2006, but resumed an upward trajectory to about 3 million birds in 2008 (Alisauskas et al. 2012).

Jeffries et al. (2003) attributed this virtual population explosion to at least 4 main factors: (1) the expansion of agriculture areas and increased yields associated with the application of nitrogen-based fertilizers; (2) the establishment of refuges, mostly between 1930 and 1970, which protected birds from harvest; (3) a decline in harvest rates; and (4) climate change. The latter may have been responsible for a southward shift in breeding distribution by the 1970s to below 60° N latitude, where the climate is less severe (Abraham et al. 1996). Yields from planted crops have increased substantially over the past 50 or so years (1950–2002) in the Central and Mississippi Flyways, due to the heavy application of nitrogen fertilizers and the introduction of high-yield crop varieties. Such increases are about 3.5-fold for corn, 2.5-fold for rice, 2-fold for wheat, and 6.5-fold for soybeans (Abraham et al. 2005). Waste corn in particular is an especially important food for migrating and wintering Lesser Snow Geese (Frederick and Klaas 1982, Alisauskas and Ankney 1992b), as well as for other migratory waterfowl and Sandhill Cranes (*Grus canadensis*) in midcontinent North America (Krapu et al. 2004a). In addition, pulse crops (legumes) have been planted in recent years on areas of the Great Plains where the growing season is adequate, particularly in Saskatchewan and North Dakota, and geese are now opportunistically using the associated waste grain (Pearse et al. 2010). In 2005–6, 2.3 million ha of pulse crops were planted in Canada (mostly in Saskatchewan) and 0.5 million in the United States (mostly in North Dakota; Abraham et al. 2012). Along with such an abundant and predictable food source, there is an extensive network of 1.5 million ha of national wildlife refuges in the Central and Mississippi Flyways, in addition to numerous state and private protected areas. Many of these sites were established when major increases in corn production were taking place; they thus of-

fered geese significant safety from hunters, as such refuges often contained large open-water areas or wetland complexes that were in close proximity to agricultural fields. Together, the rise in grain production and the establishment of these refuges were the perfect recipe for enlarging the winter carrying capacity for Snow Geese, and thus were a major factor for the increase in the MP Lesser Snow Geese as well as Ross's Geese.

As for actual numbers of geese, unfortunately there is no yearly breeding survey for Snow Geese over the vast majority of their breeding grounds. Aerial photo surveys in 1997 estimated the total continental population of Snow Geese at 6.7 million, with approximately 5 million breeding adults and the remainder made up of nonbreeding 1- and 2-year-olds (Cooke et al. 2000, Mowbray et al. 2000). The accuracy of estimates of population size is difficult to determine, however, as not all breeding or wintering grounds are surveyed. In addition, as the numbers of Snow Geese continue to increase, and as they expand their distribution on both breeding and wintering areas, population estimation becomes even more difficult (see New Estimation Methods).

The U.S. Fish and Wildlife Service evaluates long-term trends in winter populations of Snow Geese / light geese through the Midwinter Survey. This survey shows that light goose numbers have increased 230%, from 1.1 million (in the late 1950s) to 3.6 million (1997–2001). The 2001–10 average was 3.6 million: 46.5%in the Mississippi Flyway, 26.5% in the Central Flyway, 16.1% in the Pacific Flyway, and 11.0% in the Atlantic Flyway (the latter mostly Greater Snow Geese). The Midwinter Survey revealed that the MP, by far the largest of the 4 populations, had experienced an annual growth rate of 3.4% from 1969 to 2004; from 2002 to 2011 the rate was 2%/year, with a 2011 estimate of 3.2 million (U.S. Fish and Wildlife Service 2011). The WCFP increased 8.3% annually from 1972 to 2004, and 10% annually from 2002 to 2011, with a 2011 estimate of 196,100 birds from this population in the United States. The combined WIP and WAP, the source of nearly all the Snow Geese wintering in the Pacific Flyway, was 863,800 in 2010, which was 4% lower than in 2009, although the population has increased by 8% annually from 2001 to 2010 (U.S. Fish and Wildlife Service 2011). Most geese in the combined population are from breeding colonies in the western Canadian Arctic, as the 2010 breeding colony on Wrangel Island was only 140,000 adults.

The Greater Snow Goose population was believed to be only 3,000 at the beginning of the twentieth century, but it increased to about 25,000 in 1965, and to over 800,000 by 2000 (Ankney 1996, Bechet et al. 2004). Since 1965, the Canadian Wildlife Service has conducted an annual photographic survey during spring migration in southern Québec, and the U.S. Fish and Wildlife Service surveys them via the Midwinter Survey in the Atlantic Flyway. Average population sizes (2000–2010) were 868,118 for the Québec Survey and 398,237 for the Midwinter Survey. Like the MP Lesser Snow Geese, the Greater Snow Goose population has increased markedly, averaging (from Québec data) 149,780 in the 1970s, 246,970 in the 1980s, 604,660 in the 1990s, and 868,118 from 2000 to 2010. The spring estimate has increased by 5%/year from 2001 to 2010 (U.S. Fish and Wildlife Service 2010).

New Estimation Methods

Abraham et al. (1996) suggested that the Midwinter Survey underestimates the light goose population, because large, dense concentrations are undercounted; the MP alone may number about 6 million. Boyd (2000) found that visual counts underestimated Snow Goose flocks that exceeded 2,000 birds, with flocks >4,000 requiring a correction factor of 1.6 in comparison with photo counts. Eggeman and Johnson (1989) noted inconsistency issues with the Midwinter Survey, including variations in both survey efforts and timing, as well as a lack of clear sample design; Heusmann (1999) also discussed this issue. Rusch and Caswell (1997) and Abraham and Jefferies (1997) found accuracy difficulties in the Midwinter Survey that were

specific to light geese, in that the large flocks of wintering light geese often are underestimated by visual counts, and coverage of all areas used by the geese is difficult, especially since the population has expanded into new wintering areas. Furthermore, most of the Midwinter Survey routes meld Snow Geese and Ross's Geese together as light geese, which obviously restricts an estimation of population size by species. Lastly, breeding-ground surveys are not undertaken every year; nor do they cover all breeding colonies when they are conducted. Hence there is significant need for better methods of estimating the population size of arctic-nesting geese.

Alisauskas et al. (2009) addressed this issue by using the Lincoln Estimator to provide population estimates for several arctic goose populations, including MP Lesser Snow Geese. Such an approach is not new, and dates back to Lincoln's original suggestion that marked ducks (band returns) could be used to estimate continental population size (Lincoln 1930). That estimator, now often referred to as the Lincoln-Petersen Index, is elegantly simple and widely used to estimate the size of closed wildlife populations. For waterfowl, population size can be estimated using the size of the harvest for a given year and the ratio of the total number of banded birds harvested during their first year after banding to the total number banded, which estimates the harvest rate. The subsequent math of the relationship is simple: population size in a given year equals total harvest divided by harvest rate (Alisauskas et al. [2009] discuss assumptions with this method).

The Lincoln Estimator approach yielded dramatically higher estimates of Lesser Snow Geese in the MP, compared with traditional surveys, such as the Midwinter Survey and surveys of nesting colonies. The bias-corrected 1998 estimate was a stunning 18.7 million adults, compared with 3.8 million estimated from 1997–98 photographic surveys of colonies (Kerbes et al. 2006). Assessments of the August population (1999–2006) suggested that the MP contained more than 20 million adults and 5 million goslings by 2006. Growth rates for the MP were estimated at 10.3%/year from 1971 to 1998, but only 2.8%/year from 1998 to 2006, following the implementation of special harvest initiatives.

Harvest

Unlike their dramatically increasing population, the harvest of Snow Geese has not concurrently grown over the last 4 decades. With large year-to-year fluctuations and no overall trend, the harvest in the United States averaged 429,000 geese from 1965 to 1974, 499,000 from 1975 to 1984, 366,000 from 1985 to 1994, and 651,000 from 1995 to 2000. From 1999 to 2008 the harvest averaged 639,800 Snow Geese, of which 73.1% were white morphs and 26.9% were blue morphs. Most of the blue morphs were harvested in the Mississippi Flyway (57.3%) and the Central Flyway (42.1%), with only 0.5% in the Atlantic Flyway and 0.1% in the Pacific Flyway. Of the total harvest (both morphs), 47.4% were taken in the Central Flyway, 36.0% in the Mississippi Flyway, 8.9% in the Pacific Flyway, and 7.7% in the Atlantic Flyway (the latter mainly Greater Snow Geese).

In the Central Flyway, 68.0% of the Snow Geese were harvested in Texas, 17.3% in the Dakotas, and 8.4% in Kansas and Nebraska. In the Mississippi Flyway, most were shot in Arkansas (42.2%) and Louisiana (32.8%), with lesser numbers taken in Missouri (13.1%). Snow Geese in the Pacific Flyway were harvested primarily in California (71.6%), Washington (13.3%), and Oregon (10.3%); most in the Atlantic Flyway were taken in Delaware (32.0%), Maryland (22.8%), and Pennsylvania (15.1%).

In Canada, the harvest of Lesser Snow Geese averaged 180,823 birds from 1999 to 2008, with a high of 229,400 in 2008. Over this time period, 68.6% were harvested in Saskatchewan, 21.3% in Manitoba, 6.2% in Alberta, 1.9% in Québec, and 2.0% in the remaining provinces. The harvest of Greater Snow Geese over this same time period averaged 76,305 birds, with a high of 120,666 in 2008.

Nearly all (98.2%) were harvested in Québec. At a continental level, the annual harvest of light geese (Snow Geese and Ross's Geese combined) during the regular fall and winter hunting seasons peaked at 912,557 birds in 1998, with an average of 428,630 from 1962 to 1997 and 471,654 from 1962 to 2007 (Johnson et al. 2012).

Special Harvest Initiatives. To control the rapid population growth of both Snow Geese and Ross's Geese, which are collectively referred to as light geese or white geese, in spring 1999 the U.S. Fish and Wildlife Service instituted a special conservation order that was designed to increase the harvest of light geese in the Mississippi and Central Flyways. In particular, the conservation order allowed the harvest of both goose species beyond 10 March (the date limit imposed by the Migratory Bird Treaty Act of 1918), with no bag or possession limits, as well as the use of unplugged shotguns and electronic calls. Studies have shown that electronic calls are 5 times more likely to attract Snow Geese within a hunter's shooting range, and that the size of flocks flying within shooting range is 1.8 times larger with electronic versus traditional calls; collectively, this produces a hunter kill that is 9.1 times greater with electronic calls (Olsen and Afton 2000). A Final Rule, published in the *Federal Register* in November 2008, made the harvest regulations permanent in the Central and Mississippi Flyways; the Atlantic Flyway was eligible to implement them as well. The Final Rule authorizes states to effect a conservation order to allow the harvest of light geese outside of traditional hunting seasons, due to their overabundance.

In spring 1999, the Canadian Wildlife Service followed suit with special conservation measures that authorized the spring harvest of light geese in selected areas of Québec and Manitoba, allowing the use of electronic calls and, in some cases, bait (with a permit). Such measures were also implemented in Nunavut and Saskatchewan during spring 2001, and in southeastern Ontario in 2012, the latter due to the large numbers of Greater Snow

Geese now staging there in the spring. Ross's Geese were excluded from spring conservation harvests in Canada, however, because a court ruled in 1999 that sufficient evidence had not been presented to show that they were overabundant and that they had contributed to the habitat damage caused by Snow Geese.

Special harvest initiatives in the United States and Canada have clearly increased the harvest of light geese in the midcontinent, with an average overall harvest of 1,300,244 during the period of conservation action (1998–2007), which was an increase of 154% from the prior 10-year average (Johnson et al. 2012). From an estimated harvest of 398,455 during the first season of operation in spring 1999, the harvest resulting from these special initiatives averaged 691,896 from 2000 to 2010, which was more than half (54%) of the total harvest. Most (98.7%) of the light geese harvested under the auspices of special harvest initiatives were taken in the United States: 60.2% of this amount in the Mississippi Flyway and 39.8% in the Central Flyway. Johnson et al. (2012) reviewed the figures for light geese associated with the special harvest initiatives in both the United States and Canada, and state surveys estimated that from 1999 through 2008, 6,612,513 light geese were harvested under the conservation order, of which 59% (3.89 million) were taken in the Mississippi Flyway and 41% (2.72 million) in the Central Flyway. The top 5 states were Missouri, Arkansas, Louisiana, South Dakota, and Nebraska, which (combined) accounted for 73.4% of the total harvest since 1999. Studies indicate, however, that these harvest figures for light geese may be overestimates.

Using a different approach (the Lincoln Estimator), Alisauskas et al. (2011) calculated the harvest of MP Lesser Snow Geese in the United States and Canada, which suggested that state surveys to estimate conservation-order harvests had a quite high bias. They estimated an average harvest of 604,061 birds during the regular season from 1998 to 2006, with another 318,187 (77% adults) harvested through the conservation order,

compared with an average of 659,260 for state-estimated conservation-order harvests (a difference of >207%). According to Alisauskas et al. (2011), 34.5% of the overall harvest (37% of the adults and 28% of the juveniles) was attributable to special harvest provisions. Some 97% of the harvest outside of the regular season occurred in the United States. The question remained, however, as to whether these initiatives were meeting the objective of reducing the midcontinent population of light geese in general and Lesser Snow Geese in particular.

Rockwell et al. (1997) conducted an analysis with the best data available at that time to predict that a reduction in adult survival to 71%–73% would cause the Midcontinent Population of Lesser Snow Geese to decline by 15%/year. Their subsequent estimates predicated that harvests would have to increase from 305,000 adults and young (1985–94 average) to about 915,000 adults and young annually for 3–7 years to reduce the MP by 50%. Cooke et al. (2000) later revised these predictions (using updated survival estimates) and suggested that an annual harvest of between 1.5 and 3.4 million MP geese (5–10 times the current harvest) was needed to reduce populations by 10%–15% annually.

In comparison with these target estimates, Alisauskas et al. (2011) concluded that the MP had continued to grow during the years of the conservation order relative to prior years, but at a reduced rate. The annual population growth rate (the Lincoln Estimator) was 14.4% from 1990 to 1998 (before special harvest measures), compared with 5.0% afterward (1998–2006). Their analysis also noted a decline in annual survival from about 89% to around 83% among Snow Geese nesting in southern areas (south of 60° N latitude), but that group only includes about 10% of the MP. In contrast, there was no change in the survival of birds from northern nesting colonies (north of 60° N latitude), where annual survival was about 87% from 1989 to 2006. High survival rates were consistent with low harvest rates, which—in the northern cohort—were only 2.4% during 1989–97 (prior to the conservation order), and 2.7% in 1998–2006. For birds from southern colonies, the rate was 3.1% from 1989 to 1997 and 3.7% from 1998 to 2006. The higher harvest rate of the southern population was correlated with their earlier fall migration, exposing them to harvest pressure sooner, while the late migration of the much larger northern population resulted in a greater ratio of geese to hunters, and a resultant lower harvest rate.

The total annual harvest did increase through the special initiative measures, but it did not exceed 0.75 million adults in any of the years assessed between 1989 and 2006, with the take of both adult and young geese exceeding 1 million in only 2 of the 9 annual harvest periods analyzed. In addition, the harvest of adult birds was lower than expected, as was the harvest rate, which never exceeded 4.8%. A subsequent estimation of population size, applying the Lincoln Estimator to harvest and harvest rates (Alisauskas et al. 2009), suggested that the MP has been seriously underestimated; estimates of the August population exceeded *15 million adults and 5 million goslings* from the start of the conservation order through 2006 (or >20 million adults, depending on the model input). The number of birds in 3 major populations of light geese—MP Lesser Snow Geese, Greater Snow Geese, and Ross's Geese—all exceed the population estimates when the conservation order was first implemented. Alisauskas et al. (2011:179) concluded, "We are confident that the abundance and population growth rate of midcontinent Snow Geese (as well as Ross's [Geese] and Greater Snow Geese) currently exceeds the ability of existing numbers of hunters to exert harvest pressure that is necessary to impose sufficient additive mortality and thus effectively influence population growth." It thus remains to be seen when the MP might begin to exceed the carrying capacity of its arctic breeding habitats and result in a measurable decline in that population's growth rate.

In the most recent summary evaluation of the effect of special harvest initiatives on Lesser Snow Geese and Ross's Geese, Dufour et al. (2012) noted the inability of hunter harvests to lower population levels. They presented an example where the hunting mortality (kill rate) required to reduce adult survival of MP Lesser Snow Geese to below 80% (Rockwell et al. 1997) would necessitate increasing the current kill rate of about 2.5% to at least 9%. A similarly large increase in the kill rate (from 2% to 12%) would be needed to affect the Ross's Goose population. Dufour et al. (2012:25) concluded that "our results offer little support for the suggestion that spring harvest in Canada and conservation order harvest in the United States has had a negative impact on Lesser Snow Goose productivity or recruitment." A summary conclusion of the 2012 Arctic Goose Habitat Working Group (Leafloor et al. 2012:3) is telling: "In the absence of drastic population control measures, continued increases in population size of Midcontinent Lesser Snow Geese and Ross's Geese are expected, and are likely to lead to more destruction of arctic wetland habitats used by geese and other species."

There is some indication, however, that the special harvest initiatives implemented for Greater Snow Geese have at least temporarily arrested the growth of that population, and they may have even caused a decline since such measures began in spring 1999 on staging areas in southern Québec (Reed and Calvert 2007, Dufour et al. 2012). Calvert and Reed (2005) found that the special harvest provisions reduced adult survival from 83% at the initiation of these measures in 1998/99 to 73% afterward (until 2002), with the kill rate of adults increasing by 130%. Juvenile survival, however, did not change. Population models predicted an annual growth rate of 7.8% without special harvest initiatives, compared with an 8.0% annual decline with these measures in place. The Canadian harvest of Greater Snow Geese averaged 110,028 birds from 1998/99 to 2010/11, of which 33.2% were taken via special harvest initiatives. The U.S. harvest averaged 46,740 over this period during the regular season, and 41,070 with special harvest provisions implemented during 3 of those seasons (2008/9 to 2010/11). Unlike the situation for the MP, hunting may reduce the Greater Snow Goose population, both because of the large number of hunters and the goose-hunting tradition in the mid-Atlantic states, and because the initial population size for these geese was not as large.

Lastly, although special harvest measures in both the United States and Canada can affect light goose populations directly by increasing adult mortality, hunter disturbance is an indirect effect that can reduce reproductive output (fecundity). The hunting disturbance associated with a special conservation hunt of Greater Snow Geese (15 Apr–31 May) on their spring staging grounds in southern Québec reduced the breast muscle of surviving females by 5%–11% and their abdominal fat reserves by 29%–48% (Féret et al. 2003). While surviving birds continued their northward migration, they presumably arrived on their breeding colonies with fewer stored reserves, which probably would reduce their breeding success and thus would represent an additional sublethal effect of the spring hunt beyond the actual harvest of geese. During spring in southcentral Nebraska, the lipid content of Lesser Snow Geese was 25% less in areas open to hunting, compared with closed areas, but it was similar among areas after the cessation of hunting (Pearse et al. 2012). In contrast, Greater White-fronted Geese also had lower (24%) lipid levels in hunted versus nonhunted areas, but that difference persisted after the hunting season; hence such disruption was an unintended consequence of spring Snow Goose hunting. Alisauskas (2002) suggested that spring hunting seasons farther north, especially in the northern prairie states and southern staging areas of the Prairie Provinces (e.g., southern Manitoba), can have a 2-fold effect on the breeding population of light geese by both decreasing the number of potential breeders and upsetting the spring accumulation of fat reserves, which can reduce breeding success.

Hunter interest in special harvest initiatives may be waning somewhat, however. In the United States, the number of hunters pursuing light geese through the conservation order in the Central and Mississippi Flyway states from 1999 to 2006 ranged from 41,163 in 1999 to 75,727 in 2000 (Johnson et al. 2012). There were only 3,571 more hunters in 2008 for all 18 participating states, compared with the first year (1999), when only 10 states joined in. In the Central Flyway, hunter numbers averaged 33,399, but declined to 19,844 in 2008 from the peak of 49,047 in 2000. Numbers of hunters in the Mississippi Flyway have been more stable, ranging between 23,401 and 28,888 (1999–2008). The average number of hunters/state (1999–2007) has ranged from merely 55 in Indiana to 11,729 in Texas.

The conservation order has resulted in an estimated 2.53 million hunter-days between 1999 and 2008, with 58% taking place in the Central Flyway and 42% in the Mississippi Flyway, but these numbers seemingly peaked between 2002 and 2004 and are now trending downward (Johnson et al. 2012). Overall hunter success averaged 11.7 birds/hunter (from 8.5 in 2000 to 16.3 in 2007). Within individual states, hunter success varied from 1.2 birds/hunter in Indiana (2000) to 52.7 in Arkansas (2006), with Arkansas hunters consistently reporting the highest average success (38.2 birds/hunter).

BREEDING BIOLOGY

Behavior

Mating System. Snow Geese are monogamous and generally mate for life, with the pair bond maintained throughout the year (Cooke et al. 1981). Pair separation is rare, although a female was documented as having established a new pair bond within several days after the death of a mate on the breeding grounds (Abraham et al. 1981). Most pair bond breakups result from the death of a member of the pair (Cooke and Sulzbach 1978). Mating primarily occurs during the northward migration, particularly at rest stops, with pair bonds first forming when individuals are in their second winter/spring (Prevett and MacInnes 1980).

At the La Pérouse Bay colony (1970–81), most Lesser Snow Geese were paired at 2 years of age, yet only about 30%–60% bred as 2-year-olds, compared with about 75%–>90% for 3-year-olds and

>90% for 4-year-olds (Cooke et al. 1995, Cooch et al. 1999). Snow Geese do not pair as yearlings. During a 3-year study (1997–99) of Lesser Snow Geese breeding on Wrangel Island, <5%–20% formed pair bonds as 2-year-olds, versus 38%–50% for 3-year-olds, 81% for 4-year-olds, and higher still for birds 5 years old and older (Ganter et al. 2005). The Snow Geese on Wrangel Island may pair later than those in the Low Arctic, which could be related to the harsher weather in the High Arctic. On Bylot Island, Greater Snow Geese began to breed as 2-year-olds (25%), with the percentage increasing to 57% for 3-year-olds, and 100% for 4-year-olds (Reed et al. 2003b). The probability of starting to breed was not affected by snow cover or hatch date, but it was considerably reduced during years when lemming populations (*Lemmus sibiricus, Dicrostonyx groenlandicus*) had crashed, as arctic foxes (*Alopex lagopus*) then switch to alternate prey and become more focused on goose colonies (Bêty et al. 2002).

There appears to be a cost associated with breeding for the first time, however. At La Pérouse Bay (1973–90), first-time breeders were 22.8% less likely to breed the following year, and those first breeding as 2-year-olds were 45.2% less likely to breed the following year (Viallefont et al. 1995). The majority (68.3%) of the females in the population first bred as 3-year-olds. If there was an equal cost of breeding or not breeding as a 2-year-old, a predicted 54.8% of all 3-year-olds would breed each year, which is much lower than the actual percentage. Hence successfully nesting 2-year-olds are less common among the subsequent cohort of successful 3-year-olds than are nonbreeding or unsuccessful 2-year-olds. On average, the lifetime production of goslings by individuals first nesting as 2-year-olds was similar to those first starting as 3-year-olds, in large part because birds first breeding as 2-year-olds are less likely to then breed again as 3-year-olds. This finding led to the conclusion that available resources limited nesting at an early age in Lesser Snow Geese, as only the best-conditioned individuals nested as 2-year-olds.

Lesser Snow Geese exhibit a strong tendency to select a mate of the same color morph (assortative mating), a behavior first described by Cooch and Beardmore (1959) and subsequently well studied in the La Pérouse Bay colony. Long-term research (1968–91) at La Pérouse Bay revealed that 90.4% of females reared by white-morph parents returned with white mates and 78.2% reared by blue-morph parents returned with blue mates (Cooke et al. 1995). Such assortative mate selection in Snow Geese reflects imprinting of the goslings on the parental color type as well as on sibling color (Cooke and McNally 1975, Cooke et al. 1976), which led to the prediction that mixed pairs would eventually disappear from the population. But the assortative system is not perfect, as 9.6% of the goslings from white parents chose blue mates, and 21.8% of the goslings from blue parents choose white mates (Cooke 1987). From 1968 to 1991 at La Pérouse Bay, the percentage of mixed pairs averaged 16.6%; all-white pairs, 64.3%; and all-blue pairs, 19.1% (Cooke et al. 1995).

Mixed pairs persist at colonies like La Pérouse Bay, so biologists then asked whether mixed pairs exhibited increased fitness. Findlay et al. (1985) examined several components of reproductive success and found few differences between pure and mixed pairs, although recruitment rates were higher among mixed pairs for 2 out of 7 cohorts, and nest failure for mixed pairs was also consistently lower. Although their results indicated an advantage in mixed pairing, assortative pairing predominates, probably because the color phases were allopatric until the early part of the twentieth century. Thus color-phase cues for species recognition may still act as an isolating mechanism against mixed pairs (Cooke 1987). Cooke et al. (1995) provide a thorough review of this aspect of Snow Goose mating systems, noting that early taxonomists (prior to 1930) were correct in considering the blue-phase and white-phase Snow Geese as distinct taxa, because they looked quite different and were allopatric (separated on both their breeding and wintering areas).

Sex Ratio. The sex ratio (males:females) is basically equal during all stages of the life cycle of Snow Geese, from hatching to adulthood (Cooke and Harmsen 1983, Cooke et al. 1995). An analysis of 23 years of data (1969–91) at the La Pérouse Bay colony, however, has revealed a decline in the percentage of males at fledging: from about 51% in the early years, when the food supply was abundant, to about 48% in more recent years, as the food supply became scarcer (Cooch et al. 1997). Sex ratios at hatching did not change; hence the mortality of males had increased between hatching and fledging. The change in sex ratios only occurred among younger adults (<5 years old), however, probably because they were less likely to adapt to alterations in their food supply. Nonetheless, although the decrease in surviving male goslings was about 20% among younger parents, the overall decline at the population level was only about 3%.

Site Fidelity and Territory. Lesser Snow Goose females have high rates of nest-site fidelity. At the La Pérouse Bay colony, 22% of 2,127 females banded as goslings were located in the colony 3 years later (Cooke et al. 1995). As this 22% is close to the estimated survival rate for juveniles as a whole, the return rate includes virtually all surviving females. In contrast, <0.3% of marked males returned to their natal site. At La Pérouse Bay, 73% of the marked females returned to within 500 m of their previous nest site, and 43% nested within 100 m (Cooke and Abraham 1980). In response to rapid population growth and concurrent deteriorating habitat conditions at the La Pérouse Bay colony, due to overcrowding, Ganter and Cooke (1998) noted only slight changes in breeding site fidelity among adult birds. The mean age of breeding birds, however, increased in what were formerly the central parts of the colony, and young birds apparently settled on the periphery of the colony, even when space was available in the center. This settlement pattern by young birds caused a long-term shift in the colony's location.

As adults, males follow their mates in returning to breeding colonies, because Snow Geese mate for life (although males do engage in extra-pair copulations). At La Pérouse Bay, the return rates of color-marked Lesser Snow Geese after the first year were 73%–79% for females and 50%–66% for males (Cooke and Sulzbach 1978), but the male rate was 68%–82% when calculated from pair bonds in the colony. Such return rates approximate survival rates, and thus indicate that nearly all of these birds are faithful to their breeding colonies. In comparison with that of Lesser Snow Geese, site fidelity is much less among Greater Snow Geese, which, on Bylot Island, use different locations for their breeding colonies, based on available resources and on the colonies' proximity to Snowy Owls. The proportion and location of isolated nests also varied from year to year (Lepage et al. 1996). Young from the previous year (yearlings) have also returned with the adults, but most leave the nesting territory soon after incubation begins (Prevett and MacInnes 1980).

Pairs establish territories that vary in size depending on colony density; these territories are smaller in the interior than along the perimeter of the colony. Once established, territories are vigorously defended from predators and from intruding males seeking extra-pair copulations. Dunn et al. (1999) noted that extra-pair copulations were commonplace, though, forming 47% of the copulations observed during 2 years of study (1993–95) at the Karrak Lake colony; most (76%) were attempted when females were on their nests. Yet extra-pair copulations were not very effective at producing offspring, with just 2%–5% deemed successful (through DNA testing) at Karrak Lake, and 2.4% at La Pérouse Bay (Lank et al. 1989b).

Mineau and Cooke (1979) argued that the function of territories was to defend the female, not to defend the space per se (see the Ross's Goose account). On Bylot Island, male Greater Snow Geese spent 32% of their time being alert during the egg-laying period, which is when females are fertile (Gauthier and Tardif 1991). Females, however, use a territory for foraging during recesses from incubation. Foraging time during incubation recesses averaged 60.9%–80.0% for females monitored at the Karrak Lake colony in 1993 and 1995 (Gloutney et al. 1999). Greater Snow Geese on Bylot Island spent an average of 75.5% of their time feeding during recesses (Gauthier and Tardif 1991).

Courtship Displays. Courtship displays and the formation of pair bonds primarily occur during the spring, although some happen in late winter. Courting males closely follow females while enlarging their body contours and adopting an exaggerated *erect* posture (Mowbray et al. 2000). Females assume a more passive role in the courtship process. As in all species of geese, maintenance of the pair bond is reinforced by the *triumph-ceremony*, which is also performed by other family members (Prevett and MacInnes 1980). Pairs carry out the *triumph-ceremony* when they meet, especially after an aggressive interaction with an intruder. The display is performed simultaneously by the pair and involves stretching their necks forward and upward while orienting laterally and calling loudly.

Copulation occurs in shallow water or on land, with the female adopting a *bowed* posture, with her neck lowered, often accompanied by *head-* or *bill-dipping* (Mowbray et al. 2000). Actual *treading* lasts about 5 seconds. After copulation, the pair rises up vertically, with extended heads and necks, while vocalizing and flapping their wings. The final sequence of the postcopulatory display is *preening* and *bathing* (Cooke et al. 1995). Most copulations between pair members occur during the prelaying and laying period.

Nesting

Nest Sites. The female selects the nest site. Snow Geese, however, nest in colonies that can vary in size from only a few hundred to >100,000 birds. Lesser Snow Geese usually locate their nesting colonies on lowland grassy tundra plains within a few kilometers of the ocean, but colonies also occur along broad shallow rivers near the coast, as well as on islands in shallow lakes that are as far as 60 km inland (e.g., Karrak Lake in Nunavut). Lesser Snow Geese along the Cape Churchill Peninsula on the western coast of Hudson Bay can nest as far as 15 km inland, but mostly nest within 5 km of the coast (Rockwell and Gormezano 2009).

Nest density varies with habitat type and colony size. At Karrak Lake in 1976, the density on islands was 23 nests/1,000 m² in rock habitat, compared with 14 in heath, 14 in mixed habitat (low-growing vascular plants, lichens, and mosses on relatively dry gravel-filled soil), 6 in tussock habitat, and 2 in moss (McLandress 1983a). Density on mainland sites was 16 nests/1,000 m² in mixed habitat and 11 in tussock. Barry (1967) reported that on the Anderson River delta, nests were as little as 3–4.5 m apart in dense areas within the central part of a colony, but about 15 m apart on a ridge 30–60 m wide. Lesser Snow Geese on Wrangel Island and Greater Snow Geese on Bylot Island are known to nest in association with Snowy Owls (Tremblay et al. 1997, Baranyuk 2000). On Bylot Island, Tremblay et al. (1997) reported a colony of 236 nests at a density of 3.9/ha; Kerbes et al. (1999) observed nest densities of 4.4–42.8/ha.

At the Karrak Lake colony, the nest density of Lesser Snow Geese was negatively correlated ($r = -0.24$) with the amount of total vegetation (Alisauskas et al. 2006b). The percentage of grass was highest where no nesting (either by Lesser Snow Geese or Ross's Geese) occurred, intermediate in areas occupied for 1–10 years, and lowest in areas in use for 11–20 years. The proportion of heather (*Cassiope tetragona*) was lowest and that of moss highest in the oldest parts of the colony.

On Bylot Island, Greater Snow Geese nest on well-drained westward slopes, as well as on hilltops where vegetation occurs (Lemieux 1959b). They do not always use the same colony sites each year, with the deciding factor thought to be the snow-cover distribution. Greater Snow Geese tend to nest in less compact colonies than Lesser Snow Geese (Gaston et al. 1986).

In Lesser Snow Geese, only the female constructs the nest, with her mate in close attendance. The nest itself is a built-up scrape requiring several years to reach final form (Cooch 1958). Initial nests can be little more than scrapes in moss or gravel that are enlarged by the female as eggs are laid. The female deposits a small amount of down in the bowl of the nest, some with the first egg but most with the third and fourth eggs (Barry 1967). Although the nest is primarily lined with down, the lining can include material from the depression. In other situations, moss, willow, and grass are added to old nests to form fairly substantial raised structures than can reach 100–200 cm in diameter; they vary in morphology, though, depending on where the nest is constructed. At Karrak Lake, as nest locations went from heath, to rock, to mixed habitats, to moss (i.e., from more protected to less protected habitats), the outer diameter, wall thickness, circumference, rim height, and nest mass generally increased (McCracken et al. 1997). Total nest mass averaged 564 g in heath habitats, compared with 1,379 g in moss.

Clutch Size and Eggs. Clutches of Lesser Snow Geese normally range between 1 and 7 eggs. Among 2,773 nests monitored during an 11-year period (1973–84) at La Pérouse Bay, the most frequent clutch size was 4 eggs (34.9%), followed by 5 (27.8%), and 3 (17.6%; Rockwell et al. 1987). Females breeding for the third time or more had larger average clutch sizes (4.6 eggs) than those breeding for the second time (4.0) or the first (3.5; Cooke et al. 1981). Clutch size also increases with the female's age. At La Pérouse Bay (1974–87), the average clutch size (adjusted for eggs added via brood parasitism and laying date) was 3.0 eggs for 2-year-olds, 3.5 for 3-year-olds, 3.8 for 4-year-olds, and 3.9 for 5-year-olds (Cooch et al. 1989). There are no detectable age effects on clutch size after a goose is >5 years old (Rockwell et al. 1983). Larger clutch sizes have been associated with early snowmelt. During an 8-year study (1958–65) on the Anderson River delta, the smallest average clutch size (3.2 eggs) occurred during the year when snowmelt was latest (Barry 1967). Clutch size at La Pérouse Bay (1973–91) was also negatively correlated with laying date ($r = -0.55$; Cooke et al. 1995).

On Bylot Island (1957), clutch size of Greater Snow Geese averaged 4.6 eggs ($n = 118$; range = 2–9 eggs); clutches initiated earliest averaged 6.5 (Lemieux 1959). The number of eggs laid also decreased in proportion to the number of days nest initiation was delayed. During a later 7-year study (1991–97) on Bylot Island, clutch size averaged 3.9 eggs ($n = 2,253$), with late-nesting females laying fewer eggs than early-nesting females (Lepage et al. 2000). The seasonal decline in mean clutch size varied from −0.11 to −0.26 for each day of delay.

The eggs of Snow Geese are elliptical ovate and white, but they later become stained a yellowish hue from the nest material (Palmer 1976a). At Karrak Lake, egg measurements (1 each from 83 clutches) averaged 79.6 × 53.1 mm (Slattery and Alisauskas 1995). At La Pérouse Bay, the eggs of females >3 years old ($n = 457$) averaged 79.7 × 52.9 mm, compared with 80.5 × 52.4 mm for 2- and 3-year-olds ($n = 33$; Robertson et al. 1994). On Bylot Island, egg measurements for Greater Snow Geese (1 each from 869 clutches) averaged 80.0 × 52.7 mm (Mowbray et al. 2000). Egg mass (1 each from 276 clutches) averaged 119.3 g.

Egg mass varies markedly within clutches and among individuals. At La Pérouse Bay, the average mass of 22,562 eggs was 124.5 g, but the smallest hatched egg was 86 g, compared with 166 g for the largest (Cooke et al. 1995). Egg mass is highly repeatable and heritable among females, which accounts for 80% of the variation among individuals (Lessells et al. 1989). In other words, females that initially lay large eggs continue to do so in future years, and females laying small eggs continue to maintain that size.

The egg-laying rate is about 1 every 33 hours, although that pattern can vary (Schubert and Cooke 1993). Cooch (1958) found that females usually laid 1/day for the first 3 eggs, skipped a day, and then completed their clutches. The laying period is synchronized within the colony, as seen during studies at La Pérouse Bay (1973–79), where the last geese began to lay an average of 11.9 days after the first geese (Findlay and Cooke 1982a, 1982b). Females laying and hatching eggs in synchrony experienced reduced nest predation; those nesting before or after the peak suffered greater predation by Herring Gulls (*Larus argentatus*) and arctic foxes. Synchronous nesting mediates the problem, as predators are overwhelmed by a superabundance of prey. On Bylot Island (1991–97), Greater Snow Geese initiated 90% of all clutches within an 8-day period (Lepage et al. 2000).

Incubation and Energetic Costs. Females begin incubation after the last egg is laid. Only the female incubates, but the male stands guard very close to the nest. The average incubation period is 23.6 days, but it is shorter for clutches with fewer eggs (Cooke et al. 1995). The incubation period averaged 23.1 days for 201 nests observed by Cooch (1958), and 22.4 days for 48 nests monitored by Ryder (1969b), with an average of 1.4 days for all eggs in a clutch to hatch.

The reproductive effort and ultimate success of Lesser Snow Geese depend heavily on their reserves of proteins, lipids, and other nutrients acquired prior to arrival in the Arctic and then used during the formation and subsequent incubation of the clutch. Such reserves are generally accumulated on spring staging areas, as reported during an 8-year study (1983–84, 1988–93) in the corn-producing region of southern Manitoba, which is a major staging area for the MP (Alisauskas 2002). Geese gained an average of 9.8 g/day in body mass from mid-April to early May ($n = 398$), with no detectable yearly variation, although gains in lipids varied annually from 6.0 to 17.5 g/day ($n = 393$).

In their landmark study at the McConnell River colony in the Northwest Territories, Ankney and MacInnes (1978) documented the use of stored nutrient reserves by breeding Lesser Snow Geese. Females ($n = 78$) averaged 2,950 g when they arrived at the colony, of which 492 g were lipids, 228 g were

protein, and 41 g were calcium. During laying, their average body mass then decreased by 429 g (14%); protein, by 34 g (15%); lipids, by 122 g (25%); and calcium, by 7 g (17%). Late in the incubation period, females were basically emaciated, having lost 42% of their original body mass, almost all (89%) of their lipid reserves, and substantial (35%) amounts of protein. A few females actually starved to death at their nests. After their clutches hatch, females feed heavily to recover body condition, molt, and then prepare for fall migration.

In contrast to Lesser Snow Geese and other species of geese, female Greater Snow Geese use very little of their lipid reserves to produce a clutch, which is probably a consequence of their long and energetically costly migration from spring staging areas along the St. Lawrence River to their breeding grounds in the High Arctic. These geese encounter little food while traveling northward across Québec and Labrador, while Lesser Snow Geese find rice, corn, and wheat during their migration through the midcontinent.

During a 2-year study (1989–90) on Bylot Island, at 73° N latitude, most female Greater Snow Geese collected shortly after arrival had only moderate fat reserves and did not begin egg laying until >1 week later (Choinière and Gauthier 1995). None had developing egg follicles. In contrast, Lesser Snow Geese at McConnell River arrived with well-developed egg follicles and began laying eggs 2–5 days later (Ankney and MacInnes 1978). Female Greater Snow Geese thus use a comparatively long prenesting delay to feed intensively and acquire nutrients to use for reproduction; Gauthier and Tardif (1991) noted that prelaying females fed for 75% of a 24-hour day. Hence, unlike Lesser Snow Geese, the nutrients that Greater Snow Geese need for clutch formation are acquired on the breeding grounds. Their lipid reserves decreased during incubation, however, from 374 g during laying to 253 g postlaying (−33%). Protein reserves, on the other hand, decreased during egg laying, which suggested that some reserves were deposited in the eggs.

During incubation, female Greater Snow Geese spent an average of 93% of their time on the nest, but that percentage declined to 91% about 6 days before the hatch, and then increased to nearly 100% during the last 4 days (Reed et al. 1995). In the last half of the incubation period, females averaged 5–7 recesses/day; they remained close to the nest but fed for >90% of the time. Recesses averaged 15–16 minutes, but increased to >30 minutes during the last 1–4 days, although recesses were infrequent then. Females captured during late incubation had only lost about 17% of their body mass; their reserves were clearly not depleted.

Nesting Chronology. When snow conditions allow, Lesser Snow Geese often begin nesting within a few days after their arrival on the breeding grounds. At La Pérouse Bay, the mean egg-laying dates from 1973 to 1992 ranged from 19 May to 11 June, with peak nest initiation averaging 11.0 days after arrival (range = 7–13 days) during the 10 years when the peak could be determined (Cooke et al. 1995). Initiation of the first nests varied between 12 May and 8 June. Laying was highly synchronized, occurring over a 10-day period, with a standard deviation of only 2.2 days; about two-thirds of all nests were initiated within a 5-day period and >90% were begun within 4 days of the mean initiation date. Differences in the timing of nest initiations are largely determined by the pattern of snow disappearance. At Karrak Lake (1991–99), about 900 km to the north of La Pérouse Bay, the mean egg-laying date was 25 May–1 July; on Bylot Island (1989–99), about an additional 725 km north, Greater Snow Geese initiated clutches (n = 3,444) between 6 and 20 June (Lepage e al. 2000, Mowbray et al. 2000).

The evidence at hand also indicates that laying in the nesting colonies of the western and southern Arctic begins a week or more sooner than in the colonies of the eastern Arctic. Climate changes within the western and central Canadian Arctic (1961–90) have resulted in earlier snowmelts, which allow Lesser Snow Geese to begin nesting sooner (Cohen et al. 1994). In the southern Hudson Bay region, snowmelt during the 1970s occurred an average of 2 weeks earlier than in previous and successive decades (Abraham and Jefferies 1997).

Nest Success. Lesser Snow Geese usually experience high nest success, except during years of unusually severe weather. Of 4,839 nests at La Pérouse Bay in Manitoba (1973–91), only 8.1% of the nests were categorized as total failures (range = 1%–25%; Cooke et al. 1995). A high degree (27%) of clutch loss occurred prior to incubation, with 90% of the mortality from predation or abandonment happening before laying the second or third egg. Of 4,326 nests with <7 eggs but first located at the 1-egg stage, 26.8% failed during laying, of which 73% of those failures occurred when only 1 egg was in the nest; >90% failed when 2 or fewer eggs were in the nest. Early-nesting birds also suffer more preincubation nest failure than later-nesting birds. In contrast, nest failure once incubation began was usually under 15% and often <5%, but it was unusually high during 2 years when a large quantity of caribou (*Rangifer tarandus*) in the colony destroyed sizeable numbers of nests.

During his 8-year study (1958–65) on the Anderson River delta, Barry (1967) found that only 6.2% of 3,929 nests failed completely. Nest success at Karrak Lake in Nunavut (1991–99) ranged from 47% to 88% (Mowbray et al. 2000). At the McConnell River, MacInnes et al. (1990) noted a positive relationship between early snowmelt and nest success, and early nests at La Pérouse Bay had greater success than mid- to late nests (Cooke et al. 1984). Jackson et al. (1988) observed that nest success was directly correlated with vegetative cover, with nests located in tall willows (*Salix* spp.) experiencing much higher success rates than those in shorter willows or in areas with no willows.

On Bylot Island (1991–97), the nest success of Greater Snow Geese (n = 2,068) averaged 63.3% (Lepage et al. 2000). Early nests and late nests were about 10% less successful than nests initiated closer to the mean date. Greater Snow Goose nests located near Snowy Owl nests on Bylot Island were more successful than those farther away, and the former tended to be initiated earlier (Tremblay et al. 1997). The presence of the owls, as well as Rough-legged Hawks, provides additional protection for the geese, as the raptors aggressively defend their own nests. Goose nesting success was lower (23%–42%) during a year when Snowy Owls were absent, compared with when they nested with the geese. A colony of 236 nests located over a 1.5 km² area of lowland tundra was centered only 66 m from a Snowy Owl nest and experienced a nest-success rate of 89.8%. Predation by arctic foxes was the main cause of nest failure. Few habitat features influenced nest success, although nests in pond habitats had lower success than nests in wet meadows or moist tundra. Isolated nests were also more successful on hillsides than in lowlands, while colonial nests were more successful in lowland wet meadows where tall willow (*Salix lanata*) occurred, compared with on hillsides or in moist tundra.

In addition to arctic foxes, other important predators of Snow Goose nests are Herring Gulls, Glaucous Gulls (*Larus hyperboreus*), and jaegers (*Stercorarius* spp.). Predators that cause more minor losses include grizzly bears, black bears (*Ursus americanus*), gray wolves (*Canis lupus*), Common Ravens (*Corvus corax*), and even Sandhill Cranes (Abraham et al. 1977, Cooke et al. 1995, Tremblay et al. 1997). Global climate change is also introducing a new predator of Snow Goose nests, polar bears (*Ursus maritimus*). The earlier breakup of ice floes in spring is forcing polar bears to the mainland sooner, where they then potentially overlap with nesting Snow Geese, as is occurring along the Cape Churchill Peninsula on western Hudson Bay (Rockwell et al. 2011). Such predation will reduce reproductive output, as happened in 2009 when there was complete reproductive failure on the Cape Churchill Peninsula, in part due to heavy losses caused by a variety of predators, including polar bears. Models also indicate, however, that the onshore arrival of polar bears and nesting by Snow Geese will not always overlap, so such new predation will affect but not eliminate the Snow Goose population in this part of the Arctic.

Brood Parasitism. The colonial nature of Snow Goose nesting promotes relatively high levels of brood parasitism. Large clutches of up to 21 eggs have been observed as a result of brood parasitism (Cooke et al. 1995). At La Pérouse Bay (1969–86), brood parasitism affected an average of 22% of the nests (range = 12%–31%; Lank et al. 1989a). Brood parasitism accounted for 5.3% of the goslings hatched annually (range = 1.8%–9.3%). The rate of parasitism was low during years when nest sites were readily available, and high during years when fewer sites were clear of snow and water. Parasitism was not prevalent during years of high nest failure during laying, which indicated that parasitism is a response to a lack of nest sites and not to nest failure. Brood parasitism was deemed a salvage reproductive strategy used by females with a low probability of successfully nesting.

The actual act of parasitism can occur when potential parasitic females and their mates approach an occupied nest, which is defended by the nesting pair (Lank et al. 1989b). Parasitic females prefer to lay eggs in or adjacent to occupied and defended nests, rather than undefended nests. While the 2 males engage in aggressive interaction, the parasitic female approaches the nest and attempts to dislodge the resident female and deposit an egg in the nest. A parasitic egg may also be laid just outside the nest and then rolled into the nest by the attending female, to make the actual location of the nest less obvious to predators. Other parasitized nests occur when a female takes over an abandoned nest that contains only 1 or 2 eggs, which may occur in up to 6%–8% of the nests (Cooke et al. 1995).

As with other waterfowl exhibiting intraspecific brood parasitism, this strategy is not as effective as incubating one's own nest. At La Pérouse Bay, parasitism did not affect the hatching of host eggs, but the hatching success of parasitic eggs was lower than that of host eggs, primarily because of poor timing by parasitizers in relation to the initiation of incubation by the host female (Lank et al. 1990a). Davies and Cooke (1983) found that parasitic eggs had little chance of hatching when added to nests 4 days after the host female began incubation, because they are not incubated in synchrony with the host clutch. During a 13-year study (1973–86) of brood parasitism at the La Pérouse colony, parasitic eggs formed 6.6% of all eggs and exhibited a mean hatching efficiency of 62.7%, versus 69.2% for host eggs (Lank et al. 1990). The hatchability of parasitic eggs (excluding an outlying year), however, declined 6.5% for each 1% increase in the rate of parasitism during a given year.

Renesting. There is no evidence of renesting by Snow Geese.

REARING OF YOUNG

Brood Habitat and Care. Preferred brood habitats include sedge meadows, tidal saltmarshes, shallow lakes and ponds, and upland tundra, which provide ample foraging opportunities for the developing goslings and their parents (Mowbray et al. 2000). At La Pérouse Bay, Cooke et al. (1995) observed that families of Lesser Snow Geese use coastal saltmarshes, where their principal foods were a stoloniferous grass (*Puccinellia phryganodes*) and a rhizomatous sedge (*Carex subspathacea*). Broods often travel great distances in search of food, especially at La Pérouse Bay, where the expanding number of Lesser Snow Geese has caused most broods to disperse 15–50 km to the east and southeast of the bay in search of better forage (Cooch et al. 1993). Goslings from dispersed broods were 7.3% heavier and had longer culmens (3.1%), heads (2.6%), and tarsi (1.9%) than goslings remaining at La Pérouse Bay. Snow Goose families (pairs with 1 or more goslings) exploiting food patches at La Pérouse Bay were always dominant over pairs without goslings, and the rate of these dominance interactions was significantly higher on experimentally created patches of high food biomass (Mulder et al. 1995). Geese also fed longer

in high biomass patches, compared with control areas (19.2 vs. 2.9 min/visit), and they defended these areas from intruders. Such actions probably allow these geese to increase their food intake rate, decrease foraging costs, and reduce predation risk, due to heightened vigilance by the parents and stronger cohesion of the family unit.

At Karrak Lake in Nunavut (1994–95), broods surveyed about 40 days after hatching were somewhat homogeneously distributed over the 5,000 km² study area, with larger concentrations occurring closer to the coast, about 70 km from the colony (Slattery and Alisauskas 2007). At the flock level (5–1,500 birds), geese avoided lichen heaths, used other terrestrial habitats as available, and preferentially selected freshwater sites. Individual habitat selection was for lowland areas, such as wet sedge meadows, hummock graminoid tundra, and freshwater locales (70% of birds observed); upland habitats were avoided. Broods probably selected lowland habitats because of the greater availability of food and easier predator avoidance, compared with upland areas. Most geese used freshwater sites, demonstrating that assessments of brood-rearing habitat must include areas beyond the coastal saltmarshes, which are traditionally considered the primary brood-rearing habitat in this region for Lesser Snow Geese and Ross's Geese.

On Bylot Island (1989–90), radiotracking of 20 Greater Snow Goose families over a 70 km² area of wet tundra and uplands revealed 3 categories of habitat users: sedentary geese, shifters, and wanderers (Hughes et al. 1994b). Initial movements from the nest to brood-rearing areas varied from <1 to 5 km, but home ranges then averaged 680 ha for sedentary families, 1,660 ha for shifters, and 1,820 ha for wanderers. The home ranges of sedentary families included more ponds and lakes, while shifters and wanders utilized more upland habitat; such differences potentially accounted for variations in the body mass of goslings captured near fledging. Sedentary families also nested earlier, which indicated that they were the more experienced breeders.

Brood movements associated with radiomarked females on Bylot Island were assessed during a 5-year study (1997–2001) by Mainguy et al. (2006). Total movement ($n = 41$ females) averaged 25.6 km and did not differ among years, but it varied among individuals (from 2.6 to 52.5 km), depending on the area selected for brood rearing. Most of this movement (92%; $n = 22$) involved the distance traveled within the first 6 days after the hatch. Broods in 2000–2001 traveled an average of 32.3 km from the main nesting colony to brood-rearing areas, which were upland habitats dominated by mesic tundra and widely scattered wetlands with predator-safe areas, such as ponds and lakes. On the southern plain of Bylot Island, Massé et al. (2001) calculated that wetlands made up about 11% of a 1,600 km² study area. Suitable forage plants for adults and goslings were most plentiful in stream (95% occurrence) and wet polygon habitats (60%), but scarce in lake polygons (7%), polygon channels (5%), and lakes (1%). Plant cover was high in all habitats (≥90%), except for lakes (50%).

In 1981, during the brood-rearing period at Jungersen Bay (northern Baffin Island), Greater Snow Geese used 3 major habitat types: tidal marshes dominated by *Carex subspathacea* and *Puccinellia phryganodes*; wet moss-covered meadows with up to 5 cm of standing water that were dominated by *Carex stans*, *Dupontia fisheri*, *Calamagrostis neglecta*, and *Arctagrostis latifolia*; and the edges of ponds with bands of vegetation 1–2 m wide, dominated by *Carex stans*. The 3 most important plant species used by the geese were *Puccinellia phryganodes*, *Carex subspathacea*, and *C. stans*. In comparing foraging quality between sedge and grass habitats for broods on Bylot Island, Manseau and Gauthier (1993) found that grass meadows had a higher nitrogen and lower fiber content than sedge meadows, probably because goslings fed selectively on *Eriophorum* in the grass habitat.

Goslings typically leave the nest within 24 hours after hatching, which takes 15–36 hours and tends to be highly synchronized (Mowbray et al.

2000). Both parents guard the brood, but only the female broods the young. Brooding time ranges from 31.7%–37.5% of a 24-hour day immediately after hatching to only 9.8%–10.8% by day 16; the amount of time is less in warmer weather (Lessells 1987). Family groups remain together, but they join loose assemblages of 20–100 families for the 4- to 6-week brood-rearing period (Cooke et al. 1995). The parents and their young remain together through the first winter and often do not break up until after arrival on the breeding grounds the following year.

Brood Amalgamation (Crèches). In a study by Williams et al. (1994), brood amalgamation resulted in 5% of the goslings being reared by parents other than their own, with 13% of the families accepting 1 or more adopted goslings. Most (79%) families adopted only a single gosling. The timing indicated that the cost of fostering an extra gosling was low, because 46% of the adoptions occurred 15 days after hatching, when parents can recognize their own goslings. There was no evidence that adopted goslings selected high-quality parents, and there were no differences between adopting and nonadopting adults in terms of age, initial brood size, or body condition 5–6 weeks after hatching. Moreover, adoption appears to be beneficial to all parties. For parents, vigilance and brooding time do not increase with larger broods, but such broods may foster faster gosling growth rates and provide a higher dominance rank. Winter survival of the young may also increase, as return rates were higher in adopting versus nonadopting adults.

Development. Snow Goose goslings begin foraging soon after hatching and grow rapidly thereafter during the perpetual daylight that characterizes summer in the Arctic. Goslings may spend up to 90% of each day foraging (Owen 1980). Male goslings reportedly grow more rapidly than females (Ankney 1980), and goslings from larger-sized broods grow faster than those within smaller broods (Cooch et al. 1991a). At the McConnell River in the Northwest Territories, goslings from the heaviest eggs had a greater average weight than those from light eggs (94.9 g vs. 73.1 g) and survived starvation longer (112 hours vs. 93 hours), but the potential advantage between egg mass and body mass decreased as the goslings grew, so that the body mass of goslings was not different at fledging (Ankney 1980). At La Pérouse Bay, gosling growth was more rapid and survival much greater for those that dispersed away from the colony to forage, compared with those that remained within the colony (Cooch et al. 1993).

On Akimiski Island, located on the western side of James Bay (53° N latitude), Lesser Snow Geese averaged 66.8 g on day 1 (4% of adult weight), grew to 1,129 g by day 31 (64% of adult weight), and reached 1,315 g near fledging (43 days), which was 74% of adult weight (Badzinski et al. 2002). Body parts important for gathering food were especially large at hatching. The culmen averaged 17.2 mm on day 1 (32% of adult size); the tarsus, 31.0 mm (38% of adult size). In contrast, on day 1 the wing length was only 13% of adult size; the small intestine, 24% of adult size. The authors suggested that selection favors goslings able to process food (development of the bill, the digestive organs, and the leg muscles) and thus take advantage of the long arctic days and short-term availability of high-quality plant foods. Aubin et al. (1986) reported that Lesser Snow Goose goslings within the McConnell River colony in the Northwest Territories fledged at an average of 72% of adult weight. Ankney (1980) reported fledging at 42 days at the McConnell River colony.

The growth and development of Greater Snow Goose goslings is of particular interest, both because of their larger size (compared with Lesser Snow Geese) and because their main colony on Bylot Island is among the northernmost goose-breeding areas on the continent (73° N latitude), with a very short brood-rearing season. The goslings grow very rapidly, however, with mean body mass increasing from 80 g at hatching to 2,332 g by the time they reach their fall staging areas along the St. Lawrence estuary, 110 days later (Lesage and Gauthier 1997). The growth constant ($k = 0.093$) for the goslings is among the highest reported for precocial birds. Leg muscles grow especially rapidly, achieving 50% of asymptotic mass 11 days before body mass reaches the same point, and 95% of asymptotic mass at fledging (43 days). Goslings are especially lean during their growing period, with the accumulation of lipids delayed to the point where they form only 1% of body mass at fledging; hence young birds begin fall migration with very low lipid reserves. Greater Snow Geese fledged at about the same age as species of smaller geese, but they did so at 68% of adult body mass, compared with 79% for Lesser Snow Geese and 89% for Cackling Geese (*Branta hutchinsii minima*; Sedinger 1986). Sedinger concluded that selection for rapid growth is exceptionally strong in arctic-nesting geese, because short summers force goslings to grow quickly and depart early.

A subsequent study of this same population found that late-hatched goslings may adjust their growth patterns in response to reduced food availability (Lesage and Gauthier 1998). Although early-hatched goslings had more body protein than late-hatched goslings, the mass of organs associated with food acquisition (legs, esophagus, intestines, liver) was similar between the 2 groups. Late-hatched goslings had much smaller breast muscles, however. In attempting to maintain a necessarily high nutrient intake, late-hatched goslings thus appear to favor the development of organs associated with food acquisition at the expense of other organs.

RECRUITMENT AND SURVIVAL

Recruitment of Snow Geese depends on the breeding propensity of adults, clutch size, hatching success, nest success, gosling survival, and the survival of fledged young, all of which can very annually in response to habitat and demographic conditions. At La Pérouse Bay (1973–91), clutch size at hatching was 3.5–4.5 eggs, but the number of goslings leaving the nest averaged 3.1–4.3, a decline of 4.4%–11.4%; brood size at fledging averaged 1.6–3.5 goslings (Cooke et al. 1995). Losses among broods are highest during their first week after hatching, with the main predators of goslings being gulls, jaegers, Common Ravens, arctic foxes, red foxes (*Vulpes vulpes*), wolves, and polar bears. Total brood loss, averaging 8.5%, was especially variable (range = 1%–62%), but 75% of the broods that did hatch reached fledging. Other reproductive parameters associated with the recruitment of Lesser Snow Geese at La Pérouse Bay also exhibited significant annual variation: preincubation nest failure (17%–45%), total nest failure (1%–25%), clutch size (3.6–4.8 eggs), goslings leaving the nest (3.1–4.3), and the production index (0.5–3.3 goslings). On Bylot Island (1990), mean brood size declined significantly, from 3.17 ($n = 414$) early in the season to 2.70 ($n = 1,060$) late in the season (Hughes et al. 1994a).

At the La Pérouse Bay colony, Cooch et al. (1989) demonstrated the effects of increasing population size on recruitment, as the colony grew from just over 2,000 pairs in 1970 to about 9,000 in 1984. Average clutch size varied from 3.7 to 4.9 eggs/year (1973–87); clutch size adjusted for brood parasitism declined significantly—0.05 eggs/year, or 0.72 eggs (16% of the initial mean)—and was independent of a female's age and experience. Mean clutch size was also negatively correlated with colony size ($r = -0.71$), as well as with the size of the winter population in the Mississippi and Central Flyways combined ($r = -0.554$). Because Lesser Snow Geese accumulate reserves for clutch formation before they begin nesting, this decrease in clutch size may reflect increasing competition for food during spring migration. Gosling survival declined in areas of high densities because adults remove large quantities of plant food, which leaves much less available during the brood-rearing period (Cooch et al. 1993).

Alisauskas (2002) provided a broader study of recruitment in Lesser Snow Geese in his examination of the relationship between arctic climate, spring nutrition, and recruitment in the MP, based on age ratios in the harvest in the Central and Mississippi Flyways (1962–99) and spring nutrient reserves in southern Manitoba (1983–84, 1988–93). Age ratios (immatures:adults) were inversely correlated with arctic weather, as indexed by average snow depths in May and June and mean temperatures in June. Fall age ratios were also correlated with body mass, as well as with lipid reserves during staging in April. Recruitment was affected by the arctic climate (69%), spring lipid reserves (20%), and the particular flyway that was used (11%). Nonetheless, after accounting for these effects, the age ratio in this population exhibited a long-term (1962–99) decline of 0.017/year; body mass also decreased (1983–93). As at La Pérouse Bay, this study implicates the expanding Lesser Snow Goose population in the decline of recruitment, but it happens over a broader scale than at La Pérouse Bay.

Recruitment of young is also affected by a female's age, as seen during studies (1973–81) of marked females at La Pérouse Bay (Ratcliffe et al. 1988). A greater percentage of hatchlings were recruited by females that were 5–7 years old, compared with younger (2–4 years) and older (8–13 years) females. When calculated as the number of recruits observed within 4 years of their hatching year, 50% were produced by 5- to 7-year-old females, 40% by 2- to 4-year-olds, and 10% by 8- to 13-year-olds. About 35% of all recruits were produced by fewer than 20% of the females.

Reproductive success at La Pérouse Bay (1973–88) increased among females aged 2–6, due to significant increases in clutch size and gosling survival and a reduced probability of total nest or total brood loss (Rockwell et al. 1993). Clutch size grew by an average of 14.4% for females between 2 and 6 years of age, but it did not increase thereafter, to at least age 15. After 6 years, however, reproductive success declined significantly, due to reduced egg hatchability and decreased gosling and brood survival. The lower hatchability was attributed to senescence. This decline in reproductive output among older birds may reflect their reduced physical capacity to provide adequate parental care, but it may also be influenced by the philopatry of females to areas where habitat deterioration and a resultant loss of forage occurred, due to overpopulation. In addition, replacement mates for older females may be younger, more inexperienced males.

The best measure of an individual's contribution to the next generation is measured by its lifetime reproductive success, although annual or age-specific reproductive success can be misleading if there are tradeoffs among survival, breeding propensity, and reproductive success (Cooke et al. 1995). Modeling the lifetime reproductive success of Lesser Snow Geese at La Pérouse Bay revealed that 75% of the fledglings had a lifetime reproductive success of zero, which included individuals dying before their first breeding attempt, as well as those surviving to 2 years of age and attempting but failing to breed successfully at any point along the reproductive cycle. Among successfully breeding birds, lifetime success declined gradually, as expected, given the high probabilities of adult survival and breeding. In the model, the most successful female produced over 40 fledglings. The mean estimated lifetime reproductive success was 2.1 recruits, which is slightly above a replacement value of 2.0.

Adult Snow Geese have high survival rates, which is a key factor contributing to their population explosion. Moreover, an analysis of 30,000 band recoveries for 230,000 Lesser Snow Geese banded in the MP during 1950–88 indicated that few differences existed between survival rates for males and females (Francis and Cooke 1992). There were indications that males might have a slightly higher survival rate than females, but estimates can be biased by the emigration of males to migration routes where recovery rates are lower. The mortality rates of adults are primarily influenced by hunting, while mortality for young birds mostly occurs before they leave the breeding grounds.

Survival rates can vary annually, but they can also exhibit patterns over the long term. At La Pérouse Bay (1970–88), survival rates for adults rose from about 78% in 1970 to nearly 88% by 1987 (Francis et al. 1992a). This increase coincided with a decline in hunter harvests, which therefore may have contributed to greater survival during that period. In contrast, the survival of juveniles decreased from about 60% in 1970 to approximately 30% in 1987, despite the reduction in hunting pressure. This decline indicated that juveniles were experiencing increased natural mortality, most likely either before they departed the breeding grounds or during fall migration. Greater juvenile mortality appeared to be correlated with a slower growth of goslings and reduced body size at fledgling, due to deterioration of the habitat on brood-rearing areas. A more recent analysis of recoveries of birds banded in La Pérouse Bay indicates that adult survival continues to increase. Cooke et al. (2000) estimated the annual survival for both males and females to be 94% (1990–94). Adult survival estimates from Lesser Snow Geese banded in the Queen Maud Gulf averaged 94% for males and 92% for females.

Survival rates also vary with age and breeding status, as reported by Francis et al. (1992b) in their analysis of recoveries and recaptures (1952–87) from the 350,000 Lesser Snow Geese that were banded at colonies in Canada, migration stopover points in the Dakotas and Missouri, and on their wintering grounds in Louisiana and Texas. The annual first-year survival of goslings banded on 5 breeding colonies varied from 10% to 70% of adult survival rates, but averages for the 5 colonies ranged from 25.0% at the northernmost colony at Baffin Bay (66° N latitude) to 61.3% at the southernmost, on Cape Henrietta Maria (55° N latitude). Most mortality occurred on the breeding area or early in migration. Young birds that survived to reach migration stopovers or wintering grounds were more vulnerable to hunting than were adults (recovery rates of 1.3–7.5 young vs. 0.5–3.4 adults), but their survival was only slightly lower. The greater hunting vulnerability and lower survival of young geese continued through their second year. Among adults, nonbreeders and failed breeders were much less vulnerable to hunting than were successfully breeding adults, but they did not have a higher survival rate. There was also evidence that experienced breeders had higher survival rates.

Mean survival rates from the various locations were 42.4% for young geese and 81.6% for adults at La Pérouse Bay (1970–88); 47.8% for young and 82.7% for adults at Cape Henrietta Maria (1969–79); 65.8% for young and 82.4% for adults in the Dakotas (1961–76); 77.2% for young and 78.8% for adults in Missouri (1961–70); 79.6% for young and 84.2% for adults in Texas (1970–88), and 92.4% for young and 81.1% for adults in Louisiana. Cooke et al. (1995) reported first-year survival of females at La Pérouse Bay (1973–91) as 20%–68%, compared with 15%–60% for first-year males. Adult survival was 68%–100%. During a study of Greater Snow Geese on Bylot Island (1991–96), first-year survival averaged 40%, and annual survival of adult females averaged 83.3% (Menu et al. 2000).

The longevity records for Snow Geese are 27 years and 6 months for a Lesser Snow Goose recovered in Texas (sex unknown), and 22 years and 9 months for a male Greater Snow Goose recovered in New Jersey. Both birds were banded as adults.

FOOD HABITS AND FEEDING ECOLOGY

Snow Geese are herbivores, with various above- and belowground plant parts and agricultural grains comprising nearly all of their diet throughout their annual cycle. The preferred foraging habitats of Lesser Snow Geese include tidal marshes, brackish marshes, braided creek estuaries, shallow freshwater wetlands, sedge (*Carex* spp.) meadows, and agricultural fields.

As in other species of geese, Snow Geese ingest large amounts of food that are retained for a short period of time. Lesser Snow Geese staging during late summer on the Beaufort Sea coastal plain in northeastern Alaska consumed an average of 666 g/bird (wet weight) of tall cottongrass (*Eriophorum angustifolium*) during a minimum 10-hour day, which was equivalent to 29% of the

body mass of a 2–3 kg goose (Hupp et al. 1996). On average, food was retained for only 1.37 hours, and 48% of the organic matter in the bases of plant stems was metabolized. During the 21-day staging period at this site, roughly 300,000 Snow Geese were conservatively estimated to have removed >900,000 kg of cottongrass (dry weight).

Feeding occurs in large flocks, with the geese gleaning, grazing, grubbing for plant roots, and picking or plucking fruits (Mowbray et al. 2000). Tundra plants, sedges, and grasses favored by Snow Geese can generally recover and even thrive if they are only lightly to moderately grazed. Moreover, sedges and alkaligrass (*Puccinellia phryganodes*) decline and are replaced by other plants when grazing by Snow Geese ceases (Abraham and Jefferies 1997). The bill is especially adapted for grubbing, with a horny area along the edge of each mandible (the grinning patch) that produces a viselike grip for extracting the underground tubers and rhizomes of marsh plants (Bolen and Rylander 1978). Grubbing occurs especially during the spring, when food sources are scarce, and again during the late summer, when plants are dying back and most of their nutritional value is found in the roots. It is also an important foraging method in coastal marshes during the winter, when most of the nutritional value of marsh plants is stored in the rhizomes and tubers.

Extensive grubbing for plant roots and stems has been extremely destructive to tundra breeding grounds as Snow Goose populations have increased in number, with large concentrations of geese known to remove the vegetation from an entire area over the course of the breeding season. Grubbing by large flocks can also denude marsh vegetation and create large openings. At the McConnell River, grubbing by Snow Geese has exposed underlying glacial gravels (Kerbes et al. 1990). At La Pérouse Bay, Williams et al. (1993) reported that the intact vegetative community on the breeding grounds located on saltmarsh swards decreased by 12% from 1985 to 1992, while the remaining areas showed roughly a 50% decrease in productivity from 1979 to 1991. Because overgrazing degrades coastal saltmarsh tundra areas, Lesser Snow Geese are forced to move to inland meadows and upland habitats in search of food (Cooke et al. 1995). Large numbers of geese, both parents and young, have been known to travel south and east to coastal areas along Hudson Bay in search of food (Cooch et al. 1993).

Breeding. Important foods for breeding adults include the leaves and roots of grasses, sedges, rushes (*Juncus* spp.), forbs, willows (*Salix* spp.) and other tundra shrubs (Owen 1980, Cooke et al. 1995). Prior to incubation, Lesser Snow Geese at La Pérouse Bay feed intensively on watersedge (*Carex subspathacea*), goosegrass (*Calamagrostis deschampsioides*), and red fescue (*Festuca rubra*), but then switch to the young shoots of saltmarsh grasses and sedges later in the breeding season (Ganter and Cooke 1996). Preferred forage plants on the coastal tundra of Hudson and James Bays were sedges and cottongrass. Favored grasses included Fisher's dupontia (*Dupontia fisheri*), pendantgrass (*Arctophila fulva*), beach rye (*Elymus arenarius*), and alkaligrass (Abraham and Jefferies 1997).

On Bylot Island, the diet of breeding Greater Snow Geese was 100% vegetative and did not differ between males and females (Gauthier (1993a). Maydell oxytrope (*Oxytropis maydelliana*) roots and rhizomes, alpine bistort (*Polygonum viviparum*), grasses, and sedges formed 40% of their diet, with the remainder (60%) consisting of the young leaves of grass, Scheuchzer's cottongrass (*Eriophorum scheuchzeri*), arctic willows (prostrate willows), and watersedge (*Carex aquatilis*) stems. Grasses, willows, cottongrass, and oxytrope were high in protein content (≥20%), although grasses, willows, cottongrass, and sedges (stems) were low in acid detergent fiber (<25%). The emerging leaves of willows were potentially the most nutritious food. The diet of Greater Snow Geese also differed by habitat type and breeding status, with the diets of prelaying and laying birds using upland tundra consisting of 45% belowground plant parts and 55% aboveground plant parts. In contrast, birds during early incubation in upland habitats ate the emerging leaves and catkins of arctic willows (44.4%), alpine bistort bulbs (27.5%), and the leaves of various grasses (17.8%); <30% of their diet was belowground plant parts. Prelaying and laying birds feeding in sedge meadows primarily consumed the basal stems of water sedges (62.0%) and the emerging leaves and catkins of arctic willows (22.9%). At Jungersen Bay in northwestern Baffin Island, the principal forage items were alkaligrass and sedges (Giroux et al. 1984).

Young goslings feed on fruits and flowers, horsetail shoots, and midge (Chironomidae) larvae (Palmer 1976a, Owen 1980). According to Gauthier et al. (1995), Fisher's dupontia and cottongrass are important gosling forage foods on arctic breeding grounds. Juveniles collected along the St. Lawrence River area consumed more bulrush stems (40%), spikerushes (*Eleocharis*), and rushes (*Juncus*) than did adults (Giroux and Bédard 1988). During 2000–2001, Audet et al. (2007) noted that the chief foods of Greater Snow Goose goslings on Bylot Island consisted of grasses (primarily *Arctagrostis latifolia*), smartweeds (Polygonaceae), rushes (Juncaceae), cresses (Cruciferae), and horsetails (*Equisetum* spp.).

Migration and Winter. On their major staging areas in the Arctic and on the extensive coastal marshes in Louisiana and Texas, Lesser Snow Geese still forage exclusively on plant parts. During spring on the Mackenzie River delta in the Northwest Territories, Lesser Snow Geese first fed on sedges, grasses, and horsetails and then moved inland, where they consumed cranberries (*Vaccinium* spp.), curlewberries (*Empetrum* spp.), salmonberries (*Rubus spectabilis*), and cottongrass (*Eriophorum* sp.) tubers (Barry 1967). During late summer and early fall, staging Lesser Snow Geese on the coastal plain of the Arctic NWR (*n* = 75) fed extensively on the belowground stems of tall cottongrass and the aboveground shoots of northern scouringrush (*Equisetum variegatum*), which together formed up to 94% of their diet (Brackney and Hupp 1993). Within the Fraser River delta in British Columbia, wintering Lesser Snow Geese almost exclusively ate three-square (American) bulrush rhizomes (Burton et al. 1979).

The most important feeding area for migrating Snow Geese from the Hudson Bay colonies is a strip of coastal marsh 1–10 km wide, extending about 2,000 km from Rupert Bay in Québec to near Churchill, Manitoba. During the fall in 1975 and 1976, the food habits of 273 Lesser Snow Geese collected in the James Bay area revealed that 9 plant species in the Equisetaceae, Juncaginaceae, Gramineae, and Cyperaceae totaled 90% of the foods eaten (Prevett et al. 1979). At 3 locations in the southwestern part of James Bay, bulbs of arrowgrass (*Triglochin palustris*) were the most preferred food. During spring (1976, 1978–80) along the western coast of James Bay, the diet of 148 Lesser Snow Geese consisted of the rhizomes of sedges (*Carex* spp.; 42.2%), the shoots of goosegrass (*Puccinellia phryganodes*; 20.7%), and arrowgrass bulbs (13.8%); the latter was the most heavily selected food in both spring and fall.

In an associated study of body weight and nutrient-reserve dynamics of Lesser Snow Geese staging during spring and fall in 1976 on the southern part of James Bay, Wypkema and Ankney (1979) noted that the geese fed heavily on arrival in the spring, which subsequently increased their body weight and protein reserves; the latter amounted to 27% of the reserve used later on the breeding grounds by males, and 20% of that used by females. The protein reserve for females was enough to lay about 1 more egg (0.8) in an average clutch. The lipid reserves of both adults and juveniles on southern James Bay increased significantly during the fall, which lengthened their theoretical maximum flight range, so this accumulation was an important part of fall migration. Juveniles also forage in the fall to increase their structural size, as they are not fully grown when leaving the breeding grounds.

Alisauskas et al. (1988) discussed the feeding ecology of Lesser Snow Geese wintering in 3 different habitat types: (1) the coastal marshes of Louisiana; (2) inland rice prairies in Texas and Louisiana; and (3) agricultural land in Iowa, Kansas, and Missouri, where waste corn was widely available. In coastal Louisiana, their January–February diet was predominantly (61%–90% aggregate percentage) the subterranean parts (rhizomes and tubers) of marshhay cordgrass (*Spartina patens*), seashore saltgrass (*Distichlis spicata*), Olney's bulrush (*Scirpus olneyi*) and saltmarsh bulrush (*S. robustus*). In contrast, geese feeding in rice stubble mainly ate grasses (28%) and forbs (70%), but when winter rains loosened the soil, the birds began grubbing for rhizomes. Corn formed 99%–100% of the diet of geese collected from habitat type (3), although these geese also used other agricultural fields, such as winter wheat, soybeans, and milo (sorghum). Nutritionally, diets from rice fields were the highest in protein, but fiber content was greatest in the diets of birds collected in coastal marshes. Meeting the daily existence energy requirements of the geese (167 kcal/bird) would necessitate consuming 2.4 times more food in coastal Louisiana and 4.3 times more in the rice fields, compared with cornfields.

The foraging methods for a diet dominated by rice are different from one consisting primarily of corn, and Alisauskas (1998) found morphological differences among geese from diverse habitats. Those collected from marsh habitats had the largest body size; those from rice prairies were intermediate, but the most variable; and those to the north, relying on corn resources, were smallest. Additionally, geese in marsh habitats had thicker bills, longer skulls, and longer culmens than geese using cornfields.

Almost everywhere in the West and Midwest, crops are important sources of food for Lesser Snow Geese. Since the early 1970s, changing crop patterns have allowed several hundred thousand Lesser Snow Geese to linger into January along the Missouri River in Iowa, Nebraska, Kansas, and

Missouri. In California, Lesser Snow Geese consume barley, wheat, and rice grains and graze on the green shoots of pasture grasses and cereal crops. Waste corn gleaned in harvested fields is particularly important, but tubers and rootstocks of bulrushes and cattails are still eaten at the Bear River marshes in Utah, the Klamath basin in California, and Summer Lake in Oregon (Bellrose 1980). Snow Geese on the Bear River marshes feed extensively on the succulent tips and tubers of alkali bulrush (*Scirpus paludosus*). During the spring (8 Mar–18 Apr), approximately 250,000 Lesser Snow Geese were observed feeding in an 80 ha plowed cornfield near Havana, Illinois, that had been subjected to considerable wind damage before harvest (Frank Bellrose). The waste corn provided forage for the entire group for at least a week, while thousands continued to feed for a month.

Greater Snow Geese staging along the St. Lawrence estuary traditionally spent the majority of their time grubbing for rhizomes of three-square (American) bulrush. Extensive feeding by this expanding population, however, has adversely affected bulrush vegetation, especially in sanctuaries, which has led to more birds using saltmeadow cordgrass (*Spartina patens*) marshes farther toward the mouth of the estuary, as well as adjacent agricultural areas planted in corn and hay (Bédard et al. 1986, Giroux and Bédard 1987). During winter along the mid-Atlantic coast, Greater Snow Geese roost and feed in saltmarshes and managed impoundments located on several national wildlife refuges, but they forage extensively in surrounding fields of waste corn.

MOLTS AND PLUMAGES

Detailed accounts of molts and plumages can be found in Palmer (1976a), Cooke at al. (1995), and Mowbray et al. (2000). Unlike all other waterfowl, Lesser Snow Geese are unique in that goslings, immatures, and adults occur in color phases, or morphs (white or blue), with intermediate intergrades when the morphs interbreed. There are as many as 7 identifiable plumage types among adults (Cooke and Cooch 1968), ranging from pure white with a minimum of 7 black primaries and little or no gray on the wing coverts (Type 1), to the pure blue morph with gray underparts and mostly pale or white plumage between the legs and the tail. Like all other geese, as well as the whistling-ducks, adult Snow Geese have 1 plumage and 1 associated molt/year.

Goslings begin to acquire juvenal plumage about 2–3 weeks after hatching and complete the plumage around 6 weeks of age, at which time they can fly. The juvenal white morph is pale gray on the sides of the head and grayish on the crown, the nape, the sides, and the back; the underparts are white. In contrast, the juvenal blue morph is similar to the adult, but the head and the neck are a slate brown, with a white chin spot. The mantle and the scapulars are browner; the brown underparts are paler, with more white on the belly. Juvenal plumage is replaced by first basic about 12 weeks after hatching and is usually completed during winter, although there is considerable individual variation. The molt associated with first basic replaces all the body feathers except the juvenal remiges. The white morph has grayish upperparts and whitish underparts, all of which become whiter through the winter. The blue morph scarcely changes over the winter, although some white feathering may appear on the head and the neck (Ogilvie 1978).

Successful breeders acquire their definitive plumage when their goslings are about 3 weeks old; they also molt later than yearlings, nonbreeders, and failed breeders. The molting period is such that the family attains flight at about the same time.

CONSERVATION AND MANAGEMENT

The Snow Goose population explosion during the latter half of the twentieth century has dramatically affected not only the behavior and distribution of Snow Geese, but also the foraging habitat on breeding areas in the Arctic. Impacts of the burgeoning goose population on northern ecosystems have been well documented at several staging areas and nesting sites used by MP Lesser

Snow Geese and Ross's Geese that breed in Queen Maud Gulf, Southampton Island, and the Hudson Bay Lowlands of Nunavut, Manitoba, and Ontario. All contain plant communities and associated soils that are adversely affected at a large spatial scale (Abraham et al. 2012). Tundra habitats in and around nesting colonies of geese have been impacted most severely, as the pressures from large colonies and their foraging methods of grubbing and pulling have resulted in a tremendous loss of vegetative biomass in these areas. Further, the fidelity that Snow Geese exhibit toward their breeding areas ensures that they nest in the same arctic locales for decades, although as nesting areas in coastal tundra are degraded, some nesting birds have gradually shifted farther inland and impacted freshwater wetlands. Moreover, once an area has been severely degraded, it takes only a small number of geese to sustain the damage and thus inhibit recovery. At 9 study areas in southern Hudson Bay, up to 43% of the vegetation loss occurred since 1996, with no evidence of intertidal vegetation recovery within the region (Jefferies et al. 2006). The problem was initially described in a summary report by Abraham and Jefferies (1997), and in an update to that report, Abraham et al. (2012:12) stated, "The overall conclusion is there is no evidence of a decline in the rate of habitat degradation either in intertidal habitats or inland freshwater habitats along southern Hudson Bay since the initial report was written."

Good, easy-to-read accounts of the problems overabundant Snow Geese create on their arctic breeding grounds are contained in Rockwell et al. (1996), Abraham et al. (2005), and Abraham et al. (2012). They note that excessive grubbing and pulling destabilizes the thin arctic soil layer, which allows melting snow and spring rain to erode the earth and reveal glacial tills and gravels. In some situations, the geese will also create depressions that subsequently fill with water and expand each year as the birds feed along the edges. The loss of soil-stabilizing tundra sedges and grasses also leads to greater evaporation of soil moisture,

which draws inorganic salts to the surface and thus increases soil salinity to as much as 3 times that of sea water. Such adverse soil conditions then inhibit the growth of preferred forage grasses, as evidenced by studies at La Pérouse Bay where exclosures erected on exposed sediment in 1982 were still devoid of vegetation in 2004 (Abraham et al. 2005). McLaren and Jefferies (2004) found that patches of exposed sediment only 20 cm in diameter could become hypersaline over a single growing season and thus limit the establishment of graminoids. On a landscape scale, such results are catastrophic. Some 65% of the 54,800 ha intertidal saltmarsh located between the Maguse River in Nunavut (on the western coast of Hudson Bay) and Attawapiskat, Ontario (on the western coast of James Bay), is no longer productive, while the remaining area in this saltmarsh is heavily utilized by the geese (Abraham et al. 2012). There are no indications of large-scale recovery of overgrazed plant communities at any site.

In addition, as habitat conditions deteriorate in the intertidal zone, the geese move inland to forage within nearby freshwater wetlands dominated by the sedge *Carex aquatilis*, the base of which is a rich source of carbohydrates and protein. Continued feeding in these areas, however, can cause the sedges to die and expose the underlying peat, which can then lead to various detrimental changes in plant cover, including exposure of glacial tills and marine clays. *Carex aquatalis* is also not a good food for the developing goslings, so family groups need to forage over broader areas. The loss of aboveground plant cover can be extensive wherever families concentrate in numbers, such as on Bylot Island (Greater Snow Geese) and on Lesser Snow Goose breeding areas in the central Arctic (Abraham et al. 2012).

At the Karrak Lake colony in Nunavut, overgrazing by Lesser Snow Geese and Ross's Geese virtually removed all grasses and sedges in lowland areas of the colony, which then contributed to an order-of-magnitude reduction in the abundance of voles and lemmings there, compared with lowland areas outside the colony (Samelius and Alisauskas 2009). At its extreme, overgrazing by geese leads to dried ponds with cracked earth and a "forest" of dead willow stubs around the periphery. Lastly, once the tundra loses its layer of protective vegetation, the subsequent recovery can take decades, and the damage may be permanent in the most affected areas (Kerbes et al. 1990, Jefferies et al. 2003). This destruction of arctic habitats has been likened to a "Snow Goose ghetto" and is dramatically depicted in the video *Snow Geese in Peril*, developed by Ducks Unlimited, along with an accompanying book, *Snow Geese—Grandeur and Calamity on an Arctic Landscape* (Batt 1998).

In addition to geese, the deterioration of these habitats also affects other species. Few invertebrates can survive in the resultant new system, which reduces habitat for shorebirds. The loss of soil moisture causes the death of willows and other shrubs that provide nesting cover for tundra-nesting species such as Yellow Rails (*Coturnicops noveboracensis*), Savannah Sparrows (*Passerculus sandwichensis*), Northern Shovelers, and American Wigeon (Rockwell et al. 1996), although habitat alteration by geese may need to be quite widespread and severe before population declines of other tundra-nesting birds occur over large areas (Sammler et al. 2008).

The degradation of forage and nesting habitat has other impacts on Snow Geese when concentrations get too high. At La Pérouse Bay, mean clutch size from 1973 to 1987 declined by 0.72 eggs, or 16% of the initial mean clutch size (Cooch et al. 1989). In addition, the average mass of Snow Goose goslings decreased by about 16% between 1976 and 1988, and the mean size of adult females declined by 15% (Cooch et al. 1991b). Such data indicate that overcrowding on at least some breeding colonies is having a density-dependent effect that affects reproductive output. In response to such overpopulation and habitat degradation, goose families at La Pérouse Bay are also dispersing from their traditional feeding areas, and goslings from these dispersed broods on average are 7.3% heavier than

those that remained on traditional sites (Cooch et al. 1993).

On a much broader scale, Alisauskas (2002) noted a long-term decline of −0.017/year in the age ratio (immatures:adults) of the MP Lesser Snow Geese from 1962 to 1999, which reflects a widespread reduction in productivity. There was less variation in this decline after 1980, which may reflect an accelerating reduction in productivity as overcrowding increased. Adults are more adept at finding food and avoiding hunters and, thus, are able to maintain high survival rates in relation to young geese, so the total population of Snow Geese is now adult dominated.

In light of all this, the major management concern for Snow Geese (and Ross's Geese) is contending with the problems associated with overpopulation. In 1997 the Arctic Goose Habitat Working Group conducted a comprehensive review of this issue, which resulted in the publication of *Arctic Ecosystems in Peril* (Batt 1997). The group's recommendations called for a 50% reduction of the midcontinent population of light geese by 2005, and a reduction in the annual population growth rate (λ) to 0.85–0.95 (a 5%–15% reduction in total numbers per year) from the current growth rate of about 1.05 (5% growth per year). To achieve the proposed reduction in population growth, the group also recommended an increase in the harvest rate to about 3 times the existing level, which would require 2.2 million of these geese to be taken annually. Suggested control methods included extending the hunting seasons in both autumn and spring; permitting larger bag numbers; expanding shooting hours; liberalizing the use of electronic calls, bait, and live decoys; increasing harvest opportunities on reserves and refuges; attracting more hunters by issuing a single standard Snow Goose hunting license for all states, provinces, and countries; and permitting more egg collecting by native peoples on the tundra breeding grounds. The intent of these actions was to reduce adult survival, which is a key driver of population growth.

Despite both implemented and proposed harvest measures, however, high populations of Snow Geese (and Ross's Geese) continue to damage tundra ecosystems, due to their long life span and low rate of natural mortality, even though productivity (as indexed by age ratios) has exhibited a long-term decline. What is most alarming is that such large-scale alterations in arctic ecosystems affect other wildlife using the same areas, and this degradation may continue until light goose populations begin to regulate their own numbers (Alisauskas et al. 2011). Nonetheless, harvest management of light geese is likely to continue to be an area of major conservation concern in the coming decades, as increasing harvest opportunities and their associated kill rate is probably the only viable option to control the light goose population. Johnson and Ankney (2003) summarized other possible, more direct measures to control the burgeoning population of light geese in North America (e.g., trapping on arctic breeding grounds), but they emphasized that increasing the harvest by hunters was the most desirable solution for reducing the overabundant population of light geese. Such efforts are not yet affecting the survival and abundance of the MP, however, which is by far the largest component of the continental Lesser Snow Goose population.

A key area needing more research is the estimation of carrying capacity in the central and east-ern Canadian arctic regions, as this capacity appears to exceed the current size of the MP there, which continues to grow, even if at a reduced rate (Alisauskas et al. 2011). Preferred freshwater wetlands and their associated lowland tundra habitats in the western half of the Queen Maude Gulf Bird Sanctuary appear to be largely unexploited by Snow Geese and Ross's Geese (Abraham et al. 2012), which indicates that there could be room for continued expansion of breeding populations in the central Arctic. In contrast, nearly all of the intertidal marshes associated with the Hudson Bay Lowlands have been severely degraded by foraging geese.

A plausible scenario for these arctic habitats is that the supply of agricultural grains (e.g., waste corn) on wintering grounds in the United States, and the increasing acreages of new crops used by Snow Geese farther north in Saskatchewan and elsewhere, will continue to benefit and perhaps even add to an already high carrying capacity, which presumably will continue to support large numbers of Lesser Snow Geese and Ross's Geese during migration and winter in the midcontinent for the foreseeable future. Unfortunately, this situation translates into large numbers of geese becoming available to continue their degradation of arctic breeding areas.

Ross's Goose

Chen rossii (Cassin 1861)

Left, adult; *right*, juvenile

The Ross's Goose is the smallest of the 3 forms of white geese in North America, where it is endemic. About 95% of the population breeds in the Queen Maude Gulf Migratory Bird Sanctuary in the central Canadian Arctic, primarily at Karrak Lake. Ross's Geese nest colonially with Lesser Snow Geese (the two species are collectively referred to as "light geese") to form the largest goose-nesting colony in the world. Their breeding grounds were not discovered until 1938, however, on a small lake southeast of the Perry River. The population, estimated at only 5,000–6,000 as recently as the 1930s, has proliferated to about 1 million today. Along with this increase in numbers, their breeding range has expanded into parts of the eastern Canadian Arctic. However, along with a burgeoning Lesser Snow Goose population, the increased numbers of Ross's Geese are damaging their fragile arctic breeding ecosystem. Their primary wintering range was once almost exclusively in the Central Valley of California, but wintering populations have now expanded into the Gulf Coast states of Texas and Louisiana, as well as into Arkansas, New Mexico, and the northern highlands of Mexico. Ross's Geese nest within

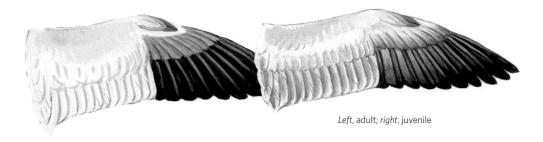

Left, adult; *right*, juvenile

3–5 days of arrival at their arctic breeding grounds, prompted by the short breeding season they encounter, but nest success is very high (often >80%). The Ross's Goose is closely related to the Lesser Snow Goose, and hybrids are not uncommon. The species was named in honor of Bernard R. Ross (1827–1874), the chief factor (trader) of the Hudson's Bay Company at Fort Resolution, on Great Slave Lake in the Northwest Territories.

IDENTIFICATION

At a Glance. Ross's Geese are small white geese with black wing tips in adults of both sexes. This species can be confused only with the Snow Goose.

Adults. The sexes are identical, with all-white plumage except for the black primaries. Ross's Geese are small, with short necks. The legs and the feet vary from bluish pink to an intense pink-red color, with black nails. These geese have a short, stubby, pink bill that lacks the black tomia (the inner cutting edges) forming the grinning patch in Lesser Snow Geese. Both males and females often have warty protuberances between the nostrils and the base of the upper bill; these are more pronounced in older geese, possibly serving as an indicator of social status (McLandress 1983b). The irises are blackish gray. A blue form (morph) is extremely rare and may be the result of hybridization with the blue form (morph) of the Lesser Snow Goose. McLandress and McLandress (1979) recorded only 3 blue morphs among 38,825 Ross's Geese in California.

Juveniles. Juvenile Ross's Geese are a pale slate gray that is noticeably lighter in color and not as exten-

sive as the plumage in juvenile Snow Geese. Juveniles do not lose this gray body color, especially on the head, until well into the following summer after they hatch. The legs and the feet are dark gray to greenish gray, with tan webs and toes. The bill is black to blackish gray, with the first pink coloring appearing at about 10 weeks of age (Dzubin 1965). The bill can also be a grayish blue, with a pale grayish-pink nail. The bill is generally all pink by mid-November, with males tending to show this pinkish color before the females. The irises are grayish brown, with olive-gray upper eyelids.

Goslings. The color pattern of Ross's Goose goslings is quite variable, but there are 2 basic morphs: yellow and gray. Ryder (1967) identified 7 color categories, and Nelson (1993) added an eighth: light yellow, yellow green, dark yellow, black yellow, dark gray, pearly gray, white, and medium gray. Different color morphs can appear in the same brood, although color differences are not evident after 3–4 weeks. In 2 studies at Arlone Lake in the Perry River region of the Northwest Territories, goslings in 95 out of 155 (61.3%) broods were a single color morph (monochromatic); 60 broods (38.7%) contained a mixture of 44.5% yellow goslings and 55.5% gray (Hanson et al. 1956, Ryder 1967). In 334 monochromatic broods, 30.5%–32.6% were yellow morphs and 67.4%–69.5% were gray morphs. The genetic basis of these color morphs is not known (Cooke and Ryder 1971).

Voice. Ross's Geese are less vocal than Snow Geese, but their calls are higher pitched (Sutherland and McChesney 1965). Migrants give a rapid, high-pitched flight call (*keek, keek, keek*) that is well

Left:
Ross's Goose; note the characteristic short and stubby bill.
Ian Gereg

Below:
Ross's Geese in flight.
RyanAskren.com

known among hunters. Other calls are given during displays and aggressive encounters.

Similar Species. The Ross's Goose can only be confused with the white color form (morph) of the Snow Goose. At a distance, the Ross's Goose appears to be similar to the Snow Goose, but up close the Ross's Goose is much smaller. It has a distinctly smaller bill that lacks the grinning patch of the Snow Goose, a faster wing beat, and a more high-pitched call. Ross's Geese are often found in mixed flocks with Snow Geese, which permits easy identification in a side-by-side comparison between the 2 species.

Weights and Measurements. Ross's Geese are the smallest of the 3 varieties of white geese, about two-thirds the size of the Lesser Snow Goose. Bellrose (1980) reported the following average body weights for Ross's Geese: 1,814 g for adult males ($n = 47$), 1,615 g for adult females ($n = 32$), 1,588 g for immature males ($n = 44$), and 1,474 g for immature females ($n = 36$). Wing length averaged 394 mm for adult males, 376 mm for adult females, 373 mm for immature males, and 361 mm for immature females. Ryder and Alisauskas (1995) reported an average bill length of 43.4 mm for adult males ($n = 54$) and 40.1 mm for adult females ($n = 47$). Tarsus lengths averaged 85.2 mm for the males and 81.0 mm for the females.

DISTRIBUTION

Both Ross's Geese and their sister species, Lesser Snow Geese, are endemic to North America. Trauger et al. (1971) documented 32 hybrids in a sample of 5,471 Ross's Geese and 6,489 Lesser Snow Geese (white phase) that were examined between 1961 and 1968. From these data, they then estimated that 4.7% of the continental population were Ross's × Lesser Snow Goose hybrids and expressed concern about potential genetic swamping. For Ross's Geese banded in the central Canadian Arctic, however, Kerbes (1994) determined that the incidence of hybridization was only 1.9% (56 out of 2,943 geese).

The primary breeding ground for Ross's Geese is the Queen Maud Gulf region of Nunavut in Canada, and the breeding location for this species was the last to be discovered among all geese in North America. Bent (1925:185) remarked that after a Ross's Goose has left its wintering grounds in California, "whither it goes when it wings its long flight northeastward across the Rock Mountains in the early spring no one knows, probably to remote and unexplored lands in the Arctic regions." Discovery of the breeding grounds did not occur until June 1938, and Gavin (1947) was the first to describe a nest from a colony of about 100 pairs on an island in a small lake 22.5 km southeast of the Perry River delta (now called Discovery Lake), which passes through what is now the Queen Maude Gulf Migratory Bird Sanctuary (MBS) in the central Canadian Arctic. Surveys in 1939–41 located 2 more nesting colonies on other lakes in the Perry River area, with collectively about 600 pairs on the 3 lakes. In June 1949, Hanson et al. (1956) discovered a fourth colony of about 260 pairs about 40 km south of the Perry River delta at what is now known as Arlone Lake. Subsequently, larger colonies were located by Ryder (1969a), Alisauskas and Boyd (1994), and Kerbes (1994).

Breeding. The Queen Maud Gulf region in the central Canadian Arctic contains an estimated 95% of all breeding Ross's Geese. Early photographic surveys of known colonies estimated roughly 34,000 Ross's Geese in 1966, and their numbers increased rapidly and steadily to 77,300 in 1976, 90,800 in 1982, 188,000 in 1988, and 567,000 in 1998 (Kerbes 1994, Kelley et al. 2001). These estimates are minimal, however, because the Queen Maude Gulf MBS covers a huge (54,000 km²) area; some colonies were probably missed, although Kerbes (1994) visited 92 light goose colonies. Nonetheless, the majority (91%) of Ross's Geese nest at only 5 colonies in the Queen Maud Gulf MBS (Alisauskas and Rockwell 2001). By far the largest colony is found at Karrak Lake, where an estimated 709,000 adult Ross's Geese nested in 2010 (U.S. Fish and Wildlife

Breeding Distribution ■
Wintering Distribution ■

numbers in Arkansas and Louisiana, and even into Mississippi. Prior to 1970, however, nearly all Ross's Geese wintered in the Central Valley (Bellrose 1980), where they apparently form 2 distinct subpopulations: those in the Sacramento Valley to the north (primarily in national wildlife refuges and the Gray Lodge Wildlife Management Area), and those in the San Joaquin Valley to the south (McLandress 1979). Wintering Ross's Geese in California numbered 106,412 in 1977, 214,722 in 1989, 250,537 in 2000, and 219,805 in 2005 (Weaver 2005), a dramatic increase from a 1956–60 average of only 13,000 (Bellrose 1980). Of the 2005 total, most (89%) inhabited the Sacramento Valley, with 3% in the San Joaquin Valley and 5% in the Imperial Valley. A few also winter in the lower Colorado River valley.

To the east, wintering Ross's Geese in New Mexico occur primarily at and around the Bitter Lake National Wildlife Refuge (NWR) in the Pecos Valley and the Bosque del Apache NWR in the Rio Grande valley. Surveys of the middle Rio Grande valley in New Mexico during December 2000 tallied 55,600 light geese, of which 21% of the adults were Ross's Geese (Drewien and Benning 2001). Sullivan (1995) estimated 66,800 wintering Ross's Geese in Texas in December 1994: from Sabine to Galveston Bay (5%), coastal Galveston to Lavaca Bay (21%), interior rice fields adjacent to that coastal area (54%), Lavaca Bay to the Rio Grande (2%), the Panhandle region (18%), and northeastern Texas (<1%).

On the interior highlands of Mexico, among a total of 229,288 light geese (Ross's Geese and Snow Geese) in 1998 and 310,204 in 1999, Drewien et al. (2002) observed 49,200 Ross's Geese in 1998 and 58,200 in 1999. Ross's Geese were most abundant (86%–93%) in the northern part of the state of Chihuahua, with some (7%–13%) in northern Durango. Important wintering sites in Chihuahua were 2 reservoirs at North Casas Grandes, and 2 lagoons and a reservoir in Ascension Valley. Primary feeding sites for these geese were fields of corn, oats, and sorghum. Ross's Geese also winter from the states

Service 2011), up from 224,000 in 1995 and 395,000 in 2000 (Alisauskas and Rockwell 2001).

As the Ross's Goose population exploded, nesting colonies have become established largely eastward of Queen Maude Gulf, in association with traditional colonies of Lesser Snow Geese: on Baffin, Southampton, and Banks Islands; on the western coast of Hudson Bay, at the McConnell River in Nunavut and at La Pérouse Bay in Manitoba; and at Cape Henrietta Maria in Ontario, where James and Hudson Bays meet (Didiuk et al. 2001). Helicopter surveys indicate more than 10,000 Ross's Geese on Baffin Island during some years, and as many as 2,250 pairs may be found as far south as Cape Henrietta Maria. A new colony near the McConnell River in Nunavut numbered about 90,000 in 2005, up from >70,000 in 2003. In 1983, well outside this normal breeding range, a single Ross's Goose nest was found on Howe Island in Alaska, along with the nests of 100–200 pairs of Snow Geese, but no Ross's Geese have been observed there in subsequent years (Johnson and Troy 1987).

Winter. The major wintering areas for Ross's Geese are the Central Valley of California, New Mexico, Texas, and interior Mexico, with ever-increasing

Ross's Geese and Lesser Snow Geese; note the much larger bill and characteristic black "grinning patch" of the Lesser Snow Geese.
Tim J. Moser

of Chihuahua to Zacatecas (Turner et al. 1994, Howell and Webb 1995). They are rare south to Jalisco, as well as along the Gulf Coast in the state of Tamaulipas (Saunders and Saunders 1981).

Migration. Not much is known about staging areas in the subarctic and boreal zones (Alisauskas 2001). To the south, however, Ross's Geese stop on the Athabasca River delta in northwestern Alberta, and then move south to shallow lakes in the Sullivan Lake region in southeastern Alberta and southwestern Saskatchewan, between the North and the South Saskatchewan Rivers (Dzubin 1965, Melinchuk and Ryder 1980). In the 1960s and 1970s, 90% of Ross's Geese used this route. From Alberta and Saskatchewan the birds move farther south; Freezeout Lake in Montana is a major staging area. The birds then cross the Continental Divide to reach their winter destinations in the Central Valley of California.

As the Ross's Goose population expanded, more and more birds used an eastern migration route, traveling southward along the western coast of Hudson Bay to eastern Saskatchewan and Manitoba, and then south through the Dakotas to win-

tering areas in Arkansas, Louisiana, Texas, New Mexico, and the northern highlands of Mexico. Prevett and MacInnes (1972) first noted the increase of Ross's Geese at migration stopovers in the central United States and at wintering areas along the Gulf Coast during 1967–70, and they predicted that numbers of Ross's Geese at these areas would increase. This eastward shift in migration movements is illustrated by an examination of 2,152 recoveries of Ross's Geese banded at Queen Maude Gulf from 1989 to 2001 (Alisauskas et al. 2006a). Recoveries in Canada (30%) were primarily from Saskatchewan (85%); recoveries in the United States (70%) were distributed across the Pacific (49%), Central (39%), and Mississippi (12%) Flyways. In contrast, 96% of Ross's Geese banded at Queen Maude Gulf previously (during the 1960s) were recovered in the Pacific Flyway (Melinchuk and Ryder 1980).

Ross's Geese are early fall migrants. They begin to arrive at the Athabasca River delta in early September, with most flocks arriving during the last 2 weeks of the month (Dzubin 1965). They similarly show up in southeastern Alberta and southwestern Saskatchewan during the first week in September, but the major influx does not occur until the third week. Dzubin (1965) reported that departures from the Alberta/Saskatchewan staging area began after the first week in October and were largely completed by mid-October. By the late 1990s, however, Ross's Geese remained on staging areas in southern Saskatchewan until the end of October, mixed with flocks of Lesser Snow Geese (Alisauskas 2001). Early arrivals of Ross's Geese reach Freezeout Lake in Montana by the first week in October, with their numbers increasing to late October or early November. They then travel over the Continental Divide to arrive in the Klamath basin in California by mid-October, and they remain there for a few weeks before moving southward into the Sacramento Valley by late October.

In late February or early March, Ross's Geese begin to move north through the Sacramento Valley to the Klamath basin, where they pause for 2–4 weeks before migrating to the Malheur NWR area in eastern Oregon, where they usually appear in late March or early April (Bellrose 1980). By late April they arrive in the Freezeout Lake region of Montana and southeastern Alberta (Dzubin 1965). Peak migration in western Saskatchewan occurs after 1 May, with sizeable numbers still occurring there as late as 15–20 May (Alisauskas 2001). During a 3-year study (1966–68) on their breeding grounds at Karrak Lake, Ross's Geese first arrived on 30 May 1966, 10 June 1967, and 4 June 1968; peak numbers occurred over the next 3–4 days. At Arlone Lake (1963), Ross's Geese first showed up on 5 June, in small flocks of 2–50 individuals. Peak arrival was 8 June, but individuals continued to arrive until the end of the month.

MIGRATION BEHAVIOR

There are no specific migration ecology studies of Ross's Geese. They are, however, very gregarious during migration, often gathering in flocks of many thousands at staging areas. They arrive on their nesting grounds in family units, with the yearlings remaining with their parents until incubation begins (Ryder 1967).

Ryder (1967) analyzed the spring migration of Ross's Geese at Arlone Lake relative to temperature. The prevailing isotherm on their wintering grounds in central California was 4.4°C in January. The geese then moved northward, arriving at each subsequent location just after the 0°C isotherm (32°F), which ensured that snow and ice had started to melt, making food available along the migration route. They arrived on their breeding grounds when temperatures there were 1.1°C–3.3°C.

Molt Migration. Ross's Geese do not appear to move long distances from their breeding sites to undergo molting. The first Ross's Geese to enter the molt are the subadults that have accompanied their parents back to the breeding grounds. The subadults leave their parents when the latter begin the incubation period, and they initially move to communal areas (Palmer 1976a). Subadults then travel to molting areas by the time the adults are hatching goslings. The molting subadults are usually flightless by the second week of July; they are then joined by failed breeders, and eventually by successful breeders and their young.

At Arlone Lake, Ryder (1967) noted that Ross's Geese slowly but continuously left nesting islands and moved as family units or small flocks to inland lakes and river courses to molt. Some traveled as far south as the Perry River estuary. By 9–10 July, 80% had departed the nesting islands. Three weeks after hatching, these molting flocks are as large as 200 birds; single families are seldom observed. Ross's Geese remain in these large flocks until molting is completed.

HABITAT

Ryder (1969a) described the general habitat of much of the Queen Maude Gulf MBS, which he divided into 3 physiographic regions: western upland, eastern upland, and central lowland. The central lowland (which covers most of the sanctuary) is a flat plain of sand and silt that overlays Precambrian rock; it slopes about 0.48 m/km northward over roughly 135 km to Queen Maude Gulf. Wet meadows and marsh tundra dominate the area, which is characterized by frost-heaved tussocks of tussock cottongrass (*Eriophorum vaginatum*), sedges (*Carex* spp.), dwarf birch (*Betula glandulosa*), dwarf Labrador tea (*Ledum decumbens*), and cloudberries (*Rubus chamaemorus*). Elevated, well-drained sites are characterized by a mixed association of lichens, mosses, and vascular plants. McLandress (1983a) provided a more detailed description of habitats used by nesting Ross's Geese and Lesser Snow Geese in the Karrak Lake colony (see Nest Sites).

During fall migration on the Athabasca River delta, Ross's Geese fed and roosted in the associated marshes (Dzubin 1965). During spring migration in northeastern California (1990), Ross's Geese foraged in grain fields in early March (67%) and switched to alfalfa fields from late March to mid-April (43%–64%). They heavily (78%) grazed wet meadows in mid-March (McWilliams and Raveling 1998). Wintering birds make substantial use of agricultural crops during the day and return to roost in shallow wetlands or reservoirs at night. In Texas (1991–92), Ross's Geese were 11 times more common in rice prairies than coastal marshes, but the sample size was small ($n = 112$ geese; Harpole et al. 1994).

POPULATION STATUS

Ross's Geese were believed to be near extinction in the early 1900s, and there were only 5,000–6,000 as recently as the 1930s (Ryder and Alisauskas 1995). According to Lloyd (1952), the number of Ross's Geese within the core breeding area of Queen Maud Gulf was <5,000 in the early 1950s but increased to 18,000 by 1966–68. Dzubin (1965) estimated a total population of 30,000 in 1965. Since these early estimates, the Ross's Goose has become an abundant species, with its population and distribution expanding on both breeding and wintering areas.

Aerial photographic reconnaissance in the Queen Maude Gulf MBS by Kerbes et al. (1983) revealed 77,300 Ross's Geese nesting in 30 scattered colonies in 1976, which expanded to 188,000 in 1988 and about 982,000 by 1998 (Canadian Wildlife Service Waterfowl Committee 2010). Since initial estimates in the 1950s, the Ross's Goose population south of Queen Maude Gulf grew at 13.1%/year from 1952 to 1988 (Ryder and Alisauskas 1995). The largest colony is at Karrak Lake, which is the only colony annually surveyed. The population at Karrak Lake was estimated to be 12,000 in 1965 (Ryder 1969a), grew to 225,000 in 1993, and reached 1.1 million by 2005 (Slattery and Alisauskas 2007). In 2010, an estimated 709,000 adult Ross's Geese nested at Karrak Lake. The population at Karrak Lake has increased 7%/year from 2001 to 2010 (U.S. Fish and Wildlife Service 2011). The eastern Canadian Arctic population of Ross's Geese is surveyed less periodically and was estimated at 52,000 in 1998 (Kelley et al. 2001).

Harvest

Trends in the geographic distribution of the Ross's Goose harvest reflect the eastward expansion of their winter range, as they were initially harvested in the Central Flyway in 1974, the Mississippi Flyway in 1982, and the Atlantic Flyway in 1996 (Kelley et al. 2001). The Ross's Goose harvest in the United States averaged 74,490 from 1999 to 2008: 64.2% in the Central Flyway, 18.1% in the Mississippi Flyway, 17.4% in the Pacific Flyway, and only 0.2% in the Atlantic Flyway. Most of the Central Flyway harvest occurred in Texas (74.2%); in the Mississippi Flyway, primarily from Arkansas (47.2%), Louisiana (26.0%), and Mississippi (15.8%); and in the Pacific Flyway, almost all were from California (93%). The harvest of Ross's Geese in Canada from 1999 to 2008 averaged 25,560 birds, with a high of 54,941 in 2001; 80.0% were harvested in Saskatchewan, 10.3% in Manitoba, 8.5% in Alberta, and 1.2% in the remaining provinces.

Starting in 1999, the U.S. Fish and Wildlife Service introduced a special conservation harvest season for white geese (Snow Geese and Ross's

Geese), with the aim of stemming population growth by increasing harvest mortality. This conservation season, or spring hunt, was established in the Mississippi and Central Flyways and allowed hunters to use unplugged shotguns and electronic calls; there was no harvest limit. In 2001, Alisauskas et al. (2006) estimated that 69% of the Ross's Goose harvest occurred during the regular season (1 Sep–31 Jan), 22% during an extended season (1 Feb–10 Mar), and 9% after March 10.

BREEDING BIOLOGY

Behavior

Mating System. Ross's Geese appear to be monogamous. Little information is available on the timing of pair-bond formation, the duration of the pair bond, or when breeding first occurs, but presumably Ross's Geese are similar to Lesser Snow Geese in these aspects. There is evidence that some females breed when they are 2 years old, but most probably do not breed until ≥3 years old (Ryder 1969b). In Louisiana, 19-month-old captive birds began pair formation during February (Palmer 1976a). Most pairing must occur during spring migration, because nesting at Karrak and Arlone Lakes is initiated within 4–5 days after arrival, with peak initiation averaging 5 days after the first nests are started (Ryder 1967, 1972).

Sex Ratio. The sex ratio (males:females) was 0.97:1 (49.2% males) for 4,932 Ross's Geese banded at the Queen Maude Gulf MBS in 1989–94 (Ryder and Alisauskas 1995), and 1.1:1 (52.4% males) for 1,564 adults banded during their flightless period on the Perry River (Melinchuk and Ryder 1980). McLandress (1983c) reported a sex ratio of 1.1:1 for 505 adult Ross's Geese trapped in Saskatchewan during fall migration, and 1.0:1 for 608 adults shot by hunters in California. The ratio from 190 hatchings at Arlone Lake in the Northwest Territories was 0.88:1 (Ryder 1972).

Site Fidelity and Territory. Drake and Alisauskas (2004) examined philopatry and the dispersal of Ross's Geese at Queen Maude Gulf, based on re-sightings of neck-banded geese at 5 of the largest known colonies, including Karrak Lake, during 1999–2003. Resighting probabilities varied among colonies: from as low as 6.9% at 3 colonies to 61.2% at Karrak Lake. Given the estimated annual survival rate of 63%–68% for females, fidelity to at least some colonies is very high. Annual movement probabilities among the colonies, however, ranged from 2% to 14% for females and 12% to 38% for males, which was substantially higher than expected. The authors' figures demonstrate the importance of dispersal, which affects the distribution of breeding birds, the potential influence of immigration on colony-specific growth rates, and gene flow among colonies.

The concept of territoriality in colonially nesting geese is somewhat controversial, with the debate focused on whether an actual space is defended or whether the female is defended, especially given the close proximity of nests (see Nest Sites). Ryder (1967) reported that 2 pairs at Arlone Lake defended areas of 5 m² and 10 m², with undefended areas between the territories around the nests. He later proposed that territories in colonially nesting geese evolved to a size that protected the pair from nonsexual harassment by other males and provided some food resources so that the defending male did not have to leave the site for long periods (Ryder 1975). More recently, Gloutney et al. (2001) did not believe that territories on Karrak Lake contained much available food, because an average of 2.8 neighboring pairs nested within a 10 m radius, and overcrowding had degraded the habitat.

In Snow Geese, Mineau and Cooke (1979) argued that territories could not protect food resources, because defended areas overlapped. Furthermore, other Snow Geese, including nonneighboring males, enter these areas without challenge during the latter stages of incubation, when the female is no longer fertile. The authors reasoned that territorial behavior served to protect the nesting female from unwanted copulations by other males and keep the nest free of eggs laid by unfamiliar females. Thus what initially appeared to be a defense of a space (territory) might actually be a defense of the female.

In a 2-year study (1993 and 1995) at Karrak Lake, 11 out of 20 (55%) observed copulation attempts were by extra-pair males, most (82%) of which were made while the female was on the nest during the laying period (Dunn et al. 1999). The extra-pair attempts were consistently resisted by the female, and the pair-bond male always returned to attack the intruding male, which inevitably was a neighbor. Despite female resistance, however, 50% of the extra-pair copulations appeared to be successful (based on cloacal contact), which equated to 38% of all successful copulations. Hence extra-pair copulations, both attempted and successful, were clearly common occurrences among Ross's Geese nesting at Karrak Lake. Extra-pair copulations did not lead to high levels of paternity, however, as DNA fingerprinting revealed that 2 out of 83 (2.4%) goslings in 2 out of 24 (8.3%) families were the result of extra-pair copulations. These results supported behavioral observations that males guard their mates when the females are fertile, but otherwise the males seek extra-pair copulations with females that are not well guarded. Nonetheless, extra-pair copulations appear to be an inefficient reproductive tactic, mainly because there are a large number of males seeking such copulations with females that may no longer be fertile.

Courtship Displays. There is no specific study of pair formation and pairing displays in wild Ross's Geese, but aggressive displays have been described (Ryder and Alisauskas 1995), with the pair performing the *triumph-ceremony* characteristic of geese and swans. Otherwise, Ross's Geese exhibit low levels of intraspecific aggression, with the *forward* posture the most aggressive threat display given by both males and females. The *forward-threat* involves running at an intruder with the neck stretched forward, and its intensity is increased by opening the mouth and hissing. Males and females together will give a *diagonal-neck-threat* posture while chasing conspecifics and Lesser Snow Geese. This display is performed with the neck stretched upward and the furrowed neck feathers erected and vibrated, with the bill sometimes open. Males and females also give a *bow-threat* posture to displace conspecifics on breeding and wintering areas (McLandress 1983b). Johnsgard (1965) noted mutual *head-dipping* as a precopulatory display, but postcopulatory displays were relatively weak.

In Louisiana during the winter, <7% of adult and <4% of juvenile Ross's Geese were in family groups, compared with 10%–22% of adult and 12%–15% of juvenile Snow Geese (Jónsson and Afton 2008). In addition, Snow Geese won 70% of their social encounters with Ross's Geese. The Snow Geese maintained their family groups longer than did Ross's Geese, which probably contributed to the former's dominance during winter.

Nesting

Nest Sites. Ross's Geese nest on open areas of low tundra or on islands in shallow lakes, often in association with Lesser Snow Geese. At the Queen Maude Gulf MBS, nesting colonies of Ross's Geese and Lesser Snow Geese in the 1960s were exclusively located on islands in shallow lakes, yet in 1990–91 greater numbers were found nesting on the mainland areas (Alisauskas and Boyd 1994). The massive population explosion of both species, coupled with the lack of unoccupied islands in small lakes, probably forced both species to select other nesting habitats. At Karrak Lake, Ross's Geese primarily nest in lowland habitats, while Lesser Snow Geese nest in upland habitats (Didiuk et al. 2001). Nest densities of Ross's Geese at Karrak Lake were negatively correlated with the proportion of total vegetation ($r = -0.38$), grass cover ($r = -0.42$), and lichen cover ($r = -0.32$; Alisauskas et al. 2006b).

Ryder (1967) formulated 4 nesting habitat classifications, based on his early studies (1963–64) at Arlone Lake in the Northwest Territories: (1) "birch" indicated variably sized patches of dwarf birch ≤1 m above the substrate; (2) "rock," piles of boulders or single rocks next to the nest; (3) "open," areas of little or no cover, with a moss substrate; and (4) "mixed," a combination of dwarf birch and/or willows, grasses, sedges, and other vascular plants. The dimensions of 122 nests varied among these 4 habitat types: outside diameter averaged 39.1–49.0 cm; inside diameter, 15.5–15.7 cm; nest-wall thickness, 23.1–33.3 cm; and depth, 5.8–8.1 cm. Based on the outer diameter, nests were largest in moss habitats, followed by rock, mixed, and then birch habitats; the differences reflected the availability of nest material and the nest's exposure to weather. Some nests are reused in subsequent years, but it is not known if reuse is by the same pair.

Nest densities at Arlone Lake were 102/1,000 m² in mixed habitat, 35 in rock habitat, 37 in birch habitat, and 9 in open habitat; total density was highest (220) in mixed habitat on an island. Preferred nest sites were occupied first; later-nesting females used less suitable sites interspaced among the previously selected ones. During a 3-year study (1966–68) of Ross's Geese nesting on an 18.6 ha island in Karrak Lake, most nests (n = 597) were located in rock habitat (49.8%–57.0%), with 10.2%–21.8% in mixed rock with dwarf birch (Ryder 1972). The density of successful nests within a 6 m radius was 3.0–3.2, and 9.9 within a 12 m radius. The mean distance to the closest neighbor was 4.6 m.

McLandress (1983a) studied nest-site selection at Karrak Lake during 1976, using habitat classifications modified from those of Ryder (1967): (1) "rock," on the tops of drumlins, characterized by gravel and various-sized rocks with almost no vegetation; (2) "mixed" (the most common habitat), along drumlins at lower elevations, with a variety of low-growing vascular plants, as well as lichens and mosses; (3) "heath patches," mainly (dwarf) Labrador tea and white heather (*Cassiope tetragona*), with a few dwarf birch and willows in wind-sheltered locations along the slopes of rock ridges; (4) "sand tussock" (uncommon), low-lying areas of frost-heaved pingos of sand, often ringed with sedges or other vascular plants; and (5) "moss" (predominant in all low-lying areas), a wet mat of mosses overlaying 5–10 cm of peat. The nest densities of Ross's Geese were highest in heath patches (64/1,000 m²), followed by sand tussocks (42), mixed (40), moss (18), and rock (16). Of 301 nests, the majority were located <3 m from another Ross's Goose nest (51.0%) or Lesser Snow Goose nest (12.9%), and fewer were ≥6 m from another Ross's Goose nest (26.1%) or Lesser Snow Goose nest (10.1%). On an island in Karrak Lake (1966–68), nests were an average of 15.0 m apart; unsuccessful nests were closer together (13.5 m) than successful nests (15.4 m; Ryder 1972).

McCracken et al. (1997) provided detailed measurements from nests at Karrak Lake during a 1994 study comparing the nest morphology of Ross's Geese with that of Lesser Snow Geese. Nests varied among 3 habitat types (heath, mixed, and moss), but they increased in average mass from 585 g in heath, to 755 g in mixed, to 936 g in moss. Similarly, the average outer diameter progressed from 37.6 cm in heath, to 51.6 cm in mixed, to 55.4 in moss. Nest-wall thickness went from 12.7 cm in heath, to 19.7 cm in mixed, to 21.7 cm in moss. In general, nests were smaller in habitats that provided more cover from wind and precipitation. Lesser Snow Geese built nests in rock habitats or the upper parts of eskers, but such sites were not used by Ross's Geese, probably because the latter is a smaller species, and these sites are energetically less favorable, compared with more sheltered heath and mixed habitats; Ross's Geese are about two-thirds the size of Lesser Snow Geese, which translates into a daily basal metabolic rate that is 1.18 kcal/g higher. Ross's Geese constructed relatively larger and more insulated nests than Lesser Snow Geese, which most likely minimized energy expenditure during incubation, quickened embryo development, and lessened the cooling of embryos during nest recesses. This nest construction probably also reflects an adaptation to the smaller egg size and higher metabolic rate of Ross's Geese, compared with Lesser Snow Geese.

The female is responsible for nest building, although the male may provide minimal assistance. Nests are constructed and maintained by stripping and grubbing plant materials immediately surrounding the nest site, including such items as leaves and twigs from dwarf birch, willows, and Labrador tea, as well as mosses, grasses, and even old dried scats (Ryder and Alisauskas 1995). Down is added near the end of egg laying or into the incubation period. For 55 nests at Arlone Lake, 2% had down in the nest before the last egg was laid, 82% had down when the clutch was completed, and 16% did not have down until after the clutch was completed (Ryder 1967).

Clutch Size and Eggs. Clutch size at Karrak Lake averaged 3.3 eggs in 5,182 nests monitored from 1991 to 2000 (Alisauskas 1991); in a 1966–68 study there, clutch size averaged 3.9 eggs (range = 3.6–4.0 eggs) for 597 marked nests (Ryder 1972). Over all years, most clutches contained either 4 (42.9%) or 3 (18.9%) eggs; 12.1% were 2-egg clutches, and 2.5% were 6-egg clutches. There were also a few 7-egg ($n = 2$), 8-egg ($n = 1$), and 9-egg ($n = 1$) clutches. Successful clutches initiated early in the nesting season were larger than clutches initiated later (4.0–5.8 eggs vs. 2.6–3.9).

At Arlone Lake, clutch size ranged from 1 to 8 eggs, with an average of 3.7 eggs in 1963 ($n = 769$) and 3.6 in 1964 ($n = 906$) (Ryder 1967). Clutches with >6 eggs most likely were deposited by more than 1 female, and clutches with only 1 egg probably experienced unrecorded predation. Clutches started early in the season were larger than those started later (4.1–5.0 eggs vs. 2.9–3.1), even though the difference in initiation was only 2–4 days.

Fresh Ross's Goose eggs are smooth, mostly subelliptical, and a nonglossy white or light cream in color, but they become stained in the nest (Ryder and Alisauskas 1995). Eggs measured 73.0 × 65.4 mm ($n = 398$), with a mean volume of 83.1 cc ($n = 135$) and a mass of 89.2 g ($n = 27$). Mean fresh egg mass is 5.4% of the female's mass during laying, so a 4-egg clutch is about 20% of the female's body mass. The shell thickness of single eggs from 8 clutches at Karrak Lake averaged 0.3 mm. Egg laying often begins before nest construction is complete and then proceeds at a rate of 1.2/day (range = 1.0–2.3 eggs), determined from 460 clutches monitored by Ryder (1969b). Most clutches are completed in 4–6 days. Skipped days during egg laying are directly related to unfavorable weather. During normal seasons a day is usually skipped between depositing the second and third eggs for 3- or 4-egg clutches (28%–52.5%), and the third and fourth eggs of 4-egg clutches (29%).

Incubation and Energetic Costs. Ryder (1972) reported an average incubation period of 21.8 days for 254 nests at Karrak Lake. Most (79.9%) were incubated for 21–22 days. At Arlone Lake, the incubation period for 45 clutches averaged 22.0 days (range = 19–25 days; Ryder 1967). In 1993, 7 females at Karrak Lake, monitored via Super 8 (8 mm film) cameras, averaged 99.2% of their time on the nest during incubation, with a decline of 0.06%/day as incubation progressed (Jońsson et al. 2007). Nest recesses ranged from 3 to 43 minutes (average = 13.8 minutes; $n = 48$); nest recesses averaged only 0.4 hr/day ($n = 17$; Jońsson 2005).

This high incubation constancy and reliance on stored lipid reserves may have evolved because foraging opportunities are limited during incubation, especially for a colonially nesting species like the Ross's Goose. Individuals in the best condition may also remain on the nest the longest. At Karrak Lake in 1994, the incubation constancy of Ross's Geese and Lesser Snow Geese ($n = 90$) averaged 91.8%, with no difference between the 2 species (LeShack et al. 1998); Jońsson et al. (2007) also noted this similarity between the geese. Gloutney et al. (2001) found that the recess length of 6 females averaged 15.6 minutes, during which time feeding was the dominant (82%) activity. Mean daily foraging time was 1.97 hr/day during recesses, but not much (5.7 g) food was actually consumed.

Energetically, arctic-nesting geese commonly arrive on their breeding areas when the snow cover is still extensive, which limits food availability. In addition, Ross's Geese nest within a few (<5) days of arrival, and thus have very little time to exploit local food resources before they begin to lay eggs, although they can and do obtain food during recesses from the nest. At Karrak Lake, Ross's Geese apparently consume very little from the time of arrival until they depart the colony with their goslings (Gloutney et al. 1999). Female body weight declined 21% (from 1,288 g during laying to 1,012 g during incubation), with a 73% decrease in abdominal fat (from 29.5 g to 7.9 g) and a 28% reduction in protein from the breast muscles (from 112.5 g to 81.4 g). Females averaged 9.17 hr/day foraging during the preincubation period, when they primarily consumed mosses (50.8%), chickweeds (*Stellaria* spp.; 25.4%), and sedges (12.5%). During nest recesses, the average food intake was only 2.91 g/hr (5.7 g total); mostly (82.5%) mosses, which are of low nutritive value, were consumed. Thus, although females spend a significant amount of time foraging during nesting, their nutrient intake is minimal, and reliance on stored reserves appears paramount. The authors explored 4 hypotheses to explain why Ross's Geese may forage on poor-quality foods: (1) the use of any food source was beneficial, but intake rates were low on the degraded habitats at Karrak Lake; (2) foraging is a mechanism of territorial defense; (3) the geese ingest a minimal amount of food so as to maintain gut microflora; and (4) the geese are searching for calcium and not food energy. Calcium obtained from eggshells produced in previous years may be a more important resource to Ross's Geese than organic foods, as data indicated that 56% of the 41.8 g of calcium needed to produce a 4-egg clutch came from eggshells.

Nesting Chronology. Ross's Geese began nesting within 4–5 days after their arrival at the Arlone and Karrak Lakes colonies, with peak initiations occurring 5 days later (Ryder 1967, 1972). Ross's Geese within a colony start their nests in synchrony, as seen during a 2-year study (1993 and 1995) at Karrak Lake, when all laying was initiated within a span of 16–19 days (Dunn et al.1999). The synchrony index for Ross's Geese was 66%, compared with 70% for Lesser Snow Geese and an average of only 32% for 22 species of songbirds. In an earlier 3-year study (1966–68) on an 18.6 ha island at Karrak Lake, the interval between the initiation and the completion of clutches was 15 days during the relatively early season of 1966, versus 9 days in 1967 (55% less), and 18 days in 1968 (Ryder 1972).

The nesting season is short, due to the Arctic location of the breeding grounds, but nesting chronology nonetheless varies with weather conditions. At Karrak Lake, the first nests were begun on 3 June 1966 (an early season), 16 June 1967, and 8 June 1968. Corresponding peaks in nest initiations were 2–5 days later: 6 June 1966, 18 June 1967, and 13 June 1968 (Ryder 1972). Peak periods of egg laying were 7–8 June 1966, 18 June 1967, and 15 June 1968. At Arlone Lake in 1963, nesting was delayed 5 days by high winds (Ryder 1967).

Nest Success. Unless there is severe weather during nesting, the nest success of Ross's Geese is impressive, due to the high degree of incubation constancy and attendance at the nest. Hanson et al. (1956) reported only 6 nest failures (from gull predation) among 260 nests monitored on islands in Arlone Lake during 1949. Ryder (1967) found that nest success at Arlone Lake was 97% (n = 93) in 1963 and 83% (n = 59) in 1964. Of 581 eggs deposited in these nests during those two years, 88.0% hatched (93.7% in 1963 and 79.2% in 1964). Most egg losses were due to predation (58.6%) or dead embryos (20.0%).

Nest success can vary annually. At Karrak Lake, it averaged 83% from 1966 to 1999, but ranged between 68% and 92% (Alisauskas and Rockwell 2001). In an earlier Karrak Lake study (1966–68), nest success (n = 597) was 67.7%–88.3% (Ryder 1972). The lower rate occurred in 1968 and was attributed to a large percentage of inexperienced geese nesting for the first time. Hatching success of 2,235 individually marked eggs during this study was 70.3% (range = 60.6%–80.3%), with most (23.9%) of the unsuccessful eggs lost to predation, although the annual predation rate on eggs varied markedly (13.1%–33.2%). Nest predators were Parasitic Jaegers (*Stercorarius parasiticus*), Long-tailed Jaegers (*S. longicaudus*), Glaucous Gulls (*Larus hyperboreus*), and Herring Gulls (*L. argentatus*). Egg loss by other causes (such as infertility) was 5.8%. Ryder and Alisauskas (1995) attributed the loss of both eggs and nesting adults to predation by arctic foxes (*Alopex lagopus*), gray wolves (*Canis lupus*), grizzly bears (*Ursus horribilis*), and wolverines (*Gulo gulo*). At Arlone Lake (1964), arctic fox predation resulted in the complete abandonment of a major nesting island (Ryder 1967).

Brood Parasitism. Ross's Geese and Lesser Snow Geese nest closely together, in the open, so instances of intra- and interspecific brood parasitism are not surprising. In 1988 at the Queen Maude Gulf MBS, 107 out of 2,148 (5.0%) Ross's Goose and Lesser Snow Goose nests contained at least 1 egg from both species (Kerbes 1994). At Arlone Lake (1963–64), Ryder (1967) reported 9 parasitized nests that contained 10–29 eggs. The effects of brood parasitism have not been investigated, however, for either the parasite or the host.

Renesting. Renesting has not been documented in Ross's Geese and is doubtful, given their short breeding season in the Arctic.

REARING OF YOUNG

Brood Habitat and Care. Many families of Ross's Geese, along with those of Lesser Snow Geese, travel all the way to the coast of Queen Maude Gulf, a distance of about 70 km from the Karrak Lake colony; the mean dispersal distance was 45 km for Ross's Geese (Slattery 1994). Some broods reached the coast in <17 days. During these movements, the goslings feed in lowland habitats, especially those along drainages that lead to the coast. In an aerial assessment of habitat use late in the brood-rearing period (about 40 days posthatching), approximately 70% of both Ross's Geese and Lesser Snow

Geese selected lowland habitats (wet sedge meadows, graminoid tundra hummocks, and freshwater sites) and avoided upland habitats (Slattery and Alisauskas 2007). The lowland choices probably offer greater food availability and easier predator avoidance. The geese were distributed fairly homogeneously between Karrak Lake and the coast, with some clumping observed near the latter; the study also indicated, however, that most light geese were found on freshwater habitats that could be >120 km from the coast. Flock sizes observed during this assessment averaged 82 birds (range = 5–1,500 birds); about 75% of the flocks contained <100 geese, but these flocks only accounted for 45% of all geese tallied.

Both parents lead their goslings from the nesting colony within 1–2 days after hatching, at which time the goslings walk and run well (Ryder and Alisauskas 1995). The parents also zealously guard their young. Ryder (1967) noted that, in response to an intruder, the female leads the goslings away while the male stands guard with outspread wings and mouth agape. The young remain with their parents for up to a year, although family groups of Ross's Geese break up more readily than those of larger goose species (McLandress 1983b).

At Karrak Lake, when males were experimentally removed from 17 pairs of Ross's Geese early in the incubation period (days 1–8), there were no differences between widowed and paired females in nest success, nor in the body weight of the female at hatching or at posthatch recapture, although widowed females spent more time alert while on the nest (22.9% vs. 8.6%) and less time with their bills tucked (9.1% vs. 23.7%). Paired females were recaptured almost twice as far from the nesting colony, however, as widowed females (50.9 vs. 27.3 km). The study suggested that male removal during incubation had little effect on nest success, perhaps because territories were established, laying was completed, and extra-pair copulations were unlikely. Since paired females were recaptured much farther from the colony, the presence of the male may allow pairs to travel greater distances to better brood-rearing areas. Male parental care after hatching is also important, because the male is the primary defender of the brood, while the female spends most of her time feeding (to replenish the body weight and the reserves she lost during laying and incubation).

Brood Amalgamation (Crèches). While there are no specific studies of brood amalgamation in Ross's Geese, Ryder (1967) found that family groups at Arlone Lake combined after leaving the breeding colonies, with aggregations of as many as 200 birds occurring 3 weeks after hatching. Single families were seldom seen.

Development. Goslings average 64.3 g at hatching (Ryder and Alisauskas 1995). MacInnes et al. (1989) noted that Ross's Geese grew more rapidly than Lesser Snow Geese, and that Ross's Geese had more genes contributing to that rapid growth. The body-weight growth constant was 0.0788 for Ross's Geese, compared with 0.0723 for Lesser Snow Geese, an 8.2% difference. Growth rates were also higher for culmen length (10.0%) and height (13.4%), as well as tarsus length (10.5%).

Slattery and Alisauskas (1995) observed that 1-day-old Ross's Goose goslings had proportionately more body protein and larger gizzards than 1-day-old Snow Goose goslings, which may have allowed the former to process food more rapidly. Juvenal plumage is acquired by 5–6 weeks, with flight attained in about 42 days (Palmer 1976a). As further evidence of their rapid growth rate, Slattery (1994) reported that Ross's Goose goslings averaged 80% of adult structural size (culmen, tarsus, and midtoe) in only 3 weeks. Similarly, Ryder (1967) noted that the bill, the tarsus, and the midtoe were within the range of adult size by the time the goslings were 10 weeks old.

The increased growth rate for Ross's Geese apparently shortens their development time, so they reach flight stage earlier; their smaller final size also produces a flying individual in a shorter time span (MacInnes et al. 1989). This growth pattern is clearly adaptive, given the extreme conditions characteristic of the arctic breeding habitats used by Ross's Geese. In 1963–64, Arlone Lake was still completely ice covered on 1–2 June, and ice was still rising off the lake bottom as late as mid-July (Ryder 1967).

RECRUITMENT AND SURVIVAL

Ryder (1967) noted an immediate drop in brood size after hatching, as goslings were lost from abandonment and predation. In 1964, brood size averaged 2.9 at his Arlone Lake study area, 2.7 at fall staging areas in Saskatchewan, and 1.7 during the winter in California. Alisauskas and Rockwell (2001) estimated annual survival probabilities for Ross's Geese banded and recovered during 1961–99. Most (13,083) adults were banded north of 56° N latitude, as were 12,915 out of 15,252 juveniles; the resultant survival estimates were 86.6% for adults and 54.0% for juveniles. The stochastic rate of population growth was estimated at 1.0904. This rate declined to 1.0811 with a 1% reduction in adult survival, which equated to an 0.85% reduction in the population growth rate.

Longevity for Ross's Geese is 22 years and 6 months, recorded from a female banded as an adult and recovered in California.

FOOD HABITS AND FEEDING ECOLOGY

Ross's Geese are exclusively vegetarians, with their primary food items consisting of grasses, sedges, legumes, and cultivated grains. The small bill of this goose species is well adapted for grazing on short grasses and sedges. Foraging habitats include agricultural fields, shallow wetlands, wet meadows, marshes, and native-prairie grasslands. During migration and over the winter, Ross's Geese generally feed in fields during the day and return to wetlands at night to roost. They commonly forage in association with Lesser Snow Geese, often concentrating in flocks of many thousands.

Breeding. On their breeding grounds at Arlone Lake, Ryder (1967) reported that the diet of 26 Ross's Geese (adults and goslings) consisted of the roots, leaves, stems, and spikelets of grasses, sedges, and birches (*Betula* spp.). At Karrak Lake in 1993, Ross's Geese (n = 13) averaged 9.2 hr/day feeding before they began incubation (Gloutney et al. 2001); the average foraging bout was 24.2 minutes, with 79% of the time spent feeding. Food intake, however, was only 12.4 g/day (1.35 g/hr). The geese mainly ate mosses (50.8%), although they also fed on more nutritious foods, such as chickweeds (25.4%), and sedges (12.5%). In addition, they consumed eggshells (9.7%) from prior years, which probably provided some of the calcium needed for eggshell formation. In contrast, incubating birds were restricted more to their nest sites and thus primarily (82.5%) consumed mosses. The authors suggested that the use of such low-quality foods and the small amounts consumed reflected the significant amount of vegetation removed by the dramatically increased populations of both Ross's Geese and Lesser Snow Geese at Karrak Lake over the previous 25 years, from 17,000 birds in 1965–67 to 480,000 in 1995.

Migration and Winter. When first arriving on major migration staging area in central Saskatchewan during the fall, Ross's Geese heavily used temporary wetlands containing mats of spikerush (*Eleocharis* sp.) and sedges for feeding and resting (Dzubin 1965). Disturbances from hunting quickly forced the geese to use larger lakes located in game preserves, from which they then engaged in twice-daily feeding flights to waste grain fields (primarily wheat and barley). Spring migrants in the Central Valley of California, along with other species of geese, used winter wheat, barley, and alfalfa fields (Heitmeyer et al. 1989). During the winter in California, Ross's Geese ate the leaves of various species of grasses, rice seeds, cottongrass (*Eriophorum* sp.), millets (*Echinochloa* spp.), sedges, bulrush (*Scirpus* sp.) leaves and stems, and spikerush and saltgrass (*Distichlis* sp.) shoots (Ryder and Alisauskas 1995). The diet of Ross's Geese wintering in New Mexico and the state of Chihuahua in Mexico consisted of corn and other cultivated crops (Drewien and Brown 1992).

During 2 winters in southwestern Louisiana (2002/3 and 2003/4), Ross's Geese spent an average of 53.3%–57.1% of their time feeding, 20.0%–23.9% alert, 16.0%–19.5% inactive (e.g., resting and preening), and 2.9%–7.2% in locomotion (Jónsson and Afton 2009). Feeding time was highest for juveniles in families of 4 or more geese (72.2%); it was also high for juveniles in families of 3 birds (51.4%), and for lone juveniles (52.3%) without adults. In contrast, the feeding time for adult parents with 2 or more juveniles was only 28.9%, while their alert time was 43.4%. Feeding time for a lone adult or for a pair without juveniles was 49.8%–50.9%. Juvenile geese probably spend more time eating than adults, because they forage less efficiently and are less able to avoid competition from adults. Juveniles are also not yet fully grown during winter, and thus increase their feeding time to meet their higher energy requirements. When adults are present in family groups, their high alert time probably allows juveniles to spend more time feeding and resting. Pecking rates (bouts/min) were higher for Ross's Geese (average = 1.4) than Lesser Snow Geese in the same area (average = 1.0); the former also fed more (53.3%–57.1% vs. 45.4%–46.3%).

These findings are consistent with the prediction that the smaller-bodied Ross's Goose needs to spend more time eating than the larger-bodied Snow Goose, in part because the former reach their lower critical temperature (about 6°C vs. 2°C) sooner. During winter 2002/3 in Louisiana, however, both species averaged a 3.8% decrease in feeding time for each 1°C increase in temperature, which is not consistent with this prediction. Changes in ambient temperatures in Louisiana may affect both species similarly, because the difference in lower critical temperature between the species was small, compared with within-day fluctuations in temperatures associated with the warm winter climate that usually prevails in southwestern Louisiana.

MOLTS AND PLUMAGES

Birds in juvenal plumage, which is acquired at 5–6 weeks of age (latter half of Aug), have white heads with a grayish crown, nape, and sides of the neck. The rest of the body is whitish gray (Palmer 1976a, Ryder and Alisauskas 1995). They are able to fly at about 42 days old. First basic replaces juvenal plumage from December/January through April and is worn until July. Birds in first basic largely exhibit white feathering by mid-December to late January, and they retain the juvenal remiges until a full body molt occurs in the summer, when the birds are about 1 year old. Definitive basic also appears in summer (Jun–Jul) and is a complete molt of the body and wing plumage. Birds are flightless for about 28 days during the associated wing molt, which occurs after the body molt has begun. Breeding birds do not become flightless until 15–20 days after the peak of hatching (Ryder 1967), and they regain flight at about the time the goslings are ready to fly.

CONSERVATION AND MANAGEMENT

The population explosion of Ross's Geese, like that of Lesser Snow Geese, has negatively affected tundra vegetation, although this impact is less studied in Ross's Geese. It is also difficult to separate the damage by species, since both nest together in large colonies, although Ross's Geese at Karrak Lake primarily nest in lowland habitats, while Lesser Snow Geese nest in uplands. One difference between the 2 species is in bill structure. The smaller bill of the Ross's Goose is less effective at grubbing, but it allows them to crop vegetation at even lower levels, which can slow or even prevent recovery of tundra plants in degraded areas (Didiuk et al. 2001).

At Karrak Lake, Alisauskas et al. (2006b) documented the long-term effects of the increasing light goose population on the vegetation, where they damage stems both from the removal of vegetation by foraging geese and from their use of vegetation for nest building and nest maintenance. The proportion of total vegetation declined as the duration of nesting by geese increased, with the percentage of vegetation on areas with no history of nesting geese averaging about twice that of places with ≥20 years occupancy (91% vs. 44%). The percentage of grass cover was greatest in locales with no documented goose nesting (24%), compared with those having 1–10 years of occupancy (12%), 11–20 years (2.2%), and >20 years (7.8%). The percentage of lichens was also greatest (21%) in areas without goose nesting and declined with colony age, reaching only 1.3% in the oldest part of the colonies. In contrast, the percentage of moss was greatest (31%) in the oldest colony areas and lowest (19%) in portions of the colony where geese had not nested. Similarly, the percentage of white heather was lowest (3.9%) in the oldest parts of the colony. Overall, the proportion of damaged habitat—determined by the sum of exposed substrate, exposed peat, and *Senecio congestus*—was greatest (38.6%) in the oldest parts of the colony, compared with 8.5% in unoccupied areas. The percentage of exposed substrate was highest (14%) at the center of colonies, compared with only 1.4% in parts of the colony where geese had not nested.

During the brood-rearing period, Slattery (2000) used exclosures to demonstrate that the degradation of plant biomass by both Ross's Geese and Lesser Snow Geese occurred in a continuum with the distance from the colony, but resulted in the removal of about 50% of all the aboveground vegetation over a 5,000 km² area. Given that Ross's Geese made up about 50% of the total light goose population, they potentially removed 25% of all the aboveground vegetation over an extensive area. (See the Snow Goose account for information on measures taken to reduce the impacts of light geese on arctic ecosystems.)

Brant (Atlantic and Black)

Branta bernicla (Linnaeus 1758)

Left, Atlantic Brant; *right*, Black Brant

Brant are small maritime geese with a breeding distribution that spans arctic coastal areas throughout the Northern Hemisphere. There are two subspecies of Brant recognized in North America: the Atlantic Brant (*Branta bernicla hrota*) and the Black, or Pacific, Brant (*B. b. nigricans*). These subspecies are further divided into 4 populations (2008/9 population estimates are in parentheses): in the eastern Arctic, the Atlantic Brant (151,300) and the Eastern High Arctic Brant (38,900), and in the central and western Arctic, the Black Brant (147,400) and the Western High Arctic Brant, sometimes referred to as Gray-bellied Brant (16,200). Some authorities (e.g., Livezey, 1997) separate these birds into 2 species of Brant: the Black, or Dark-bellied, Brant (*Branta nigricans*) and the Pale-bellied, or Atlantic, Brant (*Branta hrota*). Most of the Black Brant population breeds on the Yukon-Kuskokwim (Y-K) delta in Alaska, where these birds have been well studied, in contrast to eastern Arctic Brant populations (especially the Eastern High Arctic Brant), which breed as far north as any species of bird and thus are more inaccessible. Brant are monogamous and mate for life, producing an average clutch of 3.6 eggs/pair that is incubated for about 25 days. Goslings attain flight in 45–50 days. Early-hatched goslings have faster growth rates and a better chance of survival than late-hatched young, and family groups stay together through migration, winter, and into the early stages of spring migration. Brant are accomplished flyers, with most Black Brant mainly flying nonstop from their major staging area at Izembek Lagoon on the Alaska Peninsula to wintering grounds in Baja California, >5,000 km away. Brant are vegetarians, depending heavily on eelgrass (*Zostera marina*) during migration and over the winter on the Pacific coast; they also eat sea lettuce (*Ulva* spp.) on the East Coast.

Brant adult

Brant juvenile

IDENTIFICATION

At a Glance. Brant are small dark maritime geese, characterized by small bills; long dark wings; and black heads, necks, bills, breasts, primaries, tails, and legs. Small white bars form a white neck ring, or "necklace," around the middle of the neck, which is usually absent in immatures. The extensive white uppertail coverts are distinctive in flight. Two subspecies are recognized in North America: the Atlantic Brant and the Black (or Pacific) Brant. The Atlantic Brant has a pale belly that contrasts sharply with the black breast; the necklace does not meet in front. The Black Brant has a dark belly and breast, with little contrast between them; the neck ring is more extensive and meets in front.

Adults. The sexes are similar in appearance in both subspecies of Brant. Adults have a white crescent necklace on each side of the upper neck, which tends to reach up higher in males (Boyd and Maltby 1980). In both subspecies, the bill, the head, the neck and the chest are black; the back and the wings are a dark grayish brown. The tail coverts form a distinctive white band at the base of the black tail. The legs and the feet are black. The irises are dark brown. In Atlantic Brant, the breast and the belly are grayish white and contrast sharply with the black chest. The sides and the flanks are white, barred with gray brown that becomes more sharply defined posteriorly. Black Brant differ in having a black breast and a black belly as far back as the legs. The sides and the flanks are strongly barred with gray and white. The white necklace is more extensive and meets in the front on Black Brant, but does not join together on Atlantic Brant.

In flight, a Brant looks more like a duck than a goose because of its relatively small size, rapid wing beat, and short neck. Brant are the swiftest of all geese in flight. They usually fly in irregularly shaped flocks or in lines that are trailing or abreast, usually close to the water. A key identification feature is the whitish-gray belly of the Atlantic Brant, which is very apparent in flight, in contrast to the blackish belly of the Black Brant.

Juveniles. Juveniles have the same basic coloration as the adults, except that their dark grayish-brown greater and middle wing coverts are tipped with white, which presents a strong edging effect on the back that is usually discernable until March. The white necklace does not appear until midwinter. The legs, the feet, and the bill of young geese are black or dark gray. The eyes are dusky brown.

Goslings. Goslings are grayish white, with a dark cap and mask about the eyes, and a white V between the breast and the belly (Nelson 1993).

Flock of
Atlantic Brant
in flight.
GaryKramer.net

Atlantic Brant (*left*) and
Black Brant (*right*); note
the difference in the
white neck markings and
the flank coloration.
Ian Gereg

Black Brant goslings are darker than Atlantic Brant goslings, although individual variation is common.

Voice. Brant calls are hoarser and less resounding than the calls of Canada Geese. Palmer (1976a) reports a guttural *cronk* given by adults in flight or at rest, as well as a comparatively quiet *cut-cut-cut cronk* or *cut-cut-cut* call given in flight as an intraflock call. A mild alarm call involves a louder, more drawn-out *crrronk*, singly or repeated, given by lone birds separated from flocks or by Brant on their nesting territory. A shorter explosive *cruk* is another alarm call, given once or rapidly repeated, that is accompanied by head and neck movements.

Similar Species. Brant are unlikely to be confused with any other goose species. All subspecies of Canada Geese have a distinct white chin strap as

Atlantic Brant in a characteristic goose *threat* posture.
Ian Gereg

opposed to the white necklace of Brant, and the Brant's white rear is more extensive and conspicuous. The calls of the two taxa are also very different.

Weights and Measurements. Males are larger than females, averaging about 10% heavier in weight and 6% larger in linear measurements. Data on body weights and linear measurements for both subspecies of Brant are summarized in Reed et al. (1998b). Body weights of breeding Atlantic Brant (Jun–Jul) averaged 1,293 g for males ($n=27$) and 1,224 g for females ($n=41$). Weights during the winter (Jan) averaged 1,446 g for males ($n=54$) and 1,236 g for females ($n=40$); during the spring, weights averaged 1,515 g for males ($n=20$) and 1,398 g for females ($n=21$). Culmen length averaged 33.6 mm in males ($n=27$) and 31.9 mm in females ($n=40$). Body weight of breeding (Jun) Black Brant averaged 1,165 g for males ($n=48$) and 1,025 g for females ($n=152$); winter weights (Jan) averaged 1,519 g for males and 1,396 g for females. Culmen length averaged 33.9 mm in males ($n=152$) and 32.2 mm in females ($n=278$).

DISTRIBUTION

Taxonomic authorities recognize 3 subspecies of Brant: the Brent Goose (*Branta bernicla bernicla*), the Black Brant (*B. b. nigricans*), and the Atlantic Brant (*B. b. hrota*). All are circumpolar in their breeding distribution, situated along the coasts of the Arctic Ocean and extending southward into connecting bays and seas.

The Brent Goose, or Dark-bellied Brent Goose, is the nominate subspecies and breeds in western Siberia, mainly along the coastline of the eastern Taymyr Peninsula, which juts out into the Laptev and Kara Seas of the Arctic Ocean. They also are found on nearby Kolguev Island, Severnaya Zemlya (Archipelago), and the Franz Joseph Land Archipelago (Johnsgard 1978, Boyd 2005). These Brant largely winter along the coast of northwestern Europe in Denmark, the Netherlands, the United Kingdom, and France, although they sometimes extend east to the Lena River delta, where they mix with Black Brant; the boundary between the 2 groups is unknown (Ward et al. 1993, Boyd 2005). Black Brant occur farther to the east, from the New

Western High Arctic
Eastern High Arctic
Pacific Brant
Atlantic Brant

Breeding Distribution
Year-round Distribution
Wintering Distribution

delta (Stickney and Ritchie 1996, Sedinger and Stickney 2000). These groups appear to be quite discrete, as Barry (1967) noted that only 6 of the 10,000 Brant banded as young on the Y-K delta were ever recaptured on the Anderson River delta; likewise, only 3 of the 3,000 Brant captured and banded on the Anderson River delta were recaptured or appeared elsewhere in Alaska. The Y-K delta has long been recognized as an important breeding area for Black Brant (Spencer et al. 1951, Hansen and Nelson 1957); this population also extends east along the northern coast of the Northwest Territories to Banks Island, Victoria Island, and the coastal islands in Queen Maud Gulf.

In addition, Black Brant breed in northeastern Siberia, from the Lena River delta eastward to the Anadyr River basin and the Chukchi Peninsula, as well as on Wrangel Island (Reed et al. 1998b). Johnsgard (1978) considered these Brant to be a separate subspecies, *orientalis* (Pacific Brant). In contrast, Delacour and Zimmer (1952) used *nigricans* to identify a dark-bellied subspecies called Lawrence's Brant, and *orientalis* for all other Black Brant. Lawrence's Brant had a dark gray belly that was paler than that of Black Brant, and apparently the differentiation was familiar to the waterfowl hunters in the area (Boyd 2005). Nonetheless, the identity of Lawrence's Brant is based entirely on 3 specimens in the American Museum of Natural History (Delacour and Zimmer 1952) and thus is probably not a valid subspecies, but rather an intergrade between the Atlantic and the Black Brant subspecies. Manning et al. (1956) examined 66 Brant collected in the Arctic from Alaska to Greenland and concluded that 12 specimens showed color intergradation between Atlantic and Black Brant.

Siberian Islands and the Jana River delta to Wrangel Island and the Anadyr [Chukchi] Peninsula. Wetlands International recognizes these Brant as forming 2 groups: *nigricans* (Black Brant), which winter in Kamchatka and Japan, and *orientalis* (Pacific Brant), which winter in Korea and China (Boyd 2005). Johnsgard (1978) recognized *orientalis* as a separate subspecies, but whether these Black Brant are reproductively isolated from those in Alaska is a continuing source of debate (Boyd 2005). In North America, within the two subspecies present, 4 populations of Brant are recognized: Black Brant, Western High Arctic Brant, Atlantic Brant, and Eastern High Arctic Brant.

Black Brant (Pacific Black Brant)

Breeding. Most (>80%) Black Brant nest along the coast of the Y-K delta in Alaska, with smaller colonies located at the Nugnugaluktuk River on the Seward Peninsula, Prudhoe Bay and the Colville River delta on Alaska's North Slope, and the Anderson River delta in the Northwest Territories (Sedinger et al. 1993). In general, the Arctic Coastal Plain supports about 500–1,500 breeding pairs of Brant, with the highest nest densities found on the Colville River delta and the Sagavanirktok River

Winter. Black Brant in North America winter locally along the Pacific coast from Alaska south to Mexico, where they are found on lagoons on both the west coast of the Baja California peninsula and the mainland coasts of the states of Sonora and northern Sinaloa. A large percentage of the

Female Black Brant with her newly hatched goslings.
Maynard Axelson

Black Brant population once wintered along the Pacific coast of Washington, Oregon, and California, as well as in Baja California, but a shift to what are now their primary wintering grounds in Mexico began in the late 1950s (Smith and Jensen 1970, Ball et al. 1989). This trend continued, with the wintering population of Black Brant in the United States declining from 44% in 1957 to only 7% by 1975 (Kramer et al. 1979). Reasons for the decline are speculative but involve harassment and other human disturbances, such as hunting (Chattin 1970, Smith and Jensen 1970). Black Brant from breeding grounds in Russia have also been observed on wintering grounds in Baja California (Ward et al. 1993). An occasional Black Brant reaches the Atlantic coast (Veit and Petersen 1993).

From 2001 to 2008, an average of 122,163 Black Brant were recorded during Midwinter Surveys in Mexico. Most were located in the coastal lagoons of Baja California (San Quintín, Scammon's, San Ignacio, and Magdalena), and 86% of the Black Brant surveyed from 1998 to 2007 occurred in these lagoons (Mallek and Wortham 2008). In 2008, 103,299 Black Brant were surveyed in Mexico, 81% in the lagoons on the Baja Peninsula and 19% in coastal lagoons on the mainland (Tiburón, Obregón, Agiabampo, Topolobampo, Bahía Santa María, and Pabellón). The total distance traveled to Mexico varies from about 4,800 km for Brant from the Y-K delta to 8,900 km for Brant migrating from Victoria Island (Reed et al. 1998b).

Lindberg et al. (2007) found that Black Brant wintering in Mexico exhibited strong fidelity to particular bays and lagoons, with the probability of site fidelity between winters being >87%, although there was movement both within and among winters. Movements between winters were related to issues of habitat quality, while within-winter movements were more closely tied to migration patterns.

North of Mexico, from 2001 to 2008, an average of 19,775 Black Brant wintered in Washington, 370 in Oregon, and 4,827 in California (Trost and Sanders 2008). Concentration areas in Washington include Skagit, Padilla, Samish, Fidalgo, and Willapa Bays; in Oregon, Coos and Tillamook Bays (Ball et al. 1989, Reed et al. 1989); in California, Humboldt Bay (Moore and Black 2006a); and in British Columbia, Boundary Bay (Reed et al. 1998c). Since 1992, Black Brant numbers have in-

Juvenile Black Brant, with the characteristic juvenile white barring on the wings.
GaryKramer.net

creased in the Boundary Bay and Robert's Banks area of the Fraser River delta, with 1,264 estimated in 2007/8 (Canadian Wildlife Service 2008). Brant wintering at Boundary Bay arrive in early November (Reed 1997).

Migration. During the fall, most Black Brant from North America stage at Izembek Lagoon, near the tip of the Alaska Peninsula, generally arriving from mid- to late August; they are probably joined there by most of the Black Brant breeding in Siberia (Dau 1992). Reed et al. (1989) found that the average date of arrival at Izembek Lagoon by radiomarked adult Black Brant from widely separated breeding areas reflected the distance birds had to travel. Average arrival dates were 18 September for Black Brant from the Y-K delta (900 km away) and the Mackenzie delta (3,500 km away), and 3 October for those from Victoria Island and the Western High Arctic Brant population from Melville and Prince Patrick Islands (all roughly 4,500 km away). The mean duration of their stay at Izembek was 49 days (range = 29–68 days). Within the lagoon, there was also notable segregation, with Black Brant using the western portion and Western High Arctic Brant using the eastern portion. Small groups of several hundred Black Brant stop in the Puget Sound area, temporarily joining with the Western High Arctic Brant population, as well as staging at Humboldt Bay in California, before heading farther south to winter in Mexico (Moore and Black 2006a).

Izembek Lagoon is significant to fall-staging Black Brant, because Izembek is where these Brant accumulate energy reserves for their long nonstop migration to wintering areas in Mexico (see Migration Behavior). During their stay at Izembek, body weight increased 10.0 g/day for males and 7.8 g/day for females (Dau 1992).

Mason et al. (2006) noted that the mean date of

departure from Izembek Lagoon was 6 November, which coincided with decreasing air temperature, day length, and frequency of the low tides needed for foraging. Dau (1992) reported that peak departure from 1959 to 1988 was between 22 October and 22 November, with an average departure date of 4 November. In California, large numbers of Black Brant on migration have been observed near the Channel Islands (Garrett and Dunn 1981); south of Humboldt Bay, migrants have been spotted from ships or from promontories that jut into the ocean (Einarsen 1965).

Kasegaluk Lagoon, located on the northeastern coast of Alaska, adjacent to the Chukchi Sea, is also an important fall staging area, where Black Brant most likely are attracted by its large quantities of green algae (probably *Ulva fenestrata*). Fall numbers of Black Brant using the lagoon were 40% (1989), 15% (1990), and 49% (1991) of the estimated total population of Black Brant in those years (Johnson 1993). A large percentage of these Brant may come from breeding populations in arctic Alaska, Canada, or Russia. Black Brant also migrate through this lagoon in July, apparently en route to molting areas on the Beaufort Sea (Derksen et al. 1979).

Unlike fall migration, where there is a single major staging area (Izembek Lagoon), spring migrants use several key staging areas along the Pacific coast of the United States and British Columbia. Spring migration north from Mexico begins in late January, although Kramer et al. (1979) noted that peak departure from Bahía de San Quintín did not occur until mid-March. All bays along the Pacific coast that provide suitable habitat have influxes of northbound migrants from February to early May that result in populations many times larger than those of late fall. Several are important concentration points: Morro, Bolinas, Tomales, Drakes, and Humboldt Bays in California; Coos, Yaquina, Tillamook, and Netarts Bays in Oregon; Willapa Bay, Grays Harbor, Dungeness Bay, and the Puget Sound area, particularly Padilla and Samish Bays, in Washington; and Boundary Bay in

extreme southwestern British Columbia, as well as other sites in the Strait of Georgia.

According to Moore and Black (2006a), who analyzed trends in numbers of Brant using Humboldt Bay during 1931–2000, approximately 15,000–24,000 Black Brant staged there in March–April 1999 and 2000. Lee et al. (2007) used mark-recapture data from 1,061 Black Brant at Humboldt Bay to study northward migration (Jan–May) in 2000–2001. Migration began in late December and ended in mid-May. Peak numbers of Black Brant occurred in mid-March and represented 13% of the Pacific Flyway population. The mean stopover duration was 26 days and was inversely related to age: older birds arrived earlier and stayed for less time than did younger birds. Humboldt Bay was visited by 28% of the Flyway population in 2000 and 58% in 2001.

In Washington, the use of Willapa Bay as a spring staging site for Black Brant is directly related to the availability of foraging areas with extensive beds of eelgrass. The number of migrating Black Brant spotted in this bay declined by 63% over a 31-year period (1960–90), which corresponded to a 31% loss of eelgrass beds (Wilson and Atkinson 1995); the destruction of these beds is largely attributed to an increase in their use for commercial oyster production. In British Columbia, Hagmeier et al. (2008) estimated that between 28,927 and 33,181 Black Brant staged on the Fraser River delta and nearby areas in the Strait of Georgia in spring 1999, and 21,621–25,405 in spring 2000, corresponding to 18%–26% of the Pacific Flyway population.

The majority of use days by Brant at 2 major staging areas in Washington (Willapa and Dungeness Bays) occurred during the spring. Peak arrival was the last week of April, with only a few birds remaining by mid-May (Wilson and Atkinson 1995). Gilligan et al. (1994) noted spring flights of migratory Black Brant passing off the coast of Oregon at a rate of over 500/hr during early April. Angell and Balcomb (1982) reported that the majority (100,000–125,000) of the Pacific coastal migrants staged within Puget Sound during the spring. From their final staging areas in Washington and British Columbia, it is a comparatively short flight for Black Brant across the Gulf of Alaska to Izembek Lagoon, where the mean arrival date was 14 April (Mason et al. 2006).

Not all migrants return to Izembek in the spring, however, as at least part of the population bound for eastern Arctic breeding grounds passes through the interior of Alaska, via the Yukon basin and the Porcupine River (Irving 1960). Kessel (1989) reported that over 40,000 Black Brant fly over western Alaska each spring. Black Brant arrived on the Y-K delta in early or late May, depending on the weather (Mickelson 1975), while those migrating farther north along the coast reached the Arctic Coastal Plain of Alaska and the Anderson River delta in late May (Barry 1967). Birds that continued to Victoria and Banks Islands, did not arrive until early June (Manning et al. 1956, Parmelee et al. 1967).

Outside North America, Black Brant nest in small numbers in eastern Siberia, from the delta of the Lena River to the basin of the Anadyr River and the Chukotka [Chukchi] Peninsula, as well as on the New Siberian Islands, the De Long Islands, and Wrangel Island (Dement'ev and Gladkov 1967, Portenko 1981, Ward et al. 1993), although Ward et al. (1993) found that Black Brant on Wrangel Island were primarily molt migrants from Alaska. Ground surveys revealed <100 breeding pairs, compared with the 1,000–2,000 breeding pairs previously reported by Uspenski (1965). Band recoveries indicate that most of the Black Brant breeding in Siberia migrate to North America and its associated wintering areas, rather than to Asia. Nonetheless, 4 Black Brant marked in Alaska during the breeding season were observed during winter along the coasts of Hokkaido and Honshu Islands in Japan, where they probably made an overwater flight of 3,850 km from their staging area at Izembek Lagoon (Derksen et al. 1996). Japan is the only Asian country known to winter Black Brant (in small number).

Western High Arctic Brant

Breeding. The Western High Arctic population of Black Brant breeds in the Queen Elizabeth Islands, primarily on Melville and Patrick Islands, but also on Eglinton Island and adjacent smaller islands in the Arctic Ocean. Maltby-Prevett et al. (1975) and Boyd and Maltby (1979) provided most of the information on Western High Arctic Brant breeding in these areas, which included distribution and abundance surveys and marking with neck collars. These Brant are genetically unique, with a 0.74% divergence in mitochondrial DNA (which is used in comparing how closely related species and populations within a species are, and in distinguishing among them) from other Black Brant populations in North America, from which they may have been reproductively isolated for about 400,000 years (Shields 1990). Their plumage is also different, with a medium to dark gray belly that contrasts with the dark breast, leading to them being referred to as Gray-bellied Brant (Boyd and Maltby 1979, Boyd et al. 1988). Plumage characteristics indicate an average of 74% Gray-bellied Brant in the population (Boyd and Maltby 1979, Reed et al. 1989).

Winter. Most of the Western High Arctic Brant winter in Puget Sound and the Strait of Georgia, which encompass an embayment of the Pacific Ocean along the Canada / United States border. An especially important site is Padilla Bay in Washington, where Reed et al. (1989) recorded about 15,000 Western High Arctic Brant during winter 1987/88 that were largely segregated from other Black Brant populations. They also winter in the Fraser River delta in British Columbia and in estuaries along the Alaska Peninsula, which Dau and Ward (1997) attributed to the warmer temperatures and increased food availability there. A small number move through California to winter as far south as Mexico (Boyd and Maltby 1979).

Migration. Western High Arctic Brant migrate west along the coastline, passing through the Beaufort Sea, continuing along the Alaskan coast, and then heading south to join other populations of Black Brant on staging areas within Izembek Lagoon (Reed et al. 1989). They depart Izembek for Puget Sound in Washington, although a few individuals continue farther south (Boyd and Maltby 1979). Most Western High Arctic Brant migrate about 6,700 km, compared with 8,400 km for the small number that travel as far as Mexico.

Atlantic Brant (Light-bellied Brant)

Breeding. In North America, Atlantic Brant breed within the mid- to lower arctic regions of Canada, around the Foxe Basin, with major colonies on Southampton Island and northwestern and southwestern Baffin Island (Boyd 2005), as well as others on Prince Charles Island, Air Force Island, and North Spicer Island (Abraham and Ankney 1986, Gaston et al. 1986).

Winter. Atlantic Brant winter along the Atlantic coast, stretching from Massachusetts down to central North Carolina, with concentrations on Long Island (in New York State) and in New Jersey. Stragglers occur as far north as Maine and as far south as Florida (Kirby and Obrecht 1982). On rare occasions, individual Atlantic Brant have strayed to the Pacific coast (Unitt 1984). From 2006 to 2009, an average of 152,522 Atlantic Brant were estimated from the Midwinter Waterfowl Surveys: 48% in New Jersey, 39% in New York (Long Island), 7% in Virginia, and <2% in each of the remaining Atlantic coastal states (Klimstra and Padding 2009).

Migration. Most Atlantic Brant depart Foxe Basin in early September for staging areas on James Bay, where they accumulate fat reserves. The majority then fly nonstop to the Atlantic coast, a travel distance of 2,500–3,000 km, arriving from late October to early November. In most years, however, some of these Brant will stop briefly near the confluence of the Ottawa and St. Lawrence Rivers in Québec, on Lake Ontario in New York, and occasionally elsewhere (Reed et al. 1998b).

In the spring, Atlantic Brant depart their wintering areas in April and May (Sibley 1993) and fol-

low 2 major routes back to their breeding grounds. The first route goes along the coast of New England and the Maritime Provinces into the Gulf of St. Lawrence. From the St. Lawrence estuary they travel directly to James Bay and then proceed along the eastern coast of Hudson Bay to breeding areas in the Foxe Basin. A second, shorter route is an overland flight to the confluence of the Ottawa and St. Lawrence Rivers (Lewis 1937, Palmer 1976a, Vangilder et al. 1986, Erskine 1988).

Eastern High Arctic Brant

Breeding. The population of Atlantic Brant breeding in the eastern High Arctic of North America are the northernmost breeding birds in the world, surpassing even the breeding range of the Greater Snow Goose (*Chen caerulescens atlantica*). Eastern High Arctic Brant largely breed on the eastern and central Queen Elizabeth Islands, principally on Melville, Bathurst, Ellesmere, Axel Heiberg, and Devon Islands (O'Briain et al. 1998, Boyd 2005).

Outside North America, most Brant in this population breed in the Svalbard Archipelago (Spitsbergen), although Eastern High Arctic Brant may also breed in northwestern Greenland (Arctic Goose Joint Venture Technical Committee 2008). The Svalbard population (along with the Greenland birds) winter in Denmark and at Lindisfarne in the northeastern United Kingdom (Denny et al. 2004). In Russia, Dement'ev and Gladkov (1967) reported that these Brant are found as far east as the Franz Josef Land Archipelago, where they appear to form a distinct breeding group (Boyd 2005). The various groups of Eastern High Arctic Brant totaled about 6,600 birds in 2001 (Denny et al. 2004). The few Light-bellied Brant breeding in northwestern Greenland (Boertmann 1994) may belong to either the Eastern High Arctic Brant population or the Spitsbergen population (Reed et al. 1998b).

The greatest nesting densities in the northern Canadian Arctic are found on the southern coast of the Bell Peninsula, near East Bay of Southampton Island, on western Baffin Island, on North Spicer Island, and on the southern coasts of Prince

Charles and Air Force Islands. Within East Bay, observers recorded a conservative estimate of 15.7 birds/km² (Gaston et al. 1986). Aerial surveys of the Foxe Basin have identified 5 major Eastern High Arctic Brant sites: Cape Dominion, Prince Charles Island, Air Force Island, North Spicer Island, and the northwestern coast of the Baird Peninsula (Reed et al. 1980).

Winter. Eastern High Arctic Brant winter almost exclusively in Ireland, although a small number are found on the Channel Islands and in northwestern France (Maltby-Prevett et al. 1975). It was not known that Eastern High Arctic Brant wintered in Ireland until 1971, when 2 adults banded on Ellesmere Island were later recovered in Ireland (Maltby-Prevett et al. 1975).

Migration. James Bay is an important fall staging area for Eastern High Arctic Brant, which then fly to staging areas on the western coast of Greenland in September. These Brant later fly across Greenland to stage in Iceland, followed by a 1,300 km flight over the North Atlantic to their wintering grounds in Ireland, where they arrive in late September–early October, with >75% staging at Strangford Lough. They remain there for about 6–8 weeks, feeding on eelgrass, before making a large-scale dispersal around the Irish coast (Maltby-Prevett et al. 1975, O'Briain and Healy 1991). The total distance traveled is about 4,500 km.

In the spring, Eastern High Arctic Brant depart Ireland during late April and stage in Iceland for about a month. They then leave Iceland for their flight over the Greenland ice cap, arriving on their breeding areas in early June (Alerstam et al. 1990). On Bathurst Island, records from 1968 to 1979 indicated that most Eastern High Arctic Brant arrived during the first 10 days of June; the earliest arrival was 28 May 1977 (O'Briain et al. 1998).

MIGRATION BEHAVIOR

During their fall migration, Brant make the longest flights of all waterfowl. Black Brant may fly as far as 4,200 km just to reach their Izembek La-goon staging areas, and then continue for another 5,300 km to Bahía de San Quintín, their northernmost wintering area on Baja California (Dau 1992). Atlantic Brant customarily fly from their breeding grounds in the Foxe Basin to the southern part of James Bay, a distance of 1,600 km, and then travel an additional 1,300 km to their New Jersey wintering grounds.

Einarsen (1965:21) observed Black Brant and Snow Geese flying in severe winds at Puget Sound and remarked that they "maintain an almost uniform and level flight above the water as they bore steadily against the wind. Their rapid wing beats overcome wind resistance and they will pass a point in a fraction of the time necessary for the slower [snow] geese." He clocked Black Brant speeds at 98 km/hr.

Images from radar tracks of Black Brant revealed that they leave en masse from Izembek Lagoon on an east-southeastern course of 131° (Dau 1992). The birds usually departed during a 2- to 3-hour period in the early evening when there were moderate-velocity northerly tailwinds that followed the passage of a low-pressure system. Black Brant had attained an altitude of 1,149 m within 75 km of their departure point, and some flocks had reached 1,220 m within a distance of 131 km, suggesting that they gained altitude to encounter more favorable wind speeds. They potentially maintained this altitude until arrival on their wintering grounds in Baja California, principally Bahía de San Quintín. During 23 years of a 30-year period (1959–88), Black Brant departed in falling snow or freezing rain, as well as when winds were flowing from 317° with a velocity of 40 km/hr at 850 millibars (about 1,400 m).

Their probable migration route from Izembek Lagoon to Bahía de San Quintín averages 5,300 km, but a direct-route migration distance averages 4,400 km. Black Brant complete this migration in about 54 hours, flying nonstop at an average ground speed of 99 km/hr. Kramer et al. (1979) recorded that Black Brant wintering at Bahía de San Quintín arrived 45–60 hours after the first major departure from Izembek Lagoon, which would indicate a flight speed of 75–100 km/hr. Whatever their speed, this flight has a considerable energetic cost, as males lose an average of 33% of their body weight and females lose 31% (Dau 1992).

Eastern High Arctic Brant undergo a strenuous migration to and from their wintering grounds in Ireland. In the spring, Alerstam et al. (1990) found that birds departing their staging area in Iceland left in small flocks of 40–45 birds and flew at low altitudes (<100 m) until no longer in sight from land. Subsequent studies equipping Eastern High Arctic Brant with satellite transmitters found that spring migrants leaving Ireland to fly over the Greenland ice cap gained altitude, but at a very slow (0.01–0.06 m/sec) rate. In contrast, the climb rate for the much smaller Knot (*Calidris canutus*) was 1.1 m/sec (Gudmundsson et al. 1995). Eastern High Arctic Brant also decreased their forward progress as they crossed the Greenland ice cap, which reaches heights of about 3,000 m above sea level. Such migration behavior suggested that the birds came down on the ice to rest and recover from the steep climb.

Molt Migration. Large numbers of both failed and nonbreeding Black Brant from Alaska, Canada, and Siberia undergo a molt migration to the Arctic Coastal Plain of Alaska, where they arrive from late June through mid-July (King and Hodges 1979, Derksen et al. 1982, King and Derksen 1986, Taylor 1995). The major molting site in the area is the 2,000-km² Teshekpuk Lake region, which lies within the National Petroleum Reserve of Alaska, an area characterized by many large tundra lakes and very little elevational relief (Derksen et al. 1979). The importance of the Teshekpuk Lake region to molting Black Brant cannot be overemphasized, as banding data suggest that nearly all of the nonbreeding Black Brant north of the Bering Strait in Alaska, Canada, and Siberia come to this area to molt. Bollinger and Derksen (1996) reported that molting Black Brant in the Teshekpuk Lake region originate from 10 nesting colonies in Canada and Alaska, including Gray-bellied Brant from as far away as Melville Island in the western Canadian High Arctic. Of 6,910 Black Brant banded during this study (1987–92), 76% were adults, and 57% were males; 61% of the adult females were failed breeders; and 91% of all recaptures were <6 years old. Black Brant also exhibited a very high (95%) degree of site fidelity, but this rate varied inversely with their traveling distance to the Teshekpuk Lake area, probably due to the costs of longer migrations.

Derksen et al. (1979) reported 14,000–33,000 molting Black Brant in the Teshekpuk Lake region during surveys from 1976 to 1978. More recently, a 2007 survey of 209 lakes in this area recorded 27,000 molting Black Brant (Mallek 2007). Preferred molting habitats in the Teshekpuk Lake area are along the shores of deep open-water lakes (Derksen et al. 1979). The birds also congregate on the mossy, peat-covered shorelines of shallow tundra lakes on Wrangel Island in Russia (Ward et al. 1993). Molting Black Brant were present at Teshekpuk Lake from 26 June through 19 August, when they regained the ability to fly, they then traveled to staging areas along the coast (Taylor 1995).

HABITAT

During the breeding season in the Low Arctic, Black Brant will nest colonially near the upper edges of saltmarshes or river deltas, where there is usually an abundance of low-growing graminoid vegetation, such as *Puccinellia phryganodes*, *Carex subspathacea*, and *Dupontia fisheri*. At the Tutakoke colony on the Y-K delta, Black Brant nested several hundred meters from coastal mudflats, in wet sedge meadows dominated by *Carex ramenskii* (Flint and Sedinger 1992). In contrast, Black Brant and Atlantic Brant nesting in the High Arctic encounter fewer and less extensive saltmarshes, so nesting is more dispersed, at times up to 30 km inland, with reports of nests on gravel islets (Reed et al. 1998b, Boyd 2005).

During the fall and spring migrations, Brant are found in sheltered marine habitats at staging areas, such as Izembek Lagoon on the Alaska Peninsula, that are characterized by extensive growths of aquatic vegetation, especially eelgrass. Ward et al. (1997a) determined that about 44%–47% (15,000–16,000 ha) of the 34,662 ha Izembek Lagoon (49 km long × 3–11 km wide) was covered with eelgrass, which constituted the largest stand of eelgrass in the world. Large (6,000–7,000 ha) contiguous tracts of eelgrass occurred in the central and southern sections of the lagoon, and the extent of that eelgrass in the lagoon had remained relatively stable over an 18-year period (1978–95). The next-largest beds of eelgrass on the Pacific coast are found in Boundary Bay in British Columbia, but these are only about a third the size of those at Izembek Lagoon (Ward et al. 1997a). Black Brant and Atlantic Brant also use habitats with sea lettuce (*Ulva fenestrata*) and other green algae, such as additional species of *Ulva* and *Enteromorpha*.

Eastern High Arctic Brant breeding on Bathurst Island choose areas characterized by extensive sedge meadows interspersed with numerous small ponds, as well as rolling uplands with sparser herbaceous vegetation, such as *Saxifraga oppositifolia* (O'Briain et al. 1998).

In the winter, Black Brant and Atlantic Brant are also found in protected coastal habitats, such as bays and estuaries, where there is an abundance of eelgrass and green algae. In Mexico, Black Brant have not been recorded south of Bahía Santa María, which is probably the southern limit of eelgrass growth (Kramer and Migoya 1989). This habitat preference contrasts markedly with all other goose species in North America, which instead make extensive use of agricultural habitats during the winter months. Along the Atlantic coast, however, some Brant graze on lawns and golf courses (Smith et al. 1985).

POPULATION STATUS

Black Brant. Most (>80%) Black Brant nest on the Y-K delta in Alaska, where the breeding population decreased by >60% during the early 1980s, with predation by arctic foxes (*Alopex lagopus*) and the local subsistence harvest most likely contributing to this decline (Sedinger et al. 1993). Since 1992, aerial surveys of the 5 major nesting colonies on the Y-K delta (Kokechik Bay, Tutakoke River, Kigigak Island, Baird Peninsula, and Baird Inlet Island) provide highly comparable estimates of breeding-population size (Anthony et al. 1995). The total number of nests located in all colonies in 2008 (9,995) was 40% lower than in 2007 and about 43% below the long-term (1992–2007) average, which yielded a declining trend estimate of about 3%/year (Wilson 2008). Sedinger et al. (1993) estimated fewer than 400 breeding pairs at colonies outside the Y-K delta. Aerial surveys in 2008 on the Arctic Coastal Plain estimated 2,669 breeding Black Brant (Dau and Larned 2008). A 2005 aerial survey of Victoria Island and nearby areas recorded 1,117 Black Brant (Conant et al. 2006).

The Midwinter Survey estimated 147,400 Black Brant in 2008; since 1970 the survey figures have ranged from a low of 98,500 in 1987 to a high of 186,100 in 1981 (U.S. Fish and Wildlife Service 2009). Black Brant populations in general have averaged 126,700 from 1970 to 2008, with numbers above 100,000 each year (except for 98,500 in 1987) (U.S. Fish and Wildlife Service 2009).

Western High Arctic Brant. The status of this population is monitored via Midwinter Waterfowl Surveys on their principal wintering grounds in Washington State: 16,200 were estimated in 2009, which was 76% more than in 2007. Population estimates have increased by an average 7%/year from 2000 to 2009. Since 1970, this population of Brant has ranged from a low of 2,100 in 1983 to a high of 16,900 in 1995 (U.S. Fish and Wildlife Service 2009).

Atlantic Brant. The Atlantic Brant population has fluctuated dramatically in response to environmental conditions on both their breeding and wintering areas. On Southampton Island, Atlantic Brant significantly reduced their breeding efforts during late-arriving springs (Barry 1962). Elsewhere, in winter 1976/77, severe cold froze feeding areas along the entire Atlantic coast, from Maine to Pamlico Sound in North Carolina, which resulted in the starvation of about 60,000–80,000 Atlantic Brant from an estimated fall population of 120,000, reducing their numbers to perhaps the lowest level ever recorded (Nelson 1978, Myers et al. 1982). Preferred habitats were so affected that Atlantic Brant were observed feeding on dead grass along highway embankments and roadsides (Nelson 1978), and substantive feeding programs were implemented for Atlantic Brant and other affected waterfowl (Myers et al. 1982). Brant numbers had also been reduced along the Atlantic coast in the early 1930s, following a precipitous decline in eelgrass (Cottam and Munro 1954, Rasmussen 1977).

The Midwinter Survey estimated 151,300 Atlantic Brant in 2009, which was 7% lower than in 2008, but no general trend was discerned from 2000 to 2009 (U.S. Fish and Wildlife Service 2009). The population averaged 190,000 from 1960 to 1969, but only 84,000 from 1970 to 1979, a 44% decline. Since 1970, the population has ranged from a low of 40,800 in 1973 to a high of 184,800 in 1992. Populations generally were smaller (40,800–97,000) from 1972 to 1981, with the exception of 127,000 in 1976. The maximum number of Atlantic Brant ever recorded during the Midwinter Survey was 265,500 in 1961, and the minimum was 40,800 in 1973 (Klimstra and Padding 2009).

Eastern High Arctic Brant. Due to their High Arctic breeding distribution, the Eastern High Arctic Brant population has fluctuated even more dramatically than that of Atlantic Brant. During a 22-year study (1968–89) of Eastern High Arctic Brant on Bathurst and Seymour Islands, there were 3 years (1974, 1986, 1988) when Brant did not attempt to nest (O'Briain et al. 1998). The population of Eastern High Arctic Brant is currently estimated from counts on staging areas in Iceland and on their wintering grounds in Ireland. In Ireland, population numbers have increased from <10,000 in the late 1960s to >33,000 in winter 2004/5. The

2007 survey estimated 38,900 birds (U.S. Fish and Wildlife Service 2009).

Harvest

The harvest of Atlantic Brant in the eastern United States averaged 26,200 birds from 1999 to 2008, ranging from a low of 17,600 in 2004 to a high of 33,400 in 2002 (Klimstra and Padding 2009). The 2008 harvest totaled 27,200 birds: 29% in New Jersey, 28% in New York, 20% in Virginia, and 7% in Maryland. There were an estimated 8,300 active Brant hunters in the Atlantic Flyway in 2008, from Virginia (25%), New York (18%), New Jersey (18%), and North Carolina (16%). The harvest in Canada was 1,065 in 2008; most (95.2%) occurred in Québec (Raftovich et al. 2009).

The harvest of Black Brant in the Pacific Flyway states averaged 3,746 from 1999 to 2007, ranging from 1,700 in 2002 to 5,513 in 2006 (Trost and Sanders 2008). Most of the harvest occurs in California (50.0%), with 26.5% in Alaska, and 17.6% in Washington. Subsistence harvests in the Y-K delta averaged 3,208 Black Brant from 1985 to 2005, although there was no survey in 2003 (Wentworth 2007a). From 2001 to 2005 these harvests averaged 4,553 birds. This figure was considered more accurate, because of survey participation by more villages near prime Brant habitat. Wolfe and Paige (1995) summarized the subsistence harvest of Black Brant by region in Alaska for the early 1990s; important harvest areas (other than the Y-K delta) were the North Slope (3,000), Seward Peninsula and the northwestern Arctic (2,700), the eastern Aleutian Islands (1,200), and Bristol Bay (300). The authors' statewide estimate for the subsistence harvest was 11,000 birds.

Black Brant are also harvested on their primary wintering grounds in Mexico, but harvest surveys in that country have not occurred on a regular or complete basis. Most of the harvest in Mexico occurs at Bahía de San Quintín; Kramer et al. (1979) reported a harvest of 1,741 Black Brant during winter 1974/75 and 6,532 in 1975/76 (including a 29% crippling loss). Other summary data from periodic Black Brant harvest surveys in Mexico from 1974/75 to 2000/2001 reported an average annual harvest of 1,408 (Pacific Flyway Council 2004).

Native peoples in Canada also regularly harvest Brant for subsistence purposes. In eastern Canada, the harvest of Atlantic Brant from 1973 to 1977 by native hunters averaged 6,400/year within the James Bay area in Québec (Reed 1991). From 1974 to 1976, native hunters took an average of 200 Atlantic Brant/year in Ontario, in the western James Bay and Hudson Bay region (Prevett et al. 1983). The Inuvialuit Harvest Study in northwestern Canada (based on interviews with hunters) reported an average harvest of 489 Black Brant from 1988 to 1997, but these numbers are probably underestimates (Inuvialuit Harvest Study 2003, Pacific Flyway Council 2004).

BREEDING BIOLOGY

Behavior

Mating System. Brant are monogamous and pair for life, unless a member of the pair dies (Owen 1980). Courtship activity occurs on their wintering grounds between mid-January and April (Einarsen 1965), but when actual pairs form is not well known. Pairing probably occurs in the second year, most likely 1–2 years before sexual maturity (Barry 1962, Reed et al. 1998b). Brant arrive paired on the breeding grounds, although they remain in flocks until the snow melts from their nest area (Barry 1962). From observations of marked Black Brant along the coasts of British Columbia and Washington, Reed (1993) reported that most family groups remained intact through fall migration, winter, and the first stages of spring migration. Any overt expression of familial bonds, however, is infrequent, which may facilitate feeding in flocks without disruption from conspecifics. Family bonds may be retained into spring so juveniles can learn the location of spring staging areas, many of which are not used during fall migration. An earlier report that Brant families dissolved during fall staging on the Alaska Peninsula was probably erroneous, because of the difficulty in recognizing the infrequent and subtle forms of family association within very large flocks of birds (Jones and Jones 1966).

Most Brant breed as 3-year-olds, but favorable seasons encourage perhaps 10% of the 2-year-old birds to nest (Barry 1967). Not all adults breed every year. Barry (1962) estimated that 60% of approximately 2,500 Atlantic Brant on Southampton Island did not nest following a late spring in 1957. At the Tutakoke colony on the Y-K delta in Alaska, the breeding probability of females was 67% for 2-year-olds, 68% for 3-year-olds, 84% for 4-year-olds, and 90% for female Black Brant >5 years old (Sedinger et al. 2001). The probability of breeding for males was 51% for 2- to 4-year-olds and 78% for those >5 years old. The lower probability of male breeding probably reflected the dispersal of males whose mates had died.

Abraham et al. (1983) reported that Atlantic Brant on Southampton Island mated "assortatively," because individuals chose mates with a similar necklace pattern. The authors believed that early learning of familial phenotypes such as the necklace pattern led to future assortative mating. Further, such a system could restrict interbreeding by affecting pair formation on wintering areas where these Brant mixed with other populations.

Sex Ratio. The sex ratio (males:females) of Black Brant has only been reported for birds captured during the molt. In Alaska, 57% of 7,949 Black Brant were males (Bollinger and Derksen 1996). In contrast, males constituted 45% of 1,095 birds captured on Wrangel Island (Ward et al. 1993).

Site Fidelity and Territory. Brant have a strong tendency to return to the same nest area each year. Lindberg et al. (1998) reported that 34%–100% of the females ($n = 4,076$) and 1%–29% of the males ($n = 4,071$) banded as goslings at 6 Black Brant breeding colonies on the Y-K delta from 1986 to 1991 subsequently returned to their natal area, thus demonstrating a remarkable affinity for their original nest sites. Ward and Flint (1995) noted return rates of 57%–83% for color-banded female Black Brant on the Y-K delta. Another study (Lindberg and Sedinger 1997) found that most females nested within 50 m of the previous year's nest. Barry (1967) noted that 40% of the adults neckbanded at the Anderson River delta returned a year later. On Southampton Island, 38% of the nesting Atlantic Brant adults returned to breeding sites despite very poor habitat conditions (Barry 1962). Of the Eastern High Arctic Brant marked as breeding adults on Bathurst Island (1984–87), 48% were seen again in subsequent years (O'Briain et al. 1998).

The home range of Black Brant during the breeding season varies, based on food availability and competition. In high-density colonies on the Y-K delta, there have been as many as 57 Black Brant nests in a 50-m radius, while only 4 may occur in low-density areas (Welsh 1988). Nest territories in high-density areas may be just 0.6–1.8 m apart.

Atlantic Brant newly arrived at their arctic breeding grounds on Southampton Island wait a few days on the snow, after which they make long flights to search for nest sites starting to appear on the snowfields (Barry 1962). Territories are defended by both sexes prior to incubation, and thereafter by the male until the eggs hatch. Under crowded conditions, territorial interactions are vigorous and continuous, and only the most aggressive birds maintain a territory (Barry 1962). Territory size varies in relation to habitat and competition. Palmer (1976a) reported that 66 Atlantic Brant nests on Southampton Island were an average of 33.8 m apart. Barry (1967) noted a single territory of about 200 m in diameter that was centered on a small lone hummock, in contrast to a single 6 × 12 m hummock that contained 7 Atlantic Brant nests and 2 Snow Goose nests. The few yearlings still in family groups are tolerated on territorial margins for the first few days of incubation and then are driven away. These yearling Brant and other nonbreeders then gather on undefended communal grounds (Barry 1962).

Courtship Displays. Most sexual activity and the resultant pair formation occur on winter and spring staging areas, but there is no detailed description of the pair-formation process. Brant arrive paired on the breeding grounds. The *forward-threat* posture of Brant is similar to that of other *Branta* species and precedes a direct attack (Johnsgard 1965). Brant enhance this display, however, by raising their neck feathers so the white necklace appears to surround the otherwise black head and bill, an effect that is imposing at Brant level (Boyd 2005). Johnsgard (1965) noted that Brant will vibrate the white feathers of the necklace, a behavior that can be seen at close range. A precopulatory display consists of mutual *head-dipping.*

Brant also exhibit a distinctive but infrequent *3-bird-flight* that is fast and low to the ground and usually involves a second male attempting to break up the pair (Barry 1967, Boyd 2005). Such flights could also assist the chasing male in judging the breeding status of the female and, therefore, her vulnerability to extra-pair copulation. Waterfowl are most susceptible to successful extra-pair copulations when females are actively developing egg follicles (Cheng et al. 1982, 1983), and male waterfowl may assess female status by searching for those with a drooped abdominal profile, as Sorenson (1994) observed during her studies of extra-pair copulation in White-cheeked Pintails (*Anas bahamensis*). Barry (1962) noted that female Atlantic Brant on Southampton Island were easy to identify in the field from the abdominal bulge caused by ovarian follicles developing within 5–7 days of their arrival.

Brant readily engage in extra-pair copulations, perhaps because they often nest colonially, which affords more opportunities for males to copulate with females who are not their mates. At the Tutakoke colony on the Y-K delta, 7 out of 28 (25%) observed Black Brant copulations were known to be extra-pair copulations (Welsh and Sedinger 1990). Farther north on Southampton Island, however, Barry (1962) never saw any extra-pair copulations among Atlantic Brant, despite 24 continuous hours of daylight.

Nesting

Nest Sites. The female selects the nest site. In the Low Arctic, Black Brant usually nest in grassy vegetation near the coast, but they also will nest on islands or mudflats with scattered patches of grass (Barry 1967, Mickelson 1975). On the Y-K delta, Black Brant once nested in a large contiguous colony that extended approximately 160 km along the coast (Spencer et al. 1951), but they now are found in separate (albeit large) colonies; elsewhere they nest in smaller colonies or solitarily (Sedinger et al. 1993). Barry (1967) described nest locations of Black Brant on the Anderson River delta as being 0−0.9 m above the normal high-tide line or above standing water (*n* = 250 nests), 0.3−30 m from snow cover (*n* = 108 nests), and 0−30 m from standing water (*n* = 108 nests).

At the Tutakoke colony on the Y-K delta, an area of about 12 km², nesting Black Brant are nonetheless patchily distributed, with nest densities ranging from 0.01 to 280/ha (Flint and Sedinger 1992). Eisenhauer (1977), however, reported 200 Brant nests within 500 m and 4,000 within 5 km of an observation tower at a colony in the Tutakoke area. On Banks Island in the Northwest Territories, the majority (80%) of Black Brant colonies and solitary nests were found on islands in large ponds or inland lakes, and 67% of the nesting locations supported 10 or fewer nests (Cotter and Hines 2001). The remainder of the colonies were located on the mainland and centered around active Snowy Owl (*Bubo scandiacus*) nests. Most of the island colonies also had 1 or more pairs of nesting gulls (*Larus* spp.). The mean distance between nests was 5.5 m on islands (*n* = 260), compared with 23.5 m in Black Brant colonies with Snowy Owl (*n* = 38). On Southampton Island, the average distance between 66 Atlantic Brant nests was 33.8 m, and nest density on 1 study site was 4.1/ha (Barry 1960).

In contrast, only 6−7 nesting pairs of Eastern High Arctic Brant occurred in a 10-km section of a river valley on Bathurst Island (O'Briain et al. 1998). These Brant do not encounter extensive grassy habitats for nesting and thus locate their nests farther inland, often on gravel bars of rivers or other unvegetated rocky sites. On Seymour Island, Eastern High Arctic Brant built their nests on gravel ridges, frequently near a large rock. Nests on nearby Bathurst Island were situated on low gravel islands along braided streams. All of the nests were very exposed, with no concealing cover, although nearby rocks may have provided protection from the wind (O'Briain et al. 1998).

The female builds the nest, which initially is a depression formed as she scrapes away vegetation and substrate until reaching the permafrost (Barry 1967). Nearby sedges then are molded around the scrape, and down is added as egg laying and incubation progress. Down is more abundant in the nests of Brant than in those of any other species of waterfowl, with the possible exception of Common Eider (Thompson and Raveling 1988). Brant down also appears to have the same quality as that of eiders, being quite cohesive and adhering to the nest depression in a single mass, thus resisting being blown away by the wind (Palmer 1976a). Cottam et al. (1944) described the Brant nest as the most beautiful of all waterfowl nests, with a grass foundation and a symmetrical ring of pure down, 36−46 cm in diameter. Barry (1960) reported averages for nest dimensions of Atlantic Brant: depth, 5.6 cm (*n* = 10), inside diameter of the down, 18.8 cm (*n* = 28), outside diameter of the down, 30.5 cm (*n* = 29), and outside of the entire nest, 45.7 cm (*n* = 18). Palmer (1976a) also listed nest dimensions: depth, 5.6 cm (*n* = 10); diameter of the cavity, 30.5 cm (*n* = 29); and external diameter of the nest mound, 45.7 cm (*n* = 18). Females will reuse nests from previous years (Mickelson 1975).

Clutch Size and Eggs. An average clutch weighs 320 g, which is 23% of the female's prelaying body weight (Ankney 1984). Average lipid reserves of female Atlantic Brant on Southampton Island declined by 36%, from 124 g in prelaying birds to 79 g in postlaying birds, which was sufficient to meet the lipid requirement to produce a clutch of eggs. In contrast, protein in the females declined by 20% (from 99 g to 79 g), which was only enough to meet 70% of the protein needed to produce a clutch of eggs (Ankney 1984).

From 1980 to 2002 on the Y-K delta, the average size of 22,853 Black Brant clutches was 3.6 eggs, ranging from a low of 2.5 in 1981 to highs of 4.1 in 1987 and 1992 (U.S. Fish and Wildlife Service 2002a). Mickelson (1975) reported an average clutch size of 3.3 eggs (range = 2.76–4.1 eggs) from 1969 to 1972 on the Y-K delta, and Flint et al. (1995) stated that clutch size in late incubation ranged from 3.92 to 4.16 eggs for 1,461 nests monitored on the delta from 1987 to 1989. Sedinger et al. (1998) noted that clutch size of Black Brant on the Y-K delta increased with age, averaging 3.4 eggs for 2-year-olds and 4.4 for 5-year-olds.

On the Anderson River delta (1959–65), Barry (1967) reported an average of 3.9 eggs for 700 Black Brant nests, which ranged from 2.6 eggs in 1959 to 4.5 in 1961. Clutch size varied from 1 to 10 eggs, and most frequently was 4. Large clutches are rare, and they probably represent brood parasitism. From 1958 to 1965, outside his immediate study area, Barry (1967) reported an average clutch size of 3.3 eggs for 3,308 nests, ranging from 2.8 eggs in 1964 to 4.0 in 1960. Clutch size was reduced during years when nesting habitats were not clear of snow, which delayed nest initiation and caused ovarian follicles to become atretic and reabsorb.

Lindberg et al. (1997) measured the clutch size of 1,751 Black Brant nests in the Tutakoke colony on the Y-K delta from 1987 to 1993. Clutch size varied significantly among years, from a low of 4.1 eggs in 1993 to a high of 4.4 in 1987. Contrary to earlier studies, the mean clutch size was greater in late springs (4.4 eggs, $n = 804$) than in early springs (4.2 eggs, $n = 947$), and it was positively correlated to the date of 100% snowmelt. In late springs, 45% of the clutches contained 5 eggs, compared with only 34% in early springs. The increase in clutch size may have occurred because Black Brant ac-

cumulated more endogenous (stored) nutrient reserves on staging areas during late springs, as nest-initiation dates (6–12 days after peak arrival) did not vary between early or late springs. A more likely explanation, however, is that fewer younger or small-bodied Brant breed during late springs, which would result in a lower number of smaller clutches and, therefore, an increase in the overall mean clutch size (Eichholz and Sedinger 1999). In contrast, a 1992–93 study of Black Brant on Banks Island in the Northwest Territories found that mean clutch size was larger (3.8 eggs, $n = 397$) during 1993, when June temperatures were milder and the snow melted earlier, compared with 3.5 eggs ($n = 148$) in 1992, when spring occurred later (Cotter and Hines 2001).

On Bathurst and Seymour Islands (1969–89), the clutch size of Eastern High Arctic Brant ($n = 24$) averaged 4.5 eggs (range = 2–6 eggs; O'Briain et al. 1998). Average clutch size for Atlantic Brant on Southampton Island was 4.0 eggs in 1953 ($n = 203$), 3.8 in 1956 ($n = 444$), and 3.6 in 1957 ($n = 206$), a poor reproductive year (Barry 1962). More recently, Abraham and Ankney (1986) reported a mean clutch size of 3.5 eggs ($n = 851$) for Atlantic Brant on Southampton Island.

Brant eggs are subelliptical to long elliptical. Egg color is creamy white or light buff, but becomes pale olive with age (Barry 1960, Palmer 1976a). There is little variation in egg size among the populations: on Southampton Island, Atlantic Brant eggs averaged 72.9 × 47.2 mm ($n = 521$; Barry 1960); those of Black Brant on the Y-K delta in Alaska, 72.4 × 47.2 mm ($n = 376$; Palmer 1976a); and those of Eastern High Arctic Brant on Bathurst Island, 71.4 × 47.3 mm ($n = 12$; Reed et al. 1998b). The mean volume of 3,478 Black Brant eggs measured on the Y-K delta was 84.0 cm^3 (Flint and Sedinger 1992). Egg size on the Y-K delta increased with clutch size and the female's age, and decreased with the laying date, year, and position in the laying sequence (Flint and Sedinger 1992). The range of egg sizes was similar for clutches of 3 (70.0–94.3 cm^3), 4 (69.0–94.8 cm^3), and 5 eggs (72.6–99.6 cm^3). The laying rate averaged about 0.8 eggs/day (Alisauskas and Ankney 1992).

Incubation and Energetic Costs. At the Tutakoke colony on the Y-K delta in 1992 and 1993, incubation of 88 Black Brant nests averaged 25 days (range = 23–29 days; Eichholz and Sedinger 1998). There was no relationship between incubation length and ambient temperature, mean egg size, or nest attentiveness, but the duration of incubation decreased with later nest initiation. The reduction of incubation length by late-nesting females may decrease the cost of nesting late by limiting the exposure time of eggs to predators and improving the growth rates and survival of goslings (Sedinger et al. 1995b). Elsewhere, Barry (1967) reported an average of 24 days for incubation by Black Brant on the Anderson River delta, and O'Briain et al. (1998) noted 23 days as the average for Brant nesting in the Canadian High Arctic. On Southampton Island, Barry reported that incubation time for Atlantic Brant averaged 24 days.

Only the female incubates, but the male stands guard nearby and accompanies the female during feeding recesses. The female covers the eggs with down before she leaves. Compared with other species of geese, female Brant take longer recesses from incubation, spending only an average of 89.6% of their time on the nest; Cackling Geese (*Branta hutchinsii minima*) remain on their nests 93.6% of the time, and Emperor Geese, 95% (Thompson and Raveling 1987, 1988). Welsh (1988) reported that nest attentiveness in Brant averaged 85%. Brant take an average of 6.7 recesses/day, averaging 22 minutes each and 148 min/day (Thompson and Raveling 1988). To compensate for these long absences from incubation, Brant cover their eggs with large quantities of down, providing better insulation for their eggs than the practices other geese use with their nests. When exposed to windy conditions, eggs in Brant nests cooled about 15% less than those of Cackling Geese (*minima*) and Emperor Geese (Thompson and Raveling 1988).

Both sexes of Brant lose significant body mass during incubation, but their nutrient reserves are not sufficient to allow them to fast during the entire incubation period. On Southampton Island, Atlantic Brant females lost an average of 34% of their lipid reserves and 20% of their protein reserves during egg laying (Ankney 1984). The females began incubation with relatively low (7% of body mass) lipid reserves, which only supplied about 22% of the energy required for incubation; hence 78% of this energy came from exogenous sources, obtained while on feeding recesses. Males lost an average of 65% of their fat reserves (Ankney 1984).

These fluctuations in nutrient reserves are closely related to the regulation of incubation behavior in individual Black Brant, which was studied by Eichholz and Sedinger (1999) at the Tutakoke colony on the Y-K delta. Their results also found that mean nest attentiveness for Black Brant was low: 87% (range = 79%–91%) over a 24-hour period and 85% (range = 82%–92%) during daylight hours (04:00–24:00). In general, Black Brant took nest recesses to feed when ambient temperatures were higher (at midday and on sunny days), which minimized egg cooling. There was considerable (12%) variation, however, in nest attentiveness among females, which indicated tradeoffs between the energy they invested in egg production and the energy needed for incubation. Given 2 females with the same investment of energy in the production of eggs, the more attentive female during incubation has a larger nutrient reserve remaining after egg laying.

Clearly the small size of Brant precludes their arrival in the Arctic with sufficient reserves to meet the requirements of both egg production and incubation, as well as normal daily metabolic maintenance, so females must leave their nests for longer periods than other species of geese. Brant appear to resolve this problem by nesting in habitats that provide food in close proximity to nesting sites (Ankney 1984). On Southampton Island, Atlantic Brant nested immediately above the high-tide line in saltmarsh habitat dominated by *Puccinellia* and

Carex. This area was the last to become snow free, but then it contained a flush new growth of high-quality food for female Brant nesting nearby.

Nesting Chronology. At the Tutakoke colony on the Y-K delta, nest initiation occurred within 6–12 days after arrival for 3,556 Black Brant nests monitored between 1987 and 1993, but the initiation date varied among years (Lindberg et al. 1997); the mean initiation date was 2 June during late springs and 26 May in early springs. There was also a positive relationship between the nest-initiation date and the date of 100% snowmelt, which differed by 10 days between early and late springs. Nest initiation was also more synchronized in late versus early springs. Near Onumtuk on the Y-K delta, egg laying commenced on 19 May in both 1969 and 1970, but because of the late arrival of spring in 1971 and 1972, egg laying did not start until the first week of June (Mickelson 1973).

On Bathurst and Seymour Islands, the earliest known nest of Eastern High Arctic Brant appeared on 10 June, and the latest on 22 June; the peak was 16 June ($n = 15$; O'Briain et al. 1998). On Victoria Island, Black Brant began egg laying on 9 June and continued to mid-June (Parmelee et al. 1967). During a 3-year study (1953, 1956–57) on Southampton Island, Atlantic Brant laid their first eggs between 16 and 23 June, 7–16 days after their arrival (Barry 1962). The mean date of nest initiation for Black Brant on Banks Island in the Northwest Territories was 12 June during 1993, when there were mild June temperatures and an early snowmelt, versus 20 June in 1992, when spring occurred later (Cotter and Hines 2001).

Nest Success. Nest success in Black Brant is highly variable, due to the influences of extreme spring weather, flooding, and predation by arctic foxes. On Southampton Island, a late spring and the resultant lingering snow cover in 1957 resulted in 60% of the adult Atlantic Brant not breeding (Barry 1962). Many females reabsorbed developing follicles, which contributed to a decreased clutch size or even complete breeding failure. On the central Arctic Coastal Plain of Alaska, Black Brant nested every year at their primary sites, but less frequently at secondary or solitary sites (Stickney and Ritchie 1996). During 10 years (1974–77 and 1984–89) of a study in the Canadian High Arctic, Atlantic Brant did not attempt to breed during 3 years of below-normal spring temperatures (O'Briain et al. 1998). Thus severe spring weather can prevent Brant from breeding in High Arctic habitats, although Brant have bred every year in Low Arctic habitats, such as the Y-K delta (Lindberg and Sedinger 1997).

On the Y-K delta, arctic foxes are the major nest predators, and they were primarily responsible for the dramatic decline in the nest success of Black Brant that was observed on the delta during the 1970s and 1980s (Raveling 1984, King and Derksen 1986). Nest success of Black Brant at 4 major colonies on the Y-K delta decreased from an average of 73.7% in 1964–73 to 39% in 1981–85, largely due to arctic fox predation (Anthony et al. 1991). Nest success was especially low at the Tutakoke colony: 2% in 1984 and 7% in 1985. Excessive predation by arctic foxes may have occurred because low populations of tundra voles (*Microtus oeconomus*) on inland areas dramatically reduced breeding opportunities for the foxes, which then moved into coastal areas and preyed heavily on Black Brant colonies (Anthony et al. 1991). Walrus (*Odobenus rosmarus*) carcasses on the coast may also have attracted foxes into areas used by nesting Brant. Raveling (1989) found that arctic foxes caused the loss of 55% and 85% of the nests in 2 small (10–37 pairs) Black Brant colonies, and 31%–32% of the nests in 2 large colonies composed of thousands of individuals. During years when Brant populations are low, they may be especially vulnerable to arctic fox predation, because they lose the effect of predator swamping conferred by nesting in large colonies.

On the Anderson River delta, nest success was 28% ($n = 229$) in 1991 and 81% ($n = 103$) in 1993, with Glaucous Gulls (*Larus hyperboreus*) and Parasitic Jaegers (*Stercorarius parasiticus*) being the main nest predators during both years (Armstrong 1996). Sedinger (1990) reported that Parasitic Jaegers were opportunistic predators, consuming Brant eggs when nests were left unattended. On Banks Island in the Northwest Territories, the nest success of Black Brant was 66.2% in 1992 and 77.4% in 1993; 17.2%–19.4% of the nests were destroyed by predators and 3.2%–16.6% were abandoned (Cotter and Hines 2001). On Seymour Island in 1976–77, 10 out of 12 (83%) nests of Eastern High Arctic Brant hatched successfully (O'Briain et al. 1998).

Brant nests can also become susceptible to storm tides, because they are often located very close to the ocean. On the Anderson River delta, Barry (1967) reported a nesting season where about 3% of the Black Brant nests were destroyed by storm tides; he also mentions a report from local residents that a storm tide in 1953 destroyed the entire Brant nesting effort. On the Y-K delta, Hansen (1961) also noted the loss of Black Brant nests to storm tides.

Brood Parasitism. Brant occasionally lay eggs in the nests of Emperor Geese (Petersen et al. 1994), but intraspecific brood parasitism is rare (Bellrose 1980). During their 2-year (1992–93) study of Black Brant on Banks Island in the Northwest Territories, Cotter and Hines (2001) believed a clutch of 7 eggs represented a parasitized nest, but such nests were rare among the 673 they located during their study. The authors found single clutches of 8 and 17 eggs in 1992; as well as 5 clutches with 8 eggs, 3 with 9, and 1 with 10 in 1993. Parasitic nests may be more common for late-nesting Brant, which could account for the larger clutch sizes seen in late nests (Lindberg et al. 1997).

Renesting. Brant do not renest after the complete loss of a full clutch or brood, and they are probably physiologically incapable of doing so (Barry 1962). They may build a second nest near the first, however, and lay the remaining eggs of a clutch if part of it was laid but destroyed (Barry 1962, Palmer 1976a).

REARING OF YOUNG

Brood Habitat and Care. The many tidal creeks and sloughs characteristic of Black Brant breeding habitat on the Y-K delta facilitate movements to brood-rearing areas, which can easily be 10–15 km from nesting colonies (Lindberg and Sedinger 1998). At the Tutakoke colony, brood-rearing areas were in saltmarsh communities dominated by *Carex subspathacea* and *Puccinellia phryganodes* (Sedinger et al. 1995b). Adult females exhibited fidelity to brood-rearing areas at Tutakoke during studies of marked birds from 1987 to 1993 (Lindberg and Sedinger 1998). This fidelity was variable, however, ranging from 49% to 100%, with no obvious rationale. Movements among brood-rearing sites also occurred, and were best explained by the distance between sites, although the relationship differed among years and among sites. Nonetheless, Black Brant clearly gain advantages from fidelity to their brood-rearing areas: improved feeding efficiency, reduced aggression with other Brant, and increased knowledge of local predators.

The eggs in the clutch hatch at about the same time, with 24–48 hours usually required from the pipping of eggs to the drying of young, although it may take up to 72 hours during stormy weather (Barry 1967). According to Palmer (1976a), it takes 24–60 hours for the goslings to hatch, dry, and depart the nest. When the young are dry, the male assumes the dominant role in family life, leading the goslings from the nest, accompanied by the female in the rear, so the goslings are bracketed by their parents as they travel to brood-rearing areas. The family maintains a small mobile territory that is vigorously defended, but it becomes smaller after about 2 weeks (Barry 1967).

Cell walls make up about 50% of the plant foods used by Brant goslings (and other geese). The walls are primarily composed of relatively indigestible structural carbohydrates and lignin (Sedinger and Raveling 1984, Sedinger et al. 1989). In comparison with animal foods, these plant materials are also low in protein and deficient in some amino acids (Sedinger 1984); hence geese must compensate by ingesting large amounts of food (Sedinger and Raveling 1988). At the Tutakoke colony, studies of Black Brant brood behavior from 1987 to 1993 found that goslings foraged for 70%–74% of 20 hours of daylight on the arctic tundra (Sedinger et al. 1995a), while adult males foraged 28%–40% of that time, and adult females foraged 43%–59%. The adults also spent considerable time in alert behavior: 34%–48% for males and 13%–27% for females. This division of labor is consistent with all other such studies of geese: the female forages more to recover from the rigors of egg laying and incubation, while the male becomes the principal guardian of the family (Lessells 1987, Sedinger and Raveling 1990). Alert behavior was positively correlated with brood size for males, but not for females. Increased alert behavior may occur because larger broods spread out over more of the landscape and thus require more attention to achieve the same level of security as smaller broods. Alternatively, high-quality males may be more vigilant, and more vigilant males may thus become associated with larger broods.

Brood Amalgamation (Crèches): Broods seldom mix, as Bregnballe and Madsen (1990) observed for Atlantic Brant on Svalbard. They recorded only 3 instances where a gosling followed the wrong family, and then only for about 10 minutes. Mickelson (1975) reported that Black Brant on the Y-K delta occasionally formed amalgamated broods. The mean brood size increased until about 15 days after the hatch, which is the age at which goslings probably recognize their parents (Ramsay 1951). Such amalgamation most likely was caused by the departure of some females, whose goslings were then adopted by other Atlantic Brant. It seems plausible that brood amalgamation in Brant might be more common in the large nesting colonies characteristic of the Y-K delta, where sizeable numbers of Black Brant goslings create many op-

portunities for mixing and adoption, compared with lower-density nesting populations in the High Arctic.

Development. Goslings average 43.6 g at hatching and grow rapidly thereafter, reaching a weight of 315–398 g in 14 days, which is when pinfeathers appear in the tail and the wings, and 967 g in 32 days, by which time the wings and the body feathers are well developed (Palmer 1976a). Captive-reared Black Brant goslings from the Y-K delta averaged 60.5 g approximately 24 hours after hatching (Morehouse 1974). Their growth rate averaged 18.1 g/day during the first week ($n = 20$), reached a maximum of 46.7 g/day during week 3, then declined to 17.6 g/day in week 4, before increasing again to 27.1 g/day in week 5, followed by 8.6 g/day in week 6, 4.3 g/day in week 7, and –2.9 g/day in week 8. Maximum (1,837 g) weights of young Black Brant were reached by the middle of November, but then declined markedly to 1,271 g by early April. On most birds, remiges first appeared on day 10, and retrices on day 8. Primaries grew faster than secondaries, with the ninth primary growing most rapidly during weeks 2 (7.3 mm/day) and 3 (6.9 mm/day); at 54 days it averaged 190.0 mm; and its final length was 216.0 mm.

Sedinger and Flint (1991) calculated the asymptotic body mass of Black Brant when they were 1,237 days old, which estimated that goslings increased their body mass 28-fold from hatching to fledging. Atlantic Brant on Southampton Island reached flight stage in about 45–50 days (Barry 1962). Einarsen (1965) reported that Black Brant goslings were capable of flight in 49 days.

Due to the short arctic growing season, there is strong selective pressure for Brant to nest early, which affects the growth of their goslings. Sedinger and Flint (1991) found that late-hatched Black Brant goslings on the Y-K delta were smaller than early-hatched goslings, with weight, tarsus, and culmen all declining significantly in relation to hatch date. A second study of only female goslings found that such relationships clearly affected overall fitness, as larger goslings survived at a higher rate than smaller goslings (Sedinger et al. 1995b). Larger goslings were also more likely to breed as 2- or 3-year-olds than were medium-sized or small goslings; 71%–72% of the large goslings were observed breeding, compared with only 50%–53% of the small to medium-sized goslings. Sedinger et al. (1995b) did not know when differential mortality of the goslings occurred, but a later study by Ward et al. (2004) found that monthly survival of Black Brant goslings during early fall was 20%–24% lower during the year with the latest hatch dates and slowest gosling growth. Most mortality occurred during the first 2 months after banding as fledglings. Hence arctic geese are especially adapted to breed early, because delayed breeding results in both a lower survival rate and decreased future fecundity of the offspring (Sedinger and Flint 1995). Barry (1962) poignantly recorded the cost of a late hatch in Atlantic Brant on Southampton Island: following a late reproductive year in 1956, he found 21 well-preserved young Brant frozen in the ice the following spring, all fully developed except that their flight feathers were 4–5 days short of allowing the birds to fly.

The density of the nesting population also affects the growth and development of Brant goslings, as was recorded at the Tutakoke colony from 1986 to 1995, when the colony increased from 1,100 to >5,000 pairs of Black Brant (Sedinger et al. 1998). Gosling mass at 30 days in 1986 decreased from 764 g in males and 723 g in females to 665 g in males and 579 g in females for the 1994 cohort. These declines of 13.0% for males and 20.0% for females were directly correlated with the number of broods.

RECRUITMENT AND SURVIVAL

Bellrose (1980) reported mean brood sizes for Black Brant on the Y-K delta, determined by biologists over several years: Class I = 3.22 goslings (n = 517 broods), Class II = 2.41 (n = 410), and Class III = 2.30 (n = 109). Brood sizes on the Y-K delta during most years from 1980 to 2002 averaged 3.0 in Class I (n = 3,656) and 2.8 in Class II (n = 1,062; U.S. Fish and Wildlife Service 2002b). The studies discussed below provide detailed insight into the ecology of recruitment and survival in Brant.

Annual estimates of Black Brant productivity have been conducted each fall since 1963 at Izembek Lagoon. The data gathered include the numbers of adults and juveniles, and the average number of juveniles/family. In 2005, a sample of 25,361 Black Brant contained 33.3% juveniles, and family size averaged 2.89; the percentage of juveniles was 83% higher than in 2004 and 42% above the long-term (42-year) average. From 1963 to 2005, the percentage of juveniles ranged from a low of 4.6% in 1974 to a high of 41.8% in 1966; the average was 22.3% (Groves 2006).

At a large-scale population level, Sedinger et al. (2006) demonstrated the effect of El Niño events on the distribution and reproductive performance of Black Brant. Fewer of these Brant wintered in Mexico during every El Niño event since 1965, and fewer Brant were observed breeding following each El Niño event since breeding surveys began in 1985. After the strong El Niño event in winter 1997/98, 23% of the ≥5-year-old Brant and 30% of the 3-year-old Brant did not breed the following year, in comparison with non–El Niño years. Such findings are consistent with the life-history theory that predicts longer-lived species will preserve adult survival at the expense of reproduction.

For Atlantic Brant, productivity is monitored by the age ratios in birds observed primarily in New York and New Jersey. From 1978 to 2007, the percentage of juveniles averaged 19.0%, ranging from lows of 1.5% in 1999 and 3.7% in 1986 to a high of 41.0% in 1979 (U.S. Fish and Wildlife Service 2007). In 2007, the percentage of juveniles was 31.1%, which was 28.0% higher than in 2006 and 63.7% greater than the long-term average. Kirby et al. (1986) used band-recovery data from 1956 to 1975 to estimate the annual survival of adult Atlantic Brant, which averaged 78.5% (range = 59.1%–98.4%). These rates were higher than those for other species of geese, but at the low end for the ranges reported for other populations of Brant. More recent preliminary results (2001–5) estimated Atlantic Brant adult survival of 81% for birds from Southampton Island and 73% for Baffin Island (Arctic Goose Joint Venture Technical Committee 2008). Survival of juveniles from Baffin Island was estimated at 37%.

Since 1960/61, the Eastern High Arctic Brant population has been assessed by surveys on their principal wintering grounds in Ireland, where the percentage of juveniles has ranged from a low of near zero to a high of 30% (Arctic Goose Joint Venture Technical Committee 2008).

Flint et al. (1995) estimated the survival of goslings associated with marked adult female Black Brant on the Y-K delta from 1987 to 1989. Of 61 radiotagged adult female Brant, 82% fledged at least 1 gosling (brood success); gosling survival to fledging ranged from 56% in 1989 to 79% in 1987. Most (82%) gosling mortality occurred during the first 15 days of the brood-rearing period, with losses due to predation from gulls and arctic foxes and to harsh weather conditions.

Ward et al. (2004) examined first-year survival of juvenile Black Brant by marking 7,442 goslings captured during 1990–93 about 1 month after hatching at major breeding sites in Alaska: the Y-K delta (Tutakoke River and Kokechik Bay) and the Arctic Coastal Plain. These Brant were then observed on their fall staging area (Izembek Lagoon) and on wintering areas on the Baja Peninsula in Mexico. In all years, survival was 5%–27% lower in early fall (15 Jul–1 Oct) than late fall (1 Oct–15 Feb), despite the fact that the Brant had completed a substantial migration flight in late fall to their primary wintering areas in Baja California, roughly 5,000 km away. Most juvenile mortality thus occurred prior to arrival at Izembek Lagoon (their primary staging area), and ≥60% of juvenile mortality was related to nonhunting causes. The lowest (68.1% for males, 69.1% for females) early-fall survival rate occurred in 1992 at Tutakoke River, which was a year of late hatch and slow gosling growth. Survival estimates of goslings were higher (93.7%–95.5%) on the Arctic Coastal Plain during early fall than at either site on the Y-K delta (85.2%–90.3%). Brant goslings on the Arctic Coastal Plain grow faster and are about 30% larger than Tutakoke goslings (Sedinger et al. 2001), and thus probably survive better, because both growth rate and size are directly linked to the first-year survival of goslings (Sedinger at al. 1995b). Subsequent work by Sedinger et al. (2004) demonstrated that Black Brant goslings reaching a larger size late in their first summer have a higher probability of breeding as well as a greater first-year survival.

Sedinger et al. (2002) found that the survival of adult female Brant on the Arctic Coastal Plain (90%) was only slightly higher than that for Black Brant on the Y-K delta (85% at Tutakoke River, 86% at Kokechik Bay). Hence the longer migration and increased exposure to subsistence harvest experienced by Eastern and Western High Arctic Brant did not reduce their chances of survival relative to Black Brant on the Y-K delta. The authors then speculated that the reduced productivity of Arctic Brant must be compensated for by increased survival elsewhere in the annual cycle (e.g., first-year survival) if the population was to be sustainable. Sedinger et al. (2002) concluded that the variation in life-history traits between these Brant populations was primarily caused by the different environmental conditions confronting each group. In another study, Ward et al. (2004) suggested that the lifetime reproductive success between these 2 different Brant populations was probably equal but achieved by different means: low reproductive investment but higher juvenile survival and recruitment for Brant on the Arctic Coastal Plain, versus a higher reproductive investment but lower juvenile survival and recruitment for Black Brant on the Y-K delta.

Ward et al. (1997b) estimated the annual and seasonal survival rates of marked adult Black Brant on the Y-K delta. Annual survival did not vary among years, averaging 84.0% from 1986 to

1993 (range = 77.0%–90.4%). The mean monthly survival rates, however, were lowest (90.8%) during late spring migration (15 Apr–1 Jun), the period of greatest subsistence harvest on the breeding grounds, and highest (100%) during winter (1 Jan–1 Mar), when sport harvest was greatest. (Sport harvest of Brant had declined by >60% over the 15 years prior to their study, and thus had little effect on adult mortality.) Other calculated monthly survival rates were 98.4% when nesting (1 Jun–15 Jul), 94.4% in early fall (15 Jul–15 Sep), 98.7% in late fall (15 Sep–1 Jan), and 98.8% in early spring (1 Mar–15 Apr).

Sedinger et al. (1997) captured Black Brant during the adult molt on the Y-K delta from 1986 to 1992; they estimated an annual survival of 78.2% for initially captured adults and 72.6%–90.0% for recaptured adults. Gosling survival in the first year ranged from 73.8% in 1986 to 26.0% in 1991, but the inclusion of winter observations increased these estimates by an average of 30%, which demonstrated that permanent emigration had a significant effect on survival estimates. An important result of their study was that they did not detect any effects of handling on subsequent Black Brant survival.

The most extensive analysis of Brant survival to date is that of Sedinger et al. (2007), who analyzed 53 years of data (1950–2003) from Black Brant banded on the Y-K delta. Survival rates increased from 70% in adult males and 71% in adult females during the 1950s to 88% for both sexes in the 1990s. The survival rates recorded in the 1990s were among the highest ever estimated for Black Brant; these rates did not increase in the 2000s, however, despite reductions in the sport harvest. An analysis of the subsistence harvest from 1985 to 2003 found little impact on the birds' overall survival. The authors concluded that although harvest was largely additive to other sources of Black Brant mortality prior to the 1990s, low recruitment probably caused their continued population decline in the 1990s and 2000s.

Hines and Brook (2008) followed the analysis of Sedinger et al. (2007) with a study of long-term trends in survival rates of Black Brant from the western Canadian Arctic. They examined band-recovery data from 1962 to 2001 to estimate historical (1962–67 and 1975–79) and recent (1991–2001) survival rates. For adult Brant on the mainland, survival averaged 71.9% for the 1960s and 69.3% for the 1970s, but improved significantly (to 88.4%) during 1991–2001. There was a statistically significant increase in survival of juvenile Brant over time, with annual survival averaging 21.6% in the 1960s, 46.0% in the 1970s, and 69.4% from 1991 to 2001. On Banks Island, 250 km northeast of the mainland population, adult survival rates averaged 94.5% from 1992 to 1994.

The longevity records for Brant in the wild are 22 years and 7 months for an Atlantic Brant banded and recovered in New York, and 27 years and 6 months for a Black Brant banded in Alaska and recovered in Washington. There are 11 reports of Black Brant living >15 years in the wild.

FOOD HABITS AND FEEDING ECOLOGY

All Brant are almost entirely herbivorous and have extremely restrictive diets, relying mainly on eelgrass as their primary food source, especially during the nonbreeding season, a time when no other species of goose relies more heavily on a single native plant (Reed et al. 1998b, Ganter 2000). The distribution of Black Brant outside of the breeding season is largely superimposed on the distribution of eelgrass (Cottam et al. 1944). Atlantic Brant were forced to alter their food habits drastically after the near-complete disappearance of eelgrass in 1930–31, due to a "wasting disease" associated with the mycetozoa *Labyrinthula*, which led to the abrupt and near-complete destruction of eelgrass beds along the entire Atlantic coast (Cottam et al. 1944, Cottam and Munro 1954, Rasmussen 1977). Prior to 1932, an analysis of the stomach contents of Atlantic Brant (*n* = 28) revealed a diet of 85% eelgrass and 12% widgeongrass (*Ruppia maritima*), while stomach contents examined from 1932 to 1941 (*n* = 80) contained only 9% eelgrass, versus

16% widgeongrass and 50% algae, mainly sea lettuce (Cottam et al. 1944).

Brant feed on nutritionally poor foods that are not readily digested. The protein content of eelgrass, for instance, is typically below 15%, in contrast to about 25% for the more terrestrial graminoids typically used by geese (Moore and Black 2006b). Further, Brant only metabolize about 41% of the hemicellulose and protein available in cell walls, which is probably associated with the length of time food is retained in the gut; this interval may be limited in Black Brant, as well as in other species of small geese (Sedinger et al. 1989). Brant compensate for these deficiencies by feeding for long periods of time and ingesting large amounts of food. On their primary wintering grounds in Baja California, Black Brant spent 78% of all daylight hours foraging (Kramer et al. 1979). Black Brant grazing during the fall and winter in Boundary Bay in British Columbia removed about 50% (262 tons) of the aboveground biomass and 43% (100 tons) of the belowground biomass of exotic seagrass (*Zostera japonica*; Baldwin and Lovvorn 1994a).

Preferred foraging areas of Brant include mudflats and shallow water within barrier beaches or intertidal estuaries, eelgrass and algae beds, short graminoid saltmarshes, and agricultural grasslands and grain fields (Reed et al. 1998b). Brant feed by tipping, dabbling with the head immersed underwater, pecking, and grazing. Brant are primarily vegetarians, with the reported consumption of animal foods such as snails, fish eggs, and shrimp thought to be incidental, as a result of feeding in eelgrass beds. Black Brant staging at Humboldt Bay in California during the spring preferred to feed close to tidal channels and in the deepest areas permitted by tides; the biomass and nutrient content of eelgrass were greatest there, and depletion from prior grazing was probably less than elsewhere in the bay (Moore and Black 2006b).

Brant are social, gregarious birds, often feeding in large scattered flocks, or rafts. They have also been observed feeding in association with Tundra Swans, Emperor Geese, and diving ducks, such as goldeneyes and scaup (Einarsen 1965).

Breeding. On their northern breeding grounds, where eelgrass does not occur, Atlantic Brant feed primarily by grazing on short grasses and sedges found under kelp drifts and near their nesting sites (Palmer 1976a). Phrygian alkaligrass (*Puccinellia phryganodes*) and Hoppner's sedge (*Carex subspathacea*) are especially preferred by breeding Black Brant (Barry 1967, Derksen and Ward 1993, Reed et al. 1996), the latter probably because of its high protein content (Sedinger and Raveling 1984). Other important foods consumed during the breeding season include Fisher's dupontia (*Dupontia fisheri*), tufted hairgrass (*Deschampsia caespitosa*), watersedge (*Carex subspathacea*), slender arrowgrass (*Triglochin palustre*), marestail (*Hippuris vulgaris*), sheathed pondweed (*Potamogeton vaginatus*), mosses, and saxifrage (*Saxifraga* spp.; Reed et al. 1998).

Adult Black Brant on their breeding grounds on Prince Patrick Island in the Northwest Territories consumed mostly tundra grasses and moss (Handley 1950). On the Anderson River delta, leafy pondweed (*Potamogeton foliosus*) is an important food source for Black Brant, and berries are occasionally consumed on the tundra (Mickelson 1975).

Migration and Winter. On the Pacific coast, the large and luxuriant eelgrass beds at Izembek Lagoon in Alaska attract large numbers of Black Brant each spring and fall. All along the coast (from Izembek Lagoon to lower Baja California), sizeable concentrations of Black Brant occur on those bays with the most extensive beds of eelgrass, which are heavily used during migration and in the winter (Cottam et al. 1944, Kramer et al. 1979). Off the coasts of Washington and Oregon, sea lettuce is also an important food source for Black Brant (Einarsen 1965). When disease ravaged the eelgrass beds along the California coast in the early 1940s, Brant sought food in unlikely places, including grain fields, pastures, and golf courses, where they clipped grass for food (Moffitt and Cottam 1941). When eelgrass beds became scarce in Oregon bays during the spring, Brant left bay waters several times a day to forage in adjacent meadowlands (Batterson 1968).

Penkala (1975) examined the stomach contents of 801 Atlantic Brant collected along the New Jersey coast during the winters of 1972/73 and 1973/74 and noted 11 foods that they utilized. The most important items eaten (in terms of frequency of occurrence) were sea lettuce (65.5%), eelgrass (41.5%), widgeongrass (17.5%), and salt-marsh cordgrass (*Spartina* spp.; 21.5%). Sea lettuce constituted 80.5% of the food content in the stomachs of Atlantic Brant in which it was found. Consumption of sea lettuce, compared with eelgrass, did not appear to have an adverse effect on Brant weights over the winter. In Nova Scotia, the chief food items for Atlantic Brant included sea lettuce and Irish moss (*Chondrus crispus*; Tufts 1986). A shift in the food habits of wintering Atlantic Brant appears to have occurred since the severe winter of 1976–77, when nearly all of their traditional marine and estuarine habitats froze (Nelson 1978), and the Brant began to use upland habitats, such as lawns and golf courses (Smith et al. 1985). Cultivated grass accounted for 65.6% of the food in the stomach contents of 34 Atlantic Brant collected during the mid-1980s in New York (Smith et al. 1985). In contrast, eelgrass made up <0.5% of the food found in Atlantic Brant collected during winter in New Jersey (*n* = 38) and New York (*n* = 34), although it formed 34.1% of the food in the stomach contents of Brant from Virginia (*n* = 11). Sea lettuce totaled 42.7% of the food found in Atlantic Brant from Virginia, 35.6% in those from New Jersey, and 15.8% in those from New York. In New Jersey, cordgrass (*Spartina alterniflora*) was an important winter food (49.7% of the stomach contents). Atlantic Brant appear quite capable of engaging in field-feeding behavior on croplands located well away from the coast, as has been recorded for Dark-bellied Brant in England, which, since the mid-1970s, have readily foraged on agricultural crops (wheat and barley), and grasses (Round 1982).

Animal foods listed in food-habits studies of Atlantic Brant include various insects, mollusks, crustaceans, and worms (Palmer 1976a). Rarer reports have noted Brant using more atypical foods. Sheppard (1949) stated that a flock of 30 Brant destroyed two-thirds of a 5.3-ha carrot field during their migration through Ontario. Crowberries (*Empetrum nigrum*) were the preferred food during fall migration in Greenland (Salomonsen 1950). In coastal New England, foods consumed by Brant included mussels (*Mytilus edulis*) and frozen quahogs (*Mercenaria mercenaria*; Forbush 1925).

MOLTS AND PLUMAGES

Adult Brant have only 1 plumage (basic) that is identical between males and females and molted annually via the prebasic molt. The descriptions of molts and plumages are primarily from summaries in Palmer (1976a) and Reed at al. (1998), which should be consulted for further details.

The downy plumage of goslings is replaced by juvenal plumage in 42 days, at which time they can fly. Juvenal plumage is characterized by a brownish to sooty head, neck, and breast, and the necklace can either be absent or not as distinct as in definitive basic. The sides of the body have an uneven or blotchy pattern. The mantle feathers, including the scapulars, have distinct white outer margins. Most of the upper coverts also have conspicuous white endings, as do the secondaries and, sometimes, the inner primaries. This plumage is worn briefly, and by early winter it has largely been succeeded by first basic, except that the juvenal wing is retained until the following summer. Juvenal plumage differs from both first basic and definitive basic and is readily distinguished from them, an attribute used to determine fall age ratios.

First basic is acquired through the fall or early winter, and the associated molt can be protracted in some individuals. This plumage resembles definitive basic in its color and pattern, although various feathers are intermediate in their shape and tapering. First basic "bleaches" before the next prebasic molt, although it is not worn as long as succeeding basic plumages.

Definitive basic starts in the second summer and includes a complete body and wing molt. Non-breeders and failed breeders begin this molt from mid-June to early July, while Brant with broods do not begin molting their flight feathers until the second week of July. Most of the nonbreeders/failed breeders complete the wing molt by late July or early August; those with broods finish by the second week of August. The flightless period lasted about 3 weeks on the Anderson River delta (Barry 1967), 20–22 days on Bathurst Island (O'Briain et al. 1998), 22–25 days on Melville Island (Boyd and Maltby 1980), and 23–24 days at Teshekpuk Lake in Alaska (Taylor 1995).

Taylor (1995) found that adult Black Brant at Teshekpuk Lake, a major molting area for Black Brant on the Arctic Coastal Plain, molt and regrow (to 70% maturity) all of their remiges and 25% of their back, rump, breast, and side regions during a 3.5-week period. Lipid reserves declined 71%–88% during that time, forming only 2%–4% of fresh body mass by late in the molt (Taylor 1993). Sub-adults molted with greater intensity than adults; their back, rump, breast, and side feathers consisted of ≥50% blood quills, versus ≥25% in adults. Brant departed the molting area completing only 11%–21% of all feather growth. Most Brant began molting their primaries in the first week of July; they were capable of short flights when the ninth primary averaged 62% of mean final length, and of sustained flight when this primary was at 70%. Taylor's work supports the hypothesis of Gates et al. (1993) that arctic-nesting geese complete the remigial molt first and prolong the body molt until well after departure from the molting area. Successfully breeding adult Brant began to molt about 14 days after their young were hatched (Barry 1962).

CONSERVATION AND MANAGEMENT

Both the Atlantic Brant and Black Brant populations are cooperatively managed by Canada, the United States, and Mexico, and associated management plans are in place for both subspecies (Atlantic Flyway Council 2002, Pacific Flyway Council 2004).

In terms of population size, the goal for Atlantic Brant is to maintain or exceed 124,000 birds in Midwinter Survey estimates, and to retain or improve current population surveys as well as harvest management strategies. The second major goal of the plan is to preserve existing habitat on the birds' breeding, migration, and wintering areas. Other objectives address human uses of Atlantic Brant, including sport and subsistence hunting, as well as viewing and photography; minimizing nuisance and depredation issues; and conducting relevant research. The research base for Atlantic Brant is considerably smaller than that for the Black Brant, so studies that increase knowledge of Atlantic Brant biology will improve the certainty with which biologists make management decisions regarding the habitats and populations of this subspecies (*B. b. hrota*). Research priorities include estimations of annual production and survival, and associated habitat issues. Other priorities call for understanding the birds' wintering ecology better, improving the existing sea lettuce survey, determining spring body condition, and developing a population model.

For the *nigricans* subspecies of Black Brant, the population goal is to have 162,000 birds (as estimated by the Midwinter Survey): 150,000 Black Brant and 12,000 Western High Arctic Brant (Pacific Flyway Council 2004). Management priorities call for continued surveys and population assessments, as well as harvest assessments and management. The plan also focuses on monitoring and protecting habitats on Black Brant breeding, wintering, and migration sites by fostering efforts to protect or acquire critical habitats through coordination with various Joint Ventures of the North American Waterfowl Management Plan and other habitat protection initiatives. Nine specific research needs are identified in the plan, including studies of the effects of disturbance, of winter ecology, and of the ecology of habitats associated with Brant at Izembek Lagoon.

Black Brant face specific issues associated with petroleum exploration and extraction in Alaska, on the Arctic Coastal Plain as well as near the Beaufort and Chukchi Seas. Johnson (1993) demonstrated the crucial importance of Kasegaluk Lagoon in the Chukchi Sea as a fall staging area for Black Brant, yet this area is targeted for oil and gas leases. The critical Black Brant molting area at Teshekpuk Lake is located within the National Petroleum Reserve (Derksen et al. 1979), as are important breeding areas on the Arctic Coastal Plain (Stickney and Ritchie 1996).

Human disturbance is also a significant management concern for Black Brant, as they are very sensitive to such disturbances, especially hunting (Smith and Jensen 1970, Henry 1980, Moore and Black 2006a). At Teshekpuk Lake, molting Black Brant (as well as other geese) were highly bothered by low-flying, single-engine aircraft (Derksen et al. 1979). Single-engine airplanes at altitudes below 610 m created the maximal response from the birds, but no response occurred when aircraft were above 1,500 m. At Izembek Lagoon, staging Black Brant were disturbed an average of 0.8 times/hr, and aircraft were the most frequent (0.4/hr) cause of disturbance. These Brant reacted to 67% of all disturbances and took flight in response to 49% (Ward et al. 1994). At Humboldt Bay, an important spring and fall staging area in California, nonhunting human disturbances caused Black Brant to take flight an average of 0.6 times/hr from November to May (Henry 1980). Of the disturbances in the bay, 29% were from aircraft, but 39% were from humans (mostly clammers).

Lastly, global climate change is likely to affect Brant as well as other waterfowl breeding in the Arctic. A significant paper by Ward et al. (2009) demonstrated that the winter distribution of Black Brant along the Alaska Peninsula has shifted northward over the past 42 years (1963–2004). The number of Black Brant wintering on the peninsula has also increased by approximately 7%/year, based on aerial surveys between 1986 and 2004. Such modifications may be directly linked to global climate change, as mean temperatures on the Alaska Peninsula have increased about 1°C over this 42-year interval, which resulted in a 23% reduction of days with freezing temperatures, and a 34% decline in the number of days when ice cover prevented Black Brant from accessing food resources. The number of days with extreme ice cover averaged 41 days before 1977 (a period of 14 years) compared with only 27 days after 1976 (a 28-year span). Such weather trends corresponded to states in the Pacific Decadal Oscillation (a pattern of climate variability), which provided strong support for the contention that changes in Black Brant distributions in Alaska were linked to climate warming.

The population of Black Brant wintering in Alaska also increased as the number of November days with strong northwesterly winds decreased. This correlation is significant, because changes in the occurrence of tailwinds, in combination with the birds' increased survival due to milder winters, can potentially cause a sudden change in the wintering strategy of Black Brant, shifting from long-distance migration to a short-distance detour migration, or even to winter residency (Purcell and Brodin (2007).

Canada Goose

Branta canadensis (Linnaeus 1758)

Cackling Goose

Branta hutchinsii (Richardson, 1832)

Giant Canada Goose

Vancouver Canada Goose

Cackling Goose (*B. h. minima*)

Western Canada Goose

Taverner's Cackling Goose

typical wing for all subspecies

Interior Canada Goose

Richardson's Cackling Goose

Lesser Canada Goose

Dusky Canada Goose

Atlantic Canada Goose

Aleutian Cackling Goose

Canada Goose flock descending for landing.
Joshua Stiller

Canada/Cackling Geese are among the most well-recognized species of wildlife in North America. Once considered a single species with a debated number of subspecies, in 2004 the American Ornithologist's Union split the Canada Goose into 2 species: the Canada Goose and the Cackling Goose. As a species, the Canada Goose is composed of 7 subspecies of large to medium-sized geese, while the Cackling Goose species contains 4 smaller subspecies. The group includes the largest goose in the world, the Giant Canada Goose (*Branta canadensis maxima*), where large males can weigh up to 4.5 kg. In contrast, the Cackling Goose (*Branta hutchinsii minima*) of the Yukon-Kuskokwim (Y-K) delta weighs 1.3–1.5 kg, not much larger than a Mallard. The group also contains 2 of the most notable success stories of wildlife management in North America: the Giant Canada Goose and the Aleutian Cackling Goose (*Branta hutchinsii leucopareia*). Giant Canada Geese were once thought to be extinct by the 1950s, but a flock of 200 birds was rediscovered in 1962 at Silver Lake in Rochester, Minnesota. Reintroduced into their former range but also widely introduced into new areas, Giant Canada Geese now number over 1 million in the Atlantic Flyway and 2 million in the Mississippi Flyway. The Aleutian Cackling Goose

was reduced to only 200–300 birds on remote islands in the Aleutian Archipelago and declared endangered in 1967. Subsequent conservation efforts have led to a population of 79,500 in 2009. Canada/Cackling Geese mate for life, and family groups stay together through the winter and into spring. They occupy an array of habitats, from arctic tundra to semiarid areas in the Great Basin (Intermountain Region), with all populations taking advantage of agricultural crops during migration and winter. Giant Canada Geese are considered a nuisance in many urban areas, but several management approaches offer potential solutions to conflicts with humans.

IDENTIFICATION

Canada/Cackling Geese are among the most widely recognized species of wildlife in North America. Virtually everyone in the North Country knows the hopeful feeling transmitted from those first resonant honkings heard late at night in the seeming heart of winter: spring is about to arrive. As Aldo Leopold so eloquently wrote in *Sand County Almanac*, "One swallow does not make the summer, but one skein of geese, cleaving the murk of March thaw, is the spring."

Canada Goose and gosling on a muskrat house. *Wendy Macziewski*

The Canada Goose was first described as a single species by Linnaeus, based on a bird collected near Québec City, Québec, in 1758. Ornithologists and taxonomists subsequently noted distinct size, color, and morphological differences among specimens, as well as markedly separate breeding and wintering ranges, and these differences led to the designation of multiple subspecies. There has been controversy, however, as to how many subspecies should be recognized: 12 were recognized by Delacour (1954) and Johnsgard (1978), 8 by Palmer (1976a) and Owen (1980), and only 6 by Livezey (1997). Hanson (2006) examined morphological characters of more than 1,800 Canada Goose skins collected throughout North America and recommended a split into 6 separate species and over 200 subspecies, but these designations should be viewed with caution (Banks 2007). The existing classification will probably

see future revision, but it is the system on which this account is based: the species and subspecies recognized in the fourth supplement (2004) to the seventh edition of the American Ornithologists' Union's (AOU) *Check-list of North American Birds* (AOU 1998, Banks et al. 2004).

The current taxonomy of the Canada Goose is grounded in modern-day techniques of molecular biology that have yielded new information not available in earlier studies, which all relied on morphological characteristics. Pierson et al. (2000) noted distinct genetic differences between 2 groups that were considered to be small-bodied Canada Geese: the Aleutian Canada Goose and the Cackling Goose. Scribner et al. (2003b) examined mitochondrial DNA (which is used to compare genetic relationships among species and populations within a species) from 11 populations of Canada Geese and found that large- and small-bodied

Aleutian Cackling Geese. *GaryKramer.net*

birds in western North America were highly divergent and represented distinct groups.

In July 2004, evidence from genetic studies led the AOU to split the then Canada Goose into 2 separate species: the Canada Goose and the Cackling Goose (Banks et al. 2004). The AOU maintained the subspecies previously designated under the original Canada Goose species, divided among the 2 species. The Canada Goose species (*Branta canadensis*) includes the larger subspecies: Giant (*B. c. maxima*); Western, or Moffitt's (*B. c. moffitti*); Vancouver (*B. c. fulva*); Dusky (*B. c. occidentalis*); Interior (*B. c. interior*); Atlantic (*B. c. canadensis*); and Lesser (*B. c. parvipes*). The Cackling Goose species (*Branta hutchinsii*) contains the smaller subspecies: Taverner's (*B. h. taverneri*); Aleutian (*B. h. leucopareia*), which includes the extinct Bering Canada Goose (*B. canadensis asiatica*, as the split had not yet been made between

B. canadensis and *B. hutchinsii*); Richardson's (*B. h. hutchinsii*); and Cackling (*B. h. minima*). The British Ornithologists' Union's Records Committee also accepts this separation into 2 species, but instead uses the names "Greater Canada Goose" and "Lesser Canada Goose" rather than Canada Goose and Cackling Goose. I have chosen here to lump the Canada Goose and the Cackling Goose within a single account, with separate information provided for all 11 extant forms.

At a Glance. All sex and age classes of Canada/ Cackling Geese are similar in appearance, with the male only slightly larger. The black head and neck are punctuated by a white band under the chin (the "chin strap"), which makes these birds unmistakable, even though the subspecies vary greatly in size. A white neck ring (at the base of the neck) is often characteristic of adult Aleutian Cackling

Geese (*leucopareia*), and occurs rarely in the other subspecies. The U-shaped white rump band is also diagnostic in flight. Both species are noted for their V-formation flight, although they sometimes fly in trailing lines.

Adults. The sexes are similar, with easy identification using the criteria above. All subspecies have the following in common: a black head and neck, with a white cheek patch that usually covers the throat; a gray-brown to dark brown back and wings; a mouse-gray to dark brown breast and sides, which are a lighter shade than the back in all but the Pacific coastal subspecies; a white belly, flank, and undertail coverts; a black tail and rump, separated by a white V-shaped bar formed by the white uppertail coverts; and black legs, feet, and bill.

The Giant subspecies has the longest neck in proportion to its body size, while the Cackling Goose (*minima*) has the shortest neck. Although the color of the wings and the body varies considerably among individuals within a single subspecies, there are differences that can assist in identification. The lightest-colored subspecies are the Giant and the Western; increasingly darker subspecies are the Richardson's, the Lesser, the Taverner's, and the Interior. Darkest of all are the large-sized Vancouver, the medium-sized Dusky, and the tiny Cackling (*minima*). In general, the darker subspecies are in the West and the lighter subspecies in the East.

Several subspecies of Canada/Cackling Geese stand out because of their size, plumage, or distribution: the Giant, the Western, the Atlantic, the Interior, the Vancouver, the Dusky, the Aleutian, and the Cackling. Other subspecies may require careful measurements of the bill (culmen), the folded wing, the central tail feathers, and the tarsus for identification. Pearce and Bollinger (2003) noted, however, that it is not possible to properly distinguish subspecies of Pacific Flyway Canada Geese by culmen length or plumage color, due to the extensive overlap in these characteristics. They recommended that morphological measurements—such as bill width at the nail, bill width at the base, head length, or midwing length, following data in Johnson et al. (1979)—be used to differentiate subspecies.

Juveniles. By fall, juveniles attain plumage that is only subtly different from the plumage of adults. Hence juveniles are generally recognized in the hand by their notched tail feathers, as well as by the presence of a juvenile penis and/or the bursa of Fabricius.

Goslings. Goslings of both species and all subspecies have a yellowish or greenish-yellow base color, with a well-defined, round dark crown cap (Nelson 1993). The base color is yellower and the pattern color lighter in goslings from the midcontinent and eastward to the Atlantic coast.

Voice. Calls range from the resonant *uh-whonk* of the males of larger-sized subspecies to the yelping high-pitched *unc* of the smaller ones. Female calls are shorter, a higher-pitched *hrink* or *hrih*. In general, the larger the subspecies, the longer the duration of the call, the lower the pitch, and the more sonorous the quality. Giant Canada Geese call less frequently than other subspecies, but they have the most prolonged call at the lowest pitch. The diminutive Cackling Goose (*minima*) calls the most frequently, and it has the highest pitch and the briefest notes.

Similar Species: Canada Geese have a distinctive appearance that might conceivably be confused only with Brant. Brant have an all-black head that lacks the characteristic white cheek patch of Canada Geese. In addition, Brant are a marine species, almost entirely confined to seacoasts.

Weights and Measurements. Adult weights and measurements are presented for each species under Distribution, with the subspecies arranged from largest to smallest.

DISTRIBUTION

Canada Geese and Cackling Geese are widespread in North America, breeding in every Canadian province and all 48 contiguous states, although most nest in Canada (Rusch et al. 1995). The breeding range was originally thought to be restricted to between 35° and 70° N latitude, but translocations and introductions have expanded their breeding range to the southern United States (Mowbray et al. 2002). Their winter range extends from southern Canada into northeastern Mexico, with smaller numbers wintering in Baja California and states in Mexico's central plateau, extending from Sonora south to Veracruz and Chiapas (Saunders and Saunders 1981, Howell and Webb 1995). Outside North America, Canada/Cackling Geese breed in Greenland and have been introduced throughout northern Europe, Russia, and New Zealand. Vagrants are seen farther west, in the Hawaiian Islands, Japan, and Siberia (American Ornithologists' Union 1998); and to the east, in the Bahamas, the Caribbean islands, and Bermuda (Amos 1991, Raffaele et al. 2003). Below is a brief synopsis of the distribution, biology, and physical appearance of each of the subspecies of Canada/Cackling Geese.

Canada Goose (*Branta canadensis*)

Giant Canada Goose. The Giant Canada Goose (*Branta canadensis maxima*) is the largest goose in the world, with a noticeably longer neck and larger bill than other subspecies. Body weights of Giant Canada Geese during late winter (23–27 Feb) at Silver Lake near Rochester, Minnesota, averaged 4,460 g for adult males ($n = 26$), 3,770 g for adult females ($n = 31$), 3,980 g ($n = 26$) for immature males, and 3,340 g for immature females ($n = 21$; McLandress and Raveling 1981). The bill length is the longest of all the subspecies, averaging 60.7 mm in adult males ($n = 27$), 57.3 mm in adult females ($n = 26$), 60.3 mm in immature males ($n = 65$), and 56.5 mm in immature females ($n = 48$; Hanson 1997).

The white chin patch extends farther toward the crown, and most individuals have a white band across the forehead between the crown and the bill. Giant Canada Geese are often called "resident geese," as they are regularly found year round in urban areas (such as parks, golf courses, and in-

dustrial parks) throughout much of the continental United States and southern Canada. Farther north, Giant Canada Geese nest from Saskatchewan to northern Ontario, the Hudson Bay Lowlands, Akimiski Island in James Bay, and southern James Bay. These birds historically wintered in southern Canada south through the Dakotas and the Great Lakes states, as well as in the central and southeastern United States (Rusch et al. 1996, Nelson and Oetting 1998).

Once a common breeding bird in the central portions of the Mississippi and Central Flyways (from southcentral Manitoba and eastern North Dakota south to northern Arkansas), the Giant Canada Goose was extirpated from most of its range during the 1930s, and thought to be extinct by the 1950s (Rusch et al. 1996). Then, in January 1962, Harold Hanson of the Illinois Natural History Survey discovered a flock of about 200 birds at Silver Lake, in a park in Rochester, Minnesota (Hanson 1997).

Giant Canada Geese are especially amenable to restocking efforts, because they are relatively nonmigratory, are easily propagated under confinement, readily use artificial nest structures, and have high rates of nest success and brood survival, even in areas with substantial densities of humans and modified landscapes. Most introductions have occurred in the Atlantic and Mississippi Flyways, but Giant Canada Geese have been introduced in nearly all of the lower 48 states and eastern Canada. They were even introduced as far south as Louisiana in 1960, with birds originating from Wisconsin; additional birds from Saskatchewan and Missouri were added to this population in 1962 (Chabreck et al. 1974). Some of the introductions of this subspecies around the United States also included the Western Canada Goose (*moffitti*) and mixtures or cross-hybrids of Giant Canada Geese and Western Canada Geese (Rusch et al. 1996).

Western Canada Goose. The Western, or Moffitt's, Canada Goose (*Branta canadensis moffitti*) is also known as the Great Basin Canada Goose. This

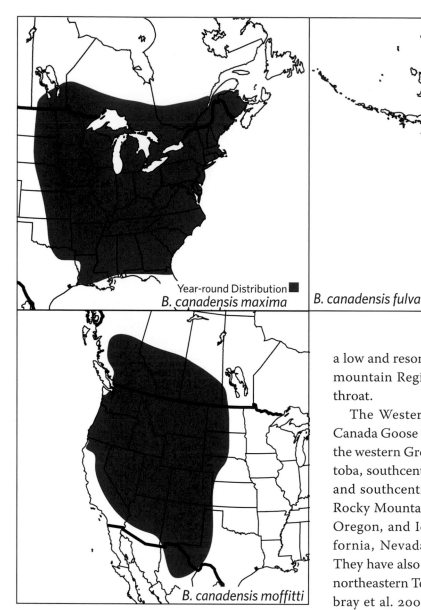

Year-round Distribution ■

B. canadensis maxima

B. canadensis fulva

B. canadensis moffitti

subspecies is only slightly smaller than the Giant Canada Goose (*maxima*), and some authorities consider them to be conspecifics (Palmer 1976a). Mowbray et al. (2002) reported an average weight of 4,017 g for males (*n* = 18) and 3,450 g for females (*n* = 3); bill length averaged 49.6 mm for males and 45.7 mm for females. Western Canada Geese have a light breast that contrasts with the black neck. They also have a slow wing beat, and their call is

a low and resonant *ahh-onk*. Birds from the Intermountain Region may have a dark stripe on the throat.

The Western Canada Goose is the common Canada Goose in the West, generally ranging from the western Great Plains north into western Manitoba, southcentral Saskatchewan, central Alberta, and southcentral British Columbia. West of the Rocky Mountains they extend from Washington, Oregon, and Idaho south to northeastern California, Nevada, northern Utah, and Colorado. They have also been introduced in Oklahoma and northeastern Texas (Krohn and Bizeau 1980, Mowbray et al. 2002). They nest in a variety of habitats near open water, including lakes, reservoirs, and marshes. Western Canada Geese winter over much of their breeding range, but they are also migratory, traveling to southern California, Arizona, New Mexico, southern Texas, and northern Mexico.

Vancouver Canada Goose. The Vancouver Canada Goose (*Branta canadensis fulva*) is the largest of the darker subspecies. Ratti et al. (1977) reported the average weight of males as 3,690 g, with an average bill length of 51.2 mm (*n* = 175). The average

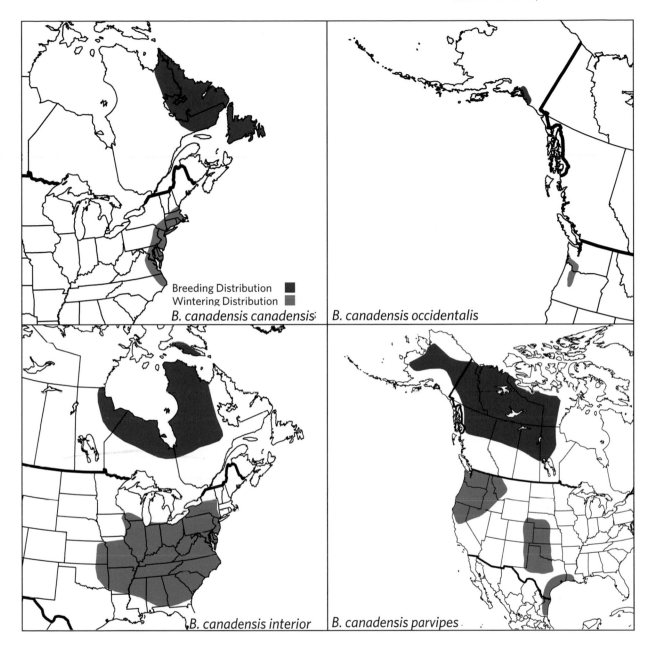

Breeding Distribution

Wintering Distribution

B. canadensis canadensis

B. canadensis occidentalis

B. canadensis interior

B. canadensis parvipes

weight of a combined sample of males and females was 3,043 g (*n* = 134). The Vancouver Canada Goose is similar in appearance to the Dusky Canada Goose (*occidentalis*), but with the upperparts and underparts more homogeneously colored; also, the bill is slightly smaller. They rarely have a white neck ring.

This unique subspecies nests in temperate rainforests from Glacier Bay through the Alexander Archipelago in southeastern Alaska, south along the Alaskan coast and associated islands, to British Columbia and northern Vancouver Island, where their core range is in the Nimpkish Valley (Ratti and Timm 1979, Lebeda and Ratti 1983). They are considered to be local residents, with <2% of the population migrating as far south as the Willamette Valley in Oregon (Ratti and Timm 1979). McKelvey and Smith (1990) found that 92% of the

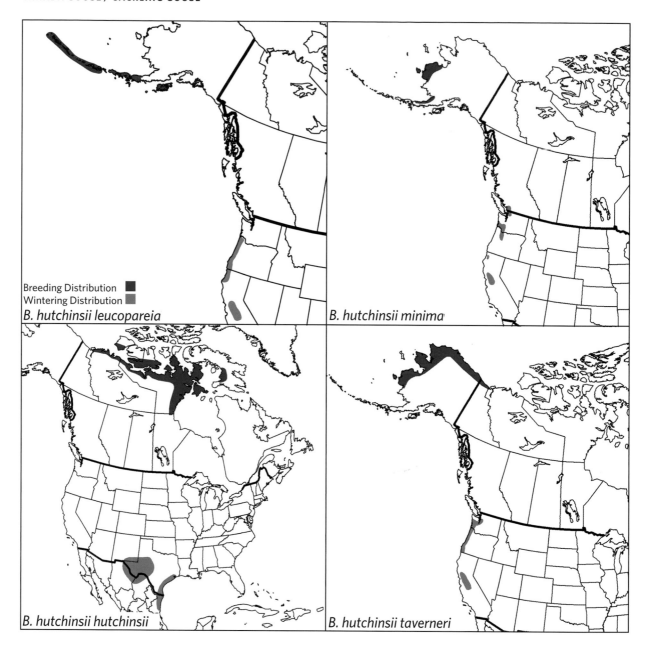

Breeding Distribution ■
Wintering Distribution ■
B. hutchinsii leucopareia

B. hutchinsii minima

B. hutchinsii hutchinsii

B. hutchinsii taverneri

recoveries of this subspecies banded in the Vancouver area occurred locally.

Lebeda and Ratti (1983) noted that Vancouver Canada Geese on Admiralty Island nested within dense forest habitats, with 1 nest located 9 m off the ground in a spruce snag. Adults often used trees as perch sites, especially during incubation, which is a unique behavior among geese. The Vancouver Canada Goose is among the least known

of all the subspecies, with fewer than 10 nests described as recently as 1973.

Dusky Canada Goose. The Dusky Canada Goose (*Branta canadensis occidentalis*) resembles the Vancouver Canada Goose (*fulva*), but is smaller. Males measured by Johnson et al. (1979) averaged 3,233 g (*n* = 130), and females, 2,640 (*n* = 131); bill length averaged 46.3 mm for males and 43.5 mm for

Canada Goose landing.
RyanAskren.com

females. The Dusky Canada Goose is darker than the Vancouver Canada Goose (*fulva*), with overall dark chocolate-brown underparts that appear to blend with the black neck. A white neck ring is rarely present. The white chin patch is reduced and occasionally absent. This subspecies breeds almost exclusively on the Copper River delta of Alaska, but it also occurs in the Gulf of Alaska (Middleton Island) and Prince William Sound (Cornely et al. 1985b, Campbell 1990).

The population of Dusky Canada Geese has declined markedly, from an estimated 20,000–25,000 in the 1970s, to 12,000–15,000 in the 1990s (Bromley and Rothe 2003), to an average of 10,900 from 2000 to 2009. The 2009 population estimate of 6,700 birds was the lowest on record since 1986 (U.S. Fish and Wildlife Service 2009). This decline has been attributed, in part, to the 1964 Good Friday earthquake, as the epicenter was only 138 km from the Copper River delta. This powerful quake

uplifted the Copper River delta by 1.8–3.4 m and subsequently changed the habitat from marsh to scrub-shrub, which, in turn, allowed an influx of previously uncommon predators, such as coyotes (*Canis latrans*) and brown bears (*Ursus arctos*). Many studies have estimated a decline in apparent nest success, from an average of 70% in the 1960s and 1970s, to 30% in the 1980s, and only 22% in the 1990s (Bromley and Rothe 2003). Nest success was <5% in 1993 and 1994 (Campbell 1990). Grand et al. (2006) reported mean annual nest-success rates of 21%–31% from 1997 to 2000, during which time Bald Eagles (*Haliaeetus leucocephalus*) destroyed 72% of the 895 nests attacked by predators, and brown bears decimated 13% (Anthony et al. 2004). Bald Eagles are also significant gosling predators, and thus play a major role in limiting goose productivity (Fondell et al. 2008).

Dusky Canada Geese migrate about 2,500 km, primarily along the Pacific coast, to arrive on wintering areas in late September, completing that migration in an average of 11 days (Bromley and Jarvis 1993). Their wintering grounds are located in southcentral Washington, the floodplain of the lower Columbia River in western Washington and Oregon, and the Willamette Valley of western Oregon (Cornely et al. 1998, Pacific Flyway Council 2008). They share wintering habitat with over 250,000 Canada Geese (4 other subspecies) and Cackling Geese (*minima*), which has greatly complicated harvest management designed to protect the Dusky Canada Goose population amid large wintering concentrations of geese (Bromley and Rothe 2003). Departure for spring migration occurs in early April.

Interior Canada Goose. The Interior Canada Goose (*Branta canadensis interior*), also known as the Hudson Bay Canada Goose, or Todd's Canada Goose, is relatively large and similar to the Atlantic Canada Goose (*canadensis*), but the body is a darker grayish color, with fine barring and whitish feather tips. Moser and Rolley (1990) reported average weights as 4,472 g for males (*n* = 22) and 4,188 g for females (*n* = 18); bill length averaged 52.9 mm in males and 49.1 mm in females. The dark coloration on the back continues without interruption and blends with the black neck. The neck is long and slender; the bill is sloped; and a white neck ring is seldom present.

The Interior Canada Goose is the most prevalent of all the Canada Goose subspecies; they make up >40% of the total number of Canada Geese and occur in 4 of the 18 management populations (Boyd and Dickson 2005). The broad breeding distribution of this subspecies encompasses taiga and coastal tundra habitat (Atlantic Flyway Council 2008b). Malecki and Trost (1990) reported that 90% of the Interior Canada Geese they observed breeding in northern Québec occupied an area on the Ungava Peninsula and along southern Ungava Bay. Most nesting was concentrated (1.6 pairs/km²) along the coasts, which formed only 7% of their survey area but contained 56% of the breeding pairs. This subspecies also breeds in southwestern Greenland, as well as on Baffin Island, on Akimiski Island (in James Bay), and west through the coastal lowlands in northern Ontario to just below South Indian Lake in Manitoba (Mowbray et al. 2002). Interior Canada Geese intergrade with Atlantic Canada Geese (*canadensis*) on the Ungava Peninsula, with Giant Canada Geese (*maxima*) in Québec and southern Ontario, and with Western Canada Geese (*moffitti*) in western Ontario and Manitoba (Palmer 1976a, Nelson and Oetting 1998). Malecki et al. (2000) reported over 5,000 Canada Geese nesting along the western coast of Greenland, and later genetic studies suggest that they may be Interior Canada Geese (Scribner et al. 2003a).

The wintering grounds of this subspecies are widespread: in southern Ontario and Québec in Canada, and extending from southeastern Minnesota south through eastern Iowa, Missouri, and Arkansas, and east from theses states to the Atlantic coast. Along the Atlantic coast, their winter range extends through the New England states south to South Carolina, with major concentrations on the Delmarva Peninsula (most of Delaware and parts of Maryland and Virginia), as well as in the Finger Lakes Region of New York, New Jersey, and additional areas in Virginia. Although they historically once wintered in the southern Atlantic Flyway, Interior Canada Geese no longer occur in Georgia and Florida during the winter (Mowbray et al. 2002, Atlantic Flyway Council 2008b, U.S. Fish and Wildlife Service 2009).

The Atlantic Population of Interior Canada Geese (see Management Units), which breeds primarily (80% of the population) on the Ungava Peninsula, experienced a dramatic decline: from 118,000 breeding pairs in 1988 (0.53/km²), to 90,000 in 1993, to 29,300 (0.13/km²) in 1995 (>75% in less than a decade), while numbers estimated between 1986 and 1995 on the Midwinter Survey in the Atlantic Flyway states decreased from 900,000 to 650,000 birds (Atlantic Flyway Council 2008b). This drastic reduction in the population resulted in the cessation of sport harvest within the Atlantic

Flyway in 1996 (Hindman et al. 1996); limited sport harvest resumed in Canada and most of the United States in 1999, when the estimated number of breeding pairs in the Atlantic Population increased to 77,500. By 2002, the number of breeding pairs had rebounded to 164,800 (0.74/km²) and many harvest restrictions were lifted, except in their southern winter range, where the season remains closed in North Carolina (Hindman et al. 2004). The 2000–2010 average was 159,494 breeding pairs; the estimate for 2010 was 154,028 pairs, with 0.69 pairs/km² (Harvey and Rodrigue 2010).

Atlantic Canada Goose. The Atlantic Canada Goose (*Branta canadensis canadensis*) is light-colored overall and is the common subspecies in northeastern North America. Average weights reported in Bellrose (1980) were 3,992 g for adult males ($n = 4,175$) and 3,447 g ($n = 3,452$) for adult females. Bill length averaged 56.0 mm for adult males ($n = 7$) and 53.9 mm for adult females ($n = 7$). This subspecies resembles the Interior Canada Goose (*interior*), but Atlantic Canada Geese have paler white underparts that contrast strongly with the black neck. They rarely have a white neck ring.

This subspecies breeds in bog and fen habitat, mainly in eastern Québec (Ungava Bay), Labrador, and Newfoundland, and north to southeastern Baffin Island (Mowbray et al. 2002). They intergrade with Interior Canada Geese to the west on the Ungava Peninsula. Atlantic Canada Geese also are regularly found on Anticosti Island and the Magdalen Islands in the Gulf of St. Lawrence, and occasionally to the south in Maine and Massachusetts. Atlantic Canada Geese winter from Nova Scotia to North Carolina (rarely South Carolina), but they are most abundant in southern New England and Long Island (Atlantic Flyway Council 2008a).

Erskine (2000) separated the North Atlantic population of Atlantic Canada Geese into 2 groups: (1) a Newfoundland stock that breeds in Newfoundland and winters primarily along the Atlantic coast of Nova Scotia; and (2) a Labrador stock that breeds in eastern Québec and Labrador, stages within the southern Gulf of St. Lawrence and on Prince Edward Island, and winters in the New England and mid-Atlantic states, where they mingle with members of the Atlantic and Southern James Bay Populations. This mixing confounds assessments of population status on the wintering grounds, although Atlantic Canada Geese tend to use coastal areas more than other subspecies, such as the Interior Canada Goose (U.S. Fish and Wildlife Service 2005). Less than 20% of the Newfoundland birds winter within the United States (Erskine and Payne 1997), while those breeding in Labrador winter primarily in southern Massachusetts, eastern Long Island, and Rhode Island (Malecki et al. 2000). Observations of 1,850 neck-banded birds and recoveries of 642 leg-banded Atlantic Canada Geese led Hestbeck and Bateman (2000) to conclude that wintering Atlantic Canada Geese have remained farther north over the past several decades. From 1991 to 1996, their winter distribution was primarily (77%) in New England, with 22% in the mid-Atlantic States and 1% in North Carolina.

Lesser Canada Goose. The Lesser Canada Goose (*Branta canadensis parvipes*) is similar in color to but smaller in size than the Western Canada Goose (*moffitti*), and about the size of the Lesser Snow Goose (*Chen caerulescens caerulescens*). Mowbray et al. (2002) reported average weights of 3,266 g for males ($n = 70$) and 2,854 g for females ($n = 59$); bill length averaged 42.4 mm for males and 40.6 mm for females. The breast is pale brown; the white cheek patches usually extend under the head.

Lesser Canada Geese nest within the tundra and taiga of interior Alaska, the Yukon, the Northwest Territories, and Alberta; on Victoria, Jenny Lind, King William, Baffin, and Southampton Islands in the Canadian Arctic; and in parts of Queen Maude Gulf and Hudson Bay. They winter from Washington and Oregon east to Colorado, New Mexico, Nebraska, Kansas, Oklahoma, northern Texas, and possibly western Missouri, and south along the Gulf Coast from Louisiana to the state of Tamaulipas in Mexico (Johnson et al. 1979, Mowbray et al. 2002). Due to the overlap in body size, the Lesser Canada Goose is difficult to distinguish from the Taverner's Cackling Goose (*taverneri*) in places where both subspecies mix on their wintering grounds (Johnson et al. 1979). These two subspecies also mingle with other subspecies of Canada/Cackling Geese, especially on wintering areas, so there are no reliable population estimates by subspecies. Most genetic studies of mitochondrial DNA indicate no interbreeding between Taverner's Cackling Geese and the larger-bodied subspecies Canada Geese, although Baker and Marshall (1997) noted genetic similarities between these closely associated species/subspecies.

Based on morphological measurements, both Lesser Canada Geese and Taverner's Cackling Geese stage in interior Alaska on agricultural lands (mainly those planted in sorghum) close to Fairbanks, near Creamer's Field and Delta Junction (Eichholz and Sedinger 2006). Most Lesser Canada Geese then migrate through central British Columbia and winter in Washington and Oregon, primarily east of the Cascades, and reverse this migration in the spring (Eichholz and Sedinger 2006). Some migrate to southeastern Alaska and then follow a coastal route to wintering areas.

Cackling Goose (*Branta hutchinsii*)

Taverner's Cackling Goose. The Taverner's Cackling Goose (*Branta hutchinsii taverneri*), also known as the Lesser Cackling Goose, resembles the Lesser Canada Goose (*parvipes*) but is smaller and generally darker overall. Johnson et al. (1979) reported average weights of 2,607 g for males ($n = 60$) and 2,421 g for females ($n = 61$). Bill length averaged 37.8 mm for males and 36.1 mm for females.

The separation of Taverner's and Lesser Canada Geese into 2 subspecies has been controversial (Palmer 1976a), although Taverner's Cackling Geese generally breed in tundra habitats, and Lesser Canada Geese breed in interior forests (Johnson et al. 1979). Taverner's Cackling Geese are also recognized by their unique mitochondrial DNA haplotypes, a series of DNA sequences used to distinguish among populations and species (Pearce et al. 2000). They usually lack a contiguous white chin strap. Some individuals are lighter in color, resembling Aleutian Cackling Geese (*leucopareia*) but with a rounder head and larger size. The breeding range of Taverner's Cackling Geese is in the interior tundra of western and northern Alaska, including the Y-K delta and the North Slope (Johnson et al. 1979, Jarvis and Bromley 1998). A recent analysis by Eichholz and Sedinger (2006) indicates that most of these geese winter west of the Cascades, in the lower Columbia River valley of Washington and Oregon and the Willamette Valley of western Oregon, but some also winter in the Central Valley of California (Johnson et al. 1979, Simpson and Jarvis 1979).

Aleutian Cackling Goose. The Aleutian Cackling Goose (*Branta hutchinsii leucopareia*) also includes the extinct Bering Canada Goose (*asiatica*), which once bred off the coast of Russia's Kamchatka Peninsula, on the Kuril and Commander Islands. Adult Aleutian Cackling Geese usually have a large white ring at the base of the black neck. This ring is thicker in front, with a brown marking and sparse white feathering on the back of the neck. Aleutian Cackling Geese are noticeably larger than Cackling Geese (*minima*). The former also have a paler gray-brown breast, as well as a white chin patch separated by a black stripe under the chin. The neck and bill are short. Mowbray et al. (2002) reported average weights of 1,946 g for males ($n = 36$) and 1,704 g for females ($n = 46$). Bill length averaged 36.6 mm for males and 35.1 mm for females.

Aleutian Cackling Geese breed within the Aleutian Islands, where they are known to nest on rocky slopes and cliffs among large colonies of nesting seabirds, including gulls (*Larus* spp.) and jaegers (*Stercorarius* spp.; Hatch and Hatch 1983). They winter along the Pacific coast south to central California.

Aleutian Cackling Geese were listed as federally endangered by the U.S. Fish and Wildlife Service in

Female Cackling Goose (*Branta hutchinsii minima*) on her nest. *Tim J. Moser*

1967, and subsequent recovery efforts have successfully increased their numbers. Their decline began, in part, when both arctic foxes (*Alopex lagopus*) and red foxes (*Vulpes vulpes*) were introduced on about 190 islands in Alaska during the height of the fur trade (1915–36), although some introductions occurred as early as 1750. The foxes decimated the goose population; no geese were seen from 1938 to 1962, until a small (200–300) group was discovered on remote Buldir Island in the Aleutians. A second wild population of no more than 200 birds was later discovered on Chagulak Island, also in the Aleutians (Bailey and Trapp 1984); the subspecies also apparently survived on Kiliktagik Island in the Semidi Islands group, located in the Gulf of Alaska. Overhunting throughout the range of this subspecies also contributed to the population crash (Springer and Lowe 1998).

Aleutian Cackling Geese began their recov-ery after the elimination of foxes from 35 nesting islands, to which family groups of these geese were then translocated from Buldir Island. Populations were subsequently established on Agattu, Amchitka, Amukta, Anowik, Little Kiska, and Nizki-Alaid Islands (Byrd 1998). Thereafter the population began to grow slowly, with 790 birds counted in 1975 on their principal wintering grounds in California, and about 2,700 estimated in California in the early 1980s. The Aleutian Cackling Goose was subsequently downgraded to threatened in 1991, when the population had reached 7,000, and removed from the Endangered Species List in 2001, after the population had reached 34,290 in 2000 (Byrd et al. 1991, Pacific Flyway Council 1999a, Collins and Trost 2009). The population was 79,500 in 2009 (U.S. Fish and Wildlife Service 2009), and these geese are now hunted.

Aleutian Cackling Geese migrate east through

the Aleutians, and depart from there in September for a trans-Pacific flight to wintering areas along the Pacific coast and inland to the Central Valley of California, where they arrive throughout October and into November and early December (Woolington et al. 1979). During the Midwinter Survey of 2009, 45,500 Aleutian Cackling Geese were estimated in the San Joaquin Valley. Their primary spring staging area is the coastal section of farmland just north of the Castle Rock National Wildlife Refuge (NWR), near Crescent City in northern California. In 1997, however, a large percentage of Aleutian Cackling Geese began using dairy pastures near Humboldt Bay in California as their new spring staging area, with 19,000 estimated there in early March 2002 (Black et al. 2004). Successful translocations of Western Canada Geese (*moffitti*) to the Humboldt Bay area may have encouraged the use of this new locale for spring staging by Aleutian Cackling Geese (Griggs and Black 2004).

Richardson's Cackling Goose. The Richardson's Cackling Goose (*Branta hutchinsii hutchinsii*), also known as the Baffin Island Cackling Goose, is only slightly larger than the diminutive Cackling Goose (*minima*). Mowbray et al. (2002) noted average weights of 2,180 g for males (*n* = 129) and 1,920 g for females (*n* = 125). Bill length averaged 39.0 mm in males and 37.7 mm in females. Richardson's Cackling Geese are the palest of the pale subspecies, and the smallest of the eastern subspecies of Canada/Cackling Geese. They can appear almost silvery at a distance, and some birds show a pale white collar at the base of the neck.

Nesting occurs on the coastal tundra from the Mackenzie River delta in the Northwest Territories east through coastal Nunavut and the Melville Peninsula, as well as on Victoria, Jenny Lind, King William, Baffin, and Southampton Islands. Richardson's Cackling Geese intergrade with Lesser Canada Geese (*parvipes*) in an overlap zone around Hudson Bay. Richardson's Cackling Geese winter in western Texas south to the northern cen-tral highlands of Mexico, as well as along the Gulf Coast from Louisiana to Mexico, as far south as the state of Veracruz (Mowbray et al. 2002).

Cackling Goose. The Cackling Goose (*Branta hutchinsii minima*) is the smallest and darkest subspecies in the Cackling Goose species, not much larger than a Mallard. Johnson et al. (1979) reported average weights from birds primarily measured on the Y-K delta in summer as 1,546 g for males (*n* = 152) and 1,312 g for females (*n* = 152). Bill length averaged 29.7 mm in males and 28.1 mm in females.

The breast of the Cackling Goose is brown or bronze and may have a purplish sheen. The head is rounded, with a sloping forehead and small stubby bill; the white cheek patch continues underneath the chin. There is usually no white neck ring, although some individuals have a thin incomplete white ring at the base of the black neck. Members of this subspecies produce a high-pitched call and have rapid wing beats during flight.

They nest exclusively on the Y-K delta in western Alaska and winter in the lower Columbia River valley of Washington and Oregon and Oregon's Willamette Valley (Pacific Flyway Council 1999b).

Gill et al. (1996) reported that some Cackling Geese have staged on Ugashik Bay and Cinder Lagoon on the northern side of the Alaska Peninsula, and others stage on Nunivak Island. They arrive at staging areas in southwestern Alaska in late September, peak in mid-October, and leave by late October or early November. Smaller numbers stop at the Copper River delta in southern Alaska, and the Queen Charlotte Islands and Vancouver Island in British Columbia (Pacific Flyway Council 1999b). Historically, Cackling Goose migration focused on the Klamath basin in northern California, which was once a major stopover for almost the entire population, before they proceeded to their wintering grounds in the Central Valley of California. Now, however, the majority remain in Washington and Oregon. An average of only 1,843 Cackling Geese were estimated during fall surveys in these

2 states from 1979 to 1985, but the average was 141,375 from 1995 to 1998 (Collins and Trost 2009). Spring migrants depart wintering areas in mid- to late April, arrive at Cook Inlet near Anchorage from late April to early May, and then progress to the Y-K delta via a coastal route or a flight over the Alaska Range (Mowbray et al. 2002).

Migration

Chronology. The timing and extent of migration varies greatly among populations of Canada Geese and Cackling Geese. Across arctic and subarctic areas, geese are prone to leave their breeding areas in late summer or early fall, long before weather conditions dictate a need to migrate, yet birds from these same populations return to subarctic or arctic breeding grounds while snow covers the ground and the first open water is just appearing. Giant Canada Geese and Western Canada Geese are notorious for their late-fall departures from breeding areas. Many remain through the winter as long as snow and ice conditions permit, and then move only short distances farther south (Hanson 1997).

During the last 20 years, 2 factors have affected the chronology and distribution of arctic and subarctic subspecies of Canada/Cackling Geese east of the Rocky Mountains: (1) the development of a large population of temperate-breeding geese, and (2) an increase in the availability of waste corn and open water, the latter provided by areas such as power plant cooling lakes, which have supplied geese with open water in subzero weather far north of where they formerly could exist.

As resident populations have increased, migration patterns have altered, with additional short-stopping by Canada/Cackling Geese bound for traditional wintering grounds farther south (e.g., the Mississippi Valley Population). The presence of resident geese attracts migrants to cooling lakes and waste grain resources, and even to grass lawns in parks and golf courses, as well as to pastures. This short-stopping has greatly delayed, and, in some instances, eliminated the arrival of Canada/Cackling Geese on traditional wintering grounds,

a factor that has made their harvest increasingly difficult to manage.

Fall. Arctic breeders, which travel the longest distances, initiate fall migration first, generally beginning in late August through September, reaching the northern United States between late September and early October. Those groups wintering farthest south will arrive by mid-November; most will reach their final destination by mid-December (Mowbray et al. 2002).

Cackling Geese (*minima*) leave the Y-K delta in October and fly directly to the Alaska Peninsula (Ugashik Bay), where they remain for up to 3 weeks, accumulating the energy necessary to complete a 2,800 km migration to wintering areas in the Klamath Valley in Oregon and California (Sedinger and Bollinger 1987). The geese complete this migration in 72 hours, during which time individuals lose between 400 and 600 g (23%–33%) of their body mass. Cackling Geese (*minima*) staging at Ugashik Bay accumulate energy reserves for migration by feeding 53%–84% of the time, primarily on nutritious tubers of *Triglochin palustris*.

Adult female Interior Canada Geese in the Mississippi Valley Population that were tagged with radiotransmitters during 1984–86 ($n = 457$) left their breeding areas in northern Ontario from late September to mid-October (Tacha et al. 1991). Some birds arrived in Wisconsin (Horicon and Grand River marshes) by mid-September, with peak numbers accumulated by mid-December in 1984, but by early to mid-November in other years. Early migrants reached southern Illinois (the Crab Orchard NWR and the Horseshoe Lake NWR) by late September, and their numbers rose through early November. Major increases in goose numbers in southern Illinois occurred around mid-December, however, coincident with decreases in Wisconsin. Early arrivals at wintering areas suggested that some birds flew nonstop.

Satellite tracking of Canada/Cackling Geese is revolutionizing our understanding of migration in ways not possible from conventional radiotelemetry techniques. Malecki et al. (2001) satellite-tracked 34 adult female Interior Canada Geese from the Atlantic Population on breeding areas in northern Québec in 1996 and 1997, which then provided significant new information on migration and the wintering ecology of this population. Those marked near the coast of Hudson Bay and the northern Ungava Peninsula left their breeding areas in late September and migrated through western Québec, southeastern Ontario, and central New York to wintering areas in the Chesapeake Bay / Delaware Bay area of the mid-Atlantic coast. In contrast, those marked in the southern Ungava Bay region migrated through central Québec, western Vermont, Massachusetts, and Connecticut, and eastern New York, primarily arriving on wintering areas in the lower Hudson River drainage. Both groups reached their wintering areas by mid- to late October. Migration was highly synchronized, with over 50% of the birds migrating >1,125 km during the fall in <7 days, and some making this flight in only a single day.

Spring. Canada/Cackling Geese spend as little as 2 weeks in many of their wintering areas before starting their spring migration in late January. Midcontinent populations are especially regulated by the snow and ice cover, following the advancing edges of snowmelt that makes waste grain and open water available (Wege and Raveling 1983). The spring migration of Canada/Cackling Geese is often associated with the 2°C isotherm, which produces heavy snow and cold air. These conditions will stop spring migration and can even send geese in reverse; hence the spring migration is more prolonged than fall migration (Bellrose 1980). By late April, however, early migrants have already reached their breeding grounds from James Bay northward through Ontario and Québec.

Among radiomarked Interior Canada Geese from the Mississippi Valley Population that departed from wintering refuges in Illinois, over 98% of all locations were between the Illinois River valley and the big bend of the Wisconsin River, Lake Shelbyville (Illinois), and the western coast of Lake Michigan. Within that corridor, however, the Kaskaskia River valley was identified as a major spring staging area. The duration of spring migration from wintering areas to breeding grounds averaged 54 days (Tacha et al. 1991). For satellite-tracked Canada Geese in the Atlantic Population, spring migration from the mid-Atlantic coast began in late February, with the birds remaining in central New York and southern Ontario during March and April, followed by a major movement north that occurred in early May, with arrival on the breeding grounds 1–2 weeks later (Malecki et al. 2001).

Radiomarked Giant Canada Geese wintering at Silver Lake near Rochester, Minnesota, began their spring migration during early to mid-April and arrived at Marshy Point, Manitoba, 6–14 days later (Wege and Raveling 1983). Their principal departure occurred when major rivers in southern and southwestern Minnesota were ice free and meltwater was abundant in western Minnesota and eastern North Dakota.

MIGRATION BEHAVIOR

The 11 subspecies of Canada/Cackling Geese vary greatly in their migration routes as well as in the distances traveled. In general, the small-bodied subspecies tend to nest farther north and migrate farthest south (Sedinger and Bollinger 1987), while the large-bodied subspecies nest farther south and may only migrate short distances, if at all. Such distributions appear to be correlated with heat loss and body size. Larger subspecies, such as Giant Canada Geese, have a heat-loss rate about 40% less than that of Lesser Canada Geese; hence the former's bigger size allows them to winter farther north, because they can better withstand cold temperatures (Lefebvre and Raveling 1967).

Lesser Canada Geese and Richardson's Cackling Geese undertake the longest migrations, traveling up to 4,800 km from Baffin Island to the coast of

Mexico (Moser et al. 1991). As migrants, Canada/Cackling Geese are extremely flexible in their behavior, which enables them to alter their patterns to adjust to changing environmental conditions. Reverse migrations during the fall or spring are common responses to unfavorable weather conditions (Raveling 1976). Other populations, particularly the Mississippi Valley Population and Eastern Prairie Population, have dramatically altered their migration chronology and winter distribution in response to changes in food distribution.

Migration is most often initiated at dusk, but it can begin at any time, and Canada/Cackling Geese fly during both day and night (Palmer 1976a, Bellrose 1980). At Marshy Point in Manitoba, radio-marked Canada/Cackling Geese departed on their fall migration under weather conditions with trailing surface winds, a decreasing mean temperature, and increasing barometric pressure (Wege and Raveling 1983). Flocks are highly variable in size, but the small subspecies tend to migrate in bigger flocks than the large subspecies, and late-season flights are apt to contain larger flocks than early-season ones. In central Illinois, 74 flocks of Interior Canada Geese ranged in size from 23 to 300 birds, and averaged 96 (Bellrose 1980). Flock sizes observed in spring migration through the Ottawa, Ontario, area ranged from 6 to 300 geese and averaged 66 (Blokpoel and Gauthier 1980).

Canada/Cackling Geese almost always migrate in a V formation, although the angle of the formation can vary from 24° to 124° (Heppner et al. 1985). A primary advantage of the V formation is the energy it saves in long-distance flights. Each bird flies slightly above the bird in front of it, and updrafts created at the wingtips of the individual in front provide lift to the individual behind. Birds flying in formations can save up to 51% of the energy cost of solo flights (Badgerow and Hainsworth 1981), although the average is about 10% (Badgerow 1988). Other calculations show that 25 geese in a V formation can reduce induced drag by 65% and increase their range by 70% (Lissaman and Shol-

lenberger 1970). The lead individual must change position during a migratory flight, however, because of the substantial energetic disadvantage of being in front (Hainsworth 1987).

The V formation also facilitates visual communication among flock members. The visual field of view for a Canada/Cackling Goose is 135°, with a binocular overlap of 20° and a blind spot at the back of the head, which extends 29° on either side of the midline. Hence, as long as the angle of the formation is >29°, each bird in the formation can see every other bird in the line, including those behind it, and visual communication is of obvious importance in maintaining flock integrity (Heppner et al. 1985). The visual benefits of the V formation were used by military aircraft in World War II. In short-distance local flights, however, Canada/Cackling Geese will fly in horizontal or vertical lines.

The migration altitude of Canada/Cackling Geese varies greatly with size, weather conditions, and the distance between departure and arrival points. Several hundred pilot reports received at traffic control centers of the Federal Aviation Administration indicate that during fall migration, most flocks of Canada/Cackling Geese flew at an altitude of about 600 m; 64% were at heights between 229 and 1,067 m. During spring migration, maximum altitudes were higher. The highest was 2,134 m, and a few were over 1,219 m, but 77% of the geese were sighted between 229 m and 914 m (Bellrose 1980). Wege and Raveling (1984) found that radiomarked Canada/Cackling Geese largely migrated nonstop between Lakes Manitoba and Winnipeg to Silver Lake (near Rochester, Minnesota), flying at heights of 100–600 m. To reduce migration costs, geese decreased air speeds as tailwind speeds increased, but ground speeds did not change. The birds also used visual ground cues to correct for course changes caused by wind drift.

Wege and Raveling (1984) reported that flight speeds averaged 65 km/hr in the air (range = 32–101 km/hr) and 83 km/hr on the ground (range = 49–

110 km/hr). Owen (1980) listed the average migration speed of geese at 64 km/hr, but noted speeds of up to 148 km/hr with a tailwind. Gill et al. (1996) found that during fall migration, Canada/Cackling Geese took advantage of the steady winds that preceded the back or western side of low-pressure systems. During spring migration, the geese most frequently traveled north on the back side of high-pressure systems.

Molt Migration. In general, successfully breeding Canada Geese molt on the breeding grounds with their broods, while failed breeders and non-breeding geese migrate to traditional molting sites. These molt migrations always involve northward movements to large bodies of water at a time that coincides with peak nutrient availability in new plant growth (Salomonsen 1968, Owen and Ogilvie 1979, Madsen and Mortensen 1987, Abraham et al. 1999a). A northward migration during this period also allows for more feeding time, because the days are longer (Derksen et al. 1982), and individuals undertaking a molt migration avoid competition with successful breeders and their goslings (Sterling and Dzubin 1967).

Canada/Cackling Geese vary greatly, however, in the distances traveled to undertake the molt. Abraham et al. (1999a) reported that Canada/Cackling Geese originating from populations in 26 states and 6 Canadian provinces underwent molt migrations to James and Hudson Bays, a migration distance that ranged from a few km to >1,500 km. Canada Geese nesting in midcontinent North America, mostly Giant Canada Geese, undergo a molt migration of at least 1,000 km to the Thelon River and other areas in the Northwest Territories (Sterling and Dzubin 1967). Some Western Canada Geese travel from northeastern California and undergo a molt migration of >1,000 km to the Northwest Territories (Rienecker 1985). Davis et al. (1985) noted >30,000 large Interior Canada Geese from the Eastern Prairie Population pass Cape Churchill and the McConnell River delta on a northward molt migration.

Among Giant Canada Geese at Crex Meadows in Wisconsin, 97% of the nonbreeders and 90% of unsuccessful breeders left the area to molt, with some traveling >1,300 km to molt in the Hudson Bay Lowlands in northern Manitoba, but then returned from late August to late October (Zicus 1981b). A portion of the Giant Canada Geese fitted with satellite- or radiotransmitters at 27 locations in southern Michigan underwent northward molt migrations of >900 km to areas around James and Hudson Bays, after which they then returned to their former nesting areas, although others did not undergo a molt migration (Luukkonen et al. 2008). In the Atlantic Flyway, satellite-tracked geese from the AFRP underwent molt migrations of 564–1,941 km to northern Québec (Sheaffer et al. 2008); however, 56% of the birds monitored did not attempt a molt migration.

The increasing numbers of Giant Canada Geese that undertake molt migration to areas such as James Bay and northern Ontario are problematic, because they occupy brood-rearing areas earlier than locally nesting geese and their offspring (Abraham et al. 1999a). The presence of large numbers of molting Giant Canada Geese, as well as nesting Lesser Snow Geese, compete with Interior Canada Geese breeding on Akimiski Island in Nunavut, where they have affected gosling growth (Ankney 1996).

HABITAT

The variation in habitat use by Canada/Cackling Geese in North America may exceed that of any other bird (Boyd and Dickson 2005); the 11 subspecies of Canada/Cackling Geese make any generalizations about habitat requirements virtually impossible. Breeding birds use vastly diverse habitats, ranging from arctic tundra on the Y-K delta in Alaska (Cackling Geese [*minima*]), to semidesert sagebrush habitat in the Intermountain Region (Western Canada Geese), to temperate rainforests along the coast of British Columbia and southeastern Alaska (Vancouver Canada Geese). Giant Canada Geese successfully breed in urban areas throughout the majority of the lower 48 states, and they also readily use agricultural areas (Lee et al. 1984).

Because they are primarily grazing herbivores, Canada/Cackling Geese are always associated with grasses and sedges as a food supply during the breeding season, and often during migration and winter as well, although nonbreeding birds can be heavily dependent on agricultural crops. The habitats breeding birds select always have high-quality sources of vegetation that are essential for the growth and development of goslings and the replenishment of nutrient reserves depleted in adults during reproduction. Breeding Canada/Cackling Geese primarily select areas with islands that are available for nests. Outside the Arctic, breeding grounds are commonly associated with reservoirs and large lakes, where the geese will commonly nest on islands and forage on nearby grasses and in agricultural fields.

Nesting Richardson's Cackling Geese on the Kent Peninsula in the Northwest Territories use hummock and mudflat habitats before snowmelt (Carrière et al. 1999). After snowmelt, however, they avoid hummocks and prefer mudflats, pond margins, and ponds. In the Thelon River valley in Nunavut, Richardson's Cackling Geese nest on cliffs and steep rock slopes of the Clarke River, in colonies of several hundred breeding pairs (Norment et al. 1999). Interior Canada Geese on the Ungava Peninsula in Québec use coastal lowlands marked by small ponds and ephemeral wetlands (Malecki and Trost 1990). Interior Canada Geese in the Mississippi Valley Population nesting along the coast of Hudson Bay near Winisk, Ontario, choose islands and sites that are the first to be free of meltwater, while pairs with broods select tidal salt marshes (Bruggink et al. 1994). Elsewhere in the Hudson Bay Lowlands, Raveling and Lumsden (1977) noted that fen ponds and bog-pond systems were especially important for nesting geese once these sites became ice free. On the northeastern coast of Hudson Bay, near Nunavik, Interior Canada Goose goslings use wet sedge meadows most heavily, because that habitat contains the most nutritious plants (Cadieux et al. 2005).

In the Flathead Valley of Montana, 90% of 479 Western Canada Goose nests were located on islands (Geis 1956). Giant Canada Geese in Rhode Island use inland and coastal ponds, rivers, marshes, and reservoirs during the nesting season, all of which are characterized by a wide variety of vegetation (Allin 1980). Both nesting and brood-rearing areas contain abundant browse, and often occur on grassy areas adjacent to ponds and lakes. Postbreeding birds congregate on larger bodies of water to molt, while wintering birds accumulate near agricultural crops of winter rye, wheat, oats, and the waste from harvested corn, or in pastures.

The principal winter habitat of Canada/Cackling Geese is simpler to describe: refuges (Raveling 1978a). Refuges are especially important to protect geese from overharvest. Large reservoirs and impoundments are also widely used by wintering Canada/Cackling Geese. Of the wintering geese from the Rocky Mountain Population, 64% were associated with reservoirs and adjacent farmlands (Krohn and Bizeau 1980).

POPULATION STATUS

The increase in the continental population of Canada/Cackling Geese during the last 5 decades is among the more spectacular accomplishments of wildlife management, rivaling the comeback of white-tailed deer (*Odocoileus virginianus*) and wild turkeys (*Meleagris gallopavo*). By compiling population estimates from different surveys, Moser and Caswell (2004) estimated an average spring population of 7.1 million Canada/Cackling Geese in North America during 2000–2002.

Management Units
All of the Canada/Cackling Goose subspecies exhibit strong fidelity to their traditional breeding, migration, and wintering grounds, which results in distinct populations that are very useful for management. Such cohesion and fidelity led biologists to designate 17 Management Units (or Popu-

lations) for Canada/Cackling Geese across North America, each with relatively little overlap during breeding, migration, and wintering (U.S. Fish and Wildlife Service 2009). Some populations contain a single subspecies (e.g., the Dusky Canada Goose, the Cackling Goose (*minima*), the Aleutian Cackling Goose), although most are composed of several subspecies that breed, migrate, and winter in specific areas. More mixing of the various populations now occurs on the wintering grounds, however, so the focus has been shifted to identifying and managing populations based on their breeding areas. Below is a brief description of the population status for each of the Management Units.

North Atlantic Population (NAP). The NAP is largely composed of Atlantic Canada Geese that primarily breed in Labrador and in Newfoundland, but also in parts of eastern Québec. They winter along the Atlantic coast from the Maritime Provinces south to Long Island Sound, and occasionally are found farther south. During winter they mix with other subspecies of Canada Geese, although the NAP tends to use more coastal habitats.

The NAP was not recognized as a separate population in the Atlantic Flyway until 1996; the official population estimate for the NAP is now derived from those geese surveyed during springtime along Strata 66 and 67 of the Waterfowl Breeding Population and Habitat Survey (WBPHS). The estimated population from 1996 to 2007 averaged 66,000 breeding pairs: 36,000 in Labrador and 29,000 in Newfoundland. The 2009 survey estimated 53,700 pairs and 179,700 individuals (U.S. Fish and Wildlife Service 2009).

Atlantic Population (AP). The AP is composed of the most northern and eastern breeding population of Interior Canada Geese. The AP breeds north of 48° N latitude, along the coasts of the Ungava Peninsula in northern Québec, particularly (80% of the total breeding population) on the northeastern shore of Hudson Bay, but also in the interior of the Ungava Peninsula (Atlantic Fly-

way Council 2008b). The wintering range for AP geese extends from southern Ontario, east through southernmost Québec to the New England states, and south to South Carolina.

The Midwinter Survey was the primary means of assessing the status of the AP until the mid-1990s, after which it became confounded by the dramatic increase in the Giant Canada Goose population in the Atlantic Flyway. The primary means of assessing the population status of the AP is now an annual survey on their breeding grounds in northern Québec. The 2009 estimate was 176,100 breeding pairs, and the 2000–2009 average was 160,050 pairs (U.S. Fish and Wildlife Service 2009). The population goal is to achieve and maintain an index of 225,000 breeding pairs (Atlantic Flyway Council 2008b). Midwinter Survey data averaged 811,098 AP geese from 2006 to 2009: 45% in Maryland, 18% in New Jersey, 14% in Virginia, and 3% in New York (Klimstra and Padding 2009).

Atlantic Flyway Resident Population (AFRP). The AFRP is largely made up of a mix of subspecies, including *B. c. maxima, B. c. moffitti, B. c. interior*, and *B. c. canadensis* (Atlantic Flyway Council 2011). These birds were introduced into the Atlantic Flyway through the release of captive flocks when the used of live decoys for hunting was outlawed in 1935, and by translocation efforts of wildlife agencies from the 1950s to the 1990s to establish local populations, primarily in urban areas. Individuals in the population undertake a limited migration and are thus considered "residents" in the lexicon of waterfowl managers. The AFRP inhabits the region from southern Ontario east through southern Québec and the Canadian Maritime Provinces (New Brunswick, Nova Scotia, and Prince Edward Island), and then south from Maine to northern Florida.

The AFRP has increased steadily and dramatically, with state breeding-waterfowl surveys recording a tripling in size, from roughly 300,000 in the late 1980s to over 1 million in the late 1990s.

From 2003 to 2009, the AFRP averaged 1,092,214 birds (U.S. Fish and Wildlife Service 2009). According to Sheaffer and Malecki (1998), 45% of the Atlantic Flyway residents bred within New York, Pennsylvania, and New Jersey during the late 1980s.

Much of the harvest of AFRP geese occurs during special September seasons, before the arrival of the migrant populations, although the former are also harvested during the regular season. During the 2008 September season in the Atlantic Flyway, 238,400 were harvested: 33% in New York, 30% in Pennsylvania, and 7% in North Carolina (Klimstra and Padding 2009).

Southern James Bay Population (SJBP). The SJBP, formerly known as the Tennessee Valley Population, is largely composed of Interior Canada Geese that breed on Akimiski, Twin, and Charlton Islands, as well as the adjacent coastal lowlands in Nunavut, and on the James Bay coast, from Ekwan Point in Ontario south to Moosonee and east to the Ontario/Québec border, and then north along the James Bay coast in Québec to 53° N latitude (Abraham and Warr 2003). About 20% breed on Akimiski Island in James Bay, although it forms only 3% of the SJBP breeding range (Arctic Goose Joint Venture Technical Committee 2008).

The SJBP winter range extends from southern Ontario and Michigan south to South Carolina, Georgia, Alabama, and Mississippi. Within the Mississippi Flyway, numbers of SJBP geese from 1996 to 1998 (Midwinter Survey) averaged 110,594: 27% in Michigan, 27% in Ohio, 18% in Tennessee, and 13% in Ontario (Abraham and Warr 2003). After 1997, however, the Midwinter Survey no longer separated populations in the Mississippi Flyway, due to widespread mixing with Giant Canada Geese, which resulted in overestimating the SJBP (Leafloor et al. 1996a). An analysis of direct recoveries from 29,000 Canada Geese banded on Akimiski Island from 1971 to 1987 revealed that most birds migrated through southwestern Ontario, Michigan, Indiana, Ohio, and northwestern Pennsylvania, to winter in southern Michigan, Ohio, Kentucky, Tennessee, and northern Alabama. Only 19% of the recoveries were in the Atlantic Flyway, primarily in Pennsylvania (Trost et al. 1998). Important staging and wintering areas are the Shiawassee NWR in Michigan, the Mosquito Creek Wildlife Area and the Ottawa NWR in Ohio, the Sloughs Wildlife Management Area in Kentucky, and the Wheeler NWR in Alabama (Abraham and Warr 2003).

From 2005 to 2009, a spring aerial breeding survey estimated an average of 17,267 SJBP geese on Akimiski Island (12,668 as breeding pairs), and another 81,263 on the adjacent lowlands of southern James Bay (66,818 as breeding pairs), for an average total population of 98,530; 81% were observed as breeding pairs (Klimstra and Padding 2009). From 1990 to 2009, total numbers for the SJBP have ranged from 77,345 in 1993 to 160,430 in 2006. From 1990 to 2009, however, the population has declined by 23%, and that on Akimiski Island has decreased by 57%. Some of this decline is attributed to Giant Canada Geese molt migrants, which compete with SJBP goslings for food in northern Ontario and western James Bay, resulting in high gosling mortality (Abraham et al. 1999a). There is also significant habitat loss on Akimiski Island, caused by the overabundant Lesser Snow Goose (*Chen caerulescens caerulescens*) population (Abraham et al. 1999b).

Mississippi Valley Population (MVP). The MVP is a well-known population that consists principally of Interior Canada Geese. This population breeds over a large area, stretching west from the Hudson Bay Lowlands in Ontario north of the Albany River into northeastern Manitoba near York Factory, where they meet the Eastern Prairie Population (Hanson and Smith 1950, Vaught and Arthur 1965, Craven and Rusch 1983). During the 1989–96 surveys, Leafloor and Abraham (2000) found nest densities were greatest (3.0/km²) along coastal tundra, decreasing to 1.4/km² in lowlands within 80 km of the coast, and only 0.6/km² in interior lowlands >80 km from the coast.

The MVP leaves its breeding grounds from late September to mid-October and follows migration corridors along both sides of Lake Michigan. An analysis of band recoveries of Interior Canada Geese marked on breeding areas in both Hudson and James Bays from 1995 to 1998 found that 32% were recovered in Wisconsin (on the western side of Lake Michigan) and 21% in Michigan (on the eastern side of Lake Michigan; Fritzell and Luukkonen 2004). Major winter concentrations are in Wisconsin, Illinois, and Michigan.

The MVP was the subject of considerable controversy when changing agricultural practices in Wisconsin began to short-stop birds, delaying their arrival at traditional wintering areas in Illinois. Prior to the 1950s, most MVP geese flew almost nonstop from James Bay to wintering grounds in southern Illinois. By the early 1960s, however, goose management at the Horicon NWR attracted peak fall counts of 100,000 MVP geese to east-central Wisconsin, which increased to >200,000 by 1974 and led to significant crop-damage issues with area farmers. Subsequent management from 1976 to 1980 discouraged goose use of the Horicon area and reduced their peak fall numbers to about 100,000 by 1980 (Craven et al. 1985). A reversal of management strategy in 1980 began another cycle of increasing goose populations in the Horicon area, which reached 257,000 by fall 1984, with subsequent surveys reporting about $1.6 million in crop damage. Revenue from the throngs of people coming to the area to view the spectacle provided by the geese, however, was $2.2 million in 1986 (Rusch et al. 1985, Heinrich and Craven 1992). The issue of short-stopping continues today, as the early winter population at Horicon averaged 274,000 from 1996 to 2000. Departure of the MVP from Wisconsin varies greatly with the snow and ice cover on corn stubble and on water. In 1994 and 1995, 80% of the MVP geese observed in the Midwinter Survey were found in Wisconsin, and only 20% in Illinois. In contrast, from 1971 to 1975 only 7% of the wintering geese were in Wisconsin (Gamble 2002).

Breeding-ground surveys of MVP geese have averaged 327,360 breeding adults from 2000 to 2009, but the estimates have varied from a low of 239,600 in 2009 to 402,600 in 2007 (U.S. Fish and Wildlife Service 2009). The total population in 2009 was estimated at 518,200. The Midwinter Survey reported an average of 840,119 Canada Geese in the Mississippi Flyway from 2006 to 2009 (not all are MVP geese): 18% in Missouri, 17% in Iowa, 15% in Ohio, 14% in Illinois, 12% in Wisconsin, 9% in Minnesota, and 6% in Michigan.

Eastern Prairie Population (EPP). The EPP also consists of Interior Canada Geese. This population nests in the Hudson Bay Lowlands of Manitoba, mainly from about 80 km north of Churchill to about 25 km east of York Factory, where it overlaps the western border of the MVP and portions of the Tallgrass Prairie Population. Few EPP geese nest south of 56° N latitude (Malecki et al. 1980, Sheaffer et al. 2004).

The EPP initiates fall migration in early September, moving through southcentral Manitoba, western Minnesota, and Iowa to traditional wintering areas, primarily in Missouri and Arkansas, that are generally west of the Mississippi River (Vaught and Kirsch 1966, Samuel et al. 1991). Historically, however, this population's winter range extended southward through Arkansas to coastal Louisiana and Texas (Vaught and Kirsch 1966). During the mid-1980s the majority of EPP geese migrated directly to the Swan Lake NWR in Missouri to winter, but now as many as 50,000 stop at Minnesota's Lac qui Parle State Park on the upper Minnesota River. Over the past several decades, increasing numbers of EPP geese have delayed their migration to traditional wintering grounds, with 64% of the population reported from Minnesota and Iowa in December 1991 to 1995, compared with just 16% from 1981 to 1985, and only 5% from 1971 to 1975 (Humburg et al. 1985, Humburg et al. 2000). There can also be significant annual variation in their migration to wintering areas. Sheaffer et al. (2004) noted that 25% of marked EPP geese had arrived on southern wintering areas by 1 November, and 55% by 1 December in 1985, while only 10% had arrived by 1 November in 1986.

Breeding-ground surveys of EPP geese have averaged 145,220 breeding birds from 2000 to 2009, increasing by an average of 3%/year. Breeding population estimates have varied from a low of 122,200 in 2001 to 169,200 in 2009, with 279,900 total birds estimated for 2009 (U.S. Fish and Wildlife Service 2009). Densities on the breeding grounds were more than 10 times higher in coastal areas than in the interior (Humburg et al. 1998).

Mississippi Flyway Giant Population (MFGP). The MFGP is composed of Giant Canada Geese reestablished or introduced in all Mississippi Flyway states. Their distribution ranges from the eastern edge of the Great Plains to the western edge of the Atlantic states: from southern Manitoba and Ontario south to northern Louisiana, Mississippi, and Alabama. This population has increased dramatically, from 855,300 in 1993 to 1,906,200 in 2009, a rise of 2%/year from 2000 to 2009 (U.S. Fish and Wildlife Service 2009).

Western Prairie Population and Great Plains Populations (WPP/GPP). The WPP breeds around the myriad small lakes that dot the boreal forest of northeastern Saskatchewan and northwestern Manitoba. Most of these birds are Interior Canada Geese, but an unknown number are Western Canada Geese and Giant Canada Geese. In contrast, the GPP is composed of large Western Canada Geese (the result of restoration efforts from Saskatchewan to Texas), but their breeding range is centered in the Prairie Pothole Region of southeastern Saskatchewan, southwestern Manitoba, and the Dakotas. The range of the GPP overlaps somewhat with the MFGP in the east and the Hi-Line Population to the west. Nieman et al. (2000) separated the WPP and GPP in eastern Saskatchewan and western Manitoba at 50°30′ N latitude, and the authors noted that the southern group of breeding WPP geese has increased by 10-fold over the past 3 decades (1970 to 1999). Both populations

mix on their migration and wintering areas, and they are surveyed and managed jointly during the winter.

The WPP migrates into the Missouri River valley of the Dakotas, where significant numbers winter at the Lake Andes NWR and the Fort Randall Reservoir in southeastern South Dakota, while many others continue down the Missouri River to the Squaw Creek NWR in northwestern Missouri. Smaller numbers continue on to Texas. In contrast, the GPP migrates south through the Dakotas and Nebraska into Kansas, Oklahoma, and northern Texas. The WPP and the GPP mix during migration along the Missouri River in the Dakotas and on reservoirs from Kansas to Texas in the winter. They also intermingle along the Platte River in Nebraska, where Vrtiska and Lyman (2004) noted a 10% annual increase in numbers of wintering WPP and GPP geese from 1960 to 2000.

Winter surveys have estimated an average of 577,380 WPP/GPP Canada Geese from 2000 to 2009, decreasing by an average of 2%/year. Counts have varied from a low of 415,100 in 2005 to 710,300 in 2002. In 2009 a spring population of 922,900 was estimated in that portion of the WBPHS that included the range of the WPP/GPP; this estimate has increased by 2%/year since 2000 (U.S. Fish and Wildlife Service 2009).

Tallgrass Prairie Population (TGPP). The TGPP is composed of Richardson's Cackling Geese and smaller numbers of Lesser Canada Geese; hence, under the new Canada Goose nomenclature, this population is a mixture of the 2 species: the Canada Goose and the Cackling Goose. The TGPP generally nests east of 105° w longitude and north of 60° N latitude: on Baffin, Southampton, and King William Islands, and on the mainland in the eastern Queen Maud Gulf region and along the Hudson Bay coast north of the Maguse and McConnell Rivers (U.S. Fish and Wildlife Service 2009). The TGPP migrates southeast across Manitoba and western Ontario, and then south through the Dakotas, Nebraska, Kansas, Oklahoma, and eastern Texas, to its wintering grounds on the Gulf Coast of Texas and as far south as Tampico, Mexico. Major staging grounds along this route include the Devils Lake region of North Dakota, the Sand Lake NWR in northeastern South Dakota, Cheyenne Bottoms and the Quivira NWR in southcentral Kansas, the Salt Plains NWR in northern Oklahoma, and the Tishomingo and the Hagerman NWRs on Lake Texoma in Oklahoma and Texas.

On their wintering grounds, the TGPP mixes with other populations of Canada/Cackling Geese, which makes estimations of the former's population size difficult. Winter surveys, however, have estimated an average of 431,360 birds from 2000 to 2009, increasing by an average of 5%/year. Counts have varied from a low of 149,100 in 2001 to 680,300 in 2007 (U.S. Fish and Wildlife Service 2009). On their Baffin Island breeding grounds, the population of adult and subadult Richardson's Cackling Geese and Lesser Canada Geese averaged 100,000 from 1993 to 1999 (Dickson 2000).

Shortgrass Prairie Population (SGPP). The breeding range of the SGPP is west of the 100th meridian, on Victoria and Jenny Lind Islands in the Canadian Arctic, and on the adjacent mainland from the Mackenzie River east through the Northwest Territories to Queen Maud Gulf and south to northern Alberta and Saskatchewan. This population contains Lesser Canada Geese and Richardson's Cackling Geese. Recognition of these subspecies, however, is virtually impossible in the field, which confounds population assessment and management (Arctic Goose Joint Venture Technical Committee 2008).

Hines et al. (2000) conducted an intensive survey (1989–93) of breeding Canada/Cackling Geese affiliated with the SGPP in the Inuvialuit Settlement Region, extending from Alaska east along the Arctic Coastal Plain to Coronation Gulf, including Banks Island and the western third of Victoria Island. They recorded 17,974 (0.68/km²) geese on the mainland coastal regions, 61,220 (0.58/km²)

on Victoria Island, and 675 (0.02/km²) on Banks Island. All areas were not surveyed, but the total estimated population was 90,000–95,000.

Band-recovery data and observations of neck-banded birds indicate that SGPP geese migrate to a staging area between Hanna, Alberta, and Kindersley, Saskatchewan, arriving from early September through October, where they remain until freezeup in mid-November to mid-December (Grieb 1970). This staging area is also noted for its concentrations of Lesser Snow Geese and Greater White-fronted Geese (*Anser albifrons frontalis*). These birds then pass through the Platte River near Ogallala, Nebraska, to winter largely in southeastern Colorado, northeastern New Mexico, the Oklahoma Panhandle, and the Texas Panhandle. Most birds from the western part of the population migrate to the Peace River area of northwestern Alberta and the Alberta/Saskatchewan border, after which they proceed to wintering areas, although a small number winter in the Central Valley in California (Hines et al. 2000).

Most SGPP geese once wintered in the Texas Panhandle, but band recoveries indicated that a larger percentage occurred farther north in the 1990s than in the 1970s (Hines et al. 2000). In the 1970s, 67% of the band recoveries were in Colorado; 2 decades later, recoveries there had increased to 82%, with a reduction from 27% to 6% in Texas and New Mexico.

Winter surveys estimated an average of 192,040 SGPP geese from 2000 to 2009, increasing by an average of 3%/year. Estimates have varied from a low of 149,100 in 2001 to 680,300 in 2007. In 2009, a spring population of 134,100 was estimated in the Northwest Territories by the WBPHS, a tally that has increased by 6%/year since 2000 (U.S. Fish and Wildlife Service 2009).

Hi-Line Population (HLP). The HLP consists of two subspecies: the Giant Canada Goose and the Western Canada Goose. It resulted from diverse translocations establishing new breeding flocks. The HLP of large Canada Geese nests in southeastern Alberta, southwestern Saskatchewan, eastern Montana and Wyoming, and Colorado. Many reservoirs and stock ponds were constructed in this region since the reintroductions, which has led to a 10-fold increase in the HLP, from less than 20,000 birds in the 1960s to over 60,000 estimated in 1995 (Nieman et al. 2000).

The HLP migrates only a short distance to wintering grounds that extend from northcentral Colorado southward into central New Mexico. The Midwinter Survey shows that most HLP geese winter in the reservoirs east of the Front Range in northern Colorado; others are found in Montana and Wyoming, and lesser numbers in Nebraska and New Mexico.

The Midwinter Survey estimated an average of 228,980 HLP geese from 2000 to 2009, decreasing by an average of 1%/year. Counts have varied from a low of 180,200 in 2007 to 270,700 in 2000. In 2009, a spring population of 298,400 geese was estimated in Alberta, Saskatchewan, and Montana by the WBPHS; this estimate has decreased by 2%/year since 2000 (U.S. Fish and Wildlife Service 2009).

Rocky Mountain Population (RMP). The RMP primarily consists of Western Canada Geese, which nest in central Alberta, western Montana, and the Intermountain Region (Idaho, Wyoming, Nevada, Utah, and Colorado) south to eastcentral Arizona and northwestern New Mexico. The major nesting range for the RMP is from southern Alberta to northern Utah. The Intermountain birds have a discontinuous breeding distribution, resulting from scattered water areas providing breeding habitat among the region's mountain ranges and arid basins (Krohn and Bizeau 1980). The RMP winters from central and southern California to central Arizona, and as far north as southern Alberta. Known molting concentrations are found on reservoirs and lakes in southern Alberta, southwestern Montana, Wyoming, and northern Utah.

The breeding population averaged 162,950 from 2000 to 2009, with the largest numbers nesting in southern Alberta and central Montana. In 2009

the spring breeding population totaled 128,400, and it has decreased by 1%/year since 2000 (U.S. Fish and Wildlife Service 2009).

Pacific Population (PP). The PP consists of Western Canada Geese and hybrids between Interior Canada Geese and Giant Canada Geese (Nelson and Oetting 1998, Smith 2001). Unlike the RMP, this population is composed of both resident and migratory segments. The PP nests and winters west of the Rocky Mountains, with its breeding range extending from northern British Columbia and Alberta south through the Pacific Northwest to central California and western Nevada. A large portion of this population exists as a consequence of reintroduction efforts in the 1960s and 1970s, some as a result of the salvage of goose eggs from the construction of the John Day Reservoir on the Columbia River (Nelson and Oetting 1998). In general, reintroduced birds do not migrate, except during severe winters, when some move to the southern portion of their range.

Most of the PP is surveyed in northern Alberta during the spring WBPHS (Strata 76–77), along with the Spring Survey in Washington, Oregon, California, and Nevada (U.S. Fish and Wildlife Service 2009). The total for the PP was 127,000 birds in 2009, of which 68,100 were estimated in Alberta.

Vancouver Canada Goose Population (VCGP). Although the VCGP is not designated as a distinct Management Unit by the U.S. Fish and Wildlife Service (2009), this population of Vancouver Canada Geese is resident from Glacier Bay south along the Alaskan coast and associated islands to British Columbia and northern Vancouver Island.

Dusky Canada Goose Population (DCGP). The DCGP is composed solely of Dusky Canada Geese, and it is among the smallest populations of geese in North America. The DCGP breeds almost exclusively on the Copper River delta in Alaska and winters in Washington and Oregon. In 2007 the official population survey of the DCGP was changed from a wintering estimate to a direct count on the breeding grounds. The population has subsequently averaged 8,667 birds from 2007 to 2009, but the lowest number ever recorded was the 6,700 that were counted in 2009.

Cackling Goose Population (CGP). The CGP contains only Cackling Geese (*minima*), which nest exclusively on the Y-K delta and winter in Washington and Oregon. Since 1998 the size of the CGP has been based on the relationship between spring surveys of adults on the Y-K delta, and population estimates during the fall on their winter range. Population estimates averaged 68,581 birds from 2000 to 2009, with a low of 50,187 in 2002 and a high of 84,699 in 2008. This population has increased by 1%/year since 2000. The number of indicated pairs (singles + pairs) tallied on the Y-K delta from 2000 to 2009 averaged 47,379 (Collins and Trost 2009).

Lesser and Taverner's Population (LTP). The LTP consists of Lesser Canada Geese and Taverner's Cackling Geese. The Taverner's Cackling Goose is the largest subspecies of small-bodied Cackling Geese, while the Lesser Canada Goose is the smallest subspecies of large-bodied Canada Geese. These 2 subspecies nest throughout Alaska and winter in Washington, Oregon, and California. In general, Lesser Canada Geese nest in the boreal forest, from upper Cook Inlet north to the edge of the Brooks Range, west to the transition zone of boreal forest/coastal tundra, and east to the Yukon River tributaries in the Yukon. In contrast, Taverner's Cackling Geese nest farther north, on the coastal tundra of the northern and western coasts of Alaska (Johnson et al. 1979).

These two subspecies mix with other Canada/Cackling Geese throughout the year, especially on their wintering areas; hence there are no reliable estimates of the separate populations. The current estimate of the LTP comes from the strata of the WBPHS that are predominantly occupied by these subspecies (Strata 1–6, 8, and 10–12). The information from this survey is used as a population index, which numbered 68,000 geese in 2009 and has increased by 3%/year since 2000 (U.S. Fish and Wildlife Service 2009).

Aleutian Cackling Goose Population (ACGP). Aleutian Cackling Geese are the sole component of this distinct population. The original breeding distribution of the ACGP is not well known, but historical records suggest that most of the larger Aleutian Islands and the Kuril Islands in Asia were once within its original distribution range. These geese winter along the Pacific coast to central California. The ACGP is a conservation success story; the 2009 population estimate, calculated from observations of neck-banded Aleutian Cackling Geese in California, was 79,500, with the population increasing by 10%/year over the previous 10 years (U.S. Fish and Wildlife Service 2009).

Harvest

Canada/Cackling Geese are widely harvested in all 4 flyways, with the substantial numbers of birds taken reflective of the large size of many of the populations. From 1999 to 2008, the harvest of Canada/Cackling Geese in the United States averaged 2,573,562 geese, including both the regular season and special early and late seasons. Most (37.6%) were harvested in the Mississippi Flyway, with 26.2% in the Atlantic, 24.2% in the Central, and 11.9% in the Pacific Flyways. From 1999 to 2008, the harvest of Canada/Cackling Geese in Canada averaged 658,735 birds, with a high of 735,005 in 2008. Over this time period, 23.5% were harvested in Ontario, 22.5% in Saskatchewan, 19.0% in Alberta, 14.4% in Manitoba, 12.4% in Québec, and ≤3.0% in the remaining provinces.

There are also harvest data for some individual Management Units, with a few examples presented here. Although these numbers are confounded because of mixing from other populations, most of the harvest of NAP Canada Geese occurs in the Atlantic provinces of Canada, which averaged 39,800 birds from 2001 to 2005 (Atlantic Flyway Council 2008a). In the United States, Sheaffer

(2005) used band-recovery data to calculate an average harvest of 8,100 adult NAP geese from 2000 to 2004. Harvest estimates for the AP are also confounded by other Canada Goose populations that share the same wintering grounds, but the Canada Goose harvest in the Atlantic Flyway states increased from 113,000 in 1999 to 520,600 in 2005, and harvests in AP areas in Ontario and Québec have grown from 77,000 in 1999 to 164,800 in 2006 (Atlantic Flyway Council 2008b). Sheaffer (2005) estimated that harvest rates for adult AP geese during 2002–4 ranged from 6.3% to 8.6%.

BREEDING BIOLOGY

Behavior

Mating System. Canada/Cackling Geese are monogamous and pair for life, but they will form a new pair bond after the loss of a mate (Jones and Obbard 1970, Raveling 1988). Raveling (1988) found that 93% of 73 neck-collared Giant Canada Goose pairs remained together until death, although 5.5% separated and obtained new mates while their former mates were still alive. MacInnes and Lieff (1968) observed a 1%–2% separation rate for Richardson's Cackling Geese at the McConnell River in the Northwest Territories. Zicus (1984) noted the separation of a pair of Canada Geese and the subsequent formation of 2 new pairs at Crex Meadows in northwestern Wisconsin.

Pair-bond formation is poorly studied, but it probably happens during winter and early spring (Raveling 1969). Pairing can occur among yearlings after they return to their breeding areas, and Sherwood (1965) observed unpaired older geese in Missouri establishing pair bonds in a matter of hours. In Washington, about 20% of territorial pairs of Western Canada Geese did not nest (Ball et al. 1981).

Breeding age varies among the subspecies, but a significant number begin nesting as 2-year-olds, with peak breeding propensity reached as 4- to 5-year-olds (Mowbray et al. 2002). Among the larger subspecies, 46% of the Giant Canada Geese

at Marshy Point in Manitoba first bred as 2-year-olds, with 85% breeding as 3-year-olds (Cooper 1978). Similarly, Craighead and Stockstad (1964) observed that 31% of the wild Western Canada Geese in the Flathead Valley of Montana first bred as 2-year-olds, with 100% breeding as 3-year-olds; 14% of their captive birds did not breed until age 3. The reported percentage of nonbreeders is higher among captive populations, because nonbreeders are easier to detect and 2-year-old captives rarely nest (Cooper 1978). Among the smaller subspecies, MacInnes and Dunn (1988) observed Lesser Canada Geese and Richardson's Cackling Geese first breeding as 2-year-olds (33%–75%). They also stated that a few (2%–4%) males bred as 1-year-olds. In Missouri, Brakhage (1965) found that 42% of the yearling Giant Canada Geese males were paired, versus only 14% of the females. In general, however, breeding does not occur until the second or third year, because family groups remain together during the first winter. The percentage of the population that is breeding also increases with age. Among Interior Canada Geese near Cape Churchill in Manitoba, the breeding effort was 7%, 15%, 40%, 100%, 95%, and 94%, respectively, for females that are 2, 3, 4, 5, 6, and 7 or more years old (Moser and Rusch 1989).

Sex Ratio. The sex ratio (males:females) of prefledging birds is close to 50:50 (Mowbray et al. 2002), but differential mortality among adults can lead to unequal sex ratios in some populations. At Lake Ellesmere in New Zealand, 45.5% of 14,379 banded adults were males (Imber 1968). Imber also calculated sex ratios based on band-recovery data for the Swan Lake flock in Missouri and found some variation occurring with age, as the percentage of males ranged from 45.0%–55.3% for geese 1–2 years old or 9–10 years old.

Site Fidelity and Territory. There is a strong tendency for female Canada/Cackling Geese to return to the same nesting area year after year, and females commonly will reuse the same nest site. In the Flathead Valley of Montana, 45% of the nests were either on the same site or within 7.6 m of it (Geis 1956). On islands in Dowling Lake in southeastern Alberta, at least 28% of 111 clutches were observed in old nests (Vermeer 1970a). Among Cackling Geese, 42.9% of the marked females used the previous year's nest under conditions of heavy snow cover, but only 19.4% reused old nests when snow cover was light (Petersen 1990).

Canada/Cackling Geese are also highly territorial, but the size of the defended site varies greatly with the density of the breeding population, the age of the birds, the nature of the surrounding cover, the chronology of breeding, and some subtle factors that are not readily apparent. On the Y-K delta, the territory size of Cackling Geese ranged from 0.9 to 1,755 m², with the size of the territory varying with the size, shape, and vegetation of the pond associated with the nest site, the aggressiveness of individual males, and the nesting chronology (Mickelson 1975). Sherwood (1966) also reported annual variations in territory size for individual pairs of geese, but he also believed that they had proprietary access to the prior years' territory that other geese recognized and rarely challenged. Vegetative cover and male aggressiveness also may influence territory size, with the most aggressive ganders maintaining the largest territories (Brakhage 1965, Wood 1965, Ewaschuk and Boag 1972).

Courtship Displays. Pair-formation displays of Canada/Cackling Geese occur during winter and spring, as in other species of *Branta*, for which Owen (1980) provided an excellent summary. In the closely related Barnacle Goose (*Branta leucopsis*), pair formation occurs in 4 chronological stages: (1) *mate-searching*, (2) *herding*, (3) *mock-attacks* (including wing displays), and (4) prolonged *triumph-ceremonies* (Black and Owen 1988). Males engage in *mate-searching* by walking through flocks while displaying an array of *forward-neck-extensions*, *head-pumping*, and *erect* postures, accompanied by vocalizations such as those used in the *triumph-ceremony*. *Herding*, or "*shepherding*" (Owen 1980), occur as a male lo-

cates a particular female and attempts to separate her from the flock. *Mock-attacks* occur when a male runs up to 15 m from the female and exhibits *threat* postures toward "imaginary aggressors," although these attacks usually do not elicit a response from other birds. Black and Owen (1988) believed a *mock-attack* signified the initiation of a trial liaison that could lead to a long-term bond.

In Canada/Cackling Geese, males increase their aggressive activities with displays that are largely self-descriptive: *bent-neck*, *forward*, *erect*, and *head-pumping* postures. Male aggression often includes loud vocalizations accompanying *herding*, which leads to a *mock-attack*, and then the *triumph-ceremony*. Raveling (1970) and Akesson and Raveling (1982) provided a good description of the behaviors of breeding Canada Geese, including *aggressive-approach*, *retreat*, *calling*, *head-tossing*, *head-pumping*, and the *triumph-ceremony*; additional information can be found in Blurton Jones (1960), and Klopman (1962, 1968).

The *triumph-ceremony* is a diagnostic and important display by Canada/Cackling Geese and serves to maintain or strengthen pair bonds and assert the dominance of pairs in aggressive interactions with conspecifics and other intruders. This ceremony occurs throughout the year and is performed not only by pairs, but also by family members (Raveling 1970, Akesson and Raveling 1982). It happens most frequently in the spring, however, before the pairs disperse to their breeding territories (McLandress and Raveling 1981a). The posture of the *triumph-ceremony* is similar to the aggressive displays of geese, but here both male and female birds usually participate, orienting themselves laterally instead of head-on, as in aggressive displays. Both then stretch out their necks and call simultaneously. The *triumph-ceremony* is also performed when a pair reunites, especially after an aggressive interaction with a rival.

Precopulatory displays are exhibited by both sexes and are characterized by mutual and synchronized *head-dipping* that somewhat resembles *bathing* movements. As the display progresses, the female lowers herself into the water in preparation for mounting. Both sexes join in a postcopulatory display, with the male lifting his wings and both sexes calling loudly (Collias and Jahn 1959, Johnsgard 1965).

Nesting

Nest Sites. The female leads her mate on a search for a particular nest site, which she selects; it often can be a previously used nest (Collis and Jahn 1959, Hanson and Browning 1959, Cooper 1978). Good visibility over the surrounding terrain is a major factor affecting the nest-site selection of Canada/ Cackling Geese, because a pair of geese is formidable enough to drive off most nest predators if they are detected in time (Samelius and Alisauskas 2001). Hence muskrat houses and islands are commonly used as nest sites. At Flathead Lake in Montana, 90% of 479 nests of Western Canada Geese were located on islands. In southeastern Michigan, muskrat lodges were the most commonly used nest site by Giant Canada Geese (Kaminski and Prince 1977). Of 201 Canada Goose nests in northern California, 38.8% were located on muskrat houses and 37.8% were on islands (Miller and Collins 1953). At Ogden Bay in Utah, Western Canada Geese nesting on islands preferred those that were well spaced, to avoid territorial conflicts; tall; close to open water; and in shorter vegetation, to enhance visibility (Reese et al. 1987).

In the Arctic, Canada Geese commonly nest on hummocks (Raveling and Lumsden 1977) and will use islands when available (Petersen 1990). Richardson's Canada Geese on Victoria Island in Nunavut, in the Canadian High Arctic, often used small islands near ample forage (Jarvis and Bromley 2000). On the Y-K delta, 80% of 808 Cackling Goose (*minima*) nests were located on small islands (Mickelson 1975). Canada Geese also usually nest close to open water (Kaminski and Prince 1977). Miller and Collins (1953) found that 61% of 201 Canada Goose nests in refuges in northern California were within 2.7 m of open water; only 1 nest was >46 m away.

Canada Geese, particularly the Giant and Western subspecies, also use a greater diversity of nest sites than all other species of waterfowl. Western Canada Geese have been found nesting in marshes on mats of bulrushes (*Scirpus* spp.), on the tops of muskrat houses (Martin 1964), and on haystacks (Steel et al. 1957), cliffs (Geis 1956), and dikes (Miller and Collins 1953). Geis (1956) also found Western Canada Goose nests in abandoned heron, egret, and Osprey (*Pandion haliaetus*) nests. In Alberta, Canada Geese have occasionally nested in trees, primarily in old platform nests constructed by Ferruginous Hawks (*Buteo regalis*) and Swainson's Hawks (*B. swainsonii*; Schmutz et al. 1980). Nesting in trees and on cliffs has been commonly reported for both small- and large-bodied Canada/Cackling Goose subspecies (Geis 1956, Lebeda and Ratti 1983, Norment et al. 1999). Giant Canada Geese and Western Canada Geese are especially prone to use artificial nesting platforms, which were widely erected during restoration programs (Brakhage 1965, Szymczak 1975). In the only published study on the nesting ecology of the poorly known Lesser Canada Goose, nest sites on the Tanana River near Fairbanks, Alaska, were located on gravel islands and shore habitats, most commonly among driftwood logs associated with patches of alders (*Alnus* spp.) and willows (*Salix* spp.; Ely et al. 2008).

Unless a previously used nest is selected, new nest scrapes are rounded out on mats of vegetation or on the domes of muskrat (*Ondatra zibethicus*) houses. Only the female constructs the nest, by reaching out from the saucer-shaped nest depression to gather vegetation for the base and rim, with nest improvement continuing as egg laying progresses. Down is plucked from the breast and added to the nest lining at about the time the second, third, or fourth egg is laid, and more down is applied until incubation commences (Brakhage 1965, Vermeer 1970a, Mickelson 1975). Nest dimensions vary from an inside diameter of about 16 cm in the *minima* subspecies of Cackling Geese (Mickelson 1975) to 25 cm in the Giant Canada Goose (Kossack 1950).

Clutch Size and Eggs. Dunn and MacInnes (1987) provide the most comprehensive data summary and analysis of clutch size differences among the various subspecies of Canada/Cackling Geese. Their analyses found that clutch size tended to increase with female body weight, but body weight only explained 18% of the variations in clutch size. Clutch size declined at higher latitudes, but latitude only accounted for a small (16%) amount of the variation in clutch size. A multiple regression model with latitude, longitude, altitude, and body weight explained 33% of the variation in clutch size, but other factors (e.g., weather) that were affected by latitude also correlated with clutch size. In addition, clutch size is smaller in late-nesting females (Rohwer and Eisenhauer 1989) and in younger birds (Aldrich and Raveling 1983).

Clutch size in 34 studies of 11,786 Canada/Cackling Goose nests, including several subspecies, averaged 5.1 eggs (Bellrose 1980). Completed clutches ranged from 1 to 12 eggs, but those containing >8 probably represented eggs laid by more than 1 female; almost 90% contained 4–7 eggs. In general, average clutch size declined with subspecies size: 5.3 ($n = 6,366$) for Western Canada Geese and 5.2 ($n = 2,982$) for Giant Canada Geese, compared with 4.6 ($n = 522$) for Interior Canada Geese and 4.3 ($n = 1,038$) for Cackling Geese (*minima*). At the Hanford Reservation along the Columbia River in southwestern Washington, clutch size averaged 5.5. for 2,688 Western Canada Goose nests monitored from 1953 to 1970, with 90% of all clutches containing between 4 and 7 eggs (Hanson and Eberhardt 1971). Most estimates of clutch size, however, do not account for partial nest predation, where only 1–2 eggs from a clutch might be destroyed. In northern Ontario, partial nest predation on Interior Canada Goose nests averaged 7.6% during a 6-year study (1985–90; Bruggink et al. 1994).

Dunn and MacInnes (1987) reported representative clutch size data from smaller to larger subspecies: 4.7 ($n = 3,504$) for Cackling Geese (*minima*) on the Y-K delta in Alaska, 5.6 ($n = 188$) for Aleutian Cackling Geese on Buldir Island in Alaska, 4.3 ($n = 580$) for Richardson's Cackling Geese at the McConnell River in the Northwest Territories, 5.2 ($n = 1,497$) for Dusky Canada Geese on the Copper River delta in Alaska, 4.1 ($n = 2,355$) for Interior Canada Geese at Cape Churchill in Manitoba, 5.6 ($n = 3,816$) for Western Canada Geese at Hanford Reach in Washington, and 5.6 ($n = 477$) for Giant Canada Geese at Marshy Point in Manitoba. Cline et al. (2004) reported an average clutch size of 5.5 for 2,132 Giant Canada Goose nests in northeastern Illinois (2000–2002).

Canada/Cackling Goose eggs are generally creamy white, smooth, and elliptical/subelliptical, but they often become stained from nest vegetation during incubation (Palmer 1976a, Bellrose 1980). As expected from the differences in size among the subspecies, egg mass ranges from an average of 101 g ($n = 4,417$) in the *minima* subspecies of Cackling Geese (Mowbray et al. 2002) to 144 g ($n = 374$) in Dusky Canada Geese (Bromley and Jarvis 1993), 159 g ($n = 74$) in Interior Canada Geese (Raveling and Lumsden 1977), and 168 g ($n = 1,896$) in Giant Canada Geese (Cooper 1978). Egg size varies from 9.0 × 6.0 cm in Giant Canada Geese (Hanson 1997) to 7.5 × 4.9 cm in the *minima* Cackling Goose subspecies (Mickelson 1973). The general egg-laying rate is about 0.7/day (Alisauskas and Ankney 1992a).

Incubation and Energetic Costs. Only the female incubates the eggs, but the time period is typically shorter for the smaller arctic-nesting subspecies. Incubation for Cackling Geese (*minima*) on the Y-K delta averaged 26 days ($n = 45$), ranging from 24 to 31 days (Mickelson 1975), while Western Canada Geese in southeastern Alberta averaged 26.8 days ($n = 140$), with a range from 25 to 28 days (Vermeer 1970a).

Incubation constancy is related to body size, with the smaller subspecies taking more recesses than larger subspecies, because the former store smaller reserves of energy and thus must leave the nest more often to forage (Thompson and Raveling

1987). The incubation constancy of Cackling Geese (*minima*) averaged 93.6%, with females averaging 3.5 recesses/day of 26 minutes each (92 minutes total; Aldrich 1983). In contrast, incubation constancy was 98.5% in Giant Canada Geese that averaged only 1.4 recesses/day for 15 minutes (20 minutes total; Cooper 1978). Richardson's Cackling Geese averaged 4 recesses during incubation, with breaks ranging from 15 to 30 minutes each; 41% of the recess time was spent foraging (Jarvis and Bromley 2000).

Canada/Cackling Geese, like other geese, use a large percentage of their body reserves for egg laying and as an energy source during incubation. Northern-nesting species transport these reserves to the breeding grounds from spring staging areas, as reported in a landmark study of the *minima* subspecies of Cackling Geese (Raveling 1979a). Females arrived on the Y-K delta in May and June with an average body mass that had increased by 46% (595 g) over weights recorded in April at spring staging areas in California. Lipid reserves grew by a dramatic 209% (360 g), and protein rose by 21% (62 g). Females had gained 1.8 times more body mass, 2.4 times more lipids, and 1.4 times more protein than did males. Females then used these reserves during the reproductive period, becoming emaciated by the end of incubation, having lost 42% (795 g) of their body mass since they arrived 43 days earlier. Their lipid levels were nearly exhausted, decreasing from 532 g at arrival to 33 g; protein was at its lowest (249 g) level in the annual cycle. Lipid reserves in particular provided 85% of the energy for incubation (Raveling 1979b). In contrast, the body mass of males had not decreased significantly by the end of incubation, although their lipids had declined by 86%, probably because males use considerable energy defending the nests from predators and the females from other males.

Female Interior Canada Geese in northern Ontario used 99% of their fat reserves, 10% of their protein reserves, and 24% of their mineral reserves during prenesting, egg laying, and incubation (Gates et al. 1998). Males lost 93% of their stored fat, 10% of their protein, and 16% of their mineral reserves over the same period. Dusky Canada Geese on the Copper River delta in Alaska increased their lipid and protein reserves during the prelaying period, because they experienced a 52% decline in lipids during spring migration (Bromley and Jarvis 1993). Most (93%) of the energy required during egg laying was acquired by feeding on the breeding grounds, but most (76%) of the energy used during incubation came from stored reserves.

Nesting Chronology. The larger subspecies (Giant Canada Geese and Western Canada Geese) breed farther south than other continental species of geese; hence they are among the very first waterfowl to nest in the spring, beginning as early as late February into March. In northwestern Missouri, Giant Canada Geese began nesting between 15 and 20 March each year from 1961 to 1964 (Brakhage 1965). Over a 6-year study (1977–82) at the Eufaula NWR in Alabama/Georgia, the nesting season of resident geese (mostly Giant Canada Geese) usually began in late February and averaged 95 days (Combs et al. 1984).

In contrast, arctic-nesting Canada/Cackling Geese undergo an extensive migration, often arriving at snow- and ice-covered marshes, ponds, and lakes that are just commencing to thaw. In some years, Richardson's Cackling Geese, Lesser Canada Geese, and groups within Interior Canada Geese and Taverner's Cackling Geese may not nest until mid-May or even early June (Dunn and MacInnes 1987). Nest initiation, however, typically occurs within 10 to 13 days of peak arrival, the period of time necessary for the development of egg follicles (Raveling 1978b). Snow cover did not restrict access to nest sites within a few days of the peak arrival of Dusky Canada Geese at the Copper River delta in Alaska, with nest initiation commencing between 22 April and 16 May in 1977–79 (Bromley and Jarvis 1993).

If weather conditions cooperate, nest initiation can be highly synchronized in Canada/Cackling Geese. Dusky Canada Geese on the Copper River delta initiated 48% of all nests over a 4-day period in 1977. Most (89%) Cackling Geese (*minima*) initiated nesting within 10 days of arrival on their Y-K delta breeding grounds (Raveling 1978), and Aleutian Cackling Geese began nesting on Buldir Island in Alaska from 25 to 30 May, with all clutches initiated in 11–16 days in 1975 and 1976 (Byrd and Woolington 1983). At Winisk in northern Ontario, Interior Canada Geese arrived an average of 26 days before the onset of nest initiation (Bruggink et al. 1994). The earliest nests were initiated 1–6 days before peak spring runoff, with a median initiation date of 10 May in "early years" and 16–19 May in "late years." Peak nesting by Interior Canada Geese from the SJBP on Akimiski Island in James Bay occurred within the first 2 weeks of May (Gleason et al. 2003). Ely et al. (2008) noted that Lesser Canada Geese along the Tanana River near Fairbanks, Alaska, initiated nesting from 27 April to 20 May, with peak initiation occurring from 3 to 8 May. Most (89%) Cackling Geese (*minima*) on the Y-K delta initiated nesting within 10 days of their arrival on the breeding grounds (Raveling 1978). Based on data collected on the Y-K delta between 1982 and 2009, however, the average hatching/nesting date for Cackling Geese (*minima*) is occurring 0.33 day earlier each year, or about 9 days earlier in 2009 than in 1982 (Fischer et al. 2009).

Nest Success. The nest success rates of geese are generally higher than those of ducks, because a pair of geese can successfully defend a nest from an array of predators. Nest success rates from 15 studies of Giant Canada Geese averaged 69.7% for 4,272 nests. Bellrose (1980) reported an average of 70.1% from multiple studies and a variety of subspecies involving 14,048 nests. Nest success also can vary annually, however, as occurred among Dusky Canada Geese during a 3-year study (1977–79) on the Copper River delta: 85% in 1977, 54% in 1978, and only 17% in 1979 (Bromley and Jarvis 1993).

Interior Canada Geese on Akimiski Island in Nunavut experienced nest success rates of 65%–89% (Mayfield estimates) from 1993 to 1999 (Leafloor et al. 2000). On islands in the Columbia River, nest success was 70% for 3,824 Western Canada Goose nests monitored from 1950 to 1970 (Hanson and Eberhardt 1971). In a detailed analysis of success among 1,852 Dusky Canada Goose nests found from 1997 to 2000 on the Copper River delta in Alaska, mean nest success ranged from 21% to 31%, but the survival probability of individual nests ranged from 7% to 71% across years and nest-initiation dates (Grand et al. 2006). On the Y-K delta, the nest success of Cackling Geese (*minima*) averaged 79% from 1985 to 2009, ranging from a low of 42% in 1985 to a high of 94% in 1987 and 2000 (Fischer et al. 2009). The nest success of Aleutian Cackling Geese on Buldir Island in the Aleutians averaged 91% for 123 nests located in 1975–76, which is among the highest ever recorded for any Canada/Cackling Goose (Byrd and Woolington 1983). The apparent nest success for 18 Vancouver Canada Goose nests was 55.6% (Lebeda and Ratti 1983).

The nest success of Canada/Cackling Geese is especially high when they nest on islands. In west-central Illinois, the success rate for 242 Giant Canada Goose nests was 92.8% on islands, compared with 42.9% for nests on the shore (Perkins and Klimstra 1984). On the Y-K delta, nest predation was 80% for 100 Cackling Goose (*minima*) nests located along the shore, compared with only 40% for 385 nests on islands (Thompson and Raveling 1987).

The hatching success of eggs is generally high for Canada/Cackling Geese. Among the large subspecies, hatching success was 86%–90% for Western Canada Geese in Montana (Geis 1956), and 67% for Giant Canada Geese in Manitoba (Cooper 1978). For small subspecies, hatching success was 75% for Aleutian Cackling Geese (Byrd and Woolington 1983), and 65% for the *minima* subspecies of Cackling Geese (Mickelson 1975).

Predators of Canada/Cackling Goose nests range from striped skunks (*Mephitis mephitis*), coyotes, American Crows (*Corvus brachyrhyn-*

chos), Black-billed Magpies (*Pica hudsonia*), and black rat snakes (*Elaphe obsoleta obsoleta*) in the southern breeding areas to arctic foxes, red foxes, Common Ravens (*Corvus corax*), jaegers, and Glaucous Gulls (*Larus hyperboreus*) in the northern areas. On the Y-K delta, an experiment that used domestic chicken eggs to simulate Cackling Goose (*minima*) nests found predation rates of 48% for 96 nests, 78% of which were destroyed by avian predators and 22% by mammals (Vacca and Handel 1988). Predation rates were also higher in nests that were not covered with down (and thus exposed to view), compared with those that were down covered (61% vs. 35%). Smith and Hill (1996) reported predation by polar bears (*Thalarctos maritimus*) on Canada Goose nests on Akimiski Island in Nunavut. Bald Eagles and red foxes preyed on the nests of Lesser Canada Geese along the Tanana River near Fairbanks, Alaska (Ely et al. 2008).

Brood Parasitism. Extremely low rates of brood parasitism have been observed in smaller subspecies (Richardson's Cackling Geese and Lesser Canada Geese) that nest in colonies at the McConnell River (MacInnes et al. 1974). During this 7-year study (1965–71), only 4 nests were found to contain >7 eggs, and these were presumed to be parasitized nests.

Renesting. Canada/Cackling Geese rarely attempt to renest if their eggs are lost much more than 1 week after incubation begins (Smith et al. 1999). On the Copper River delta in Alaska in 1999–2000, 72% of the female Dusky Canada Geese attempted renests if the first nest was lost prior to the midlaying period, while only 30% renested when the nests were lost during early incubation (Fondell et al. 2006). An average of 11.9 days passed between the first nest and the second attempt; on average, the second nest was 74.5 m from the original nest and contained an average of 0.9 fewer eggs. On islands along the Columbia River near Hanford, Washington, a substantial number of Western Canada Geese that deserted their nests early in the season renested 1–2 weeks later (Hanson and Eberhardt 1971). Renesting is less common in the High Arctic, because of the shorter growing season.

REARING OF YOUNG

Brood Habitat and Care. In general, brood habitat contains abundant plant food in the form of short grasses and sedges along pond or river shorelines that allow easy access to water, or coastal arctic areas with extensive mudflats and barrens (Mowbray et al. 2002). On the Y-K delta, Cackling Goose (*minima*) broods used 4 major habitat types: (1) shallow ponds and mudflats, (2) sloping pond shorelines with plentiful vegetation, (3) river and slough meadows dominated by grasses and sedges, and (4) relatively dry tidal-meadow flats dominated by sedges (Babcock and Ely 1994).

The key requirement of brood habitat is nutritious, high-protein plants (Sedinger and Raveling 1984), on which goslings will intensively feed (Collias and Jahn 1959). On the Y-K delta, Cackling Goose (*minima*) goslings fed for 50%–70% of daylight hours, or 11–13 hours/day feeding (Sedinger and Raveling 1988). Farther north in the Arctic, Richardson's Cackling Goose goslings averaged 15–16 hr/day feeding (Lieff 1973). In Washington, Western Canada Goose goslings fed for only 7–8 hr/day (Eberhardt et al. 1989a). During a 5-year study (1992–96) on the Y-K delta, Cackling Goose (*minima*) goslings fed 64%–81% of the time, versus 30%–51% for their parents (Fowler and Ely 1997). Adult males devoted more (35%–44%) time to being alert than adult females, although the females still spent 26%–34% of their time in this mode. In contrast, goslings were rarely (1.6%–2.6%) alert.

The female will brood young at the nest for 1–2 days, after which both parents lead the goslings to feeding areas. The new growth of green vegetation that is characteristic of brood habitats can either be virtually adjacent to nest sites or several kilometers away. Near Churchill, Manitoba, Interior Canada Geese traveled 1.5–2.3 km from nest sites to brood-rearing areas (Didiuk and Rusch 1998). Western

Canada Geese on the Columbia River in south-central Washington used an average of 8.8 km of the river and a mean home range of 983 ha to rear their broods (Eberhardt et al. 1989a). The broods used the river as an escape route, as 66.7% of their terrestrial locations were within 5 m of water and only 7% were farther than 50 m. At Crex Meadows in Wisconsin, Zicus (1981a) observed that 86% of 70 Giant Canada Goose broods remained on the same marsh for the entire brood-rearing period. In Ohio, daily travel by Giant Canada Goose broods seldom exceeded 0.4 km (Warhurst et al. 1983).

Brood Amalgamation (Crèches). Brood amalgamation is not uncommon in Canada Geese, especially on densely occupied brood-rearing areas. In Missouri, Brakhage (1965) reported that broods of Giant Canada Geese began to mix after the goslings were 5–7 days old. Such groups (crèches) were accompanied by 2–5 productive pairs and a variable number of nonbreeding females. The largest crèche contained 100 goslings, accompanied by 21 adults and subadults. In Wisconsin, most brood amalgamation occurred when the goslings were 2–3 weeks old (Zicus 1981a). In reporting the effect of brood amalgamation on gosling survival, Warhurst et al. (1983) noted that of 103 Giant Canada Goose goslings joining another brood, 90% survived to fledging.

Development. Growth and development among the various subspecies of Canada/Cackling Geese varies somewhat, but a detailed study of Cackling Geese (*minima*) by Sedinger (1986) revealed the general pattern. Recently hatched goslings (2–3 days old) weighed an average of 61.4–61.8 g and contained a small lipid reserve that would satisfy energy requirements for 1 day. The gizzard was the muscle weighing the most (11%–12% of total body weight) at hatching. As in all arctic geese, goslings grew rapidly after hatching (the Gompertz growth equation), and fledged in 48–49 days at 87%–89% of adult body weight. The leg muscles grew rapidly until the goslings were 30–35 days old, and these muscles were 71%–72% of adult size after 34 days. In contrast, the breast muscles began rapid growth when the goslings reached 15 days old, with the quickest growth in these muscles occurring after 40 days (near fledging). The size of the breast muscle in fledged goslings was between 62% (males) and 74% (females) that of an adult.

Among the larger subspecies, Western Canada Goose goslings fledged when they were between 49 and 56 days old in California (Moffitt 1931), and at 70 days in southcentral Washington (Eberhardt et al. 1989a). Interior Canada Goose goslings on Akimiski Island in James Bay fledged in about 63 days (Hanson 1997).

RECRUITMENT AND SURVIVAL

The survival of goslings to fledging is generally higher than that in ducks, because both parents attend the brood and are capable of defending goslings from predators. Mortality nonetheless can be substantial. Bellrose (1980) summarized an array of studies across several subspecies and noted that survival to fledging ranged from 49% to 70%. Most gosling mortality occurred in the first week after hatching, often during the trek from nesting to brood-rearing areas. An average of 4.7 goslings/nest left 5,959 successful nests, and 4.0 reached flight stage. About 70% of the breeding pairs were successful, resulting in a net production of approximately 2.8 young/pair.

In northern Ontario, the mean estimate of gosling survival for Interior Canada Geese ranged from 35% to 49% over 4 years (1986–88, 1990; Bruggink et al. 1994), while survival of Interior Canada Goose goslings on Akimiski Island in James Bay averaged 59% from 1993 to 1999, ranging from 49% to 67% (Leafloor et al. 2000). Along the Columbia River in Washington, gosling survival among Western Canada Geese was 49%; mortality was highest in the first 2 weeks, compared with the following 8 weeks (Eberhardt et al. 1989). Among the smaller subspecies, survival to fledging by Cackling Goose (*minima*) goslings on the Y-K delta was 88% (Mickelson 1975). Ely (1998) reported survival rates in Cackling Geese (*minima*) ranged from 52% to

78%, excluding total brood loss. Survival of Richardson's Cackling Geese and Lesser Canada Geese at the McConnell River in the Northwest Territories was 86% (MacInnes et al. 1974). Giant Canada Geese nesting in urban habitats in Connecticut had a gosling survival rate of 76% (Conover 1998).

On the Columbia River in southcentral Washington, 15 out of 27 (55.6%) female Western Canada Geese tagged with radiotransmitters successfully fledged at least 1 gosling. The 70-day survival rate for a gosling was 49.1%, but it was significantly lower (65.9%) the first 14 days after hatching versus the last 56 days (74.7%). Most total brood losses (10 out of 12) occurred during the first 14 days.

There is also a strong relationship between age and reproductive performance, as Raveling (1981) documented during in his landmark work on Giant Canada Geese at Marshy Point on Lake Manitoba. Of the 2-year-old females, 25% raised broods averaging 2.3 young, while 31% of 3- to 4-year-olds had broods averaging 2.9 young. In contrast, however, 58% of geese >4 years old raised broods averaging 3.7 young. These older females were only 26% of the breeding population but produced 50% of the young, and older females were also twice as likely to raise at least 1 gosling to flying age than were the younger age classes. Regardless of age, however, individuals successfully raising a brood were more than twice as likely to successfully raise a brood in the subsequent year.

Survival rates of postfledging Canada/Cackling Geese are variable, from lows of 48% for Western Canada Geese (Rexstad 1992) to 90% for Richardson's Cackling Geese (Alisauskas and Lindberg 2002). There are no broad-based geographic patterns, however, probably because postfledging Canada Geese have low rates of natural mortality, although survival rates are strongly influenced by harvests (Rexstad 1992). In the Atlantic Flyway, Hestbeck (1994) found that the years (1984–88) with higher (27.2%) average annual harvest rates corresponded to lower (72.6%) survival rates, as well as to population declines of wintering Canada Geese (from 913,000 to 472,000). Conversely,

populations increased from 612,000 to 967,000 from 1963 to 1974, when lower (17.0%) average harvest rates corresponded to higher (82.0%) survival rates. Similarly, annual survival of neck-banded EPP Interior Canada Geese averaged 65.1% during years when harvests were restrictive (1987–89), compared with 59.5% in more liberal harvest years (1990–93; Sheaffer et al. 2004).

There are, however, a plethora of survival studies of Canada/Cackling Geese from various geographic ranges, a few of which are reported here. In an extensive study of Cackling Geese (*minima*) from 1982 to 1989, annual survival averaged 54.1% for adult males (range = 42.1%–62.2%), 59.5% for adult females (range = 50.4%–69.2%), 49.1% for immature males (range = 16.0%–67.5%), and 50.4% for immature females (range = 38.2%–65.0%; Raveling et al. 1992). The mean annual survival rates of adult females and immatures of both sexes were lower in years when sport hunting was permitted (31.4%–51.6%) than when it was not (57.0%–64.1%). From 1996 to 2000 in Nebraska, the survival rates of Canada Geese (primarily Giant Canada Geese) averaged 68.8% for adults and 61.1% for juveniles (Powell et al. 2004). In Utah, the survival rate of Western Canada Geese from 1965 to 1984 averaged 46.3% for adults and 48.2% for immatures (Rexstad 1992). Eichholz and Sedinger (2007) estimated the annual survival rate of Lesser Canada Geese from 1994 to 1998 as 68% for adults and 49% for immatures. These rates were low in comparison with those for other subspecies, and there may have been an overharvest of Lesser Canada Geese during the period of the authors' study. Annual survival rates of Interior Canada Geese leg-banded at the Horicon NWR from 1974 to 1980 averaged 78.6% for adults and 68.5% for immatures (Samuel et al. 1990). In the Atlantic Flyway, the survival rate of Canada Geese (mixed subspecies and age classes) banded during winter from 1983 to 1986 averaged 77.3% (Hestbeck and Malecki 1989).

The longevity records include a wild female Canada Goose, banded in Ohio and shot in Ontario, that lived 33 years and 3 months. The U.S.

Geological Survey database also reports Canada Goose longevity records of 28 years and 5 months for 1 goose (sex unknown), and 3 others (2 males, 1 sex unknown) that were >24 years old when recovered. A Cackling Goose (*minima*) banded in California and shot in Alaska was 18 years and 4 months old (sex unknown).

FOOD HABITS AND FEEDING ECOLOGY

Canada/Cackling Geese are almost exclusively herbivorous, although there are rare reports of animal foods in their diets, such as the ingestion of alkali flies (*Ephydra hians*) during the fall by Canada Geese at Mono Lake in California (Jehl 2004). In general, however, Canada/Cackling Geese are most often associated with terrestrial grazing of vegetation, although the long necks of the large and medium-sized subspecies allow them to regularly tip up and feed on aquatic vegetation (Owen 1980).

Canada/Cackling Geese have adapted and benefited from the agricultural activities of humans, more so than any other species of waterfowl in North America. Giant Canada Geese and other waterfowl began feeding in wheat and other agricultural fields around Whitewater Lake in southeastern Manitoba almost as soon as farmers commenced agricultural activities there in the early 1880s (Bossenmaier and Marshall 1958). Geese foraging on leafy aquatic vegetation can easily spend 7–8 hr/day foraging for food, especially during winter (Owen 1980), but foraging is greatly reduced when the geese feed on agricultural foods. Wintering Cackling Geese (*minima*), for example, satisfied their daily energy requirements in about 2 hours when feeding on agricultural grains, but increased their foraging time to 8–9 hours (80% of the day) when feeding on poor-quality green vegetation (Raveling 1979b). This adaptation to grain and cereal foods on the wintering grounds has no doubt allowed Canada/Cackling Geese (and other species of geese) to dramatically increase their numbers since the 1960s (Ankney 1996). Grain fields that are attractive to geese generally are large and open, with an undisturbed body of water nearby that is large enough to provide a feeling of security.

The bill shapes of various subspecies can affect feeding ecology: the long narrow bills of large subspecies like the Giant Canada Goose are adapted for shearing vegetation and stripping grass seeds (Hanson 1997), while the smaller bills of Cackling Geese are better at rapid and precise pecking (Owen 1980).

Breeding. During the breeding season, the principal nutritional requirement of Canada/Cackling Geese is a high-protein diet that can satisfy their growth and plumage-development needs. Hence, during the spring and summer, geese primarily select actively growing green leaves that are high in crude protein content but low in fiber, as Sedinger and Raveling (1984) observed in Cackling Geese (*minima*) on the Y-K delta. The birds selected arrowgrass (*Triglochin palustris*) leaves, which contained more protein and less fiber than other plant foods that were available.

Cadieux et al. (2005) detailed the foods used by Interior Canada Geese breeding on the northeastern coast of Hudson Bay near Nunavik, Québec. Their study was especially insightful in contrasting food use between adults and goslings. During the first 4 weeks of the brood-rearing period, adults ($n = 25$ females, 27 males) primarily consumed graminoids (>65%), especially species with the highest nitrogen concentrations (*Carex aquatilis* and *Eriophorum* spp.). Graminoids were also important to the goslings ($n = 59$), but they consumed a greater (68%) amount of other species, perhaps due to their foraging inexperience. As the nitrogen content of graminoid plants declined late in the brood-rearing period, however, adults shifted their diet (>40%) largely to berries (mostly *Empetrum nigrum*). Goslings consumed fewer berries (24%) but maintained their use of nitrogen-rich species (mostly graminoids, at 53%), presumably to complete their growth. On brood-rearing areas near Winisk, Ontario, Interior Canada Geese fed on nitrogen-rich plants, primarily watersedge (*Carex subspathacea*) and alkaligrass (Bruggink et al. 1994). The principal food of adult Vancouver Canada Geese nesting on Admiralty Island in southeastern Alaska was American yellow skunk-cabbage (*Lysichitum americanum*), which formed nearly 24% of their diet (Lebeda and Ratti 1983).

On the Y-K delta in Alaska, Cackling Goose (*minima*) goslings preferred the high nitrogen levels and low fiber content of arrowgrass (*Triglochin* spp.). As they grew, their diet switched to sedges, grasses, and four-leaved marestail (*Hippuris tetraphylla*), followed by crowberries and blueberries (*Vaccinium* spp.) just prior to fledging (Sedinger and Raveling 1984).

Migration and Winter. On migration and in winter, Canada/Cackling Geese switch more of their diet to cultivated grains, as well as browsing on the leaves of clovers and noncultivated grasses, and various parts of other plant species. During 1981 and 1982 in southern Illinois, important foods were wheat (26%), corn (21%), blunt spikerush (*Eleocharis obtusa*; 6%), and smaller quantities of white clover (*Trifolium repens*), bluegrasses (*Poa* spp.), Johnsongrass (*Sorghum halepense*), nodding white smartweed (*Polygonum lapathifolium*), and Pennsylvania smartweed (*Polygonum pensylvanicum*; Havera 1999). In a study of the feeding activities of Interior Canada Geese on the Crab Orchard NWR in southern Illinois, Bell and Klimstra (1970) noted that cornfields attracted 41% of the birds; small grains, 24%; pasture, 22%; soybeans, 9%; and wheat stubble and clover, 4%.

Near Rochester, Minnesota, McLandress and Raveling (1981) found that Giant Canada Geese fed heavily on corn during the winter but shifted to a diverse diet in spring that largely included bluegrass (Kentucky bluegrass; *Poa pratensis*). In general, corn provided carbohydrates and Kentucky bluegrass supplied protein. In southern Illinois and southeastern Missouri, Interior Canada Geese in the MVP fed on agricultural crops and pasture forage during the fall and winter, including corn, soybeans, winter wheat, alfalfa, clover, fescues (*Festuca* spp.), bluegrasses, and sunflower seeds (Gates et al. 2001). Spikerushes (*Eleocharis* spp.) were the only notable plants consumed in wetlands.

Although agricultural crops are unquestionably the mainstay of Canada/Cackling Geese on their migration and wintering grounds, food-habits studies reveal that many geese consume noncultivated foods. On the migratory-staging areas of James Bay, the chief spring foods of Giant Canada Geese were sedge seeds, marestail (*Hippuris vulgaris*), needle spikerush (*Eleocharis acicularis*), and slender arrowgrass (*Triglochin palustris*) bulbs; winter foods included bur-reeds (*Sparganium* spp.), grasses, crowberries (*Empetrum nigrum*), and cranberries (*Vaccinium* spp.; Reed et al. 1996a). In the Maritime Provinces, eelgrass (*Zostera marina*) is a major food of Atlantic Canada Geese, although they also eat saltmarsh grasses (*Puccinellia americana* and *Spartina* spp.; Erskine 2000). Historically, Atlantic Canada Geese wintering in Currituck Sound in North Carolina exclusively used native marsh plants such as pondweeds (*Potamogeton* spp.; 49%), widgeongrass (*Ruppia maritima*; 19%), and naiads (*Najas* spp., 11%), to the exclusion of agricultural grains.

In Alaska, Cackling Geese (*minima*) staging in the spring on Cook Inlet forage extensively in meadows of *Puccinellia* and *Triglochin*, as well as *Carex ramenskii* (Pacific Flyway Council 1999b). During the fall at the Nisutlin River delta in the Yukon, Canada Geese primarily ate pondweed (*Potamogeton richardsonii*) when low water levels allowed access to these aquatic plants. When the water levels were moderate to high, the geese fed on horsetails (*Equisetum palustre, E. fluviatile*), buttercups (*Ranunculus reptans*), and marsh spikerush (*Eleocharis palustris*; Coleman and Boag 1987).

MOLTS AND PLUMAGES

Like all geese, adult Canada/Cackling Geese have only 1 plumage (basic) that is molted annually via the prebasic molt. Palmer (1976a) and Mowbray et al. (2002) should be consulted for further details.

Juvenal plumage is similar to adult plumage, but the feathers are somewhat darker, fewer, shorter, narrower, and more rounded at the tips. This plumage covers goslings by the age of 5–6 weeks, but it is not fully developed until fledging. The tail feathers are notched. Juvenal plumage, except on the wings, is immediately replaced by first basic, and the latter is largely acquired by October (often earlier), although acquisition can continue into the winter. This plumage also activates new feather follicles and thus contains more feathers than juvenal plumage. The timing of the tail molt associated with first basic is variable, being completed between 5 months of age and the following spring.

Yocom and Harris (1965) provided a detailed description of juvenal plumage development, based on their study of 12 captive-reared Western Canada Geese. The first traces of primary and tail feathers appeared between 16 and 25 days, and the first noticeable feathers appeared between 26 and 33 days. Goslings were half feathered between 34 and 40 days and mostly feathered between 41 and 46 days, at which time the first honking sounds were heard. The last down was visible between 47 and 55 days, when the primaries were 75% grown and the white tail band was evident. The goslings were fully feathered but still flightless between 56 and 65 days, although a trace of down remained on the back of the head. The birds flew well at 65–70 days.

Definitive basic is acquired in the summer and includes the wing molt, which involves a flightless period of 29–35 days. Successful breeders begin this molt late in the brood-rearing period, while nonbreeders may molt as much as a month earlier. Parents regain flight at about the same time as their goslings.

Gates et al. (1993) provided a detailed description of the molts and plumages of Interior Canada Geese. The wing molt was most intense in late July, about 35–40 days after the goslings hatched; at that time, primary growth was 50% complete in adults and 40% complete in juveniles. Nearly all of the wing molt and about 50% of the body molt occurred before the geese migrated in late September, and most of the body molt occurred after the geese regained flight. The entire prebasic molt spanned at least 5–6 months for both adults (Jul–Dec) and juveniles (Aug–Dec), with >60% of the geese showing some molting through March; hence the molt in these birds required at least 8 months. Only the nesting season (May–Jun) was entirely free of molt activity. In other words, the birds complete the wing molt and most of the body molt during late summer and early fall, when the weather is favorable, but can still complete their molt in fall and winter if necessary. Such a strategy appears to minimize nutritional conflicts, because the molt process itself is spread out over 8 months (or more) in some individuals.

CONSERVATION AND MANAGEMENT

A primary management dilemma is how to control the numbers of local geese in urban and suburban areas where human conflicts arise, while preserving the diversity and populations of the northern-nesting subspecies of Canada Geese and Cackling Geese. The spectacular success of the reintroduction (and introduction, in some cases) of Giant Canada Geese in the lower 48 states has now led to an unforeseen quandary. These geese have taken full advantage of the human-altered landscape, from cultivated lawns associated with retention ponds in suburban developments, to power plant cooling lakes that offer open water during the winter months, to rural areas with large expanses of cultivated crops that provide additional food sources during fall migration and over the winter. This combination of ample food and large amounts of suitable habitat, in conjunction with a reduced harvest in urban settings and almost no natural predators for adults, are obvious factors in the increase of locally breeding Canada Goose populations.

A portion of the public certainly enjoys Canada Geese in urban areas, but the effects from large numbers inhabiting places such as parks, golf courses, and swimming areas has led to significant conflicts with humans. Grazing geese leave substantial amounts of excrement that can cost up to $60/bird to remove (Allan et al. 1995), and they can also damage landscaping by overgrazing. Harassment by aggressive geese during the breeding season has also created problems for humans, and geese grazing on or near airports can cause serious flight-safety issues. In 1995, Canada Geese caused the crash of a U.S. Air Force Boeing 707 jet departing Elmendorf Air Force Base in Alaska, killing the entire 24-person crew aboard the $184 million aircraft. On 15 January 2009, Canada Geese caused the failure of both engines on a US Airways Airbus A320 minutes after takeoff from LaGuardia Airport in New York, forcing a "miracle" landing on the Hudson River, where, fortunately, all 155 people on board were rescued.

To complicate matters further, harvest management of different populations that overwinter in the same geographic area is difficult, because it is generally not possible to distinguishing local geese from arctic- and subarctic-nesting birds during the harvest. To overcome these problems, many states have established early and/or late seasons that target temperate breeders, as these are periods when arctic and subarctic goose populations are either not present or occur only in small numbers. In the Atlantic Flyway, Canada Goose seasons in September are timed to target resident geese, before the arrival of northern-nesting migrants (Malecki et al. 2001).

Nonetheless, many resident geese are largely invulnerable if they remain in urban areas where hunting is not permitted. The relocation of nuisance birds from nonhunting areas with high densities of resident geese to locales where harvest is possible has met with some success. In New York, resident Canada Geese were translocated from urban habitats to state-owned wildlife management areas where hunting was permitted. Holevinski et al. (2006) reported that 23.8% of the adults and 22.9% of the juveniles were subse-

quently harvested. Only 8.8% of the translocated adults returned to the original capture site. In the lower Fraser River valley in British Columbia, the majority of Canada Goose band returns were from birds moved there from urban areas (Smith 2000). Egg addling can be effective not only in reducing reproduction, but also in reducing survival rates of adults by making them unsuccessful breeders, which can sometimes induce a molt migration and subsequently expose them to harvest during fall migration. Luukkonen et al. (2008) found that the survival rate of molt migrants from southern Michigan was low (60%), because they were exposed to hunting, compared with a 93% survival rate for successful breeders that remained in areas where hunting was restricted. Encouraging molt migrations of Giant Canada Geese may not be an optimal solution, however, because these migrants subsequently compete for food with Interior Canada Geese on northern breeding grounds (Abraham et al. 1999a).

Harvests that target specific components of the population, particularly nesting females, can be especially effective at controlling local Giant Canada Goose numbers, as Coluccy et al. (2004) demonstrated in Missouri. Their sensitivity models indicated that adult survival had the largest effect on population growth (−37.5%), followed by subadult survival (−28.6%). In contrast, nest success was only −19.0%. Hence, given their estimated statewide population of 64,222 Giant Canada Geese, about 11,699 nests would need to be destroyed annually over a 10-year period to stabilize the population. In contrast, the annual removal of about 10% of the adults and 26% of the juveniles would achieve the same population objective. Lastly, removal of 32% of the nesting females (49,333) would stabilize the population.

In *Managing Canada Geese in Urban Environments*, Smith et al. (1999) provide a comprehensive review of both lethal and nonlethal techniques to manage overabundant geese. Such methods include the use of chemical repellents, hazing (harassment through noisemakers, dogs, spotlights, lasers, etc.), the removal of nesting materials, the use of reproductive inhibitors, egg destruction through oiling or shaking, and various approaches to the harvest. In general, labor-intensive methods, such as the removal or destruction of nests or eggs, are a costly means of control and often do not reduce the overall goose population. Cooper and Keefe (1997) estimated the cost of nest and egg manipulation at $6.38/egg. It should also be noted that all subspecies of Canada Geese are federally protected by the Migratory Bird Act, so a permit is necessary for most control activities, especially disturbances of nesting geese.

Numerous landscape changes can be made to alter the habitat and behavior of resident Canada Geese and thus reduce their use of urban areas. These approaches include eliminating favorable nesting habitat, such as straight shorelines, islands, and peninsulas in man-made wetlands; removing goose-nesting structures; constructing new ball-fields well away from water bodies; creating barriers with fences, grid wires, shrubs, tall trees, or rocks to prohibit access to water bodies; decreasing the number and quality of grazing areas by reducing the amount of mowing; and planting less desirable grasses and legumes (Smith et al. 1999).

Specific to the Pacific Flyway, harvest management to protect the small population of Dusky Canada Goose is complicated by the fact that they now winter in association with >250,000 Canada/Cackling Geese whose large numbers are causing crop depredation issues in the lower Columbia River Valley in Washington and Oregon and the Willamette Valley in Oregon. Currently, Canada Goose hunting is allowed only as long as the harvest of Dusky Canada Geese is less than an annual quota of 250 birds, which has dramatically increased the survival of adult birds in this subspecies (Bromley and Rothe 2003). This restrictive approach is clearly working, because Dusky Canada Geese made up only about 1% of the overall Canada Goose harvest in the Washington/Oregon wintering area since 1996 (Pacific Flyway Council 2008).

In general, Canada/Cackling Goose populations present problems of overabundance, which in many respects is a tribute to the conservation efforts directed toward these birds. The status of Aleutian Cackling Geese changed from endangered to now being widely harvested in the Pacific Flyway. The Giant Canada Goose, a subspecies that was considered extinct by the 1950s, exploded to a 2009 population exceeding nearly 2 million geese in the Mississippi Flyway, An "embarrassment of riches" indeed, but perhaps no apology is necessary from the waterfowl community.

Muscovy Duck

Cairina moschata (Linnaeus 1758)

The Muscovy Duck is poorly known as a wild species, found only in lowland habitats of the New World tropics, but its range has recently (1980s) extended into the lower Rio Grande valley of Texas. The English-language common name "Muscovy" may well refer to the transport of this species to England and France in the sixteenth century by the Muscovy Company. They are large ducks: adult males can reach 4,000 g and are twice as big as females. Adult males have jet-black plumage with distinct white upper- and underwing patches. Females are similar but duller, and distinctly smaller when both sexes are observed in a group. Their mating system may be polygynous, although wild birds are often observed in pairs. They nest in cavities but also take readily to nest boxes. Clutch sizes for normal nests range from 8 to 15 eggs, but brood parasitism occurs among females. Muscovy Ducks are generalist feeders, using both plant and animal foods, and they will commonly eat agricultural foods when available, especially corn. Their population status is poorly known, but their numbers have probably declined, due to wetland habitat loss, timber harvests, egg collecting, and overhunting. Many other aspects of their life history also remain unknown or poorly known, and a detailed investigation of the biology and ecology of Muscovy Ducks is very warranted.

IDENTIFICATION

At a Glance. Muscovy Ducks are large blackish ducks, with a greenish-purple gloss on the upperparts. In flight, the large white upper- and underwing patches contrast markedly with their otherwise black plumage. In the United States, their population is restricted to the lower Rio Grande valley in Texas.

Adult Males. Adult males are the largest ducks in North America. Their overall plumage is blackish, with a metallic greenish-purple cast on the upperparts. All adults also have bright white upperwing and underwing coverts on the leading edge of the wings, which are usually visible when the birds are at rest. The amount of white is probably proportional to age (Leopold 1959). A short crest of feathers atop the head and the back of the neck can be erected during courtship and aggression. Their general body shape, broad wings, and large tail give them a chunky appearance. The unusually short legs and the feet are black, with long sharp claws that facilitate perching. The bill is black at the base, and has fleshy white mottling, a blackish band extending across the middle, and a pink spot near the tip. Males also have a blackish to dark red knob at the base of the bill, and bare facial skin between the bill and the eyes that is adorned with red caruncles of varying sizes and numbers. The irises are yellowish brown.

Adult Females. Females are only about half the size of the males. Their plumage is similar to that of the males, but duller. They lack the crest on the head, the knob on the bill, and the bare facial skin. The legs and the feet are dull blackish. The end of the bill is pale pinkish. Occasionally females have no white on the wings (Leopold 1959).

Juveniles. The overall body coloration of young Muscovy Ducks is similar to that of the adults, but it is a dull brownish black. Except for the green tertials and secondaries, the wings are entirely or mostly black (a small white spot on the secondary coverts may be present). Adult plumage is not attained until 2 years of age, although the white wing patches are usually acquired during the first winter. The bill is gray.

Ducklings. Ducklings are boldly patterned with yellow and dark brown, and have prominent dorsal spots and wing patches (Nelson 1993). The forehead, the throat, and the sides of the head are yellow, with a conspicuous supraorbital stripe. The feet are a dull yellow, patterned with grayish brown.

Voice. Muscovy Ducks make few vocalizations. Although seldom heard, females give a quiet *quack* or croak when frightened and produce soft but shrill calls to communicate with their ducklings. Males may hiss and give repetitious puffs during courtship (Fischer et al. 1982).

Similar Species. Domesticated Muscovy Ducks may be all white; black splotched with white; or even white with traces of buff, blue, or silver. Domestic forms also have a pronounced red facial area, with protuberances that often become quite large and grotesque. They are frequently much larger than wild birds (Donkin 1989).

Weights and Measurements. Few data are available, but body weights of adult males in Venezuela averaged 3,077 g ($n = 8$), and females, 1,689 g ($n = 12$; Gómez-Dallmeier and Cringan 1989). Palmer (1976b) found males that weighed up to 2,500 g and were about twice as large as females. Bill length averaged 67.9 mm for males, and their wing chord, 385 mm ($n = 9$). Female bill length averaged 51.4 mm, and their wing chord, 307 mm ($n = 4$). Leopold (1972) reported the weights of birds in Mexico as 1,990–4,000 g for males and 1,100–1,470 g for females, but he gave no sample sizes.

DISTRIBUTION

Muscovy Ducks are found exclusively in the tropical and subtropical regions of the Western Hemisphere, and Donkin (1989) provides an exhaustive description of their range. They are permanent residents in lowland areas from northern Mexico to Argentina. Leopold (1959) gives the northern range of Muscovy Ducks in Mexico as the central area in the state of Nuevo León on the east coast, and central Sinaloa on the west coast. Markum and Baldassarre (1989) monitored nest box use by Muscovy Ducks near San Fernando, in the state of Tamaulipas, about 100 km south of Brownsville, Texas. Brush (2009) noted breeding records from 2003 to 2006 that were farther south in Tamaulipas, near Ciudad Victoria.

In the United States, Muscovy Ducks are a recent and rare breeder, appearing only along the Rio Grande in Texas. The first recorded wild nest was found there in 1984, in a nest box erected for Black-bellied Whistling-Ducks that was 1.6 km west of Bentsen-Rio Grande Valley State Park (Brush and Eitniear 2002). Texas records state that wild Muscovy Ducks are now occasional residents in Hidalgo, Starr, and Zapata Counties, and they occur less frequently in Webb County. In Texas, Muscovy Ducks are found in heavily forested areas of the river away from human settlements; these wild birds are extremely wary, unlike the domesticated varieties.

Muscovy Ducks are most abundant, however, in lowland tropical forests south of the Tropic

Year-round Distribution ■

the Wind God, Ehecatl. The bird portrayed in the mythical bird serpent motif of the Aztec god-hero Quetzalcoatl, the Plumed Serpent, is the Muscovy Duck, and its prestige among native peoples even today is reflected in their common name for this species, "Royal Duck" (Whitley 1973). According to Phillips (1922), Muscovy Ducks were also popular in Africa and the Pacific islands, but less so in Europe in the 1900s than in centuries past (the species had been imported there as early as 1550). Apparently the Muscovy Duck is the duck that Columbus was referring to as having been domesticated by the native West Indians, who had "ducks as large as geese" (Phillips 1922). The English-language name "Muscovy" may well refer to the transport of this species to England and France by the Muscovy Company, which was incorporated in London in 1555 (Hoffmann 2005).

As a domestic bird, Muscovy Ducks are extensively bred on farms in the warm regions of the world. Within the United States, feral populations have become established in Texas, Florida, and other locations along the Gulf Coast, as well as on ponds in urban parks throughout many other states. The *Florida Breeding Bird Atlas* shows them being widely distributed in cities and towns throughout the state, as a result of feral domestication. An average of 3,340 Muscovy Ducky have been recorded on Florida Christmas Bird Counts conducted from 2004/5 to 2008/9. A few Muscovy Ducks in that state have hybridized with feral Mallards (Pranty et al. 2006).

of Cancer and north of the Tropic of Capricorn. In Mexico, Muscovy Ducks are found on coastal slopes and tropical lowlands, extending from the states of Sinaloa to Chiapas on the west coast, and from central Nuevo León and Tamaulipas south to the Yucatán Peninsula on the east coast (Howell and Webb 1995). A large number have been reported within the lower Tempisque basin of Costa Rica during the dry season (Stiles and Skutch 1989). In South America, Muscovy Ducks are considered to be residents in an area extending from Columbia and Venezuela (Rodner et al. 2000) to Buenos Aires, Tucuman, and Santa Fe in Argentina (Barnett and Pearman 2001). South of Ecuador, the range of Muscovy Ducks lies east of the Andes Mountains (Hoffmann 2005).

The Muscovy Duck is among the oldest domesticated species of fowl in the world. Early Spanish explorers noted that prior to their own arrival, native peoples in Peru and Paraguay had already domesticated this duck. In pre-Columbian times, Aztec rulers often adorned themselves in rich cloaks made from the feathers of Muscovy Ducks, because the Muscovy was considered the "nahual"—the magical totem animal, or alter ego, of

MIGRATION BEHAVIOR

The warm climate and seasonal stability of their forest habitats probably has resulted in a lack of programmed migration behavior for Muscovy Ducks. When inland wetlands are impacted by drought, however, they are known to move to coastal lagoons and swamps.

Molt Migration. Muscovy Ducks do not undergo a molt migration.

HABITAT

Muscovy Ducks inhabit wooded lowland rain-forests near swamps, slow-moving streams, and coastal mangrove thickets (Bolen 1983, Hoffman 2005). They are adept at flying through the forest to reach feeding areas in the early morning and evening, after which they perch in a favorite tree (Leopold 1959). They also roost in trees at night. Muscovy Ducks are cavity nesters and thus require large mature trees, due to the size of the females. Nest boxes built to accommodate Muscovy Ducks have an entrance-hole diameter of 21 cm (Wood-yard and Bolen 1984), in comparison with 8–10 cm for Wood Ducks (Bellrose 1980).

POPULATION STATUS

Very little is known about the population status of Muscovy Ducks, because they are not surveyed anywhere within their range. Surveys would also be difficult to conduct, because these ducks prefer forested habitats. Nonetheless, Wetlands International (2002) suggested that their population is declining, with numbers ranging anywhere from 100,000 to 1,000,000 birds. In Mexico, Saunders and Saunders (1981) stated that the clearing of bottomland forests and overhunting were responsible for a drastic decline in Muscovy Duck numbers in many areas. Leopold (1959) observed that a decrease in their populations in Mexico occurred even where bottomland forests were intact, and he considered the cause of their decline to be excessive hunting. Bolen (1983) reported that overhunting and the use of eggs by local people continue to take a toll on the wild populations in Mexico and Central America. In Belize, Muscovy Duck populations have declined in areas with no management action (Jones 2003).

In the early 1980s, Ducks Unlimited of Mexico erected >4,000 nest boxes for Muscovy Ducks in northern Mexico (Markum and Baldassarre 1989), which apparently promoted population growth that facilitated a range expansion into the lower Rio Grande valley of Texas (Cruz-Nieto

1991, Brush and Eitniear 2002). Currently, Muscovy Ducks can be observed in remote parts of the Rio Grande valley, where their numbers have slowly increased since 1984 (Lockwood and Freeman 2004). The highest (22–28) numbers of wild birds have been found along the section of the Rio Grande in Starr County, between Falcon Dam and Roma (Brush and Eitniear 2002), and the single largest flock of Muscovy Ducks recently seen in Texas consisted of 28 birds observed in August 1998 (Brush 2005).

Harvest

Specific harvest estimates are not available, but overharvest appears to be a key factor in the population decline of Muscovy Ducks. In addition, the loss of rainforest habitat in many areas has eliminated the large trees that are necessary for cavity nesting.

BREEDING BIOLOGY

Behavior

Mating System. Despite claims that no pair bonds exist (Johnsgard 1978), which implies polygyny, captive birds (Sibley 1967) and wild birds in Mexico (Rojas 1954) usually are observed in pairs during the breeding season. Captive birds breed when they are 1 year old, but the breeding age in the wild is unknown.

Sex Ratio. Sex ratios for Muscovy Ducks are unknown.

Site Fidelity and Territory. There is no information on the fidelity of Muscovy Ducks to their breeding sites, but females are probably faithful to nesting cavities and nest boxes. Muscovy Ducks are non-territorial in the breeding season but otherwise are aggressive toward each other (Hoffmann 2005).

Courtship Displays. Displays are not well studied in the wild, but the supposedly promiscuous breeding behavior of Muscovy Ducks results in a low level of preliminary courtship and pair-formation displays. A receptive female attracts the most aggres-

Muscovy Duck pair; note the large knob at the base of the bill on the male (*foreground*).
Ian Gereg

sive male, who has achieved dominance in pitched combat with other males frequenting a pond or other small water area (Stai 1999). The courtship displays of Muscovy Ducks are relatively simple in comparison with those of other ducks. Obvious male displays are *crest-raising, tail-shaking, back and-forth-head-movements,* and *partial-elevation-of-the-wings-over-the-back,* while making soft hisses (Johnsgard 1978, Fischer et al. 1982). Females generally lack any social displays, and they may not even have an *inciting* display. Copulation is on the water. It may occur forcefully by the much larger male, or be solicited by the female as she assumes a *prone* position; the female *bathes* after copulation. Males do not defend the nest or care for the young. More study is needed, however, on the courtship behavior of wild Muscovy Ducks.

Males are highly aggressive toward each other, as Fischer et al. (1982) observed at Palo Verde National Wildlife Refuge in Costa Rica. Most (90%) encounters involved the male displays mentioned above, but 10% were dramatically more aggressive. In those instances, the males faced each other and then exhibited a synchronous *vertical-hovering-flight* of 1–3 m, during which time they appeared to strike each other with their wings. Such flights were repeated 2–6 times, until 1 of the 2 males submitted and left the area.

Nesting

Nest Sites. Muscovy Ducks prefer tropical lakes, lagoons, mangrove swamps, marshes, and slow-moving wooded streams for nesting, but Markum and Baldassarre (1989) found that they used nest boxes erected in open habitats without canopy cover. These ducks have nested in coastal brackish wetlands, but they prefer freshwater wetlands. Nests are usually constructed in tree cavities and hollows, but Muscovy Ducks may occasionally nest on the ground, in dense vegetation near water (Madge and Burn 1988). The walls of a limestone cave provided an unusual nest site for Muscovy Ducks in northeastern Mexico (Eitniear et al. 1998). Phillips (1922) reported nest sites located 3–20 m above the ground, frequently in cavities but also commonly between palm leaves. Bolen (1983) noted

Muscovy Duck female.
Ian Gereg

cavity nests located at heights between 15 and 20 m. Markum and Baldassarre (1989) observed 13 nests in boxes 0.8–3.5 m above the ground, which represented the range of heights available among 407 nest boxes monitored in Tamaulipas.

This large duck requires a fairly big cavity to accommodate its size, and the loss of primary forests has reduced the availability of natural nest cavities. Fortunately, Muscovy Ducks will use nest boxes placed on suitable wetlands. The dimensions of the nest boxes monitored (but not constructed) by Markum and Baldassarre (1989) were 42 cm wide and deep, with sides tapering from 62 cm at the front to 55 cm in the rear (Woodyard and Bolen 1984). A nest base of 2–5 cm of sawdust and/or dried grass was added to each box.

Clutch Size and Eggs. Johnsgard (1978) noted that clutch size was between 8 and 15 eggs, but not many clutches have been reported from wild birds. Leopold (1959) recorded clutches of 8–14 eggs, with an average of 10. Woodyard and Bolen (1984) located 9 eggs in each of 2 normal nests. Markum

and Baldassarre (1989) found clutches of 9–15 eggs (average = 12.6 eggs) in 9 normal nests.

Muscovy Duck eggs are glossy white, although some have a slight green or buff sheen (Markum and Baldassarre 1989). The dimensions of 110 eggs from 10 nests in the state of Tamaulipas averaged 62.3 × 44.6 mm, which was smaller than the 67 × 44 mm average reported from Mexico by Leopold (1959). The average weight of 31 eggs from 4 nests was 66.4 g, with a normal laying rate of 1 egg/day (Markum and Baldassarre 1989).

Incubation and Energetic Costs. Under domestic conditions, Phillips (1922) stated that the incubation period was 35 days. In the wild, however, Markum and Baldassarre (1989) reported an incubation period of 30–31 days for 7 successful females. Nothing is known about incubation energetic costs or nest attendance by the female.

Nesting Chronology. Muscovy Ducks, like other resident waterfowl in the semitropical areas of Mexico and farther south, initiate nests over a broad period of time. The rainy season may trigger egg production, but Muscovy Ducks reportedly breed during any month of the year (Hoffmann 2005). Cruz-Nieto (1991) stated that nesting in Mexico was from May to September, although Eitniear et al. (1998) noted nesting in December. Leopold (1959) observed indications of breeding in July and August in Mexico, and he collected a female with an enlarged ovary as late as 3 October. In Tamaulipas, the nesting season lasted about 135 days, from 24 April to 6 September (Markum and Baldassarre 1989). Elsewhere, Muscovy nests have been reported from February to May in the Caribbean islands, March in Peru, June in Panama, July and November in Venezuela, and November in Bolivia (Madge and Burn 1988).

Nest Success. Little is known about the success of and predation on wild Muscovy Duck nests. In nest boxes, Markum and Baldassarre (1989) reported that 10 out of 13 nests (77%) hatched, similar to the 75% for 4 nests studied by Woodyard and Bolen (1984). The 3 unsuccessful nests were deserted: 2 probably because of observer disturbance, and 1 because of flooding. Ducklings are equipped with sharp claws and a hooked bill that help them climb out of nest cavities (Hoffmann 2005).

Hatchability was 68%: 73% in 7 normal nests, and 59% in 3 parasitized nests (Markum and Baldassarre 1989). Raccoons (*Procyon lotor*) are probably the chief predators of Muscovy Duck females and eggs, as they are for other cavity-nesting ducks. Eggs are also eaten by small mammals and snakes, while ducklings (and occasionally adults) are taken by tiger-fish (*Hoplias malabaricus*), American alligators (*Alligator mississippiensis*), and domestic cats (*Felis catus*; Phillips 1922).

Brood Parasitism. As with other cavity-nesting waterfowl, brood parasitism has been suspected. It was first documented in Muscovy Ducks using nest boxes in Tamaulipas in Mexico (Markum and Baldassarre 1989), where 4 parasitized nests averaged 17.7 eggs (range = 15–21 eggs). Each of the 4 nests was successfully incubated by a Muscovy Duck, although 2 of the nests also contained 2 eggs each from Black-bellied Whistling-Ducks. Woodyard and Bolen (1984) also found eggs of Black-bellied Whistling-Ducks in 2 out of 4 Muscovy Duck nests.

Renesting. Renesting presumably occurs among Muscovy Ducks, as it does in other waterfowl, but it has only been documented once. In Tamaulipas, 1 of the 2 marked females that deserted their nests was subsequently observed to successfully incubate a second clutch (Baldassarre and Bolen 1989).

REARING OF YOUNG

Brood Habitat and Brood Care. Only the female rears the brood, but otherwise virtually nothing is known about brood habitat or the brood-rearing period. Presumably, Muscovy Ducks use heavily vegetated emergent marshes, especially when the ducklings are young. Hoffmann (2005) reported strong bonds among brood members after fledging, even after the female has departed.

Brood Amalgamation (Crèches). No information is available.

Development. Almost nothing is known about the growth and development of wild Muscovy Ducks. Woodyard and Bolen (1984), who hand-reared 24 ducklings from 3 clutches laid by semidomesticated Muscovy Ducks in Mexico, reported on plumage development. They noted that the first feathers appeared on days 28–46 (midpoint = 35) and the last down disappeared on days 55–67 (midpoint = 63); birds were "feathered-flightless" at 68 days or more, which indicates a long period until fledging. Baicich and Harrison (1997) stated that ducklings were capable of flight at 70 days. In contrast, Hoffmann (2005) noted that their wing feathers grow slowly, with fledging at 3.5 months (about 105 days), which increases the exposure of the young to predation.

RECRUITMENT AND SURVIVAL

In Tamaulipas, 96 Muscovy Ducks were produced from 10 successful nests in nest boxes, but no other data were collected on duckling survival (Markum and Baldassarre 1989). Donkin (1989) simply noted that mortality among ducklings was "apparently high," reporting predators to be caimans (Alligatoridae), crocodiles (Crocodylidae), and some fish. Large wading birds are also likely to be duckling predators. There are no data on adult longevity or survival.

FOOD HABITS AND FEEDING ECOLOGY

Muscovy Ducks are versatile generalist feeders that use both open marshes and grain fields. They readily use agricultural fields—especially corn, but also rice—and in wetlands they will eat an array of seeds, tubers, and invertebrates (Stiles and Skutch 1989). Foraging is accomplished by tipping in shallow water, dabbling at the water's surface, and (less frequently) grazing along grassy shorelines (Madge and Burn 1988). Muscovy Ducks obtain food by various means: straining organic matter, grazing, gleaning, and grasping, the latter a useful technique when feeding on small crabs (Goodman and Fisher 1962).

In the state of Veracruz in Mexico, the food volume in 15 adult Muscovy Ducks collected by Woodyard and Bolen (1984) consisted mainly of the seeds of waterlilies (*Nymphaea* sp.; 51.7%) and mangrove (*Avicennia nitida*; 14.1%). Animal matter formed 34% of the food volume: 24.5% soldier fly (Stratiomyidae) larvae, and 7.4% Syrphidae, and the remainder from some 15 other inverte-brate taxa that were also identified, with most only present in trace amounts. In contrast, the stomach contents of 7 adults collected in Tamaulipas contained 100% corn by volume, with trace amounts of animal matter. Leopold (1959) observed Muscovy Ducks feeding at daybreak in cornfields along the lower Río Grande de Santiago in the state of Nayarit, and in sesame as well as cornfields in eastern San Luis Potosí. Leopold (1959) also cited reports that Muscovy Ducks eat termites, breaking open the mounds with their bills.

MOLTS AND PLUMAGES

Aside from Woodyard and Bolen's (1989) descriptions of early plumage development until near fledging, nothing is known about molts and plumages in Muscovy Ducks.

CONSERVATION AND MANAGEMENT

The Muscovy Duck is among the world's least studied waterfowl, despite its widespread distribution in the New World. A detailed ecological study would yield new information on virtually all aspects of its life history. Where Muscovy Ducks can be protected from hunting, the use of nest boxes is clearly a viable management technique that has led to increased numbers and range expansion (Markum and Baldassarre 1989, Brush and Eitniear 2002). Protection of lowland tropical wetlands is also essential throughout the range of Muscovy Ducks.

Wood Duck
Aix sponsa (Linnaeus 1758)

Left, hen; *right*, drake

The Wood Duck is among the most recognized species of ducks. It is indigenous to North America, except for a small population in Cuba. The brightly colored male is unmistakable. Near extinction by the early 1900s, the recovery of this species ranks among the most successful wildlife conservation stories anywhere. Breeding populations now number about 3 million birds in 2 main groups. The vast majority range through the Atlantic coastal provinces and states from Newfoundland to Florida, and then westward across southern Canada south to the Great Plains (where they have expanded their range in recent decades) and the Gulf Coast. A much smaller population occurs in the Pacific Northwest and California. A cavity-nesting species, these ducks readily use nest boxes, which provide an important management tool. They prefer wooded and scrub-shrub wetland habitats throughout their annual cycle but also readily occupy emergent wetlands. Wood Ducks are primarily vegetarians, but females use large amounts of invertebrates during the breeding season to meet the demands of egg production. Acorns are a preferred fall and winter food, for which they will forage in upland areas. Wood Ducks are among the most well-studied species of waterfowl in the world, and *Ecology and Management of the Wood Duck*, by Bellrose and Holm (1994), is the authoritative treatise on them. The species name *sponsa* refers to "bridal dress" in the male.

Wood Duck. *Left*, hen; *right*, drake

IDENTIFICATION

At a Glance. Adult males are among the most readily recognized ducks in North America. The large crest, the white throat and chin patch, the red base of the bill, and the red irises are all distinctive. Birds in flight are characterized by broad wings, a large head (pointed down), and a long rectangular tail. The females are grayish, but the head shows a readily visible white eye ring with a trailing teardrop. Males and females are readily differentiated by their wings alone: the tertials are purple or black in males but bronze in females. The steel-blue secondaries are edged with single white bars in the males, but the white bars are broader and shaped like teardrops in the females. Wood Ducks bob their heads up and down in flight, like no other duck.

Adult Males. The flamboyant colors of the males are readily apparent in birds at rest. The large crested head shows many hues of purple and green, with 2 white parallel lines extending from the base of the bill and from the back of the eye to the rear of the crest. The white of the chin and the throat sweeps upward in U-like prongs onto the sides of the head. The red irises and base of the bill complete the vivid head color. The burgundy chest, flecked with white, is separated from the bronze sides by "fingers" of black and white. The bronze sides are vermiculated with fine black lines. The glossy purplish black of the back and the tail is in marked contrast to the white of the breast and the belly. The legs and the feet are a dull straw yellow. At close range, the short bill has a black tip and

ridge, white on the sides, and a red base bordered with a narrow yellow line.

Adult Females. The adult female is among the most attractive of the female puddle ducks. The pronounced white eye ring with a trailing teardrop on an otherwise sooty-gray, slightly crested head, and the white chin and throat are all discernible at considerable distances. At close range, the brownish crest is glossed with green. The chest, the sides, and the flanks are gray brown, with disconnected lines of white dashes. The belly is white. The back is an olive brown, with a shimmer of iridescent green; the undertail coverts are white, like the belly. The bill is a dark blue gray. The irises are brown black.

Juveniles. Birds in juvenal plumage resemble adult females, but their bellies are a streaked and mottled brown rather than white. Juvenile males display the 2 white prongs extending upward on the sides of the head. Young females may or may not show white eye rings, depending on their age; these rings are less apparent than in adults.

Ducklings. Ducklings show a distinct, well-marked stripe from the eye to the occiput. Otherwise their down has a dark pattern that contrasts with a yellowish base color (Nelson 1993).

Voice. The loud, squealing *wee-e-e-ek*, *wee-e-e-ek* alarm call by the female is often the first indication of the presence of these birds in their preferred forested wetland habitat. No other duck has a more distinctive and easily remembered call as the female Wood Duck. In addition, each sex uses as many as 10 different calls in response to threats and in feeding, nesting, or courtship activities. The

Wood Duck pair; note the droop in the abdomen of the female (*right*), due to an egg in the oviduct. *GaryKramer.net*

Male Wood Duck, performing a comfort movement. *RyanAskren.com*

male's call is a *twee, twee* uttered so softly as to be rarely heard (Bellrose and Holm 1994).

Similar Species. Females in flight can be confused with female American Wigeon, because of their similar size and white bellies. Wood Ducks, however, have tails that are long and square, and American Wigeon show white on the upperwing coverts.

Weights and Measurements. Bellrose and Holm (1994) provided extensive information on weights and measurements of Wood Ducks. In Mason County in Illinois (1950–68), body weights of

273

Closeup of a male
Wood Duck feeding
in duckweed.
RyanAskren.com

adults during spring (7 Mar–1 May) averaged 693 g for males ($n = 171$) and 651 g for females ($n = 149$). Summer weights (11 Jul–4 Sep) averaged 662 g for males ($n = 182$) and 589 g for females ($n = 204$), with weights during the flightless period averaging 655 g for males ($n = 35$) and 617 g for females ($n = 73$). Weights during early winter in Mississippi and Louisiana (1979–83) averaged 671 g for males ($n = 400$) and 625 g for females ($n = 246$). Delnicki and Reinecke (1986) reported midwinter body weights of Wood Ducks in Mississippi as 686 g for males ($n = 199$) and 640 g for females ($n = 640$). Average wing measurements from an array of studies ranged from 20.8 to 24.0 cm in males and 18.8 to 23.1 cm in females. Culmen length averaged 2.8–3.8 cm for males and 3.1–3.8 cm for females.

DISTRIBUTION

Except for a small population in western Cuba, the Wood Duck is indigenous to North America, where there are 2 generally distinct breeding ranges. The first extends from the Great Plains to the Atlantic, reaching from southern Canada east to Nova Scotia, and possibly Newfoundland, and stretching south from that entire area to the Gulf Coast and along the Atlantic coastal provinces and states to Florida and Cuba. The second (Pacific) population extends from southern British Columbia south to central California and east into Idaho and Montana. The presence of both suitable nesting cavities and associated wetlands are the primary factors in delineating the breeding range of Wood Ducks. They winter in the southern half of their breeding range. Much of the information on distribution and numbers reported here are from the landmark work by Bellrose and Holm (1994) and references cited therein.

Breeding. The Wood Duck is the most abundant breeding species of waterfowl below about 40° N latitude (Haramis 1990). Ransom et al. (2001) used mitochondrial DNA analyses to support findings from banding data, noting that a significant genetic structure (and hence a close relationship) existed among 8 separate breeding populations. The best comparative information, however, is from Bellrose and Holm (1994), who estimated the numbers and distribution of Wood Ducks in North America from 1981 to 1985. In the Atlantic Flyway, their figure was 1,073,795 breeding Wood Ducks: 21.5% in

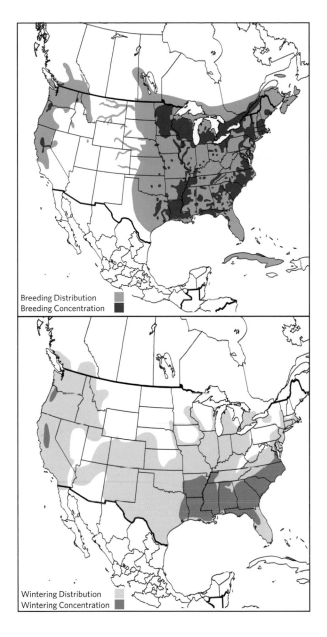

Breeding Distribution
Breeding Concentration

Wintering Distribution
Wintering Concentration

fied on the 1 km² areas that characterize the survey region. This survey covers 11 eastern states, except Maine and Vermont. It reported an average population of 387,577 Wood Ducks from 2000 to 2009. Of that total, 29.3% occurred in Pennsylvania, 23.7% in New York, 12.8% in Virginia, and 6.7% in New Hampshire.

Dennis (1990) provided estimates of the Wood Duck population in Canada, which averaged 117,118 from 1972 to 1985: 79.3% in Ontario, 16.4% in Québec, but only 1.8% in New Brunswick. Lepage and Bordage (2010) reported an average of 13,000 Wood Ducks in Québec from 2004 to 2007. Wood Ducks breed south of the St. Lawrence River in Québec (between Montréal and Québec City), as well as reaching west into Ontario along the Ottawa River between Ottawa and Montreal. Elsewhere in Ontario, they are regular breeders south of a line from North Bay to Sault Ste. Marie, but they occasionally venture as far as 160 km to the north (Harry Lumsden).

Bellrose and Holm (1994) estimated that 282,240 Wood Ducks bred from Texas to Alabama in the Gulf Coast states, where they occurred across most of the states, except in the coastal marshes. Most (47.6%) were found in Louisiana, with 29.0% in Mississippi and 14.8% in Alabama. In Mississippi, high densities occurred in the Yazoo basin and along the Mississippi River. In Texas, Wood Ducks were largely restricted to the forested areas in the eastern third of the state. Just north of the Gulf Coast, Arkansas harbored the most Wood Ducks (117,503) along the Mississippi Alluvial Plain.

Just over 2.4 million Wood Ducks were estimated in the midwestern states, from Minnesota south to Missouri and east to Tennessee and Ohio. Significant numbers occurred in the lake states of Minnesota (342,474), Wisconsin (236,982), and Michigan (139,360; Bellrose and Holm 1994). Large numbers in Wisconsin were found in the eastcentral region (south of Green Bay) and the northwestern lake district. In Illinois, high densities (93,455) occurred along the Illinois River and its tributaries, and along the Mississippi River.

Ontario, 9.7% in Maryland, 8.5% in Virginia, 8.1% in New York, and 7.8% in North Carolina. Farther south, they reported 84,188 Wood Ducks in North Carolina, 72,013 in South Carolina, and 55,083 in Georgia, as well as 32,788 in Florida, but broods have not been observed south of the Loxahatchee National Wildlife Refuge (NWR), southeast of Lake Okeechobee.

The Atlantic Flyway Breeding Waterfowl Plot Survey also provides good information on population distribution, because all waterfowl are identi-

Wood Duck pair at a nest box. *GaryKramer.net*

Farther west, Wood Ducks were historically very rare in the Central Flyway states, except for a few river drainages in Nebraska, Kansas, southeastern Oklahoma, and extreme eastern Texas. Since the mid-1970s, however, Wood Ducks have extended their range across the Great Plains, moving north into South Dakota and central North Dakota, and west from Nebraska to the Rocky Mountains (Ladd 1990). They first pioneered westward along the major rivers, followed by expansions along the tributary streams. Bellrose and Holm (1994) estimated breeding populations of 13,100 in North Dakota, >8,600 in South Dakota, 4,000–5,000 in Nebraska, and 10,100 in Oklahoma. Wood Ducks are not abundant breeding birds in the Prairie Provinces of Canada, with Dennis (1990) noting 1972–85 averages of only 868 birds in Manitoba, 13 in Saskatchewan, and 30 in Alberta.

In the Pacific Flyway, Bellrose and Holm (1994) reported 59,904 Wood Ducks, with major concentrations of 3,000 in British Columbia, 26,000 in California, 17,000 in Oregon, and 7,900 in Washington. Dennis (1990) listed 1,224 birds for British Columbia, and Campbell et al. (1990) observed that most inhabited the lower Fraser and Creston Valleys. In California, the largest segment was found in the Central Valley, with a secondary distribution in the Coast Range north of San Francisco (Bartonek et al. 1990). In Oregon, the majority occurred west of the Cascades, where they were particularly abundant in the Willamette River valley from Eugene to Portland and in the lower Columbia River valley, especially on Sauvie Island (Bartonek et al. 1990). Wood Ducks are rare breeders east of the Cascade Mountains.

Winter. There are 3 distinct winter ranges for Wood Ducks: the northern zone, where birds are generally absent except where open water and food are present; the central zone, where they commonly winter, except when harsh whether forces them south; and the terminal southern zone, where freezeup is not a factor (Bellrose and Holm 1994). Although small numbers of Wood Ducks in the Atlantic Flyway occasionally winter as far north as southern Ontario, most winter from Maryland south along the Atlantic Coastal Plain and the Piedmont into Florida. In general, the northern border of Virginia marks the northern winter range for the Atlantic states, except along the coast. The central zone includes southeastern Virginia and North Carolina, while the terminal southern zone covers South Carolina, Georgia, and Florida. Recoveries from Wood Ducks banded in the northern zone suggest that 73% winter in the southern zone, and 27% in the central zone. Winter population numbers are sparse because of the difficulty of surveying Wood Ducks in their forested habitats, but an estimated 267,000 winter in North Carolina, 299,000 in South Carolina, 243,000 in Georgia, and 167,000 in Florida (Frank Bellrose).

In the Mississippi Flyway, Wood Ducks can winter as far north as Minnesota, Wisconsin, and Michigan, but overwintering is generally sporadic north of about Nashville, Tennessee (Bellrose and Holm 1994). In contrast, large numbers winter in terminal southern zone states, where populations were estimated at 321,000 in Arkansas, 188,000 in Louisiana, 338,000 in Mississippi, and 184,000 in Alabama (Frank Bellrose). The Midwinter Survey does not reliably tally Wood Duck numbers, with an average estimate of only 30,878 from 2000 to 2010.

Small numbers of Wood Ducks winter from southwestern British Columbia through western Washington and Oregon. Most Wood Ducks breeding north of California, however, migrate into the Sacramento Valley and adjacent areas for the winter. Naylor (1960) estimated that at least 90% of the Pacific population wintered in California. Small numbers winter as far south as central Mexico, with rare records in the Yucatán Peninsula (Howell and Webb 1995).

Migration. Migration routes for Wood Ducks are not well defined, because about one-third of their breeding range also overlaps with their primary wintering range. Hence migration distance is short for many populations and virtually nonexistent for resident ones, particularly in the southern states (e.g., Arkansas, Louisiana, Alabama, and Mississippi). Compared with populations to the east, the Pacific population has a better-delineated migration route, with Wood Ducks moving from southern British Columbia through western Washington to the lower Columbia River, the Willamette Valley in Oregon, and into the Sacramento River valley and associated drainages in California (Naylor 1960, Bellrose and Holm 1994).

Recovery data from Wood Ducks banded in the northeastern states show a general movement south to southwest: to the Atlantic Coastal Plain in the East, and to the Mississippi River valley in the Midwest. South of the Appalachian Mountains, substantial movement occurs southwest to the Florida Panhandle and then west to Louisiana and eastern Texas. Conversely, appreciable numbers of Mississippi River valley migrants move east to South Carolina, Georgia, and Florida (Bellrose and Holm 1994).

During fall in New England, some Wood Ducks initiate migration through October, most migrate in November, and a few overwinter in the area (Bellrose and Holm 1994). In the middle Atlantic states, numbers rise and fall as flights from farther away arrive and depart, although many migrants remain all winter in the southern part of the region. Wood Duck numbers in the southern Atlantic states rise steadily from September through December as ingress occurs from farther north. Wood Ducks begin their departure from the Great Lakes region in late September, with migration continuing through October and November. On Pacific coastal breeding areas north of California, departures for the Central Valley in California begin in early September and continue for about 6 weeks, with peak numbers reaching the valley in October.

In the spring, migrant Wood Ducks initiate departure from their southernmost wintering grounds early in February, at the same time as the local birds begin to nest (Bellrose and Holm 1994). Hence the lengthening photoperiod stimulates 2 different reactions: breeding in the resident group, but departure in the migrants. Migration is more rapid in the spring, accelerating as the Wood Ducks progress northward. In the upper Midwest, spring migration extended over 6 weeks, compared with 10 weeks in the fall (Bellrose and Holm 1994). Spring weather, however, has a greater influence on their northward passage, altering their migration chronology in the interior states by as much as 2 weeks. The first Wood Ducks arrive in the Midwest by early March, and they reach their northernmost breeding grounds by mid-April, with other birds arriving over a period of about 2 more weeks.

MIGRATION BEHAVIOR

Among North American waterfowl, the Wood Duck is the only species with both a large migratory population and a nonmigratory one. Wood Ducks that migrate primarily do so at night and are considered partial, short-distance migrants, although the average distance from the northern United States to the ducks' principal southern wintering areas is 1,392 km (Nichols and Johnson 1990). In general, Wood Ducks migrate in small (11–63) flocks, at altitudes of up to 244 m and at flight speeds averaging 53 km/hr, a speed that allows them to cover the greatest distance with the least energetic cost (Bellrose and Holm 1994). In late summer, some Wood Ducks migrate northward before undertaking a southward migration. Most of this dispersal involves adult and immature males, as well as a few immature females.

Calculations by Bellrose and Holm (1994) indicate that 75% of the Pacific Flyway population is nonmigratory, as well as 40% of the birds in the Atlantic Flyway, and 30% in the Mississippi Flyway. An analysis of recoveries (1960–87) for Wood Ducks banded in the southern and southeastern United States indicated that younger birds migrated longer distances than adults (Hepp and Hines 1991), which may predispose them to arrive later on their breeding grounds in the spring

(Semel and Sherman 2001). There was no difference in the distances traveled by males versus females (Hepp and Hines 1991).

Molt Migration. Studies conducted at Nauvoo Slough in Illinois indicated that while most females remained on their nesting areas for the molt, the majority of the males undergo a northerly migration to regions where they probably molt (Bellrose and Holm 1994). In northcentral Minnesota, Gilmer et al. (1977) observed that a minimum of 59% of 45 radiomarked females molted locally, remaining on the breeding area until the end of August; 48% of the radiomarked males also stayed there during the flightless period. In northern Alabama, the mean dispersal date for 62 radiomarked females was not until 9 November, with only 4 departing before 1 November; hence both males and females remained on these breeding areas to molt (Thompson and Baldassarre 1989).

HABITAT

Wood Ducks are capable of using virtually all types of freshwater wetland habitats, but they avoid brackish and saltwater marshes, as well as large expanses of open water and turbulent streams. Streams are an especially important breeding area because of the varying habitats they provide, such as oxbows, ponds, or other areas cut off from the main channel (Bellrose and Holm 1994). In general, preferred Wood Duck breeding habitats have about a 50%–75% coverage of either shrubs and water-tolerant trees or dense stands of tall emergent vegetation, such as bur-reed (*Sparganium americanum*), arrow arum (*Peltandra virginica*), American lotus (*Nelumbo lutea*), and similar plants (McGilvrey 1968, Hepp and Bellrose 1995). Beaver (*Castor canadensis*) ponds are also important breeding habitats (Arner and Hepp 1989). In Alabama/Georgia, 47 radiomarked females at the Eufaula NWR used beaver ponds, temporary wetlands, managed impoundments, and lake habitats (Hartke and Hepp 2004). Managed impoundments and lake habitats were the most preferred, but results suggested that breeding females could satisfy the demands of egg production while using a variety of wetlands.

Postbreeding and wintering Wood Ducks prefer forested wetlands. In northern Alabama (1984–85), 82 Wood Ducks marked with radiotransmitters and tracked from August through December favored wooded swamps and wetlands during the day, with seasonally flooded hardwoods being the only habitat they chose after 15 December. Open water was not used, although it formed >82% of the available habitat (Thompson and Baldassarre 1988). At Mingo Swamp in southeastern Missouri, Wood Ducks inhabited flooded live forests more than expected during all seasons, with >60% of all Wood Ducks present when the forests became flooded (Heitmeyer and Fredrickson 1990). Pin oak (*Quercus palustris*) forests were preferred more than overcup oak (*Q. lyrata*) ones. Green tree reservoirs (bottomland hardwood forests) were used extensively in early fall, when live forests were dry.

Microhabitat use by wintering Wood Ducks occupying green tree reservoirs in Mississippi was generally characterized by less "openness" than that chosen by wintering Mallards (Kaminski et al. 1993). During the winter, Wood Ducks used habitats with 7.3%–10.4% openness, 9.9%–13.7% understory cover, and 39.3–43.6 cm water depth. Areas containing good food resources averaged 3.0–4.7 ducks/m² for those with acorns, and a much greater 47.3–53.1 ducks/m² for those with invertebrates.

Nocturnal roosting of Wood Ducks has been documented in many areas of the eastern United States, with 1,000–3,000 birds commonly reported from a single roost (Bellrose and Holm 1994). A roost by the Mississippi River in northeastern Iowa contained 5,400 Wood Ducks (Hein and Haugen 1966a). During that study, 15 roosts were located along a 160 km stretch of the river, with an average distance of 10.5 km between roosts. All roosts were in emergent vegetation, where the birds were easily approachable to within 15 m. In northern Alabama, stands of American lotus were the favored roosting habitat during August and September, but wooded swamps were the most preferred locale by November, as decreasing water levels and leaf senescence made the lotus plants unavailable (Thompson and Baldassarre 1988). In southern Illinois, 11 Wood Ducks radiomarked during autumn (Sep–Nov) preferred a buttonbush (*Cephalanthus occidentalis*) swamp, where 75% of their daytime locations and 99% of their nighttime ones were recorded (Parr et al. 1979). In Louisiana, 44 roost sites from various locales across the state were largely characterized by an overstory of cypress (*Taxodium distichum*), tupelo gum (*Nyssa aquatica*), willow (*Salix* sp.), and ash (*Fraxinus* sp.), with an understory of water elm (*Planera aquatica*), buttonbush, and swamp privet (*Forestiera acuminata*; Tabberer et al.1971).

Roosts are usually located in the vicinity of the surrounding wetlands that are occupied by Wood Ducks during the day. These areas (and roosts) are used year after year, for as long as that habitat persists (Bellrose and Holm 1994). In Illinois, the daytime locales for most Wood Ducks were within 2 km of their roosts. In Iowa, the farthest roosting flight was 12.9 km (Hein and Haugen 1966a). The functions of roost sites appear to be related to finding food and loafing areas, regulating the local population density during fall migration, and perhaps providing courtship opportunities (Bellrose and Holm 1994).

Roost flights commonly take place from mid-summer until the birds depart for their fall migration. In general, evening roost flights commence shortly before sunset, peak after sunset, and continue until dark. Observations of 294 Wood Duck roost flights on the Mississippi River in northern Iowa revealed that evening flights averaged 45 minutes in mid-August but decreased steadily to only 8 minutes by early November (Hein and Haugen 1966b). Early August flights began about 25 minutes before sunset and ended around 25 minutes after sunset, while early November flights began about 20 minutes after sunset and

lasted for roughly 10 minutes. Departures from roost sites in mid-August began about 25 minutes before sunrise, with all birds leaving before sunrise. Wood Duck roost fights were thus controlled by light intensity: a threshold of about 0.5 foot-candles triggered the morning flight, and below 200 foot-candles triggered the evening flights, with lower light values operating as the days shortened. In North Carolina, 112 observations made at roost sites noted that 8% of the birds arrived as singles, but most arrived in flocks of >2 birds (Perry). Overall, 72% of the birds arrived on their roosts after sunset, but during the November–January hunting season, 96% arrived after sunset.

Surveys of roost sites have been used to provide indices of local populations (Hein and Haugen 1966a, Tabberer et al. 1971). A subsequent study using radiomarked Wood Ducks in southern Illinois, however, revealed that roost counts were unreliable indicators of population size, because some birds either did not roost on certain days or never left the roost (Parr and Scott 1978). In addition, the roosts were not always discrete, as changing food supplies caused the numbers of Wood Ducks to vary at a given roost, and different weather and light conditions made it difficult to count all the birds. As a result, 15–22 simultaneous counts in an area were deemed necessary, and even then the accuracy level was only 15%.

POPULATION STATUS

The preservation of the Wood Duck is among the most successful wildlife conservation stories in North America. Passage of the Migratory Bird Treaty Act (MBTA) in 1918, between the United States and Canada, probably saved this species from near extinction. Prior to the MBTA, waterfowl seasons usually extended from September to April, bag limits were either nonexistent or so large as to be meaningless, market hunting abounded, and enforcement was almost unknown. In the early 1900s, ornithologists and conservationists were so alarmed by the virtual disappearance of Wood Ducks that several voiced fears of their extinction. Grinnell (1901:142) stated, "Being shot at all seasons of the year they are becoming very scarce and are likely to be exterminated before long." As an eminent ornithologist from Massachusetts remarked, "Only the most rigid enforcement of the law can save this, the most beautiful of North American wild ducks, from extermination" (Forbush 1913:353).

Under the auspices of the MBTA, Wood Ducks were given immediate and complete protection, in the form of a closed season. In little more than a decade after the passage of the MBTA, Phillips and Lincoln (1930:296) reported, "Once greatly reduced by summer shooting, especially in our Northern States, this fine duck has recovered everywhere with protection." By 1941, this species had sufficiently increased in number to permit a single Wood Duck to be taken in a hunter's bag in 15 states, and by 1942 this allowance extended to all states (Bellrose 1976). Today Wood Ducks are second only to Mallards as the most harvested duck in the Atlantic and Mississippi Flyways.

Population estimates of Wood Ducks are difficult, because this species is excluded from the Traditional Survey, which largely is conducted outside the range of the Wood Duck. The Atlantic Flyway Breeding Waterfowl Plot Survey, however, identifies all waterfowl to species and thus provides a good population estimate for the 11 states it covers: New York, Vermont, New Hampshire, Massachusetts, Connecticut, Rhode Island, Pennsylvania, New Jersey, Maryland, Delaware, and Virginia. Results from this survey indicated an average population of 387,577 Wood Ducks from 2000 to 2009. For Canada, Dennis (1990) estimated an average population of 117,118 birds (1972–85). Breeding population estimates (1981–85) included in Bellrose and Holm (1994) averaged 1,073,795 for the Atlantic Flyway, 1,654,570 for the Mississippi Flyway, 75,500 for the Great Plains states and Prairie Provinces, and 59,904 for British Columbia and states within the Pacific Flyway. Their grand total was 2,863,769 breeding Wood Ducks in North America. Wetlands International (2006) reported

breeding populations of 2,800,000 from eastern North America and western Cuba, 665,100 in interior North America, and 66,000 in western North America, for a total of 3,531,100.

Wood Duck populations have increased and clearly expanded their range northward and westward since the 1970s. This proliferation, in part, was due to the expansion of beaver populations, which create wetlands heavily used by breeding Wood Ducks (Cringan 1971, McCall et al. 1996). Bellrose and Holm (1994) calculated that from 1959 to 1990, Wood Duck populations in the Atlantic Flyway states increased by 10.3%/year, and by 8.8%/year in the Mississippi Flyway. An analysis of data from the Breeding Bird Survey also shows an increasing population trend. Populations in the Atlantic Flyway increased by an average of 1.7%/year from 1966 to 2008, by 2.7%/year from 1989 to 2008 (20 years), and by 3.3%/year from 1999 to 2008 (10 years). In the Mississippi Flyway, increases were 2.6%/year from 1966 to 2008, 3.2%/year from 1989 to 2008, and 3.6%/year from 1999 to 2008 (U.S. Fish and Wildlife Service 2009).

Harvest

The harvest in the United States averaged 1,201,409 Wood Ducks from 1999 to 2008: 59.7% in the Mississippi Flyway, 29.1% in the Atlantic Flyway, 7.0% in the Central Flyway, and 4.2% in the Pacific Flyway. Harvests ranged from 1,076,202 in 2006 to 1,554,347 in 1999. In the Mississippi Flyway, 16.7% of the harvest occurred in Louisiana, 15.4% in Minnesota, 11.8% in Arkansas, and 10.2% in Wisconsin. A special early Wood Duck season is conducted during September, allowing locally produced birds to be harvested before northern migrants arrive. In 1973, this special season was offered to all Mississippi Flyway states except Minnesota, Wisconsin, Michigan, and Iowa; in 1977, it was expanded to include the Atlantic Flyway states of Virginia, North Carolina, South Carolina, Georgia, and Florida (Bowers and Martin 1975, Bowers and Hamilton 1977). Harvests during the special Wood Duck season from 1999 to 2009 averaged 32,788 in the Mississippi Flyway (Tennessee and Kentucky) and 2,698 in the Atlantic Flyway (Florida). In Canada, the 1999–2008 harvest of Wood Ducks averaged 69,125, with a high of 83,806 in 1999. Over this time period, 71.0% were harvested in Ontario, 21.0% in Québec, 4.0% in New Brunswick, 1.6% in Manitoba, 1.6% in Nova Scotia, and 0.8% in the remaining provinces.

BREEDING BIOLOGY

Behavior

Mating System. Wood Ducks are monogamous, with the females usually breeding as yearlings. In South Carolina (1982–87), 82% of 51 females nested as yearlings, but drought conditions in 1 of those years delayed breeding by some females until their second and even third year (Hepp et al. 1989). In Massachusetts, only 58% of 88 females bred as yearlings (Heusmann 1975), which probably is due to a shorter breeding season. Grice and Rogers (1965) noted that 2%–24% of the returning Wood Ducks in Massachusetts (1951–56) were 2 years old when they nested for the first time. Wood Ducks not nesting until their second year may reflect delayed sexual maturity as a result of late hatching, since yearlings from late-hatched nests or from northern breeding areas may not form pair bonds until late in the following spring or during their second year (Bellrose and Holm 1994).

Some pair bonds are created as early as mid-September, with pairing continuing through the fall and winter. By late February, probably 90% of the females are paired (Bellrose 1980). At Mingo Swamp in southeastern Missouri, the percentage of paired females increased from 70.9% in fall, to 89.7% in winter, and to 97.5% by early spring (Heitmeyer and Fredrickson 1990). Armbruster (1982) reported that the percentage of paired females in southeastern Missouri was 60%–67% in October–November, 45%–46% in January–February, 83% in March, 91% in April, and 95% in May. Early pair bonds appeared tenuous, however, and were constantly tested by conspecifics until nest-searching activities commenced in March. Male Wood Ducks

remain with their mates longer than most ducks, usually until the eggs are pipped, but late-nesting females are deserted by their mates at earlier stages of incubation (Bellrose 1980). In northern Minnesota, Gilmer et al. (1977) found that males left their mates about 6.5 days into incubation, perhaps because of the shorter nesting season at northern latitudes (Haramis 1990).

Sex Ratio. Sex ratios are male dominated. Hunters' bags (1966–2000) contained an average of 1.6 adult males for each adult female and 1.5 immature males for each immature female. Across a broad distribution of banding sites during winter (1967–69), the sex ratio (males:females) of 5,504 Wood Ducks was 1:0.8.

Site Fidelity and Territory. Wood Ducks exhibit a phenomenal fidelity to their breeding areas. In 16 studies summarized by Bellrose and Holm (1994), the return rates of Wood Duck females averaged 52.2%, and in 7 of those reports the percentage recaptured in the same nest box averaged 36.0%. After accounting for mortality rates, this homing rate represents almost every surviving adult female, as well as a moderate proportion of the yearlings (Bellrose 1980). Homing rates are greater for adults than for immatures, as was seen in Illinois, where 49.1% of the adult females but only 6.5% of the immature females returned to the same nesting area the subsequent year (Bellrose et al. 1964). During an 8-year study (1979–86) at the Savannah River Site (a nuclear facility) in South Carolina, the mean annual survival rate of 55% for 181 females in nest boxes was similar to the survival rates calculated from band-recovery data, which indicates that virtually all surviving females returned to that breeding area (Hepp et al. 1987a). In southeastern Missouri, Hansen (1971) reported that 56% of 88 successful females returned and used the same nest boxes, and 6% used boxes >0.8 km away. In contrast, 10 unsuccessful females moved an average of 2.7 km, compared with an average of 0.24 km for successful females.

Relative to natal philopatry, the return rate of

1,459 females marked as ducklings in South Carolina during 1982–87 averaged 5.2% (Hepp et al. 1989). Of 67 females captured during their first nesting attempts, 58 were 1 year old, 8 were 2 years old, and 1 was 3 years old; none were older. Most (60%) females banded as ducklings did not return to their precise natal areas, but rather dispersed a short distance away (mean = 1.6 km).

Males lack the great homing ability of females, as reported by Grice and Rogers (1965) during their classic study of Wood Ducks in Massachusetts. Of 210 males bait-trapped at Great Meadows, only 19 (9%) were banded there in previous years, compared with 48% of 217 females. Doty and Kruse (1972) noted that only 3 of the 93 pen-released males returned the following year.

Males are not territorial per se, but the male of a pair will defend a moving space around the female (Grice and Rogers 1965, Bellrose 1980). The home range of 47 males radiomarked at Eufaula NWR in Alabama/Georgia averaged 367 ha during the combined prelaying and egg-production periods (Hartke and Hepp 2004). Home-range size, however, was inversely correlated with the percentage of managed impoundments and lake-influenced wetlands. Fall (Sep–Nov) home-range sizes of 11 radiomarked Wood Ducks in southern Illinois averaged 90.6 ha, with some birds moving up to 10 km/day (Parr et al. 1979). At Nauvoo Slough in Illinois, 28–73 pairs remained within an area of about 25 ha (Bellrose and Holm 1994). In north-central Minnesota, 50% of 31 radiomarked females remained within 0.5 km of their nest sites and 70% remained within 1 km, which suggested that a land unit of about 3 km² was used by pairs during the breeding period.

Male Wood Ducks do not appear to have fixed waiting sites, as do many prairie-nesting ducks, but instead swim over extensive areas, using a variety of floating logs, stumps, and bank sites for resting. Females on recess from their nests thus call loudly upon landing on the water, so as to draw the attention of their mates, as the males may be close by or as far as 1.6 km away. The unusually loud calls of female Wood Ducks may have evolved to attract their mates in what is often a visually restricted habitat (Bellrose 1980).

Courtship Displays. Armbruster (1982) provided significant insight into the courtship behavior of Wood Ducks, based on her observations in southeastern Missouri. Display bouts usually occurred in the early morning and around sunset, and they always involved vocalizations. Courting parties averaged 11.4 birds, and the sex ratio (males:females) was 2.2:1, with the proportion of males increasing throughout fall and spring. Courtship displays were first observed on 21 September, by which time all males had attained alternate plumage. Temporary associations of males and females often became courting parties, where pair bonds were initiated but continually tested. As the pairs formed, the length of the courtship bouts decreased, from 1.9 minutes/sequence in the fall to 1.0 minute in the spring. Birds in groups displayed and vocalized more in the fall than in the spring, and groups displayed and vocalized more than pairs.

Bellrose and Holm (1994), Hepp et al. (1994), Palmer (1976b), and Johnsgard (1965) reviewed the displays of Wood Ducks, and Hepp and Bellrose (1995) summarized the array of calls associated with both male and female courtship. In males, the *display shake* is analogous to the *grunt-whistle* of species in the genus *Anas* and considered to be the most elaborate in the male's repertoire. Males in this display erect the crest, extend and lower the head, and then raise the breast to expose the white belly, after which they return to a normal posture on the water. *Turn-the-back-of-the-head* was the most common display observed by Korschgen and Fredrickson (1976), and it apparently serves to strengthen the pair bond, as does *preen-behind-the-wing*. A *wing-and-tail-flash* is unique to Wood Ducks and Mandarin Ducks (*Aix galericulata*). Performed by pairs and unpaired males, the display involves rapidly raising the closed wing and tail while orienting broadside to the female. A *rush* display is common in courtship parties and used by both sexes, perhaps to indicate that more intense courtship displays are to follow. *Inciting* is given by both sexes and involves lowering the head over the shoulder while the bill makes pointing movements toward another bird. Paired males use this display against unpaired males, and females use it to indicate mate preference. A *coquette-call* given by females is a loud *terwee* first heard in August, especially at nighttime roost sites. The call attracts males and reinforces pair bonds, as well as maintains contact between the sexes during nest searching by the female (Bellrose and Holm 1994). Copulation occurs on the water and is preceded by several displays from both sexes. The postcopulatory display involves *bathing* by the female as the male gives several displays.

Nesting

Nest Sites. The male follows the female while she searches for a nest site, which is either a natural cavity or an artificial nest box. Very rare reports exist of nests on the ground (Zipko and Kennington 1977, Mason and Dusi 1983), on a muskrat (*Ondatra zibethica*) house (McIlquham and Bacon 1989), and even a leaf/twig nest in a tree (Hall 1969), but most (95%) Wood Ducks nest in natural cavities (Bellrose 1990). Unlike other cavity-nesting waterfowl, however, a significant amount of information exists on cavity characteristics and use patterns by Wood Ducks, from which 3 major patterns have emerged: (1) most cavities are in live trees, (2) large trees contain more cavities than small trees, (3) and some tree species are more prone to cavity formation than others.

In a major review of Wood Duck nest-cavity characteristics, Soulliere (1990) noted that cavity density was between 0.1 and 5.5/ha, with 90% of the cavities occurring in living trees; ≥60% of these cavities originated from broken branches or heart rot. Overall, 85% of Wood Duck cavity nests were in live trees. Dead trees seldom remain standing long enough to provide dependable nesting sites for these ducks year after year, but a few species with particularly durable wood (e.g., baldcypress,

Taxodium distichum) may be unusually persistent. Trees containing suitable cavities generally are >30 cm in diameter at breast height (dbh), and commonly near 60 cm dbh. The types of cavity-forming trees important for nesting vary from north to south and include aspens (*Populus* spp.), maples (*Acer* spp.), American basswood (*Tilia americana*), elms (*Ulmus* spp.), sycamore (*Platanus occidentalis*), American beech (*Fagus grandifolia*), oaks (*Quercus* spp.), willows (*Salix* spp.), tupelo (*Nyssa* spp.), and baldcypress.

The cavity density (5.5/ha) reported by Prince (1968) was from virgin hardwoods in a floodplain forest in New Brunswick, where the average cavity tree was 230 years old. In Minnesota, mature quaking aspen (*Populus tremuloides*) stands that were an average of 68 years old contained 57% of 28 cavities inspected (Gilmer et al. 1978). Mature northern hardwood stands, ranging in age from 100 to 120 years old, also were important cavity-production sites, which averaged 15.3/ha; 7.1/ha were in sugar maples (*Acer saccharum*). In floodplain forests in westcentral Illinois (1994–95), suitable cavities averaged 2.12/ha, with 74.3% of the available cavities in silver maples (*Acer saccharinum*), 8.2% in eastern cottonwoods (*Populus deltoides*), 8.0% in willows, and 6.7% in green ashes (*Fraxinus pennsylvanica*; Yetter et al. 1999). In Indiana (1984–85), American beeches, red maples (*Acer rubrum*), and sycamores produced 72% of the suitable cavities but composed only 28% of the total basal area of tree cover. Wood Ducks nested in only 7%–9% of those cavities that occurred at minimum densities of 0.08–0.13/ha (Robb and Bookhout 1995). The Wood Ducks ($n = 14$) used cavities in forest stands that averaged 65 ha (range = 3–420 ha). In Mississippi, cavity density ranged from 0.19 to 0.23/ha in 2 mature bottomland hardwood forests (Lowney and Hill 1989). At 1 site, sycamores and American beeches contained 60% of the cavities but totaled only 2.6% of the available trees.

Wood Ducks select cavities that range in height from about 2–15 m above the ground, and they seem to prefer the higher sites (Bellrose and Holm

1994). Soulliere (1990) also noted that the greatest use occurred within cavities highest up in trees and those with smaller and more visible entrances. At the Muscatatuck NWR in Indiana, Wood Ducks selected natural cavities with smaller entrances and volumes, and those with vertical-facing entrances, more than other cavity types (Robb and Bookhout 1995). Several studies demonstrate that cavities excavated by Pileated Woodpeckers (*Dryocopus pileatus*) create nest sites preferred by Wood Ducks. In westcentral Illinois, Wood Ducks used a greater proportion of Pileated Woodpecker cavities when available, probably because the small entrances provide security from some nest predators (Yetter et al. 1999). Cringan (1971) suggested that the increased population of Wood Ducks in Ontario was in part due to an increase in the numbers of Pileated Woodpeckers there. Conner et al. (2001), however, found competition for nest sites between these 2 species and observed instances where Pileated Woodpeckers successfully defended their cavities from Wood Ducks. In 2 instances, male Pileated Woodpeckers were observed entering nest cavities and forcibly ejecting female Wood Ducks.

Bellrose and Holm (1994) reported the mean dimensions of Pileated Woodpecker cavities as 40.6 55.9 cm deep and 19.1–23.1 cm in diameter, with entrance holes measuring 11.9–13.0 cm vertically and 9.9–11.1 cm horizontally. Wood Ducks often select the smallest entrance through which they can move easily, but they have used cavities as large as 30 × 61 cm (Bellrose et al. 1964). Bellrose and Holm (1994) reported an average entrance size of 11 × 20 cm. In contrast, the recommend dimensions for a standard Wood Duck nest box are a 61 × 30.5 cm front with an elliptical 10.2 × 7.6 cm entrance hole, which prevents entrance by raccoons (*Procyon lotor*) weighing ≥4.5 kg (Bellrose 1980).

In Minnesota, 26% of the cavity nests were located at least 1.0 km from water (Gilmer et al. 1978); in Illinois, Wood Ducks used cavities at upland sites that were an average of 1.4 km from water (Ryan et al. 1998). In Indiana, the distance to water was greater for successful nests (122 m) compared with unsuccessful nests (50 m), probably because raccoons spent more time foraging near water. The cavities were 0.0–1.2 km from the brood habitat (Robb and Bookhout 1995). Fox squirrels (*Sciurus niger*) and raccoons were important competitors for cavities, and nest success (22%) was limited by raccoon predation.

Although nest boxes are widely used for the management of Wood Ducks and other cavity-nesting waterfowl, they provide a miniscule number of nest sites in comparison with natural cavities. Soulliere (1990) used an average estimate of 0.6 cavities/ha to calculate that around 60 million natural cavities were probably present in the approximately 97 million ha in the eastern United States. In their unique study of cavity availability and ecology in the Mississippi River floodplain and adjacent upland forests of southern Illinois, Nielsen at al. (2007) demonstrated that the highest densities of cavities were in sycamores (0.50/tree) and beeches (0.41/tree); cottonwoods contained only 0.05/tree. Sycamores also had the highest (89%) 10-year cavity persistence, and cottonwoods the lowest (54%). The most prevalent causes of cavity loss were tree fall (50.0%), cavity-floor deterioration (37.5%), and a narrowing of the entrance (12.5%). Tree mortality from flooding caused the loss of 42.7% of the 75 cavity trees in westcentral Illinois, which potentially limited Wood Duck production in the area for several decades (Yetter et al. 1999). Forest-stand projections for southern Illinois by Nielsen et al. (2007), however, indicated that cavity abundance would increase by 34% during the first 10 years after the current level, and 44% after 50 years. Their study called for an evaluation of the cost-effectiveness of nest box programs in light of these projected increases in the abundance of natural cavities.

Before laying, the female rounds out the bottom material of the nest box or cavity into a saucer-like depression and covers the first eggs either with litter from the bottom of the nest cavity or sawdust placed in the nest box. A few down feathers are added after the sixth or seventh egg, supplemented with small amounts daily until the clutch nears completion, at which time large amounts of down are plucked from her breast and added to the nest. Leopold (1951) reported about 3,300 cm³ of down in early nests, but often <1,600 cm³ in renests.

Clutch Size and Eggs. Evaluation of the normal clutch size of Wood Ducks is confounded by the occurrence of parasitized nests, but most studies develop criteria to discriminate between parasitized and nonparasitized nests: a nest with ≥16 eggs generally is considered a parasitized nest (Bellrose 1980). Data from nest box studies reported the average clutch size of normal Wood Duck nests as 10.3 eggs for 82 nests during a 3-year study (1965–67) in Oregon (Morse and Wight 1969), 10.5 for 158 nests during a 7-year study (1973–79) in New York (Haramis and Thompson 1985), and 11.4 for 534 nests during a 9-year study (1966–74) in Missouri (Clawson et al. 1979). Average clutch size from natural cavities in Mason County in Illinois (1938–59) averaged 10.9 eggs (range = 6–16) for 85 nests (Bellrose and Holm 1994). In general, clutch size is independent of a female's age but positively correlated with her body mass (Hepp and Kennamer 1993). In contrast to normal nests, parasitized (dump) nests averaged 16.4 eggs for 120 nests in Oregon (Morse and Wight 1969), 20.2 for 524 nests in Missouri (Clawson et al. 1979), and 22.5 for 105 nests in New York (Haramis and Thompson 1985).

Wood Duck eggs are elliptical to subelliptical and vary in color from white or creamy white to dark tan (Hepp and Bellrose 1995). In South Carolina, egg dimensions (*n* = 105) averaged 51.3 × 39.4 mm (Hepp et al. 1987b). Fresh egg mass averaged 44.2 g and consisted of 53.1% albumen, 36.4% yolk, and 9.6% shell. In Missouri, eggs averaged 42.6 g, of which 53.5% was albumen, 35.8% yolk, and 10.7% shell (Drobney 1980). The eggs contained 69.5% water, 14.1% lipids, and 13.5% protein. Using average values for egg mass, clutch size, and body weight, Bellrose and Holm (1994) estimated that an average clutch was 78% of the female's body mass. The egg-laying rate is 1/day (Drobney 1980), with females capable of producing large numbers of eggs during a single breeding season. Captive females raised at the Max McGraw Wildlife Foundation in Illinois between 1977 and 1987 laid an average of 22.8 eggs/season over a 10-year period, with a high of 33.3 (Bellrose and Holm 1994).

Incubation and Energetic Costs. Haramis (1990) noted an average incubation period of about 30 days throughout the breeding range of Wood Ducks. In Illinois, the incubation period for 218 nests averaged 30.8 days, but ranged from 28 to 37 days (Bellrose and Holm 1994). In New York, Haramis and Thompson (1985) reported an average of 28.8 days (range = 27–30) for 16 nests. In Alabama, the incubation period of 40 radiomarked females averaged 31.8 days (range = 29–36 days; Folk and Hepp 2003). Semel et al. (1988) found that the hatchability of eggs was inversely correlated with population density, the frequency of brood parasitism, and the number of eggs laid/nest. In Missouri, the hatching rate of 4,505 eggs averaged 78% from normal nests versus 63% of 6,672 eggs from parasitized nests (Clawson et al. 1979). During a 12-year study (1976–87) of nest box use at the Max McGraw Wildlife Foundation in northern Illinois, an average of 62.5% (range = 38.4%–76.9%) of 6,941 eggs laid in 448 nest starts hatched (Semel et al. 1988). During a 7-year study (1973–79) at the Montezuma NWR in New York, 78% of 1,430 eggs laid in boxes were unhatched when nesting space became saturated by 1977, at which time 73% of all nests were parasitized (Haramis and Thompson 1985).

Females delay incubation until the clutch is near completion. This behavior, which is adaptive for species with a large clutch size and precocial young, maintains the viability of the oldest eggs in the clutch (those laid first) while still assuring that they hatch synchronously with the remainder of the clutch (Arnold et al. 1987). Some embryonic development may occur prior to the onset of incubation, however, due to increased attention by females to the eggs that were laid first. The phenomenon known as intraclutch developmental asynchrony (IDA) was defined by Kennamer et al. (1990) as the difference (in days) between the most-developed and least-developed eggs in a clutch, which averaged 2.2 days for Wood Ducks in South Carolina. Developmental synchrony is probably limited, however, because hatching success declined when IDA was >3 days.

In South Carolina, the nest attentiveness of 3 females averaged 50% of the day during the later stages of incubation (Kennamer et al. 1990). Females were also attentive at night, however, which probably allowed more diurnal foraging time to meet the nutritional demands of egg production. In Alabama, the incubation constancy of 40 females averaged 81.3%, with the females taking an average of 2.2 recesses/day (Folk and Hepp 2003). A nest recess in both the morning and the afternoon was taken on most (60%) days. Hepp and Bellrose (1995) noted that morning recesses were shorter than evening recesses (86.5 minutes vs. 138 minutes). Incubation constancy declined with nests produced later in the breeding season, with the increased use of preferred habitats, and with the distances females traveled from the nest. These patterns indicated that Wood Ducks nesting at southerly locations are not energetically constrained during incubation, but such patterns may not occur where environmental conditions are more severe.

Hepp et al. (2006) demonstrated that how females actually incubate eggs affects both egg hatchability and the resultant ducklings. Among eggs incubated at 34°C, 36°C, and 37°C, hatching success was least for eggs incubated at the lowest temperature, and embryos incubated at lower temperatures used a greater quantity of their egg protein. In contrast, duckling mass increased with rising incubation temperatures. Hence incubation temperature, and decisions by females to maintain an optimum temperature, can modify both the incubation period and the resulting condition of the ducklings.

Drobney (1980) thoroughly detailed the energetic reproductive dynamics of female Wood Ducks during his landmark study in bottomland hardwood habitats in southeastern Missouri. Total caloric costs for a 12-egg clutch were 5,996 kJ, but those for a single egg were incurred for only 6 days during the laying cycle, because females spread the costs of clutch production over an 18-day period, not just the 12 days required for laying. Females underwent hyperphagia in the spring to deposit large quantities of lipids that were later used during egg laying. Stored (endogenous) lipids were used to meet 88% of the costs of clutch production and its associated biosynthesis (e.g., growth of the oviduct), with their lipid reserves decreasing by 76%, from 134 g to only 31 g. The use of stored lipids also gives females enough time to meet their protein and mineral requirements by foraging on aquatic invertebrates, which requires a substantial effort.

Incubation is energetically costly for female Wood Ducks, as they lose body mass during the process. The body mass of 152 females monitored over 3 breeding seasons (1986–88) in South Carolina averaged 578 g early in the incubation period, compared with 553 g late in incubation (Hepp et al. 1990). In 1 of the 3 years, females that were heavier at the end of incubation survived better than lighter birds; thus incubation can indeed be costly, at least for some individuals. There was a positive association between clutch mass and incubation length for parasitized nests during each of the 3 years, which indicated previously undocumented costs of intraspecific brood parasitism.

Male Wood Ducks also use substantial nutrient reserves to cope with the time and energy constraints of reproduction. A study by Hipes and Hepp (1995) at the Eufaula NWR on the Alabama/Georgia border was the first to document the energetic costs of reproduction in males. The males lost 79% of their lipid reserves during reproduction, an amount nearly identical to that of the females. Most (86%) of this loss occurred between pair formation and egg laying, which probably reflected the reduced feeding time of males (34% vs.73% for females), since they spent more time in alert behavior and guarding their females.

Nesting Chronology. Wood Ducks nest over a broad range of latitudes, and nest initiation dates vary widely. Bellrose and Holm (1994) noted a delay of 4.1 days for each degree of latitude in the Atlantic Flyway and 3.4 days in the Mississippi Flyway. Haramis (1990) summarized nest-initiation data from an array of studies across the range of Wood Ducks and noted an 86-day difference between nest initiations from south (South Carolina) to north (New Brunswick): in South Carolina (33.5° N latitude), the earliest nest was recorded on 20 January and the nesting season averaged 157 days; in Missouri (37.0° N latitude), the earliest was around 21 February and the season averaged 181 days; in Massachusetts (42.5° N latitude), the earliest was 23 March and the season averaged 106 days; in New Brunswick (46° N latitude), the earliest was about 15 April and the season averaged 82 days. Weather will also affect nest initiation, as Grice and Rogers (1965) observed during their 6-year study (1951–56) in Massachusetts, where the start of the nesting season differed by up to 22 days.

As expected, the peak periods of nest initiation also vary with latitude. Populations nesting below 31° N latitude exhibit peaks during February, while peaks occur during March at 31°–37° N latitude (Bellrose and Holm 1994). During their 9-year study (1966–74) in Missouri, Clawson et al. (1979) noted the largest peak for nest initiations was in late March, which also saw initiation of the bigger parasitized nests. A smaller peak occurred in late April.

Nesting chronology also varies with the age and the body condition of the female. In South Carolina, adults nested 11–19 days before yearlings, and heavier females nested before lighter females, regardless of age (Hepp and Kennamer 1993). Also, females returning to the same nest site in South Carolina nested earlier than either inexperienced females or females that switched nest-site locations (Hepp and Kennamer 1992).

Wood Ducks are the only North American species of waterfowl that commonly produces 2 broods in a given year, no doubt due to the long breeding seasons associated with southern latitudes. At the Eufaula NWR on the Alabama/Georgia border, the incidence of second broods was 7 out of 101 (6.9%) successful nests in 1985 and 16 out of 139 (11.5%) successful nests in 1986; 4 females produced double broods in both years (Moorman and Baldassarre 1988). The length of the nesting season was 149 days in 1985 and 170 days in 1986, with the interval between the hatch of the first clutch and the initiation of the second clutch ranging from 15 to 51 days in 1985 and 17 to 72 days in 1986. The mean minimum age of double-brooded females was 2.6 years. The distance between their first and second nests averaged 328 m in 1985 and 473 m in 1986, although 1 female moved 2,390 m in 1985 and 7,650 m in 1986. Excluding parasitized nests, hatching success did not differ between the first and second broods, with the second broods accounting for 5.2% of total duckling production in 1985 and 9.3% in 1986. Clutch size, however, was always lower for the second clutch. Moorman and Baldassarre (1988) used the data reported for second broods to develop a linear regression equation, which indicated that the incidence of second broods increased by 1.2% for each 1° decrease in latitude.

In South Carolina from 1982 to 1986, 21 out of 219 (9.6%) successful nesting attempts were second nests (Kennamer and Hepp 1987). The minimum age of double-brooded females was 2.3 years, but 1 yearling produced a second brood. Mean clutch size averaged 13 eggs for first nests and 10 for second nests. The body weight of females was also greater for first nests than second nests. Farther north, in southeastern Missouri, the incidence of second broods during a 13-year study (1962–74) of Wood Ducks was lower (3.8%; Fredrickson and Hansen 1983). During a different 13-year study (1976–86) in the northern San Joaquin Valley of California, 56 out of 1,540 (3.6%) nesting attempts were second broods (Thompson and Simmons 1990). The time interval between clutches averaged only 26 days, which indicates that females either lost or abandoned their broods before initiating second clutches. Most (80%) females first laid second clutches when they were ≥2 years old, but 2 females did so as yearlings.

Nest Success. Nest-success estimates are variable, particularly given the large number of such studies, but a useful approach is to compare success from natural cavities with that from nest boxes. Bellrose and Holm (1994) summarized Wood Duck nest success data from 9 such studies and reported an average success of 40.5% from 512 nests (range = 31.2%–50.0% for studies with sample sizes ≥24). In southern Illinois, the nest success of 44 radiomarked females using natural cavities averaged 64%, and the rates were not statistically different between floodplains (80%) and upland forests (59%; Ryan et al. 1998).

Nest success in nest boxes is considerably higher than in natural cavities, because most boxes are equipped with predator guards. Bellrose and Holm (1994) summarized data from a very large number of nest-box studies across the United States, noting an average box use of 40.3% and an average success rate of 63.9%. Box use specific to national wildlife refuges was somewhat higher (51.8%), as was success (68.1%). In South Carolina, females that used a nest box successfully tended to return to that box, compared with unsuccessful females (Hepp and Kennamer 1992). Such females, however, did not have better nest success, were not more likely to have the nest parasitized, and did not survive better than females that switched boxes. In contrast, females that switched boxes tended to improve their nesting success.

At the Max McGraw Wildlife Foundation in Illinois, Wood Ducks started 448 nests during a 12-year period (1976–87): 206 (46%) were considered parasitized nests, 204 (46%) were normal, and 38 (8%) were drop nests (nests used by females to drop eggs in, although no female incubates the nest). The hatchability of the 6,941 eggs laid in those 448 nests averaged 62.5% (range = 38.4%–76.9%;

Semel et al. 1988). Nest box occupancy varied from 47% to 85%, and nest success, from 50% to 90%. In Mason County in Illinois, long-term (1939–74) data from 6,369 available nest boxes revealed that 40.0% were used, 59.4% successfully (Bellrose and Holm 1994). During a 5-year study (1966–70) in southeastern Missouri, 125–270 boxes were available each year. Nest box use averaged 44.6%; and 51% of the 438 nest attempts in boxes were successful (Hansen 1971).

In natural-cavity nests in southern Illinois, raccoons and black rat snakes (*Elaphe obsoleta obsoleta*) were believed to be the principal nest predators (as is common in studies of other cavity-nesting species), along with gray squirrels (*Sciurus carolinensis*) and fox squirrels (Ryan et al. 1998). Long-term (1939–74) data from nest boxes in Mason County, Illinois, revealed that raccoons caused the highest rate of nest loss (10.3%), followed by fox squirrels (9.5%), European starlings (*Sturnus vulgaris*; 5.5%), bullsnakes (*Pituophis catenifer sayi*, 5.0%), and woodpeckers (2.0%; Bellrose and Holm 1994). As of yet, no strategies have been devised in designing nest boxes that would reduce losses from starlings, woodpeckers, and other avian competitors. In Missouri, Hansen (1971) found that black rat snakes caused the greatest (15%) nest losses, followed by desertion (15%), raccoons (5%), and starlings (4%). In addition to their role as major egg predators, black rat snakes killed 3 incubating females during the study. Another female was found sitting on top of a large black rat snake, which, in turn, was coiled over the last 2 eggs in the clutch. The eggs had been incubated for about 19 days, but all were gone 3 days later.

Brood Parasitism. Intraspecific brood parasitism is common and well studied in Wood Ducks. Aside from using DNA techniques, intraspecific brood parasitism is identified by a laying rate of >1 egg/day, ≥16 eggs, and/or the presence of nonterm eggs. Estimations of parasitism based on clutch size, however, can be misleading. Morse and Wight (1969) found that 51.7% of 120 parasitized clutches in Oregon contained <15 eggs. Bellrose and Holm (1994) summarized results from 17 studies of brood parasitism, noting a mean clutch size of 21.7 eggs from successfully hatched parasitized nests. In Massachusetts, yearling females were less frequently parasitized than experienced breeders (30.8% vs. 56.0%), in part because they nested later in the season, at which time parasitism declined (Rohwer and Heusmann 1991). In central Illinois, the clutch size of parasitized nests declined as the season progressed, from an average of 16.7 eggs prior to 15 April ($n = 216$), to 16.0 from 15 to 30 April ($n = 133$), 13.8 from 1 to 15 May ($n = 114$), 11.9 from 16 to 31 May ($n = 53$), and only 9.7 from 1 to 15 June ($n = 12$; Bellrose and Holm 1994).

Parasitism is also more common in nest boxes than in natural cavities, probably because natural cavities are more difficult for multiple females to locate. In their review of clutch sizes for nests in natural cavities ($n = 28$), Semel and Sherman (1986) found that only 29% were clearly parasitized (as evidenced by clutch sizes of 19–31 eggs), compared with 95% of the nests in boxes they monitored in southeastern Missouri. Of nests in natural cavities in Illinois ($n = 91$), only 7% contained >16 eggs (Bellrose and Holm 1994). During a 13-year study (1967–79) of Wood Ducks nesting in boxes in Massachusetts, 43.3% of all nests were parasitized (Rohwer and Heusmann 1991).

The location of nest boxes, however, influences the incidence of brood parasitism. During a 12-year study (1976–87) by Semel et al. (1988), parasitism occurred in 49.5% of the boxes erected singly but in highly visible locations, and in 49.5% of the boxes erected in highly visible groups. In contrast, parasitism was only 29.8% in boxes erected singly in visually occluded habitats. This study also documented the effects of brood parasitism, as the hatchability of eggs in parasitized nests (16–44 eggs) was 57.5%, versus 67.3% for eggs in normal nests (7–15 eggs). Mean clutch size was 15.7 eggs in single but visible boxes, 16.3 in visible but clumped boxes, but only 12.4 in well-hidden boxes. Hatchability was 82% in well-hidden boxes, compared with 74% in visible boxes. The study concluded that reduced parasitism and increased hatchability occur when boxes are placed in habitats and at densities that mimic the natural conditions in which Wood Ducks evolved.

Intraspecific brood parasitism is reported from every study where Wood Ducks are nesting primarily in boxes, with the incidence and effects of parasitism increasing as the population size increases. During a 7-year study (1973–79) at the Montezuma NWR in New York, the percentage of parasitized nests increased from 14% in the first year of the study to 73% by the fourth (Haramis and Thompson 1985). Nesting efficiency (number of ducklings exiting boxes ÷ number of eggs × 100) had declined to a low of 22% by the fifth year, compared with 39%–80% in other years. The study impoundment was not flooded during the last 2 years of the authors' observations, which reduced the density of breeding pairs. Subsequently, the percentage of dump nests declined to 20%–30% and nesting efficiency rose to 50%–60%. In other studies evaluating the effects of brood parasitism, Heusmann (1972) found that the recapture rates of banded ducklings in Massachusetts were nearly equal for those from normal nests (26.0%) and from parasitized nests (26.4%). In Oregon, parasitized nests contributed 31.5% more ducklings to the population. All studies, however, reported reduced egg hatchability from parasitized nests.

Behaviorally, intraspecific brood parasitism has been documented but seldom actually seen, except for detailed observations of marked females in southeastern Missouri (Semel and Sherman 1986). Their study examined the development of 21 clutches, 20 of which increased by ≥2 eggs/day, for a parasitism rate of 95%. Despite this high level of parasitism, however, 46% of nearby boxes were unused. Morse and Wight (1969) also concluded that dump nesting among Wood Ducks in Oregon was a common occurrence and did not appear to be correlated with competition among females for nest sites. The authors found that 19 (76%) of the occupied boxes contained ≥19 eggs, and 9 nests

(43%) were abandoned before incubation. Most (76%) nests were initiated by 1 female, and the heaviest parasitism occurred during the latter half of the laying period. At least 4 different females typically contributed to each parasitized nest, but as many as 7–8 were also recorded depositing eggs in a single nest. On average, parasitized nests increased by 1.8 eggs/day, but they could grow by as many as 7–8 eggs/day. The females attempted to evade brood parasitism by avoiding boxes when conspecifics were present and by aggressively excluding intruders from their nests.

Interspecific brood parasitism occurs between Wood Ducks and Hooded Mergansers. Hansen (1971) reported 15 mixed clutches of Hooded Mergansers and Wood Ducks during a 5-year study (1966–70) in southeastern Missouri: 9 were incubated by Wood Ducks, 5 by Hooded Mergansers, and 1 was abandoned. Clutches were always incubated by the species that contributed the most eggs to the clutch, with no more than 3 Wood Duck eggs in any Hooded Merganser nest, but with 8 Hooded Merganser eggs in 1 Wood Duck nest. During a 5-year study (1981–85) of nest-box use in Minnesota, Zicus (1990b) reported that 5.2% of the nests containing Hooded Merganser eggs were incubated by Wood Ducks or Common Goldeneyes. Additionally, Hooded Merganser eggs made up 12.3% (19 out of 155) and 20% (6 out of 30) of the eggs that were incubated by Wood Ducks and Common Goldeneyes, respectively.

Renesting. Female Wood Ducks commonly renest if their first clutch is destroyed, especially in the southern part of their range, where the nesting seasons are very long. Nonetheless, in Massachusetts 1 female renested 3 times (Grice and Rogers 1965). Also, renesting females select a different cavity. In Illinois, 26 renesting females moved an average of 312 m to renest, with 1 female moving 12.1 km (Bellrose and Holm 1994). The time interval between renests depends on the stage of incubation at the time of nest destruction and the physical condition of the female; if a nest is destroyed during lay-

ing, the female may continue to lay at a new site. In central Illinois, the mean interval between the first and the second nest was 14.4 days ($n = 16$), with a new clutch delayed by 0.7 days for each day of incubation prior to the nest loss (Bellrose and Holm 1994). Of marked renesting females in central Illinois, adults accounted for 69.2% of renests but formed only 55% of the nesting population. In Massachusetts, Grice and Rogers (1965) reported a renesting interval of 0.33 days for each day the initial eggs had been incubated: most hens required at least 11 days, but 1 needed only 5 days. In southeastern Missouri, Hansen (1971) found the renesting interval of 6 females to be a minimum of 9 days.

REARING OF YOUNG

Brood Habitat and Care. The brood-rearing habitat for Wood Ducks is similar to their breeding habitat. McGilvrey (1968) recommended a mixture of shrubs (30%–50%), herbaceous emergents (40%–70%), and trees (0%–10%). Brood-rearing females in southeastern Missouri used scrub-shrub habitats (Heitmeyer and Fredrickson 1990). Beaver ponds are especially important brood-rearing habitats for Wood Ducks. In South Carolina, large (1.51–3.80 ha) beaver ponds were chosen more often than small (0.03–0.50 ha) ponds, and wetlands without beaver ponds were used primarily to travel from 1 beaver pond to another (Hepp and Hair 1977). Broods were seldom observed using nonvegetated, open-water areas of beaver ponds. On the Holston River in Tennessee, 34 radiomarked females with broods preferred aquatic bed and emergent cover types (Cottrell et al. 1990).

The female broods her ducklings for about 24 hours before she leads them from the nest site with a soft *kuk, kuk, kuk* call. She usually calls 2–4 hours after sunrise and after she has scanned the landscape for danger. The ducklings respond with peeping calls and immediately begin to leap upward toward the nest entrance, stopping there momentarily before springing outward to the water or ground below. They land unhurt, despite jumps that can approach 20 m (Bellrose 1980). Brood care by females averaged 31 days in Minnesota (Ball et al. 1975), 35 days in Michigan (Beard 1964), 39 days in South Carolina (Hepp and Hair 1977), and 56 days in Massachusetts (Grice and Rogers 1965). Late-hatched broods are usually deserted by the hens when the young are about a month old.

Females often travel long distances after departing the nest area. Of 9 radiomarked females with broods in South Carolina, 6 were highly mobile within 24–48 hours after hatching, moving an average of 1.4 km to brood-rearing habitats (Hepp and Hair 1977). Females used waterways such as creeks, lakes, and temporary wetlands to move broods to beaver pond habitats. On a riverine system in Tennessee, 34 radiomarked females with broods moved an average of 1.1 km from their nest sites to brood-rearing areas (Cottrell et al. 1990). During the first 8 weeks after hatching, the home range for these Wood Ducks averaged 46.1 ha (30.8 ha of wetland and 15.3 ha of associated shoreline).

Brood Amalgamation (Crèches). Wood Ducks have not been observed to form crèches.

Development. Body weight of day-old ducklings ($n = 43$) averaged 23.7 g (range = 19–28 g); it did not differ by sex (Hepp et al. 1987b). Lipids averaged 32.5% of a duckling's dry mass. Captive-reared ducklings attained asymptotic body mass in 7–9 weeks, at which time the males averaged 511 g and the females, 468 g (Brisbin et al. 1986, 1987). Maximum growth occurred at 19–20 days. In studies of captive Wood Ducks fed diets of 5%, 10%, 15%, and 20% protein until 60 days of age, growth was significantly reduced on the lower-protein diets (Johnson 1971). Duckling survival to 60 days was also least on the lowest protein diet, compared with the highest (11%–20% vs. 30%–63%).

RECRUITMENT AND SURVIVAL

Bellrose and Holm (1994) summarized brood survival from an array of studies across North America. Brood size at flight stage averaged 5.7, with an average of 52.1% of the ducklings lost since hatching. Total brood loss is not uncommon; in Minnesota, Wood Ducks raised an average of 41% of their ducklings to flight stage, but 5 out of 21 (23.8%) radiomarked females with ducklings lost their entire brood (Ball et al. 1975). Most mortality occurs within the first 2 weeks after hatching: 90% in Maryland (McGilvrey 1969), 86% in Minnesota (Ball et al. 1975), and 74% in Mississippi (Baker 1970). In 18 radiomarked broods in eastcentral Texas, 63% of the duckling mortality occurred during the first 2 days, and 83% in the first 10 (Ridlehuber et al. 1990). At Nauvoo Slough in Illinois (1984–87), early-hatching broods lost an average of 4.95 ducklings/brood by 4 weeks of age, with most (95%) mortality occurring in the first 2 weeks (Bellrose and Holm 1994). Bellrose and Holm (1994) also reviewed studies relating habitat quality to brood survival. Low (32%–48%) mortality rates were reported from impoundments and high-quality stream habitats where females could disperse their broods but not incur the hazards of overland travel. Conversely, mortality rates of 59%, 81%, and 92% were reported from poor-quality streams in small upland marshes.

The most extensive study of Wood Duck brood survival was conducted in Mississippi and Alabama by Davis et al. (2007). In Mississippi, the average survival of 300 radiomarked ducklings on the Noxubee NWR was 21% during 1996–99; brood survival was 64% ($n = 91$ broods). On the Tennessee-Tombigbee Rivers and Waterway System in Alabama (1998–99), the survival of 129 radiomarked ducklings averaged 29%; brood survival was 71% ($n = 38$). Interday distance traveled was an important predictor of duckling mortality at both sites, with the greatest number of deaths occurring among ducklings moving to scrub-shrub habitats. The survival of ducklings weighing 30 g at hatching was almost double that of ducklings weighing 24 g. Ducklings in smaller broods survived better than those in larger broods, perhaps because small broods were less detectable by predators; smaller broods were also easier for females to care for during inclement weather. In this study the survival of ducklings in late-hatched broods was greater than in early-hatched ones, a finding that contradicts a near-universal phenomenon of reduced survival for waterfowl ducklings in late-hatching broods. Very early nesting in the Deep South, however, causes broods to be hatched when vegetative cover is sparse and alternative prey is not yet available for predators. Avian predation was greatest for early-hatched broods, because there was little herbaceous cover. Fish predation was also highest on early-hatched broods. In general, an array of duckling predators were identified; Red-shouldered Hawks (*Buteo lineatus*), Great Blue Herons (*Ardea herodias*), and cottonmouth snakes (*Agkistrodon piscivorus*) were especially important (Davis et al. 2009). The authors also categorized cause-specific mortality for ducklings, combined across years and sites: 46% avian ($n = 155$), 23% aquatic predators ($n = 79$), 11% snakes ($n = 21$), 5% mammals ($n = 18$), 2% exposure related ($n = 7$), and 13% unknown ($n = 44$).

Nichols and Johnson (1990) used band-recovery data to estimate the survival of adult and immature Wood Ducks from 6 banding reference areas across North America. Survival averaged 55.6% for adult males, 50.6% for adult females, 47.5% for immature males, and 43.2% for immature females. In South Carolina, the survival rates of 181 females captured in nest boxes during an 8-year study (1979–96) averaged 55% (range 41%–71%; Hepp et al. 1987a). During a 13-year nest box study (1967–79) in Massachusetts, Rohwer and Heusmann (1991) noted that brood size had no effect on the annual survival of females, which averaged 52.8% over all age classes. Survival rates declined by 6.1%/year of breeding experience, however, a finding that could reflect the rigors of breeding efforts but requires further study.

Longevity records for Wood Ducks are 22 years and 6 months for a hatching-year male banded in Oregon, and 17 years and 7 months for an after-hatching-year male banded in Tennessee. Records for females are 17 years and 7 months for a hatching-year female banded in Iowa, and 13 years and 5 months for an after-hatching-year female banded in Massachusetts.

FOOD HABITS AND FEEDING ECOLOGY

Wood Ducks use a wide variety of plant seeds, fruits, and invertebrates, with their preferred foraging areas being flooded timber, scrub-shrub swamps, and shallow marshes. Wood Ducks feed in pairs or small groups, generally by tipping up or dabbling at the water's surface. In southeastern Missouri, 88% of 155 Wood Ducks collected for a food-habit analysis were surface-feeding in water with a mean depth of 19–40 cm (Drobney and Fredrickson 1979). Shallow water (<30 cm) was deemed especially important in providing access to invertebrate foods for breeding Wood Ducks. These ducks will occasionally dive for food, with both males and females observed diving for the acorns of pin oaks in water averaging 3 m deep (Briggs 1978). Their small narrow bills permit items such as acorns to be easily picked up. An extremely distensible esophagus allows Wood Ducks to easily ingest acorns measuring up to 1.9 × 5.7 cm, with 30 small acorns (probably from pin oaks) recovered from a single bird (Bellrose and Holm 1994).

Breeding. The food selection of breeding Wood Ducks in relation to breeding status was examined from spring to fall in southeastern Missouri (Drobney and Fredrickson 1979). Invertebrate consumption by 60 females increased from an aggregate percentage of 54% in prelaying females to 79% during laying, and then declined to 43% during incubation. The consumption of invertebrates is necessary to meet the demands of egg production, and insects dominated the spring diets of both females (46%; $n = 60$) and males (25%; $n = 55$). Adult and larval beetles (Coleoptera; 13%–33%) and various groups of flies (Diptera; 3%–16%) consistently ranked as the top 2 insect taxa eaten by females. Vegetatively, 28 species of plant foods were used, but only 4 were of major importance: the seeds from silver maples, watershield (*Brasenia schreberi*), elms (*Ulmus* spp.), and buttonbushes.

Young ducklings feed almost exclusively on animal matter. At John Sevier Lake in eastern Tennessee, the diets of 57 ducklings showed a decline in the percentage (by volume) of animal matter eaten as they grew older: 70.4% when they were <1 week old, 28.4% at 2–3 weeks old, and only 4.0% once they were ≥6 weeks old (Hocutt and Dimmick 1971). Diptera (50%) were by far the most important food item for young ducklings. At the Noxubee Wildlife Refuge in Mississippi, invertebrates—particularly mayfly (Ephemeroptera) and dragonfly nymphs—formed 85% of the diet of 37 ducklings (Baker 1971). Ducklings began using more plant material by 4 weeks of age, and tiny fish made up 15% of the foods they consumed.

Migration and Winter. Plant foods dominate the diets of Wood Ducks during the fall and winter, but acorns are preferred whenever available, for which these birds will readily forage in upland habitats. Wood Ducks are very opportunistic foragers; the food habits of 200 birds collected year round in South Carolina showed that 99 different plant foods were used during the study period (Landers et al. 1977). Bellrose and Holm (1994) reviewed virtually all the fall and winter food-habitat studies of Wood Ducks through the early 1990s, and they should be consulted for further details.

During midwinter in the Mississippi Alluvial Valley of western Mississippi, acorns (50.3% aggregate percentage), particularly from Nuttall's oak (*Quercus nuttallii*; 31.7%) and water oak (*Quercus nigra*; 14.7%), dominated the diets of 94 Wood Ducks (Delnicki and Reinecke 1986). In feeding trials, Barras et al. (1996) noted that captive female Wood Ducks preferred acorns from willow oaks (*Quercus phellos*) over those from 3 other oak species, even though willow oak acorns contained less fat and carbohydrates. In comparison with other species, however, acorns from willow oaks had small widths, thin shells, and the greatest ratio of meat-to-shell mass. Acorns also dominated the diets of Wood Ducks (*n* = 32) collected over 2 hunting seasons on managed impoundments in South

Carolina (57.2% aggregate percentage; Landers et al. 1976). Seeds from an array of emergent vegetation were also consumed, as was corn (16.0%).

In a study by Landers et al. (1977) in South Carolina, acorns were important (20%–32% aggregate percentage) food items from November through January, but a variety of other food items were also eaten. Acorns composed 40.4% of the winter diet in 1973, but only 3.7% in 1974, following a general mast failure in the area. Soybeans were the ducks' other major (32.8%) food item. Animal foods formed 23% of the diet in March. The seeds of white waterlily (*Nymphaea odorata*) were eaten in large quantities in August and September. Asiatic dayflower (*Aneilema keisak*) seeds—an especially nutritious food, averaging 21.3% protein and 49.0% carbohydrates—were the most consistently used food during late fall and winter. The data from this South Carolina study supported the need for habitat diversity in satisfying the nutritional requirements of Wood Ducks during their annual cycle.

MOLTS AND PLUMAGES

Adult Wood Ducks have 2 plumages/year, basic and alternate. Palmer (1976b), Kirby and Fredrickson (1990), Bellrose and Holm (1994), and Hepp and Bellrose (1995) should be consulted for further details.

In Massachusetts, Grice and Rogers (1965) reported that the first retrices appeared at 20 days, with the primaries emerging from their feather sheaths at 40 days. Body feathering is complete at 55 days, and almost all juvenal plumage is achieved in 70 days. Both sexes are primarily grayish, but the white chin and throat markings are visible on males in juvenal plumage at 45 days, and their irises begin to turn red at 60 days. Most (70%) young attain flight at 60 days, with nearly all becoming flight capable by 70 days, although the remiges are not fully grown until about 100 days. Bellrose and Holm (1994) reported fledging at 8–10 weeks (56–70 days).

Wood Ducks do not appear to have a first basic plumage (Bellrose and Holm 1994), but rather molt from juvenal directly to first alternate, which involves most feathering except for the remiges. First alternate is acquired by both sexes in about 100–150 days and resembles definitive alternate. Definitive basic involves molting all the feathers; hence there is a flightless period, lasting about 3 weeks. Nonbreeding yearling males are the first to molt, beginning as early as mid-April, while adult males begin molting soon after breeding, with almost all males in definitive basic by July. Late-nesting and renesting males and females, however, may delay their molt until late July. Some males regain alternate plumage by mid-September, but most do so between then and mid-October. Females begin to acquire definitive basic in late winter, a plumage involving variable amounts of feathering, but not the juvenal wing, which is retained until after nesting. This molt is then completed after nesting / brood rearing, during which time females also undergo a flightless period. Females acquire much of their alternate plumage when the basic feathers are replaced.

CONSERVATION AND MANAGEMENT

A discussion of Wood Duck management must begin with nest boxes. Production from nest boxes contributes only 4%–5% of the juveniles to the continental population (Bellrose 1990), but the judicious use of boxes can establish and increase local populations, although seldom in sufficient numbers to enhance populations at a state level (Bellrose and Holm 2004). Massachusetts and California, however, are exceptions. Heusmann (2000) reported that Massachusetts main-tains >7,000 nest boxes, from which 4,300 young were fledged, compared with 5,500 harvested. Under the auspices of the California Waterfowl Association, a Wood Duck nest box program was launched in 1991, with 2,293 boxes hatching 5,401 ducklings (2.4/box). By 2009, the program had grown to 5,244 boxes that hatched 34,574 ducklings (6.6/box). Over the 1991–2009 period, the box program yielded 542,107 ducklings (6.3/box), which is significant, given that the 1999–2009 Wood Duck harvest in California averaged about 30,000 birds. Hence Wood Duck boxes clearly have a place in management programs, and they also provide opportunities for research and education. Bellrose and Holm (1994) reviewed all aspects of nest box construction and placement, and they also addressed the use of metal nest boxes.

As Bellrose (1980) pointed out long ago, however, habitat is the crucial element in successful waterfowl management, as all other limiting factors are relatively transitory. Over-exploitation was correctly identified as the primary cause in the near demise of Wood Ducks, but the long-term effect of habitat loss was decimating populations as well. Clearing mature timber in the eastern deciduous forest removed untold nesting cavities, and overharvesting beaver reduced breeding and brood-rearing habitats throughout the range of the Wood Duck. Hence forest management practices that retain old-growth timber, coupled with the protection of nearby wetland habitats, is the long-term key to successful management of Wood Ducks.

Gadwall

Anas strepera (Linnaeus 1758)

Left, hen; *right*, drake

The Gadwall is a fairly well-studied, medium-sized dabbling duck. The species is widely distributed in North America, but its principal breeding range is the Prairie Pothole Region (PPR). Gadwalls primarily winter along the Gulf Coast in Louisiana and Texas, but some winter as far south as Mexico. They are also a widespread breeding species in Eurasia. Gadwalls are almost entirely herbivorous during the winter, consuming large amounts of often poor-quality submerged vegetation. This foraging strategy necessitates spending a large portion of their time feeding, and breeding Gadwalls, especially females, consume more aquatic invertebrates to meet the demands of egg production. They are also an exception among female puddle ducks in using stored protein to offset the costs of egg production. Gadwalls preferentially nest on islands and in dense cover, which leads to higher nest success than for other species of prairie-nesting ducks. Unlike many other species of dabbling ducks, Gadwall populations have increased in

Gadwall. *Left,* hen; *right,* drake

the PPR and expanded their range in both the eastern United States and parts of the West.

IDENTIFICATION

At a Glance. The Gadwall male is primarily gray, with a contrasting brownish head, black rump, and black undertail coverts. Females are brownish. In flight, both sexes show a distinct white speculum, and the white belly and underwing coverts are sharply defined by the darker breast, side, and flank feathers.

Adult Males. The overall body plumage is gray, with a grayish-brown buffy head, as well as a black rump and black tail coverts that contrast with the ashy-gray tail. There are also fine white and black vermiculations on the sides and the flanks that become much darker and crescent shaped over the breast. The belly is white. At close range, the light gray tertials form a noticeable patch at the rear of the folded wing; the long tawny-orange scapulars are also noticeable. The legs and the feet are yellowish or dull orange; the bill is dark gray. The irises are brown.

In all plumages, the wings in both males and females provide definitive identification characteristics. The inner third of the speculum has a white patch, while the outer area contains either a gray patch (immatures) or a black patch (adults). Forward of the speculum, the greater coverts show varying amounts of black, which is most pronounced in adult males. Males also show chestnut feathers on the shoulder, a coloring that is more noticeable in adults than in immatures.

Adult Females. Females are brownish tan in color overall, with a distinctly steep forehead and a flattened crown. Most of the body feathers are broadly edged with buff. The breast and the belly are white. A close look reveals a dark crown, a faint eye stripe, and a whitish throat. Females usually show a trace of chestnut on the shoulder. The feet are pale orange or yellow. The bill is orange, with a dark center and black spots on the lower edge. The irises are brown.

Juveniles. Immatures are similar to adult females, but generally browner, and the trace of chestnut on the shoulder is barely discernible. The varying degree of black on the greater coverts is least apparent in immature females. Juveniles are also more colorful than adult females, with brighter beige sides and a lighter grayish head, and a slightly greater contrast between the head and the body. The feet are pale orange to yellow. The irises are dark brown.

Ducklings. Ducklings have a pale creamy-yellow base color; the back is mottled, with smaller shoulder spots and larger rump spots (Nelson 1993). The bluish-gray / bluish-olive-gray upper mandible and pinkish-yellow lower mandible are also characteristic. Oring (1968:361) described ducklings at hatching as "cream-buff below and sepia above (except for four light dorsal spots), some individuals being yellower than others."

Voice. The call of the female is a *quack,* similar to but less strident than that of a Mallard. Males are quieter, but they utter a loud *kack, kack* when

Gadwall courtship flight; the lead bird is the female. *Ronnie Maum*

alarmed. The burp is a common vocalization in males, a deep reed-like *araeaeb*, uttered singly or in a series of 2–3 (LeSchack et al. 1997).

Similar Species. In flight, the white speculum distinguishes Gadwalls from any other species. American Wigeon in flight also show extensive white on the wings, but this coloration is on the coverts, not the secondaries. On the water, female Gadwalls can be confused with female Mallards, American Wigeon, and Northern Pintails, but the latter 2 species have bluish, not orange, bills. Female Mallards also have an orange bill, but they are larger, brownish, and lack the spotting on the edges of the bill.

Weights and Measurements. Bellrose (1980) reported average body weights of 966 g for adult males (*n* = 37), 835 g for adult females (*n* = 45), 857 g for immature males (*n* = 204), and 776 g for immature females (*n* = 200). Wing length averaged 27.2 cm for adult males (*n* = 22), 25.7 cm for adult females (*n* = 6), 26.7 cm for immature males (*n* = 58), and 24.9 cm for immature females (*n* = 56). Body mass of Gadwalls wintering in southwestern Louisiana from November through March averaged 982 g for adult males (*n* = 133), 887 g for adult females (*n* = 69), 897 g for juvenile males (*n* = 10), and 846 g for juvenile females (*n* = 31; Gaston et al. 1989).

DISTRIBUTION

Gadwalls have a broad distribution over much of the middle latitudes of the Northern Hemisphere (the Holarctic), breeding in both North America and Eurasia between about 40° and 60° N latitude, and wintering between 20° and 40° N latitude (Fox 2005a). An analysis of mitochondrial DNA (which helps to clarify evolutionary relationships

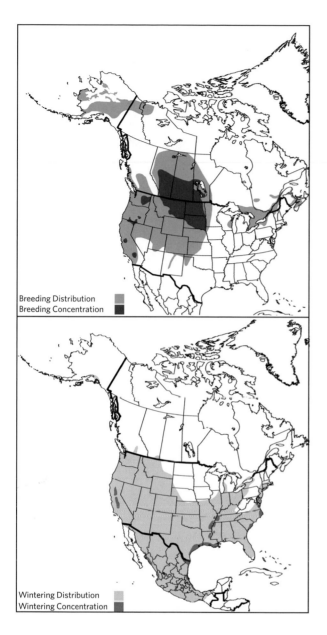

Breeding Distribution
Breeding Concentration

Wintering Distribution
Wintering Concentration

DNA structure (used in assessing relationships between populations) in North American Gadwalls (Peters and Omland 2007), so that future distinctions between populations will be less clear. Gadwalls are closely related to the Falcated Duck (*Anas falcata*) of Asia, as well as the wigeons (Livezey 1997).

In North America, Gadwalls principally breed in the PPR, although smaller breeding populations occur in virtually every other western state, from Washington and Oregon (east of the Cascades) south to California, and then eastward into northern Nevada, Utah, Colorado, Nebraska, western and central Kansas, and parts of the Rio Grande valley in New Mexico (LeSchack et al. 1997). Scattered breeding populations also occur in Wisconsin and New York (Peterjohn 1989), especially on state and federal wildlife refuges in westcentral New York (McGowan and Corwin 2008). In Canada, they breed eastward along the Great Lakes and up the St. Lawrence River valley. Breeding populations occur sporadically elsewhere in eastern Canada and on the Atlantic coast, with breeding as far north as Prince Edward Island and as far south as coastal Virginia (LeSchack et al. 1997). The major wintering area for Gadwalls is along the Gulf Coast, south to the states of Oaxaca, Veracruz, and Tabasco in Mexico, and the West Indies, as well as along the mid-Atlantic coast south to Florida. About 60% of the global population for this species occurs in North America.

The Coues's Gadwall (*Anas strepera couesi*), an extinct subspecies, formerly occurred in the Republic of Kiribati (on Washington and New York Islands) in the central Pacific Ocean, about 1,600 km south of the Hawaiian Islands (Johnsgard 1978). The 2 specimens collected in 1874 were similar to the mainland form but much smaller, about the size of teal, and they had more lamellae in the bill and somewhat different soft parts (Palmer 1976a). Livezey (1997) placed the Gadwall in the genus *Mareca* and used the species name *americana*.

In addition to their occurrence in North Amer-

among species) indicates that Gadwalls from Eurasia probably colonized North America during the Late Pleistocene (about 81,000 years ago), but the confidence interval (8,500–450,000 years ago) on that estimate was very large (Peters et al. 2008). Gadwalls from Alaska and Washington were well differentiated from other populations, but female-mediated gene flow, as well as historical and contemporary population and range expansions, have probably caused an overall weak mitochondrial

Left:
Gadwall pair.
Chadd Santerre

Below:
Female Gadwall; note the exposed lamellae.
Ian Gereg

ica, Gadwalls are widespread breeding ducks in Europe and Asia, with about 60,000 in northwestern Europe, 75,000–150,000 elsewhere in Europe, 130,000 in northeastern Africa and central/southwestern Asia, 500,000–1,000,000 in northeastern Asia, and 300,000 in southern Asia (Wetlands International 2006). Populations are small (100–500 pairs) in Iceland, England (introduced), the Netherlands, Czechoslovakia, and Greece, and sporadic (<10 pairs to occasionally 100 pairs) in Ireland, Scotland, Norway, Sweden, Finland, Denmark, Poland, Switzerland, Spain, Italy, and North Africa (Cramp and Simmons 1977). In Asia, major winter concentrations occur on the Black Sea and the eastern Mediterranean Sea, with smaller populations elsewhere. Gadwalls have bred as far east as Japan, although not commonly.

Breeding. Gadwalls breed over a broad area in western North America, but their principal range is in the mixed-grass prairie region of the PPR, with prairie parklands being of secondary importance. Dramatically fewer numbers breed to the north and east, including in Alaska (Bellrose 1980). Breeding also occurs south and west of the PPR, as well as sporadically in the Midwest and the eastern United States and Canada.

Gadwalls have expanded their breeding range northward, however, as noted by Palmer (1976a), and there are breeding records in Alaska from the Copper River delta, Juneau, and the Alaska Peninsula (Kessel and Gibson 1978). The northern limit of their breeding range is near Yellowknife in the Northwest Territories, at 62° N latitude. Small breeding populations are also present at Hay-Zama Lakes in northwestern Alberta, about 490 km southwest of Yellowknife, and the Athabasca River delta in northeastern Alberta, about 468 km southeast of Yellowknife (Fournier et al. 1992). Canning and Herman (1983) noted the range expansion of Gadwalls in the Pacific Northwest, west of the Cascades in Washington.

In the PPR, the Traditional Survey recorded 3.0 million Gadwalls in 2010: 39.6% in the eastern Da-

kotas, 30.7% in southern Saskatchewan, 13.2% in Montana and the western Dakotas, 10.8% in southern Alberta, and 5.7% in other survey regions, largely those areas north of the PPR (U.S. Fish and Wildlife Service 2010). In research over 6 years (1976–81) on 2 different 22.6 km² study areas in the PPR in North Dakota, pair density averaged 5.3/km² on the Koenig Study Area and 5.2/km² on the Woodworth Study Area; the highest (10.6/km²) density occurred on the Koenig Study Area in 1980 (Lokemoen et al. 1990a). Pair density did not appear to be strongly influenced by drought, with densities ranging from a low of 0.4–3.5/km² during a drought year (with only 31 basins on the study areas) to 3.9–4.4/km² during the wettest year (when 335 basins held water); pair density was 5.4–10.6/km² during a moderately wet year, when 129 basins were present.

From 1955 to 2010, the distribution of Gadwalls by region in the Traditional Survey was 33.6% in southern Saskatchewan; 22.0% in North Dakota; 17.1% in southern Alberta; 12.9% in South Dakota; 6.0% in Montana; 3.9% in southern Manitoba; 2.8% in the Northwest Territories, British Columbia, and northern Alberta; 0.7% in northern Manitoba; 0.1% in Alaska/Yukon; and 0.1% in western Ontario (Collins and Trost 2010). Gadwall breeding populations also occur in virtually all the western states outside the PPR, with populations from 1990 to 2010 averaging 93,414 in California, 51,108 in Oregon (just for 1994–2010), 14,877 in Washington (primarily the Columbia River basin), 8,645 in Utah (mainly the Great Salt Lake marshes), and 6,963 in Nevada (Great Basin marshes).

In eastern North America, Forbush (1925) reported that Gadwalls were a rare migrant and winter visitor in the New England states. Gadwalls were first reported breeding along the East Coast in 1939, however, after which they rapidly extended their breeding range to include more than 30 locations in the East, >2,000 km from their main breeding range in the West. Along the East Coast, where their numbers have largely increased since 1960, most breeding Gadwalls occur from

Long Island in New York down to South Carolina (Henny and Holgersen 1974). Tufts (1986) noted that Gadwalls were increasing as a breeding species in Nova Scotia. Breeding locations in the East are associated with freshwater impoundments created on the brackish marshes of national wildlife refuges and state wildlife management areas. No specific research exists to explain the range expansion of Gadwalls into eastern North America. Climate change and the construction of reservoirs and other man-made impoundments, however, are probably contributing factors (LeSchack et al. 1997). The Atlantic Flyway Breeding Waterfowl Plot Survey averaged 9,560 Gadwalls from 2003 to 2010, with the largest percentages in Pennsylvania (24.5%), New York (18.3%), and New Jersey (7.7%; Klimstra and Padding 2010). The Eastern Waterfowl Survey averaged only 315 Gadwalls from 1996 to 2010, most of which were in Québec (78.4%) and Ontario (10.8%), mainly in the St. Lawrence River valley.

Winter. The broad winter range of Gadwalls extends from southeastern Alaska east to southern Ontario and from there as far south as southern Mexico. Their principal wintering range is along the Gulf Coast, with lesser numbers occurring down the Pacific coast and the mid-Atlantic coast, southward into Florida.

The Midwinter Survey in the United States averaged 2.2 million Gadwalls from 2000 to 2010: 57.7% in the Mississippi Flyway, 33.2% in the Central Flyway, 7.8% in the Pacific Flyway, and only 1.3% in the Atlantic Flyway. Overall, 48.1% of the Gadwalls in this survey were recorded in Louisiana and 32.4% in Texas. Within the Mississippi Flyway, 83.2% occurred in Louisiana (coastal marshes) and 6.7% in Arkansas, with the remainder distributed elsewhere. In the Central Flyway, 97.5% were found in Texas, mostly along the Gulf Coast (the Laguna Madre). The 2010 survey reported 178,911 Gadwalls in the Pacific Flyway: almost all (93.2%) were in California, primarily in the San Joaquin Valley (51.8%) and the Sacramento Valley (46.8%; Collins

and Trost 2010). Only a small number winter north of Washington State, with a maximum winter population of 136 Gadwalls observed within the Strait of Georgia in British Columbia (Butler et al. 1989). The few Gadwalls in the Atlantic Flyway were mainly in the Carolinas (54.5%) and Virginia (25.3%); <100 were recorded from states north of New Jersey.

The average number of Gadwalls estimated during winter surveys from 1978 to 2006 along the east coast of Mexico was 23,686, with a steady decline in their occurrence from north to south in the survey areas: 33.7% in the Rio Grande delta, 31.3% in the lower Laguna Madre, 14.0% in the Tamesí/Pánuco River deltas, 9.1% in the Tamiahua Lagoon and the coastal area to Veracruz, 8.1% in the Alvarado Lagoons, 3.3% in the Tabasco Lagoons, and only 0.6% in the Campeche-Yucatán Lagoons. Summary data for all of Mexico from 1961 to 2000 averaged 68,468 Gadwalls; from 1981 to 2000 the average was 71,489 (Pérez-Arteaga and Gaston 2004).

Gadwall numbers on the west coast and Baja Peninsula of Mexico averaged 16,067 birds during surveys conducted at 3-year intervals from 1982 to 2006 (Conant and King 2006). Recent numbers have been smaller, however, declining from 11,000 in 2000, to only 1,130 in 2003, and slightly higher (to 3,173) in 2006. Pabellón, Topolobampo, and Agiabampo Lagoons, as well as Bahía de Santa Bárbara, are important wintering sites on the mainland west coast of Mexico. Some Gadwalls winter on marshes and reservoirs in the interior highlands of Mexico, with an average of 14,927 estimated during Midwinter Surveys from 1948 to 2006 (Thorpe at al. 2006). The lower west coast of Mexico averaged only 5,017 Gadwalls over this period.

Migration. Gadwalls are moderately early fall migrants, mostly departing their prairie breeding areas in September. At Delta Marsh in Manitoba, Oring (1964) noted their departure by mid-September, with nearly all birds gone by late September. Gadwalls are absent from their breeding areas in Alberta by early October (Semenchuk 1992). Gadwalls in British Columbia begin their migration in September, with most gone by early October (Butler and Campbell 1987). Janssen (1987) noted peak arrival in Minnesota by mid-October, as did Peterjohn (1989) in Ohio. Gadwalls appear in Iowa in late October and early November (Dinsmore et al. 1984). Large influxes of Gadwalls do not reach southern states (e.g., Arkansas and Alabama) until mid-October through November (James and Neal 1986, LeSchack 1993).

Gadwalls are among the last ducks to arrive in sizeable numbers on breeding areas in the spring. Bellrose (1980) noted that across their wintering range, the departure of Gadwalls commenced in early February. Gadwalls began to leave the Tennessee River valley in late February, with peak departure in March; most were gone by April (Bellrose 1980, LeSchack 1993). Gadwalls (mostly unpaired males) first arrived in early March at the Ogden Bay Waterfowl Management Area in Utah (1956–57), but they were generally among the last ducks to arrive at that site (Gates 1962). Peak passage occurred during a brief period in mid-April and was composed almost exclusively of mated pairs. Spring migration was completed by mid-May. Dinsmore et al. (1984) noted peak arrival in Iowa in mid-April, with birds arriving on nesting areas in North Dakota in late March and early April (Dwyer 1974, Hammond and Johnson 1984). At Delta Marsh in Manitoba (1962–64), the first Gadwalls arrived in mid-April, with peak arrival in early May.

Bellrose (1968, 1980) outlined migration corridors for Gadwalls east of the Rocky Mountains, noting that these birds primarily migrate through the plains states and the Mississippi Alluvial Valley en route to their major wintering grounds along Louisiana's Gulf Coast. Most Gadwalls follow the midplains migration corridor, which extends from central Saskatchewan southeast for roughly 2,700 km to the coastal marshes of Louisiana. Large wildlife refuges and reservoirs provide important habitats along the migration routes, probably because of the presence of submerged aquatic vegetation, a preferred Gadwall food (McKnight and Hepp 1998). Large concentrations of Gadwalls appear in the fall at the Lacreek National Wildlife Refuge (NWR) in South Dakota, the Valentine NWR in Nebraska, the Great Bend Wildlife Area in central Kansas, and the Squaw Creek NWR in Missouri. Large numbers of Gadwalls moving to the mid-Atlantic coast use the Reelfoot Lake / Tennessee River valley area of Tennessee, which is probably also a staging area for birds migrating to Louisiana. Gadwalls migrating to coastal Texas use the western plains migration corridor, which originates in southwestern Saskatchewan and southeastern Alberta. Gadwalls also stage at the Great Salt Lake marshes in Utah, from which they migrate to the Central Valley of California and the Texas coast.

Gadwalls are very capable of extensive travel over open ocean, with birds reported on the island of Hawaii and Midway Island in the central Pacific (Palmer 1976a). Yocom (1964) reported 100–200 Gadwalls during winter 1959/60 on Kwajalein Atoll in the Marshall Islands, about midway between Hawaii and New Guinea. The Coues's Gadwall, an extinct island subspecies from Washington and New York Islands in the mid-Pacific, probably resulted from Gadwalls that wandered well off course during migration.

MIGRATION BEHAVIOR

Gadwalls primarily migrate at night (Bellrose 1980), in small flocks of ≤100 birds (Duebbert 1966). Otherwise there is no information on the migration behavior of Gadwalls.

Molt Migration. Males and unsuccessful breeding females leave the breeding areas to gather in molting concentrations within suitable habitats. Near Delta Marsh in Manitoba, molting Gadwalls began to accumulate on large marshes and lakes by mid-June, but the distances traveled from breeding to molting areas is not known (Hochbaum 1944, Oring 1969). At the Lower Souris NWR in North

Dakota, Duebbert (1966) reported that nearly all males were in molting flocks by 1 July. Successful breeding females molt near their brood-rearing areas (LeSchack et al. 1997). Preferred habitats for molting are large lakes and marshes bordered by dense vegetation, which provides cover (Oring 1964, Hohman et al. 1992a). In Colorado, most (67%–86%) recaptured Gadwalls were encountered on the same reservoir where they were originally captured, indicating strong fidelity to their molting areas (Szymczak and Rexstad 1991).

HABITAT

The largest numbers of breeding Gadwalls occur in the mixed-grass prairies of the Dakotas and the Prairie Provinces of Canada (Sousa 1985). Breeding pairs of Gadwalls in North Dakota ($n = 746$) primarily inhabited seasonal wetlands (54.6%) and semipermanent wetlands (35.4%); only 3.6% were on temporary wetlands (Kantrud and Stewart 1977). Breeding pairs in South Dakota during 1973–74 ($n = 248$) used stock ponds (20.9%–25.0%), dugouts (8.2%–10.8%), and semipermanent wetlands (39.6%–47.8%; Ruwaldt et al. 1979). In the PPR in Canada, the percentage of Gadwall nests initiated in brush habitats was 49% in prairies and 9% in parkland regions (Greenwood et al. 1995). Odd areas—patches of cover <2 ha in size with an array of features, such as rock piles, upland vegetation borders around wetlands, and vegetation along fences between croplands—were also well used in parklands (79%) and prairies (41%).

During the winter, Gadwalls inhabit reservoirs, beaver ponds, other ponds, and coastal marshes, all containing abundant submerged vegetation (Paulus 1982, McKnight and Hepp 1988, Gaston 1991). During migration, Gadwalls use reservoirs and larger wetlands in the fall, and smaller wetlands in the spring (Kantrud et al. 1989, Pederson et al. 1989).

POPULATION STATUS

Unlike almost all other duck populations in North America, Gadwalls have increased in number and expanded their breeding range since surveys began in the 1950s. On the Traditional Survey, Gadwalls averaged 1.8 million from 1955 to 2010, but the overall upward trend of the past 56 years can be described in 3 phases. The population averaged 745,560 from 1955 to 1964, but it was above 1 million each year from 1965 to 1991, with an average of 1,506,000 birds. From 1992 to 2010 the population was above 2 million in each of those 19 years, and above 3 million during 6 of them; the average was 2,796,300. The highest population was tallied in 1997, at 3,897,200 Gadwalls. The 2010 population was 2,976,700, which was 67% above the long-term (1955–2009) average.

Harvest

Gadwalls are an important game species in North America. The average harvest in the United States was 1,530,308 birds from 1999 to 2008: 55.8% in the Mississippi Flyway, 30.5% in the Central Flyway, 11.0% in in the Pacific Flyway, and 2.7% in the Atlantic Flyway. Of the Mississippi Flyway harvest, most were taken in Louisiana (33.3%) and Arkansas (26.4%), with 7.2% in Mississippi. From 1999 to 2008, the harvest in Canada averaged 38,969 birds, with a high of 44,069 in 2008. Over this time period, 32.0% were harvested in Saskatchewan, 31.7% in Alberta, 19.2% in Manitoba, 7.2% in Ontario, 5.6% in Québec, and 4.3% in the remaining provinces.

BREEDING BIOLOGY

Behavior

Mating System. Gadwalls are monogamous, with both sexes breeding as yearlings, although yearling males are much less likely to breed than adult males. In Manitoba, Blohm (1982) noted that in 39 pairs where the age of both sexes was known, 74% of those pairs involved adult males, compared with 26% for yearling males. In 40 other instances where only the age of the male was known, adults were present in 78% of the pairs, and yearlings in only 22%. Of the 39 known-age pairs, 51% were adult male/adult female pairs, 23% adult male/

yearling female pairs, 18% yearling male/yearling female pairs, and 8% yearling male/adult female pairs. In a flock of captive birds where all the males where yearlings, Oring (1969:48) noted that "there was very little breeding and very few pairs formed."

Pair formation in Gadwalls begins either during fall migration or on the breeding grounds, as 45% of the females ($n = 384$) observed in coastal southwestern Louisiana were paired by mid-October (Paulus 1983). Pairing then continued rapidly, with 81% of the females ($n = 736$) paired by late November. The percentage paired then increased slowly, with 90% ($n = 1,589$) paired by April. In coastal North Carolina, 91.8% of the females ($n = 49$) were paired in November and 100% ($n = 20$) in February. Courtship behavior was initially observed in late October, but displays indicated that courtship began earlier (Hepp and Hair 1983). The percentage of time spent in courtship displays was highest (2.2%) in November, then 0.9% in December, 0.4% in January, and only 0.04% in February. In Louisiana (1977–78), the time spent by males in courtship behavior was highest (1.6%) in October–November, compared with only 0.3% in December (Paulus 1984a). Most courtship behavior occurred in the morning, but courtship activities were twice as common during periods of rainfall than at other times. Paulus (1984b) also noted that only male Gadwalls with 75% or more of their alternate plumage engaged in courtship activities and/or were paired.

The duration of the pair bond is variable. In Utah, males usually deserted their mates before mid-incubation, but a few early-nesting pairs remained together until hatching (Gates 1962). The pair bonds of captive Gadwalls dissolved at varying times during the incubation period, including an instance where the bond lasted until day 23 (Oring 1969). In North Dakota, Duebbert (1966) also observed that the pair-bond duration varied widely; bonds dissolved as early as day 7 of incubation, while others lasted until the end of incubation. LeSchack et al. (1997) remarked that most males desert their mates after the clutch is completed.

Sex Ratio. The sex ratio is slightly male dominated. Males averaged 55.2% among 206 Gadwalls observed during spring 1947 in Manitoba (Bellrose et al. 1961). During spring in Saskatchewan, the sex ratio of 715 observed Gadwalls was 52.3% males (Hines and Mitchell 1983a). Most Gadwalls were paired on arrival to the Lower Souris NWR in North Dakota, where the sex ratio was fairly equal (52.4% males; Duebbert 1966). The percentage of males during winter in North Carolina was 55.6%–58.2% for 5,192 Gadwalls observed from November through March, 1978/79 and 1979/80 (Hepp and Hair 1984).

Site Fidelity and Territory. During a 6-year study (1976–81) involving large numbers of marked female Gadwalls in North Dakota, the return rates (the percentage of hens resighted on the study area the year after banding) increased significantly with age: 9% for hatching-year birds, 34% for second-year birds, 49% for third-year birds, and 67% for after-third-year birds (Lokemoen et al. 1990a). For hatching-year females that subsequently returned, 52% were not resighted until their second summer after banding. Also, females that nested successfully returned at a significantly higher rate (68%) than unsuccessful hens (38%). In Manitoba, Blohm (1979) reported a return rate of 56% for females ($n = 101$), compared with only 8% for males ($n = 284$). At the Ogden Bay Waterfowl Management Area in Utah (1956–57), the return rate of females was at least 60% (Gates 1962). In Colorado, estimates of the recovery and recapture rates of banded Gadwalls were 2.14% and 4.21% (respectively) for adult males and 1.86% and 5.80% for females, which reflected fidelity to their breeding wetlands and a nearby reservoir for molting (Szymczak and Rexstad 1991).

At Ogden Bay Marsh in Utah, Gadwalls did not establish home ranges until 17 days after their arrival (Gates 1962). The female determined the location of the home range, which was used for loafing, feeding, and nesting. The home range of 5 marked females averaged 27 ha (range = 14–35 ha). Island-nesting Gadwalls in North Dakota used pothole wetlands up to 4.8 km away; "home ranges were apparently very large, consisting of several hundred acres," although many pairs were highly mobile and did not favor specific loafing sites (Duebbert 1962:17). Aggression by males appears to be in response to defending the female, however, not defending a physical territory (Duebbert 1966).

In North Dakota, pairs established an activity center during the prenesting period, from which they exhibited two types of defense toward intruders (Dwyer 1975). Males respond to an intruding male or males with a *chin-lift* posture or *open-bill threat* display, while the pair may exhibit a *mutual-chin-lift* posture, preceded by *inciting* by the female. Intruders usually departed following these displays; otherwise the paired male initiated an actual attack. Paired males will also react to other pairs by initiating *3-bird-flights* (*pursuit-flights*) that can last up to 1 minute and extend over 1.6 km. The overall energy expenditure of paired males declined during the breeding season, from 164 kcal/day during spring arrival to 156 kcal/day during prenesting, and to 132 kcal/day during the laying period.

Courtship Displays. The courtship displays of Gadwalls are similar to other species of puddle ducks in the genus *Anas*, and LeSchack et al. (1997) provided a succinct description and drawings of the major Gadwall displays. Males give an *introductory-shake* as a preliminary display. Major displays are *head-up-tail-up*, *down-up*, *grunt-whistle*, *preen-behind-the-wing*, *turn-the-back-of-the-head*, and *burp*. The *burp* display is a low grunt given with the neck extended and the bill usually pointed at the female. The *grunt-whistle* is the least frequent of the major displays, but it is also fairly elaborate. The male rears up high off the water with the head stretched forward and the bill dipping into the water, often displacing a string of droplets as the head is then raised and the bill pressed against the breast. The *grunt-whistle* is accompanied by a loud *whistle*, followed by a low *burp*. *Inciting* is the common courtship display of the female. Both sexes exhibit precopulatory *head-bobbing*. After copulation, the male gives the *grunt-whistle*. Both sexes *bathe* after copulation, which is common in puddle ducks.

During November in North Carolina, *burping* was the most frequent (20.1%) male display, although 5.0% were *jump-flights* and 3.2% were *grunt-whistles* (Hepp and Hair 1983). Swimming was observed 47.9% of the time, but it was considered a reproductive behavior, because individuals used swimming to position themselves for actual displays. *Inciting* was the common (39.4%) female display. *Jump-flights* usually occurred late in the courtship process, during which time competition among the males was most intense. *Pursuit-flights* were common on the breeding grounds in North Dakota, with as many as 50/hr observed near islands where Gadwalls were nesting in high densities (Duebbert 1966). Prior to egg laying, *pursuit-flights* involved 3–15 birds and were usually initiated from flocks on the water when 1 pair approached too close to another pair. *Pursuit-flights* became most intense during incubation, with nearly all such flights involving the male from 1 pair chasing the female of another pair; mid-air fights were common between the 2 males. In North Dakota, Dwyer (1975) noted that the time spent in *pursuit-flights* was highest (0.7%) during prenesting, compared with the laying period (0.3%). Males spent an average of 1.4% of their time in *threat* postures during spring arrival, but only 0.7% during prenesting and 0.4% during laying, as pairs became more isolated from unmated males and other pairs.

Gadwalls exhibit considerable intraspecific aggression during the winter, which was well documented in southwestern Louisiana (Paulus 1983). Aggression was greatest (0.8%–0.9% of the time) during October and November, with Gadwalls more likely to threaten other individuals of similar social status (e.g., paired vs. unpaired). Gadwalls used *bill-threats* (89%), *chasing* (5.8%), and *biting* (5.2%) during these aggressive encounters. Pairs were dominant to unpaired birds, as pairs won

84% of the contests with unpaired males and 78% of the contests with unpaired females. Between unpaired Gadwalls, males won 41% of the encounters and females won 59%. Only 14% of the aggressive encounters were directed toward other species of ducks, mostly (77%) American Wigeon, which used the same feeding areas as Gadwalls. The study suggested that the dominance exhibited by paired birds probably afforded them better access to preferred foods and thus allowed them to be more successful in meeting nutrient requirements than unpaired birds.

Nesting

Nest Sites. Gadwalls are upland nesters, and pairs searching for nest sites fly low over upland areas, landing where the vegetation is relatively short. The female, however, probably selects the actual nest site (Blohm 1979, Bellrose 1980). They typically locate their nests in dense areas of brush, forbs, and/or grasses. At Waterhen Marsh in Saskatchewan (1972–73), islands with dense patches of western snowberries (*Symphoricarpos occidentalis*) or tall forbs such as slim nettle (*Urtica gracilis*) were preferred Gadwall nesting sites (Hines and Mitchell 1983b). Most nests were in vegetation with >25% canopy coverage, >30 cm canopy height, and a lateral-concealment rating of 3 or 4 (4 was the maximum). The dominant (64%) plant species at 451 nests were western snowberries; 10% were in smooth brome (*Bromus inermis*). In snowberry patches ($n = 151$ nests), canopy cover averaged 50.2%, height averaged 53.4 cm, and lateral concealment averaged 3.5. Shrubs were dominant at 71% of the nest sites, forbs at 17%, and grasses or sedges at 12%. Nest densities were 1/ha in uplands, 30/ha on ditchbanks, 62/ha on artificial islands, and 74/ha on a 2.2 ha natural island. The average distance from water for 21 upland nests was 56 m.

At the Ogden Bay Waterfowl Management Area in Utah (1956–57), nests ($n = 156$) were located on channel banks (48%), dikes (31%), and miscellaneous sites (21%; Gates 1962). Vegetative cover at nest sites included upland forbs and grasses (57%), spikegrass (*Distichlis stricta*; 18%), and spikerush (*Eleocharis rostellata*; 9%). In general, nesting Gadwalls preferred the densest and driest available cover close to water; 94% of all nests were within 46 m of water. At the Tule Lake and the Lower Klamath NWRs in California (1952 and 1957), Gadwall nests were located on islands (60.3%–61.9%), fields (23.1%–23.9%), and dikes (13.4%–15.9%; Rienecker and Anderson 1960). Most nests were located in stands of nettle (*Urtica californica*; 49.8%–79.9%), thistle (2.6%–22.0%), or saltbush (*Atriplex* sp.; 7.6%–17.3%).

Gadwalls will readily nest on islands when they are available. During their survey of 209 islands in the PPR (in the Dakotas and northeastern Montana) in 1985–86, Lokemoen and Woodward (1992) found that Gadwalls had the highest island-use index among all waterfowl. Gadwalls also often nest at very high densities on islands. On 2.2-ha Goose Island in Waterhen Marsh in Saskatchewan, Gadwall nest densities were 284/ha in western snowberry patches (Hines and Mitchell 1983b). These females obviously nested in close proximity, but they would not nest at distances within about 1 m of a neighboring female. On 2.8-ha Ding Island at the Lower Souris NWR in North Dakota, however, Duebbert (1966:20) noted that "many nests were found within a few feet of each other, and some were [at] less than 12 inches." In an area of especially high density, 28 nests were located within a 23-m radius. Overall, 78 nests were located on the island in 1956, and 121 in 1957 (27.8–43.2 nests/ha). In 1957 about 80% of the nests were located in tall nettle (*Urtica procera*), with most nests initiated when the plants were 15.2–25.4 cm high, compared with plant growth to 1.8–3.1 m late in the incubation period. During a 5-year study (1976–80) on a 4.5-ha island in 385-ha Miller Lake in North Dakota, Gadwall nest density ranged from 139 to 237/ha in patches of shrub cover (Duebbert et al. 1983). Over 97% of all duck nests on the island occurred within 4 patches of shrubs, totaling about 1 ha, that contained western snowberries and Woods' rose (*Rosa woodsii*).

Gadwalls will also use planted grasslands in places where such habitats are available. At the Koenig Study Area in North Dakota (1979–81), nest density was highest (60.6/km²) in seeded cover (Lokemoen et al. 1990). Elsewhere in North Dakota (1971–73), estimated nest initiations were highest (21/km²) in seeded grasslands, as was also the case (42/km²) in South Dakota (Klett et al. 1984). In fields planted with dense nesting cover in northcentral North Dakota, duck-nesting densities averaged 77/km² over a 3-year period (1971–73), with Gadwalls contributing 24% of all nests (Duebbert and Lokemoen 1976).

The female constructs the nest by scraping a depression in the soil and forming a bowl of twigs and leaves lined with pieces of vegetation and down feathers plucked from her breast (Kear 1970). Where Gadwalls nested in high density on an island in North Dakota during 1977, 49% of the nests occurred in previously used nest bowls (Duebbert et al. 1983).

Clutch Size and Eggs. Gadwalls usually lay 5–13 eggs/nest, but parasitized nests may contain as many as 20 eggs. In multiple studies of 2,545 nests from a variety of areas, the average clutch size was 10.0, with a range of 8–11 eggs/nest (Bellrose 1980). In the PPR in the Dakotas (1993–95), clutch size averaged 9.9 eggs for 1,753 Gadwall nests (Krapu et al. 2004c). At the Woodworth Study Area in North Dakota (1966–81), clutch size of 256 nests averaged 9.9 (Higgins et al. 1992). Clutch size of 82 nests located in planted nesting cover in South Dakota averaged 10.7 (Duebbert and Lokemoen 1976). At Waterhen Marsh in Saskatchewan (1972–73), clutch sizes of early nests (initiated before 20 Jun) averaged 10.4 eggs ($n = 295$) compared with late nests that averaged 8.0 ($n = 27$: Hines and Mitchell 1983a). Clutch size averaged 11.0 for 586 successful Gadwall nests on the Tule Lake and the Lower Klamath NWRs in California (Rienecker and Anderson 1960). During a 2-year study (1956–57) in Utah, the average size of all completed clutches (including renests) was 10.0 ($n = 141$), but the clutch

size of 92 initial nests averaged 11.1 (Gates 1962). In Manitoba, Blohm (1979) observed that yearling females laid an average of 1 less egg than did older females.

Clutch size of nests located on Ding Island in North Dakota averaged 9.5 in 1956 ($n = 51$) and 9.7 in 1957 ($n = 79$; Duebbert 1966). The author also noted the frequency distribution of 140 clutch sizes during incubation: 5–6 eggs (2.9%), 7 eggs (12.1%), 8 eggs (12.9%), 9 eggs (22.1%), 10 eggs (22.1%), 11 eggs (12.9%), 12 eggs (9.3%), 13–18 eggs (4.3%), and 20 eggs (1.4%). On an island in Miller Lake in North Dakota (1977), 82% of 161 Gadwall nests contained 8–14 eggs, 9% had >14 eggs, and 8% included parasitic eggs, usually from Mallards; 1% contained dwarf (runt) eggs (Duebbert et al. 1983). During all years (1976–80) of the study, 83% of 213 incubated clutches contained 8–14 eggs, with an average clutch size of 10.3.

Gadwall eggs are smooth and glossy, ovate-shaped, creamy white to grayish green in color, and measure 55.3 × 39.7 mm ($n = 100$; Bent 1923). Eggs are laid at a rate of 1/day, most often in the early morning between 05:00 and 07:00 (Duebbert 1966).

Incubation and Energetic Costs. The female incubates the eggs, with an average incubation period of 24.0 days (range = 22–26 days) for 6 clutches in captivity and 25.8 days (range = 24–27 days) for 8 wild clutches (Oring 1969). On average, females spent 85% of each day incubating eggs, with 1.9 recesses/day that averaged 118 minutes each (Afton and Paulus 1992). Females devoted 66% of their recess time to foraging.

Ankney and Alisauskas (1991) studied the reproductive energetics of female Gadwalls, because this species is among the most herbivorous of dabbling ducks and, therefore, the females were predicted to forage inefficiently for invertebrates (protein sources) needed for egg production. The lipid reserves of the females declined 0.78 g for each gram deposited in their eggs, and their protein reserves decreased by 0.16 g. In other words, 78%

of the lipids deposited in eggs came from endogenous (stored) reserves, compared with 16% of the protein. Thus female Gadwalls are an exception among temperate-nesting dabbling ducks in their use of stored protein for egg production, although American Wigeon use protein reserves for 44% of their egg production (Alisauskas and Ankney 1992a).

Yearling Gadwall females had smaller lipid reserves than adults, but neither age group initiated egg production until the females had attained a threshold/minimum level of reserves. Females at or near the end of egg laying exhibited considerable variation in the size of their energy reserves, however, which may be due to different tactics for successfully completing incubation. This pattern probably reflects the annual variation in environmental factors, such as temperature and food availability. Thus some years may favor females that invest more energy in producing eggs (deemed a "risky" strategy), compared with other years when less energy is invested in laying eggs but more is available during incubation (a "cautious" strategy).

Nesting Chronology. Gadwalls are late nesters, with a considerable delay between the time of arrival on their breeding grounds and nest initiation. At Delta Marsh in Manitoba (1963–64), the average nest-initiation date (26–29 May) was 23 days after peak arrival (Oring 1969). In Utah, the interval between spring arrival and nest initiation was 28 days (Gates 1962). Of 108 nests located in Saskatchewan, the earliest were initiated 11–17 May in 1972 and 25–31 May in 1973, with peaks occurring during the last week of May in 1972 and the first week of June in 1973, about 1 month after most Gadwalls arrived (Hines and Mitchell 1983a).

During a 4-year study (1972–75) in Manitoba, the average nest-initiation dates ranged from 30 May to 20 June, with the peak nest-initiation date averaging 1 week later for yearlings (Blohm 1979). In North Dakota (1977), nest initiation began the first week of May, and 92% of the nests were initiated before June 9 (Duebbert et al. 1983).

Lokemoen et al. (1990a) found that in North Dakota (1976–80), yearling females nested about 1 week later than 2-year-old females, and 2-year-old females nested about 1 week later than females >2 years old (Lokemoen et al. 1990a). Older females also nested over a longer period: 40 days for >3-year-old females, compared with 24 days for 2-year-old females. In southcentral Saskatchewan (1996–97), the first clutches hatched on 23–30 June, with a median hatch date of 9–11 July (Gendron and Clark 2002). All clutches hatched over a 32- to 48-day period.

In 1956–57, Gadwalls began nesting at Ogden Bay in Utah about mid-May and continued to initiate nesting until mid-July (Gates 1962). Of 241 Gadwall nests monitored in the Klamath basin in California during 1957, 95.5% hatched between 22 June and 12 July (Rienecker and Anderson 1960). On Ding Island in North Dakota in 1956–57, the nesting period spanned 7–9 weeks from first laying until the hatching of the last nests; most nests hatched between 8 and 14 July (Duebbert 1966).

Nest Success. Gadwalls have higher rates of nest success than other species of prairie-nesting ducks, because they prefer to nest on islands and in dense cover. On Ding Island in North Dakota, nest success was 85.7% for 78 nests found in 1956 and 92.7% for 121 nests found in 1957; predators destroyed only 10 nests (Duebbert 1966). The hatching success of the eggs was 80.3% ($n = 710$) in 1956 and 85.9% ($n = 1,045$) in 1957. On an island in Miller Lake in North Dakota (1976–80), the success of 811 nests averaged 88.4% (Duebbert et al. 1983). Nest success on Goose Island in Saskatchewan was 82% for 287 nests (Hines and Mitchell 1983b). In the Klamath basin in California, where most (60.3%–61.9%) Gadwall nests were on islands, nest success was 90.3% for 344 nests located in 1952 and 87.4% for 242 nests seen in 1957 (Rienecker and Anderson 1960).

In upland situations, the nest success of Gadwalls across a broad area of the PPR in the Dakotas and Minnesota averaged 12% ($n = 425$) for 1966–74,

16% ($n = 1,153$) for 1975–79, and 15% ($n = 926$) for 1980–84 (Klett et al. 1988). Unlike the 4 other species of dabbling ducks also studied, the nest success of Gadwalls was highest in cropland habitat (28.3%) and idle grasslands (23.7%); success was lowest in haylands (5.7%). Their nest success was much higher than that of Mallards using the same nesting areas (13% vs. 7%). At the Woodworth Study Area in North Dakota (1966–81), nest success (Mayfield Method) was 21.7% for 654 nests (40.4% apparent success; Higgins et al. 1992). Mammals caused 87% of all the nest losses. Elsewhere, Gadwall nest success averaged 50% for 136 nests monitored in planted grass/legume cover in South Dakota in 1971–73 (Duebbert and Lokemoen 1976). Reynolds et al. (2001) evaluated 1,919 Gadwall nests in upland areas in the PPR in the Dakotas and northeastern Montana (1992–95) and reported daily survival rates of 94.8% on both Conservation Reserve Program habitats and Waterfowl Production Areas.

The hatchability of eggs is high in successful Gadwall nests. In the Klamath basin, the hatching success was 94.2% for 3,611 eggs in 1952 and 92.7% for 2,459 eggs in 1957 (Rienecker and Anderson 1960). Egg hatchability was 87% for 1,661 eggs in successful nests on an island in Miller Lake in North Dakota (Duebbert et al. 1983). On Goose Island in Saskatchewan, 95.8% of 2,481 eggs hatched, with the hatching failures due to infertile eggs or the early death of the embryo (Hines and Mitchell 1983a). On Ding Island in North Dakota, the hatching success of the eggs was 80.3% ($n = 710$) in 1956 and 85.9% ($n = 1,045$) in 1957 (Duebbert 1966). Of 114 eggs remaining in Gadwall nests after the hatch, 77% had no embryo visible but were probably not infertile.

Losses of Gadwall nests are primarily caused by the array of mammalian nest predators that occur in the PPR, and secondarily by avian predators (Sargeant et al. 1993). Significant mammalian predators are striped skunks (*Mephitis mephitis*) and coyotes (*Canis latrans*); avian predators include American Crows (*Corvus brachyrhynchos*),

Common Ravens (*Corvus corax*), California Gulls (*Larus californicus*), and Ring-billed Gulls (*Larus delawarensis*). Of 44 Gadwall nests attacked by predators in Ogden Bay in Utah, California Gulls destroyed 56.8% and striped skunks accounted for the remaining 43.2% of the losses (Gates 1962). Ring-billed Gulls destroyed 18.6% of all Gadwall eggs laid at Farmington Bay in Utah (Odin 1957). During a 5-year study (1976–80) on an island in Miller Lake in North Dakota, however, desertion was the primary (6%–17%) cause of nest failure, probably due to intraspecific interactions among females nesting in close proximity and the harassment of incubating females by males (Duebbert et al. 1983).

Brood Parasitism. Both intra- and interspecific brood parasitism occurs in Gadwalls, especially where nests are concentrated on islands. In North Dakota, 34.4% of 64 Gadwall nests on islands were parasitized, compared with only 2.8% of 106 nests located on peninsulas (Lokemoen 1991). Gadwalls deposited eggs in 3% of all the parasitized nests on islands. A parasitized Gadwall nest contained 8 host eggs, 4 Lesser Scaup eggs, and 3 Redhead eggs. In another instance, a Mallard nest received 1 Gadwall egg and 2 Redhead eggs. Redheads parasitized 10.1% of 190 nests on Ding Island in North Dakota (Duebbert 1966). On an island in Miller Lake in North Dakota, 9% of all Gadwall nests in 1977 contained compound clutches of >14 eggs, and 8% contained parasitic eggs (usually from Mallards). No Redhead nests were located on the island in 1977, but Redhead eggs occurred in 7 Gadwall nests (Duebbert et al. 1983). Evidence of egg movement between clutches was noted in this study, as 14 nests (either Mallard or Gadwall) had 36 eggs added, and 32 nests had 55 eggs removed.

At Waterhen Marsh in Saskatchewan, at least 60 out of 355 (17%) completed Gadwall clutches were parasitized: 9% by other Gadwalls, 8% by Lesser Scaup, and <1% by both species (Hines and Mitchell 1984). Nests parasitized by other Gadwalls contained at least 7–8 parasitic eggs, with the final clutch size in 33 nests averaging 18.0 eggs (range = 15–22 eggs); the normal clutch size of 295 nonparasitized nests was 10.1. In contrast, Gadwall nests parasitized by Lesser Scaup averaged only 1.6 parasitic eggs (range = 1–5 eggs), with the final clutch size averaging 10.7 for 27 nests. Lesser Scaup also parasitized nests in shorter or sparser cover, while Gadwalls parasitized nests in cover similar to that of nonparasitized nests. With a single exception, all instances of intraspecific parasitism were found in 0.5 ha of snowberries on 2.2 ha Goose Island in Saskatchewan.

Intraspecific parasitism reduced nest success from 76% to 54%, egg success (the percentage of all eggs in the sample that hatched) from 74% to 45%, and the hatchability of eggs (the percentage of eggs incubated to term that hatch, which excluded any eggs lost to predation or breakage) from 97% to 91% (Hines and Mitchell 1984). In contrast, parasitism by Lesser Scaup had no significant effect on nest success or egg hatchability, but egg success decreased from 74% to 67%. In North Dakota, Lokemoen (1991) observed that brood parasitism reduced the host clutch size, as 24 parasitized nests averaged 8.6 eggs, compared with 9.9 in 119 nonparasitized nests. Parasitism also affected nest success, which was 82% in nonparasitized nests versus 66% in parasitized nests. In North Dakota, brood parasitism caused the abandonment of 9 out of 115 (7.8%) Gadwall and Mallard nests with abnormally large clutches (>13–14).

Renesting. Gadwalls will readily renest if their nest is destroyed. Near Brooks, Alberta, 82% of the Gadwalls tried again after the loss of an initial nest (Keith 1961). In a 2-year study (1956–57) in Utah, 74% of 23 females renested after the experimental removal of the first clutch, 27% after the removal of the second clutch, and 25% after the removal of the third clutch; the overall percentage of renesting females was 52% (Gates 1962). Completed clutches in renesting attempts averaged 7.8 eggs (n = 24), compared with 10.7 eggs for initial nests (n = 19). Of 19 marked females producing an initial renest,

4 renested a second time, and 1 renested a third time. Continuous egg laying was exhibited where nests were experimentally destroyed, with 1 female laying 22 eggs in 3 clutches over a period of 22 days, another laying 12 eggs in 2 clutches in 12 days, and a third laying 17 eggs in 2 clutches in 17 days. Otherwise, the renesting interval lengthened as the stage of incubation advanced, although such intervals were highly variable after 10 days of incubation. Each female moved to a new nest site for each clutch and resumed laying the day after her nest was destroyed. The mean distance between first nests and renests averaged 220 m and ranged between 101 and 480 m (n = 35). The success of 30 renests was 53%, compared with 51% for 61 first nests. An average of 8.1 ducklings hatched from first nests, compared with 6.6 in renests. Renesting was not a significant factor on Ding Island in North Dakota (1956–57), where Gadwalls nested in high densities and nest success was 90% (Duebbert 1966).

REARING OF YOUNG

Brood Habitat and Care. In contrast to most other puddle ducks, Gadwall broods generally prefer open-water areas of wetlands rather than areas of dense emergent vegetation (Evans and Black 1956, Mack and Flake 1980). The habitats of Gadwall broods on Garrison Reservoir in North Dakota were characterized by permanent or semipermanent wetlands (80%), with 61%–100% open water (90%), water depths ranging from 0.6 to >1.2 m (80%), and <50% emergent cover (70%; Sayler and Willms 1997). In contrast, Mallard broods chose seasonal wetlands (67%) characterized by only 0%–60% open water (57%) that averaged <0.6 m deep (67%), with ≥50% emergent cover (81%).

Observations of 1,072 Gadwall broods over 18 years (1958–63 and 1967–78) in the PPR in the Dakotas revealed that most (61%) were on semipermanent wetlands, with 18% on seasonal wetlands, and 9% on permanent wetlands (Duebbert and Frank 1984). In southeastern Alberta (1953–57), Gadwall broods were not observed on potholes <0.5 ha in size and <0.6 m deep, but they were "very abundant" on saline lakes 8.5–19.4 ha in size and >1 m deep (Keith 1961). In eastern Washington, 80% of the Gadwall broods used wetlands that were 0.4–0.9 m deep and had dense aquatic plant cover, such as pondweeds (*Potamogeton* spp.) and Eurasian watermilfoil (*Myriophyllum spicatum*). Niche overlap there was high with broods of Blue-winged Teal and Cinnamon Teal, because these latter species had similar midseason hatching dates and also fed on dense stands of submerged vegetation (Monda and Ratti 1988).

In Saskatchewan, newly hatched broods used mixed areas: cattails (*Typha latifolia*), open water, and mudflats with sparse submerged vegetation but abundant invertebrates (Hines and Mitchell 1983a). As the ducklings matured, however, their food use shifted from invertebrates, and they more frequently selected areas with submerged vegetation, increasing from 22%–32% use by Class I broods, to 38% for Class II broods, and 67% for Class III broods. Broods (*n* = 173) fed at an average distance of 62 m from emergent cover. Gadwall ducklings were also more tolerant of rough water conditions than other dabbling ducks, as they were often found feeding in waves >5 cm in height.

The female leads her young to the brood-rearing habitat within 24–36 hours after hatching and will brood the ducklings for up to 2 weeks (LeSchack et al. 1997). At Ogden Bay in Utah, 13 marked females moved their broods an average of 1.0 km from the nest site to brood habitat (Gates 1962). On the first day of hatching on Ding Island in North Dakota, Duebbert (1966) observed some females leading their broods to brood-rearing habitats across 460 m of open water, although a favored area for most broods was only 274 m away.

Brood Amalgamation (Crèches). Brood amalgamation occurs in Gadwalls, but no published information is available. Brood amalgamation was commonly observed in North Dakota, especially where many broods were found on larger water bodies; it also occurred among older broods (Class II through fledging), perhaps as the females abandoned their broods to begin the wing molt (David Brandt). Amalgamation was less common in younger broods and among those using seasonal basins, where brood densities were lower.

Development. Duckling weight at hatching averaged 30.8 g for males (*n* = 50) and 30.5 g for females (*n* = 47; Blohm 1979), after which their growth was similar to that of other puddle ducks. Based on wild-caught birds in North Dakota, Lokemoen et al. (1990b) reported the average body mass of males as 188.0 g at 17 days (*n* = 63), 406.6 g at 33 days (*n* = 131), and 616.9 g at 48 days (*n* = 50). Females averaged 186.6 g at 17 days (*n* = 40), 380.7 g at 33 days (*n* = 91), and 573.0 g at 48 days (*n* = 33). Oring (1968) described the juvenile growth of about 250 captive birds and presented a growth curve that depicted a rapid increase in body weight after the ducklings were 15 days old. All but 3 out of 50 captive birds (26 males, 24 females) were capable of flight in 50–56 days. Gadwalls in South Dakota reached flying stage in 48–52 days (Gollop and Marshall 1954).

RECRUITMENT AND SURVIVAL

At Waterhen Marsh in Saskatchewan (1972–73), the mean brood size at hatching was 10.0 (*n* = 260), and then 8.9 in Class Ia (*n* = 127), 7.72 in Class Ib (*n* = 69), 7.3 in Class Ic (*n* = 18), 7.1 in Class II (*n* = 24), and 5.4 in Class III (*n* = 9; Hines and Mitchell 1983a). Overall mortality was 45% between hatching and flight, with 45% of those deaths occurring in the first 10 days. During 16 years of research (1966–81) on waterfowl production in the Woodworth Study Area in North Dakota, Higgins et al. (1992) noted that hen success (the percentage of females producing at least 1 young that reached fledging) averaged 30%, with an average brood size of 7.3 at fledging. Duckling loss between nesting and fledging averaged 26%, with the total recruitment rate for the population averaging 76%. About 85% of duckling mortality for the 12 duck species studied occurred within the first 2 weeks after hatching.

The most intensive study of duckling and brood survival of Gadwalls was conducted in the prairie pothole habitat of eastern North Dakota during 1990–94 (Pietz et al. 2003). The 30-day brood survival rate was 84%, with 9 out of 58 (16%) radiomarked females losing their entire broods. Brood size was the best predictor of brood survival, with the risk of total brood loss declining by 24% for each additional duckling in a given brood. Duckling deaths may be more independent within a brood than in other species of puddle ducks (such as Mallards), because Gadwall broods are less cohesive, often spreading out over a wider area. The 30-day survival rate of 212 radiomarked ducklings from 94 broods averaged 43.8% and revealed that the daily mortality was twice as high when seasonal ponds were scarce, compared with when they were abundant; 30-day survival was 57% when ≥41% of their seasonal wetlands contained water, versus 32% when ≤22% contained water.

The 30-day survival rate was also highest when the minimum temperature exceeded 10°C and no rain occurred, but duckling survival was lowest during the first 7 days after hatching, whether rain fell or not. Of 87 recorded duckling deaths, 86% were attributed to predators, with American mink (*Mustela vison*) accounting for ≥68% of the deaths where the predator could be determined. Krapu et al.'s (2000) observations supported the apparent resilience of Gadwall populations during times of drought, as Gadwall brood survival was higher than that of Mallards on the same study area during the same years. In addition, Gadwall brood survival may be higher because this species nests comparatively later in the spring; their broods experience milder weather, and brood predators probably have access to a greater abundance of alternative prey. Gadwall broods may also be less susceptible to mammalian predation because they spend more time in open-water habitats, away from shore. Moreover, the study documented the importance of seasonal wetlands to duckling survival for Gadwalls, although not to brood survival, perhaps because fewer seasonal wetlands are generally available to Gadwalls, due to their later nest-initiation dates, compared with Mallards.

In southcentral Saskatchewan (1996–97), Gendron and Clark (2002) studied brood survival in Gadwalls by monitoring 73 brood-rearing females fitted with radiotransmitters. Duckling mortality was greatest during the first 14 days. Averaging their results over 2 years for 2 study sites yielded an average brood size of 8.9 ducklings at hatching and 6.5 at 30 days, a loss of 2.4 ducklings/brood. Duckling survival was lowest (81.9%) over the first 7 days after hatching. but then increased to 95.0% over days 8–14. The 30-day survival estimate was 73.1%. Five broods were abandoned early, and the ducklings presumably died. Two females were killed during brood rearing. Female condition and body size significantly influenced duckling survival at 1 study area, while duckling survival at the second site decreased with the distance traveled for early-hatched broods, but not for late-hatched broods. Survival was greater in this study, compared with others, because of the high-quality brood habitat, the presence of fewer predators, and a lack of adverse weather during the study years. Females in poor condition at the end of incubation, however, experienced lower brood survival, probably because they had to increase their foraging time to recover lost body mass and thus reduced their brood care. It is also probable that females in the poorest condition would be the most likely to abandon their broods, but that assessment was not part of the study. Duckling production was greater from larger broods, and more ducklings were produced from early-hatched clutches, but that relationship varied between sites and years.

In contrast to prairie pothole habitat, duckling survival to fledging was much lower for the broods of radiomarked Gadwall females (*n* = 24) nesting on islands in the Garrison Reservoir in North Dakota (1985–87): 0%–38% on Lake Sakakawea and 0%–41% on Lake Audubon. Gadwall females moved their broods an average of 2.5 km to brood-rearing habitats. Survival was less in years with low water levels in Lake Sakakawea and for broods produced after the median hatching date of 22 June, as well as for broods moved to off-island wetlands for rearing. Poor duckling survival on these lakes was attributed to oligotrophic water conditions; a lack of emergent cover along the shoreline; long travel distances to brood-rearing wetlands; exposure to severe weather and wave actions; and predation by mink, California Gulls, and other predators.

Szymczak and Rexstad (1991) used recoveries from 3,739 adult male and 2,509 adult female Gadwalls banded in Colorado (1975–85) to estimate a survival rate of 75% for the males and 69% for females. In North Dakota, annual survival rates were 54% for hatching-year females ($n = 184$), compared with 78% for after-hatching-year females ($n = 200$; Lokemoen et al. 1990a). Based on 87 recoveries from 1,318 Gadwalls banded in Utah, Gates (1962) estimated survival at 33% for juveniles and 48% for adults. Banding and recovery data from 7,791 adults and 23,073 immatures across North America from 1978 to 1996 yielded average survival estimates of 61% for adult males, 60% for adult females, 50% for immature males, and 46% for immature females (Frank Bellrose).

Longevity records for Gadwalls are 19 years and 6 months for a male banded as a local bird in Saskatchewan in 1962 and shot in Louisiana in 1981, and 17 years and 8 months for a male banded as a juvenile in Saskatchewan in 1952 and harvested in Mexico in 1970.

FOOD HABITS AND FEEDING ECOLOGY

Gadwalls feed in open-water areas of wetlands and on lakes and reservoirs where there is an abundance of submerged vegetation, their preferred food. Nearly 100% of the winter diet of Gadwalls can consist of submerged vegetation, but it is usually of poor nutritional quality and thus requires large amounts of feeding time. In Louisiana during the winter, Gadwalls fed by dabbling, tipping, or filtering from the surface, but not by diving. Feeding occurred throughout the day and night (Paulus 1982). During the breeding season in North Dakota, Gadwalls on saline wetlands most often fed by dabbling, tipping, surface picking, and filtering from the surface while swimming with the head and neck submersed (Serie and Swanson 1976).

Gadwalls commonly feed in deep water, where they graze on submerged aquatic vegetation, aquatic insects, or zooplankton. In comparison with 13 other species of ducks, Gadwalls wintering in coastal Texas used deeper water with submerged vegetation (White and James 1978). In North Dakota, foraging Gadwalls were collected at sites averaging 32.9 m deep (Serie and Swanson 1976). Gadwalls frequently feed together with American Coots (*Fulica americana*) and American Wigeon, but, unlike American Wigeon, they rarely graze on upland grass pastures or grain fields.

Breeding. Like other puddle ducks, Gadwalls increase their consumption of aquatic invertebrates during the breeding season to meet the demands of egg production. On saline wetlands in North Dakota (1971–72), the esophageal contents of 107 Gadwalls averaged 45.9% animal matter and 54.2% plant matter between 17 April and 25 August (Serie and Swanson 1976). The principal animal foods were crustaceans (primarily Anostraca and Cladocera; 29.6%) and insects (16.3%). The most important plant foods were filamentous algae (Chlorophyceae; 25.3%) and widgeongrass (12.7%). Females ate an average of 48.2% animal matter, compared with 40.4% for males. The percentage of animal foods consumed by females varied markedly over the breeding season, however, from 47.7% during the prelaying period to 72.0% during laying, but only 46.3% during postlaying. Aquatic insects dominated the diet of egg-laying females (52.4%), with another 19.6% made up of crustaceans. Within these categories, midge (Chironomidae; 25.8%) larvae and aquatic beetles (Coleoptera; 16.2%) were the most important insect foods, and Cladocera (9.9%) were the most important crustaceans. Filamentous algae was the primary (18.3%–32.1%) plant material used by females throughout the breeding cycle.

Gadwalls foraged during all daylight hours and were observed feeding after dark on dense and accessible cladoceran populations and emerging midge larvae. Small crustaceans such as Cladocera were usually obtained by filtering, while surface picking or tipping was used for insects and filamentous algae. Organisms <5 mm were not eaten unless they were extremely abundant, but Gadwalls were very adept in obtaining small midges (5.5 mm) and beetle larvae (6.2 mm) from the substrate.

Gadwall ducklings fed almost exclusively (89%–100%) on invertebrates for the first 5 days after hatching, but the percentage of plant foods they consumed steadily increased to 79% at age 16–20 days, and to 95%–99% thereafter until fledging (Sugden 1973). Their diet from hatching to fledging was 90% plant material and 10% animal material ($n = 167$). About 80% of the animal foods eaten were insects: primarily midge larvae and beetles (Coleoptera), but also Cladocera (16%). Important plant foods were pondweed (*Potamogeton pusillus*; 34%) and green algae (Cladophoraceae; 19%).

In North Dakota, breeding female Gadwalls spent an average of 70%–80% of their time feeding during arrival, prenesting, and laying, while males averaged about 70% on arrival, 50% during prenesting, and 30% during the laying period (Dwyer 1975). Males fed for 27% of the time when accompanying their mates, but 36% of the time when alone.

Migration and Winter. During fall migration in Illinois (1979–81), Havera (1999) noted that the stomach contents of 22 harvested Gadwalls contained 99% (by volume) plant material, consisting mainly of brittle naiad (*Najas minor*; 29.1%), fennelleaf pondweed (sago pondweed; *Potamogeton pectinatus*; 15.5%), duckweed (*Lemna minor*; 14.5%), common arrowhead (*Sagittaria latifolia*; 9.2%), and water hemp (*Amaranthus rudis*; 7.4%). During fall migration at Ogden Bay in Utah, Gates (1957) reported that the diet of Gadwalls consisted of sago pondweed (27%), widgeongrass (17%), bearded

sprangletop (*Leptochloa fascicularis*; 11%), seeds from hardstem bulrush (*Scirpus acutus*; 8%), and horned pondweed (*Zannichellia palustris*; 8%).

Gadwalls wintering in southwestern Louisiana from November through March 1977–78 ($n = 86$) consumed 95.3% plant matter, 0.5% seeds, and 4.2% animal matter (aggregate volume). Algae composed 31.1% of their diet and was found in 69.8% of all birds collected (Paulus 1982). Other important plant foods were common widgeongrass (*Ruppia maritima*; 14.9%), dwarf spikerush (*Eleocharis parvula*; 12.8%), spiked watermilfoil (Eurasian watermilfoil; *Myriophyllum spicatum*; 9.7%), and common hornwort, or coontail (*Ceratophyllum demersum*; 6.8%). Animal matter apparently was eaten incidentally while consuming aquatic vegetation. Widgeongrass, watermilfoil, and pondweeds were the preferred foods in fall and early winter, when these species were abundant. These plants were rare by mid-January, however, so Gadwalls then fed primarily on dwarf spikerush and common hornwort for the reminder of the winter.

From October through April in 1977/78, Gadwalls wintering in southwestern Louisiana averaged 64% of their time feeding, which was similar among all cohorts of sex and pair status and indicated that their nutritional requirements were comparable (Paulus 1984a). Gadwalls spent more time feeding at night (69.3%) than during the day (61.1%), with their peak feeding activity usually occurring when temperatures were lowest and thus thermoregulatory costs were highest. Gadwalls averaged 68% of their time feeding when ambient temperatures were <14°C (their lower critical temperature), compared with 55% at temperatures ≥14°C. Feeding time also increased from 44% in October to 77% in April, but it varied among habitat types: 55.0% in deep natural marsh, and 72.5% in shallow natural marsh. In addition, feeding times differed according to food choices, ranging from a low of 45.7% when foraging on poor-quality food (such as dwarf spikerush) to 75.0% when consuming higher-quality spiked watermilfoil. Predation pressure and harassment from

Northern Harriers (*Circus cyaneus*) were also much lower at night than during the day, which allowed Gadwalls to have more undisturbed feeding time. The remainder of their time budget during the winter was spent in locomotion (11%), resting (11%), alert behavior (9%), preening (5%), and agonistic and courtship displays (<1%).

Body weight and lipid reserves of Gadwalls wintering in southwestern Louisiana (1984–85) were reflective of their food choices and the region's mild environment (Gaston et al. 1989). Lipid levels in adult males (*n* = 133) increased steadily, from 11% of their body weight in November to 20% in March; the lipids of adult females (*n* = 69) increased from 10% of their body weight in November to 19% in January, but then decreased slightly (to 18%) in March. There was no significant decline in midwinter lipid reserves during this study, which might have been the result of the mild winter encountered by these Gadwalls, although midwinter lipid reserves could decline during colder winters. Hepp (1985) observed that Gadwalls may increase their foraging speed and rate of dipping in response to shorter days and colder temperatures, behaviors that may have contributed to the absence of a midwinter lipid decline among wintering Gadwalls in Louisiana.

From November to February on Guntersville Reservoir in Alabama, the diet of Gadwalls (*n* = 75) was 99.2% vegetation. Eurasian watermilfoil was by far the most important food, forming 58%–99% of the wet mass consumed during all months (McKinght and Hepp 1998). Large plant stems were always avoided by Gadwalls, but they preferred leaves in December and January and small stems in December. The nutrient content of available foods changed over the winter, but the nutritional composition of Gadwall diets remained relatively constant, since they selectively fed on the higher-quality parts of milfoil as well as other foods. Such dietary shifts potentially explain how Gadwalls survive on a diet dominated by milfoil. During January, Gadwalls consumed invertebrates at 8 times their availability in the environment. January was also the only month when their food preferences were related to nutritional content, as evidenced by a positive correlation between preference rank and amounts of protein.

Wintering Gadwalls (*n* = 31) using managed tidal impoundments in South Carolina (1972–74) consumed 90.8% plant material (Landers et al. 1976). The most important foods (as a percentage of total volume) were seeds from flatsedge (*Cyperus odoratus*; 28.0%), the seeds and rhizomes of redroot (*Lachnanthes caroliniana*; 20.4%), and widgeongrass (18.3%), although widgeongrass was eaten most frequently (71.0%).

MOLTS AND PLUMAGES

Molts and plumages are fairly well described for Gadwalls in Oring (1968), Palmer (1976a), Paulus (1984b), and LeSchack et al. (1997). Oring (1968) provides a detailed description of the plumage sequence in Gadwalls based on observations of a captive Gadwall flock over a 2-year period.

Ducklings begin to acquire juvenal plumage when they are 7 days old, but this plumage is not readily noticeable until about 2 weeks of age, and it is not fully developed until approximately 56 days. Fledging occurs in 50–56 days, but late-hatching ducklings are capable of flight at a younger age than early hatchlings (Oring 1968). The birds are brownish at this time, with the crown and the nape dark but the rest of the head pale buffy. The back is dark; the upper breast is sepia; many feathers are buffy edged. The lower breast and the belly are white or buff and streaked or spotted with black. The vent usually has brown markings. Males molt into first basic during mid-August through December, which replaces some of the body feathering but not the remiges or retrices. First basic is similar to definitive basic in females. First alternate quickly follows first basic and is a replacement of all feathers except the juvenal remiges and retrices; this plumage is retained until the following summer. First alternate resembles definitive alternate in both males and females, but the vent usually lacks any brown markings.

Males acquire definitive basic on the breeding grounds, beginning as early as May but usually in June and July. At Delta Marsh in Manitoba, most captive males were molting body feathers during the first week of June (Oring 1969). Of the wild males Oring observed, 10%–15% were flightless by 23 July, and 50% were flightless by 5 August. All feathering is replaced, so there is a flightless period of about 25 days, although the remiges are not completely regrown until 35–40 days. Nearly all males will acquire new retrices and remiges by September. Males in this plumage resemble females in definitive alternate. Females acquire definitive basic from January to late May, but the heaviest molting occurs during March and April. Most of the body feathers are replaced at this time, but successfully breeding females do not molt their remiges until about 6 weeks after their ducklings hatch. This plumage is similar to definitive alternate.

Males acquire definitive alternate beginning in late summer, but this plumage extends into winter for the tertials and retrices. In Louisiana, males had completed definitive alternate by late January. Females acquire definitive alternate in association with their breeding cycle and the wing molt; some females begin to acquire this plumage when their ducklings are 2–3 weeks old. In Louisiana, Paulus (1984a) noted that adult females replaced their feathers from November through March, with the majority of this replacement occurring during the prealternate molt in fall (Nov) and the prebasic molt in spring (Mar). Chabreck (1966) reported an unusually late wing molt during October among females in the coastal marshes of Louisiana in 1962 ($n = 18$) and 1964 ($n = 32$). These females probably were late in rearing their broods and thus delayed the wing molt until their arrival on the wintering grounds.

CONSERVATION AND MANAGEMENT

As with other species of North American puddle ducks, habitat loss in the PPR is the primary factor that will affect Gadwall populations (see the Blue-winged Teal account). In contrast, their wintering habitat has been less affected, and even enhanced, by the creation of large reservoirs. Managed waterfowl impoundments also provided important wintering habitat for Gadwalls in South Carolina (Landers et al. 1976) and Louisiana (Paulus 1982), although spring drawdowns removed important foraging habitat. Paulus (1984) noted the importance of natural wetland habitats for Gadwalls wintering in southwestern Louisiana and emphasized the need for protecting these areas.

American Wigeon

Anas americana (Gmelin 1789)

Top, hen; *bottom*, drake

The American Wigeon is a midsized dabbling duck breeding only in North America. Often called the "Baldpate," the colorful head of the male is characterized by a distinctive whitish crown and forehead that conferred this alternative common name. The breeding range of American Wigeon is extensive, with its principal area extending from tundra habitats in Alaska and western Canada southward through the Prairie Pothole Region (PPR) of provinces and states. A small breeding population occurs in eastern North America at locations around the Great Lakes and eastward, largely along the St. Lawrence River. Wintering birds primarily occur along the Gulf Coast and in the Central Valley of California, but they also range south into Mexico. The bill of the American Wigeon is very goose-like and indicative of their eating habits: they are primarily grazing herbivores, especially during winter, when virtually 100% of their diet consists of leafy aquatic vegetation. They commonly graze in upland habitats on agricultural crops (such as wheat and clover) and various species of grasses. The breeding population averaged 2.6 million from 1955 to 2010 in the Traditional Survey, with a 2010 estimate of 2.4 million. American Wigeon are important game ducks in North America, with an average harvest of 760,000 in the United States from 1999 to 2008.

American Wigeon hen

American Wigeon drake

IDENTIFICATION

At a Glance. The white forehead and crown of American Wigeon males are distinctive, as is the blue-gray bill. Females are brownish. Both sexes show an elliptical white belly that is sharply outlined by the brown chest and sides. Males in flight show large white patches on the upperwing, but these wing patches are indistinctly grayish in females and immatures.

Adult Males. Males show a whitish crown and forehead that give the bird its alternative common name, "Baldpate." A broad green streak extends from the eye to the back of the head, arcing downward and contrasting with an overall buffy head that is heavily speckled with gray. The pinkish-brown breast and the similarly colored sides, both marked with fine wavy black lines, are sharply separated from the black undertail coverts by the white femoral tract and the belly feathers. The white wing patch is very conspicuous in flight, but it also is sometimes visible as a narrow elliptical mark on birds at rest. The speculum is black, with a greenish stripe. The legs and the feet are blue gray. The bill is bluish gray, with a black tip. The irises are dark brown.

Adult Females. Adult females have a roundish gray head that is heavily streaked with black; it contrasts with the gray-brown back and the rust-colored sides and chest. The legs and the feet are blue gray. The small stout bill is grayish to slightly bluish gray, with a black tip and a slight black border at the base, which is not present on Eurasian Wigeon (*Anas penelope*). The irises are dark brown. Both adult and immature females have an indistinct gray-tinged shoulder patch, with most of the whitish greater coverts having black margins. A tinge of green sometimes occurs in the dark speculum, which is black in adults and brownish gray in immatures.

Juveniles. Juveniles are similar to adult females, except that immatures lack a white bar on the greater secondary coverts of the wings.

Ducklings. Ducklings have a yellowish-brown secondary base color, with a partial or complete light eye ring (Nelson 1993). The typical eye stripe and ear spot of most *Anas* ducklings is either faint or lacking in American Wigeon. The short tapered bill is rather goose-like, and young ducklings have a characteristic upright posture.

Voice. The frequently heard call of male American Wigeon in flight is the *slow whistle*, which consists of 3 whistling, piping notes, described as *whew-*

whew-whew, with the middle note loudest (Wishart 1983a). Females give an *inciting* call variously described as *kaow kaow* and a low *errr err*.

Similar Species. In North America, the females are most similar to female Gadwalls and Northern Pintails, but the smaller bill and the contrast between the grayish head and the brownish breast are distinguishing features of the female American Wigeon. Female Gadwall are also differentiated by their orangish, rather than grayish, bill.

Weights and Measurements. Bellrose (1980) reported average body weights of 821 g for adult males ($n = 84$), 767 g for adult females ($n = 68$), 794 g for immature males ($n = 358$), and 708 g for immature females ($n = 373$). Wing length averaged 26.4 cm for adult males ($n = 61$), 24.6 cm for adult females ($n = 34$), 25.7 cm for immature males ($n = 62$), and 24.3 cm for immature females ($n = 74$). Body weights of American Wigeon collected throughout the year in British Columbia, Saskatchewan, and Manitoba averaged 792 g (range = 635–1036 g) for adult males ($n = 65$) and 719 g (range = 513–872 g) for adult females ($n = 68$; Wishart 1979). Bill length averaged 40.5 mm for males ($n = 45$) and 35.1 mm for females ($n = 28$; Wishart 1983a).

DISTRIBUTION

There are 3 species of wigeon: the Chiloe Wigeon (*Anas sibilatrix*), the Eurasian Wigeon, and the American Wigeon. The Chiloe Wigeon inhabits the southern third of South America, from southern Brazil, northern Argentina, and central Chile to Tierra del Fuego and the Falkland Islands. The Eurasian Wigeon is widespread, breeding from Iceland east across Eurasia, largely north of 48° N latitude, almost to the Arctic Ocean. Eurasian Wigeon winter from central Eurasia south nearly to the equator in western and eastern Africa, Jordan, Iraq, all of India, Myanmar (Burma), southern China, and Japan (Johnsgard 1978). Breeding of Eurasian Wigeon in North America is suspected but not confirmed (Edgell 1984, Fournier and Hines 1996); hence this species is treated briefly here.

Eurasian Wigeon (*Anas penelope*)

Eurasian Wigeon are now a fairly common fall migrant and wintering bird in North America, with lesser numbers reported during spring migration. Sightings have occurred in all provinces and states (except Mississippi), and there are 2 records from northern Mexico (Edgell 1984). Eurasian Wigeon generally are found along the Pacific and Atlantic coasts during the fall and winter, and they have been observed in the interior of North America during the spring. Most sightings are of single birds, with 2, 3, and 4 being progressively less common, although Hasbrouck (1944) listed flocks of 21 and 20–30. Eurasian Wigeon are common spring and fall migrants, however, in the westernmost islands of the Aleutian Archipelago, where they regularly arrive in spring flocks of >10, with a maximum of 35 observed on 22 May 1976 (Gibson 1981). Band-recovery data indicate that Atlantic coastal birds originate in Iceland, while Pacific coastal birds are probably migrants from birds breeding on the Kamchatka-Anadyr area in Russia (Edgell 1984).

In adult Eurasian Wigeon males, the russet-red head and neck, and the gray sides and back, readily distinguish them from American Wigeon. Adult females occur in 2 color phases, gray and red, but they are more difficult to separate from female American Wigeon; hence most Eurasian Wigeon sightings are of males. The axillaries are finely speckled with dark gray in the Eurasian Wigeon, while those of the American Wigeon are almost pure white. There are also recognizable differences in the greater coverts. Distinctions are harder to spot between immature birds from the 2 species. These 2 species also readily hybridize (Merrifield 1993), and mixed pairs are observed during the winter.

Hasbrouck (1944) summarized 596 records of Eurasian Wigeon in North America through 1940, noting that the first record was that of a bird killed on 3 December 1842 on Long Island in New York and later purchased in the Fulton Market. His findings revealed that 67% of the sightings were on the Atlantic coast, 20% on the Pacific coast, and the

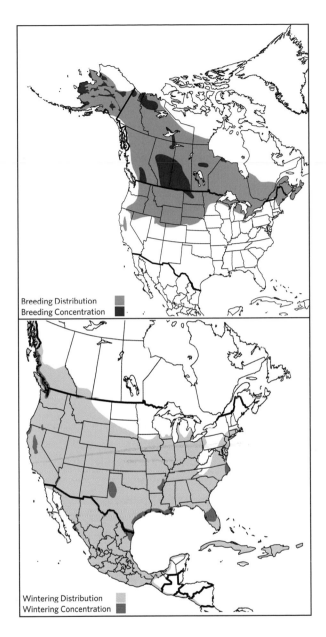

Breeding Distribution
Breeding Concentration

Wintering Distribution
Wintering Concentration

mainly concentrated in the Puget Sound area and California, while most of the Atlantic coastal sightings were between Massachusetts and Chesapeake Bay. Spring and summer records also occurred on both coasts, but the percentage of birds declined from the winter sightings: 39% in the Pacific Flyway, 16% in the Atlantic Flyway, 10% in the Mississippi Flyway, and 35% in the St. Lawrence River valley. Edgell's data showed a dramatic increase in Eurasian Wigeon sightings since the mid-1960s, with records from Christmas Bird Counts on the Pacific coast increasing 20-fold from 1965 to 1981. Edgell estimated that around 1,000 Eurasian Wigeon wintered in North America in 1981. Breeding is not confirmed, but the presence of wintering and summering pairs, as well as returning pairs to winter sites in British Columbia, strongly suggests that the Eurasian Wigeon has established breeding status in North America.

The Eurasian Wigeon is now common enough in North America to occur in harvest surveys. From 1999 to 2008, the harvest in Canada averaged 111 birds, with a high of 285 in 2002. Over this time period, 25.5% were harvested in Nova Scotia, 21.5% in British Columbia, 19.7% in Newfoundland, 12.2% in Alberta, 7.9% in Québec, and 13.2% in the remaining provinces.

American Wigeon

The distribution of American Wigeon is widespread. They are among the most northerly breeding dabbling ducks in North America, with a breeding range that extends from the tundra south through the Great Basin marshes, and east across Canada and the northern United States to Nova Scotia and Maine. They breed farther north than all other dabbling ducks, except Northern Pintails (Johnson and Greer 1988). American Wigeon winter from British Columbia east to New England and south to Central America and the West Indies.

Breeding. The principal breeding range of American Wigeon is in the tundra and boreal forests of Alaska and western Canada, and then south

remainder in the interior, largely during the spring. Edgell (1984) completed a second major update of Eurasian Wigeon records, covering the years 1948–81; he reported 1,143 sightings (1,500 individuals) in the fall and winter and 492 (560 individuals) in the spring and summer. In marked contrast to Hasbrouck's summary, 55% of the fall and winter sightings were in the Pacific Flyway, compared with 38% in the Atlantic Flyway. During the fall and winter, the Pacific coastal sightings were

Male American Wigeon.
GaryKramer.net

American Wigeon
courtship flight; note
the female (*center*).
GaryKramer.net

through the PPR. The Traditional Survey recorded 2.4 million American Wigeon in 2010, with the largest percentage (43.4%) tallied in Alaska/Yukon, and 24.6% in the Northwest Territories, southcentral British Columbia, and northern and central Alberta. Fewer numbers occurred farther east and south: 3.0% in northern Saskatchewan, northern Manitoba, and western Ontario; 13.6% in the southern prairies of Alberta, Saskatchewan, and Manitoba; 6.8% in Montana and the western

Female American Wigeon with her brood in William Hawrelak Park, Edmonton, AB, Canada. *Raymond Man-Kei Lee*

Dakotas; and 8.5% in the eastern Dakotas (U.S. Fish and Wildlife Service 2010).

The regional distribution of American Wigeon from 1955 to 2010 in the Traditional Survey was 34.6% in the Northwest Territories and northeastern British Columbia ; 21.1% in Alaska/Yukon; 15.9% in southern Saskatchewan; 10.6% in southern Alberta; and <5% in the other survey regions (Collins and Trost 2010). Of the total number of American Wigeon counted in Alaska in 2009 (794,915), the largest percentages were found on Yukon Flats (30.1%), the Yukon-Kuskokwim (Y-K) delta (15.3%), and the Tanana-Kuskokwim lowlands (13.2%; Mallek and Groves 2009).

West of the PPR in the United States, American Wigeon breed in the Intermountain wetlands from central Washington and Oregon (east of the Cascades), east throughout Idaho, Montana, and Wyoming, and south through northern Nevada and Utah, northwestern Colorado, and the Sandhills of Nebraska. They are widespread in California, from the northern plateau through the Central Valley (Mowbray 1999). From 2001 to 2010, surveys have tallied 4,600 American Wigeon in Washington, 6,466 in Oregon, and 2,045 in California (Collins and Trost 2010).

There are also various small populations in the midwestern states of Minnesota, Wisconsin, and Michigan, and then eastward along western Lake Erie in Ohio (Mowbray 1999). A few isolated breeding colonies occur in eastern Ontario, southwestern Québec, and New Brunswick (Godfrey 1966). In the Maritime Provinces, they breed locally in northeastern New Brunswick, the New Brunswick / Nova Scotia border, and eastern Prince Edward Island (Erskine 1992). Goudie (1985) noted American Wigeon breeding as far north as Newfoundland. There are scattered breeding records from New York (McGowan and Corwin 2008) east to Massachusetts (Veit and Petersen 1993) and other New England states.

Winter. The Midwinter Survey in the United States averaged 1.1 million American Wigeon from 2000 to 2010, of which 58.1% were in the Pacific Flyway, 27.6% in the Central Flyway, 10.8% in the Mississippi Flyway, and only 3.5% in the Atlantic Flyway. In the Pacific Flyway, 72.5% occurred in California

Male Eurasian Wigeon.
RyanAskren.com

(mostly the Central Valley), and 19.5% in Washington. In the Central Flyway, 95.7% occurred in Texas, both along the Gulf Coast (the Laguna Madre) and in the Playa Lakes Region of the Texas Panhandle. In the Mississippi Flyway, 75.5% occurred in Louisiana, mostly on the coastal marshes. The few American Wigeon in the Atlantic Flyway primarily occurred in North Carolina (33.4%) and Florida (32.4%). Overall, 42.1% of all American Wigeon occurred in California, 26.4% in Texas, and 11.3% in Washington. The Christmas Bird Count averaged 276,027 American Wigeon in the United States from 2000/2001 to 2009/10, of which 38.0% were in California.

The numbers of American Wigeon wintering in Louisiana have changed dramatically in response to hurricanes, which created more open-water areas within the dense coastal marsh vegetation. During the mid-1950s, peak concentrations of American Wigeon in Louisiana coastal marshes were about 100,000, but from 1967 their numbers increased steadily, reaching 1,062,000 in 1970 (Bellrose 1980). Since 1970, however, the winter population in that state has fluctuated greatly, to a low of only 12,000

in 2009, which most likely reflects variations in the food resources in coastal Louisiana.

In Mexico, an average of 67,318 birds was tallied along the east coast during annual surveys conducted from 1978 to 1982, and then at 3-year intervals from 1985 to 2006. Most occurred in the lower Laguna Madre (46.5%) and the Rio Grande delta (18.5%), with smaller percentages (4.9%–8.8%) on survey areas farther south, to the states of Oaxaca, Veracruz, and Tabasco. American Wigeon numbers counted on the west coast and Baja Peninsula of Mexico averaged 59,333 during surveys conducted at 3-year intervals from 1982 to 2006 (Conant and King 2006). Recent numbers, however, have been smaller: 25,700 in 2000, 15,500 in 2003, and 25,800 in 2006. Pabellón Lagoon, Topolobampo Lagoon, and Isla Tóbari are important wintering sites on the mainland west coast of Mexico. In 2003, 3,124 American Wigeon were tallied in the central highlands, and 57,702 in 2006, along with 25,435 in the interior highlands in 2003 and 58,932 in 2006 (Thorpe et al. 2006). On the lower west coast of Mexico, their numbers were 5,420 in 2003 and 4,728 in 2006.

Migration. Fall migration from northern Alaska begins in early to mid-August, with departures farther south on the Seward Peninsula peaking during the first half of September (Kessel 1989). Migration in British Columbia begins in mid- to late August, with peaks in the interior occurring by mid-October (Campbell et al. 1990). Fall migration from the northern regions of the Prairie Provinces begins in September, with most American Wigeon gone by the end of October, when lakes freeze over (Semenchuk 1992). Census data from 25 national wildlife refuges from 1976 to 1985 revealed that substantial numbers arrive in Washington in August and September, and in Oregon and California in September and October (Hitchcock et al. 1993). Peak numbers in Washington and Oregon typically occur in November and December, while populations in California peak in December through February.

In the northern states east of the Rocky Mountains, American Wigeon are plentiful until mid-October, but their numbers then begin to decline rapidly (Bellrose 1980). Most have departed by early November. These intermediate migration areas between breeding and wintering grounds begin to receive small numbers of American Wigeon in early October, with peak numbers arriving between mid-October and mid-November, but these areas are largely devoid of American Wigeon by December. They arrive at Long Point on Lake Ontario by mid-September, peak in late October, and are gone by mid-November (Knapton 1992). In Illinois, peak numbers in the northeastern region of the state occur from 13 October to 9 November, and in the southern part from 27 October to 9 November (Havera 1999); most are gone by December.

In the Texas Panhandle (1970–71), American Wigeon begin appearing at the Muleshoe National Wildlife Refuge (NWR), near the New Mexico border, by 10 September, with most arriving on 27 October (Soutiere et al. 1972); peak numbers (at times >200,000) usually arrive at the Buffalo Lake NWR in late December. American Wigeon begin to reach Florida in late September, and their numbers continue to increase until February (Stevenson and Anderson 1994).

Spring migration generally begins in February, with the peak on intermediate areas occurring during the first 2 weeks of April. Most birds reach their breeding areas by mid-May (Bellrose 1980). American Wigeon depart the Texas Panhandle between 10 February and 1 March (Soutiere et al. 1972), and they leave Florida between mid-February and early March (Stevenson and Anderson 1994). On the West Coast, the number of wintering American Wigeon on national wildlife refuges in central and southern California decrease sharply in March; in Washington and Oregon, their numbers start to decline in January, with most gone by April (Hitchcock et al. 1993). American Wigeon reach peak numbers in Missouri from late March to early April (Robbins and Easterla 1992), and at the Illinois River in Illinois from mid-February through mid-March (Havera 1999). They arrive in Moose Jaw, Saskatchewan, between 11 and 19 April (Wishart 1983a). American Wigeon first appear in Alberta and at Cook Inlet and the Copper River delta area of Alaska in mid-April, with peak numbers occurring by 10–15 May (Hitchcock et al. 1993). Arrival in Nova Scotia is mid-March, with a peak in early April (Tufts 1986).

Of 122 recoveries from 1,223 American Wigeon banded in Alaska between 1980 and 2010, 107 (88%) were recovered in the Pacific Flyway outside Alaska: 44.9% in California, 15.9% in Washington, 10.3% in Oregon, and 8.4% in British Columbia. Outside the Pacific Flyway, the largest number of recoveries (5, or 4.7%) was in Texas, and 1 was in Florida. Most American Wigeon in the Pacific Flyway migrate from Alaska to Puget Sound along the coastal corridor. Bandings of this species on the Yukon Flats breeding area, however, indicate that important corridors also exist through the interior of British Columbia and Alberta.

Virtually all of the American Wigeon breeding

in British Columbia migrate directly southward to winter in Washington, Oregon, and California (Bellrose 1980). Of 799 band recoveries from 6,644 American Wigeon banded in the Northwest Territories, British Columbia, and Alberta, the largest percentage (35.9%) occurred in California. Recoveries outside of California were spread over a broad area that included 34 other states: 11.3% in Texas (Central Flyway), and 5.3% in Louisiana (Mississippi Flyway). Outside the United States, 9 recoveries were in Mexico, and 1 was in Cuba.

Many of the American Wigeon entering the northern areas of the Mississippi Flyway travel southeast, bound for areas in the Atlantic Flyway south of Long Island. Some of these birds, however, will winter in eastern Arkansas and western Tennessee (Bellrose 1980). Long Point Bay in Ontario serves as a migratory staging area for American Wigeon traveling between prairie breeding areas and wintering grounds along the Atlantic coast. Petrie (1998) noted that the number of American Wigeon staging there from 1991 to 1997 averaged about 400,000 in the fall and 30,000 in the spring.

MIGRATION BEHAVIOR

American Wigeon are primarily diurnal migrants, but they sometimes migrate at night, particularly during the fall (Palmer 1976a). Flocks are generally smaller (5–15 birds) during spring migration than in the fall (>100; Mowbray 1999), and flocks in the early fall are male biased, which reflects the departure of males from molting areas before the females and immatures (Rienecker 1976). American Wigeon are among the earliest of fall migrants, next to Blue-winged Teal and Northern Pintails. Wigeon begin to move south into adjacent states from Canadian breeding areas as early as mid-August, but much larger numbers initiate migration in the last half of September (Bellrose 1980).

Molt Migration. Postbreeding males and unsuccessful females undergo molt migration. After the pair bond breaks, they initially concentrate in small groups on large wetlands near the breeding area, but within a week they will head to bigger lakes and marshes (Wishart 1985). Molting areas for these birds may be either within a few kilometers of breeding marshes or a considerable distance away. In contrast, successful females will molt on their breeding area. Concentrations of birds at molting sites increase during June and July but decline thereafter, as flight is attained and fall migration begins.

HABITAT

Breeding American Wigeon use shallow freshwater wetlands and nest in nearby upland brush–grass habitats. Their primary breeding habitat is in the shortgrass prairie and open parklands of western Canada, but also the mixed grass and shortgrass area of the PPR of the United States. They are common in the coastal tundra habitats of Alaska, such as the Y-K delta. In parkland habitat near Lousana, Alberta, breeding American Wigeon preferred small (<1 ha) semipermanent water areas surrounded by hayfields or ungrazed woodlands (Smith 1971). On the Woodworth Study Area in North Dakota (1965–75), occupancy by pairs was highest (54.4%) on wetlands with >2,000 m of perimeter (Higgins et al. 1992). Occupancy of wetlands with <2,000 m of perimeter was only 11.5%.

Wintering American Wigeon use habitats with an abundance of submergent and emergent vegetation. At the Eufaula NWR on the Alabama/Georgia border, wintering birds used sandbars and islands along the Chattahoochee River that were characterized by sparse herbaceous and grass growth, as well as managed impoundments measuring 2–5 m deep (Turnbull and Baldassarre 1987). The open-water sites of impoundments were dominated by watermilfoil (*Myriophyllum* spp.) and American lotus (*Nelumbo lutea*), and the primary emergents were arrowheads (*Sagittaria* spp.), swamp smartweed (*Polygonum hydropiperoides*), and pondweeds (*Potamogeton* spp.). Wigeon devoted more time to feeding in impoundments (63%–72%) than in river sites (46%); they spent much more time resting in the latter (27% vs. 7%–10% in impoundments). On the southern High Plains of Texas, wintering American Wigeon used playa lakes but engaged in twice-daily field-feeding flights to nearby cornfields (Baldassarre and Bolen 1984).

POPULATION STATUS

The American Wigeon breeding population in North America averaged 2.6 million from 1955 to 2010 in the Traditional Survey area, with the 2010 estimate of 2.4 million being 7% below the long-term average (U.S. Fish and Wildlife Service 2010). There is no discernable trend over this period, but the population has fluctuated from a low of 1.7 million in 1986 to a high of 3.8 million in 1959. The survey averaged 3 million in the 1970s, 2.5 million in the 1980s, 2.5 million in the 1990s, and 2.4 million from 2000 to 2010. During the dry years from 1985 to 1994, the survey averaged 1.7 million American Wigeon, 33% below the long-term average. During drought conditions in the Prairie Provinces, however, numbers increased in Alaska and the northwestern Yukon (Johnson and Grier 1988, Hodges et al. 1996). Following the drought (1995–2001), the population averaged 3.3 million and included 4 consecutive years with estimates above 3 million (1997–2000), with this rise in numbers attributed to an increase in nesting habitat due to the wet conditions. American Wigeon populations are less affected by habitat degradation in the PPR, because a major part of their breeding range is farther north and west, including most of Alaska. The Eastern Survey averaged 18,005 birds from 1990 to 2010, with survey numbers ranging from a high of 58,100 in 1998 to a low of 5,100 in 1992. The Midwinter Survey shows no discernable long-term trend, with populations in the United States averaging 1.3 million in the 1970s, 911,000 in the 1980s, and 962,000 in the 1990s.

Harvest

American Wigeon are an important game species in North America. The average harvest in the United States was 759,347 from 1999 to 2008: 49.8% in the Pacific Flyway, 25.1% in the Central Flyway, 20.7% in the Mississippi Flyway, and 4.4% in the Atlantic Flyway. Of the Pacific Flyway harvest, the majority were taken in the West Coast states: 44.5% in California, 18.8% in Washington, and 18.3% in Oregon.

From 1999 to 2008, the harvest of American Wigeon in Canada averaged 38,219, with a high of 47,852 in 2005. Over this time period, 21.3% were harvested in British Columbia, 19.0% in Alberta, 17.5% in Ontario, 16.2% in Saskatchewan, 12.5% in Manitoba, and 13.5% in the remaining provinces.

BREEDING BIOLOGY

Behavior

Mating System. American Wigeon are monogamous, with almost all females breeding their first year; many yearling males, however, will not find mates (Wishart 1983a). Pair formation is initiated before their arrival to wintering areas. Various studies across the wintering range of this species (North Carolina, Alabama/Georgia, Yucatán) found that courtship behavior averages <1% of all activity during the fall and winter, and it does not change much among these months (Hepp and Hair 1983, Turnbull and Baldassarre 1987, Thompson and Baldassarre 1991).

Along the coast of North Carolina, 84% of the females were paired by November (Hepp and Hair 1984). In western Texas, the percentage of females associated with males increased from 21% to 27% in November and December to 45% in January, 67% in February, and 81% in March (Soutiere et al. 1972). Courting groups were observed as early as 21 January and as late as 13 March. In the Yucatán Peninsula of Mexico, however, the percentage of paired females was much lower than that observed farther north: 6.0% ($n = 44$) in December, 11.0% ($n = 82$) in January, 22.0% ($n = 480$) in February, and 32% ($n = 311$) in March (Thompson and Baldassarre 1992). The lower percentage of paired females may have occurred because lone females won 46.0% of their aggressive encounters with males, so male American Wigeon wintering in the Yucatán Peninsula were of equal or even lower social rank than the females. Thus females may choose not to pair with them, instead pairing with more dominant (= higher quality) males wintering farther north (see Courtship Displays).

In British Columbia (1974–75, 1976–77), the percentage of paired birds increased steadily, from 10%–15% in December to 81% by March, although only 20% were paired in January in 1 of the winters, compared with 50% in the other; about 60% were paired in February of both years (Wishart 1983b). Of 40 known pairs, 26 (65%) were adults, 10 were juvenile females with adult males, 2 were adult females with juvenile males, and 2 were juveniles. Of 56 males of known pair status, 36 adults were paired and 8 were unpaired, whereas only 8 juveniles were paired and 4 were unpaired. Paired males were structurally larger and heavier than unpaired males, although unpaired males had larger lipid and total energy reserves, perhaps because paired males expended more energy forming and maintaining pair bonds. Paired males had larger reserves than unpaired birds in late winter and early after their arrival on their breeding areas, perhaps because they spent more time feeding and less time moving about than unpaired birds. Such findings indicated that males need to attain a threshold condition before they can expend time and energy on pair formation, and thus balance the benefits of early pair formation with the energetic costs of maintaining a pair bond afterward. The pairing date of males was also significantly correlated with their completion of the prealternate molt and their age. Lastly, some male/female associations can be ephemeral, especially early in the pair formation process on the wintering grounds (Soutiere et al. 1972).

In Idaho, 60% of the males separated from their mates during the first week of incubation, and the remainder left during the second week (Oring 1964). In Saskatchewan, males averaged 26.6 days paired with their mates from the date of the first egg laid until separation; 1 male, however, attended a female with her young brood (Wishart 1983a).

Sex Ratio. The sex ratio in this species is male dominated. During the winter in British Columbia, 59.6% of the 3,360 American Wigeon observed in 1974–75 were male as were 52.4% of the 14,806 birds observed in 1976–77 (Wishart 1983b). At the Buffalo Lake NWR in Texas, the sex ratio of 1,205 American Wigeon banded between 1961 and 1966 averaged 65.5% males (Soutiere et al. 1972). In the Yucatán Peninsula of Mexico (1986–88), the sex ratio of 364 birds was 66.8% males (Thompson and Baldassarre 1992). Bellrose et al. (1961) reported an average of 54.6% males among 10,963 American Wigeon observed during the spring at various locations across North America.

Site Fidelity and Territory. Arnold and Clark (1996) reported that 39.2% of the marked adult females (*n* = 51) and 37.5% of the marked yearling females (*n* = 24) returned to the same nesting areas in southcentral Saskatchewan over a 10-year period (1983–92). During a 3-year study (1976–78) in Saskatchewan, Wishart (1983a) documented a return rate of 44.4% for adult females (*n* = 9) and 8.1% for adult males (*n* = 37). In Manitoba, 69.6% of the adult females (*n* = 23) returned to the same breeding grounds the following year; in New York, the return rate for 74 adult females was 53% (Anderson et al. 1992).

In Saskatchewan, 6 breeding pairs initially used an average home range of 25.3 ha, but territories established prior to the initiation of egg laying consisted of 1–3 potholes and adjacent uplands and averaged 7.8 ha (Wishart 1983a). Most of the wetlands and nearby uplands were used exclusively by 1 pair.

Courtship Displays. Johnsgard (1965), Soutiere et al. (1972), and Wishart (1983a, 1983b) described the displays of American Wigeon, which are not much different from those of other dabbling ducks. In general, males compete for females by means of aggression and ritualized displays, with group flights also playing a significant part of the courtship process. Mowbray (1999) provided a detailed review of all vocalizations.

In males, *swimming-shake* is a major display,

usually given in close proximity to the female in an attempt to attract her attention. The display accentuates the male's plumage markings, such as the black and white bands on the rump as well as on the head and bill. The *burp* involves stretching the head and neck into an *alert* posture and erecting the crown feathers while giving a *slow-whistle* vocalization. The display is often repeated 2–3 times by the courting individual, but it is also given concurrently by other males in a courting group. Soutiere et al. (1972) described a *jump-flight*, which is a display performed by a male lagging behind a courting group. The male then flies over the group in a short (3–5 m) flight, lands in front of the female, and gives the *turn-the-back-of-the-head* display. This display frequently elicits *inciting* from the female and often ends in a *courtship-flight*. *Inciting* females give a series of low guttural notes described as *kaow kaow* and *errr err*. Other described displays are *bathe*, *wing-flap*, *wings-up*, and *chin-lift*. Precopulatory *head-pumping* is given on the water by both pair members, but it is usually initiated by the female. The male continues *head-pumping* during copulation, which forces the female to elevate her tail. Males (n = 27) averaged 23.4 seconds on top of the female during copulation. After copulation the male assumed an *alert* posture 93% of the time, while performing *wing-flap* and *bathe* (96%) displays.

In Texas, courting groups were observed as early as late January; they involved a single female and as many as 5–14 males (Soutiere et al. 1972). Females in these situations displayed *chin-lifting* and *inciting*, as well as *threatening* postures toward males. Males performed an *introductory-shake* (*swimming-shake*) and *turn-the-back-of-the-head*, as well as *threat* displays toward other males. *Jump-flights* ensued as the intensity of courtship activities increased. Observed copulation sequences occurred between pairs that separated from the main flock.

In addition to being a part of displays among breeding birds, aggressive interactions are commonly observed among wintering birds, most of which are intraspecific. In coastal North Carolina, 87% (n = 99) of the aggressive interactions were intraspecific, with the males dominant to the females when their social status was similar (Hepp and Hair 1984). In a coastal estuary on the Yucatán Peninsula of Mexico, 86% (n = 143) of the aggressive encounters were again intraspecific (Thompson and Baldassarre 1992). *Threat* was the most common (74.5%) form of aggressive display and was characterized by a low posture in the water, with the head and neck stretched low and the bill open. *Chasing* was the next most common (16.7%) aggressive display, which involved actual rushing at the target birds and often nipping at the tail. The interactions were male versus male (41.2%), female versus male (26.5%), male versus female (16.7%), and female versus female (15.7%). The initiator won 99.0% of these aggressive encounters when the participants were of equal rank (e.g., male vs. male). Of the interspecific aggressive encounters (14%), most (12.6%) were toward Northern Shovelers.

Nesting

Nest Sites. Females presumably select the nest site, which is typically well concealed and often far from water. Of 55 nests located on the Woodworth Study Area in North Dakota (1966–81), 35% were in brush, 31% in forbs, 27% in grass, and 4% in marsh (Higgins et al. 1992). Most (49%) nests were well concealed in an erect and closed-vegetation cover type, with 82% half or totally concealed from above and 83% at least three-quarters concealed from the sides. In the PPR of North Dakota (1980–88), 39% of 232 American Wigeon nests were located in snowberries and 56% were in habitats characterized by a grass understory and a snowberry overstory (Kruse and Bowen 1996). Their nesting preference was for vegetation >25 cm tall, and they avoided vegetation <15 cm tall. In central Alaska, mixed sedges (*Carex* spp.) and grassy meadows with low shrubs were used as nesting areas (Kessel 1989). Of 21 nests in British Columbia, most (86%) were located in dense grass under shrubs or in shrubs under trees (Campbell et al. 1990).

In the Flathead Valley of Montana (1937–39), 45 nests averaged 90 m from water, with the farthest at 320 m (Girard 1941). In the Yukon, Bent (1923) reported nests located up to 800 m from open water. In southeastern Alberta, 23 nests were an average of 21.8 m from water, and the average light penetration in nesting cover was 47% (Keith 1961). On islands in Lake Newell in Alberta, where sagebrush (*Artemisia cana*) and open meadows were the predominant cover type, 18 nests were an average of 32.8 m from water (Vermeer 1970b). In southcentral Saskatchewan (1976–78), 24 out of 30 (80%) American Wigeon nests were located in dense stands of shrubs, such as western snowberries (*Symphoricarpos occidentalis*), wolfwillows (American silverberry; *Elaeagnus commutata*), and low-growing prairie roses (Arkansas rose; *Rosa arkansana*); 5 were in dense herbaceous cover, and 9 were in pure stands of western snowberry (Wishart 1983a). The most significant herbaceous covers were alfalfa (*Medicago sativa*) and yellow sweetclover (*Melilotus officinalis*). Nests (*n* = 26) averaged 42.5 m from water, with the farthest nest located 205.1 m away. Vermeer (1970b) noted that the islands used by nesting American Wigeon in both Alberta and Saskatchewan often had nesting colonies of California Gulls (*Larus californicus*), Ringed-billed Gulls (*Larus delawarensis*), or Common Terns (*Sterna hirundo*).

The nest is constructed in a slight depression, lined with grass or other nearby herbaceous vegetation and light gray down. Females add down feathers to the nest throughout egg laying and incubation (Bent 1923, Bellrose 1980).

Clutch Size and Eggs. In Alaska, 19 nests averaged 7.3 eggs (range = 5–10; Kessel 1989). In Ontario, Peck and James (1983) reported an average clutch size of 8.0 eggs (range = 3–11) in 21 nests. In British Columbia, the average was 8.2 (range = 3–11) for 32 nests. In southcentral Saskatchewan, clutch size averaged 8.6 eggs in 29 nests, with older adult females laying an average of 1.4 more eggs/clutch than first-year females (Wishart 1983a). In

the Flathead Valley of Montana (1937–39), clutch size averaged 9.6 for 45 nests (Girard 1941).

American Wigeon eggs are ovate, creamy white, and measure 53.9 × 38.3 mm (*n* = 81; Bent 1923). In Saskatchewan, Mowbray (1999) reported data on eggs measured by Wishart that averaged 52.93 × 37.46 mm for 242 eggs from 29 clutches. The average egg mass (*n* = 237) from 29 clutches was 42.7 g, about 6% of a female's body mass. A complete clutch of 9 eggs averaged 388 g, which was >50% of an adult female's body mass. As with other puddle ducks, American Wigeon eggs are laid at a rate of 1/day (Wishart 1983a).

Incubation and Energetic Costs. In Saskatchewan, the incubation period of 5 nests was 25–28 days, during which time the females averaged 87.8% of the day incubating (Wishart 1983a). These females averaged 1.8 recesses/day of 94 minutes each, for a total daily duration of 176 minutes, during which they spent an average of 45% of their time feeding. The paired male accompanies the female when off the nests, but most males have deserted their mates by the end of incubation. Hochbaum (1944) reported incubation at 23 days for 1 clutch hatched in an incubator, and Scott and Boyd (1957) observed 22–25 days for eggs from captive birds.

There are no specific studies of reproductive energetics in breeding American Wigeon. However, the basic pattern is probably similar to that of Gadwalls, which have somewhat parallel food preference for low-quality, leafy aquatic vegetation. Female Gadwalls are an exception among temperate-nesting dabbling ducks, in that about 16% of the protein used in egg production comes from stored reserves (Ankney and Alisauskas 1991). Alisauskas and Ankney (1992a) calculated that female American Wigeon, however, used protein reserves for 44% of their egg production, the highest level reported for any duck species. They rely heavily on green vegetation during the period of egg production, and green vegetation has a low-protein content in comparison with animal matter. Hence, because American Wigeon are not able to meet all of the protein requirements for egg production from foods obtained in their diet, their stored reserves decline substantially.

Nesting Chronology. American Wigeon are relatively late nesters in comparison with other puddle ducks. In British Columbia, the earliest nesting was reported on 30 April, with peak nesting occurring in mid-June ($n = 43$; Campbell et al. 1990). In southeastern Alberta, Keith (1961) reported an average nesting date of 11 May for 21 nests. During a 2-year study (1974–75, 1976–77) in southcentral Saskatchewan, the first nests were initiated on 11 May, with peak initiations occurring between 26 May and 2 June (Wishart 1983b). Nesting began an average of 35 days after arrival, but drought delayed nesting by 2 weeks during 1 year ($n = 30$). In Alaska, nesting was reported on 10 May at Minto Lakes and 23 May on Yukon Flats (Bellrose 1980). At Yellowknife, Northwest Territories, American Wigeon began nesting on 18 May, slightly over a week after arrival, but nesting did not peak until 15 June (Murdy 1964). Nests were initiated over a 50-day period at Minto Flats and 28 days at Yellowknife. The largest sample of nests ($n = 232$) comes from the Lostwood NWR in North Dakota (1980–88), where the average initiation date was 28 May, and 90% of the nests were initiated between 12 May and 11 June (Kruse and Bowen 1996).

Nest Success. The nests of American Wigeon are notoriously difficult to locate, so nest success data are not plentiful. On the Woodworth Study Area in North Dakota (1966–81), apparent success of 58 nests was 43.1%, and Mayfield success was 22.4% (Higgins et al. 1992). Apparent success of 45 nests in the Flathead Valley of Montana (1937–39) was 75% (Girard 1941), compared with only 33% for 18 nests located on islands in Lake Newell in Alberta (Vermeer 1970b). In southeastern Alberta, nest success was 39% for 18 clutches located between 1952 and 1965, but within this time span, the rate

was 58% for 12 nests discovered during a pre-drought period versus 0% success for 6 nests found during the drought (Smith 1971). Bellrose (1980) reported 57% success from nests in several locations in Alaska.

American Crows (*Corvus brachyrhynchos*), gulls (*Larus* spp.), striped skunks (*Mephitis mephitis*), Franklin's ground squirrels (*Citellus franklinii*), red foxes (*Vulpes vulpes*), and short-tailed weasels (*Mustela erminea*) have all been identified as predators of American Wigeon nests. At the Woodworth Study Area in North Dakota, red foxes were the major predators, accounting for 18 out of 25 (72%) nest predations (Higgins et al. 1992).

Brood Parasitism. Brood parasitism is not common in American Wigeon, most likely because their nests are well concealed and they do not nest at high densities. Palmer (1976a), however, cited incidents of brood parasitism by White-winged Scoters, Lesser Scaup, and Northern Shovelers.

Renesting. There are virtually no data on known renests, but renesting undoubtedly occurs. Wishart (1983a) observed 1 female that laid her first egg on 3 June and initiated a second clutch of 8 eggs 4 days later, at a site about 300 m away.

REARING OF YOUNG

Brood Habitat and Care. Like Gadwalls, American Wigeon generally use open-water wetlands more frequently than other dabbling ducks, largely because their diet is dominated by submerged vegetation as the ducklings grow older (see Food Habits and Feeding Ecology). Wigeon ducklings do not compete with Gadwall ducklings, however, despite their similar habitat use, diets, and feeding methods.

American Wigeon broods ($n = 35$) on the Seney NWR in Michigan were observed on an 8.1-ha marsh characterized by open water interspersed with sparse to medium-dense stands of sedge (*Carex lasiocarpa*) and abundant submerged vegetation; the water depth averaged 0.5 m (Beard

1964). In the Alberta parklands, broods primarily used smaller (0.8–2.0 ha) potholes that varied from closed to open (Smith 1971). In southeastern Alberta (1953–57), broods (*n* = 39) used the larger (8.5–20.8 ha) wetlands and avoided small (<0.5 ha), shallow (<0.6 m) potholes (Keith 1961). In South Dakota, broods also preferred larger (1.0–1.6 ha) potholes to small (0.2–1.0 ha) ones (Evans and Black 1952). In Saskatchewan, Wishart (1983a) noted that females often moved their ducklings >0.5 km overland to reach larger, deeper potholes.

In the PPR of southern Alberta, young ducklings (Class I) used wetlands characterized by open water (25%–35%) and submerged plants (41%–61%); older ducklings (Class II–III) used habitats with 83%–95% submerged plants (Sugden 1973). Young ducklings (Class Ia and Ib) fed on the surface (75%–79%), after which bill dipping (33%–62%) and head ducking (8%–43%) predominated. Young ducklings were particularly dispersed when feeding, with Class I broods often scattered up to 4 ha over a given pond. Feeding ducklings tended to avoid the shore and use areas of open water with submerged vegetation.

The female accompanies her brood until the ducklings are almost fully grown, and she occasionally remains with the brood until her young can fly (Bellrose 1980). At the Seney NWR in Michigan (1950–51), young ducklings fed by dabbling and gleaning from the surface, and "their feeding was so vigorous that the noise of their bills sucking in the water could be heard for a considerable distance" (Beard 1964:498). The ducklings stretched and jumped to reach insects attached to emergent vegetation. On 1 occasion, 2-week-old ducklings were observed diving for food. During morning observation periods, ducklings averaged 15 minutes sleeping and 44 minutes resting. In this study the females abandoned their broods 6–7 weeks after hatching.

Brood Amalgamation (Crèches). Brood amalgamation has not been specifically studied on American Wigeon, but it does occur. In British Colum-

bia, broods of 15–19 ducklings most likely reflect brood amalgamation, probably because females usually abandon their broods before fledging and depart the breeding area to begin molting (Campbell et al. 1990). At Yellowknife, Northwest Territories, crèches were observed in 0.4% of American Wigeon broods, and they were more common as broods got older (Toft et al. 1984). In Saskatchewan, however, Wishart (1983a) noted that females aggressively prevented other ducklings from joining their broods. Beard (1964:515) also stated that the American Wigeon female was very aggressive "toward any and all ducklings that attempted to join her brood"; she also commonly drove away adult intruders.

Development. The average weight of 8 captive-reared ducklings was 24 g at 1 day, 54 g at 1 week, 139 g at 2 weeks, 259 g at 3 weeks, 382 g at 4 weeks, and 433 g (*n* = 5) at 5 weeks (Southwick 1953). Reported fledging times were 44 days at Yukon Flats in Alaska (Lensink 1954), 45–58 days in Manitoba (Hochbaum 1944), and 47 days in South Dakota (Evans et al. 1952). There are no other specific growth data in the literature.

RECRUITMENT AND SURVIVAL

Bellrose (1980) summarized brood-size data from many regions in North America: Class I broods averaged 6.3 ducklings (*n* = 1,564); Class II broods, 6.8 (*n* = 993); and Class III broods, 6.5 (*n* = 283). In the Dakotas, brood size averaged 7.1 for Class I (*n* = 46), 6.6 for Class II (*n* = 19), and 7.7 for Class III (*n* = 3; Duebbert and Frank 1984). During a 17-year period (1965–81) at the Woodworth Study Area in North Dakota, brood size averaged 6.6, which was a reduction of 23% from the average size of a successful clutch (Higgins et al. 1992). More specifically, brood size (*n* = 58) in this study averaged 7.3 for Class I, 5.6 for Class II, and 6.0 for Class III. Brood sizes in the Alberta parklands (1953–65) averaged 7.5 in Class Ia, 6.0 in Class IIa, and 7.0 in Class IIIa (Smith 1971). In British Columbia, Campbell et al. (1990) found an average brood size of 7.0 ($n = 534$), with 57% containing 5–8 ducklings/brood. Some were probably amalgamated broods, however, with 15–19 young. From 1961 to 1965 at Yellowknife, Northwest Territories, brood size averaged 6.5 for Class Ia ($n = 7$) and 5.8 for Class I ($n = 224$), a decrease of 10.7% (Toft et al. 1984). Brood size was greater (6.7, $n = 101$) for early nesters, compared with late nesters (4.9, $n = 45$), and declined at a rate of 0.04 ducklings/brood/day. These authors proposed that this reduction in brood size for later-nesting American Wigeon, as well as for other species they studied, reflects an adaptation to changing conditions during the breeding season that favor smaller clutch sizes, and thus smaller brood sizes, for later-nesting females. Females nesting later than average incur the maximum cost/offspring but risk a greater loss of their young at fledging. Hence, if these females produce fewer young/clutch, they may have less of an investment during a given season but produce more young over a lifetime.

From a population standpoint, Leitch and Kaminski (1985) noted that only 30% of all American Wigeon pairs produced broods on a grassland Saskatchewan study area from 1950 to 1975 ($n = 2,594$ pairs). In the Alberta parklands (1953–65), 39% of all pairs reared broods averaging 7.0 young, thus producing 2.7 young/pair (Smith 1971). At the Woodworth Study Area in North Dakota, production averaged 0.31/pair (Higgins et al. 1992). An analysis of banding data across North America, corrected for harvest bias, revealed an average recruitment of 1.34 immature females/adult female from 1974 to 1997 (Frank Bellrose).

In California, an analysis of 32,097 American Wigeon banded during the winter (1951–71) estimated average survival rates of 66% for adult males and 58% for adult females (Rienecker 1976). In southcentral Saskatchewan, a mark-resighting analysis during the breeding season revealed average annual survival rates of 63.5% ($n = 25$) for adult females and 57.1% ($n = 24$) for juvenile females, with survival negatively correlated with wetland habitat conditions in the previous year

(Arnold and Clark 1996). Survival rates were 65.0% where nesting females were protected by predator-proof fences, compared with only 21.8% in unprotected areas. Survival rates based on an analysis of 19,537 banded American Wigeon across North America from 1978 to 1996 averaged 64% for adult males, 51% for immature males, 49% for adult females, and 42% for immature females (Frank Bellrose). Overall survival was 51.5%.

Longevity records for American Wigeon are 21 years and 4 months for a male of unknown age banded in Washington and shot in Nebraska, and 15 years and 5 months for an after-hatching-year male banded and later harvested in California. Rienecker (1976) calculated a much shorter average life span for American Wigeon: 2.3 years for males and 1.7 years for females.

FOOD HABITS AND FEEDING ECOLOGY

One of the most unique aspects of American Wigeon is their short and narrow bill structure, which is similar to that of a goose and thus highly adapted for grazing on upland and aquatic plants (Wishart 1983a). Their bill structure limits the efficiency of filter feeding, so American Wigeon probably would not compete well with other dabbling ducks in this respect. Thus, unlike other dabbling ducks (except Gadwalls), American Wigeon consume large amounts of leafy aquatic vegetation, as well as seeds, and commonly graze in upland settings on agricultural crops, such as clover and green wheat. Food use for this species varies geographically, but important dietary items are the stems and leafy portions of submerged aquatic plants, such as pondweeds (*Potamogeton* spp.), widgeongrass (*Ruppia maritima*), eelgrass (*Zostera marina*), exotic eelgrass (*Zostera japonica*), sea lettuce (*Ulva lactuca*), duckweed (*Lemna minor*), wild celery (*Vallisneria americana*), and watermilfoil (*Myriophyllum exalbescens*). Other major food items are the seeds and leaves of panicgrasses (*Panicum* spp.), wild rice (*Zizania aquatica*), cultivated rice, bulrushes (*Scirpus* spp.), and algae (Mowbray 1999). Foraging times are usually lengthy, because leafy aquatic vegetation is nutritionally deficient. During the winter in Alabama, American Wigeon spent an average of 45%–71% of their time feeding on poor-quality leafy vegetation, such as watermilfoil (*Myriophyllum spicatum*; Turnbull and Baldassarre 1987).

During the winter and on migration, American Wigeon commonly feed together with Gadwalls, Redheads, Greater Scaup, Lesser Scaup, and American Coots (*Fulica americana*). During fall migration on lakes in Oklahoma, American Wigeon and Gadwalls were the most frequently observed waterfowl in association with American Coots (Eddleman et al. 1985). The dietary overlap with American Coots was 75%–99% for American Wigeon and 88%–99% for Gadwalls. Foods brought to the surface during foraging dives by American Coots were often robbed by American Wigeon and Gadwall; hence American Coots made otherwise unobtainable foods available to these latter 2 species (Knapton and Knudsen 1978, Ryan 1981). Although food use was similar between these 3 species during migration in Oklahoma, interspecific interactions between the ducks and American Coots were minimized, because the ducks tended to migrate ahead of or just behind cold fronts and during the day, while American Coots migrated after cold fronts and at night.

Breeding. Consumption of animal matter by American Wigeon increases sharply during the breeding season, especially by females. From arrival through the incubation period in Saskatchewan (1976–78), the diet (aggregate percentage) of 34 females averaged 31.6% animal material, and that of 34 males, 13.3% (Wishart 1983a). Female diets averaged 15.8% ($n = 9$) animal matter upon arrival, 36.5% ($n = 19$) during prelaying and laying, 37.5% ($n = 6$) during incubation, and 22.6% ($n = 15$) during brood rearing. Caddisflies (Trichoptera) were the most important (12.4%) animal food consumed by females over this period, although other animal matter included dragonflies (Anisoptera), damselflies (Zygoptera), aquatic beetles (Cole-

optera), midges (Chironomidae), flies (Diptera), mollusks (Mollusca), and crustaceans (Crustacea). Both males ($n = 15$) and females ($n = 5$) resumed their diets of 99%–100% vegetation by the molting period and continued this during fall migration and winter. Important plant foods during the breeding period were upland species, such as the seeds of oats and barley, lambsquarters (*Chenopodium album*), and various grasses, as well as aquatic plants, such as watermilfoil, duckweed, and water buttercup (*Ranunculus subrigidus*). The seeds of bulrushes and sedges were also consumed. In the Northwest Territories, the diets of 10 adult Wigeon averaged 69% plant material and 31% animal matter; ducklings ($n = 16$) averaged 66% animal food (Bartonek 1972). In Alberta, 129 ducklings consumed 97% animal matter for the first 10 days, and then switched to 89% plant matter by 20 days of age (Sugden 1973).

Migration and Winter. The diet of American Wigeon consists almost entirely of plant matter during migration and in the winter. Submerged aquatic plants are preferred, but agricultural crops are also readily eaten. When they arrived in September at Boundary Bay in the Fraser River delta of British Columbia, American Wigeon fed almost exclusively in intertidal habitats bounded by the upper limit of exotic eelgrass distribution during the day. With the arrival of winter storms and heavy rains in November, they switched to nocturnal feeding in nearby agricultural habitats (Baldwin and Lovvorn 1984a, 1984b; Lovvorn and Baldwin 1996). Of 45 birds collected at Boundary Bay from October through March, their diet was mostly (93.3%) intertidal vegetation: exotic eelgrass (84.8%), sea lettuce (5.8%), and eelgrass (2.7%). Exotic eelgrass may have been used more than native eelgrass because it occurred at higher elevations in the bay and thus was available for a longer time during the fall, when it was also most abundant. Plant material was also the virtually exclusive (99.6%) dietary component of 30 American Wigeon sampled from upland habitats, with major food items being the leaves and roots of unidentified grasses (65.8%) and clover (13.6%), and the seeds of knotweed (*Polygonum* spp.; 6.8%) and wild millets (*Echinochloa* spp.; 3.3%).

Lovvorn and Baldwin (1996) summarized studies reporting the tremendous food biomass in the intertidal zone of Boundary Bay, a major stopover area for migrating birds in the Pacific Flyway. Dry mass estimates for exotic eelgrass were 1.4 tons of leaves and 2.1 tons of rhizomes. The biomass of gastropods (snails) totaled 272 tons, and bivalves (clams), 86.5 tons. Amphipod biomass was 16.4 tons. The juxtaposition of these food resources (in the intertidal zone) with those in nearby farmlands in the Fraser River delta area supported about 75% of the American Wigeon population (determined by radiotracking 49 birds). The farthest distance they traveled from intertidal areas was 6 km. Ice cover on the delta, however, forces dabbling ducks into alternative sites during about 13% of all winters.

During the winter (Nov–Feb) in the Humboldt Bay area of California, plant material was also the dominant (99.7%) food of 140 American Wigeon collected by Yocom and Keller (1961). Eelgrass was the most important (81.0%) item, although clover (6.0%) and spikerushes (*Eleocharis* spp.; 3.5%) were also consumed. Vegetative matter was the primary (94.5%) food of 31 American Wigeon collected during the winter on managed tidal impoundments in South Carolina (Landers et al. 1977). The most important items were widgeongrass (34.9%), Carolina redroot (*Lachnanthes caroliniana*; 21.2%), bulrushes (12.6%), and sedges (*Cyperus* spp.; 21.7%). On the Pacific coast of Mexico, the gizzard contents of 42 American Wigeon contained 99.7% (24 items) plant material, including 22.1% widgeongrass, 18.5% spikerushes, 18.3% domestic rice, and 17.6% spiny naiad (*Najas marina*; Saunders and Saunders 1981). During fall migration in Illinois, Havera (1999) noted that the food habits of 31 American Wigeon were also dominated (98.6%) by plant material, including corn (27.4%), naiads (*Najas* spp.; 16.9%), and muskgrass (*Chara* spp.;

11.6%). In Saskatchewan, males (*n* = 32) and females (*n* = 19) almost exclusively (99%–100%) used plant material during fall migration and winter (Wishart 1983a).

MOLTS AND PLUMAGES

The molts and plumages of American Wigeon are described by Southwick (1953), Palmer (1976a), Wishart (1983a, 1985), and Esler and Grand (1994a); these are summarized and detailed in Mowbray (1999).

Juvenal plumage is completed approximately 6 weeks after hatching, with the first feathers occurring in about 2 weeks. Individuals are generally grayish to grayish brown at this time, with little green in the speculum, but there can be some white on the wing coverts. First basic is acquired immediately after juvenal plumage and includes the body feathers but not the wing feathers, which are retained until the following summer. Acquisition of this plumage can begin as early as 7 weeks after hatching and proceeds rapidly, although most individuals do not replace the retrices until December or January. First basic in males is generally a dark plumage, with off-white on the head and the neck and pinkish tan on the breast, the sides, and the flanks. Females are similarly dark, with sharp markings.

First alternate in males involves most feathers except for those on the wings. It is largely complete by October in adult males in Saskatchewan. Some, however, do not fully acquire this plumage until they are on their wintering grounds in January. By April, all males complete first alternate, which resembles definitive alternate. A light to heavy molt occurs on the breast, the tail coverts, the back, and the scapulars from February through April, but it does not alter plumage coloration and is considered a continuation of the prealternate molt (Wishart 1985). Females acquire first alternate with timing similar to that of males, and the resultant plumage resembles definitive alternate.

Definitive basic is a complete body and wing molt, and males resemble females when in this plumage. Males begin their acquisition of definitive basic in June, when their mates are incubating. Molting males will concentrate on large wetlands with other males once pair bonds dissolve. The wing feathers are not replaced until almost all of the body plumage is complete, and this part of the molt involves a flightless period of 35 days. In Saskatchewan, most males were flightless from 15 to 31 July. In Saskatchewan, successfully breeding females began to acquire definitive basic in July.

The acquisition of definitive alternate in males begins in mid-August, after the remiges are fully grown, and it is mostly complete by October, although certain individuals can retain some basic feathers into January. On the coast of British Columbia, Wishart (1985) observed that males underwent a light to heavy molt of the tail coverts and the breast, back, and scapular feathers. Bellrose (1980) noted that there was great variation among individuals in the acquisition of definitive alternate, with some males still wearing 10%–30% of their basic plumage, and all males not acquiring full definitive alternate until mid-February. In Saskatchewan, females acquire definitive alternate from August into September, but they do not molt from October to January. Females then begin a light molt in January, with the incidence intensifying in February and March. This molt continues until their arrival on the breeding grounds, but it is not noticeable when egg laying takes place.

CONSERVATION AND MANAGEMENT

American Wigeon are vulnerable to the loss of wetland and upland habitats, which are conservation issues affecting all breeding waterfowl in the PPR (see the Blue-winged Teal account). Their wintering habitat is more secure, although habitat loss and/or disturbance occur on specific staging and wintering sites. Lovvorn and Baldwin (1996) recommended that a number of suitable sites in the Puget Sound area on the Pacific coast be available for wintering American Wigeon, because these birds move from their primary sites during periods of adverse weather.

One episode of poisoning was reported in Washington and Oregon, where wintering American Wigeon died after exposure to Diazinon AG 500, which they ingested while grazing on golf courses (Littrell 1986, Kendall et al. 1992). In October 1986, 85 American Wigeon died after foraging on one fairway in Bellingham, Washington, on the same day when Diazinon was applied and irrigated.

American Black Duck

Anas rubripes (Brewster 1902)

Left, drake; *right*, hen

The American Black Duck is endemic to North America, where its principal breeding range is eastern Canada and the northeastern United States. The species' primary wintering range is along the Atlantic coast, from the Maritime Provinces to the Carolinas, but these ducks also winter at inland sites in the Mississippi Flyway, mainly in Kentucky and Tennessee. They are part of the "Mallard complex," which, in North America, also includes the Mallard, the Mexican Duck, and the Mottled Duck. A large, dark brownish–black puddle duck, the plumage of American Black Ducks is similar between males and females, but the sexes can be readily differentiated by bill coloration. Breeding birds occupy the array of freshwater and saltmarsh wetlands that occur throughout their range, but wetlands created or altered by beaver (*Castor canadensis*) are heavily used in forested habitats. Most nests are on the ground in all habitats, but some are in trees, and even on duck blinds in Chesapeake Bay. The population, as indexed by the Midwinter Survey, has declined dramatically (>50%) between the 1950s and 1980s, which prompted a significant management and research focus on this species. In 2010, the Midwinter Survey tallied 223,500 American Black Ducks, 7% below the most recent (2000–2009) 10-year average of 241,100, and well below the population goal of 640,000 breeding birds. Various factors have been identified as affecting American Black Duck populations, including overharvest, loss and alteration of habitats, competition from Mallards, and hybridization with Mallards. All have received considerable research attention and debate. Hybridization with Mallards, however, is especially serious because there are no habitat or behavioral factors that separate Mallards from American Black Ducks, and no intervening management action can be undertaken.

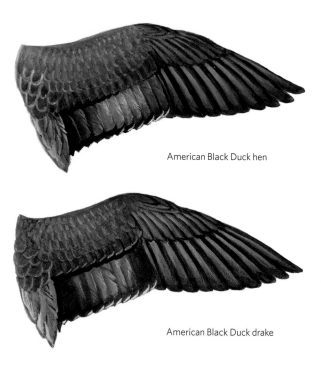

American Black Duck hen

American Black Duck drake

IDENTIFICATION

At a Glance. Both sexes of American Black Ducks have a similar plumage, which is comparable to that of female Mallards but is a much darker blackish brown. In flight, the silvery underwing linings contrast sharply with the dark body plumage. The speculum is a bluish purple to violet, bordered by black. The sexes are easily differentiated by bill color: yellow in the male, but generally olive green with black markings in the female.

Adult Males. The contrast between the male's lighter brownish-gray head and neck and the uniform blackish-brown body is quite noticeable. At closer range, the blackish eye stripe, darker cap, and a bright yellowish bill of the male are apparent. The speculum is a bluish purple to violet, bordered by black, with a thin white line at the trailing edge. The legs and the feet are brownish red to coral red, with black webs. The leg and foot color of some adult birds is a deep reddish-orange, a characteristic previously thought to denote a separate species (Shortt 1943). The irises of males are a darker brown than those of the females.

Adult Females. Female coloration is similar to that of adult males, except for the bill. In the hand (but not visible from afar), the breast feathers of the female have a V-shaped marking, compared with a U-shape in males, although males lack these U-shaped markings when in basic plumage. Females are also paler overall when viewed together with males. The legs and the feet are a fleshy orange, with black webs. In contrast to males, females have an olive-green bill, with grayish markings that merge to black on the upper mandible.

Juveniles. Juveniles are similar to adult females, but their bill color is a dark greenish olive. The feet are blackish in hatchlings, but they gradually take on a muted flesh tone (with black webs) in females and orange salmon (with black webs) in males.

Ducklings. American Black Duck ducklings resemble those of Mottled Ducks and Mallards, but the former are darker, with a larger ear spot and a much wider eye stripe (Nelson 1993). The back and the wings are dark orange yellow and olive brown, with a dusky-brown crown and forehead. Some ducklings are distinctly darker than others. The chin is a pale orange yellow that appears "silvery." The belly and the undertail are yellow, but less so than in Mallard ducklings. The feet are dark; the bill has a grayish-pink base color.

Voice. American Black Duck calls are similar to those of Mallards. Females give the familiar *quack*,

Winter flock of American Black Ducks.
Eric C. Reuter

and males most often give a low *raeb* alert call. A paired *raeb-raeb* is uttered while mating or attracting a mate. The decrescendo call of the female is a *quaegeageageageag*, usually about 6 syllables long, with the second syllable accented but the amplitude decreasing with successive notes. The decrescendo call is given mostly in the early morning and evening and functions to attract mates or conspecifics (Johnsgard 1965, Abraham 1974, Palmer 1976a).

Similar Species. This species is often confused with female Mallards, but the plumage of American Black Ducks is much darker and shows no white. In particular, American Black Ducks lack the extensive white on the tail feathers of female Mallards, and they do not have the broad white band bordering the violet-blue speculum. American Black Ducks can also be confused with Mottled Ducks, another monomorphic duck (a species where the sexes are similar in appearance) in North America that is primarily resident in Florida and along the Gulf Coast. Nonetheless, the breeding range of these 2 species does not overlap, and American Black Ducks are uncommon in the wintering range of Mottled Ducks, except in coastal South Carolina and Georgia. In comparison with Mottled Ducks, American Black Ducks have a darker plumage, less contrastingly pale versus extensive buffy edging on their body feathers, and streaking on the head and neck versus a plain and buffy coloration.

American Black Duck × Mallard hybrids are usually easily recognized by an intermediate combination of features from the 2 species: white in the tail, curled uppertail coverts, patches of green on the head, traces of white bordering the speculum, lack of a white neck ring, and/or pale grayish tertials. Kirby et al. (2000) examined 4,608 American Black Duck / Mallard wings from the Atlantic Flyway in 1977 and concluded that 13.5% were hybrids.

Weights and Measurements. American Black Ducks and Mallards are the heaviest ducks in the genus

Female American Black Duck with her brood. *Darroch Whitaker, Parks Canada*

Anas. Body weights of American Black Ducks captured along the midcoast of Maine during winter (Jan–Mar) in 1979–85 averaged 1,202 g for adult males ($n = 952$), 1,013 g for adult females ($n = 406$), 1,107 g for immature males ($n = 273$), and 962 g for immature females ($n = 343$; Krementz et al. 1989:82, average of means from table 1). Body weights of American Black Ducks captured in Maine during late summer and fall (Aug–Oct) of 1983–95 averaged 1,317 g for adult males ($n = 222$), 1,090 g for adult females ($n = 227$), 1,158 g for immature males ($n = 857$), and 1,016 g for immature females ($n = 664$; Longcore et al. 2000a). The average bill length was 54.2 mm in adult males ($n = 377$), 51.1 mm in adult females ($n = 355$), 54.5 mm in immature males ($n = 195$), and 51.3 mm in immature females ($n = 240$). Wing chord length averaged 285.0 mm in adult males ($n = 377$), 268.7 mm in adult females ($n = 335$), 280.8 mm in immature males ($n = 858$), and 265.5 mm in immature females ($n = 659$).

DISTRIBUTION

Both the breeding and the wintering distributions of American Black Ducks are confined to eastern North America, largely east of the Great Plains and south of the tundra. Their primary breeding range is in eastern Canada, including the Maritime Provinces, and south through the northern New England states. American Black Ducks winter extensively along the Atlantic coast, from the Maritimes through the Carolinas, as well as at inland sites extending to the Mississippi River and its tributaries.

Breeding. With some exceptions, in the United States American Black Ducks breed from the Mississippi River east across the northern tier of states. In Canada their primary breeding range reaches from extreme northeastern Manitoba through Ontario, eastward through Québec as far north as the taiga extends in Québec, Labrador, and Newfoundland, and throughout the Maritime Provinces. Elsewhere in Canada, they are uncommon and local in southern Saskatchewan (Smith 1996) and the Fraser River delta in British Columbia (Campbell et al. 1990).

The breeding population of American Black Ducks in Canada is assessed via the Eastern Survey, conducted since 1990 by the U.S. Fish and Wildlife Service, using fixed-wing aerial transects, and

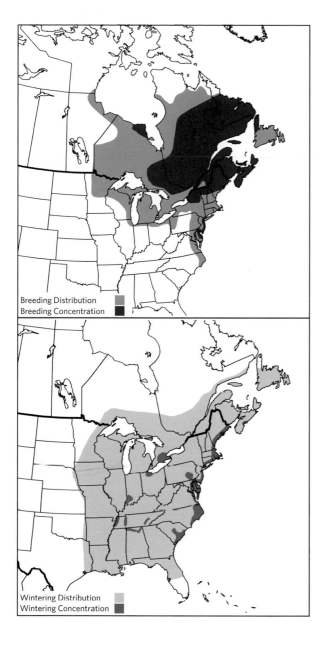

Breeding Distribution
Breeding Concentration

Wintering Distribution
Wintering Concentration

insula of Québec, as well as New Brunswick and Nova Scotia).

In the United States, American Black Ducks nest in coastal marshes from Maine as far south as Cape Hatteras in North Carolina. Historically, American Black Ducks were particularly abundant in the marshes adjacent to the eastern shore of Chesapeake Bay, but in recent years habitat degradations have affected the number of breeders, with total breeding pairs in the Chesapeake area averaging 3,658 from 1993 to 2005. (Costanzo and Hindman 2007). To the west, American Black Ducks breed in northeastern Minnesota (Janssen 1987), northern Wisconsin (Robbins 1991), Michigan, and northern Ohio, largely in the Lake Erie marshes (Peterjohn 1989). They also breed locally but irregularly in western and southern Minnesota, northeastern Illinois, central Indiana, and southwestern West Virginia. There are casual breeding records in the Dakotas, Kansas, and south to Georgia and northern Florida (Longcore et al. 2000a).

The Breeding Waterfowl Plot Survey conducted in the Atlantic Flyway states (except Maine) averaged 59,179 American Black Ducks annually between 2000 and 2010: 30.7% in the coastal marshes of New Jersey, 17% in the Adirondack region in New York, 9.3% in New Hampshire, and 7.4% in Massachusetts. The collective states bordering Chesapeake Bay (New Jersey, Maryland, Delaware, and Virginia) contained 33.2%. Maine is assessed via the Eastern Survey and contains the bulk of breeding American Black Ducks in the northeastern states, which averaged 66,463 from 1995 to 2011. Their numbers have fluctuated markedly, however, from 139,900 in 1998 to only 32,790 in 2010.

Winter. American Black Ducks primarily winter in eastern North America, east of the Mississippi River. The Midwinter Survey in the United States averaged 239,606 American Black Ducks from 2000 to 2010 in the Atlantic (88.9%) and Mississippi (11.1%) Flyways. Virtually none were recorded in the Central and Pacific Flyways. In the Atlantic

by the Canadian Wildlife Service, via helicopter surveys. The data from the combined fixed-wing and helicopter surveys averaged 474,182 American Black Ducks from 1990 to 2010: an estimated 37.6% in the western Boreal Shield (Ontario and western Québec), 24.8% in the eastern Boreal Shield (eastern Québec, Labrador, and Newfoundland), 20.7% in the central Boreal Shield (central Québec), and 16.8% in the Atlantic highlands (the Gaspé Pen-

Pair of American Black Ducks; note the yellow bill of the male (*foreground*).
RyanAskren.com

Flyway, wintering American Black Ducks concentrate in the coastal wetlands between Long Island in New York and North Carolina. The largest percentages of birds are in New Jersey (44.1%) and in the Chesapeake Bay area of Maryland, Delaware, and Virginia (23.8%). Elsewhere, 7.7% are in Maine (coastal areas), 8.6% in Massachusetts (coastal marshes), 7.7% in New York (mostly on Long Island), 3.6% in North Carolina, and only 0.5% south of that state. About 24,000 hardy American Black Ducks counted on Christmas Bird Counts wintered in Canada (1990–2004), largely along the Atlantic coast, as well as near Lake Erie and the St. Lawrence River. The Maritime Provinces harbored the greatest numbers, with almost 1,000 wintering as far north as Newfoundland.

The Mississippi Flyway averaged 26,490 American Black Ducks from 2000 to 2010, of which most (32.9%) were in Tennessee, largely on the Cross Creeks National Wildlife Refuge (NWR) and the Tennessee NWR in westcentral Tennessee, plus some on reservoirs in eastern Tennessee (Sanders et al. 1995). They also winter in the Lake Erie marshes and the large river valleys in Ohio (27.0%), and in marshes associated with Lake St.

Clair and the Detroit River in Michigan (13.2%). Other areas in the Mississippi Flyway with wintering American Black Ducks are the upper Illinois River valley, the region between the Ohio and Mississippi Rivers in Illinois, and the lower Wabash River in Indiana. Very few winter at the southern end of the flyway, with only 0.5% in Louisiana. Since about 2002, the numbers of American Black Ducks have declined rapidly on the Midwinter Survey in the Mississippi Flyway. Brook et al. (2009), however, detected a negative correlation ($r = -0.50$) between the Midwinter Survey and American Black Duck numbers estimated during the winter in Ontario (1986–2005), indicating a recent northward shift in the winter distribution of Mississippi Flyway American Black Ducks.

Unlike patterns in many other species of ducks, Diefenbach et al. (1988a) found no differences between sex and age classes in the winter distribution of American Black Ducks, based on an analysis of band-recovery data from 1950 to 1984. The authors postulated that early pair formation and predictable food supplies, especially in coastal areas, reduced age and sex segregation during the winter. American Black Ducks were also faithful to wintering areas, as measured by the percentage of recoveries from the same winter banding area. Fidelity was much higher (85% of the recoveries) for coastal areas than for inland areas (55%). There was some evidence of increased fidelity to more northern wintering areas, perhaps because the greater tidal fluctuations in northern coastal areas maintain more predictable food resources. American Black Ducks tend to remain at their wintering sites during periods of cold weather and diminished food availability, reducing their energy expenditure and waiting for conditions to improve, rather than migrating to another site. Hence the availability of quality habitat is especially important to wintering American Black Ducks (Diefenbach et al. 1988b).

On Prince Edward Island, resting was the primary activity of American Black Ducks during the fall (62.2%) and winter (76.7%), but this behavior rose to 76.9%–85.5% in response to the increasing wind chill (Hickey and Titman 1983). American Black Ducks also fed less and rested more as the tides grew stronger. Foraging activity was 17.6% in the fall and 11.9% in the winter. Among paired birds, females in late winter spent more time feeding than their mates (44.5% vs. 34.8%).

Migration. During the fall, American Black Ducks begin to reach areas of New England by late September, with peak arrival in mid-October; their passage into Michigan, Wisconsin, and Minnesota occurs 2 weeks later than in the Northeast (Bellrose 1980). They do not reach the middle states in abundance until mid-November. In the Chesapeake Bay area, migrants first show up in late September, but peak arrival is from October through November (Stewart 1962). The small numbers that winter in the southern states do not make a significant appearance until late November and early December.

Spring migration begins by early February in the southern states and mid-February in the middle states, with steady departures continuing into early April (Bellrose 1980). Spring departures from Chesapeake Bay begin by mid-February and peak in early March (Stewart 1962). The birds'

northward migration is gradual, arriving on their breeding areas in the northern states and Canadian provinces by late March and early April, as soon as open water appears. In contrast, in the Adirondack region of New York, American Black Ducks first arrive on small streams, before any other open water occurs (Chris Dwyer). In the Great Lakes states, these ducks arrive by mid-March, with numbers increasing through March and not declining until mid-April, when they head farther north (Bellrose 1980). They first appeared in Canada on 11 April in the St. John estuary, 15 April in the St. Lawrence estuary, 2 May at Hamilton Inlet in Labrador, and 26 May at Makkovik Bay in Labrador (Wright 1954). Spring migration in the United States is completed by mid-April, in the southern interior of Canada by late April, and at the northern limit of their range in Québec by late May (Longcore et al. 2000a).

MIGRATION BEHAVIOR

American Black Ducks are very cold tolerant, with many birds delaying their southward migration until the marshes freeze over. Migration occurs at night and generally involves small flocks of 12–20 individuals (rarely >30), but flocks of several thousand may depart from inland staging areas in response to the arrival of cold fronts (Palmer 1976a, Frazer 1988).

Migration is confined to North America. Breeding populations in the coastal marshes of the Maritime Provinces, the mid-Atlantic states, and Wisconsin are sedentary or move only short distances (Longcore et al. 2000a). Breeding birds from Ontario and Québec migrate the longest (1,000–1,300 km) distances. Migration corridors extend along a north–south axis and converge, funnel-like, from north to south (Bellrose and Crompton 1970), but they are not well defined because of overlapping breeding and wintering ranges and short travel distances.

Zimpfer and Conroy (2006) used band-recovery data from 1965 to 1998 to develop explanatory models of movements (16,598 recoveries) and fidelity (27,935 recoveries) to wintering areas in the United States for 3 Canadian breeding populations of American Black Ducks. Males from the western breeding area predominantly migrated to the Mississippi Flyway or southern Atlantic Flyway, whereas males from eastern and central breeding areas went to the northern Atlantic Flyway. Female movement patterns were similar. American Black Ducks also exhibited strong fidelity to their wintering areas: 99.0% for adult males, 95.5% for adult females, 92.1% for juvenile males, and 88.7% for juvenile females.

Molt Migration. Molt migration in American Black Ducks is incompletely known, but males leave their mates in late May or early June and, along with nonbreeding males, undergo a molt migration that can be a considerable distance from their breeding areas. Nonbreeding males are thought to arrive on the molting area first, followed by breeding males, while breeding females generally molt on heavily vegetated marshes near the breeding grounds (Longcore et al. 2000a). American Black Ducks in Canada migrate northward to molt on southern Hudson Bay, James Bay, Ungava Bay, and the northern coast of Labrador (Palmer 1976a), but some also head westward into Manitoba and eastern Saskatchewan (Longcore et al. 2000a). Near Cape Henrietta Maria on Hudson Bay, Peck (1972) considered American Black Ducks to be the second most abundant species, with flightless males observed in late June. Farther up the western coast of Hudson Bay, Bellrose (1980) cited surveys in 1955, 1956, and 1961 that reported large numbers of molting American Black Ducks from the Shagamu River in northern Ontario (37 km south of Fort Severn) to the Seal River in Manitoba (45 km north of Churchill). In the region of Hudson and James Bays, Ross (1984) noted important molting areas at Pen Island, river mouths in the Winisk and Severn zones, and the Hannah Bay Sanctuary in James Bay. Postbreeding American Black Ducks have occurred as far west as the Athabasca River delta in Alberta and as far north as the Thelon River and Chesterfield Inlet in Nunavut (Wright 1954).

Molting areas are located on tidal marshes and estuaries all along the Atlantic coast, from Hudson Bay to Chesapeake Bay. In Nova Scotia, Seymour (1984) reported molting American Black Ducks in saltmarshes. At Okak Bay in Labrador, most of the freshwater wetlands used by molting American Black Ducks were small (<10 ha), shallow (<1 m deep), and within 2 km of saltwater (Bowman and Brown 1992). The vegetation bordering the ponds was mostly willows (*Salix* spp.), but dwarf birch (*Betula glandulosa*) and sedges (*Carex* spp.) were also common. Some females use molting areas located away from coastal areas; these sites are generally similar to their breeding habitat and include emergent marshes, scrub-shrub wetlands, and beaver ponds.

American Black Ducks exhibit a limited fidelity to their molting areas. In the Okak Bay area of northern Labrador (1983–86), Bowman and Brown (1992) captured 320 molting male American Black Ducks, of which 29 were recaptured in subsequent years, and 2 in 3 consecutive years. Of those recaptured, 52% molted on the same pond used in a previous year, and 71% molted within 2 km of their original molting site. The overall return rate to the molting area was estimated at 10%. Only 5 females were captured during the study, however. Bowman and Longcore (1989) stated that survival was high during the flightless period, with a minimum estimate of 89%, based on 26 radiomarked males. Of the 23 ducks that were available to determine movements, 12 remained on the watersheds where they were marked and 11 moved to different watersheds. For ducks monitored until the end of the flightless period ($n = 17$), the mean distance between the capture site and the farthest location was 1.8 km. Molting birds were never observed feeding, and they lost 24% of their body weight (Bowman 1987).

HABITAT

American Black Ducks use a broad diversity of habitats during their annual cycle, occurring on virtually all wetland habitat types associated with salt, brackish, and fresh water as far north as the boreal forest. Specific habitats include estuaries, tidal marshes, tidal mudflats, freshwater wetlands, streams, and rivers, and even urban/suburban habitats such as parks. In general, most American Black Ducks breed on freshwater wetlands but winter in brackish wetlands, where they also make extensive use of tidal mudflats or feed along rocky shorelines at low tide. Large numbers breed in coastal saltmarshes along the Atlantic coast.

American Black Ducks use such a wide array of wetlands and nesting habitats that Palmer (1976a:329) remarked, "Breeding habitats and nest sites are so diverse that the presence of water, no matter how restricted, is virtually the only characteristic that is common." Breeding birds in coastal areas readily occupy coastal marshes, estuaries, and river deltas, while those breeding inland tend to choose wetlands created or modified by beaver. American Black Duck broods formed the largest (57.4%) percentage among 6 species observed using 41 beaver ponds in southcentral New Brunswick in 1969, with most (74%) of that use occurring on active ponds (Renouf 1972). A model of habitat use by breeding American Black Ducks found that the presence or absence of beaver was an important selection variable, along with the amount of surface water, the area of flooded timberlands, and their visibility when near an occupied human dwelling (Diefenbach and Owen 1989). The ducks also use emergent marshes along lakes and rivers, wooded swamps, and even open bogs within the boreal forest. Nests are dispersed, but they can occur in high densities on islands (Stotts and Davis 1960, Bélanger et al. 1998).

On inland wetlands in southcentral Maine (1977–80), radiomarked American Black Ducks ($n = 13$ females and 7 males) chose (in order of preference) persistent emergent broad-leaved deciduous forests and broad-leaved deciduous scrub-shrub (Ringelman et al. 1982a). They avoided evergreen scrub-shrub wetlands, because herbaceous vegetation in other habitats contained a higher density of the macroinvertebrates females needed

for egg production. American Black Ducks also made use of small ephemeral pools (5%–10% of the telemetry locations), some no more than 3 m² , but several thousand such pools were present on the study area. Some females (7 out of 13) used small shallow streams that were <0.5 m deep, probably to forage for invertebrates. Males and females occupied the same wetlands from prelaying until early incubation, because males stayed with the females for 90% of the time during prelaying and accompanied their mates on nest recesses during incubation.

In the western Adirondack region of New York, a heavily forested area, American Black Ducks primarily used scrub-shrub wetlands, but they were also found in deciduous forests, palustrine emergent wetlands, and unconsolidated bottom wetlands (Dwyer and Baldassarre 1994). In central Ontario, Merendino and Ankney (1994) examined habitat use by relating water chemistry and the physical characteristics of wetlands to information from American Black Duck and Mallard breeding pair surveys. American Black Ducks and Mallards shared the most fertile wetlands, and poor-quality wetlands were avoided by both species. When they were by themselves, American Black Ducks occupied less fertile wetlands in comparison with those they shared with Mallards. The 2 species appeared to potentially compete for preferred wetlands, which were characterized by a high degree of fertility, moderate areas of open water, very irregular shorelines, and small size.

Postfledging American Black Ducks ($n = 112$) in eastern Maine and southwestern New Brunswick tracked with radiotelemetry from September through early December in 1985–87 exhibited the greatest preference for palustrine emergent wetlands (Frazer et al.1990a). Riverine habitats were avoided in September, but were used more in November, once ice formed on emergent marsh habitats. At night, the ducks primarily chose emergent marshes >30 ha, but they occupied a greater variety of wetland types in the daytime. Their use of managed waterfowl impoundments on the Moosehorn

NWR in Maine was 66% during the day and 90% at night.

During migration and winter in the Atlantic Flyway, American Black Ducks make extensive use of both tidal and nontidal wetlands, such as estuaries, tidal mudflats, beaver ponds, riverine marshes, lakes, ponds, and reservoirs (Jorde et al. 1989). In the Mississippi Flyway, wetlands on and around the Great Lakes are especially important to these migrating ducks, particularly the Lake Erie marshes of Ohio, where there is an estimated 19,500 ha of wetland habitat (Bookhout et al. 1989). Away from the Great Lakes, American Black Ducks use floodplain wetlands and moist-soil impoundments during their spring and fall migration through the upper Mississippi River valley, although this migration corridor is of secondary importance compared with its use by other species, such as Mallards (Reid et al. 1989). On their wintering grounds farther south in the Mississippi Flyway (Kentucky and Tennessee), American Black Ducks feed in green tree impoundments and naturally flooded bottomland hardwood forests, moist-soil areas, and flooded croplands (Reinecke et al. 1989). Their resting and roosting areas are primarily in larger marshes, shrub swamps, beaver ponds, and oxbow lakes, but these sites also provide protection from human disturbance, as well as a location for courtship and other social behavior.

At the Chincoteague NWR on the Virginia coast (1985–86), Morton et al. (1989b) determined wintering habitat use by monitoring 22 radiomarked female American Black Ducks, comparing habitat use (percentage of locations in each habitat) with habitat availability (percentage of total area in each habitat type). The proportional use of saltmarshes (49.1%), impoundments (22.5%), and natural pools (7.4%) was significantly greater than the relative availability of these habitats, while uplands (1.0%), subtidal water (11.5%), and open water (0.3%) were used less frequently in proportion to their availability. Tidal flats (5.9%), streams (0.8%), and shrub wetlands (1.5%) were used in proportion to their availability. Refuge impoundments were occupied during the day, and saltmarshes at night, as well as subtidal areas during periods of ice cover. The mean core areas used by female American Black Ducks over the winter were larger (483 ha) for juveniles than for adults (183 ha). The time of day, tides, and icing all combined to influence habitat use.

Diurnal time-budget studies of American Black Ducks wintering at Chincoteague (1985–86 and 1986–87) revealed that birds using refuge pools spent more time (44.0% of the day) resting relative to feeding (31.8%), whereas birds on tidal marshes spent more time (52.6%) feeding than resting (11.5%; Morton et al. 1989b). Nocturnal observations collected across all habitats indicated that American Black Ducks fed considerably at night, as time budgets estimated that 42% of the nighttime was spent feeding (Morton et al. 1989a). In this latter paper, radiotelemetry locations also showed different habitat-use patterns during the day and night: 51.9% of the daytime locations were in refuge pools, compared with only 11.0% at night, while 8.5% were in tidal marsh during the day, versus 29.5% at night.

Albright et al. (1983) estimated that the daily energy expenditure of female American Black Ducks wintering in Maine ranged from 159 Kcal/bird/day at 5°C to 240 Kcal/bird/day at −20°C. American Black Ducks wintering at Chincoteague NWR spent a similar amount of energy. Due to the milder climate in Virginia, however, it was assumed that American Black Ducks there departed with larger lipid reserves than the Maine birds. Temperatures encountered by American Black Ducks wintering in Maine often were below −10°C, requiring ducks to conserve energy by resting.

Longcore and Gibbs (1988) noted that wintering American Black Ducks in Maine avoided the wind when temperatures were <0°C, seeking protection and resting in smaller (50–99) or larger (>100) flocks in the lee of landforms. These ducks also formed larger, dense flocks in response to cold temperatures, with flock sizes of 65–88 when temperatures were −4.4°C to −12.2°C, compared with groups of 25 when temperatures were between 5°C and 13.8°C. Such energy conservation behaviors probably lowered their thermoregulatory costs.

Lastly, tidal cycles strongly influence habitat use of wintering American Black Ducks, but the effects are different in northern versus more southerly areas (Morton et al. 1989a). Extreme tidal fluctuations (often >1 m) as far north as Maine and eastern Canada keep many foraging areas ice free, which enable American Black Ducks to winter in these northern areas, primarily by foraging on the bivalves and amphipods available at low tide (Jorde and Krohn 1986). In contrast, decreasing tidal amplitudes farther south increase the area of cordgrass (*Spartina* spp.) saltmarshes, where tides have less of an effect on American Black Duck feeding habits. At Chincoteague NWR, the tidal height is only about 0.6 m, and small bivalves are found there in lower densities. Aerial surveys revealed that 94% of American Black Ducks observed at Chincoteague were in the water feeding on sea lettuce (*Ulva lactuca*), rather than searching for bivalves (Morton et al. 1989b).

Habitat and Body Condition

In Maine (1974–76), both immature ($n = 34$) and adult ($n = 16$) female American Black Ducks increased their body mass and lipid reserves in late fall to provide an energy reserve during winter along the coast, although immatures at that point carried 54 fewer grams of lipids than adults (Reinecke et al. 1982). During the winter, adults used 59 g of lipids (vs. 64 g in immatures), along with 17 g of protein (vs. 25 g in immatures). The loss of protein reserves possibly lowered the ducks' daily energy requirements and increased the effective amount of their energy reserves. Nonetheless, winter was deemed the most stressful period in their annual cycle, although their winter lipid reserves were sufficient to provide adults with 8–9 days of survival time at ambient temperatures from −10°C to −20°C, and 4–5 days for immatures. Juveniles were probably at a physiological disadvantage during severe winter weather.

In contrast to the situation in Maine, where both body mass and lipids declined from high levels in the fall to low levels later in the year, American Black Ducks wintering farther south at Chincoteague, Virginia (1985–86), generally maintained their body weight and lipid reserves. Adult males ($n = 23$) had high (132–148 g) lipid reserves from early through late winter. The body weight and lipids of adult ($n = 18$) and immature ($n = 18$) females were low (58 g and 21 g of lipids, respectively) in early winter, probably because breeding activities and a later molt than males left them insufficient time to accumulate lipids. The females recovered lipids and body weight by midwinter, however, and these remained high through late winter.

The variation in body mass and lipid patterns between American Black Ducks wintering in Maine and in Virginia was explained by differences in the severity of winter weather and in their diets. Maine obviously has lower average temperatures and harsher winter weather than Virginia, which put demands on the ducks' lipid reserves. Their diet in Maine, however, is dominated by high-protein invertebrates, which do not promote lipid deposition, compared with the plant foods primarily used by American Black Ducks in Virginia. Morton et al. (1990) also argued that wintering farther south increased survival.

POPULATION STATUS

The American Black Duck population declined by >50% between the 1950s and 1980s, which engendered significant conservation concern, enough so that the American Black Duck Joint Venture was the first species joint venture to be established under the auspices of the North American Waterfowl Management Plan. Population assessment is hampered, however, because several major surveys take place, and no single survey covers their entire breeding range. Additionally, limited visibility within the forested habitats used by breeding American Black Ducks creates difficulties in extrapolating survey results to population estimates. Nonetheless, the population goal established in the North American Waterfowl Management Plan is a breeding population of 640,000 American Black Ducks.

Since 1990, the U.S. Fish and Wildlife Survey has conducted fixed-wing aerial transect surveys in eastern Canada and states in the Northeast (the Eastern Survey), while the Canadian Wildlife Service has conducted helicopter surveys in core American Black Duck breeding ranges in Ontario, Québec, and the Atlantic Provinces. These data were once analyzed separately, but they were merged in 2005 and reported as a single population estimate. In contrast to these 2 breeding surveys, the Midwinter Survey has taken place annually since 1955 and thus forms the longest dataset available to assess population trends, although this survey needs to increase its precision and reduce its bias (Conroy et al. 1988). Nonetheless, patterns seen in the Midwinter Survey data clearly reveal the extent of the decrease in the American Black Duck population.

Midwinter Survey data since 1955 reveal an almost continuous decline until the mid-1980s, when the population stabilized at 50% below the 1955 estimate. Since the mid-1990s, American Black Duck numbers have continued to decline, from 273,500 (1996–2000), to 250,000 (2001–5), to 224,700 (2006–10). In 2010, the Midwinter Survey tallied 223,500 American Black Ducks, which was 7% lower than the 10-year (2000–2009) average of 241,100. Of this total, 203,000 (86%) occurred in the Atlantic Flyway (5% below the 10-year average), and 20,400 (9%) occurred in the Mississippi Flyway (25% below the 10-year average). American Black Duck population averages in the Atlantic Flyway declined by 55% (from 403,000 in 1955–59 to 221,000 in 2000–2005). In the Mississippi Flyway, their average numbers decreased by a whopping 86% (from 213,000 in 1955–59 to 30,000 in 2000–2005). Averages for this species from 2006 to 2010 were 204,000 in the Atlantic Flyway and 21,000 in the Mississippi Flyway.

Population trends estimated by Christmas Bird Counts have depicted an even greater decline in American Black Duck numbers than the Midwinter Survey. Christmas Bird Counts during 1959–60 recorded 181,372 birds (11.4 ducks detected/party hr) in Canada and the United States, compared with 150,409 (1.6/party hr) in 1979–80, an 86% decrease in detection rates. The 2009–10 count tallied 115,331 (0.9/party hr).

The breeding population estimate for the 2010 Eastern Survey was 565,900 American Black Ducks, which was 13% below the 10-year (2000–2009) average of 650,800. The Atlantic Flyway Breeding Waterfowl Plot Survey—a ground-based survey of 1 km² in Atlantic Flyway states from New Hampshire and Vermont south to Virginia—estimated 38,200 American Black Ducks in 2010, 43% below the 1993–2009 average.

Declines are also well documented at local and regional levels. American Black Ducks were once a common breeding duck in the Chesapeake Bay area, but an analysis of 40 years of data (1966–2005) from the Breeding Bird Survey revealed a decline of 2.9%/year on the Chesapeake Bay portion of the upper Atlantic Coastal Plain, which includes coastal areas of New Jersey, Maryland, Delaware, and Virginia (Costanzo and Hindman 2007). The decrease in Maryland alone averaged 4.1%/year.

Harvest

The American Black Duck is a highly prized game species, particularly in eastern North America, one associated with a rich hunting tradition. The average harvest in the United States was 139,122 from 1999 to 2008, with a maximum harvest of 174,467 in 2000 and a minimum of 110,612 in 2004. Harvest occurs in the Atlantic Flyway (70.8%) and the Mississippi Flyway (29.1%). In the Atlantic Flyway, American Black Ducks were harvested in New York (21.4%), Maryland (14.4%), New Jersey (12.2%), Virginia (10.2%), and Pennsylvania (9.5%).

From 1999 to 2008, the harvest of American Black Ducks in Canada averaged 117,848, with a high of 174,943 in 1999. Over this time period, 30.6% were harvested in Québec, 23.0% in Nova Scotia, 14.7% in Ontario, 14.2% in Newfoundland, 11.5% in New Brunswick, and 6.0% in other Canadian provinces. There is a subsistence harvest in Canada, which was estimated at about 1,300 birds annually harvested by Cree hunters in Ontario (1967–77), and about 11,000 on the Labrador coast in 1980 (Longcore et al. 2000a).

BREEDING BIOLOGY

Behavior

Mating System. American Black Ducks are seasonally monogamous, with most birds breeding as yearlings. Pair formation begins as early as late summer and continues through fall and winter, although Johnsgard (1960a) remarked that pairs observed during October in central New York appeared to be temporary. In the Kent Island area of Chesapeake Bay in Maryland (1956–57), Stotts and Davis (1960) noted the monthly chronology of birds observed in pairs: 5.6% (26 Aug– 22 Sep), 10.4% (23 Sep–20 Oct), 15.1% (21 Oct–17 Nov), 50.0% (11 Feb–10 Mar), 58.6% (11 Mar–7 Apr), and 87.8% (8 Apr–5 May). Adults pair before juveniles, with juvenile females pairing at 6–7 months of age, and juvenile males at 7–8 months.

In an experiment with captive American Black Ducks, individuals on an ad libitum (unlimited) diet paired earlier, and their pair bonds were stronger than birds fed a restricted diet (Hepp 1986). Individuals on the ad libitum diet also had a greater peak winter weight, relative to initial weight, than those on the restricted diet (22% vs. 6.5%). Adults of both sexes paired significantly earlier than immatures: 75% of the females on the ad libitum diet were paired by the end of October, compared with 50% on the restricted diet. During the study, adult males were dominant over both adult females and young birds of both sexes (Hepp 1989). Changes in body mass during the winter were not correlated with male dominance rank (the relative status among dominant males), perhaps because dominant males expended more energy in aggressive interactions with other dominant males. Most (86%) of the aggressive encounters initiated by dominants were with other dominant males, indicating that significant energy is expended by

them during the courtship process and is a clear cost of pair formation. In the wild, Brodsky and Weatherhead (1985a) found that the courtship behavior of American Black Ducks was correlated with differences in habitat quality during the winter at 3 sites they examined in Ontario. Courtship began earliest and occupied more time at the site where food was most nutritious. Courtship at the intermediate-quality site only began after temperatures were high enough to reduce the energy costs of existence. Courtship began last and was of the shortest duration at the site with the least nutrients. Hence studies of both captive and wild American Black Ducks demonstrated that courtship activity and, therefore, pair bond formation may be delayed or even prevented at sites with limited food availability.

In coastal South Carolina, American Black Ducks began their courtship and pairing earlier than 5 other species of dabbling ducks studied by Hepp and Hair (1983). Paired American Black Ducks were common in October, so courtship behavior must have commenced before that time. In November, 96.7% ($n = 30$) of the females were paired, and 100% ($n = 45$) were paired by January. In the Cayuga Lake region of central New York, the major period of sexual display was November and December, when the percentage of paired American Black Ducks increased rapidly: from about 20% in late November to >80% by late December–January, with all birds paired by April (Johnsgard 1960a).

In Maryland, male American Black Ducks ($n = 7$) remained with their mates an average of 14.3 days (range = 7–22 days) after incubation of the first clutches began (Stotts and Davis 1960). Attendance by drakes during renesting averaged 9.1 days for 8 males monitored from 4 days before incubation began to 16 days into incubation. There were 5 reported instances where females appeared to desert their mates: at 19 and 22 days of incubation for first nests, and 1, 6, and 9 days for renests. In Maine, pair bonds terminated 19–20 days into incubation, but bonds were broken earlier if a first nest was destroyed and a renest was not promptly initiated (Ringelman et al. 1980).

On occasion, however, the pair bond in American Black Ducks may be more enduring. Barclay (1970) observed a marked pair in the spring that was still paired the following November. In Maryland, Stotts and Davis (1960) noted small numbers of pairs throughout the summer. Some of the females that remained with their mates probably were unsuccessful nesters.

Sex Ratio. The sex ratio of American Black Ducks favors the males. Bellrose (1980) used continental band-recovery and harvest data to estimate a sex ratio of approximately 55% males in the fall population. In coastal North Carolina during winter, the sex ratio was 51%–62% males (Hepp and Hair 1984).

Site Fidelity and Territory. In northeastern Nova Scotia (1973–89), 18 out of 62 (29.0%) marked adult females returned to their natal home range, with 8 nesting within 0.25 km of the wetland where they were raised; 4 females nested in 2–4 different years on the same home range (Seymour 1991). All females were paired upon arrival to the study area. Of 97 marked adult males, 23.7% returned (15.5% as paired birds and 8.2% unpaired). The return rate was 14.9% for juvenile females and 8.5% for juvenile males.

In Maine and Vermont (1956–61), 25% of 89 females trapped on nests returned to their former nesting areas in subsequent years, most to within 91 m of their previous nest sites (Coulter and Miller 1968). Homing rates were higher (25%–32%) on island sites than in sedge-meadow habitats. On Île-aux-Pommes in the St. Lawrence estuary of Québec, 7 out of 42 (22%) marked females returned to the island in subsequent years, with 1 female using the same nest in 5 consecutive years (Reed 1970).

In Maine, the home range of radiomarked American Black Ducks averaged 119 ha for females ($n = 13$) and 231 ha for males ($n = 7$). The home range of the females was larger (130 ha) during prelay-

ing and laying, compared with incubation, but the difference was not statistically significant (Ringelman et al. 1982a). Home ranges tended to be linear, with home-range length averaging 1,870–2,080 m for females and 2,608–2,903 m for males. Pairs used 4–6 wetlands during prenesting versus 1–3 wetlands during laying and incubation. The home ranges of postfledging American Black Ducks in Maine (1985–87) also tended toward linearity but varied in size according to the month, from 797 ha in September ($n = 65$) to 1,570 ha in November ($n = 105$; Frazer et al. 1990b). In the western Adirondack region of New York, the home range of 7 radiomarked birds averaged 152 ha (Dwyer and Baldassarre 1994).

Within their home ranges, American Black Ducks establish territories, which can be small wetlands or a section of shoreline along a lake, stream, or beaver pond; these domains are usually visually isolated from each other (Longcore et al. 2000a). Territories are generally not established until about 5–10 days before egg laying (Seymour and Titman 1978, Seymour 1984). On a tidal marsh in Nova Scotia, the territories of 7 marked pairs were not established until the pairs were on the marsh for 30–40 days, with their territories corresponding to tidal ponds of 0.16–3.8 ha in size (Seymour and Titman 1978). Males were on these territories almost continuously during the prelaying and laying periods, but they deserted both their mates and their territories at about mid-incubation; males occupied territories for 27–32 days, while females remained for about 45 days. On inland streams in Nova Scotia, males remained on their territories for 20–25 days (Seymour 1984).

Pursuit flights were the major mechanism for establishing and maintaining a territory. Of 132 *pursuit flights* involving marked territorial males in Nova Scotia, most began when an intruding pair, led by the female, landed on the territory (Seymour and Titman 1978). The resident male then immediately either swam toward the female or approached with a short *jump flight*. The intruding female then flew in response to the approach by the resident male. In 48 out of 132 (36%) encounters, the mates of pursued males remained on the water or lagged at least 9 m behind in the air. The median length of *pursuit flights* was 21–25 seconds, and their frequency was highest (67%) during the egg-laying period; hence the purpose of such flights may be an assessment of intruding females and their potential for extra-pair copulations rather than territorial defense by the resident male, as reported for other ducks (Baldassarre and Bolen 2006). Seymour (1984) noted that mated males attempted extra-pair copulations both before and after *pursuit flights*.

Courtship Displays. Johnsgard (1960a, 1965) described courtship displays in American Black Ducks, which are given in courtship bouts ranging from a single male to 5 or more displaying simultaneously toward a single female. There are 3 primary displays by males: *grunt-whistle*, *head-up-tail-up*, and *down-up*. The *grunt-whistle* is given somewhat early in the courtship sequence, while *head-up-tail-up* is of intermediate intensity and usually followed by *nod-swimming*, where the male will swim rapidly around the female with his head and neck outstretched on the water. The *down-up* involves dipping the bill in the water, raising the tail, and then jerking the head upward; a whistle and a *raeb-raeb* call are given when the head is at its highest point. The *down-up* is considered to be the highest-intensity display, because it formed about half of all male displays observed by Johnsgard (1960a) during the period of pair formation. The *down-up* display is prevalent when American Black Ducks are grouped with Mallards. In contrast, the *grunt-whistle* was most frequent during the first few months of pair formation. Of all *grunt-whistles* observed, 45.6% were lone displays; 65.2% of the *head-up-tail-ups* were performed as 2 or 3 birds displayed simultaneously; and 56.8% of the *down-ups* were given when 4 or more birds displayed at the same time.

The variable intensity of the male displays is in response to the intensity of the 2 primary female displays, *inciting* and *nod-swimming*. Females *incite* by moving the head and bill back and forth along the breast and flank while giving a series of low *gagg* notes. *Inciting* becomes more common later in the pair-formation period, and it generally indicates that a pair bond has been formed. The male's response shifts away from the 3 major displays to *leading*, where the male orients the back of his head toward the *inciting* female while swimming rapidly in front of her, "leading" her away from the courting group of males. Courtship activity and male competition increases in intensity with the initiation of *leading*, with males often giving short *jump-flights* to gain position in front of the female. Johnsgard (1960a) thought that of all the male displays, the *leading* display had the greatest importance in mate selection. *Leading* was most intense in December and January, when the largest number of pairs apparently were formed, occurring as much as 20–30 times/hr. Johnsgard also noted that cold temperatures, wind, and disturbance all had strong effects on the frequency of displays.

Nesting

Nest Sites. The pair flies together over suitable nesting habitat, but the female selects the nest site and is usually very wary during nest construction, which begins about 3–4 days before the first egg is laid (Stotts and Davis 1960). Females initially form a nest bowl that is about 17.8–20.3 cm wide and 3.8 cm deep, scratched out of bare ground or nestled in sphagnum moss, matted sedges, or grasses (Coulter and Miller 1968). The subsequent nest is composed of materials at the site, which vary but include grasses, twigs, conifer needles, leaves, and the like. Average nest dimensions were 29.5 cm for the outside diameter ($n = 35$) and 8.4 cm to the deepest part ($n = 33$; Jerry Longcore). In Maryland, pine needles were the most common nesting material (Stotts and Davis 1960). In New England, down was added beginning with the fourth or fifth egg (Coulter and Miller 1968). In Maryland, females normally began adding down when the clutch was half complete, with the addition of down proceeding very slowly until just before incubation began, at which time a "profuse lining" was added to the nest (Stotts and Davis 1960:145). Individual females, however, differed in the timing of when down was added to their nests.

Nest sites are extremely variable, depending on the cover available, and include tidal flats, shallow marshes, bogs, and flooded timberlands, as well as upland forests, trees, shrublands, and grasslands. Of 731 nests located during 6 years of study (1953–58) in the Kent Island area of Chesapeake Bay in Maryland, 59.5% were in wooded areas, 18.9% were on duck blinds, 16.6% were in marshes, and 5% were in cultivated areas and cultivated borders (Stotts and Davis 1960). Stotts and Davis also noted that females will reuse nest sites from previous years, especially on duck blinds. The use of duck blinds appeared to occur where other nesting cover was lacking. Nearly 70% of these nests were located on the roof, and there was a significant preference for blinds covered with eastern red cedar (*Juniperus virginiana*) or grasses. Several blinds contained as many as 4 nests.

Of the 435 nests these authors found in wooded areas, most (57.9%) were in tangles of honeysuckle (*Lonicera japonica*) and poison ivy (*Rhus radicans*). Saltmarsh cordgrass (*Spartina alterniflora*) was the most important cover for nests located in saltmarshes ($n = 121$). Small islands within their Chesapeake Bay study area contained especially high densities of nests. On 2.0 ha Bodkin Island, annual nest density over a 5-year period (1953–54 and 1956–58) was 12.3–52.8/ha, with some nests <1 m apart. The distance from the nest to water on Bodkin Island averaged 11.3 m ($n = 76$); it was 59.1 m ($n = 71$) on 52.6 ha Parsons Island. Krementz et al. (1991) found that American Black Duck nest density on Bodkin Island was 35.7/ha in 1989, despite the island eroding to only 0.4 ha in size.

On islands in the Kennebec River in Maine (1956–63), 66.7% ($n = 75$) of American Black Duck nests were under live conifers, and another 21.4% were in patches of blueberries (*Vaccinium* spp.;

Coulter and Miller 1968). In contrast, nests on islands in Lake Champlain ($n = 169$) were distributed among 11 cover types. Early in the nesting season (before 10 May), 26.6% of the nests were located among low dead herbaceous plants, particularly nettles (*Urtica* spp.) and raspberries (*Rubus* spp.); 20.9% were among fallen logs, dead tree tops, and fallen limbs; and 17.3% were in live conifer stands. Most (59.3%) nests found later in the season were in new herbaceous growth, primarily nettles. As in Maryland, nest densities were high on islands. On 2.3-ha Young Island in Lake Champlain, the nest density was 10.0–12.6/ha, with an active nest density of 4.3–5.2/ha. Five active nests occurred in a single 0.2-ha patch of raspberries. American Black Ducks tended to nest near the edges of islands, with 70% of the island nests on the Kennebec River within 3.0 m of the high-water mark, and about 33% within 0.9 m or at the edge of 1.5–6.1 m banks.

In northern bogs, the preferred nest sites (80%) were in sweetgale (*Myrica gale*), leatherleaf (*Chamaedaphne calyculata*), and sedges (*Carex* spp.). In the floodplain forest along the St. John River valley of southcentral New Brunswick, Prince (1968) located 6 American Black Duck nests in bucket-like (open) tree cavities. The nests were in mature trees averaging 209 years old, 14.6 m tall, and with a diameter at breast height of 53.6 cm. The actual cavities used were at an average height of 5.8 m. American Black Ducks, along with Mallards, also made extensive use of stumps and dead snags for nest sites in flooded timber at the Montezuma NWR in New York (Cowardin et al. 1967). The most common nest sites were on the top of stumps and in snags of large trees where the bole had broken off. Stumps that contained nests had an average diameter of 49.5 cm and were 1.9 m above water. American Black Ducks also nested in trees (82% of 188 nests) on a flooded island in Lac Saint-Louis in Québec (Laperle 1974). In the lowlands of the St. John River in New Brunswick, Wright (1954:34) remarked that "the nest frequently is placed in holes and crotches in mature hardwoods." These hardwoods flooded to a depth of 3 m, and nests at some locations were >1.6 km from dry land.

In Maryland, 1 American Black Duck nest was located in an old nest of a Common Grackle (*Quiscalus quiscula*), and 3 others were placed in abandoned Great Blue Heron (*Ardea herodias*) nests situated 21–27 m above the ground in loblolly pines (*Pinus taeda*; Stotts and Davis 1960). Cavities at the bases of trees on islands in Lake Champlain were used as nest sites early in the nesting season, as well as tree crotches and tree stubs (Coulter and Miller 1968).

On islands in Maine, 9 out of 169 (5.3%) wild American Black Duck nests were in artificial nest cylinders lined with hay (Coulter and Miller 1968). In Maryland, McGilvrey (1971) conditioned 160 captive-reared female American Black Ducks to nest in artificial cylinders (1967–69) and found that 39 (18.1%) returned to use the cylinders during the spring following their release. In Delaware, 62 out of 429 (14.4%) of the captive-reared females imprinted to use nesting cylinders returned to them the following spring (Lesser et al. 1974). None of their 368 offspring, however, returned to use the cylinders. In Massachusetts, 27 out of 245 (11.0%) conditioned females returned to use cylinders, but none of their offspring did (Heusman et al. 1979).

Clutch Size and Eggs. During the 1950s in the Kent Island area of Chesapeake Bay, the average size of 360 complete clutches over 6 nesting seasons was 9.1 eggs (range = 1–14 eggs; Stotts and Davis 1960). Clutch size varied, however, averaging from 10.5 to 10.9 eggs early in the nesting season (15 Mar–11 Apr) but only 7.5–8.0 late in the season (24 May–20 Jun). The average decline over the 6 seasons was 0.6 eggs/2-week period. The clutch size of marked females also varied with their ages. In Maryland, the size of first clutches for females to 2 years of age averaged 9.2 eggs (n = 47), compared with 9.7 (n = 17) for females >3 years old. Second clutches averaged 8.7 eggs for the first age group and 8.5 for the second. In this same area of Mary-

land, the average clutch size of 101 nests located between 1986 and 1989 varied from 9.0 to 10.2 eggs (range = 3–20 eggs; Krementz et al. 1991).

On islands in Lake Champlain, the average clutch size was 10.4 eggs for 178 nests found in 1951–56 and 9.6 for 148 nests located in 1957–63 (Coulter and Miller 1968). Other clutch-size data presented by Coulter and Miller averaged 9.0 eggs for nests located in southwestern New Brunswick and in northern, central, and eastern Maine (1938–64). Clutch size on islands in Lake Champlain and in the Kennebec River in Maine decreased from an average of 10.8–11.1 eggs (n = 63) early in the nesting season (1–20 Apr) to 8.6–9.0 (n = 15) late in the season (31 May–16 Jun). The clutch size of 31 females ≥2 years old averaged 10.5 eggs, compared with 9.4 for a sample of 117 unmarked, mixed-age females that included yearlings. On Île-aux-Pommes in the St. Lawrence estuary of Québec (1963–69), the mean clutch size of 258 nests was 9.2 (range = 5–17 eggs; Bélanger et al. 1991).

American Black Duck eggs have smooth shells and come in varied shades of white, cream white, or pale buff green. They are ovate to elliptical ovate and measure 59.4 × 43.2 mm (n = 82; Bent 1923). The mean egg mass was 56.6 g for 61 eggs collected from different clutches in the United States and Canada, with 1 egg averaging 4.4%–5.9% of the female's weight during egg laying (Longcore et al. 2000a). Eggs are laid at an average rate of about 1/day (Coulter and Miller 1968), usually around 1–2 hours after sunrise (Stotts and Davis 1968).

Incubation and Energetic Costs. The length of the incubation period reported for American Black Ducks varies somewhat among studies. In Maryland, the incubation period for 51 clutches averaged 26.2 days (range = 23–33 days; Stotts and Davis 1960). Incubation averaged 25.6 days for 13 clutches hatched in an incubator. Disturbances affected incubation in the wild, as incubation averaged 25.0–26.9 days among females flushed from their nests 1–8 times, compared with 25.5–31.0 days among females flushed 8–17 times. On the St.

Lawrence estuary, incubation averaged 28.7 days for 23 clutches (Reed 1970).

In Maine, Ringelman et al. (1982b) studied the incubation constancy of 5 females, 3 of which nested next to wetlands and 2 in uplands. The overall incubation constancy (86.7%) did not differ between the 2 groups, but the wetland nesters averaged 2.3 recesses/day of 82 minutes each, while the upland nesters averaged only 1.1 recesses/day but took a longer (183 minutes) recess. The wetland nesters usually took recesses shortly after sunrise, while the upland nesters always took recesses in the late afternoon. In addition, the wetland nesters took longer recesses, at higher air temperatures, and also following long incubation sessions. A fourth wetland-nesting female with no down in her nest apparently compensated by taking much shorter and less frequent nest recesses (1/day averaging 34 minutes). The long recess periods associated with incubation in breeding American Black Ducks may reflect the time they need to compensate for their relatively small nutrient reserves, because they forage on aquatic invertebrates, which are not highly abundant in wetlands in the northeastern United States. The low frequency of recesses in females nesting far from wetlands may serve to reduce their energy costs by restricting their flight time to once a day, which also reduces the energy cost of rewarming the eggs.

Stored lipid and protein reserves acquired away from the breeding grounds provide a small but important component of the energy needed for producing and incubating the clutch, but most of the ducks' necessary energy is acquired from foods eaten on the breeding grounds (Owen and Reinecke 1979). On Prince Edward Island, females averaged 50.5% of their time feeding during the breeding season, compared with 35.4% for males (Hickey and Titman 1983). The energy expenditure of females peaked during egg laying, at 3.5 times the basal metabolic rate (BMR), while the peak for males was 2.0 times the BMR during prenesting. Females mobilized about 46 g of lipids between prelaying and postlaying (Reineck et al. 1982). Protein content was slightly higher in laying versus prelaying females, which probably reflected the growth of the ovary and oviduct.

Nesting Chronology. In the Chesapeake Bay area of Maryland (1953–58), the average date when the first American Black Duck eggs were laid was 19 March, with the peak of initial egg laying from 18 to 30 April (*n* = 629; Stotts and Davis 1960). The average initiation date for the last clutches was 20 June. On the same study area in 1986–89, the mean dates of nest initiation were 19 April–2 May (*n* = 168; Krementz et al. 1991). The length of the nesting season during these 2 studies varied, from a low of 68 days in 1988 to 91 days in 1953.

On islands in Lake Champlain in Vermont (1957–63), American Black Ducks initiated egg laying between 3 and 9 April, versus between 1 and 23 April (in 1956–63) on islands in the Kennebec River in Maine (Coulter and Miller 1968). The latest dates for American Black Duck clutches on Lake Champlain islands were 14–15 June. The largest (19.8%) percentage of 162 nests were initiated on 11–20 April, and 16.7% were started on 21–30 April. A greater (50%) percentage of nests, however, were initiated on 1–10 April when March temperatures were above the long-term mean, compared with only 25% that were initiated between those dates when the temperatures were below the long-term mean. American Black Ducks in the Lake Erie marshes of Ohio began to nest about mid-March and continued into early June (Barclay 1970).

Farther north, American Black Ducks on the St. John River in New Brunswick laid their first eggs on 6 April, with a peak reached in May, and the few attempts to renest continuing into July (Wright 1954). On the St. Lawrence estuary, they began laying on 12–25 April, with the last clutches begun on 7–20 June (Reed 1968).

Nest Success. American Black Duck eggs have high fertility rates, as seen on islands in Lake Champlain in Vermont (1951–55), where only 14 out of 2,008 eggs (0.7%) were infertile (Coulter and Mendall 1968). On Île-aux-Pommes in the St. Lawrence estuary of Québec, just 0.9% of 1,538 eggs were infertile (Reed 1970). In Maryland, 0.3% of 336 eggs were infertile (Stotts and Davis 1960).

In the Kent Island area of Chesapeake Bay in Maryland, nest success over 4 years of study (1953, 1956–58) averaged 38% (range = 32%–63%) for 574 clutches (Stotts and Davis 1960). Nest destruction (50% loss) was primarily (34%) due to American (Common) Crows (*Corvus brachyrhynchos*) and Fish Crows (*Corvus ossifragus*), and 10.4% of the losses were by unknown agents. Another 11.5% of the nests were abandoned, mostly for unknown reasons (6.5%), although 3.5% of the desertions were caused by tides. Krementz et al. (1991) found a higher (55.3%) average nest-success rate for 170 nests located on the same study area during 1986–89 (range = 8.7%–62.5%). Daily nest-survival rates, however, did not differ between nests monitored in the 1950s and those monitored in the 1980s. Krementz et al. concluded that nest success, as well as other reproductive parameters, did not differ between the 1950s and 1980s; instead, other factors (e.g., the loss of nesting islands to erosion and an increase in Mallard populations) were responsible for the decline of American Black Ducks in the area.

Over an 11-year study (1963–73) on islands and on the mainland in the St. Lawrence estuary, apparent nest success was 42% for 590 nests (Reed 1975a): 27.7% for 83 nests on the mainland, 43.9% for 478 nests located on the main nesting island of Île-aux-Pommes, and 51.7% for 29 nests on other islands. Annual nest success was variable, ranging from 30.2% to 71.4% on Île-aux-Pommes. The probability of nest loss was much greater during egg laying than during incubation, with renesting offsetting the losses of initial nests in some years. Gulls (*Larus* spp.) were the main nest predators on islands, as about 4,500 breeding pairs of gulls occurred in the area throughout the study period. American (Common) Crows were occasional nest predators on islands, as well as on the mainland; red foxes (*Vulpes fulva*) were mainland nest predators.

Brood Parasitism. Stotts and Davis (1960) recorded 2 instances of intraspecific brood parasitism (and suspected 11 others) during their 1950s study in Maryland, for an overall parasitism rate of 1.8%. They recorded a single instance of an American Black Duck adding to a Mallard nest that contained 5 eggs. The Mallard deserted the nest after laying her ninth egg, but the American Black Duck continued laying until the final clutch reached 20 eggs. There were also single incidents of a Ring-necked Pheasant (*Phasianus colchicus*) and a Bobwhite Quail (*Colinus virginianus*) adding eggs to American Black Duck nests. Many of the American Black Duck nests in this study occurred at relatively high densities on islands, which facilitates brood parasitism because nests and females are more visible to each other, while nests and females are much more widely dispersed (and thus less visible) in an upland setting. Stotts (1987) revisited some of these earlier study sites in 1986 and reported that 7 out of 35 (20.0%) American Black Duck nests on Bodkin Island contained eggs from semidomesticated Mallards.

Renesting. American Black Ducks readily produce a second clutch when the first is lost, but they rarely renest more than once. In Chesapeake Bay, at least 16% of the ducks that lost nests or young broods renested (Stotts 1968). Renesting occurred an average of 18 days (range = 13–26 days) after the loss of the first clutch. The mean clutch size of 8 marked females that renested averaged 9.1 for first nests, versus 8.1 in second nests (Stotts and Davis 1960). A comparison of nest success between first nests and what were probably renests found a nest-success rate of 56.5% for clutches laid before 2 May (*n* = 230), compared with 47.0% for clutches laid after that date (*n* = 136).

The largest study of renesting biology for American Black Ducks occurred on island and bog habitats in Vermont and Maine, where 32 out of 102 (31%) marked female American Black Ducks renested, compared with 57% of Mallards in the same habitats (Coulter and Miller 1968). Their renesting propensity was strongly age related, as 49% of 51 females ≥2 years old renested, compared with only 1 out of 7 (14%) yearlings. Females usually only renested once in a season, but 2 females, aged at least 3 and 4 years old, both renested twice in a given year; 2 others renested each year for 3 years, producing a total of 13 clutches. Renests were located close to first nests: 48.3% within 91 m, and 16.1% between 91 and 182 m. The hatching success of renests was high (77%). The interval between the loss of the first clutch and initiation of the first renest averaged 9.7 days for nests lost between days 1 and 10 of incubation, compared with 11.7 days for nests lost between days 11 and 24. The clutch size of renests by marked females averaged 10.3 eggs for first clutches and 9.6 in second clutches.

REARING OF YOUNG

Brood Habitat and Care. Females and their broods use shallow wetlands with emergent vegetation such as reedgrasses (*Calamagrostis* spp.) and sedges (*Carex* spp.), floating aquatic plants such as cowlilies (*Nuphar* spp.), pondweeds (*Potamogeton* spp.), or scrub-shrub vegetation such as leatherleaf and sweetgale (Longcore et al. 2000a). New or reflooded beaver ponds are especially favored, because nutrients are released early in the flooding cycle, which stimulates the production of aquatic invertebrates that developing young ducklings need as food (see Food Habits and Feeding Ecology).

Females with broods normally make a single primary movement to the brood-rearing wetland. In Maine (1977–80), 8 radiomarked females and their broods traveled an average of 1.2 km, with only 1 female making a secondary move to other wetlands (Ringelman and Longcore 1982a). Small temporary wetlands (<0.5 ha) were used as stopover points. All females initiated their primary brood movement before the broods were 3 days old; 85 unmarked broods used 21 wetlands that made up 73% of all the water on the study area. Nevertheless, 43 broods (51%) were reared on just 3 wetlands, so only a small percentage of all the study-area wetlands served as brood-rearing habitat. Beaver ponds provided habitat for 18 broods (21%), while 52 broods (61%) occupied adjacent wetlands with water levels raised by the activity of beavers. New beaver flowages (<3 years old) received more use than older flowages. Emergent wetlands were used more often, while lakes and evergreen scrub-shrub wetlands were less frequented than expected. Broods were not found on wetlands classified as dead scrub-shrub, unconsolidated bottoms, or aquatic beds. Wetlands used for brood rearing were associated with large areas of flooded mountain alder (*Alnus incana*), willows (*Salix* spp.), and herbaceous vegetation; had larger areas of surface water than bypassed wetlands; and were larger in general (4.2 ha vs. 1.2 ha). Seldom-used wetlands contained large areas of open water, submerged aquatics, or ericaceous vegetation, such as leatherleaf and sweetgale.

Females with broods clearly use fertile wetlands, but they are also found on low-fertility sites with a low pH, which are common within 17% of the breeding range of American Black Ducks (Longcore et al. 1987a). In Maine, use of wetlands by American Black Duck broods was not influenced by pH or specific conductivity (Ringelman and Longcore 1982a). In New Brunswick, wetland selection by American Black Ducks and other ducks whose broods rely on invertebrates (i.e., insectivores) also was not affected by pH levels, although wetlands with the most invertebrates, and thus the most duck broods, did not contain fish (Parker et al. 1992). Dragonfly and damselfly (Odonata) larvae were especially important invertebrates on wetlands with a pH of <5.0, where these insects formed >70% of invertebrate biomass. Invertebrate biomass and diversity were greater on wetlands frequented by American Black Duck broods that were 1–36 days old, compared with wetlands that were ignored, which emphasizes the important of invertebrates to the growth of developing ducklings. In contrast, there was no difference in the abundance and diversity of invertebrates between wetlands that were used and those not used by American Black Duck broods that were 37–55 days old.

On an agricultural landscape in southern Québec, most (97%) of 134 sightings of sympatrically breeding American Black Ducks and Mallards were on waterways: 59% on streams, 19% in ditches, and 19% in mill ponds created on these streams (Maisonneuve et al. 2000a). American Black Ducks were sighted more frequently on streams (64% of sightings/brood, on average) and ditches (31%), while Mallard broods favored streams (43%) and mill ponds (37%). Extensive movements along waterways were made by broods of both species.

In coastal areas, females with broods favor brackish tidal marshes, and they will travel to them from less productive freshwater sites. In the Antigonish watershed of northeastern Nova Scotia (1973–92; Seymour and Jackson 1996), about half of the female American Black Ducks with broods annually moved either overland or along rivers from oligotrophic-mesotrophic wetlands to a large hypertrophic tidal marsh. Not only did females nesting on nearby wetlands move to the tidal marsh, but 13 broods traveled along rivers to reach it, taking 3–6 days to cover 6–12 km. Broods traveling overland ($n = 8$) moved 1.8–3.3 km. Broods in the tidal marsh were most often located in a 12-ha area between the interface of the marsh and the estuary. During 16 years of the 1973–92 study period, an estimated 73% of 1,535 broods hatched within the watershed were on sites directly accessible to the tidal marsh or on the rivers flowing into the marsh. In Chesapeake Bay, Mallard and American Black Duck females with broods ($n = 14$) moved an average of 2.3 km from their nests to a brood-rearing wetland, but they then made shorter secondary movements, averaging 1.8 km (Krementz and Pendleton 1991). American Black Ducks nesting on Île-aux-Pommes in the St. Lawrence estuary traveled at least 5.6 km to coastal marshes to rear their broods (Reed 1975).

In experiments with captive-reared birds, the salt glands of American Black Duck ducklings increased in size with exposure to increasing salinity, and their glands were always heavier than those of similar-aged Mallards and Mallard × American Black Duck hybrids (Barnes and Nudds 1991). However, maximum salt gland size was reached at 1% sodium chloride (NaCl) concentrations, indicating that osmoregulation could only occur at fairly low salinity levels. Newly hatched ducklings died after drinking water containing >1% NaCl.

In Maryland, pipping by the unhatched ducklings became very strong and regular about 48 hours before hatching (Stotts and Davis 1960). In a normal clutch, all eggs hatched within 3–4 hours, and the egg tooth fell off in 24–36 hours. The ducklings are brooded until dry, but some ducklings are dry by the time the last egg hatches.

Females then lead their ducklings from the nest within 24 hours. At first, the young remain tightly clustered under the close guidance of the females' soft quacks (Beard 1964). As the ducklings become older, however, the female allows them to separate into groups of twos, threes, and fours and to range at greater distances when feeding, but she can call the brood together and direct their course if danger threatens. At 5 weeks, the ducklings rarely assemble as a full brood and are less responsive to their mother's call. Young ducklings are capable of foraging for themselves, but the female occasionally assists by churning up bottom sediments to locate invertebrate foods (Longcore et al. 2000a). In Maine, Ringelman et al. (1982b) reported the bond between a marked female and her brood dissolved after 43 days; another female left her brood after 48 days.

Brood Amalgamation (Crèches). Brood amalgamation appears to be rare in American Black Ducks, although Tufts (1986) reported brood amalgamation in Nova Scotia when 3 females with broods quickly adopted orphaned ducklings released near them. Longcore and McAuley (2004) documented posthatch brood amalgamation by 2 females in

Maine: 1 female with 20 ducklings and the other with 18–22.

Development. Smart (1965a) reported an average duckling weight of 31.3 g ($n = 25$) within 8–36 hours after hatching. Reinecke (1979) fitted the Gompertz equation to these and other weight and body-measurement data to describe the growth and development of juvenile American Black Ducks, noting a growth constant of 0.394. Most impressive was the relationship between the growth of the flight muscles and the legs (tarsi). Tarsus length was 50% of adult size at hatching, while body weight was only 3% of that of adults. During the first 3 weeks, the pectoral (flight) muscles were only about 1% of overall body weight, and they developed slowly until 4 weeks before fledging, when these muscles gained about 180 g of tissue and increased from 2% to 20% of the duckling's body weight. In Québec, the mean weight of ducklings from hatching to 4 days was 44 g for both males ($n = 34$) and females ($n = 32$; Lepage 1973). By 58–62 days, however, the males averaged 915 g and the females, 825 g. By September, the males ($n = 19$) averaged 1,148 g and females ($n = 17$), 1,005 g.

Acid deposition can affect duckling growth and development. In an experiment by Rattner et al. (1987), captive American Black Duck ducklings (10 ± 2 days old) were reared for 10 days either on acidified wetlands (pH 5.0) or circumneutral wetlands (pH 6.8). Ducklings on the acidified wetlands grew poorly in comparison with those on circumneutral wetlands. The final weight of ducklings reared on the circumneutral wetlands increased an average of 77 g/100 g of initial weight, compared with only 7 g for those on acidified wetlands; culmen and tarsal lengths were similarly affected. Ducklings exhibiting poor growth also tended to have higher uric acid levels and lower concentrations of hematocrit, plasma protein, glucose, and cholesterol. Acidification suppressed algal and phytoplankton growth in the wetlands, which in turn restricted the abundance of invertebrates, the food needed by developing ducklings.

RECRUITMENT AND SURVIVAL

Bellrose (1980) summarized brood size data for American Black Ducks from an array of studies completed prior to 1980: 7.5 for Class 1 ($n = 171$), 6.2 for Class II ($n = 125$), and 6.7 for Class III ($n = 149$). Bellrose believed that the slight increase in the size of Class III broods over Class II broods indicated a breakup of some broods and the association of the unattached ducklings with other broods as they approached flight stage. Despite the potential addition of stray ducklings, broods in Class III were 11% smaller than those in Class I, and represented a 23% reduction over the number of ducklings leaving successful nests.

Brood sizes vary greatly both geographically and temporally, however. On 37–40 moderately to highly fertile wetlands in northeastern Maine, the average size of Class IIc–III (near fledging) broods differed between years: 4.0 in 1993 ($n = 94$) and 4.6 in 1994 ($n = 94$; Longcore et al. 1998). Brood sizes were also bigger (4.5–5.3) on 2 large impoundment complexes, compared with broods on other wetlands in forested (3.6–4.5) or agricultural (3.6–4.6) landscapes.

In southcentral Maine (1977–80), the mean Class III brood size ($n = 36$) was 5.3 (range = 3.8–6.0; Ringelman and Longcore 1982b). Young ducklings (Class Ia–IIa) had a survival rate (Mayfield Method) of 60.7%, which was significantly lower than the 70.0% survival rate for older ducklings. At least 6 females (16.7%) lost their entire broods. Also, ducklings hatched later (after 14 Jun) had a lower survival rate (10.1%) than ducklings hatched earlier (37.1%). The authors noted that the Mayfield Method provided a more accurate measure of duckling recruitment than the use of the Class III brood numbers, which overestimated recruitment by 45% in their study.

In southern Québec, brood survival to 30 days was 13.4%–20.6% (Maisonneuve et al. 2000a). Over a 16-year period during 1973–92 in the 750 km² Antigonish watershed in northeastern Nova Scotia, the habitat type used for brood rearing greatly affected fledging success (Seymour and Jackson 1996). Mean brood sizes at fledging averaged 6.72 ($n = 39$) in estuary habitat, 6.62 ($n = 45$) on lakes, 7.05 ($n = 196$) on dispersed inland wetlands, and 3.50 ($n = 54$) on the tidal marsh. Females ($n = 27$) that moved their broods from dispersed wetlands to the tidal marsh fledged fewer (3.78) young than females ($n = 54$) remaining on those wetlands (6.89). Most mortality occurred during the move to the tidal marsh, and it was probably greater if the movement occurred before the ducklings were 2 weeks old. The distance traveled may also have been a factor in duckling survival, as only 9.4% of ducklings were lost from 17 broods moving overland for <0.5 km, compared with 26.1% for 14 broods moving 1.9 km or more along rivers and overland.

Class III brood size of American Black Ducks hatched on islands in the St. Lawrence estuary from 1964 to 1969 averaged 4.5 for 219 broods (Reed 1970). In a later paper, Reed (1975) estimated brood survival on the St. Lawrence estuary from 1968 to 1971 at only 56%, and duckling survival at 34%. Of 47 marked females (1968–73), 34% fledged at least 1 duckling, 26% lost their entire broods, and the fate of the remaining 40% was unknown. Most (80%) duckling mortality occurred during age Class I, with 20% in Class II. In part, the high duckling and brood mortality in this population was attributed to the long distances they traveled over open water from nesting islands to brood-rearing areas on the shoreline.

The decline and precarious status of the American Black Duck population prompted many insightful studies of their annual and seasonal survival rates. In Maine (1977–80), 74% of 19 radio-marked females survived over the 121-day monitoring period, which included prelaying, laying, incubation, brood rearing, and postrearing. There were no differences in survival rates among the periods (Ringelman and Longcore 1983). The 3 deaths were from predation: 2 by Red-shouldered Hawks (*Buteo lineatus*), and 1 by an unknown predator. In New Brunswick (1992–94), survival during the

breeding season (15 Apr–1 Aug) was 88.9% for 31 radiomarked females (Petrie et al. 2000).

Francis et al. (1998) estimated annual survival rates using band recoveries across 6 regions during 3 time periods: 1950–66, 1967–82, and 1983–93. The average of the survival rates for these 3 periods was 58.7% for adult females, 66.1% for adult males, 56.3% for immature females, and 56.3% for immature males. The annual survival rate for the most recent period (1983–93) was 61.0% for adult females ($n = 7,653$), 66.7% for adult males ($n = 12,089$), 57.9% for immature females ($n = 25,288$), and 64.9% for immature males ($n = 32,466$).

Krementz et al. (1988) calculated annual survival rates for American Black Ducks from 1950 to 1983, using banding data from 10 major reference areas in an effort to examine the effects of hunting. The average annual mortality rate was 38.5% for adult males, 55.1% for adult females, 55.7% for immature males, and 65.0% for immature females. The percentage of annual mortality attributed to harvest was 40.6% for adult males, 32.5% for adult females, 47.7% for immature males, and 44.1% for immature females. Changes in hunting regulations, however, appeared to change the survival rates of immatures and adult males, but not adult females. Their study was significant in the debate over compensatory versus additive mortality, in which the former theory postulates that deaths due to hunting are "compensated for" by a decline in other forms of mortality (i.e., overall mortality would not be different in the absence of hunting). In contrast, if hunting mortality were additive, then overall mortality would increase beyond that seen in the absence of hunting. The authors concluded that compensatory mortality could indeed occur at major reference areas and among sex and age classes. They cautioned, however, that "managing Black Ducks under the assumption that compensatory mortality occurs for age-sex classes, at all times, and at all locations may lead to over-harvests of populations in certain reference areas" (Krementz et al. 1988:225).

Longcore et al. (2000b) studied the survival of

397 juvenile American Black Ducks radiomarked in Québec, Nova Scotia, and Vermont during autumn in 1990 and 1991. Survival did not differ between males and females, and 86% of the confirmed deaths were due to hunting. Survival during the postfledging and staging periods differed by location, with the highest (54.5%) survival rate in Nova Scotia; where the opening date of the hunting season was the latest. The lowest (39.5%) survival rate was on the Québec/Vermont border, where hunter numbers and activity were greatest. There were no differences in survival among marked cohorts from the three regions during migration and winter. Survival was 80.9%–96.5% when hunting mortalities were removed from the calculations. Such a high rate of harvest for juvenile American Black Ducks in this study led Longcore et al. (2000b:250) to remark, "Depleted local and regional breeding populations can be restored only by allowing sufficient numbers of adult breeders and progeny to return to natal habitats." In other words, overharvest before migration begins can significantly lower American Black Duck populations at local and even regional levels, because fewer individuals are available to return and breed in subsequent years.

A 1987–88 study of 103 juveniles fitted with radiotransmitters on the 980-ha Shepody Wildlife Area in New Brunswick showed the effect of local harvest on American Black Ducks (Parker 1991). Prehunting survival (mid-Jul–Sep) was 78.7% in 1987 and 97.6% in 1988. During the first 2 weeks of the hunting season (1–15 Oct), however, survival rates declined by 54% in 1987 and 43% in 1988, with 80.9%–85.0% of all hunting mortality occurring on opening day. Hence such a substantial hunting loss at a site like Shepody may offset the benefits of management designed to increase local production.

Longcore et al. (1991) provided data on mortality and harvest from their study (1985–87) of 106 hatching-year female American Black Ducks equipped with radiotransmitters on a lightly hunted area along the New Brunswick/Maine

border. Survival from August until mid-December was 59.3%, but it rose to 69.4% when hunting losses were excluded. Survival was 98.7% from August through September, but it then declined to 88.5% by 31 October and 71.8% by 30 November. Predators accounted for most (53.2%) of the nonhunting mortality, and the hunting mortality rate was 31.7%. The seasonal survival rate extrapolated to an annual rate of 26.2%, 12% lower than the estimate from Krementz et al. (1998).

The survival of 227 radiomarked females wintering in coastal marshes of New Jersey and Virginia (19 Dec–15 Feb) was 54% in 1984/85 and 72% in 1985/86, or 65% when pooled (Conroy et al. 1989a). Combining years, survival estimated from hunting risk was 84%, while survival estimated from nonhunting risk was 78%. Nonhunting mortality was caused by predation and winter stress (starvation). Of 38 nonhunting mortalities, 21 could be assigned a probable cause: most (12) of these were killed by predators, primarily red foxes and raccoons (*Procyon lotor*); 2 were killed in traps set for muskrats (*Ondatra zibethicus*); and 5 others died from starvation and/or hypothermia during a cold spell in January 1985, when 50 other American Black Ducks were found dead or unable to fly, due to the same causes.

Conroy et al. (1989a) also noted that adults had a higher survival probability than hatching-year birds (73% vs. 60%), because adults had higher survival from nonhunting risks. Adults with body weights at capture that were greater or equal to the median weight survived better than adults at weights below the median (85% vs. 61%). Body mass did not correlate with the survival of hatching-year birds, however. On the coast of Maine, Krementz et al. (1989) examined data from 1,427 American Black Ducks captured over 6 winters (1979–84) and found no relationship between late-winter body mass and annual survival for any sex/age group.

The longevity record for a wild American Black Duck is 26 years and 5 months for a male banded at an unknown age in Pennsylvania and shot in Delaware. A second bird banded as a hatching-year female in Québec was later found dead in that province after 26 years and 4 months. There are 2 other longevity records of 25 years, 6 months and 22 years, 5 months.

FOOD HABITS AND FEEDING ECOLOGY

American Black Ducks exhibit great diversity in the food items they consume, which is related to the multitude of habitats they frequent. Along the Atlantic coast, they forage on tidal mudflats and in salt, brackish, and freshwater marshes. Inland, they use marshes, lakes, managed impoundments, beaver ponds, streams, and rivers. In New Jersey the types of foods eaten were in proportion to their availability within the foraging habitat (Costanzo 1988), indicating that American Black Ducks are opportunistic feeders.

During the breeding season, American Black Ducks feed singly or as mated pairs, while wintering birds forage in small groups. They usually dabble at the surface of the water or on mudflats and tip up in shallow water, but they will occasionally dive to depths of 2.0–3.8 m for food (Brodsky and Weatherhead 1985b). The average dive duration was 5.5 seconds at a water depth of 2.0 m, compared with 9.7 seconds at 3.8 m.

Breeding. The diet (aggregate percentage) of female American Black Ducks collected on freshwater wetlands in Maine (1974–76) included 91% animal matter during the egg-laying period ($n = 9$) and 80% postlaying ($n = 9$; Reinecke and Owen 1980). The most important invertebrate foods for laying females were mayflies (Ephemeroptera; 51%), mosquito (Culicidae; 14%) larvae, and isopods (*Asellus*; 11%). Postlaying males primarily consumed snails and clams (Mollusca; 53%), as well as mayflies (15%). The diet of 11 males during the breeding period was 66% animal matter and 44% plant matter: snails and clams again were the most significant (45%) animal foods, although isopods formed 7% of their diets. Preferred plant foods (18%) were the seeds of bur-reeds and sedges, and arrowhead (*Sagittaria*) tubers. In contrast, previous studies of

the spring food habits reported >90% plant material in the diet of American Black Ducks in Maine (Mendall 1949, Coulter 1955), but these early reports included food from the gizzard in the analysis, which dramatically underestimates the importance of invertebrates (Swanson and Bartonek 1970).

The diet of ducklings is initially dominated by invertebrates, but the percentage of invertebrates in their diet declines as they become fully feathered. In Maine (1974–76), invertebrates averaged 95% ($n = 13$) of the diet in Class I (1–18 days) ducklings and 84% ($n = 12$) in Class II (19–43 days), but declined to 34% ($n = 11$) in Class III (44–63 days; Reinecke 1979). Mayflies (59%) and caddisflies (Trichoptera; 14%) were the primary foods for Class I ducklings, while clams and snails (24%) and aquatic sowbugs (*Asellus*; 21%) were the most important for those in Class II, along with aquatic beetles (11%), and flies (9%). The highest percentage of invertebrates was consumed during the ducklings' most rapid period of absolute and relative growth.

Migration and Winter. The fall and winter diets of American Black Ducks are dominated by plant material when they are in freshwater habitats and animal material when in marine habitats. Mendall (1949) first documented this pattern in his classic study of American Black Duck food habitats throughout the state of Maine from 1938 to 1946. During fall on inland freshwater habitats, the stomachs of 366 American Black Ducks contained 84.1% plant matter and 15.9% animal matter (percentage volume): 65 different plant foods were identified, which demonstrates the opportunistic foraging ecology of these ducks. The most important plant foods were sedges (Cyperaceae; 38.4%), bur-reeds (*Sparganium* spp.; 11.6%), and pondweeds (Najadaceae; 8.9%). In contrast to the diets of Black Ducks on inland waters, animal matter dominated (72.8%) the fall diets of 49 ducks from coastal waters. Various marine snails, especially periwinkles (*Littorina* spp.), were the most significant (50.0%) animal foods, although their diets also included bivalve clams (Pelecypoda; 11.8%) and amphipods (Amphipoda; 8.2%). Eelgrass (*Zostera marina*) was the primary (10.5%) plant food. In Nova Scotia, amphipod shrimp (*Orchestia grillus*) were consumed by fall migrants (Tufts 1986). During late fall (Nov–Dec) in coastal South Carolina, the gizzard contents from 32 American Black Ducks shot by hunters contained 97.2% plant material (McGilvrey 1966): the most important items were corn (13.0%) and seeds from sweetgum (*Liquidambar styraciflua*; 11.1%), green hawthorn (*Crataegus viridis*; 7.3%), and cyperus (flatsedge; *Cyperus odoratus*; 7.3%).

American Black Ducks consume more animal foods than Mallards, and this material becomes increasingly dominant in their diet over the winter. During 2 winters (1983–84) on the central coast of Maine, marine invertebrates accounted for 96% of the diet of 18 American Black Ducks collected by Jorde and Owen (1990). The most important food items (aggregate percentage) were periwinkles (24.7%), amphipods (*Gammarus oceanicus*; 21.8%), blue mussels (*Mytilus edulis*; 21.4%), soft-shelled clams (*Mya arenaria*; 6.2%), and crabs (*Pinnixa sayana*; 6.2%). Mendall (1949) reported a diet of 86.9% animal matter for 34 American Black Ducks from coastal waters in Maine. During winters in coastal New Jersey (1983–86), animal foods accounted for 92% of the diet of 40 American Black Ducks (Costanzo 1988); the most important (64%) item was the saltmarsh snail (*Melampus bidentatus*). When ice covered their foraging marshes, however, the ducks were forced to eat foods found on mudflats and tidal creeks, such as scuds (*Gammarus* spp.), clams (*Gemma gemma*), killifish (*Fundulus* spp.), and sea lettuce.

In nutritional studies of American Black Duck foods during the winter, the amphipod *Gammarus oceanicus* contained the best nutrient content in terms of gross energy, protein, fat, and true metabolizable energy, while the amount of ash in the shells of soft-shelled clams, blue mussels, and periwinkles contributed to the low metabolizable energy content in these invertebrates (Jorde and Owen 1988). Wintering American Black Ducks in

Maine forage intensively for amphipods among ledges of rockweeds (*Fucus* spp. and *Ascophyllum* spp.) until this habitat is exposed by the ebb tide; they then move to mudflats and mussel bars to eat mollusks (Jorde 1986).

During the first year of a 2-year study (2004–5) of habitats used by wintering and staging American Black Ducks on Long Island in New York, mudflats had the greatest (1,204 kg/ha) food biomass; lesser amounts came from areas with submersed aquatic vegetation (61 kg/ha) and from saltmarshes (34 kg/ha; Plattner et al. 2010). During the second year, however, freshwater habitats had the greatest (306 kg/ha) biomass, then mudflats (85 kg/ha), and then saltmarshes (35 kg/ha). These results suggested that American Black Duck numbers might be more limited by this winter habitat choice, as food densities during winter in coastal Long Island were considerably lower than resources for other dabbling ducks using inland freshwater habitats.

Elsewhere, plant material made up 94% of the food consumed by American Black Ducks in west-central Tennessee (*n* = 39; Byrd 1991). The most important items were the seeds of spikerushes (*Eleocharis* spp.; 40.8%), swamp smartweed (*Polygonum hydropiperoides*; 17.9%), and rice cutgrass (*Leersia oryzoides*; 4.3%), although the ducks also ate the stems of these plants (11.3%). On reservoirs and lakes in Ohio, Trautman (1940) reported that American Black Ducks fed on gizzard shad (*Dorosoma cepedianum*). When available, waste corn provides an important food source in the late fall and winter, with the ducks making field-feeding flights of up to 40 km to forage on it (Bellrose 1980).

MOLTS AND PLUMAGES

Palmer (1976a) has described and Longcore et al. (2000a) have summarized the molt and plumage sequences for American Black Ducks; these authors should be consulted for further details.

Juvenal plumage begins when the ducklings are 19–25 days old and is mostly complete in 34–43 days. Full juvenal plumage is acquired in 44–

60 days, at which time the primaries have fully emerged. Young birds can fly in 58–63 days.

The acquisition of first basic begins in the fall and overlaps with juvenal plumage. First basic involves a partial molt of some of the body feathers, but the juvenal wing is retained. Pyle (2005) noted that the prebasic molt in some specimens of *Anas* that he examined (n =19), including American Black Ducks, only involved 8%–19% of the head, the neck, the upper back, the breast, and the sides. Other specimens ($n = 22$), however, exhibited no evidence of molting, but instead retained their juvenal plumage. Regardless of the extent of first basic, it is quickly molted from September through November to first alternate, which replaces all feathering except the juvenal wing. First alternate is worn into summer by males, but females replace this plumage with definitive basic in late winter or early spring. Males in first alternate are virtually identical to males in definitive plumage, but the former can be identified by the juvenal wing. Females in first alternate resemble those in definitive basic, again except for the wings.

Males acquire definitive basic beginning in early June or July, which involves a complete molt of the body and wing feathers, and a flightless period of about 28–32 days. Definitive alternate for males is acquired in late summer and early fall and worn until the following summer. Only the body feathers are replaced during the associated molt leading to definitive alternate; the wing feathers from basic are retained. Females replace definitive basic with definitive alternate after brood rearing, thus acquiring this plumage later in the summer than the males.

CONSERVATION AND MANAGEMENT

The unquestionable decline of the American Black Duck population has been of great concern—and debate—among managers and researchers for decades. No universal consensus on a cause (or causes) for this decline has yet emerged, which limits effective management actions. Rusch et al. (1989), Nudds et al. (1996), and Conroy et al.

(2002) provided thorough reviews of where data and debates stood in the late 1980s to mid-1990s regarding the population ecology for this species and the potential causes of the population's downturn. These reviews recognized that overharvest, habitat loss and degradation, contaminants, competition from an increasing Mallard population, and hybridization with Mallards all had potential influence as either single or multiple agents in the decline of American Black Ducks.

Overharvest. There has been much study and debate relative to the effects of harvest on American Black Duck populations. Early work decided that this harvest acted in an additive fashion, increasing overall mortality (Martinson et al. 1968), so restrictive harvest regulations should lead to an increase in their population. Later studies overturned the additive mortality hypothesis for ducks in general and concluded that hunting mortality was compensatory; in other words, harvest did not increase *overall* mortality beyond what would occur in the absence of hunting (Anderson and Burnham 1976). Nonetheless, as a result of a 1982 lawsuit filed by the Humane Society of the United States, the U.S. Fish and Wildlife Service was forced to implement restrictive harvest regulations for American Black Ducks in 1983; Canada followed suit in 1984. After the implementation of these restrictions, the decline in American Black Duck numbers abated and stabilized: the Midwinter Survey estimate was 293,800 in 1983, and it remained at about 300,000/ year for the remainder of the 1980s. The identification of a specific cause and effect between the population size and harvest regulations remains elusive, however, probably because the causes are "multiple and complex" (Conroy and Krementz 1990) and there is a lack of truly manipulative experiments that would yield more definitive results (Nichols 1991).

Regionally, Krementz et al. (1988) presented evidence that some sex/age groups could be overharvested in certain reference areas, and Longcore et al. (2000b) found that American Black Duck survival on study areas in Québec, Nova Scotia, and Vermont was ≥80% in the absence of hunting. Locally, Parker (1991) noted a heavy harvest on a managed area in New Brunswick. There are no conclusive studies, however, linking overharvest to the decline of the American Black Duck population at a continental level. Moreover, the responses of this species' reproduction and death rates to changes in their overall population size are poorly known, which further restricts any ability to assess the effects of hunting (Nudds et al. 1996).

Habitat Loss and Degradation. Concerns over habitat degradation on breeding areas include the landscape-level clearing of land for agriculture, and forestry practices such as clearcutting. Such factors have long been postulated as key changes facilitating the movement of Mallards, a species principally from the open habitats of the prairies, into the forested habitats that form the principal breeding range of American Black Ducks. The eastward expansion of Mallards is "one of the most spectacular shifts in waterfowl distribution patterns in North America" (Johnsgard and DiSilvestro 1976:905). Heusman (1974) remarked that extensive land clearing in New England during the mid-1800s was a drastic alteration of habitat, which, along with land clearing and the construction of many farm ponds elsewhere in the eastern United States, may have created a corridor that allowed Mallards to expand eastward into the breeding and wintering range of American Black Ducks. In Maine, increases in the number of Mallards were correlated with the creation of small ponds and marshes within historical American Black Duck range (Longcore et al. 1987a). Where forest removal was extensive in Ontario, Mallards increased while American Black Ducks declined, with the former replacing the latter on the most productive wetlands first (Merendino et al.1993).

Forest clearing, however, may not always precede the movement of Mallards into traditional American Black Duck breeding areas. Ankney et al. (1987) found that American Black Duck popu-lations declined in areas that remained forested in Ontario and Québec, while those of Mallards increased. In Ontario, Merendino and Ankney (1994) observed that both Mallards and American Black Ducks favored similar wetlands. In heavily forested areas of the Adirondack region in New York, sympatrically breeding Mallards ($n = 11$) and American Black Ducks ($n = 7$) also used similar habitats, which suggests that an undisturbed forest and habitat use within that forest are not factors isolating American Black Ducks from Mallards (Dwyer and Baldassarre 1994).

Further insight into the habitat-use interactions of Mallards and American Black Ducks stems from a 1998–99 survey of 343 4-km² plots in southern Québec, along the lowlands of the St. Lawrence River valley and Lac Saint-Jean, and in agricultural areas of Abitibi-Témiscamingue (Maisonneuve et al. 2006). American Black Duck densities were higher in dairy farm and forested landscapes (>39 indicated breeding pairs/100 km²) than in cropland landscapes (8/100 km²), while Mallard densities were similar across all landscape types (30–43/100 km²). Habitat modeling indicated that the presence of American Black Ducks decreased in areas with increasing amounts of cornfields, other plowed fields, and deciduous forests, while their presence increased where the topography was undulating, with slopes of 10%–15%. These parameters had an opposite effect on the presence of Mallards. The odds of finding American Black Ducks were doubled, however, wherever Mallards were present, indicating that both species were attracted to areas with adequate habitats. These modeling results support the hypothesis that habitat changes may be a primary factor leading to the decline of American Black Ducks in southern Canada. In particular, dairy farm landscapes are of great importance for this species, and converting this landscape to one dominated by croplands represents a threat to the American Black Duck population in this region.

Other habitat changes in American Black Duck breeding areas have been the loss or erosion of important nesting islands in Chesapeake Bay (Stotts 1987). Longcore et al. (2000a) noted that urban development and agricultural expansion have reduced their habitats in southern Québec and Ontario, as well as along the St. Lawrence River. The large-scale development of hydroelectric power in the James Bay area could potentially affect molting areas (Rusch et al. 1989).

Habitat alterations on American Black Duck wintering grounds include urbanization (or urban sprawl), wetland alteration, and wetland destruction. Munro and Perry (1982) noted that the loss of submerged aquatic vegetation affected the numbers of these ducks wintering in Chesapeake Bay. In coastal Virginia, humans disturbed 14% of 179 American Black Duck flocks sampled by Morton et al. (1989a). Human disturbances caused the ducks to reduce their feeding time; the birds also flushed in 15 out of 25 (60%) disturbances. In the longer term, a predicted rise in sea levels would result in the loss of coastal marshes that provide important breeding and wintering habitats along the Atlantic coast.

There has been a contention that American Black Ducks are more salt tolerant than Mallards, so the latter would have more difficulty moving into saltmarsh breeding habitats (the habitat refugia hypothesis). Salt-tolerance studies conducted on American Black Ducks and Mallards by Barnes and Nudds (1991), however, led to the conclusion that the differences were not enough to prevent Mallards from expanding their range into estuarine habitats dominated by American Black Ducks. Belanger and Lehoux (1994) suggested that creating habitats, particularly saltmarshes, that are more suited to American Black Ducks and would exclude Mallards could be useful in increasing the productivity of pure American Black Ducks.

Contaminants. Acid deposition (acid rain) across much of their forested breeding range has been proposed as a causative agent in the decline of American Black Ducks (Haines and Hunter 1981), particularly because 17% of the breeding range

occurs in areas with a poor acid-neutralizing capacity (Longcore and Gill 1987). American Black Ducks commonly raise broods on low-fertility, low-pH wetlands, however, so additional effects from acidification seem minimal (Longcore et al. 1987a). Moreover, this species may compensate for the low numbers of invertebrates in acidified habitats by foraging on wetlands that lack fish, which compete with American Black Ducks for invertebrate foods (Bendell and McNicol 1995). Rattner et al. (1987), though, did find reduced growth in captive-reared ducklings on experimentally acidified wetlands (see Development).

American Black Ducks most certainly have ingested spent lead shot. Nationally, lead pellets occurred in 8.3% of the gizzards from 2,206 American Black Ducks harvested between 1938 and 1953 (Bellrose 1959). In Chesapeake Bay (1976–77 and 1979–80), lead levels in the livers of 128 American Black Ducks averaged 12.4 ppm, with an 18.0% incidence of ingested shot (DiGiulio and Scanlon 1984). In Merrymeeting Bay in Maine (1976–80), lead shot occurred in 6.9% of 506 American Black Duck gizzards (Longcore et al. 1982); lead pellets in bottom sediments averaged 99,932/ha at this heavily hunted site. In eastern Canada (1988–89), the median lead level in wing bones of juvenile Mallards and American Black Ducks combined ($n = 8,634$) was 5.7 ppm, with 17% at >10 ppm (Scheuhammer and Dickson 1996). Some areas where American Black Ducks experienced high levels of lead exposure were southern Ontario and the St. Lawrence estuary, from Québec City to Lake Champlain.

There are few data on what has happened since the United States began requiring steel shot for all waterfowl hunting, although this practice is having a positive effect on lead toxicity in American Black Ducks and probably other waterfowl as well. On 2 national wildlife refuges in Tennessee, exposure to lead shot, as indicated by a blood level ≥0.2 ppm, declined from 11.7% in 1986–88 ($n = 423$) to 6.5% in 1997–99 ($n = 721$; Samuel and Bowers 2000). American Black Ducks have also accumulated metal and organochlorine contaminants in areas such as Chesapeake Bay, but as yet there are no suspected adverse affects (Krementz 1991).

Competition from an Increasing Mallard Population. Competition from Mallards has been cited but not proven as a possible contributor to the decline of American Black Ducks. In central Ontario, the most fertile wetlands were used solely by Mallards or shared with American Black Ducks, while the least fertile wetlands were vacant, and relatively infertile sites contained few Mallards or American Black Ducks (Merendino and Ankney 1994). The authors concluded that competition for breeding wetlands was likely between the 2 species and may have contributed to the decline of American Black Ducks. On wetlands in Maine, however, American Black Ducks did not lose any of the interactions they initiated with Mallards; they displaced Mallards to another part of the wetland 87.2% of the time, with no change occurring in the remaining 12.8% (McAuley et al. 1998). When Mallards initiated the encounters, they forced American Black Ducks to move elsewhere in the wetland 63.3% of the time, but in 15.0% of such occasions, they themselves were displaced by American Black Ducks, and in 21.7% no change occurred. Displacement to an area outside of a given wetland was rare (16.6% of 229 interactions), and that outcome was equal between the 2 species. Regardless of some conflicting results from these studies, opportunities for competition and hybridization between Mallards and American Black Ducks increased with the release of hundreds of thousands of game-farm Mallards within traditional American Black Duck range on the East Coast, especially in New York (on Long Island) and Maryland. Maryland alone released over 300,000 Mallards between 1974 and 1988 (Heusman 1991).

Competition could potentially manifest itself in differential reproductive outputs (the reproductive hypothesis), as was seen in New York, where American Black Ducks had lower rates of nest success than sympatrically breeding Mallards and

did not renest as often, although the sample sizes in this study were small (Dwyer and Baldassarre 1993). In an agricultural area in southern Québec (1994–96), nest-initiation date, clutch size, and nest success did not differ between Mallards and American Black Ducks, but the renesting effort of Mallards was nearly twice that of American Black Ducks (Maisonneuve et al. 2000b). There was no difference in the survival rates for the 2 species. In an agricultural area in New Brunswick (1992–94), Petrie et al. (2000) found no differences in clutch size, hen success, duckling survival, or female survival.

Hybridization with Mallards. In a study during the winter near Ottawa, Ontario, male Mallards and American Black Ducks initially courted and paired intraspecifically, after which unpaired Mallards joined American Black Duck courtship groups, but only after all female Mallards had formed intraspecific pairs (Brodsky and Weatherhead 1984). Of 33 unpaired female American Black Ducks at that time, 73% paired with 85 available male Mallards, despite the presence of 57 unpaired male American Black Ducks. Such results indicated that male Mallards were competitively superior to male American Black Ducks, and that the formation of American Black Duck/Mallard pairs was a key mechanism leading to hybridization. In a subsequent mate-selection experiment with captive-reared Mallards and American Black Ducks, the female associated with a dominant male, who then monopolized her time and kept her away from subordinate males, which supported the field observations (Brodsky et al. 1988).

The issue of introgressive hybridization between American Black Ducks and Mallards is of grave concern, and warrants a thorough discussion, because the hybrids can backcross with one or both of the parent populations, and hybrid swarms may occur in locations where introgression is especially prevalent. Hence, in the absence of isolating mechanisms between the 2 species, management efforts against introgressive hybridization would be futile, and the American Black Duck species would potentially disappear.

Ankney et al. (1987) believed that introgressive hybridization, not habitat loss or overharvest, was the cause of Mallards largely replacing American Black Ducks in Ontario. The ratio of American Black Ducks to Mallards breeding in that province decreased from about 1:2 to 1:6 between 1971 and 1985, although the total number of American Black Ducks and Mallards combined had not changed much, indicating that Mallards had "replaced" American Black Ducks. In addition, the decline in American Black Ducks was steepest in areas where the greatest number of hybrids existed in relation to pure American Black Ducks, despite unchanged and even improving habitat in those areas. The authors concluded that introgressive hybridization was leading to a genetic swamping of American Black Ducks by the numerically superior Mallards, probably through mixed pairing during renesting, including extra-pair copulations. Such a scenario is plausible, because pair bonds dissolve during incubation, so the number of unpaired males is high relative to the number of renesting females, which could promote male Mallards fertilizing renesting female American Black Ducks, especially in areas where male Mallards outnumber male American Black Ducks. This possibility is supported by Brodsky and Weatherhead (1984), who noted that mixed pairs during winter only involved male Mallards with female American Black Ducks, never male American Black Ducks with female Mallards.

The conclusions of Ankney et al. (1987), however, were challenged by Conroy et al. (1989b), who argued that habitat change and overhunting could not be dismissed in controlling the decline of American Black Ducks. They also argued that conclusions about the effect of introgressive hybridization were premature, stating, "it appears that whether introgressive hybridization with Mallards contributes to the decline in Black Ducks depends on whether Black Duck habitat is deforested" (Conroy et al. 1989b:1069). Ankney et al. (1989) refuted their challenge, as forest clearing does not

appear to be a prerequisite to Mallards moving into American Black Duck habitat.

Given that introgressive hybridization occurs—and, in some areas, occurs frequently between American Black Ducks and Mallards—a meaningful research question is *why* it happens. Ankney et al. (1986) reported that the mean genetic distance was nearly indistinguishable between different American Black Duck populations (0.0007), Mallard populations (0.0010), and Mallard versus American Black Duck populations (0.0006); there was as much genetic variation within species populations as between the 2 species. Accordingly, even though taxonomic authorities recognize Mallards and American Black Ducks as distinct species, Ankney et al. (1986:708) concluded that "our *genetic data* [italics mine] do not support even subspecific status for the Black Duck." They also noted that because hybrids are common in locales frequented by the 2 species, there appears to be little, if any, behavioral premating isolation. Pair formation occurs at about the same time and in the same habitats, and courtship displays are virtually identical between the 2 species (Johnsgard 1960a). Further, because of the bright green heads and overall body coloration of male Mallards versus the duller plumage of male American Black Ducks, female American Black Ducks choose male Mallards when courted by males from both species (Brodsky and Weatherhead 1984).

The number of hybrids clearly has increased on the western edge of the American Black Ducks' range in eastern Ontario and western Québec (Ankney et al. 1987), and hybridization is occurring at varying levels throughout a large portion of the breeding range of American Black Ducks. Such hybridization does not bode well for the long-term integrity of the American Black Duck population, as there appear to be no behavioral or habitat-related isolating mechanisms between the 2 species. Ankney et al. (1986) point out that the hybridization situation is not stable, and as the number of Mallards and hybrids increase in an area, the number of pure American Black Ducks will decline. They also note that the numbers of hybrids reported are very conservative, because many of the initial hybrid offspring, and the offspring of those hybrids with pure forms, look like the parental types, especially in females. Thus a critical question is just how much hybridization has occurred.

Mank et al. (2004) provided some insight in a study that used DNA techniques to compare the genetic differentiation between American Black Duck and Mallard samples from museum specimens collected before 1940 with samples taken in 1998. Their results revealed that the genetic differentiation between the species decreased from 0.146 before 1940 to 0.008 in 1998. Their summary conclusion sheds significant light on the hybridization issue: "This is a significant reduction in genetic differentiation, and represents a breakdown in species integrity most likely due to hybridization" (395).

A final point is the well-documented effect of Mallard hybridization on the native Gray Duck (*Anas superciliosa*) of New Zealand, which potentially is a glimpse into the future of the American Black Duck. Game-stock Mallards from Europe were introduced to the South Island in 1867, and North American stock was introduced to the North Island in the 1930s (Williams 1981). The subsequent impact was stunning, as hybridization had reached 51% on the South Island by 1980–81, and the percentage of pure Gray Ducks dropped to 4.5% (Gillespie 1985). More recent genetic data revealed the widespread introgression of mitochondrial DNA (used in comparing how closely related species are, and in distinguishing among them) from Mallards to Gray Ducks throughout New Zealand, and the conclusion that "the process of speciation appears to be undergoing reversal" (Rhymer et al. 1994: 970). New Zealand apparently is doomed to harbor a hybrid swarm of Mallards × Gray Ducks, leading to the eventual loss of its endemic species. New Zealand is obviously much smaller than the American Black Duck range in North America, but unless genetically pure American Black Ducks can remain geographically or behaviorally isolated from Mallards, their ultimate fate seems to be determined, even if the time span to achieve that fate is not.

Mallard

Anas platyrhynchos platyrhynchos (Linnaeus 1758)

Left, hen; *right*, drake

The Mallard is the most abundant, most studied, and most readily recognized duck in the world, with a breeding range that extends across Eurasia and North America. In addition, it has been introduced into Australia and New Zealand. Feral populations in city parks and other urban habitats occur virtually throughout the world. The green head, chestnut brown chest, white ring collar, and curled-up tail feathers of the male make identification unmistakable. Their breeding range in North America is widespread, extending from the Aleutian Islands across virtually all of Canada except the High Arctic, and southward through most of the continental United States. Their principal breeding range, however, is the Prairie Pothole Region (PPR) of the midcontinent. Mallards winter as far north as food and open water persist, but their principal wintering range is the Mississippi Alluvial Valley (MAV). Mallards are habitat generalists, breeding on virtually every type of wetland and consuming what foods are available. Postbreeding and wintering Mallards also feed extensively on agricultural

Mallard. *Left,* hen; *right,* drake

grains such as wheat, corn, rice, and soybeans, more so than any other duck. By far the most abundant duck in North America, the Traditional Survey averaged 7.5 million Mallards from 1955 to 2010, with a 2011 estimate of 9.2 million, which was 22% above the long-term average. In 2011, there were 403,000 Mallards in the Eastern Survey area. Summing these 2 major surveys and various state surveys yields a continental breeding population estimate of 11.25 million for 2011. Mallards are the most heavily harvested ducks in North America, averaging about 35% of the total duck harvest. The harvest in the United States averaged 5.0 million from 1999 to 2008, while the harvest in Canada averaged 573,444. Taxonomically, Mallards are part of the Mallard complex, which, in North America, also includes American Black Ducks, Mexican Ducks, and Mottled Ducks. All of these species are known to hybridize with Mallards to varying extents.

IDENTIFICATION

At a Glance. Males are unmistakable, with a metallic green head and neck, a brownish breast, a white neck ring, curled-up black central tail feathers, and a yellow bill. Females are brownish overall, with an orange bill that is blotched with black.

Adult males. At rest, a white neck ring separates the metallic green head and neck of males from the chestnut brown chest, which contrasts with the gray sides, back, and belly. The rump is black, along with the upper- and undertail coverts. In flight, the white tail forms a striking V against the black upper- and undertail coverts. The central black uppertail coverts are also curled upward, a feature unique (among ducks) to the male Mallard. In both sexes, the underwings show a flash of white with each wing stroke, which contrasts with the darker outer margins of the primaries, but not nearly as much as in American Black Ducks. The violet-blue speculum is bordered in front and behind by a pronounced white stripe, which identify the male in any plumage. The legs and the feet are orange to coral red; the bill is yellow to yellowish green. The irises are brown.

Adult females. Females are mottled brownish throughout, but there is a noticeable contrast between the lighter brown of the neck and the darker breast. The crown is dark brown; there is a dark brown stripe through the eye. As in the males, females have a violet-blue speculum bordered by a pronounced white stripe at the front and behind. In flight, females are distinguished by the white-bordered speculum and whitish outer tail feathers. The legs and the feet are dull orange. The bill is usually orange, with black mottling that forms a saddle in the center below the nostril; some females have dark bills, as if completely covered by the saddle. The irises are dark brown.

Juveniles. The plumage of juveniles is similar to that of adult females. Early-hatched young, however, will begin to acquire first alternate plumage in late August, and closely resemble adults by mid-October.

Ducklings. The plumage of ducklings is yellow (the base color) and olive brown, with a very distinct dark eye stripe and ear spot (Nelson 1993). The dull orange feet have a contrasting grayish brown pattern.

Mallard pair.
Joshua Stiller

Voice. Mallards are among the most vocal of ducks. The female gives a characteristic *qua, quack, quack, quack* (decrescendo call) that is accented on the first or second note and then decreases in volume after that. Females also have an array of maternal calls associated with brood-rearing. Males give a slow, raspy *raeb* call.

Similar Species. Female Mallards can be confused with American Black Ducks, but the latter show no white anywhere on the body or the tail. Mottled Ducks are darker than female Mallards, have no white in the tail, and lack blotches on the bill.

Weights and Measurements. Mallards and American Black Ducks are the heaviest ducks in the genus *Anas*, although their body mass will vary throughout the year. Delnicki and Reinecke (1986) provided body-weight and wing-length data from a large number of Mallards measured during midwinter in Mississippi. Body weights averaged 1,246 g for adult males ($n = 1,308$), 1,095 g for adult females ($n = 453$), 1,181 g for hatching-year males ($n = 169$), and 1,040 g for hatching-year females ($n = 188$). Corresponding wing length averaged 292.8 mm for adult males, 275.5 mm for

adult females, 286.4 mm for immature males, and 270.1 mm for immature females. The mean bill length from museum specimens averaged 41.7 mm for males ($n = 96$) and 38.7 mm for females ($n = 65$; Nudds and Kaminski 1984). Whyte et al. (1986) supplied a large dataset on body weights of Mallards from autumn through early spring on the southern High Plains of Texas, and Yetter et al. (2009) included morphological and body-weight measurements from breeding Mallards in Illinois. Johnson (1961) presented data on changes in testes weights of males and ovary weights of females throughout the annual cycle, based on wild birds (81 males, 57 females) collected in southeastern Washington.

DISTRIBUTION

Mallards are the most abundant and widely distributed duck in the Northern Hemisphere, ranging from the Arctic to the subtropics in North America, Europe, and Asia. Mallards have also been introduced to Australia, New Zealand, and the Hawaiian Islands, where they have caused conservation issues, due to hybridization with native species. Humans have used Mallards for food over

371

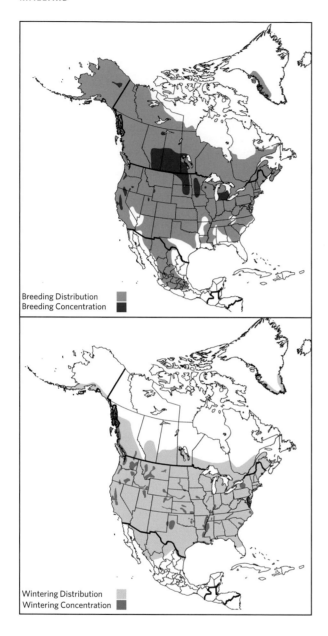

Breeding Distribution
Breeding Concentration

Wintering Distribution
Wintering Concentration

Duck, and 2 ducks on the Hawaiian Islands: the Laysan Duck (*Anas laysanensis*) and the Hawaiian Duck (*A. wyvilliana*). He combined what is sometimes recognized as the separate Greenland Mallard (*A. platyrhynchos conboschas*) with the Mallard. Greenland Mallards breed on small lakes in western and southeastern Greenland and mainly winter along the coast. They are about 10% larger, and exhibit paler plumage than the typical Mallard (Delacour 1964). Johnsgard (1978) considered all of the above taxa to be subspecies of a single Mallard species; he also recognized 2 subspecies for the Mottled Duck. The Laysan Duck of Laysan Island in the Hawaiian Islands Archipelago appears to be a very distinct Mallard form, which warrants species status (American Ornithologists' Union 1998, Rhymer 2001). Another Mallard-like duck, the Marianas Duck or Marianas Mallard (*A. p. oustaleti*)—possibly a hybrid form—formerly inhabited the western Pacific islands of Guam, Tinian, and Saipan. It has not been observed since 1981 and is now considered to be extinct.

Breeding. Mallards have the most extensive breeding range of any duck in North America, covering nearly all of the United States except for portions of the southeastern states and coastal Texas and Louisiana. Otherwise, the southern limits of the breeding range are northwestern Baja California and the states of Michocán, Morelos, and Puebla in Mexico. Their range north of the continental United States includes almost all of Canada south of open tundra, and extends northwest across Alaska to the Aleutian and Pribilof Islands in the Bering Sea. The Aleutian population is resident and appears to be genetically distinct from Asian and mainland Alaska populations (Kulikova et al. 2005). Within this broad distribution, however, the Mallards' core breeding range is the PPR, especially the prairie parklands of southcentral Canada. Their breeding densities respond to water conditions in the PPR, with overall numbers correlated with the number of wetlands containing water in May. In eastern North Dakota, the density of Mal-

the millennia; hence almost all forms of domestic ducks originated from Mallards, although Muscovy Ducks have also been widely domesticated.

The Mallard is the core taxon in the Mallard complex, which, according to Livezey (1991), consists of 14 species, largely separated into a Northern Hemisphere group, a Southern Hemisphere group, and 2 African species. In North America, Livezey accorded species status to the Mottled Duck, the American Black Duck, the Mexican lard breeding pairs from 1961 to 1980 varied from a low of 2.3/km² (a drought year) to a high of 9.5/km² in 1963, a wet year (Krapu et al. 1973). There was a positive correlation ($r = 0.53$) between pair density and wetland abundance.

The Traditional Survey estimated 9.2 million Mallards in 2011: 28.0% were in the eastern Dakotas; 22.8% in southern Saskatchewan; 10.6% in the Northwest Territories, northeastern British Columbia, and central and northern Alberta; 10.2% in southern Alberta; 9.1% in Montana and the western Dakotas; 9.0% in northern Saskatchewan, northern Manitoba, and western Ontario; 5.7% in southern Manitoba, and 4.5% in Alaska (U.S. Fish and Wildlife Service 2011). The long-term (1955–2010) distribution of Mallards in the Traditional Survey area indicated that 27.2% of the breeding population was in southern Saskatchewan; 15.0% in northern Saskatchewan, northern Manitoba, and western Ontario; 14.3% in the Northwest Territories, northeastern British Columbia, and northern and central Alberta; 14.2% in southern Alberta; 12.6% in the eastern Dakotas; 6.6% in Montana and the western Dakotas; 5.0% in southern Manitoba; and 5.0% in Alaska/Yukon.

Outside the Traditional Survey area, the central interior plateau of British Columbia contains an estimated 200,000 Mallards, most (185,000) of which are in the northern Rockies (Canadian Intermountain Joint Venture Technical Committee 2010). To the south, the average Mallard breeding populations in the western states from 2001 to 2010 were 45,466 in Washington, 90,141 in Oregon, 323,989 in California, 3,310 in Nevada, and 11, 357 in Utah (Collins and Trost 2010). To the east, 2011 estimates of Mallards were 188,000 in Wisconsin and 225,000 in Michigan. Mallards breed through the southern Great Plains as far as southern Texas, where they come into contact with Mexican Ducks. The most important breeding grounds in this region are eastern Wyoming, the Nebraska Sandhills (western and central Nebraska), and the Rainwater Basin in southeastern Nebraska. A few nest in Cheyenne Bottoms in central Kansas.

The Eastern Survey has been conducted since 1990 by the U.S. Fish and Wildlife Service (using fixed-wing aerial transects) and the Canadian Wildlife Service (using helicopter surveys). The data here are estimated from the combined fixed-wing and helicopter surveys, which averaged 396,380 Mallards from 1990 to 2010: 93.4% were in the western Boreal Shield (Ontario and western Québec), 4.0% in the central Boreal Shield (central Québec), 2.0% in the Atlantic highlands (the Gaspé Peninsula of Québec, plus New Brunswick and Nova Scotia), and 0.5% in the eastern Boreal Shield (eastern Québec, Labrador, and Newfoundland). The 2011 estimate for the entire Eastern Survey, which includes Maine and a small part of northern New York, was 403,000.

The Atlantic Flyway Waterfowl Plot Survey conducted in the eastern states (except Maine) averaged 731,365 Mallards annually between 2000 and 2010: New York (27.4%), Pennsylvania (24.0%), Virginia (8.1%), Maryland (7.4%), Massachusetts (7.0%), New Jersey (6.8%), Connecticut (4.8%), New Hampshire (4.7%), and Vermont (3.2%), with the remaining 6.6% scattered in the rest of the Atlantic Flyway.

Winter. Mallards winter throughout most of the United States, extending along the Pacific coast as far north as the Alaska Peninsula, and westward along the Aleutian Islands. Mallards commonly winter in the northern parts of the United States, and even in southern Canada, wherever open water and food resources are available, especially waste grains. Butler et al. (1989) reported as many as 51,000 Mallards annually wintering within estuaries along the Strait of Georgia in British Columbia.

The Mississippi Flyway, however, harbors the largest number of wintering Mallards. From an average 5.1 million Mallards estimated in the Midwinter Survey from 2000 to 2010, almost half (2.3 million, or 45%) were in the Mississippi Flyway. Most wintered in habitats associated with the ancient floodplain of the Mississippi River,

Left:
Female Mallard with her brood; note the eye stripe, which is characteristic of all *Anas* ducklings.
John C. Avise

Below:
Male Mallard in a comfort movement.
GaryKramer.net

also known as the Mississippi Alluvial Valley, with their core wintering grounds between Cape Girardeau in Missouri and southern Louisiana. An analysis of band-recovery data indicated that during the winter Mallards were most abundant between 34° N and 36° N latitude in eastern Arkansas and western Mississippi (Bellrose and Crompton 1970). Wintering Mallards in the MAV once exclusively used the acorns, moist-soil plants, and invertebrates made available by late-year rains that flooded its vast areas of forest. Huge losses of wetlands have occurred in the MAV, however, and perhaps only 20% of an original 10 million ha of bottomland hardwood forests remain (Baldassarre and Bolen 2006). Today Mallards depend heavily on waste agricultural crops (such as corn, rice, and soybeans), although the remaining natural wetlands still provide important foods (Kaminski et al. 2003).

From 2000 to 2010, the MAV area in Arkansas contained the largest (23.8%) percentage of Mallards wintering in the Mississippi Flyway, where present and historical populations associated with the White River in the Stuttgart area are legendary. Louisiana accounted for 17.4%, many in the Rice Belt and adjacent coastal marshes between Lafayette and the Sabine River. Elsewhere in the Flyway, 16.8% occurred in Tennessee, 16.1% in Missouri, and 6.7% to the north in Illinois.

The Central Flyway contained 33.1% of the 2000–2010 average: 40.6% in Texas, 17.5% in Nebraska, 11.9% in Oklahoma, and 10.3% in Kansas. Large numbers of wintering Mallards in Texas use the playa lakes of the Panhandle, where they field-feed on nearby waste corn. The remainder largely occur in eastern Texas. In Nebraska, many Mallards winter in open-water areas on the Platte River, and they also forage on nearby waste corn. Mallards wintering in Kansas and Oklahoma are associated with open water and consume waste grains in fields close by.

The Pacific Flyway accounted for 19.3% of the 2000–2010 average. Of these, 35.9% were in Washington, 30.5% in California, 13.7% in Idaho,

and 10.5% in Oregon. Mallards in Washington and Oregon primarily winter in the Columbia basin, although others are in the Puget Sound area of Washington and the Willamette Valley of Oregon. In California, most winter in the Sacramento Valley, but others occur in the San Joaquin Valley and the northeastern plateau. Only 2.7% of the 2000–2010 estimate was tallied in the Atlantic Flyway.

Migration. Mallards travel along several major migration corridors, with that most used by the largest number of birds extending from southeastern Saskatchewan to northwestern Illinois, and then south to eastern Arkansas, Tennessee, and Mississippi (Bellrose 1980). A second high-use corridor parallels the upper Missouri River southward to the area of Stuttgart, Arkansas. Two important corridors cross the Great Plains: one goes to eastern Texas and coastal Louisiana; the other to eastern Colorado, western Nebraska, and the Texas Panhandle. Farther west, the most important corridor extends from Alberta to the Columbia basin of Washington and Oregon, while the second most important one extends to the Snake River near Boise, Idaho. In the Atlantic Flyway, the majority of Mallards are found south of Long Island in New York, with most originating from breeding grounds in Ontario, Québec, and the Maritime Provinces.

In the midcontinent, the northern tier of states does not receive appreciable numbers of fall migrants from more northerly breeding areas until early October, with peak numbers reached late in the month and a pronounced decline after mid-December (Bellrose 1980). Small groups arrive on intermediate migration areas as early as the beginning of September, but their numbers remain low until early October, after which migrants arrive steadily until peak populations are reached either in late November or early December. On their southern wintering grounds (e.g., the Gulf Coast), Mallards begin to arrive late in October, with substantial influxes from late November through De-

Male Mallard
preparing to land.
RyanAskren.com

cember. There are usually no sharp peaks in migrant population trends during the fall, because migration is prolonged, but occasionally severe storms on the northern plains in late October or early November have created a mass departure (a grand passage) of Mallards and other waterfowl into the Central and Mississippi Flyways (Bellrose 1957, Bellrose and Sieh 1960). An analysis of band-recovery data within the Mississippi migration corridor indicated that the principal passage of Mallards occurred over a 2-month period, 15 October–15 December (Bellrose and Crompton 1970).

In the West, Mallards first arrive in the Northwest Basin, which includes the Columbia basin, by late September, with a steady build-up to a peak in late December (Bellrose 1980). Mallards also begin to reach the Central Valley in California in September and increase in number through November, to a peak in late December. Mallards tend not to arrive in the West in the distinctive pulses that can characterize the midcontinent populations. In the East, Mallard numbers begin to increase in New England from September to mid-October, but these birds arrive farther south markedly later (Bellrose 1980). Mallards initially arrive in the mid-

Atlantic states by October, with a steady increase to peak numbers in mid-December. Arrival to the southeastern coast is not until late October, with their peak in mid-December.

In spring, Mallards are again among the first of the dabbling ducks to migrate northward. Mallards traveling through the midcontinent leave their wintering areas in the South by early February, with a steady departure continuing through February and March; few remain there by April (Bellrose 1980). Mallards arrive in central Iowa during the third week of February (LaGrange and Dinsmore 1989). Departures from intermediate areas also occur during February and March, but they continue later into April than departures from traditional wintering grounds. These areas are most noted as stopovers for migrating Mallards, but they also overwinter some birds. In the Midwest, wintering Mallards are joined by migrating Mallards by the last week in February, after which their numbers increase steadily throughout March, followed by a rapid decline through April, with few remaining by the end of the month. During winter in eastcentral Arkansas (1989), 44 radiomarked female Mallards began their departure in mid-February, with

Tip-up feeding by Mallards is characteristic, as is the curling black uppertail coverts. *RyanAskren.com*

almost half of the individuals gone by mid-March; 8 were detected migrating through southeastern Missouri between 17 March and 1 April (Dugger 1997). A widespread movement of Mallard pairs into southcentral and eastern North Dakota typically begins during late March and early April but differs among years, in response to the variable weather conditions characteristic of those months (Krapu 1981, Krapu et al. 1983).

In the West, Mallards begin their northward migration from the Central Valley of California early in February (Bellrose 1980). From there up along the coast to Puget Sound, departures are gradual through mid-April. Departures from the Great Basin commence in late February and continue through March. The passage of Mallards through British Columbia occurs primarily in March, with smaller numbers seen through mid-April (Butler and Campbell 1987). In the East, spring migration commences from the southeastern coast during early February and accelerates steadily through late March; from the mid-Atlantic states, during February, with a steady decline to early April; and from New England, during February and March, with a peak in early April (Bellrose

1980). Mallards reach their breeding grounds in Ontario, Québec, and the Maritime Provinces in late April and early May.

During a 6-year study (1976–81) in North Dakota, the average arrival date was 20 April for adult females, compared with 5 May for yearlings; 90% of all resident females were seen by 14 May (Lokemoen et al. 1990a). At Delta Marsh in southern Manitoba (1946–50), Mallards first arrived between 20 March and 17 April, with peak arrival in a given year varying from 10 to 15 April until 19 to 30 April (Sowls 1955). In southeastern Alberta (1953–56), the mean date of first arrival averaged 27 March, but fluctuated between 23 March and 3 April (Keith 1961)

MIGRATION BEHAVIOR

Among dabbling ducks, Mallards and American Black Ducks are very cold hardy and thus are the latest fall migrants, although their entire migration period extends from late summer to early winter. In general, however, Mallards commonly remain north as long as open water and food are available, often though November. Mallards can even migrate during the winter; Mallards in central Illinois have been forced by snow and cold to move south as late as mid-February, only to return by 1 March (Bellrose 1980).

Mallards are short- to medium-distance partial migrants, with many populations considered to be sedentary. Migration is along traditionally used corridors that vary from 80 km wide in the Pacific Northwest to 240 km wide on the western Great Plains (Bellrose and Crompton 1970, Bellrose 1972). In central Illinois, the air speed of 9 flocks on fall migration was estimated at 91 km/hr (Bellrose and Crompton 1981). The departure dates of marked females from wintering grounds in east-central Missouri were not related to their age or condition at capture in January, but late-molting females departed later than early-molting females (Dugger 1997).

Molt Migration. Male Mallards commonly undergo a molt migration of varying distance from their breeding marshes; some females will molt on the breeding areas, while others will not. In northcentral Minnesota, all radiomarked male Mallards ($n = 27$) departed the study area when breeding activities were completed and before the flightless period; departures peaked in early June (Gilmer et al. 1977). In contrast, 26 out of 51 females raising broods remained on the breeding areas during the flightless period. Most unsuccessful females departed the breeding areas before the flightless period; they traveled independently of the males, although Oring (1964) observed that unsuccessful pairs will join molting congregations. Both sexes mainly inhabited areas of dense emergent cover during the flightless period, with females known to use large lakes and river marshes with extensive emergent vegetation, particularly wild rice (*Zizania aquatica*). The destination of Mallards leaving the study area was unknown, but most probably remained in northwestern Minnesota. Elsewhere, large numbers of molting male Mallards have been observed on large marshes, such as Delta Marsh in Manitoba (Hochbaum 1944, Sowls 1955), and lakes, such as Whitewater Lake in Manitoba (Bossen-

maier and Marshall 1958) and those on the Camas National Wildlife Refuge (NWR) in Idaho (Oring 1964).

During a 3-year study (1990–92) in the St. Lawrence River valley in New York, most radiomarked females ($n = 74$) remained in the vicinity of their breeding wetlands during the postbreeding period (Losito and Baldassarre 1995a). In contrast, 32 out of 34 radiomarked females nesting near Suisun Marsh in California departed the area before molting, with many (50%) leaving by mid-June (Yarris et al. 1994). Of 27 females located from late June through September, 25 migrated northward from the study area, and 2 remained. Molting areas were determined for 20 of those females: 9 in Oregon and 11 in California, a range of 12–536 km from the study area. Females used wetlands dominated by bulrushes (*Scirpus* spp.) and cattails (*Typha* spp.). At the Camas NWR in Idaho, molting Mallards congregated in all marshes that were larger than a few hectares in size. Dominant plants in these wetlands were cattails (*Typha latifolia*), hardstem bulrush (*Scirpus acutus*), sedges (*Carex* spp.), muskgrass (*Chara* spp.), pondweeds (*Potamogeton* spp.), duckweeds (*Lemna* spp.), naiads (*Najas* spp.), hornwort (*Ceratophyllum demersum*), and four-angled waterlilies (*Nymphaea tetragona*; Oring 1964). Mallards probably exhibit some fidelity to their molting sites, but few data are available.

HABITAT

Mallards are by no means specialists in their choice of breeding habitats, using virtually every type of freshwater wetland habitat on the continent, as well as brackish habitats along the coast. They are also quite tolerant of urban conditions and thus readily breed in parks and other urban habitats (Heusmann and Burrell 1974, Figley and VanDruff 1982). Mallards reach their greatest breeding density in the PPR, but they are also common in forested habitats to the east, stretching across Canada and the northern United States; hence these areas are logical focal points from which to discuss their breeding habitats.

In the prairies of North Dakota (1967–69), breeding pairs ($n = 811$) primarily used seasonal (59.1%) and semipermanent wetlands (29.5%; Stewart and Kantrud 1977). During a 3-year study (1988–90) involving 91 radiomarked female Mallards in central and eastern North Dakota and westcentral Minnesota, habitat use varied with their reproductive cycle (Krapu et al. 1997). In general, females occupied semipermanent wetlands during all stages of reproduction, but prenesting and egg-producing females preferred semipermanent wetlands (40.8%–48.9%), seasonal wetlands (25.2%–25.9%), and lakes (16.8%–25.2%), while incubating females largely used semipermanent (57.1%) and seasonal wetlands (23.1%). Postnesting females again primarily chose semipermanent wetlands (59.4%), with 18.3% found in lakes and 18.0% in seasonal wetlands. During long-term research (1965–75) at the Woodworth Study Area in southeastern North Dakota, the occupancy rate of wetlands by size during late May and early June surveys was 2.2%–6.0% of the wetlands between <0.09 to 0.5 ha in size; 12.4%–15.6%, between 0.51 and 1.0 ha; 15.8%–29.6%, between 1.01 and 5.0 ha; 35.3%, between 5.10 and 10.0 ha, and 55%, between 10.01 and 25.0 ha (Higgins et al. 1992). Radiomarked females ($n = 15$) in North Dakota used temporary and seasonal wetlands to a greater degree than expected (relative to the proportional distribution of habitat types on the landscape), ephemeral wetlands and permanent lakes to a lesser degree than expected, and semipermanent wetlands in relation to their availability, but the results were not statistically significant. Individual females were observed in 7–22 wetlands and spent 22%–61% of their time on the single, most-used wetland. During May in the aspen parklands of southcentral Saskatchewan (1977–80), Mallard pairs occupied wetlands in relation to their availability on the landscape (Mulhern et al. 1985). Johnson and Grier (1988) noted that Mallards tended to fill available ponds in the core areas of their breeding range and secondarily in peripheral areas.

Outside the PPR, Mallards in Ontario extensively chose wetlands created by beaver (*Castor canadensis*), but pairs were distributed evenly over all ponds (Patterson 1976). In the boreal forest of the Clay Belt ecoregion of Ontario, pairs selected beaver ponds and partially open fen wetlands over other habitat types (Rempel et al. 1997). In forested areas of northcentral Minnesota (1968–72), 24 radiomarked Mallards (12 males and 12 females) most frequently used circumneutral bogs and seasonal wetlands (Ball et al. 1975). Pairs most often chose lakeshore habitats; along lakeshores, the preferred areas were sand/gravel patches, overhanging brush, and bog mats. Among these 24 Mallards, their preference was greatest for seasonal wetlands and least for softwood swamps, but all communities were used during the breeding season.

In the heavily forested western Adirondack region of New York (1990–91), radiomarked female Mallards ($n = 14$) were tracked on a 189-km² study area containing 71% deciduous forests and 24% wetlands (Dwyer and Baldassarre 1993). They primarily used scrub-shrub wetlands (about 50%), although roughly 30% were in emergent wetlands and around 13% in deciduous-forest wetlands. Radiomarked females ($n = 97$) were also tracked on the nearby St. Lawrence River valley of New York (1990–92), where a 126-km² study area was less forested (53%) and contained fewer wetlands (11%), but was also heavily used for agriculture (35%; Losito and Baldassarre 1995a). Wetland composition on the study areas was 51.1% forested-live wetlands, 25.3% forested-dead wetlands, 12.1% scrub-shrub wetlands, and 8.2% emergent wetlands. Breeding females preferred forested-live wetlands (40.2%) over emergent wetlands (23.4%), scrub-shrub wetlands (22.1%), and forested-dead wetlands (13.3%), whereas postbreeding females chose forested-dead wetlands (34.5%), forested-live wetlands (24.9%), scrub-shrub wetlands (21.5%), and emergent wetlands (17.7%). On average, postbreeding females used fewer (2.6) wetlands of larger (192 ha) size, compared with breeding females, which used an average of 4.1 smaller-sized (101 ha) wetlands. The seasonal shift to forested-dead wetlands during postbreeding probably reflected the loss of water from the less permanent forested-live, emergent, and scrub-shrub wetland types.

As fall migration begins in the prairie areas of Canada, Mallards quickly congregate on large marshes and lakes, from which they make regular field-feeding flights to nearby croplands (e.g., fields of wheat, barley, oats, and rye), where they can cause depredation problems (Bossenmaier and Marshall 1958). Mallards continue to consume agricultural grains as they migrate southward, especially waste corn, but they also feed in natural wetlands to offset the nutritional deficiencies of a corn diet (Baldassarre et al. 1983). Mallards appear to require a greater diversity of habitats during spring migration, however. Spring migrants still use sites with residual waste grains (such as corn) in areas like the Platte River in Nebraska (Pederson et al. 1989), but they also extensively congregate in seasonal wetlands. In central Iowa (1983–84), 455 seasonally flooded farmed basins (sheetwater wetlands) provided 19,530 Mallard use days (the number of Mallards counted per day on an area, summed over the number of days they were there), compared with only 103 use days on 16 small emergent wetlands (LaGrange and Dinsmore 1989). The birds tended to choose the larger sheetwater wetlands (>2 ha), where they fed for 40% of the time. Preferred sites contained moist-soil plants or the high-energy seeds of waste corn and were farthest from disturbances. Mallards then flew ≤13 km to roost on emergent wetlands. Sheetwater and seasonal wetlands are especially important during spring, because Mallards are early migrants and often arrive at stopover sites when larger wetlands are still covered by ice.

On their principal wintering grounds in the MAV, aerial surveys during the winters of 1987/88 through 1989/90 estimated Mallard populations as high as 1.79 million during January of 1988 (Reinecke et al. 1992). During the early winter (Dec), 59%–69% chose wetlands with water levels managed to benefit waterfowl, compared with 28%–39% on unmanaged wetlands. In contrast, during the late winter (Jan) they preferred unmanaged wetlands to managed ones (52%–79% vs. 19%–43%). Croplands were used frequently (53%–85%), with 21%–39% of the birds grazing in rice fields and 11%–41% in soybean fields (except for 7% in Dec 1989). Lower-ranking sites were forested wetlands (3%–11%) and moist-soil habitats (3%–29%). Flooding increased their use of wetlands with unmanaged water regimes, as well as soybean fields and other croplands.

An analysis of band-recovery data from 1950 to 1980 in the MAV found that Mallards exhibited considerable flexibility in their winter distribution, due to the influences of temperature, water conditions, and population size (Nichols et al. 1983). When years of extreme warm and extreme cold temperatures were compared, Mallard recoveries for all sex and age classes were located farther south in cold years. The data suggested that this movement is a threshold response to low temperature, triggered by ice cover on wetlands. Also, more recoveries in the MAV were made during wet years than dry years, except for adult males. Heitmeyer and Fredrickson (1981) considered winter precipitation to be the best measure of habitat availability in seasonally flooded hardwood bottomlands like the MAV. Lastly, the percentage of young Mallards recovered in the MAV increased when populations were low. There was also evidence that greater proportions of adults wintered in the MAV when the population size was high, perhaps because adults have a competitive advantage over young birds.

Wintering Mallards on the southern High Plains of Texas spent the day on playa lakes but engaged in predawn and postsunset field-feeding flights to nearby cornfields (Baldassarre and Bolen 1984). These ducks congregate in large numbers wherever open water prevails and waste grains are available during the winter, even in areas where snowfall and low temperatures are common. In a study during 1978–80, Mallards wintering along the Platte River in southcentral Nebraska encountered average minimum temperatures of −11.1°C during the cold winter of 1980. They fed in nearby cornfields, as well as in aquatic habitats, although they increased their foraging in the fields during cold periods and reduced energetically costly behaviors such as aggression and courtship. Minimum temperatures were 31% colder in riverine versus canal habitats, so radiomarked birds moved to canal roosting sites during the coldest periods of that winter. Canals offered a more favorable microclimatic, as temperatures were milder and the banks afforded protection from the wind (Jorde et al. 1984). The Mallards then returned to riverine habitats as ambient temperatures increased.

POPULATION STATUS

Mallards are the most abundant duck species in North America. Their numbers fluctuate widely over the long term in response to patterns of precipitation and other habitat variables, such as the extent and quality of upland nesting cover. Trends in Mallard populations reflect not only the quality of available breeding habitat, but also the general status of duck populations breeding within the same geographic range, especially the PPR.

The Traditional Survey averaged 7.5 million Mallards from 1955 to 2010, but ranged from a high of 11.2 million in 1958 to just below 5.0 million in 1985. In general, the Mallard survey estimates show no definitive trend, but rather exhibit pulses of highs and lows around its long-term average (LTA) over groups of years. For example, Mallard numbers exceeded the LTA each year from 1955 to 1959 (9.7 million, on average), in all but 3 years between 1970 and 1980 (8.2 million), and in all but 3 years (2004–6) between 1995 and 2011 (8.4 million), which represents a considerable number of near-consecutive years when the population exceeded the LTA. Low points in the Mallard population occurred from 1962 to 1966 (5.7 million), 1981 to 1994 (5.9 million), and 2004–6 (7.2 million). The 2011 estimate of 9.2 million was 22% above the LTA. The North American Waterfowl Management Plan (2004) has a breeding population ob-

jective of 8.2 million Mallards in the Traditional Survey area.

The Eastern Survey area counted 403,000 Mallards in 2011, which was 5% below the LTA of 424,000 (1990–2010). The Atlantic Flyway Breeding Plot Survey estimated 586,000 in 2011, 20% below the LTA of 764,000. Elsewhere, breeding population estimates in 2011 were 68,000 in Oregon (34% below the LTA of 103,000), 315,000 in California (14% below the LTA of 364,000), 283,000 in Minnesota (26% above the LTA of 225,000), 188,000 in Wisconsin (3% above the LTA of 182,000), and 225,000 in Michigan (40% below the LTA of 374,000). Summing the 2011 numbers for all surveys yields a continental breeding population estimate of 11.3 million. The North American Waterfowl Management Plan (2004) reported a continental average of 13 million Mallards from 1994 to 2003.

Outside North America, nonbreeding Mallard populations are estimated at 4.5 million in northwestern Europe (to the Baltic Sea), 1 million in central Europe and the eastern Mediterranean, 800,000 in southwestern and central Asia, 75,000 in southern Asia, and 1.5 million in eastern Asia (Wetlands International 2006).

Harvest

Mallards are by far the most heavily harvested ducks in North America, making up about 35% of the total number of ducks in the annual harvest (Trost 1987); in an average year, hunters have harvested 20%–25% of the total Mallard population (Anderson and Burnham 1978). The Mallard harvest in the United States averaged 5.0 million from 1999 to 2008: 50.5% in the Mississippi Flyway, 22.0% in the Pacific Flyway, 18.3% in the Central Flyway, and 9.2% in the Atlantic Flyway. Of the harvest in the Mississippi Flyway, 26.8% were taken in Arkansas, 9.3% in Minnesota, 9.2% in Illinois, 8.8% in Missouri, and 7.0% in Tennessee. In the Pacific Flyway, 27.1% of the harvest was in California, 22.5% in Washington, 17.0% in Oregon, and 14.3% in Utah. For the Central Flyway, 22.4% of the harvest was in North Dakota, 18.3% in Texas, 14.2% in Oklahoma, 11.5% in South Dakota, and 11.3% in Nebraska. The Atlantic Flyway harvest primarily occurred in New York (20.0%), Pennsylvania (18.7%), Maryland (13.3%), and Virginia (10.7%).

In Canada, the harvest from 1999 to 2008 averaged 573,444, with a high of 676,376 in 2000. Over this time period, 26% were harvested in Saskatchewan, 23.6% in Ontario, 15.5% in Alberta, 13.7% in Manitoba, 12.2% in Québec, and 9.0% in the remaining provinces.

Anderson (1975) calculated a mean life span of 1.8 years for adult Mallards and 1.6 years for juveniles, based on banding data. The longevity record for a wild Mallard, however, is 27 years and 7 months for an adult male banded in Louisiana and later shot at an unrecorded location. Other longevity records are 26 years and 4 months for an adult male banded in Michigan, 23 years and 5 months for an adult male banded in Oregon, and at least 21 years and 3 months for a male banded at an unknown age in Illinois. The oldest recorded female life span was 17 years and 7 months for a hatching-year bird banded in Michigan and shot in South Carolina.

BREEDING BIOLOGY

Behavior

Mating System. Mallards are seasonally monogamous, and, along with American Black Ducks, are the earliest-pairing dabbling ducks. At Winous Point Marsh on Lake Erie in Ohio, Mallards start to establish loose pair bonds as early as August, with an estimated one-third paired by early September and 75% by late October (Barclay 1970). At the southern end of Cayuga Lake in central New York, courtship displays within a semitame flock of about 200 Mallards began on 13 September, with the estimated number of pairs increasing steadily from early October and reaching 80%–90% by late December–early January (Johnsgard 1960a). In coastal Louisiana many Mallards arrived paired: 55% of the females observed before 9 November were paired, rising to 95% by late December (John-

son and Rohwer 1998). Weller (1965) reported that 90% of the Mallards he saw in January in Louisiana were paired.

Of 15 radiomarked Mallard pairs in Minnesota, males deserted their mates an average of 4.1 days into incubation, but the range was from 5 days before the clutch was completed to 17 days into the incubation period (Gilmer et al. 1977). In Idaho, 10 out of 13 pairs separated on the first day of incubation, 1 pair remained together into the first week of incubation, another into the second week, and the third until the eggs were pipped (Oring 1964). In New York, the pair-bond duration of 9 radiomarked pairs averaged 56 days, and all pair bonds were terminated before August (Losito and Baldassarre 1996). Three pair bonds were broken when 1 member of the pair died, and 3 ended on days 9, 11, and 15 of the first nesting attempts. Of the other 3 pairs, 1 female lost her nest on day 2 of incubation but the pair bond was not broken until 44 days later. A second female produced 3 nests, all of which were destroyed 5, 8, and 11 days into incubation, but the pair bond persisted until 10 days after the loss of the third nest. The pair bond of the third female lasted 71 days, although the female was never located on a nest.

There is documentation, however, of pair bonds extending across 2 breeding seasons (Dwyer et al. 1973, Blohm and Mackenzie 1994), and Losito and Baldassarre (1996) observed a pair bond that persisted for 3 years. Such behavior appears to be rare, although Mjelstad and Sætersdal (1990) reported reuniting in 10 out of 12 marked pairs of Mallards in a resident urban population in Norway; they argued that such behavior may be normal among sedentary populations. Pairs will also commonly reunite during renesting efforts (see Renesting).

Most Mallards breed as yearlings, although some late-maturing young females probably do not, especially in those years when drought reduces water areas on the prairies (Bellrose 1980). In the St. Lawrence River valley of New York, just 41% of 44 radiomarked yearlings were found on nests (Losito et al. 1995). In westcentral Illinois (1998–2003), however, only 6 out of 143 radiomarked females did not nest, 5 of which were yearlings (Yetter et al. 2009). Adults ($n = 79$) initiated more nests (1.7/nesting female) than yearlings ($n = 48$; 1.3 nests/female), and the age distribution among nesting females was 0.6 yearlings/adult. Calverley and Boag (1977) examined the reproductive tracts of Mallards and reported nonbreeding rates of 6.1% in the parklands of Alberta ($n = 66$), but 28.6% in arctic habitats in the Northwest Territories ($n = 14$), where Mallards are displaced during drought conditions on the prairies. Although the age classes were pooled, many nonbreeders were probably yearlings.

Male Mallards commonly seek extra-pair copulations with females other than their mates when the other females are most fertilizable, such as when they are producing eggs (Cheng et al. 1982). In southwestern Manitoba, 8 out of 46 (17.4%) broods involved multiple mating events, and 9 out of 298 (3.0%) ducklings were fathered by a male that was not the female's mate (Evarts and Williams 1987). These percentages were probably underestimated, however, due to several error sources; when the authors adjusted their figures to account for the error, they estimated that at least 48% of the broods involved multiple mating events.

On the breeding grounds, the social organization of Mallards was documented on and near Ventura Marsh in northcentral Iowa during a 2-year study (1974–75) involving 134 marked males and 22 marked females (Humburg et al. 1978). Of all Mallards observed, 88% were paired shortly after their arrival in late March and early April, but 20% of the marked males were initially observed without females, which strongly suggests that a portion of the males arrive unpaired in the spring but are then capable of forming a pair bond. There was also a continual turnover of unpaired males during the breeding season; 70 marked males never observed with a female averaged only 1.3 days on the study area, compared with 17.6 days for 64 marked males seen with a female at least once. After nesting pairs were established, however, the number of Mallards on the study area was relatively constant throughout the breeding season. Over the 2 years of the study, 48% of the 134 marked males were observed with at least 1 female for an average of 10.4 days, and 11% were seen in at least 2 pair sequences, with the second averaging 6.9 days.

Ohde et al. (1983) examined the potential function of excess males in a breeding population of Mallards at Harmon Marsh in Iowa by individually marking 89 males and 47 females and then attempted to equalize the sex ratio in the population by removing 117 unmarked males and adding 45 females. The results also provided insight into the social organization and mating system of breeding Mallards. Matings were documented for 32 of the marked males, of which 16 (50%) mated with 1 female, at least 8 (25%) mated with more than 2 or 3 females, and another 8 exhibited indications of having done the same. Moreover, 5 (16%) of the mated males established regular associations with a second female while their original mates were either laying eggs or in the first few days of incubation; these males appeared to have simultaneous pair bonds with 2 females. Such behavior is consistent with a mixed reproductive strategy. The earliest-breeding males remained sexually active for the longest periods, as 9 of the last 10 males active on the breeding area were among the first 12 marked. These 9 males averaged 78 days of breeding activity, which was in sharp contrast to an earlier report (Dzubin 1969) that males deserted their mates during incubation and joined other males on molting areas.

Sex Ratio. Data are abundant on Mallard sex ratios, which vary with seasons and locations. Anderson (1975) estimated the adult sex ratio (males:females) as 1.20–1.30:1 (54.5%–56.5% males). During spring migration at 10 sites in North America, the sex ratio of 53,602 Mallards averaged 52.6% males (Bellrose et al. 1961). In northcentral Iowa, sex ratios ($n = 5,695$) shortly after Mallards arrived averaged 1.38:1 (58.0% males), but the ratio was 1.15:1 (53.5% males) on breeding areas (Humburg et al.

1978). At Harmon Marsh in Iowa, the sex ratio was 53.3% males, compared with 56.1% on a nearby sheetwater area (Ohde et al. 1983). The overall percentage was 55.8% males ($n = 3,230$). During winter in the Atchafalaya delta of Louisiana, sex ratios in 488 flocks averaged 1.17:1 (53.9% males), but monthly averages were 55.9% males in November and December (350 flocks), compared with 51.0%–51.9% males in January and February (125 flocks), and an even sex ratio during March (13 flocks). On breeding grounds in the PPR, Sargeant et al. (1984) noted that females accounted for 76% of the dabbling ducks killed by foxes. Johnson and Sargeant (1977) concluded that fox predation alone had been sufficient to alter Mallard sex ratios over time: estimated sex ratios were 1.10 (or fewer):1 ($\leq 52.4\%$ males) in pristine times, 1.17:1 (53.4% males) from 1939 to 1964, and 1.29:1 (56.3% males) from 1959 to 1964.

Site Fidelity and Territory. A summary by Anderson et al. (1992) noted that females are philopatric to breeding areas. In North Dakota (1968–71), 46% of 113 marked females returned to nest baskets in the same marshes where they were originally marked, which represented nearly all surviving females (Doty and Lee 1974). During a study of nest-basket use at Ventura Marsh in Iowa (1971–75), the homing rate of marked females was 67%–89% in all years except the first (25%), and 19 (58%) marked females homed to a specific nest basket at least once during the study (Bishop et al. 1978). In the prairie potholes of North Dakota, 46% of 113 marked females homed at least once to nest baskets in the same marsh where they were initially captured, and two-thirds were observed in the same baskets in which they were caught (Doty and Lee 1974). The homing rate of successful nesters (52%) was much greater than that of unsuccessful nesters (16%). Of an estimated 140 female ducklings, 7 (5%) returned to or near their natal marshes. Males, however, are not very philopatric to breeding areas, although some do return. Evrard (1990) reported a 2.7% return rate for 225 marked

adult males in northwestern Wisconsin; none of the 72 marked juveniles returned. In Manitoba, Titman (1983) observed a return rate of 3% for 33 marked adult males.

During a 6-year study (1976–81) in North Dakota, return rates averaged 28.8% for 52 hatching-year females. Return rates were 44.0% for 25 second-year females that nested successfully, compared with 23.3% for 30 second-year females that were unsuccessful (Lokemoen et al. 1990a). The average return rate for 150 after-hatching-year females was 33.3%, but it was greater for successful females than unsuccessful ones (52.2% vs. 17.3%). Using a resighting probability of 64.5% for marked Mallards in southcentral Saskatchewan (1982–93), Arnold and Clark (1996) calculated a return rate of 42.6% for adults and 31.4% for juveniles.

During a 7-year study (1978–84) in Poland, 198 out of 447 (44%) female Mallards captured in nest boxes were recaptured in subsequent years (Majewski and Beszterda 1990). Return rates were higher for successful females than unsuccessful ones (34% vs. 16%), with 50% of the successful females nesting in the same structure, and 84% nesting within ≤ 100 m. In contrast, only 10% of the previously unsuccessful females returned to the same structure, and 55% nested >100 m from the previous site.

Territorial defense by male Mallards is controversial, with several investigators hesitant to call the defended portion of the home range a territory (Lebret 1961, McKinney 1965, Barclay 1970). On prairie pothole habitat in southwestern Manitoba, however, Titman (1983) observed that 10 marked pairs of Mallards never ventured beyond the boundaries of a specific area from the beginning of nest-site selection to the start of incubation, a period of 13–22 days. He did use the term territory, because the area was defended for exclusive use by the pair over the entire time period. The shapes of these areas varied with the configuration of the occupied wetlands, and they ranged in size from 5.0 to 16.3 ha. Territories of neighboring pairs frequently overlapped, but pairs (especially the females) were hardly ever found simultaneously in the area of overlap, especially during the territorial period. On average, territory size was larger in habitats with breeding pair densities of 4–7.5 pairs/km², compared with habitats with densities of 22–25 pairs/km² (15.9 ha vs. 9.2 ha). Two marked males even returned to the same territory used the previous year. Such territorial behavior probably evolved in males to protect the female from attempted extra-pair copulations, provide a feeding area free from harassment, and alert the female to potential predators. Krapu (1981) noted that paired males ($n = 4$) in North Dakota arrived with a lipid reserve of 86 g, which was rapidly used up during the time period when males were establishing activity centers (territories).

Territorial defense was often manifested in the form of 3-bird-flights, which feature a male pursuing a paired female accompanied by her mate. In Manitoba, Titman (1983) noted 4 general features of 3-bird-flights: most (60%) originated on or within 0.4 ha of the territory of the chasing male; 3 birds usually (86%), but not always, participated; the flights lasted <30 seconds (60%); and the flights covered <400 m (77%). These 3-bird-flights made up 85.5% of 1,405 pursuit-flights observed, with the remainder classified as group-flights associated with attempted extra-pair copulation (rape). In 98.8% of the observed 3-bird-flights, the male chased the pair or the female of the pair; 57.5% of these chases involved a specific pursuit of the female. Only 1.2% involved 1 male chasing another male. Marked males also returned to their territories following 80 out of 82 (97.6%) 3-bird-flights. Titman concluded that for males, 3-bird-flights were compatible with a mixed reproductive strategy: the primary function served to protect their mates from extra-pair copulations, but it also allowed them to engage in their own opportunities for extra-pair copulation with fertile females or to seek new liaisons with renesting females.

In contrast to these findings, however, Hori (1963) never recorded aerial attacks of lone male Mallards in ways suggestive of territorial defense during 14 years of observation involving about 200 flights. Instead of territorial defense, the 3-bird-flights most commonly seen in the spring could represent the defense of a mate, or an assessment of another female for a potential opportunity to engage in extra-pair copulations, as Sorenson (1994) observed during her studies of this behavior among White-cheeked Pintails (*Anas bahamensis*). The chase rate in flight was 5.4 times greater for fertile females compared with unfertile females, and it was longer when a fertile female was pursued. In addition, males may have assessed female reproductivity during these flights, as a female carrying an egg in her oviduct exhibited a distinctly drooped profile. Seymour (1990) noted that such 3-bird-flights are usually most common during the egg-laying period, when females are fertile and thus vulnerable to successful extra-pair copulation. Baldassarre and Bolen (2006) also found that male ducks differentially targeted fertilizable females.

Unlike the debate associated with territory, Mallards clearly establish a home rage during the breeding season that varies in size, depending on habitat, the density of breeding pairs, and the stage of reproduction. In North Dakota, the total home range of 6 radiomarked females averaged 468 ha, compared with 111 ha during the laying period (Dwyer et al. 1979). In contrast, home-range size within forested habitats of northcentral Minnesota for 12 radiomarked females was 210 ha, but the home range was much smaller during the prenesting (135 ha) and laying (70 ha) periods. Home-range size for 12 radiomarked males in Minnesota averaged 240 ha (Gilmer et al. 1975). In contrast, home ranges of Mallards along the Mississippi River east of Bemidji, Minnesota, averaged 540 ha (range = 40–1,440 ha) for 8 females and 620 ha (range = 70–1,140 ha) for 5 males; 5 home ranges were <100 ha, but 6 were >750 ha (Kirby et al. 1985). Mallards with comparatively small and intermediate-sized home ranges consistently exhibited a high use of marshes along rivers and river channels, with the overall home-range size negatively correlated with use of river habitat ($r = -0.929$). Such data suggest a continuum of responses by Mallards to habitat conditions encountered when selecting home ranges. Almost 90% of the locations of pairs with small home ranges were in habitats providing the most cover (e.g., cattails and bog-mat shorelines), compared with only 17% for pairs with large home ranges. In heavily forested habitat in the Adirondack Region of New York, the home ranges of Mallard pairs averaged 70 ha (Dwyer and Baldassarre 1994). The home-range size of 11 radiomarked females averaged 106 ha, but varied between 12 and 313 ha.

Courtship Displays. The courtship displays of Mallards are the most well studied of all waterfowl, including publications by Lorenz (1951–53), Johnsgard (1960a), Lebret (1961), Weidmann and Darley (1971), Palmer (1976a), Drilling et al. (2002), and Baldassarre and Bolen (2006). Because many of the Mallard displays are similar to those of other species of dabbling ducks, they are described in greater detail here.

Courtship occurs in groups, with males gathering on open water for well synchronized displays in bouts lasting up to 15 minutes (and sometimes longer) that Lorenz termed "social play." During this early phase of courtship, the males perform several preliminary displays, such as *head-shake*, *head-flick*, and *swimming-shake*. Initial displays can also involve *ritualized-drinking*, often followed by *mock-preening*, where the nail of the bill is rubbed against the underside of the wing to produce a *rrr* sound that is audible up to several meters away. Males also use *jump-flights* to attract a female's attention. In addition, Mallards exhibit *courtship-flights*, which occur as a female takes flight and is followed by up to 5–15 males (Dzubin 1957). These flights are usually slow and erratic and last 5–10 minutes, during which time the female will exhibit the *inciting* display.

Preliminary-shaking commonly occurs once courtship intensifies. Males perform this display by first drawing the head inward toward the shoulders, while also ruffling their underfeathers, and elevating the body from the water's surface. The head and the body are then thrust upward in a series of shakes, with each increasing in intensity. *Preliminary-shaking* apparently serves to gain the attention of a female and indicates that more elaborate displays will follow (e.g., *grunt-whistle*, *down-up*, or *head-up-tail-up*).

The *grunt-whistle* is either displayed by a lone male or, more frequently, by groups of males. This display is performed by lowering the bill into the water and shaking it slightly from side to side. The body is then arched upward while the bill is still held low, often remaining in the water; a loud sharp whistle, followed by a deep grunt, is given just before the body's arch reaches its peak, at which time the bill touches the water and a fine string of droplets is tossed toward the target female. Males perform the *down-up* by quickly lowering the bill into the water and then jerking the head quickly upward but not raising the breast, with a whistle given when the head is highest and the bill lowest. A stream of water often rises as the bill is lifted, during which time the male gives a *raeb-raeb* call. The *down-up* is especially associated with aggression and the proximity of males to each other. Johnsgard (1960a) considered the *grunt-whistle* to be a low-intensity display given more frequently early in courtship, while the *down-up* increased in frequency and eventually formed about half of all male displays at the peak of pair formation. Of the displays observed during his courtship studies on Cayuga Lake in central New York, 58.0% of all *grunt-whistles* occurred when a single male displayed, 26.4% of all *down-ups* when males displayed in groups of 4 or more, and 60.3% of all *head-up-tail-ups* when 2 or 3 males displayed.

The *head-up-tail-up* complex forms the major social display of male Mallards and usually consists of 4 displays given in quick succession: (1) *head-up-tail-up*, (2) *turn-head-to-female*, (3) *nod-swimming*, and (4) *leading*. The male performs *head-up-tail-up* by first drawing the head backward and upward and curving the tail upward while also lifting the wings, which exposes the curled feathers on the tail for viewing from the side. The *burp*, which is a single sharp whistle, accompanies this display and only occurs at this time. The *head-up-tail-up* display lasts about 1 second and signifies an intensification of the courtship process. *Head-up-tail-up* is often immediately followed by the self-descriptive *turn-head-to-female*, and then by *nod-swimming*, where the male lowers his outstretched head and neck to the waterline and swims away from the female or moves in a circle around her. Following *nod-swimming*, the male often leads the female from the group with the *turn-the-back-of-the-head* display. During this display, the male exposes the feathers on the back of the head in such a way as to exhibit a stiffly raised central line of neck feathers, which presents the female with a striking dark line of black on the green of the head.

Females use *nod-swimming* to stimulate groups of males to display (Weidmann and Darley 1971). Females perform this display by extending the head and neck flat along the water and swimming among males, or by *steaming*, with the head held low over the water. *Inciting*, however, is the most distinctive courtship behavior of females and is basically the same display in all dabbling ducks (Lorenz 1951–53). In this display, the female swims after her mate (or potential mate) while "threatening" another male by flicking her head over her shoulder. The female gives a loud *queg geg, geg geg* while inciting, which is her only courtship call. The male responds to *inciting* by attacking the individual designated by the female or by *leading*, where the male attempts to lead the female from the courtship group by exhibiting *turn-the-back-of-the-head* as he rapidly swims away from her. Males and females associated with *inciting* displays often have formed pair bonds, but *inciting* can also occur during temporary liaisons created early in the pair-formation process.

The major precopulatory display is *mutual-head-pumping*, although males may pump their heads more vigorously than females. Nonetheless, a pair engaged in *mutual-head-pumping* signifies that copulation is imminent. Copulation in Mallards, as in all other ducks, normally occurs on the water, with the female assuming a low posture as the male then climbs onto her back and grabs the back of her head before inserting his penis. Postcopulatory display almost always involves *bathing* by the female, while the male performs *bridling* (and an accompanying call), during which the head is drawn in and the chest lifted out of the water.

Females choose mates among courting males, in part through an assessment of male ornaments: the bill color, the tail curl, the neck ring, a maroon breast, and more. Omland (1996a, b) experimentally manipulated these factors in studies of Mallard pair formation and found that only bill color significantly affected mate choice, with a higher pairing success among males with bills that were more yellow and unblemished. Despite occupying such a small part of the male body, bill color alone explained 24% of the variation in the pairing success of males, and bill color combined with molting intensity explained 39%. Apparently fleshy ornaments, such as the bill, are more accurate indicators of immediate male quality than plumage. In domestic fowl, the comb size of males is strongly correlated with blood testosterone, which reflects current physical condition. Experimentally blackening the bill of male Mallards lowered pairing success in 15 out of 19 observed pairs.

Klint (1980) conducted several experiments on mate choice by captive female Mallards that involved manipulating male plumage. Females exposed to normal males and males whose green head-coloration was removed showed a preference for the normal males (89%) in all experiments. Females subjected to males with the brown breast feathers removed generally also preferred normal males (64%). The results of removing the blue speculum were inconclusive. These experiments indicated that male plumage characteristics also influence an initial mate choice by the female.

Nesting

Nest Sites. Female Mallards begin searching for a nest site within 5–10 days after establishing a home range (Drilling et al. 2002). The process begins with the pair flying low over the marsh during the evening, with the female in the lead (Bellrose 1980). Mallards are very much generalists in their selection of nest cover and specific nest sites, which reflects the diversity of the habitats where they breed. Nesting habitat can range from grasslands and brush in the PPR to deep forests in the eastern United States, and virtually every habitat type in between. There is a tendency, however, to nest in tall dense grasses/legumes and in brushy vegetation with overhead concealment, when such habitats are available. Both upland and marsh areas are selected; hence the distance of Mallard nests from water varies greatly with the habitat type. Nests also occur on dikes and levees, small islands, and in artificial nest baskets, which also influence the distance from nests to water. In heavily cultivated areas, Mallards are often forced to nest close to a water source, because the only available cover occurs along the narrow margin of vegetation surrounding potholes and sloughs.

During a 4-year study (1982–85) of 17 high-density duck-nesting areas in the PPR in Canada (9 in parkland and 8 in prairie), the largest (39%) percentage of Mallard nests ($n = 1,885$) were located in odd areas: areas of grasses and forbs <2 ha in size, as well as rock piles, gravel pits, narrow upland vegetation borders around wetlands, and vegetation along fences between croplands (Greenwood et al. 1995). Other nest areas were rights-of-way (25%), brush (14%), woodlands (9%), wetlands (8%), croplands (5%), and grasslands (2%). Based on habitat availability, however, Mallards selected brush (81%) and rights-of-way (10%). Of 414 Mallard nests located during a long-term study (1952–65) in parklands in Alberta (Smith 1971), 34.8% were in western snowberries (*Symphoricarpos oc-*

cidentalis) or buckbrush (*Symphoricarpos orbiculatus*), 23.7% in grasslands, 10.4% in willows (*Salix* spp.), 9.2% in aspen (*Populus tremuloides*) groves, and 6.5% in wolfwillow (*Elaeagnus commutata*).

Dzubin and Gollop (1972) noted that Mallard nests in parkland habitat near Roseneath, Manitoba, were closer to water than nests located in mixed-grass prairie habitat near Kindersley, Saskatchewan; 95% of 217 nests in the parklands were within 50 m of water, compared with 498 m for 584 nest in mixed grasslands. The shorter distance in the parklands was correlated with the higher density of wetlands and the restriction of preferred nesting cover to wetland edges.

To the south, a 4-year study (1977–80) of radiomarked female Mallards in central North Dakota recorded nests ($n = 142$) in grasslands (40.1%), wetlands (16.2%), odd areas (15.5%), rights-of-way (14.1%), haylands (12.0%), and croplands (2.1%; Cowardin et al. 1985). The grassland nests were largely in native vegetation, but they were also usually associated with woody species; 68% were at sites dominated by western snowberries. Based on the availability of habitat types, the highest nesting-habitat preferences were for rights-of-way and odd areas, and the least for croplands. Woody plants were the predominant vegetative life form (36.6%); of the remainder, 27.5% were grasses and sedges (*Carex* spp.), 15.5% growing forbs, 10.6% bulrushes (*Scirpus* spp.) and cattails (*Typha* spp.), and 9.8% residual forbs. Mallards nesting during another study (1966–84) in central North Dakota exhibited a preference for planted cover (42.8%) and odd areas (17.7%), which were sites such as rock piles, haystacks, gravel pits, and shelterbelts (Klett et al. 1988). Mallards also nested in haylands (10.7%), wetlands (9.7%), rights-of-way (7.6%), idle native grasslands (6.7%), and native grasslands used for pasture or hay (4.5%). In a third North Dakota study (1979–81), Lokemoen et al. (1990a) found that Mallard nest densities were 60.6/km^2 in a seeded nesting cover of wheatgrass (*Agropyron* sp.), alfalfa (*Medicago sativa*), and yellow sweetclover (*Melilotus officinalis*); 48.5/km^2 by roadsides; 44.3/km² by the sides of canals ; 43.4/km² in odd areas; and 35.6/km² in dry wetlands.

Mallards will also readily nest in overwater vegetation, more so than any other species of dabbling duck. In southcentral North Dakota, females located 66% of 53 nests in dense overwater wetland vegetation averaging 76.7 cm tall; the nests were usually within 50 m of the wetland's edge (Krapu et al. 1979). The authors noted that other studies in the PPR also reported a high (56%–74%) percentage of overwater Mallard nests. Of the Mallard nests located in wetlands in North Dakota by Cowardin et al. (1985), 47% were in cattail vegetation and 17% were in bulrushes.

Mallards in the PPR will also commonly nest on islands, which can afford protection from mammalian predators. In northwestern North Dakota, Mallards nested on a natural 4.5 ha island within 385 ha Miller Lake, where densities reached 241–389 nests/ha (Duebbert et al. 1983). A single 0.6-ha clump of snowberries (*Symphoricarpos albus*) and western rose (Woods' rose; *Rosa woodsii*) contained 225 simultaneously active Mallard nests (Lokemoen et al. 1984). During the peak nesting period, Mallard nests, on average, were only 2.7 m from conspecifics. Mallards can also achieve high nest densities on man-made islands (Johnson et al. 1978).

Outside the PPR, Mallards (*n* = 395) nesting at the Tule Lake and the Lower Klamath NWRs in California during 1957 were primarily located on islands (38.0%), in fields (31.1%), and on dikes (28.4%; Rienecker and Anderson 1960). Mallards generally tended to prefer medium to tall cover, such as that afforded by their preferred cover types, saltbush (*Atriplex* sp.; 40.3%) and nettle (*Urtica californica*; 18.7%). On islands in Lake Champlain in Vermont, 27.9% of Mallard nests (*n* = 86) were located in low, dead herbaceous vegetation, primarily raspberries (*Rubus* spp.) and nettles (*Urtica* spp.); 18.6% were under live conifers, such as yew (*Taxus canadensis*); and 10.5% each were associated with fallen logs, dead tree tops, fallen limbs or hollow tree boles, crotches, and stubs (Coulter

and Miller 1968). Mallards, along with American Black Ducks, also made extensive use of stumps and dead snags created in flooded timber at the Montezuma NWR in New York, with the most common nest site being on the tops of stumps and the snags of large trees where the bole had broken off (Cowardin et al. 1967). The average diameter of stumps containing nests was 49.5 cm, with a 1.9 m height above water.

During a 3-year study (1990–92) in the St. Lawrence River valley in New York, 39 out of 71 (55%) nests were located in uplands and 32 (45%) were in wetlands (Losito et al. 1995). Upland nests were mostly in haylands (44%) and grasslands (31%), while wetland nests were primarily in forested-live timber (41%), scrub-shrub wetlands (28%), and emergent wetlands (22%); 8 (25%) of the wetland nests were located entirely over open water. At 4 sites in nearby southern Ontario (1997–2000), nests (*n* = 442) produced by radiomarked females were located in woody habitats more often than expected, relative to their availability (Hoekman et al. 2006a). Females generally avoided croplands and pastures. At 3 of the sites, females selected grasslands, haylands, and wetlands more often than expected, given the distribution of habitats across the landscape. At the fourth site, however, most nests were located in woody and wetland habitats, which were the dominant cover types on the landscape. Many nests were located on small hummocks within large (>10 ha) shallow wetlands, as was also observed in New York.

The female forms a nest bowl or a scrape in the litter from dead plants, with scrapes being about 18–20 cm in diameter and 3.8 cm deep (Coulter and Miller 1968). Overwater nests can either be simple bowls on mats of floating vegetation or more elaborate woven structures constructed from emergent vegetation (Krapu et al. 1979). Down generally appears in the nest between when the fourth and sixth eggs are laid (Caldwell and Cornwell 1975), but it is often added sparingly until just prior to the completion of the clutch. Females may reoccupy nest bowls from prior years, as reported from a high-density nesting population on a 4.5 ha island in North Dakota, where 73% of the nest bowls were reused in a given year (Duebbert et al. 1983).

Clutch Size and Eggs. Clutch size is affected by the nest initiation date and the age and body condition of the female. Bellrose (1980) reported an average of 9 eggs/clutch from 5,170 Mallard nests at various locations across North America. Dzubin and Gollop (1972) noted that the average clutch size from 20 North American studies (1949–57) was 8.7 eggs for 1,468 nests (range = 5.7–10.6 eggs). The size of incubated clutches from 281 nests on Miller Lake in North Dakota ranged from 3 to 24 eggs, but a clutch size of >13 (11%) was attributed to laying by multiple females (Duebbert et al. 1983).

In the parklands near Roseneath, Manitoba, clutch size averaged 8.6 eggs ($n = 111$) prior to 25 May, compared with 7.2 eggs ($n = 55$) after that date (Dzubin and Gollop 1972). Similarly, clutch size in the mixed grasslands near Kindersley, Saskatchewan, averaged 9.1 eggs ($n = 579$) prior to 15 May and 7.7 eggs ($n = 98$) afterward. In North Dakota, the clutch size of adult females declined by 0.027 eggs/day as the nesting season progressed (Lokemoen et al. 1990a). The clutch size of first nests established by captive-reared Mallards declined by about 0.1 eggs/day as the season advanced (Batt and Prince 1979). Nutritionally, in a comparison between sibling pairs of captive-reared Mallards fed a wheat diet versus an enriched diet, those on the wheat diet had a lower clutch size, laid smaller eggs, laid them more slowly, and had fewer renesting attempts (Eldridge and Krapu 1988). Diet did not, however, affect the initiation of laying, the duration of the laying period, or the seasonal changes in clutch size associated with each renest. The authors interpreted this pattern as an adaption by Mallards to the highly variable environments characteristic of the PPR, where the probability of successful reproduction declines as the breeding season progresses.

Clutch size in incubated nests in a North Da-

kota study was larger, on average, for adult females, compared with yearlings (9.0 eggs vs. 8.5; Lokemoen et al. 1990a). Krapu and Doty (1979) reported similar results for the average clutch size in the nests of 7 yearlings in North Dakota, compared with that of 46 adults (9.3 vs. 10.3). A subsequent study (1993–95) in the PPR in the Dakotas reported an average clutch size of 8.9 for 1,236 nests, with a range from 8.8 to 9.1 (Krapu et al. 2004c). Clutch size at the median date of nesting (8 Apr) was 11.3–11.4. Clutch size declined as the season progressed and lipid reserves were depleted, but not in a linear fashion, which indicated that Mallards had less trouble securing nutrient reserves for later clutches than did other species (Krapu 1981).

On a North Dakota study area, clutch size averaged 0.7 fewer eggs during a dry versus a wet year (Krapu et al. 1983). In the PPR in North Dakota and Minnesota (1988–94), however, local water conditions (as measured by the percentage of basins with water) did not affect Mallard clutch size, egg volume, or the hatching mass of ducklings (Pietz et al. 2000), perhaps because Mallards acquired the lipid reserves used in the production of early clutches prior to their arrival on the breeding grounds (see Incubation and Energetic Costs).

In the East, clutch size averaged 8.9 eggs ($n = 42$) during a 3-year study (1990–92) in New York, and it was negatively correlated with the nest-initiation date ($r = -0.40$; Losito et al. 1995). Mean clutch size was 9.2 for adults ($n = 26$) and 8.6 for yearlings ($n = 15$), but the difference was not statistically significant. On islands in Lake Champlain in Vermont, clutch size averaged 9.6 for 131 nests (Coulter and Miller 1968). Clutch size decreased with the initiation date, but this decline was not constant. In westcentral Illinois (1998–2003), clutch size averaged 9.4 for 66 initial nest attempts, with a range of 4–12 eggs (Yetter et al. 2009). Clutch size did not differ by a female's age or among the years of the study.

Egg weight is genetically controlled (heritable) and thus independent of a female's age or weight, with a repeatability measure (determined from 60 females) of 62% (Batt and Prince 1978). Duckling mass is positively correlated with egg size, with the relationship in captive-reared Mallards stronger for second clutches ($r^2 = 0.72$) compared with first clutches ($r^2 = 0.60$; Batt and Prince 1979). Rhymer (1988) also reported that duckling mass at hatching was strongly correlated with egg mass for captive-reared ducklings, averaging 25.2 g for light eggs (<48.0 g), 29.7 g for normal eggs (48.0–56.4 g), and 34.8 g for heavy eggs (>56.4 g). Egg-size variation was unrelated to a female's body size ($r = 0.17$), which again reflects the heritability component. Heavier ducklings have a lower lethal temperature (the temperature at which 50% of the individuals cannot maintain their body temperature) than lightweight or normal-sized ducklings (5°C for heavy vs. 15°C for lightweight and 10°C for normal) and thus can survive for longer periods without food. Day-old ducklings from heavier eggs also spend less energy/unit gram to maintain homeothermy.

Mallard eggs are elliptical to ovate in shape. They vary in color from light greenish buff to light grayish buff, or are nearly white, and they have very little luster (Bent 1923). Measurements of 93 eggs from different North American collections averaged 57.8 × 41.6 mm. Egg weight averaged 52.2 g, as calculated from 3 different studies in North America: 1,085 eggs of captive-reared wild-strain Mallards in Manitoba (Rhymer 1988), 613 eggs of captive wild-strain Mallards in North Dakota (Eldridge and Krapu 1988), and 455 eggs of wild Mallards in California (Pehrsson 1991). Mallards generally lay 1 egg/day, with about 1–3 days between nest building and the laying of the first egg (Coulter and Miller 1968).

Incubation and Energetic Costs. The incubation period averages 28 days, with a range of 23–30 days (Palmer 1976a). Using microprocessor data loggers, Loos and Rohwer (2004) found that female Mallards in the PPR in North Dakota and southwestern Manitoba began incubating their eggs during the egg-laying period, spending an average of 8.5 hours/day on the nest between the laying of eggs 2–10. Total time spent on the nest during this period averaged 4,057 minutes, with the amount of incubation time increasing with the number of eggs in the clutch, but it was not influenced by low temperatures, precipitation, or nest-initiation date. Caldwell and Cornwell (1975) noted that the egg temperatures of captive-reared Mallards were high enough for embryo development as early as the laying of the sixth egg in a final clutch of 10–12 eggs.

Caldwell and Cromwell (1975) also reported nest attentiveness at 94.6% during incubation, with recesses totaling 78 minutes and averaging 24 minutes ($n = 67$), which equates to 3.3 recesses/day. The percentage of recess time females spent feeding ranged from an estimated 67% (Afton and Paulus 1992) to just 38%, based on 2 radiomarked females monitored by Krapu (1981). No recesses took place at high ambient temperatures (>32°C), and only rarely during rainfall. Incubation occurred in bouts that averaged 356 minutes ($n = 56$), with a bout defined as the time spent on or in the immediate vicinity of the nest. Incubation bouts were shortest at ambient temperatures above 24°C (313 minutes) and below 6°C (324 minutes), and longest (417 minutes) between 6°C and 14°C. Females did not attend their nests at night until after the last egg was laid. Upon her return to the nest, the female shifts the position of the eggs an average of 1.2 times/hr.

Dwyer et al. (1979) noted that breeding female Mallards in North Dakota averaged 54.8% of all daylight hours feeding during egg laying, 38.3% during incubation, and only 17.6% during prenesting. In North Dakota and Minnesota, paired males spent less time feeding and more time being alert than did females. Both pair members engaged in the same behavior about 67% of the time, but females were most often feeding while males were resting or alert (Pietz and Buhl 1999). Daily activity patterns varied, however, between study sites and years, probably in relation to water conditions and food availability. Hence time-budget data on breeding Mallards are probably temporally and geographically specific.

Gloutney and Clark (1991) estimated that stored lipids accounted for 24.7% of the cost of incubation, with the remaining energy acquired during nest recesses. During 1975 at Marshy Point in Manitoba, body weights of 38 females declined 7.4 g/day during incubation, which amounted to an average of 185 g, or 18% of their initial body weight (Gatti 1983). Adults lost less weight overall than did yearlings (14% vs. 20%). Early-nesting females began incubation at the highest body weight, but they lost 295 g during incubation, or 27% (11.8 g/day) of their initial body weight, probably because they increased their nest attentiveness, and thus reduced their recesses, during the colder temperatures occurring early in the season. During the peak nesting period, females lost 17% of their body weight, compared with 11% for the late-nesting period.

Krapu (1981) documented the reproductive energetics of Mallards during a study in North Dakota. The general pattern for females was to rely on a lipid reserve acquired before arrival to the breeding grounds, and to obtain all the necessary protein and a small amount of additional lipids after arrival. Seven prenesting females averaged 109 g of lipids shortly after arrival, and they depended heavily on these lipid reserves during reproduction. An average of 63 g (57%) was used between the initiation of laying and the sixth day of incubation; all of their lipids were probably exhausted by the time hatching occurred. Body weights of 11 females averaged only 900 g during the last 5 days of incubation, which was 25% less than their body weights during prelaying. The proteins required for egg production were obtained by foraging on aquatic macroinvertebrates on the breeding areas; the use of their lipid reserves allowed the females to focus on acquiring the scarcer invertebrates to meet their protein needs.

Krapu's conclusions were reinforced in a subsequent 4-year study (1977–80) in North Dakota that assessed Mallard recruitment by monitoring 338 radiomarked females (Cowardin et al. 1985). Both the probability of producing eggs and the number of eggs produced increased with the body condition of the female, and both indices also were higher for adult females than yearlings, which indicates the higher reproductive potential of adult birds.

Nesting Chronology. Mallards are among the earliest-nesting ducks, with nest initiation occurring between 4 and 30 April over vast reaches of their breeding range (Bellrose 1980). Bellrose reported that, in general, Mallards nest over a span of 60–85 days, except in the far north; the span was 30 days at Yellowknife, Northwest Territories, and 55 days on the Saskatchewan River delta.

Nesting chronology at a given location, however, is influenced by several factors, particularly spring temperatures. In North Dakota, the average temperature in April delayed peak nesting activity by 2.2 days for each 1°C drop (Cowardin et al. 1985). During a 4-year study (1982–85) over a broad area of the PPR in Canada, temperature, precipitation, and the availability of seasonal wetlands in May were collectively correlated with nest-initiation date ($r^2 = 0.28$). The median date of nest initiation was 16 May, but it decreased by 2.3 days for each 1°C increase in the average May temperature (Greenwood et al. 1995). The nest-initiation period averaged 27 days, but it increased by 0.10 days for each 1 cm increase in May precipitation. In parkland habitat near Roseneath, Manitoba, Dzubin and Gollop (1972) noted that snow cover on nesting habitat, ice cover on wetlands, and both minimum and cumulative temperatures above freezing were apparent factors affecting nest initiation and could delay subsequent hatching peaks by 2–3 weeks.

The nest-initiation date for first nests may vary with a female's age, but the data are not conclusive. During a 4-year study (1977–80) of radiomarked females in North Dakota, Cowardin et al. (1985) were unable to detect a difference between yearlings and adults in the median nest-initiation date:

12 May–6 June for yearlings ($n = 32$), compared with 13 May–1 June for adults ($n = 92$). In another 4-year study (1978–81) in North Dakota, 10% of the Mallard nests were initiated by 26 April, with a median date of 16 May; 90% were initiated by 6 June (Lokemoen et al. 1990a). A female's age also had no significant effect on either the start or the length of the nesting season. Coulter and Miller (1968), however, suggested that yearlings nested later than adults on islands in Lake Champlain, as 75% of the known adult females ($n = 16$) nested before 24 April, compared with 35% of the mixed-aged females ($n = 20$), which, the authors believed, mostly represented yearlings.

Of 1,167 Mallard nests monitored in the Dakotas (1993–95), 50% were initiated by 18–19 May, but initiation ended largely by the summer solstice, as was the case for 4 other species of dabbling ducks monitored during the same study, although there was variation among the species (Krapu 2000). For Mallards, 75% of the nests were initiated by 2–6 June, 90% by 12–16 June, 95% by 17–21 June, and 99% by 24 June–2 July. Krapu concluded that the influence of photorefractory mechanisms constrained most nesting to the spring, because reproductive success (brood survival) is much lower later in the summer. Other factors, however—such as water conditions, food availability, and food quality—also have a strong influence on the length of the breeding season. In exceptionally wet summers, Mallards and several other species of ducks nest into mid- and late summer in North Dakota (Krapu et al. 2001). Dry years have the opposite effect. The period of nest initiation in North Dakota went from 85 days during a wet year (1976) to only 66 days (a 22% decline) during a severe drought in 1977; Mallards at 1 study site were unsuccessful at breeding and abandoned their activity centers by mid-May in 1977 (Krapu et al. 1983). In another comparison of the effects of drought on reproduction, radiomarked females ($n = 8$) monitored over 3 years of normal water conditions (1973–75) averaged 43.5 days on the breeding area, used 15.5 wetlands, and initiated

14 nesting attempts; 8 females monitored during the drought year of 1977 averaged 16.4 days on the breeding area, used 9.0 wetlands, and initiated only 1 nesting attempt.

Outside the PPR, Mallards in northcentral Iowa (1974–75) initiated nests between 5 April and 15 June in 1974, and 15 April and 10 June in 1975, with 50% initiated by 30 April in 1974 and by 5 May in 1975 (Humburg et al. 1978). During a 6-year study (1998–2003) in westcentral Illinois by Yetter et al. (2009), the mean initiation date for first nests ranged from 22 April to 6 May. The majority (75%) of all nests ($n = 195$) were initiated by 20 May, and the average length of the nesting season was 88 days (range = 78–107 days).

To the east, a 3-year study (1990–92) of Mallards in the St. Lawrence River valley of New York reported that the median date of nest initiation for radiomarked females was 8–17 May and did not differ among years (Losito et al. 1995). The nest-initiation date also did not vary between adults (10 May; $n = 44$) and yearlings (7 May; $n = 24$). On average, Mallards initiated 10% of their nests by 19 April, 50% by 10 May, and 90% by 5 June, with the central span of nesting (10%–90%) averaging 48 days. The earliest nest was located on 3 April, and the latest on 16 June. During a 7-year study (1957–63) on islands in Lake Champlain in Vermont, the average nest-initiation date of Mallards was 12 April, with a maximum annual variation of 9 days (Coulter and Miller 1968). Egg laying terminated by 16 June.

In British Columbia, Mallards in the southern Okanagan Valley began nesting around the first week of April, compared with 2 weeks later in the Nicola and Cariboo Districts to the north (Munro 1943). In areas farther north (e.g., the Athabasca delta in northeastern Alberta and Yellowknife, Northwest Territories), Mallards did not initiate nests until about mid-May, but nest initiation quickly peaked during the last 2 weeks of May (Bellrose 1980). Bellrose also reported nesting as early as 11 May on Yukon Flats in Alaska, with peak initiation between 20 May and 5 June.

Nest Success. Mallard nest-success studies are almost too numerous to count, with the results varying markedly over both time and geographic area. Highlights from studies in major areas of the Mallard's breeding range, however, depict the general pattern. A simulation model based on data from North Dakota indicated that a 15% nest-success rate was needed to maintain the Mallard population within their principal breeding range, the PPR (Cowardin et al. 1985). This rate is especially low, but it is also variable over time and geographic location across much of the PPR. Extensive details on Mallard nest success are presented for this region, because these same factors affect the nest success of other upland waterfowl in this very important breeding area.

The success of 1,572 Mallard nests located during a 4-year study (1982−85) on 17 sites in grassland and parkland regions of the PPR in Canada averaged only 12%/year, but it varied markedly (2%−29%) among the research sites (Greenwood et al. 1987). Nonetheless, only 7 out of 31 (23%) annual nest-success estimates from individual study areas were ≥15%. Nest success was strongly correlated with the amount of grass on the landscape: $r = 0.80$ in the grasslands and $r = 0.67$ in the parklands. Of the 7 nest-success estimates ≥15%, 4 were from study areas where nesting was concentrated in large native pastures, with the highest (>100) density of nests in 259 ha of pasture. Predation, mostly by mammals, destroyed 72% of the nests and caused the abandonment of 6%, but predation tended to decrease as the nesting season progressed. Nest success did not differ among Mallards and 4 other species of dabbling ducks in the study, but it did vary by habitat type: 25% in woodlands, 19% in brush, 16% in wetlands, 15% in grass, 11% in odd areas, 8% in rights-of-way, and 2% in croplands (Greenwood et al. 1995). In prairie habitat, nest success was negatively correlated with the percentage of croplands, decreasing by about 4% for every 10% increase in cropland area. Duck populations probably were not stable where croplands exceeded about 56% of the landscape.

Nest success was little better during 19 years of study over 3 time periods (1966−74, 1975−79, and 1980−84) within 5 areas of the PPR in the United States (Klett et al. 1988). The overall success of 2,399 nets monitored during these time periods was only 6%−8%, but varied within habitat types. Nest success averaged over the 3 periods was highest (24.0%) in idle grasslands, but these formed <1% of the available habitat; planted cover was next (9.7%), followed by wetlands and grasslands (8.7% each). As in prairie Canada, nest success was lowest (2.0%) in croplands, which averaged 65% of the habitat across all study areas, and 4.0% in haylands. Most (82%) nests were lost to predation, and 7% to farm machinery. The authors concluded that the wetland base in the PPR in the United States was adequate to attract large numbers of Mallards and other ducks, but that nest success was limiting their population maintenance or growth.

Higgins et al. (1992) reported a nest-success rate of only 14.9% for 367 Mallard nests monitored from 1966 to 1981 at the Woodworth Study Area in North Dakota. For radiomarked females in central North Dakota (1977−80), nest success ($n = 129$) averaged only 8.3%, with a range of 5.9%−13.1% (Cowardin et al. 1985). Nest success was highest (10.5%) in odd areas, but again was lowest (0.3%) in croplands. Predation accounted for 70% of all unsuccessful nests. Successful initial nests were closer to water than unsuccessful nests (142 vs. 212 m) and had more wetland basins within 200 m (2.1 vs. 1.6), but they were about equally distant from predator travel lanes (57 vs. 59 m). Hen success (the percentage of females ultimately hatching a clutch) averaged 15.2% but varied among years (5.9%−23.6%), due to increased renesting in wet years.

These low rates of nest success in the PPR reflect the long-term impact of the enormous conversion of upland nesting habitat to agricultural croplands, as the PPR is among the most intensively farmed regions of the world. Lynch et al. (1963) estimated that as early as the mid-1950s, 72% of all uplands in the PPR in Canada were converted to cereal grain production, with nearly all of the remainder grazed by livestock. Millar (1986) estimated that by 1982, cultivation had converted 83% of the tallgrass prairie, 84% of the shortgrass prairie, 78% of the mixed-grass prairie, and 80% of the aspen parklands across a broad swath of the PPR in Canada. A 1981–85 study of 10,000 wetland basins in the PPR in Canada found that degradation of wetland margins reached 82% in parkland habitat and 89% in grasslands (Turner et al. 1987). The dominant habitat composition on 9 parkland sites across prairie Canada (determined from 1982 aerial photographs) averaged 56% croplands (but reached 80% at 1 site), 16% grasslands, and 13% wetlands (Greenwood et al. 1995). The composition on 8 prairie sites averaged 59% croplands (with a high of 84%), 19% grasslands, and 11% wetlands. The low nest success observed during their 4-year study (1982–85) on 17 sites in grassland and parkland regions of the PPR in Canada led Greenwood et al. (1987) to conclude that "marginal farming practices" were probably the greatest threat to duck-nesting habitat, because the best habitat—native grassland—was converted to croplands long ago. In the United States, Higgins (1977) noted that as much as 84% of the landscape was cultivated in parts of eastern North Dakota.

Current rates of nest predation in the PPR reflect an interaction between habitat loss and the diverse and abundant predator community found there. Sargeant et al. (1993) identified 20 avian and mammalian predators in the region, of which 9 species of mammals and 7 species of birds commonly preyed on nests and/or adult waterfowl, as well as on ducklings. The authors observed an average of 12.2 predators on 33 study areas, ranging in size from 23 to 26 km². Among mammals, striped skunks (*Mephitis mephitis*), raccoons (*Procyon lotor*), Franklin's ground squirrels (*Spermophilus franklinii*), and red foxes are the major predators of waterfowl nests in the PPR. Raccoons particularly affect overwater nests, because they prefer to forage in and around the edges of wetlands (Fritzell 1978). In contrast, striped skunks forage more in upland areas; Greenwood et al.

(1999) found shell fragments from duck eggs in 40% of 1,248 skunk scats examined in North Dakota. Red foxes (*Vulpes vulpes*) cause more serious damage to waterfowl, in large part because they destroy not only nests, but are also capable of killing nesting females. During an extensive study (1968–73) in North Dakota, Sargeant et al. (1984) noted that Mallards made up 23% of 1,798 dabbling duck remains collected at red fox dens. A single red fox female killed 16.1–65.9 ducks/year. The study extrapolated these results to estimate that red foxes killed 900,000 breeding ducks annually in the prairie population; 76% of the dabbling ducks killed were females.

The increase in red fox numbers in the PPR was a direct result of human activities. The conversion of grasslands to agriculture created red fox habitat concurrent with the systematic removal of the principal competitors and predators of red foxes: initially gray wolves (*Canis lupus*), and then coyotes (*Canis latrans*) (Johnson and Sargeant 1977, Sargeant 1982). Historical records indicate that red foxes were sparsely distributed in the PPR, but their numbers and range expanded dramatically following control of the coyote population in the 1930s and 1950s. The relationship among the canids is fairly straightforward: gray wolves suppress the numbers of coyotes in an area, and coyotes limit the number of foxes.

Sargeant et al. (1993) noted that red foxes in the PPR were most abundant in areas with the fewest coyotes, although the latter have increased in numbers since the 1970s. Such differences in the composition of predator communities are significant, because they can explain variations in waterfowl nest success among areas. In North Dakota and South Dakota (1990–92), the nest success of ducks on 36 study areas averaged 32% where coyotes were the main canid present, compared with 17% where red foxes were the principal canids (Sovada et al. 1995). In 1 situation, nest success was 2% on an area dominated by red foxes versus 32% in an area where coyotes prevailed, even though the 2 areas were only 5 km apart. Coyotes apparently help increase nest success not only because they displace red foxes, but also because they prey on other medium-sized nest predators, such as raccoons (Sargeant et al. 1993).

The nest success of Mallards and other ducks in the PPR reflects a complex interaction among an array of factors, such as the size of the available habitat (patch size), predator abundance (and its species composition), and weather. During a 3-year study (1993–95) in the PPR in the United States, Sovada et al. (2000) noted that nest success tended to increase with the expanding patch size of planted grasslands, but success was also influenced by the nest-initiation date and year. An index of predator activity also varied among years and patch sizes of habitat. Rabies sharply reduced striped skunk populations on some study areas, and red fox activity was greatest in small patches. Drever et al. (2004) found that the nest success of Mallards and other prairie-nesting ducks did not differ over time at managed sites (where predators were excluded), but varied positively with pond density during the prior year. In contrast, at sites where predators were removed but could emigrate back to a given site, nest success fluctuated widely over time, but was not affected by pond density.

Away from upland areas in the PPR, Mallard nest success can be high for overwater nests, because red foxes, a major waterfowl predator, usually avoid entering the water. In southcentral North Dakota, the success of 35 overwater Mallard nests was 54%, compared with 14% in nearby uplands (Krapu et al. 1979). In southwestern Manitoba, Arnold et al. (1993) recorded a 43.9% hatching success of 47 overwater Mallard nests, compared with 12.2% for 49 upland nests. They also noted that a large fraction of the local population was nesting in overwater habitat, which is not searched during traditional nesting studies.

Mallards will readily use nest baskets erected in the PPR, and nest success with them can be very high. A 1966–68 study of 1,038 nest baskets erected on prairie wetlands in North Dakota,

South Dakota, Minnesota, and Wisconsin reported 397 nest attempts (389 by Mallards), with a nesting success of 83% (Doty et al. 1975). In prairie Iowa, 222 nest baskets monitored over a 6-year period (1964–69) averaged 33% use by Mallards, with a hatching success of 87% (Bishop and Barratt 1970). Although early studies advocated the use of nesting structures where upland nesting habitat was sparse, Artmann et al. (2001) demonstrated that the density of the breeding population was a more important factor in determining the use of nest structures. During their research in northeastern North Dakota, 260 structures were placed on areas with comparable amounts of wetland habitat, but the surrounding area at grassland sites was 44.8% perennial cover, compared with only 8.0% at cropland sites. Occupancy rates in nest baskets were significantly greater on grassland sites than cropland sites (17.8% vs. 3.9%), even though the grasslands had better natural nesting cover; the difference was largely a function of higher Mallard pair densities on grassland versus cropland sites (15.2 vs. 9.2 pairs/km). The authors thus concluded that erecting nest structures where Mallard pair densities are high would be the most effective way to maximize initial nest-structure use.

Mallards can also achieve very high nest-success rates on islands in lakes located in the PPR. Duebbert et al. (1983) found 1,500 Mallard nests over a 5-year period (1976–80) on a 4.5 ha island in 385 ha Miller Lake in northeastern North Dakota; nest success ($n = 1,427$) averaged 83.2%. Abandonment, averaging 15.4%, was the main cause of nest failure. An appraisal of 209 islands averaging 1.36 ha in size (range = 0.04–21.45 ha) in the Dakotas and Montana found that Mallard nest success was 76% on islands with predator control versus 58% on islands with no predator reduction (Lokemoen and Woodward 1992).

Outside the PPR, Yetter et al. (2009) conducted a 6-year study (1998–2003) of radiomarked female Mallards in Illinois. Nest success averaged 19.6% for 189 nests, but it varied from 9.8% to 33.3%.

Average success was higher among 60 yearling nests than 129 adult nests (21.6% vs. 18.7%), but the confidence intervals overlapped. There also was only weak support for nest success declining with the nest-initiation date. Of 169 nest losses, most (90.5%) were from predation: coyotes (21.9%) and raccoons (13.0%) were identified most often, but 47.9% were destroyed by unidentified predators. Nest success was highest (25.8%) in idle grasslands, which made up 14.5% of the available habitat and contained 55.7% of all nests (*n* = 167). Idle/mowed haylands also formed 14.5% of the available habitat, but nest success there was 11.4%, with this habitat containing 26.4% of the nests.

Mallard nest success was 17.6% (*n* = 67) during a 3-year study (1990–92) in the St. Lawrence River valley of New York and did not differ among years or between adults and yearlings (Losito et al. 1995). Of the 52 unsuccessful nests, most (92%) were depredated (69% mammalian, 6% avian, 25% unknown); 4% were lost to haying operations; and 4% failed to hatch (dead embryos), although they were incubated for 34–46 days. Nest success did differ among habitat types, however, being highest (38.6%) in the extensive haylands that characterized this dairy-farming area of New York, and 13.9% for nests in wetlands, most of which were forested-live trees or scrub-shrub. Wetlands housed 45% of all nests, but success there was probably lower because of higher predator populations.

In southern Ontario, Hoekman et al. (2006a) used radiotelemetry to study Mallard breeding activity over a 4-year period (1997–2000) at 4 sites where agricultural activity ranged from relatively intensive to relatively light. Of 429 nests, 74 (17.2%) were successful, 39 (9.1%) were abandoned, and 316 (73.7%) were unsuccessful. Of the unsuccessful nests, 287 (90.8%) were lost to predators, 3 (1%) were lost to other natural causes, and 26 (8.2%) were destroyed by machinery or human activity. Estimates of nest survival varied little among the 4 sites.

Brood Parasitism. Mallard nests are parasitized by several other species of ducks, but the extent of parasitism depends strongly on the type of nesting habitat and the density of the parasitic and host species. Redheads, Ruddy Ducks, Lesser Scaup, Gadwalls, Northern Pintails, Cinnamon Teal, and even Common Goldeneyes have all parasitized Mallard nests (Drilling et al. 2002).

Pelzl (1971) documented a Red-breasted Merganser parasitizing a Mallard nest in Wisconsin. Of 24 active Mallard nests located in emergent vegetation of semipermanent wetlands in North Dakota, 10 were parasitized by Redheads, 2 by Ruddy Ducks, and 1 by both species (Talent et al. 1981). The primary effect of parasitism was a reduction (−22%) in the number of Mallard eggs, from 7.2 in nonparasitized nests to 5.6 in parasitized nests. Also, the hatchability of Mallard eggs was only 43% in parasitized nests, compared with 80% in nonparasitized nests. Most of the decline in egg success (percentage of eggs that hatch from nests hatching at least 1 egg) was attributed to displacement, as 35% of the Mallard eggs in successful nests were displaced, mostly underwater.

In Utah, 11 out of 29 (38%) Mallard overwater nests were parasitized: 8 by Redheads, 2 by Ruddy Ducks, and 1 by both species (Joyner 1976). Clutches averaged 5.7 parasitic eggs and 6.8 host eggs, versus 7.2 eggs in nonparasitized nests. As in North Dakota, the main effect of parasitism was the reduced hatchability of Mallard eggs in parasitized nests (54.7% vs. 65.2%). Weller (1959) noted that Mallard nests closest to water were the most likely to be parasitized.

In North Dakota, 4 out of 16 (25%) Mallard nests located on islands were parasitized, versus only 1 out of 51 (2%) on peninsulas, with nests initiated later in the season more likely to be parasitized than early nests (Lokemoen 1991). Unlike previous studies, there was a significant difference in the number of host eggs in parasitized nests, compared with nonparasitized ones (6.0 vs. 9.6). Of 281 incubated Mallard clutches on a 4.5 ha island in North Dakota during 1977, 11% of the clutches exceeded 13 eggs, which was indicative of brood parasitism, and 4% contained known parasitic eggs from Gadwalls (Duebbert et al. 1983).

In southwestern Manitoba, overwater nests were far more likely to be parasitized than upland nests (11% vs. 0%), with most nests parasitized by Redheads (Arnold et al. 1993). Compared with nonparasitized nests, the parasitized nests averaged 1.2 fewer Mallard eggs, and egg success was 23% lower. The effects of parasitism, however, were outweighed by the much greater success of Mallard nests located overwater rather than in upland habitat (43.9% vs. 12.2%).

Renesting. Mallards persistently renest following the destruction of their first nests, and they will commonly do so following the loss of a second and even a third nest. During his studies in the Alberta parklands, Keith (1961) estimated that all unsuccessful female Mallards renested, with 63% attempting a third nest following the loss of their second. He also noted that Mallards were more persistent renesters than 4 other duck species also under study by him. Rotella et al. (1993) fitted 16 females in Alberta with implanted radiotransmitters: 3 produced 1 nest, 5 produced 2 nests, 4 produced 3 nests, 3 produced 4 nests, and 1 produced 6 nests. At Roseneath in Manitoba, Dzubin and Gollop (1972) estimated a 50% renesting rate by Mallard females that lost their initial clutches. On islands in Lake Champlain in Vermont, Coulter and Miller (1968) reported that Mallards were more persistent renesters than American Black Ducks (57% vs. 36%). Dwyer and Baldassarre (1993) found that Mallards in forested habitats in New York also were more persistent renesters than American Black Ducks, and Sowls (1955) concluded that Mallards were very persistent renesters among the species of ducks he studied in Manitoba.

Among wild Mallards nesting on islands in Lake Champlain in Vermont, 17 out of 30 (57%) marked females renested after the loss of their first clutches, with the older females being more likely to renest (Coulter and Miller 1968). Clutch size averaged 10.6 eggs for first nests and 9.6 for renests ($n = 15$). The interval between the loss of the first nest and the first renest was highly variable, but it averaged 12.5 days for 15 females. All but a single renest were initiated when the first nests were lost after 1–3 weeks of incubation. Most (65%) renests were within 91 m of the first nests, and 24% were within 91–182 m. Yetter et al. (2009) evaluated renesting by 77 radiomarked females during a 6-year study (1998–2003) in westcentral Illinois. The percentage of females renesting each year ranged between 50.0% and 85.7%, with adults more likely to renest than yearlings (75% vs. 48%). Clutch size averaged 9.4 eggs for 66 first nests, 8.8 for 19 second nests, and 10.0 for 3 third nests.

Renesting is more likely under favorable habitat conditions. In their study of Mallard population ecology, Pospahala et al. (1974) concluded that late nesting and renesting were greatest when water conditions remained favorable into July. In North Dakota, Krapu (1981) noted that the lipid content in the reserves of females about to renest averaged 74% less than the reserves of females initiating first nests, indicating that females do not store substantial amounts of lipids before renesting, but acquire what is necessary from their diet during the time the later clutch is being produced.

A propensity to renest was also seen during a 6-year study (1976–81) of captive-reared Mallards in North Dakota (Swanson et al. 1986). The birds received unlimited food, and their clutches were removed on about the second day of incubation. All 4 yearling females renested, with 2 of them producing 3 clutches. All 8 of the 2-year-old females laid 3 clutches, and 3 produced 4 clutches. All 8 of the 3-year-old females completed 4 clutches, and 3 produced 5 clutches. Average clutch size was 10.4 eggs for the first clutches, 10.0 for the second, 9.6 for the third, 8.5 for the fourth and fifth. Batt and Prince (1979) also observed a progressive decline in clutch size for captive-reared female Mallards offered unlimited food, with clutch size averaging 10.4 eggs for 151 first nests, 9.5 for 116 second nests, 8.4 for 59 third nests, and 8.2 for 13 fourth nests. The mean renesting interval was 7.1 days (range = 5–10 days), with this interval increasing by 0.18 days for each day of incubation on the lost nest. In contrast, females exposed to limited natural food had renesting intervals ranging from 6 to 24 days, and 42% exceeded the maximum of 10 days observed for females that were fed unlimited food.

In Iowa, researchers experimentally destroyed the nests of 22 marked females between days 10 and 17 of incubation to observe the mating process of females during renesting: 4 females left the study area, 3 remained on the area but did not remate, and 15 (68%) remated (Humburg et al. 1978). Of 11 known rematings, 8 females (73%) returned to their original mates and 3 (27%) changed mates. In another Iowa study, Ohde et al. (1983) found that 9 out of 14 marked females that lost nests subsequently rejoined their original mates; the authors concluded that extra males were not essential for successful renesting.

Continuous laying, which occurs when a female loses a nest during egg laying but continues to produce eggs in a new nest without interruption (renesting interval = 0), is very relevant to a discussion of renesting. In Illinois, continuous laying was reported for only 1 out of 77 (1.3%) females that lost first nests (Yetter et al. 2009). Arnold et al. (2002) documented continuous laying in 278 out of 3,064 (9.1%) radiomarked Mallards nesting in the PPR in Canada. Females laying continuously produced an average of 12.1 eggs (range = 5–18 eggs), compared with 8.9 for other females (range = 4–14 eggs). Continuously laying females averaged 25 g heavier than noncontinuously laying females. Such data support the observations of early biologists that Mallards are persistent renesters, with a large component of the population capable of laying more eggs than are found in a normal-sized clutch.

REARING OF YOUNG

Brood Habitat and Care. Like other dabbling ducks, optimal brood habitat for Mallards consists of wetlands with abundant invertebrate populations that are needed for the growth and development of the ducklings, and emergent vegetation that provides escape cover, although broods will also use open-water areas. Females with broods are also quite flexible in the habitats they occupy; they will use wetlands of all types and sizes, although their choice is clearly influenced by habitat availability. Hence management actions should strive to ensure that landscapes contain a diversity of wetland types. Protection of wetland complexes that include a sizeable amount of seasonal wetlands would provide excellent brood-rearing habitat for Mallards. Permanent and semipermanent wetlands are particularly important, because they provide brood habitat even in dry years, as well as during the later parts of the breeding season.

During a 3-year study (1987–89) in southwestern Manitoba, monitoring of 29 radiomarked females and their broods revealed no consistent trend in the wetland habitats that were used by broods, as each available wetland type was used by at least some broods (Rotella and Ratti 1992a). Mallard broods in eastcentral Saskatchewan spent the most (69%–95%) time on semipermanent wetlands (Dzus and Clark 1997). In the prairie parklands of Canada, Raven et al. (2007) monitored 201 radiomarked female Mallards and their broods during a 4-year study (1993–97) at 15 sites and compared habitat use between the early brood-rearing season (before 30 Jun) and the late brood-rearing season (after 30 Jun). Broods mostly occupied various types of semipermanent and permanent wetlands during both the early season (66.3%) and late season (76.7%), compared with seasonal wetlands (19.5% early and 15.4% late). Ephemeral, temporary, and tillage wetlands were selected even less than seasonal wetlands. Semipermanent wetlands dominated by bulrushes (*Scirpus* spp.) were the preferred (19.7%) habitat type in the early season. In the late season, these wetlands were the second choice (19.9%); permanent wetlands were used the most (22.5%). Early-season broods occupied areas with peripheral vegetation (45.0%), unvegetated areas (22.6%), areas with interspersed vegetation (21.9%), and fully vegetated areas (10.5%). In contrast, late-season use was 45.3% in unvegetated areas, 34.6% in areas with peripheral vegetation, 14.9% in areas with interspersed vegetation, and 5.3% in fully vegetated areas.

An observational study of 569 Mallard broods over 18 years in the PPR in the Dakotas recorded occurrences on semipermanent wetlands (49%), seasonal wetlands (23%), sewage lagoons (14%), and permanent wetlands (11%; Duebbert and Frank 1984). In the PPR in southcentral North Dakota, Talent et al. (1982) radiotagged females to study the habitat use, mobility, and home-range size of 25 broods: 16 during the relatively wet year of 1976 and 9 during the extremely dry year of 1977. In 1976, broods preferred seasonal wetlands dominated by whitetop rivergrass (*Scolochloa festucacea*); these areas received 46% of the use but accounted for only 20% of the available wetlands. Brood occupancy of semipermanent wetlands was less than expected (41% use vs. 47% availability), while semipermanent wetlands were the only habitat used during the dry year of 1977. In this and a subsequent study (Talent et al. 1983), females with broods also selected wetlands with high densities of midge (Chironomidae) larvae, although Dzus and Clark (1997) found that the final wetland occupied by Mallard broods in Saskatchewan did not have more midges than the first wetland they used. Talent et al. (1982) noted that most (89%) overland movements occurred during the first 2 weeks after hatching. Cumulative home-range size averaged 11.0 ha and increased rapidly until the broods were about 1 week old.

Krapu et al. (2006) used radiotelemetry to study habitat use and movements of 69 female Mallards with broods during a major drought period in North Dakota (1988–92), and then in the first 2 years of the subsequent wet period (1993–94). On average, broods occupied fewer wetlands during the dry period than during the wet one (2.2 vs. 3.6). Seasonal wetland use was also lower in the dry period (22.4%) versus the wet period (42.5%).

Semipermanent wetlands were selected more frequently during the dry period (73.2%), compared with the wet one (50.0%). Broods averaged 3.3 hours from nest departure to arrival on a wetland during the dry period, compared with only 1.0 hour during the wet period, and the maximum distance traveled from nest sites was 1.5 times greater during the dry versus the wet period (2,183 m vs. 1,484 m). The predicted number of interwetland moves by broods up to 30 days posthatching was 7.3 during the wet period, compared with only 2.0 during the dry period. Such differences in brood movements between dry and wet periods influenced why duckling survival was higher during the wet period (see Recruitment and Survival). In southwestern Manitoba, Mallard broods ($n = 29$) monitored for 30 days after hatching averaged 4.0 interwetland moves, used an average of 2.9 wetlands, and traveled an average distance of 1,548 m (Rotella and Ratti 1992a). In a mixed-grassland habitat in Saskatchewan, 1 marked female with a brood moved 4.8 km in 1 week, and another traveled 8 km in 9 days (Dzubin and Gollop 1972).

Outside the PPR, 56 radiomarked females with broods up to 30 days old in the San Joaquin Valley of California (1996–97) preferred reverse-cycle wetlands, which are seasonal wetlands dominated by upland vegetation (such as *Hordeum* spp., *Rumex* spp., and *Distichlis* spp.) and flooded from March through August to provide brood habitat. These areas made up only 19.1% of wetland availability but received 37.0% of brood use (Chouinard and Arnold 2007). Semipermanent wetlands formed 47.0% of the wetland habitat and received 34.8% use, and permanent wetlands totaled 19.7% and received 8.7% use. The distance traveled to an initial wetland averaged 662 m. The home-range size of 24 broods averaged 14.6 ha (range = 1.1–76.7 ha). On the Lower Klamath NWR in northeastern California (1989–90), 27 radiomarked female Mallards with broods generally selected seasonally flooded wetlands with a cover component, although permanent wetlands and other more-open habitat types were also chosen (Mauser et al. 1994a). Of these broods, 12 made 22 relocation movements (>1,000 m in 24 hours) during the first week after hatching ($n = 6$) and after the fourth week ($n = 16$): 7 broods made 1 move, 3 made 2 moves, 1 made 3 moves, and 1 made 6 moves. Home-range size (averaged over both years) was 93 ha.

Mallard ducklings usually leave the nest within about 12 hours after the entire clutch hatches (Dzubin and Gollop 1972), and they can survive 4.9–6.3 days without food (Krapu 1979). In North Dakota, the departure times from their nests for 69 radio-marked females with broods varied from 05:35 to about 21:30, and most departures occurred in the morning: 36% by 08:00, 57% by 10:00, and 74% by noon (Krapu et al. 2006). The ducklings follow the female in single file as she leads them to water, during which time she may vocalize as many as 200 times/min (Bjärvall 1968). Females do not always lead their broods to the nearest wetland, however. In the San Joaquin Valley, females did so on only 44% of 50 occasions (Chouinard and Arnold 2007), and in Saskatchewan, only 37% of the time (Dzus and Clark 1997). Females brood their ducklings for up to 2 weeks, as young ducklings cannot maintain their body temperature for an extended period, although they are near to being homeothermic 24 hours after hatching (Koskimies and Lahti 1964, Caldwell 1973, Ryhmer 1988). Among captive-reared, day-old Mallard ducklings of normal weight, many could not maintain homeothermy when subjected to ambient temperatures below 20°C for as short a period as 1 hour (Bluhm 1988). Brooding is especially frequent during cold weather, at night, and during precipitation.

Pehrsson (1979) provided detailed observations on duckling feeding behavior, based on a novel study that followed broods from wild eggs hatched by a domestic Mallard. Newly hatched ducklings peck at dark points and small objects as soon as they leave the nest, but this behavior quickly changes into chasing and capturing insects perched on vegetation, which was the principal feeding behavior observed until the ducklings reached Class IIb. Class I ducklings had the greatest ability to jump from the water's surface to obtain insects spotted on the vegetation, and they could observe potential prey from >1 m away. Young ducklings were also adept at catching flying insects, with a duckling sitting on the ground capable of jumping 20 cm to capture a prey item. Ducklings did not use their eyes to search for food below the water until they reached Class IIa, and they did not tip up until they reached Class IIb. In Minnesota and North Dakota, however, Class I Mallard ducklings fed with their heads submerged up to 36% of the time, and occasionally (0.2%–2.6%) they were observed tipping up (Pietz and Buhl 1999). Dive feeding was relatively rare, but it was not restricted to any particular age group. The lowest feeding rate observed by Pehrsson (1979) occurred in Class Ia ducklings (11.1 efforts/min), and the highest for Class III ducklings (22.2 efforts/min), with the increase due to interruptions about every 4 seconds for breathing by the older ducklings as they strained food below the surface. The ducklings consumed between 1.2 and 12.5 items/min. Experimentally released ducklings obtained more food on fishless lakes, since fish also compete for insects (Pehrsson 1984).

On prairie pothole wetlands in South Dakota, Class I Mallard ducklings mostly ate by surface feeding (78%) and bill-dipping (19%); Class II ducklings by head-ducking (45%) and bill-dipping (41%); and Class III ducklings by head-ducking (66%), bill-dipping (19%), and tipping up (15%; Ringelman and Flake 1980). The mean duration of activity bouts (feeding and/or swimming) averaged 51 minutes for Class I ducklings, 43 minutes for Class II, and 56 minutes for Class III. Swimming decreased and loafing increased as the ducklings grew older.

Females generally remain with the brood, and they will give alarm calls at the approach of danger. The ducklings respond either by seeking cover in dense emergents, clustering around the female, moving to the water's edge and freezing in place, or fleeing to uplands (Beard 1964, Drilling et al. 2002). Ducklings will dive to escape a direct attack. In southwestern Manitoba, data from 20 radiomarked females with broods revealed that they occasionally left their broods for as long as 2 hours (Rotella and Ratti 1992a). In Minnesota and south-central North Dakota, females ($n = 32$) left their broods for an average of >27 minutes, with a range from 2 to >80 minutes (Pietz and Buhl 1999). The length of absences was not correlated with brood age or brood size, but the purpose of these absences is unknown. Ball et al. (1975) noted that radiomarked female Mallards in Minnesota remained with their broods an average of 50.7 days, which is about the time when the young can fly. Several females, however, left their 5- to 6-week-old ducklings during the day but returned to escort them on feeding forays at night.

Brood Amalgamation (Crèches). Brood amalgamation appears to be uncommon in Mallards, but it has not been specifically examined. During a study of Mallard duckling survival in California, however, 12 out of 91 (13.2%) radiomarked ducklings joined other broods (Mauser et al. 1994b). Of 29 ducklings that survived ≥44 days, 6 (20.7%) had joined other broods, leaving their original broods at 2, 18, 18, 19, 22, and 39 days; 5 of these ducklings joined Mallard broods and 1 joined a Northern Pintail brood. Boos et al. (1989) reported 2 instances of Mallard brood amalgamation in Sweden.

Development. Average body mass within 24 hours of hatching was 31.8 g (range = 27.2–40.6 g) for 27 ducklings in Alberta (Nelson 1993), and 32.4 g in North Dakota (Lokemoen et al. 1990b). Ducklings then grow very rapidly. Southwick (1953) listed average weights for captive-reared Mallards raised from artificially incubated wild eggs: 29 g on day 1 ($n = 18$), 66 g at 1 week, 148 g at 2 weeks, 288 g at 3 weeks, 388 g at 4 weeks, 453 g at 5 weeks, and 683 g at 7 weeks. Sample sizes from weeks 1–6 were 6–9 ducklings. Sugden et al. (1981) reported duckling weights from artificially incubated wild eggs averaging 72 g at 1 week, 197 g at 2 weeks, 379 g at 3 weeks, 572 g at 4 weeks, 742 g at 5 weeks, and 871 g at 6 weeks; the maximum growth rate occurred at

2.5 weeks. Energy intake/bird increased from 63.5 kcal/day during week 1 to 334.8 kcal in week 5, but then declined to 248.7 kcal by week 8. Lokemoen et al. (1990b) noted that the maximum growth rate for wild Mallards was between 32 and 38 days.

Gollop and Marshall (1954) assembled current data from that time and reported that Mallard ducklings were down covered for their first 18 days (Class I). The initial body feathers appeared on their sides in 19–25 days (Class IIa), with the ducklings mostly feathered by 26–35 days (Class IIb), when the primaries emerged from their sheaths. The last down (on the nape, the back, or the upper rump) remained until 36–45 days (Class IIb). Ducklings were "feathered flightless" at 46–55 days (Class III), and attained flight at 52–60 days. In captive-reared birds, the first feathering appeared during week 3, when scapulars, flank feathers, and retrices became visible (Southwick 1953). By week 4, the ducklings were fully feathered across the belly but otherwise primarily downy; the secondaries were still in feather sheaths. By week 5, the scapular, flank, and tail feathers were well developed; the secondaries were visible; and the primaries were just emerging from feather sheaths. The face was also well feathered, but down persisted from the nape to the rump. By week 6, the ducklings were fully feathered, and only a small downy patch remained.

In North Dakota, Cox et al. (1998a) reported that the body mass of captive-reared ducklings ($n = 183$) at day 17 was greater for ducklings that were heavier at hatching; the differences averaged 1.7 g at day 17 for each 1.0 g of their weight at hatching. The growth ratio (the percentage of body mass attained by ducklings when last measured, relative to predicted body mass) of wild female Mallard ducklings was also positively correlated with body mass at hatching. Mean body mass at day 17, and the mean growth ratio of ducklings/brood (adjusted for body mass at hatching), were positively correlated with the numbers of aquatic insects consumed, and negatively correlated with changes in the daily minimum air temperature.

Body mass at day 17 grew by 0.29 g for each unit increase in the number of invertebrates, but decreased by 3.53 g with each unit increase in the daily minimum temperature.

RECRUITMENT AND SURVIVAL

Recruitment (the number of females added to the fall population) is a product of several demographic variables: nest success, hen success, brood survival, and average brood size at fledging. During their 4-year study (1977–80) of 338 radiomarked female Mallards in North Dakota, Cowardin et al. (1985) reported an average nest success of 8% and a hen success of 15%. Only 74% of the females that hatched a clutch, however, were later observed with at least 1 surviving duckling. Females in the spring population recruited only 0.27 young females to the fall population, from which a subsequent model predicted a 20% annual decline for the study population; a nest success of 15% and a hen success of 31% would be needed for a stable population. In addition, the April–September survival of females averaged only 80%, because females were killed on their nests, primarily by red foxes.

In a sensitivity analysis developed by Hoekman et al. (2002), nest success explained most (43%) of the population growth of Mallards in the PPR; other factors were the survival of adult females during the breeding season (19%) and the survival of ducklings (14%). In the Great Lakes states, a sensitivity analysis developed by Coluccy et al. (2008) revealed that breeding-season parameters accounted for 63% of the annual Mallard population growth rate, of which duckling survival (32%) and nest success (16%) were the most important. Female survival during the nonbreeding period accounted for the remaining 36% of the variation in annual population growth rates.

Mallards are highly adapted to successfully breed under the variable water conditions that characterize their principal breeding range, the PPR (Krapu et al. 1983). Such attributes include females returning to previously successful nest sites (philopatry), selecting safe nests sites (e.g., islands) where available, pioneering into new areas when nesting in old areas is unsuccessful, occupying a large home range to exploit its food and nesting cover, tolerating crowding when conditions are good (wet years), having the potential to lay large initial clutches (due to their ability to carry nutrient reserves to the breeding grounds), and being able to renest several times when conditions are favorable. Mallards can compensate for the occasional long periods of drought because they are fairly long lived; they can forgo breeding during dry periods and yet live long enough to return when conditions have improved. As Krapu et al. (1983: 696) noted, "these attributes explain the causes of the widespread distribution and abundance of the Mallard in the midcontinent region before intensive agricultural development and the mechanisms whereby this species has occupied and flourished in a wide array of habitats extending across most of temperate North America and Eurasia."

Outside the PPR, Hoekman et al. (2006a) determined breeding demographics of Mallards in southern Ontario during a 4-year study (1997–2000) of 224 marked females. Those remaining on the study areas built an average of 2.1 nests, and almost all (97%) initiated 1 or more nests. Nest survival, however, averaged only 13% (range = 11%–15%), and hen success averaged 37% (range = 25%–46%). Female survival during the breeding period (25 Mar–15 Jul) averaged 75% (range = 65%–84%). Of the 45 females that did not survive during the study, 56% were nesting when they died, and 16% were killed by hay-cutting machinery.

In a broader analysis, Hoekman et al. (2006b) determined population dynamics of Mallards (1992–2000) from data on radiomarked females at 4 sites in Ontario and 1 in New Brunswick. Hen success (except for 1 site) averaged 40%; duckling survival to 30 days averaged 53% for clutches hatched before 1 June, and 40% for clutches hatched after that date. Mean recruitment averaged 89% (range = 79%–98%). These eastern Canadian recruitment rates were more than twice those reported from agricultural environments in Canada's PPR (where the average from 27 sites was only 41%), and they were attributed to higher hen success, resulting from increased nest survival and a greater nesting effort.

In research (1990–92) by Losito et al. (1995) in the nearby St. Lawrence River valley of northern New York, average nest success (17.6%; $n = 67$) was similar to average hen success (18%; $n = 84$), which was very different from the findings in other studies. Hen success in New York was similar to the average reported for North Dakota (15.2%) by Cowardin et al. (1985), but nest success was not (17.6% vs. 8%). Nesting effort/female, however, was lower in New York (1.5 nests/adult and 0.7/yearling) than in North Dakota (2.3/adult and 0.9/yearling). Hence more persistent renesting in North Dakota, despite low initial nest-success rates, resulted in hen success there exceeding nest success. In New York, unlike southern Ontario, nests success was highest (39%) in haylands, probably because this habitat had extensive tracts of land with fewer predators than were found in wetlands, where nest success was only 14%.

During a 6-year study (1998–2003) in westcentral Illinois, nest success averaged 19.6% and hen success averaged 28.3% (Yetter et al. 2009). Female survival during the breeding period averaged 71.0%, but ranged between 54.6% and 100%. Estimated recruitment (female ducklings/female) ranged between 30.2% and 67.2%, with an estimated recruitment rate of 61.3% deemed necessary for a stable population.

Several factors affect brood size and attrition in Mallards, but paramount among them is the amount of water on the landscape during the brood-rearing period. Other elements include the hatching date, the distance from the nest to the nearest water after hatching, and the predator population, especially mink (*Mustela vison*). Most mortality occurs during the first 2 weeks after hatching, and total brood loss is common. During a 3-year study (1987–89) in southwestern Manitoba, the total loss among 69 broods monitored by radiotelemetry averaged 49%, and duckling survival to 30 days averaged 22% (Rotella and Ratti 1992b). Brood survival was highest (70%) during a comparatively wet year and lowest (34%) during the driest year, as were duckling survival percentages (15% vs. 8%). Survival was positively correlated with wetland density (as indexed by shoreline length), but negatively correlated with hatching date. Thus the lowest survival rates for both broods (14%) and ducklings (4%) occurred with late-hatched broods and short shoreline lengths, and the highest rates for broods (72%) and ducklings (40%) were with early-hatched broods and long shorelines. Among the broods that survived, duckling survival averaged 48% and did not differ among years.

During a 2-year study (1996–97) in the prairie parklands of southcentral Saskatchewan (St. Dennis National Wildlife Area and Allan Hills), Gendron and Clark (2002) used radiotelemetry to monitor the survival of 114 Mallard broods. Both study areas contained a high density of wetlands and had cover managed for nesting waterfowl. Only 14 females (12%) experienced total brood loss, with the greatest mortality occurring during the first 14 days, particularly in the first 7 days (25.9%). Duckling survival to 30 days was 59.5%, with an average brood size of 4.3. Both brood-survival and duckling-survival estimates in this study were among the highest ever reported for Mallards, which the authors attributed to the excellent wetland conditions that persisted throughout the study. These conditions probably influenced their findings that survival did not decrease with later hatching dates or with increasing overland travel distances. In a 4-year study (1990–93) in eastcentral Saskatchewan, overall duckling survival did not vary with the total distance traveled, but radiomarked broods surviving an average of 14 days ($n = 26$) traveled a shorter distance to water (211 vs. 310 m) than broods ($n = 9$) experiencing the loss of all ducklings (Dzus and Clark 1997).

Dzubin and Gollop (1972) reported ducklingsurvival rates from a 3-year study (1956–58) in mixed-grassland habitat near Kindersley, Saskatchewan, and 4 years of research (1952–55) in park-lands near Roseneath, Manitoba. In Saskatchewan, an estimated average of only 48% of 197 hatched clutches survived the initial move from nest site to water, which probably reflected the long distances traveled after the clutches hatched: an average of 251 m for early nests ($n = 584$ ducklings) and 321 m for late nests ($n = 108$ ducklings). In contrast, the authors noted that the average distance from nests to water in the Manitoba parklands was just 30 m at hatching. While 95% of all parkland nests were within 119 m of water at hatching, 95% of the early nests in mixed grassland were ≤498 m distant. Of the successful broods in the parklands, clutch size for early nests averaged 8.8 eggs during incubation, compared with 8.0 at hatching (–9%), and 7.2 eggs versus 6.8 (–6%) for late nests. The duckling loss between hatching and downy broods <1 week old was 20% for early broods and 10% for late broods. There was no statistical difference in brood size for early or late Class Ia, II, or III broods, but the decline from Class Ia to Class III averaged 18% for early broods and 13% for late broods, with an overall attrition rate (from incubation clutch size to Class II) of 31% for early broods and 14% for late broods. Of successful broods in the mixed grass-lands, clutch size averaged 9.4 eggs during incu-bation and 8.6 at the hatch (–9%) for early nests, and 7.9 versus 7.7 (–3%) for late nests. Here the loss between hatching and downy broods <1 week old was 27% (early broods) and 30% (late broods), the decline from Class Ia to Class III averaged 18% (early broods) and 13% (late broods), and the over-all attrition rates were 46% (early broods) and 40% (late broods).

During a 2-year study (1976–77) in PPR in North Dakota, 13 out of 25 broods (52%) monitored with radiotelemetry experienced total brood loss, with 11 out of 13 (85%) lost within 2 weeks after hatching (Talent et al. 1983). The mortality rate ($n = 171$ ducklings) was 65%; 68% of the mortality resulted from the loss of entire broods, all of which occurred within wetlands, with mink presumed to be the major predator. Brood size at fledging av-eraged 5.0 ($n = 4$). Few ducklings were lost during overland travel, and cumulative overland travel was not correlated with total brood loss among mobile broods. Of 7 broods moving ≥1.5 km overland, 5 fledged at least 1 duckling, in contrast to 7 broods moving ≤1.5 km, where only 2 survived to fledge 1 or more young. One female, a yearling, moved 5.6 km and passed through 10 different wetlands before her ducklings were 8 days old, but she only lost 1 out of 8 ducklings.

A 7-year study (1988–94) in the PPR (North Da-kota and Minnesota) reported total losses for 16 out of 56 (29%) radiomarked broods (Krapu et al. 2000). Brood loss during the first 30 days was 11.2 times more likely on areas where the percentage of seasonal wetlands holding water was <17%, com-pared with >59% elsewhere. Brood loss was also 5.2 times greater during rainy conditions than dry periods, and it increased by 5% for each 1-day delay in hatching dates between 17 May and 12 August. The authors concluded that early-hatched broods experienced higher survival rates because of the availability of seasonal wetlands, which provided excellent habitat for the aquatic invertebrates pre-ferred by young ducklings. Despite the greater survival of early-hatched broods, the poor nest-ing success of early-nesting species like Mallards nonetheless results in more late-hatched broods appearing on the prairie pothole landscape. Hence there is a distinct tradeoff between nesting early (low nest success but high brood survival) versus nesting late (higher nest success but lower brood survival).

Brood survival for Mallards, as well as for other waterfowl, is strongly affected by seasonal wetlands, not only because these habitats pro-duce abundant aquatic invertebrates, but also because they are avoided by mink, which are the major duckling predator across much of the PPR (Sargeant et al. 1973, Arnold and Fritzell 1990). Mink must have permanent water available, par-ticularly on prairie landscapes, where droughts are common (Sargeant et al. 1993). Krapu et al. (2000) concluded that areas without permanent water have low mink populations and thus higher water-fowl brood survival, as long as seasonal wetlands are abundant. As support for such a contention, a subsequent study by Krapu et al. (2004c) found that permanent water resulting from the construction of a 125 km canal in North Dakota provided optimal habitat for mink, which in turn affected the brood and duckling survival of Mallards and Gadwalls over a broad area of the surrounding landscape. Mink caused at least 65% of the identifiable predation in this area, with the 30-day survival rates of ducklings declining markedly after water filled the canal.

Orthmeyer and Ball (1990) used radiotelemetry to study the survival of 27 Mallard broods (1985–86) on impoundments at the Benton Lake NWR in northcentral Montana, which consists of 2,712 ha of uplands surrounding 2,348 ha of marshes that are divided into 8 impoundments. Nest success was high, usually >60%. Their research is of comparative interest because the habitat conditions for Mallard broods were nearly ideal: short distances from nests to water; rare secondary movement overland; excellent cover-to-water ratios on impoundments; management for invertebrates; and low predator populations, except for 2 species of gulls (*Larus*). Nonetheless, total brood loss occurred among 37% of the 27 broods they monitored, accounting for 60% of all ducklings lost. The 60-day survival rate was 39.5%, with 87% of all duckling losses occurring over the first 18 days. Survival through Class I (46.6%) was lower than through Class II (88.9%), and survival for early-hatched broods was greater than for late-hatched broods (44.2% vs. 33.2%). Overall, only about 40% of the ducklings that hatched survived to fledging, at which time brood size averaged 5.0 ducklings.

Outside the PPR, Mauser et al. (1994b) monitored brood survival for 127 radiomarked ducklings in 64 broods during a 3-year study (1988–90) on the Lower Klamath NWR in California. Total brood loss differed among the years: 81.2% in 1988, 36.8% in 1989, and 37.5% in 1990. Duckling survival in 1988 (18%) was only calculated from 0 to

10 days after hatching, while survival from 0 to 50 days was 37% in 1989 and 34% in 1990. In 1989–90, 93% of 58 duckling mortalities took place during the first 10 days after hatching, with 86% occurring before 6 days. Of the 87 mortalities from 1988 to 1990, 16 (18.4%) happened during movements from the nest to water. Most of the early deaths were caused by long-tailed weasels (*Mustela frenata*) and avian predators. Fledging success did not differ for broods from early-hatched versus late-hatched nests. The low survival rate in 1988 occurred because 70% of the seasonal wetlands were dry by late April, and 2 large permanent marshes were drained. This extensive loss of brood-rearing habitat probably exposed ducklings to high rates of predation. In contrast, refuge management kept all seasonal marshes filled with water though early June in 1989–90, which allowed the broods to disperse and probably reduced losses due to predators.

In westcentral Illinois (1998–2003), brood size averaged 8.2 ducklings at the hatch ($n = 37$) and 3.0 at day 17; 7 females (18.9%) experienced total brood loss (Yetter et al. 2009). Brood survival to 20 days ($n = 32$) averaged 75.9%, while survival of 251 ducklings from 32 broods averaged 41.3% to 20 days of age. Most (63.4%) duckling mortality occurred within the first 5 days of hatching, and 94.4% happened by day 13. Brood survival in forested habitats used by Mallards in northcentral Minnesota was determined over a 5-year period (1968–72) from data on 45 radiomarked females and 265 sightings of unmarked broods (Ball et al. 1975). On average, Mallards successfully raised 44% of the ducklings they produced, with 70% of duckling mortality occurring in the first 2 weeks after hatching.

Hoekman et al. (2004) used radiotelemetry to study the survival of 70 broods during a 4-year study (1997–2000) at 4 sites in southern Ontario. The 30-day duckling survival rate averaged 40%, but it varied from 7% to 50% across sites. The extremely low duckling survival at 1 site was apparently related to poor wetland conditions. Most mortality (77%) occurred within the first 8 days after hatching. The survival probability was 8.8 times higher for ducklings >7 days old, compared with ≤7 days old, and 1.7 times greater for early-hatched (before 1 Jun) versus late-hatched ducklings (1 Jun or later). The female's age did not influence duckling survival. Overall, early nests and the resultant ducklings contributed 61% to recruitment at 30 days.

Survival rates for Mallards have been more thoroughly studied than for any other duck species. Hence, in addition to annual survival estimates of all sex and age classes, survival rates have been partitioned within the annual cycle. Ranges of survival within sex/age groups for various combinations of years and different geographic locations were 62%–68% for adult males, 54%–59% for adult females, 48%–63% for juvenile males, and 46%–61% for juvenile females (Arnold and Clark 1966, Anderson 1975, Smith and Reynolds 1992). In his extensive analysis of 134,000 band recoveries of Mallards banded before the hunting season (mostly in 1961–70), Anderson (1975) reported survival rates of 62% for adult males, 54% for adult females, 48% for juvenile males, and 46% for juvenile females. In general, these rates did not vary among years or banding regions, although such relationships were probably masked by large sampling variances or a close relationship between survival and harvest rates. Nonetheless, Mallards banded in central breeding areas (the PPR) tended to have a higher survival rate than Mallards banded in eastern breeding areas: 68.2% vs. 62.5% for adult males, 57.9% vs. 54.6% for adult females, 55.0% vs. 49.2% for juvenile males, and 54.5% vs. 48.0% for juvenile females. Anderson concluded that regional differences in survival probably existed, even though they were not detected statistically. Average annual survival rates for Mallards banded in southern Ontario and the Central Valley of California tended to be low, while the opposite was true for Mallards banded in eastern South Dakota and southeastern Saskatchewan. An analysis of later band-recovery data (1978–96) yielded annual survival rates of 67% for adult males, 63% for immature males, 57% for adult females, and 50% for immature females (Frank Bellrose), which is within the ranges of other studies. The overall survival rate for all sex and age classes combined was 59%. Survival was highest (62%) for Mallards banded in the PPR, compared with 51% for those banded east of the pothole region, and 54% for the Great Lakes Region. Survival of Mallards banded in the western states and provinces averaged 58%.

Hestbeck (1990) reported that survival rates of adult Mallards from the midcontinental population (1962–70, 1971–78, 1979–84) were higher for those banded in the north versus those banded in the south. The mean percentage change in survival/degree of latitude was 0.494 for adult males and 0.440 for adult females. These parameters were 0.484 for juvenile males and 0.414 for juvenile females. Hestbeck found that the effects of banding latitude and time period on survival were statistically significant for adults but not for juveniles. The author believed this disparity was due to the high variability of the juvenile survival estimates, not to a process that acted differentially on adults. The cause of the link between survival and latitude was unknown, but it may be related to a similar north–south gradient in predation, to habitat degradation, to land use, and/or to agricultural practices (see Conservation and Management).

Unfortunately, little is known about the influence of age on annual survival in Mallards beyond the comparison of immature (<1 year old) and adult birds. Dufour and Clark (2002), however, provided important data on annual survival of yearling versus adult female Mallards, based on their 16-year study (1983–97) of breeding Mallards in southcentral Saskatchewan, where 145 yearlings and 145 adults were marked and released during this period. The 385-ha study area averaged 63 wetlands (May ponds), but ranged from a low of 12 in 1991 to a high of 101 in 1997. Yearling females had higher annual survival rates than adult females over the study period (58% vs. 47%), and the differences were most pronounced in years with fewer wetlands, which reflected a smaller breeding effort by yearlings and, therefore, a lower exposure to predation during nesting.

As for survival during various times of the year, the largest and most geographically diverse dataset on breeding-season survival is that of Devries et al. (2003), who reported mortality estimates from a 6-year study (1993–98) that radiotracked 2,249 breeding female Mallards at 19 different sites in the aspen parklands and mixed-grass prairie eco-regions across Alberta, Saskatchewan, and Manitoba in the PPR in Canada. Female survival averaged 76% (range = 62%–84%) during the 90-day breeding season, and was lowest when the females were nesting. The probability of survival was higher for yearling females (77.0%) than for adult females (75.7%), but the difference was slight. Survival also decreased with increasing degrees of longitude and lower percentages of wetland habitat. In other words, the results indicated that geographic location and the amount of wetland habitat interact, since female survival was markedly less in the western portions of the Canadian PPR as the percentage of wetland habitat declined. These results probably also reflected differences in the predator communities and their interactions with wetland habitats. Predator indices in the western portion of the parklands ecoregion were higher for coyotes and Red-tailed Hawks (*Buteo jamaicensis*) and lower for red foxes and mink. Hence predators other than red foxes and mink, which predominate elsewhere, are significant sources of mortality for waterfowl nesting in this ecoregion.

Another important study on breeding-season survival of Mallards is that of Brasher et al. (2006), because it is the only investigation to simultaneously evaluate breeding survival of both males and females. The authors used radiotelemetry to monitor 90 males and 272 females during 2 breeding seasons in the prairie parklands of Canada (1998 in Manitoba and 1999 in Saskatchewan). Average female survival was 84% in 1998 and 71% in 1999. In contrast, the survival of paired males was 99% in 1998 and 98% in 1999. Such results validate the long-held contention that management efforts should focus on increasing female survival rates, as the survival of males during the breeding season has little effect on the population growth of Mallards. Other breeding-season survival estimates for female Mallards in the PPR are 72% in North Dakota, 1963–73 (Johnson and Sargeant 1977); 81% in North Dakota, 1977–80 (Cowardin et al. 1985); 60% in the prairie parklands of Canada, 1981–85 (Blohm et al. 1987); and 57% for adults and 73% for yearlings in the prairie parklands of Canada, 1981–85 (Reynolds et al. 1995).

Unexpectedly, the daily female-survival rate (1990–92) during the breeding season in the St. Lawrence River valley in New York (Losito et al. 1995) was lower than that recorded by Cowardin et al. (1985) in North Dakota (1977–80), yet nest success was much higher. A possible explanation for these differences was that more females in New York were killed by avian predators, which did not concurrently affect nest survival. In North Dakota, red foxes were the primary predators for both nests and nesting females; hence low nest success corresponded with low female survival. Avian predation in that state was deemed of "limited importance" (Cowardin et al. 1985:27). In contrast, raptors killed female Mallards in New York when they were away from their nest sites, as well as during prenesting and nonnesting periods. Unlike prairie grasslands, most wetlands in the St. Lawrence River valley are forested or surrounded by forests, and this greater structural complexity probably made forested wetlands more attractive hunting areas for raptors. Such differences in habitat structure and the resultant demographic parameters illustrate why studies across the distribution range of Mallards are needed, so as to understand possible courses of management actions.

During a 6-year study (1968–74) in northern Minnesota, Kirby and Cowardin (1986) evaluated the survival of 109 radiomarked adult females within the 169-day breeding and postbreeding period. Overall survival for this period was 71.4%: 100% during nest initiation (36 days), 79.8% during incubation (28 days), 94.3% of those remaining alive after incubation during brood rearing (51 days), 94.7% of that reduced group during the molt (26 days), and 100% during premigration (28 days). Fall/winter survival was estimated at 88.1% for adults and 63.8% for young.

Several studies examined Mallard survival outside of the breeding season. Reinecke et al. (1987) estimated the winter survival of 223 radiomarked female Mallards during a 6-year period (1980–85) in the lower Mississippi River valley, which averaged 85% for adults and 81% for juveniles during the 50-day hunting season and 99% for adults and 86% for juveniles during a 20-day posthunting period. The mortality rate due to harvest for all females combined was 5.4 times their natural mortality rate during the hunting season and 2.4 times their natural rate for the full duration of winter. Only 10 (4.5%) nonhunting mortalities occurred during the study.

Hepp et al. (1986) found a relationship between the condition index of Mallards banded in the MAV during autumn (1 Oct–15 Dec) in 1981–83 and the probability of recovery during the hunting season. Of 5,610 banded birds, 234 were recovered the following winter; those in poorer condition at the time of banding were more likely to be shot during the hunting season. The relationship was true for all sex and age classes, but applied most strongly to adult males. The authors speculated that the lower survival of individuals in poor condition was related to their use of habitats where they were more vulnerable to hunting, competition with birds in better physical condition, and their greater degree of movement as they searched for "better" habitats, all of which would increase their probability of encountering hunters. In contrast, Jeske et al. (1994) did not detect a relationship between body condition and winter survival for Mallards in San Luis Valley of Colorado, but there were mitigating circumstances. Outbreaks of avian cholera are common during winter in the San Luis Valley, but mortality from avian cholera is unrelated to body condition, so cholera deaths may

have masked relationships between other causes of mortality and body condition.

Female survival during the winter in the Playa Lakes Region of Texas was also influenced by body condition, determined over 3 winters (1986/87–1988/89) by monitoring 153 radiomarked females (Bergan and Smith 1993). The mean survival rate for the 100-day period (21 Nov–1 Mar) was 77.7%, with no age or year differences, even though precipitation and food availability varied among winters. The mortality rate during the hunting season was only 1.8%, compared with a natural-mortality rate of 21%. Survivorship was correlated with winter body condition (body mass / wing length), with adults in relatively poor condition having lower survival than adults in better condition (65.7% vs. 90.7%). Similarly, the survival of immatures with a low condition index was lower than those with a high condition index (66.6% vs. 86.9%).

In eastcentral Arkansas, there were no mortalities among 92 females radiomarked after the 1988 and 1989 hunting seasons, despite monitoring during 2,510 exposure days and observing differences in habitat conditions between years (Dugger et al. 1994). The lack of evidence for winter habitat conditions affecting survival may reflect a behavior pattern by Mallards, which will leave the MAV in response to poor habitat conditions (Nichols et al. 1983). The results of this study indicated that hunting deaths and habitat conditions during the hunting season, rather than other sources of mortality and winter habitat conditions after the hunting season, were the major determinants of winter mortality among female Mallards in Arkansas.

FOOD HABITS AND FEEDING ECOLOGY
Mallards are opportunistic omnivores that are highly adaptable in their use of both natural and agricultural foods. Although some studies have indicated food preferences, the plethora of available studies of Mallard food habits demonstrate that they exploit whatever foods are available within the constraints of seasonal demands. Mallards are also highly opportunistic in their use of

foraging habitats, feeding in virtually every type of wetland, although they especially use moist-soil areas, shallow (<60 cm deep) marshes and ponds, dense emergent vegetation, rivers, lakes, and flooded timber. Mallards also forage on many cultivated crops, such as wheat, barley, rice, corn, and soybeans. Their diet is generally dominated by plant material, except during the breeding season, when females extensively consume animal matter to meet the demands of egg production. Their principal foods are the seeds and vegetative parts of aquatic plants, acorns (from *Quercus* spp.), and aquatic macroinvertebrates, but they will also opportunistically prey on tadpoles, frogs, small fish, and fish eggs. During the winter in the area below Garrison Dam in North Dakota (1996–99), the diets of Mallards were primarily composed of rainbow smelt (*Osmerus mordax*), especially during years with persistent snow and ice cover (Olsen et al. 2011).

The lamellae of Mallards are fairly widely spaced, averaging 7.96/cm ($n = 161$), compared with 10.06–13.28/cm for 5 other species of dabbling ducks and 21.48/cm for Northern Shovelers (Nudds and Bowlby 1984). Among the 7 species of dabbling ducks studied by these authors, a prey-size selectivity index, based on invertebrate foods, was 95.7 for Mallards, versus 334.1 for the most specialized species (American Green-winged Teal), and 30.0 for the most generalized species (Gadwalls).

Mallards mainly feed by tipping up and dabbling at the surface, but their foraging methods are adaptable to the conditions they encounter; diving is rare but reported. During the winter on tidal mudflats in the Atchafalaya River delta of Louisiana, Mallards fed by dabbling 47% of the time and head-dipping 47% of the time; tipping up was rare (Johnson and Rohwer 2000). The mean water depth at foraging sites was <5 cm, however, which would prohibit tipping up. When feeding on corn, Mallards use the nail on the tip of their bills to grasp the cobs and pull away kernels (Goodman and Fisher 1962).

Breeding. Swanson et al. (1979) demonstrated the importance of macroinvertebrates in the diets of female dabbling ducks during the egg-laying period, with such foods forming 70%–99% of the diet in 5 species collected in southcentral North Dakota (70% in Mallards). A subsequent paper reporting the results from 177 Mallards collected during the breeding season in the same region of North Dakota revealed similar amounts (aggregate percentage) of animal foods in the diets of paired males (38%, $n = 39$) and nonlaying females (37%, $n = 41$), but 72% in the diet of laying females ($n = 37$; Swanson et al. 1985). Important foods for laying females were insects (27.1%), gastropods (snails; 16.4%), crustaceans (12.9%), worms (Oligochaeta; 11.8%), and seeds (24.8%).

The consumption of invertebrates reflected hydrological and phenological events that occurred during the breeding season and influenced food availability (Swanson et al. 1979). Upon arrival, the primary foods available to Mallards were waste grains (corn and wheat) and residual seeds from aquatic plants such as millet (*Echinochloa crusgalli*), the latter made available by snowmelt that filled shallow wetlands but only persisted for a few days. As the ground thawed, these wetlands then dried up, but warm spring rains stimulated terrestrial worms to migrate to the surface, where they were readily available to females; worms constituted 34.9% of the females' diet in April, compared with <1% in May. Crustaceans in the diet were dominated by fairy shrimp (Anostraca), which were the first aquatic invertebrates to mature in recently flooded wetlands in the spring; they were then replaced by later-maturing clam shrimp (Conchostraca), insects, and mollusks. Snails made up <0.1% of the April diet, but 24.9% in May. As summer progressed, submerged plants appeared in the open-water areas of more permanent wetlands and provided habitat for mollusks, midge larvae, and amphipods, which were then eaten by breeding females. Such data underscore the importance of protecting complexes of wetlands in the PPR to meet the nutritional demands of Mallards and other ducks during the breeding season.

As in other ducks, young Mallard ducklings predominantly feed on aquatic invertebrates, which provide the necessary protein they need for rapid growth and development. At the Bear River Migratory Bird Refuge in Utah, the percentage of invertebrates in the diet of ducklings was 97% for Class Ia, 90% for Class Ib, and 75% for Class Ic (Chura 1961). By Class IIa, however, the dietary percentages were about 50% invertebrates and 50% plant material, mostly seeds. Adult and larval midges, water boatmen (Corixidae), and adult water scavenger beetles (Hydrophilidae) were important invertebrate foods. Terrestrial forms of aquatic invertebrates, especially adult midges, dominated (59%–87%) the diet of Class I ducklings, but these were replaced by more aquatic forms as the ducklings matured and began tipping underwater to forage. In southwestern Manitoba, ducklings consumed 91% animal matter and 9% plant matter over the brood-rearing period; animal material was especially predominant (97%) during the first 6 days (Perret 1962). The chief animal foods for the ducklings included midges, water boatmen, water scavenger beetles, and mayflies (Ephemeroptera).

Migration and Winter. The nutritional requirements of Mallards during the postbreeding and wintering periods largely focus on the acquisition of plant foods, which are rich in carbohydrates. Carbohydrates facilitate the accumulation of both premigration lipid reserves and lipid reserves needed during the winter; they also provide ready energy to meet daily metabolic demands. An examination of the esophagus and proventriculus of postbreeding Mallards collected from 54 wetlands in the aspen parklands of Saskatchewan found that noncultivated plant foods formed 69% of their diet in August ($n = 51$) and 81% in September ($n = 39$; Sugden and Driver 1980). Various nutlets and the seeds of several plant species made up 39% of the foods from the total sample ($n = 90$), while duckweeds (*Lemna trisulca* and *Lemna minor*)

accounted for 21%. Insects dominated the animal matter, with midge (Chironomidae) larvae being the most important.

Postbreeding Mallards also make extensive use of agricultural grains. On their northern breeding grounds they especially feed on wheat, barley, and oats, which they exploit via field-feeding flights from the wetlands where they roost. Mallards often cause considerable damage (depredation) to these crops, because they are heavily consumed after they are cut and swathed (allowed to dry in the fields) but not yet gathered up for harvest. In southeastern Manitoba, field-feeding Mallards in the Whitewater Lake district preferred durum wheat and barley, after these crops were cut and swathed (Bossenmaier and Marshall 1958). Durum wheat was the most favored, because ducks could obtain the highly palatable kernels very easily, while barley was of medium palatability. Common wheat was their third choice, but ranking well below durum wheat and barley. Fall migrants moving farther south primarily selected waste corn, while Mallards on the southern wintering areas heavily used rice and soybeans. At all locations, however, Mallards ate a combination of naturally occurring wetland foods and agricultural grains, albeit in proportions that varied with locations and conditions.

During the fall along the Illinois and Mississippi Rivers in Illinois, 97.7% of the volume of foods in 2,825 gizzards collected from hunters during 1938–40 consisted of plant material (Anderson 1959): primarily corn (47.4%), as well as wetland plants, such as rice cutgrass (*Leersia oryzoides*; 12.8%), coontail (*Ceratophyllum demersum*; 7.7%), and various species of smartweeds (*Polygonum*; 7.2%). During the fall in 1979–81, plant foods made up 99.4% of the Mallard diet, determined from 4,308 gizzards collected in the upper Illinois River region (Havera 1999). Corn was again the most frequently chosen item (48.0%), followed by rice cutgrass (5.5%) and an array of other moist-soil plants, such as Japanese millet (*Echinochloa frumentacea*), the tubers of chufa flatsedge (*Cyperus esculentus*), and

Pennsylvania smartweed (*Polygonum pensylvanicum*). In the Mississippi/Illinois River confluence area, 1,970 Mallard gizzards examined during the fall (1979–81) contained 99.6% plant material, although corn formed a much smaller (11.1%) part of their diet (Havera 1999). Instead, they ate larger percentages of rice cutgrass (14.4%), Pennsylvania smartweed (12.3%), curlytop ladysthumb (*Polygonum lapathifolium*; 12.2%), and an array of other moist-soil plants.

To the south, Gruenhagen and Fredrickson (1990) noted that 71 female Mallards collected during the fall on moist-soil impoundments at the Squaw Creek NWR in northwestern Missouri primarily consumed wild millet (*Echinochloa* spp.; 68.7%), along with nodding smartweed (*Polygonum lapathifolium*; 15.4%). In contrast, the tubers (2.9%) and seeds (12.8%) of arrowhead (*Sagittaria latifolia*) were the major foods for 20 females collected from arrowhead-dominated habitats, while rice cutgrass (38.9%) and millet (31.8%) formed most of the diet for 32 females collected from mixed moist-soil/arrowhead habitats. Mallards also fed in nearby agricultural fields. The diet of male Mallards (*n* = 156) collected during 3 winters in southeastern Missouri was dominated by plant foods, but the composition varied, based on the habitats where the specimens were collected and annual variations in the production of mast. Acorns from pin oak (*Quercus palustris*) and cherrybark oak (*Quercus falcata*) made up 64.5% of their diet in 1 winter but were not eaten during the other 2, when seeds from moist-soil plants predominated (46.0%–61.4%).

During November and December near the Santee NWR in South Carolina, the volume of food in the gizzards of 130 Mallards collected from hunters primarily (96.2%) contained plant foods (McGilvrey 1966a). Preferred items (by percentage of volume) were southern rice cutgrass (*Leersia hexandra*; 19.8%), hydrochloa (*Hydrochloa caroliniensis*; 12.1%), the seeds of sweetgum (*Liquidambar styraciflua*; 9.9%), and the seeds of buttonbush (*Cephalanthus occidentalis*; 7.1%). Agricultural foods were corn (6.8%) and oats (3.6%). By percentage of occurrence, Mallards most frequently ate buttonbush (61.5%) and swamp smartweed (*Polygonum hydropiperoides*; 60.8%). On tidal impoundments in South Carolina, plant foods dominated (91.6% by volume) the stomach contents of 168 Mallards collected from hunters (Landers et al. 1976). Their primary food was the seeds of dotted smartweed (*Polygonum punctatum*; 47.1%), which occurred in 61.3% of the contents in the sample. Other frequently (26.2%–32.1%) eaten plant species were the rhizomes and seeds of redroot sedges (*Lachnanthes caroliniana*), and the seeds of fall panicgrass (*Panicum dichotomiflorum*), saltmarsh bulrush (*Scirpus robustus*), flatsedge (*Cyperus odoratus*), Olney's bulrush (*Scirpus olneyi*), wild millet (*Echinochloa walteri*), and spikerushes (*Eleocharis* spp.) Corn formed only 0.7% of the total dietary volume.

On their principal wintering grounds in the MAV, bottomland hardwood forests were once the ancestral winter home of Mallards, but only about 20% of an original 10 million ha remain today (Baldassarre and Bolen 2006). Much of the original forest habitat has been converted to agricultural production, but the remaining bottomland forests, when flooded, are capable of satisfying virtually all of the food, cover, and water requirements of wintering Mallards. During winter in southeastern Arkansas (1990–91), the diet of Mallards (*n* = 52) collected while foraging in deliberately flooded bottomland hardwood forests (green tree reservoirs) consisted entirely of noncultivated plant and animal foods (Dabbert and Martin 2000). The diversity of taxa in their diet was impressive, with 17 species and 21 families/orders represented. No single taxon made up >11.1% of the total food volume, with the 3 most used items being isopods (Isopoda; 11.1%), dipterans (Diptera; 10.3%), and narrowleaf forestiera (*Forestiera angustifolia*; 9.5%). Their research demonstrated that Mallards will consume large quantities of noncultivated foods when they are made available by flooding, even though agricultural foods were within 1.6 km of the study areas and were probably also regularly eaten by Mallards.

An earlier 2-year study (1957/58 and 1958/59) in central Arkansas noted the profound effect of winter rainfall and subsequent flooding on the use of agricultural versus bottomland hardwood habitats (Wright 1961). The gullet contents (by percentage of volume) of Mallards ($n = 583$) harvested by hunters (Nov–Jan) averaged 35.5% acorns during the wet winter of 1957/58, compared with only 12.6% in 1958/59. Rice averaged 45.3% in 1957/58 and 49.5% in 1958/59. Wright noted that Mallards immediately increased their consumption of acorns when the weather changed from dry to wet. The next most frequently (11.1%) eaten foods were barnyardgrass (*Echinochloa crus-galli*) and junglerice (*Echinochloa colonum*), which were associated with agricultural fields; soybeans averaged only 6.1%.

During winter (Dec–Jan) in the MAV of west-central Mississippi, the esophageal contents of 311 hunter-killed Mallards (219 males, 92 females) examined over a 5-year period (1979–83) averaged 98.9% plant foods (by percentage of dry weight), of which 82.9% were agricultural grains (Delnicki and Reinecke 1986). Preferred vegetative materials were soybeans (41.6%), rice (41.3%), and the seeds of moist-soil plants (16.0%). The use of agricultural foods differed among years, however, with rice averaging 44.6% in 1979/80, 8.3% in 1980/81, and 63.4% in 1981/82. Soybean consumption varied inversely with that of rice. Which agricultural foods were eaten depended on water conditions, because Mallards cannot obtain rice unless they can filter-feed in shallowly flooded rice fields. Hence, during dry winters, they shifted their food choice to soybeans. The Mallards examined during this study had not eaten acorns. Hunters in the MAV commonly harvest Mallards early in the morning, when the birds arrive in bottomland hardwood forests after feeding and roosting in agricultural fields at night, so the harvested Mallards may have had little time to feed on acorns and aquatic invertebrates (R. M. Kaminski). Sheeley and Smith

(1989) also noted that hunter-killed birds analyzed for their food habits were more likely to have fed on agricultural grains.

Collectively, these studies reflect the importance of winter flooding on food use by Mallards in the MAV, which was well documented in southeastern Missouri, where Heitmeyer (2006), during winter 1982 (Dec–Feb), examined the influence of flood events on the subsequent feeding ecology and behavior of female Mallards wintering in the Mingo basin. Corn is widely grown in the area, but rice is not. Flood events clearly benefited Mallards, which redistributed into shallowly flooded (<50 cm) bottomland hardwood forests dominated by red oak species such as willow oak (*Quercus phellos*), cherrybark oak, and pin oak. During February, the Mallards also increased their foraging time (from 5.8% prior to flooding to 52.9% during flooding) and food consumption (from 30.3 g/day to 78.6 g/day). The use of wetland-associated invertebrates also changed markedly in response to flooding, from 2.9% before to 42.3% during flooding. Acorn consumption was 52.7% before and 43.3% during flooding, while the use of corn declined from 11.1% before to 0% during. Based on these data, and hydrological data for 14 rivers in the MAV from 1939 to 1940, Heitmeyer concluded that Mallards were adapted to the timing, duration, and extent of winter flooding in the MAV, and protection of the hydrological patterns associated with floodplain habitats (such as bottomland hardwood forests) was critical for Mallard populations.

Heitmeyer (1988a) also documented changes in the body composition of female Mallards wintering in the Mingo basin that were correlated with annual-cycle events and their use of food resources during 3 wintering periods: the comparatively dry winters of 1980/81 and 1981/82, and a wet winter in 1982/83. Females completed important annual-cycle events during winter in the Mingo basin: the prealternate molt in the fall, and the prebasic molt in the winter and early spring. Adults proceeded through these events sooner than immatures, but both adults and immatures completed molting events earlier during the wet winter than the 2 drier winters. Molting females were especially dependent on the protein-rich invertebrates available in bottomland hardwood habitats to meet the demands of the prebasic molt.

The importance of bottomland hardwood forests is further underscored by the fact that changes in agricultural practices can dramatically affect the availability of such foods for waterfowl. A trend in the MAV is toward planting early-maturing varieties of rice that are harvested in August and September, which makes 79%–99% of the waste grain unavailable for wintering waterfowl, because it deteriorates or germinates by early winter (Manley et al. 2004). A subsequent study concluded that overlooking such losses may have caused prior estimates of carrying capacity for wintering waterfowl in the MAV to be overestimated by 52%–83% (Stafford et al. 2006). Increases in the acreage of soybeans in the MAV also affects wintering Mallards, because waste soybeans deteriorate rapidly once they are flooded, to the point of nearly complete deterioration in about 90 days (Nelms and Twedt 1996). More importantly, soybeans are a nutritionally poor food for Mallards, as documented by Loesch and Kaminski (1989). In their study, body weights of captive-reared Mallards that were fed solely on diets of 4 agricultural grains were lowest for the group that were given soybeans. Of 8 females that died during the experiment, 5 were in the soybean group.

Outside the MAV, Jorde et al. (1983) studied the feeding ecology of Mallards wintering along the Platte River in Nebraska (1978–80), an area characterized by more rigorous winter conditions than those encountered by Mallards on more traditional wintering areas to the south. Plant foods formed most (96.8%) of their diet, which included 51.7% waste corn, and 12.6% common duckweed (*Lemna minor*). Mallards primarily exploited grazed corn stubble and cattle feedlots, particularly during periods of snow cover, when the corn fed to cattle was also available to Mallards. The ducks usually made a single field-feeding flight/day, usually during mid- to late afternoon. During the severe winter of 1979, however, the distance of feeding flights increased from 3.2 to 20 km, as Mallards sought areas where grazed stubble was more abundant. Once in the fields, Mallards spent 78% of the time feeding during the severe winter of 1979, compared with 52% in the milder 1980 winter. Time spent in the fields ranged between 39 and 217 minutes, with Mallards flying to the fields earlier and spending more time there foraging as the winter progressed.

Mallards wintering in the Playa Lakes Region of Texas also used waste corn as their primary winter food source, which they and 3 other species of puddle ducks exploited via field-feeding flights from the playas (Baldassarre and Bolen 1984). Unlike their pattern in Nebraska, however, ducks in Texas mostly undertook 2 field-feeding flights/day. A morning flight was initiated an average of 52 minutes before sunrise and lasted 23 minutes, and an evening flight began 25 minutes after sunset and lasted 37 minutes. Given a choice among fields, field-feeding Mallards and other ducks apparently minimized their foraging time by selecting fields where waste corn was most readily available; burned fields were preferred, followed by disked fields, especially those containing >60 kg waste corn/ha. In contrast, if only freshly harvested fields were available, field-feeding waterfowl consistently choose fields where waste corn was most abundant, which again would minimize their foraging time.

Mallards wintering in the Playa Lakes Region were in better condition than Mallards wintering farther north, most likely because winter weather was not as severe and thus waste corn was plentiful. Mallards examined during a 3-year study (Oct–March each year, 1979–82) by Whyte et al. (1986) arrived in the Playa Lakes Region in the autumn, when their lipid reserves, and usually their body weights, were lowest for all sex and age classes. By midwinter, however, lipid reserves had increased 40% in adult males ($n = 43$), 35% in adult females ($n = 27$), 14% in juvenile males ($n = 12$), and 31% in juvenile females ($n = 27$). Although corn is high in carbohydrates, and thus facilitates the accumulation of a needed lipid reserve, it is low in both protein content and quality (Baldassarre et al. 1983), which required Mallards and other field-feeding puddle ducks (e.g., American Green-winged Teal) to also forage on noncultivated foods in the playa lakes to supplement an otherwise nutritionally incomplete diet (Quinlan and Baldassarre 1984).

MOLTS AND PLUMAGES

Molts and plumages are well studied in Mallards. Thorough reviews can be found in Southwick (1953), Gollop and Marshall (1954), Palmer (1976a), Bellrose (1980), Young and Boag (1981), Heitmeyer (1987), and Drilling et al. (2002).

During the summer and early fall, juvenile Mallards acquire 3 plumages in succession (juvenal, first basic, and first alternate); hence feathering associated with at least 2 consecutive plumages can occur on the same individual. The acquisition of juvenal plumage begins at 18–21 days after hatching. Most feathering is complete by 26–35 days, and flight is attained in 52–60 days; sometimes this takes up to 70 days, but it can be as short as 40–43 days in Alaska. Young Mallards in juvenal plumage resemble adult females, being brownish overall, with white underwings, but subtle differences can identify males and females as the ducklings age. The notched juvenal tail feathers are replaced any time after about the start of September, depending on their hatching dates.

Once the juvenal plumage is fully developed, the next feathering is first basic, which individuals start to acquire in late summer, although early-hatched young can begin the associated molt in late July. The plumage of both males and females is similar to adult females at this time. First basic, however, is worn only briefly and involves the replacement of limited feathering on the head, the neck, the breast, the mantle, and the scapular region. There is some debate, however, as to whether first basic even exists as a distinct plumage in Mallards and other puddle ducks. Pyle (2005) noted that the prebasic molt in 19 specimens of *Anas* that he exam-

ined only involved 8%–19% of the head, the neck, the upper back, the breast, and the sides, while 22 other specimens exhibited no evidence of molting at all, but instead retained their juvenal plumage. The confusion as to whether first basic exists in puddle ducks (as well as other ducks) stems from differentiating between the actual molt and pigment deposition. During the virtually continuous molting that occurs during the summer, feathers renewed early in the juvenile period (Jul–Aug) appear to be cryptic (concealing), while those renewed later are more brightly colored, yet *all* can be part of the same molt, as a given feather follicle is only activated once during this time period. Coloration appears to be controlled by several factors, including the sex hormones, which could explain why coloring in feathers occurs later in the molting process. Pyle (2005) concluded that variations in the deposition of pigment over the protracted period of the summer molt among juvenile ducks could easily lead to the conclusion that 2 molts occurred.

Regardless of the debate over first basic, both sexes begin to acquire first alternate when they are about 65–75 days old, and they continue to acquire that plumage through the fall. All feathering is replaced except the remiges, although the juvenal wing is retained. Immature males in first alternate are difficult to distinguish from mature males in definitive alternate, with the same being true for immature females. In the hand, both can be separated from adults by the retention of identifiable juvenal wing feathers and other wing variants, such as feather vermiculations. Immature males retain first alternate into the breeding season, after which they molt into second basic, probably at about the same time as adult males. Males appear female-like at this time. Second basic is followed by second alternate, which is acquired from late summer into early fall and retained until the following breeding season. Immature females, however, acquire second basic during late winter and spring; all body feathering is replaced, but the juvenal wing is still retained. Second basic is darker and browner than first alternate but otherwise similar to definitive basic. Females begin to acquire second alternate in late summer and complete the associated molt by November.

Definitive basic in males is acquired from mid- to late summer and is a complete molt involving all body feathering and the remiges; hence the birds undergo a flightless period of about 4 weeks. The body molt is largely finished before the remiges are molted and the flightless period begins. The body plumage is only retained for a few weeks before being replaced by definitive alternate. The new remiges, however, are retained until the following summer. Females acquire definitive basic in late winter and early spring.

Males acquire definitive alternate, which involves the replacement of most body feathering and some wing feathering, in the summer and fall. The prealternate molt overlaps the flightless period associated with definitive basic, as the prealternate molt of body feathering begins while the remiges are renewed. The last feathers to be acquired in definitive alternate are the recurved central tail feathers and the tertials and tertial coverts. Females also usually begin to acquire definitive alternate during the flightless period and complete the molt in autumn. The resultant plumage is worn until late winter or early spring.

Heitmeyer (1987, 1988b) provided a thorough description of the prebasic molt in female Mallards wintering in Mingo basin in the MAV in Missouri, in which he documented the costs and associated strategies for completing this important life-history event. Females replaced all but the remiges during the prebasic molt, which amounted to 53.7 g of new plumage, about 46 g of which was protein. Females would need to ingest 84 g of protein (assuming a 55% conversion efficiency) to produce the new feathering. Protein costs were highest during the middle stages of the molt, when the largest percentage of the plumage was being replaced; protein costs for molting were otherwise spread out over 46 days. Hence a key adaptation for meeting the high protein demand associated with the prebasic

molt is to extend the cost over many days. Habitat conditions presumably would dictate just how long it would take to complete this molt, with the shortest period occurring under the best of habitat conditions (Richardson and Kaminski 1992). Females met the energetic demands of the prebasic molt by accumulating relatively large (168 g) lipid reserves, which declined in a curvilinear fashion through the middle of the prebasic molt, but then increased to an average of 188 g late in the prebasic molt, and 219 g during premigration (Heitmeyer 1988a). Although lipid reserves are primarily used to meet daily energy requirements, females compensated for their increased use of protein during this molt by feeding on protein-rich invertebrates. Hence females consumed more (15.3%–37.5%) animal matter and less fiber during the stages of the prebasic molt through to its completion (Heitmeyer 1988c).

Mallards lose body weight during the flightless period, which lowers their wing load and thus shortens the time needed to regain flight. In Poland, both male and female Mallards lost 12% of their body mass during the flightless period, which allowed them to regain flight capability in 22–29 days, at which time the remiges were 75%–83% of their final length (Panek and Majewski 1990). Remige growth in males averaged 5.5–6.7 mm/day. Pehrsson (1987) also observed such declines in body weight and condition in molting Mallards reared in captivity; short escape flights were possible at 19–26 days into the flightless period. Metabolizing stored lipids during the flightless period also allows Mallards to use habitats that afford protection from potential predators, even if food is scarce at those locations. Body condition was generally higher in females than in males, and they retained a higher body condition longer, which may reflect the females' greater capacity to endure nutritional stress, such as during egg laying and incubation.

In one of the only studies to assess survival during the flightless period, Fleskes et al. (2010) conducted a 3-year study (during August–May in 2001–2, 2002–3, and 2006–7) that monitored

181 adult females fitted with radiotransmitters on the Klamath Basin NWR complex on the Oregon/California border. The females were captured just before or early in their flightless period, and survival was monitored for 30 days thereafter. Survival for a 30-day period was 76.8%, but it ranged from 11% on a small wetland to 93% on a large wetland. Most mortality was from avian botulism (64%) and predation (26%). In contrast, the survival of 50 radiomarked females during a 26-day molting period in Minnesota was much higher (94.7%), but females there were not exposed to botulism (Kirby and Cowardin 1986).

CONSERVATION AND MANAGEMENT

Habitat conservation and restoration are major concerns in the management of most waterfowl, including Mallards. Much of this management focus is in the PPR, where the loss of historical habitat has been extensive (Cox et al. 2000). In the United States, the loss of original wetland habitat within the PPR was estimated at 89% in Iowa, 49% in North Dakota, and 35% in South Dakota (Dahl 1990). The situation in Canada is similar. An extensive 1982 survey across the major duck-producing areas in Canada documented that cultivation had claimed 83% of the tallgrass prairie, 84% of the shortgrass prairie, 78% of the mixed-grass prairie, and 80% of the aspen parklands (Millar 1986). Agricultural expansion in the PPR in Canada caused the loss of about 90% of the wetland habitat in the eastern part of the region by 1951; in the western region, quality waterfowl habitat was still available at that time (Bethke and Nudds 1995). Agricultural activity has substantially impacted wetlands since 1951, however; >824,500 ha of wetlands were lost from the Canadian PPR since 1975. These losses were believed to have caused a mean annual deficit (1975–89) in the numbers of 10 duck species, including reductions of 29.3% for Mallards, 18.2% for Blue-winged Teal, and 45.2% for Northern Pintails.

A more detailed study monitored 10,000 wetlands across the PPR in Canada from 1981 to 1985 and found that 57% of the basins and 74% of their margins had already been degraded (Turner et al. 1987). During the study, the impact on wetlands increased slightly (to 59%), with the greatest (78%) repercussions in ephemeral wetlands. The degradation of wetland margins increased to 82% in parkland habitat and 89% in grasslands. A second study, monitoring 10,500 wetlands across the PPR in Canada from 1985 to 2005, reported that the impact rate (agricultural activity altering a wetland basin or margin) had actually declined over time, but only because fewer unaffected wetlands existed on the landscape (Bartzen et al. 2010). Furthermore, the recovery rate (reestablishment with no visible agricultural impact) was always lower than the impact rate. Shallow ephemeral wetlands in agricultural fields had the highest impact rate and the lowest recovery rate. The study concluded that the "high rates and incidence of wetland impact in conjunction with low recovery rates clearly demonstrate the need for stronger wetland protection in prairie Canada" (Bartzen et al. 2010:525).

As a result of the loss and degradation of waterfowl habitat in the PPR resulting from agricultural activities, the conservation community has brought forth a number of programs to help reverse or mitigate these losses. The Small Wetlands Acquisition Program instituted by the U.S. Fish and Wildlife Service on 1 August 1958, by an amendment to the 1934 Migratory Bird Hunting and Conservation Stamp Act (Duck Stamp Act), allowed the proceeds from the sale of Duck Stamps to be used to protect waterfowl habitat in the PPR. Protected areas are known as Waterfowl Production Areas, or WPAs. From 1959 to 2006, this program had purchased 6,199 wetland parcels (totaling 262,150 ha) and acquired permanent easements on 25,582 wetlands (totaling 585,377 ha; U.S. Government Accountability Office 2007); 36% of this habitat was in North Dakota and 22% in South Dakota, both important breeding ranges for Mallards. The program had also protected 906,096 ha of grassland nesting habitat.

The Conservation Reserve Program (CRP) protected about 1.9 million ha of grassland habitat in the PPR as of 1992 (Reynolds et al. 2006). The CRP was responsible for the estimated production of an additional 12.4 million ducks (2.2 million annually) in the PPR during 1992–97 (Reynolds 2005). Collectively the CRP produced 25.7 million ducks in the PPR during 1992–2003. A model by Reynolds et al. (2006) indicated that the nesting cover provided by the CRP accounted for a 50% increase (from 12% to 18%) in the average nesting success of Mallards and a 37% increase in recruitment. The CRP has also had a positive effect on nest success for ducks in other cover types, which accounted for another 400,000 Mallards that were recruited into the fall flight annually from 1992 to 1997. Nonetheless, by 2007 approximately 1 million ha with CRP contracts expired in the Dakotas, and only about 13% of all CRP lands were predicted to remain by 2010 unless contracts were reauthorized or extended (Reynolds et al. 2006).

Management in the PPR that provides large blocks of dense upland nesting cover generally increases the nest success of ducks. Sovada et al. (2000) suggested that landscapes with a mixture of wetlands and grasslands configured in relatively large tracts are the most productive for nesting ducks, while small isolated tracts of nesting habitat have only marginal value, unless predator populations are reduced. On a lesser scale, surrounding suitable nesting habitat with electrified fencing deters predators and thus can lead to high nest success (Sargeant et al. 1974, Lokemoen and Woodward 1993). Nest baskets are readily used by Mallards and can be quite effective at increasing nest success in areas where upland nesting habitat is lacking (Doty et al. 1975). Krapu et al. (2000) noted that survival of Mallard broods was relatively high on landscapes containing the largest number of seasonal basins, which underscores the importance of maintaining seasonal wetlands as a major component of wetland complexes managed for Mallard production.

Despite massive conservation efforts, the PPR appears to be especially vulnerable to the effects of climate change, as all global climate models predict

a warming in this region (Anderson and Sorenson 2001). Models developed by Larson (1995) anticipate that a 3°C increase in temperature will reduce prairie wetlands by 22% and parkland wetlands by 56%. Assuming a doubling of carbon dioxide concentrations by 2060, Sorenson et al. (1998) then used data from the Palmer Drought Severity Index (1961–97) to predict that warming in the PPR could cause the number of wetlands to fluctuate between 0.6 and 0.8 million, instead of a mean of 1.3 million; the number of ducks would therefore range between 2.1 and 2.7 million, well below the long-term average of 5.0 million. Other simulation models suggest that a drier climate would shift the most-productive breeding habitat in the region from the center (the Dakotas and southeastern Saskatchewan) to wetter eastern and northern fringe areas that are less productive or have experienced extensive wetland drainage (Johnson et al. 2005). Unless existing wetlands are protected and drained wetlands are restored, there is currently little habitat insurance for populations of prairie-breeding waterfowl. Wetland cover types will also be altered in response to climate change. Under the scenario of a doubling of atmospheric carbon, the expected decline in water levels would reduce the amount of open water in a semipermanent wetland from 50% to only 22% in 3 years (Poiani and Johnson 1991).

On migration routes and wintering grounds, Krapu et al. (2004a) noted a marked reduction in the amount of waste corn remaining after the agricultural harvest, due to the improved efficiency of harvest combines and increased acreages planted in soybeans. Extensive fall plowing of grain stubble also greatly reduces the availability of waste corn (Baldassarre et al. 1983). Moreover, less waste rice is available to Mallards on their principal wintering grounds in the MAV. Hence, although agricultural crops provide an important food source for Mallards and other waterfowl during migration and winter, such habitats supplement but are not a substitute for high-quality natural-wetland habitats.

Habitat contamination is also an issue. Ingested lead shot has been a serious source of Mallard mortality. Early studies (1938–54) estimated annual losses of 3%–4% of the continental Mallard population from this cause (Bellrose 1959). During the 1980/81 hunting season at Catahoula Lake in Louisiana, the incidence of ingested lead shot in 218 Mallards was 69%, with lead levels of 86–131 ppm (on a dry weight basis) in their livers (Zwank et al. 1985). The probability of death from lead poisoning was estimated at 76%. A lead-poisoning die-off during 1953 was estimated to have killed 6,000 Mallards on Catahoula Lake. The use of lead shot for waterfowl hunting was banned in the United States in 1991 and in Canada in 1999, and the bans appear to be having positive effects in reducing waterfowl mortality. During the 1996 and 1997 hunting seasons in the Mississippi Flyway, for example, the incidence of ingested pellets (either lead or nontoxic) was 8.9% for Mallards ($n = 15,147$), but 68% of all pellets were nontoxic, and the ingestion of 2 or more pellets declined by 78% (W. Anderson et al. 2000). The authors also estimated that nontoxic shot had reduced Mallard mortality in the Mississippi Flyway by as much as 64%. In addition, lethal dietary toxicity values have been established for 30 organochlorine compounds, 39 organophosphates, and 15 carbamates (Hill et al. 1975). While various contaminants have occurred in the body tissues of wild Mallards, they have not been documented at levels believed to be high enough to affect population parameters.

Lastly, hunting regulations are used by managers to achieve a sustainable harvest of Mallards, for which the major tools are adjustments in the length of the hunting season and bag limits. The relationship between a manipulation of hunting regulations and the subsequent survival of Mallards is nonetheless complex. At a continental level, Anderson and Burnham (1976, 1978) did not find an increase in annual survival rates during 1960–70, years of restrictive versus liberal harvest regulations. Harvest mortality, up to a threshold point, was considered to be compensatory, with harvests not affecting overall survival in the population. In other words, harvest mortality would be counter-balanced by a corresponding decrease in natural mortality, but *only to a point*. A later analysis by Nichols and Hines (1983) generally supported the compensatory mortality hypothesis for Mallards.

In contrast, Smith and Reynolds (1992) found that survival of all sex and age classes of Mallards was higher during a period of restrictive harvest regulations (1985–88), compared with a period of liberal harvest regulations (1979–84). As the authors stated, their results most likely differed from previous studies because of differences in population abundance, weather, habitat conditions, recruitment, regulations, harvest rates, and other factors. It was possible that Mallard harvests were mostly compensatory in the 1960s and 1970s, but less so in the 1980s. On local hunting areas in southern Manitoba, Caswell et al. (1985) noted that survival rates of Mallards increased by 12.5% for adult males and 24.1% for adult females during years of restrictive versus liberal harvest regulations; there were no differences in the rates among juveniles. Greater adult female survival was attributed to a delayed opening of the hunting season, which allowed them to disperse from natal areas before they were harvested.

Collectively, studies have indicated that harvest mortality can be additive in local situations, especially for adult females, as well as additive over a broader geographic area during some years. Smith and Reynolds (1992: 315), however, concluded that "the population dynamics of Mallards are more complex than the simple depictions of the completely additive and completely compensatory models." Even if harvests were completely additive, compensatory responses could occur via density-dependent reproduction (higher production when populations are small). Kaminski and Gluessing (1986) found an inverse correlation ($r = -0.59$) between the age ratio of Mallards harvested in the MAV (1959–84) and the size of the breeding population. Reynolds and Sauer (1991) noted that annual changes in the size of the breeding Mallard population were positively correlated with an index of production rate ($r = 0.503$) and negatively correlated with an index of harvest rate ($r = -0.415$). Average harvest-rate indices did not differ between periods when breeding populations were increasing (1961–69) and when they were decreasing (1970–85), but production indices tended, on average, to be lower during 1970–85 (0.964) versus 1961–69 (1.096).

Such complexity, as well as a continued debate over compensatory versus additive mortality, eventually led to changes in the way Mallard harvest regulations were established. Prior to 1995, annual harvest regulations for Mallards and other ducks were based primarily on a given year's estimated fall population; harvest regulations were liberal in years with high populations and more restrictive in years with low populations. This approach was filled with shortcomings, however, which are reflected in sources of uncertainty that limit the ability of waterfowl managers to predict the impact of hunting on duck populations (Nichols 2000). Hence, beginning in 1995, the U.S. Fish and Wildlife Service has adopted an Adaptive Harvest Management (AHM) system in setting harvest regulations for Mallards, an approach spe-cifically designed to consider uncertainties in the response of waterfowl populations to the manipulation of harvest regulations (Williams and Johnson 1995). Implementation of the AHM is based on 4 alternative models of population response, each with different assumptions about harvest mortality (compensatory or additive) and population dynamics. Managers select a regulatory alternative, based on a given year's habitat conditions, the size of the spring breeding population, and the relative probability (model weight) currently assigned to each of the 4 models. After an assessment of the Mallard breeding population the following year, model weights are then updated (increased or decreased) in accordance with the ability of each model to predict the estimated change in population size from the previous year. Hence managers are "learning as they go" by mathematically relating the extent of agreement between the predictions based on assumptions about harvest regulations and population dynamics, and the subsequent population size. The new weights are then used to begin the process anew in setting the next year's regulations.

Mexican Duck

Anas platyrhynchos diazi (Ridgway 1886)

The Mexican Duck is among the least well known of all North American waterfowl. These ducks are part of the "Mallard complex," which, in North America, also includes the Mallard, the American Black Duck, and the Mottled Duck. Mexican Ducks are most closely related to Mottled Ducks, although the males and females are similar in appearance to a female Mallard, only darker. Mexican Ducks inhabit arid areas, where they are resident year round; their primary range is the interior plateau of central Mexico, which harbors about 98% of an estimated total breeding population of 55,000. The loss and alteration of natural wetland habitats because of intensive agriculture and livestock grazing characterize the entire range of these ducks, but they have adapted to man-made water habitats associated with these land uses. The northern edge of their range extends into southeastern Arizona, southern and central New Mexico, and the Trans-Pecos Region of southwestern Texas, where Mexican Ducks are known to hybridize with Mallards; hence many studies address morphological descriptions and taxonomic issues, especially as they relate to hybridization. Taxonomists have debated subspecies versus species status for the Mexican Duck, but recent molecular data argue for the latter. The biology of Mexican Ducks is not well known but appears to be similar to that of Mallards, although the breeding season of the former is much longer (up to 8 months). More study is needed on Mexican Ducks, especially their general biology and an assessment of the extent of their hybridization with other species. The scientific name honors Augustin Díaz (1929–93), a Mexican geographer and explorer.

IDENTIFICATION

At a Glance. Males and females are similarly colored and resemble female Mallards, but Mexican Ducks are slightly darker, although never as dark as American Black Ducks. The overall body color is a darkish brown that contrasts with the lighter neck and head. In flight, the silvery-white wing linings contrast sharply with the darker upperwing. The speculum is bluish, with a trailing white edge; the leading white edge is often indistinct or absent. Scott and Reynolds (1984) noted that all the Mexican Ducks they examined in Mexico lacked a white band in front of the speculum, although there was a trailing white band. The locales where they are observed (distribution) can also aid in their identification, especially in Mexico.

Adult males. The body of the males, especially the breast, is a rich reddish brown that contrasts with the lighter head and neck. The tail and the tail coverts are fuscous brown to black-

Male Mexican Duck.
Bert Frenz

Female Mexican Duck.
Bert Frenz

ish and lack any white feathering. The speculum is bluish, with a distinct greenish sheen. The legs and the feet are a deep orange. Scott and Reynolds (1984) noted that males of all seasonal and geographic groups had a more pearly-gray wash on the tertials than did females, and males were generally darker, particularly on the chest, which was more reddish brown. The sexes can best be distinguished, however, by the color of the bill. Of the specimens examined in Mexico, the bill of the males ($n = 52$) was usually a clear olive green, although a few were more yellow than green (Scott and Reynolds 1984).

The bill of the females (*n* = 46) was more variable in color, but it was generally a dusky orange, often diluted with olive, with discrete dark spots or a suffusion of brown on the mid-dorsal area.

The plumage of both sexes varies from the northern extent of their range in the southwestern United States to the southern end in the central highlands of Mexico (see Distribution). The southern specimens are slightly darker, with more brownish and less grayish coloration; the feather edgings are more rufescent and less buffy below (Aldrich and Bayer 1970).

Adult females. Adult females are similar to adult males, but their overall body plumage is not as richly reddish brown as in the males. The bill color of the females is also more variable than that of the males.

Juveniles. Juveniles are similar to the adults, but more streaked.

Ducklings. Mexican Duck ducklings resemble those of Mottled Ducks, but the former have a brighter brown color pattern, and (usually) a thinner and incomplete eye stripe (Nelson 1993). The dorsal spots are sometimes larger and more noticeable.

Voice. Females give a loud quacking call, similar to that of female Mallards. Males give a low raspy *rink*.

Similar Species. Mexican Ducks are most similar to Mottled Ducks, but the latter have a paler cheek, a more unicolored bill, and a more limited eye stripe that does not continue onto the crown. At close range or in hand, Mottled Ducks have black around the nares and at the base of the bill, where the mandibles join together. Distribution also aids in identification, as the range of these 2 ducks do not overlap.

Separation of female Mexican Ducks from female Mallards can be difficult without close inspection, for which Huey (1961) provides the most detailed guide. In general, the tail of female Mexican Ducks is darker than that of female Mallards, and it lacks the white outer tail feathers. The undertail coverts of female Mallards are white or near white, while those of female Mexican Ducks are dark brown with a light brown edging; Huey noted that this feature was the most striking difference between the females from these 2 taxa. Female Mallards always have a distinct white border on the leading and trailing edges of the speculum. Differentiating Mexican Duck × Mallard hybrids can be difficult, but Scott and Reynolds (1984) noted a definite greenish sheen to the speculum in all but 6 of the 98 Mexican Ducks they collected in Mexico, compared with the purplish sheen of Mallards.

Weights and Measurements. The measurements listed here are from 6 sites in Mexico (Scott and Reynolds 1984). The median body weight of males (*n* = 52) was 1,028 g (range = 849–1,243 g), and wing length was 281 mm (range = 257–297 mm). Bill length was 42.0 mm (range = 35.6–46.1 mm), and bill width was 21.0 mm (range = 18.7–22.2 mm). The body weight of females (*n* = 46) was 908 g (range = 647–1,267 g), and wing length was 261 mm (range = 248–277 mm). Bill length was 38.8 mm (range = 34.1–42.2 mm), and bill width was 19.2 mm (range = 18.0–21.0 mm).

Measurements of 18 northern male specimens averaged 273.9 mm for wing length, 53.0 mm for bill length, and 44.2 mm for tarsus length, while measurements for females (*n* = 27) averaged 254.7 mm for wing length, 51.0 mm for bill length, and 42.6 mm for tarsus length (Aldrich and Baer 1970). Measurements of 13 southern male specimens averaged 269.9 mm for wing length, 53.3 mm for bill length, and 46.3 mm for tarsus length, while those for females (*n* = 13) averaged 253.4 mm for wing length, 50.3 mm for bill length, and 42.0 mm for tarsus length.

DISTRIBUTION

The American Ornithologists' Union (AOU) formerly considered the Mexican Duck to be a separate species (*Anas diazi*) from the Mallard. There

Year-round Distribution ■
Wintering Distribution ■

is a high degree of morphological variation within the distribution range of Mexican Ducks, which led to an early designation of 2 subspecies: *A. diazi diazi* (Mexican Duck) in the south, and *A. diazi novimexicana* (New Mexican Duck) in the north. Subspecies status for the New Mexican Duck is no longer recognized, however, as the populations are contiguous and similar in body size, and the plumage differences were too slight to permit the identification of birds from northern versus southern populations (Aldrich and Baer 1970).

In 1983, the AOU declared the Mexican Duck to be conspecific with the Mallard, due to hybridization between the 2 forms, which clearly occurs at the northern edge of the Mexican Duck's range in the southwestern United States. In their respective books on species accounts of waterfowl, Johnsgard (1978) and Bellrose (1980) also considered the Mexican Duck to be a subspecies of the Mallard, although Palmer (1976a) did not. This taxonomic status of species versus subspecies for the Mexican Duck also has major management ramifications. The decision to lump what had been 2 separate species (the Mallard and the Mexican Duck) into a single species (*Anas platyrhynchos*) negated any need for special harvest management actions by

the U.S. Fish and Wildlife Service to protect the Mexican Duck, which had been listed as an endangered species in the United States in 1967.

Hybridization with Mallards occurs in several species (see the American Black Duck account), and the restricted range and small population size of Mexican Ducks allow a more detailed look at this topic. The decision to lump Mallards and Mexican Ducks into 1 species was based primarily on the work of Hubbard (1977), who used plumage indices to describe a broad phenotypic (appearance) cline, with the museum specimens he examined progressing from more Mallard-like forms in the southwestern United States to "pure" Mexican Ducks at the core of their range in the central highlands of Mexico. The associated plumage indices varied from 1.0 for "pure" Mallards north of the range of Mexican Ducks, to 20.5 in central New Mexico, 22.8 in southern New Mexico, and 34–35 for "pure" Mexican Ducks in the central highlands.

Hubbard compared plumages among specimens collected over 2 time periods in Doña Ana County in southern New Mexico, and he concluded that hybridization had increased from 73% in an 1893–1920 sample to 88% in a 1938–60 sample. This finding, however, was refuted by Scott and Reynolds (1984), in part because Hubbard's sample sizes were small ($n = 25$ and 17) and probably not random, as "odd" hybrid forms tend to be deposited in museums. There were also no recent samples. Based on a study of 181 Mexican Ducks harvested in Arizona (a more random sample), Brown (1982:116–17) stated that "the contention that Mexican Ducks in Arizona are a hybrid swarm is an erroneous oversimplification of Hubbard's (1977) work." Brown noted that the wing characteristics of the ducks in his sample revealed that only 7 were Mallard × Mexican Duck hybrids, and that few Mallards nested within the range of Mexican Ducks. He also argued that although some hybridization was to be expected at the northern edge of the Mexican Duck's range, the incidence may have been accelerated in New Mexico and southeastern Arizona following the release of 385 captive-reared Mexican Ducks at the La Joya Waterfowl Management Area, the Bosque del Apache National Wildlife Refuge (NWR), and San Simon Cienega between 1963 and 1975, as these birds were released without pair bonds and when male Mallards were present, which would increase the potential for forming hybrid pairs.

Scott and Reynolds (1984) did, however, agree that the birds in Doña Ana County could represent a "fulcral" population where phenotypes are in a state of flux (a hybrid zone). These authors also showed a cline in plumage indices, based on 98 birds collected only in Mexico, with plumage scores ranging from 34.3 in the south (Jalisco) to as high as 28.5 in northwestern Chihuahua, which is just south of New Mexico. Hence, as was seen by Hubbard (1977), there appears to be a rather abrupt break in the phenotypes of Mallard-like forms to *diazi*-like forms more or less at the United States / Mexico border. Most plumage samples were fairly uniform, although Scott and Reynolds (1984) noted more evidence of Mallard introgression from their sample at Río Conchos in southeastern Chihuahua, which may have seen immigrant birds from the Trans-Pecos Region moving into the area after large irrigation projects created habitat for Mexican Ducks.

Nonetheless, the key issues are the stability of the apparent hybrid zone in the southwestern United States, and whether it is moving southward into Mexico, where the Mallard phenotype would eventually swamp the Mexican Duck phenotype through introgressive hybridization. Scott and Reynolds (1984) concluded that large, genetically pure populations of Mexican Ducks occurred in many areas of Mexico, and that the hybrid zone was not moving south, except possibly at Río Conchos in southeastern Chihuahua. They noted that Mallards were once (pre-1920s) fairly common during the winter as far south as the Valley of Mexico, and that some of those birds may have stayed to breed. Mallards are now almost unknown in central Mexico, and scarce even in northern Chihuahua, probably because of a shift in migratory habits; they remain farther north, due to increased grain production and the construction of man-made lakes and reservoirs in the midwestern and western United States (Saunders and Saunders 1981). Howell and Webb (1995) reported that Mallards were an uncommon winter visitor in the region from Chihuahua to central Durango in northern Mexico, and they did not mention that any breeding occurs. Midwinter Surveys of Mexico's interior highlands averaged only 2,013 Mallards in 11 surveys conducted between 1980 and 2006, most of which were in the northern highlands (Thorpe et al. 2006). Hence the few Mallards reaching the interior highlands during the winter occur in the north, outside the core of Mexican Duck breeding range in the central highlands, which obviously limits opportunities for interbreeding. In Arizona, Webster (2006:7) noted: "My recent experience in eastern Cochise County and environs is not of the 'extensive' hybridization that influenced the decision to lump the two. Rather, it is of no obvious hybridization, because I have not seen an obvious Mallard during the breeding season in the areas I bird, areas in which Mexican Ducks breed at a number of localities in small to moderate numbers." During spring 2010 in southwestern New Mexico, surveys revealed that Mexican Ducks outnumbered Mallards 157 to 31 in Hidalgo County, 44 to 2 in Luna County, and 47 to 18 in Doña Ana County (Sandy Williams), which again indicates some geographic separation between Mexican Ducks and Mallards.

There may also be behavioral mechanisms that limit introgressive hybridization and/or its geographic extent. During his landmark study of Mexican Ducks in the interior highlands, Williams (1980) found that no other species of ducks (e.g., Mallards) were ever observed displaying with Mexican Ducks. In Arizona, Brown (1985) noted that most adult Mexican Ducks are seen in pairs as early as September, prior to the arrival of wintering Mallards, which limits the opportunity for mixed pair bonds. Pair bonds in Mexican Ducks are also long, and these ducks may quickly re-pair after brood rearing and the wing molt (see Mating System).

Lastly, 3 recent genetic studies conducted since the Mexican Duck and the Mallard species were combined have all considered the Mexican Duck to be a separate species, most closely relating it to the Mottled Duck, *not* the Mallard (Johnson and Sorenson 1999, McCracken et al. 2001, Kulikova et al. 2004). The American Black Duck was considered the next closest relative of the Mexican Duck and the Mottled Duck. All 3 of these monomorphic forms (Mexican Duck, Mottled Duck, and American Black Duck) are then related to the Eastern Pot-billed Duck (*Anas zonorhyncha*). Livezey (1991) also recognized species status for the Mexican Duck, based on morphological measurements. Such new findings led the AOU to reconsider species status for the Mexican Duck in 2011, but the vote failed by 6 to 5. The dissenters argued that the degree of hybridization between the 2 forms has not yet been established, and that phenotypic differences are inconclusive. In 2010, however, the International Ornithologists' Union's list of bird names recognized species status for the Mexican Duck.

At the least, subspecies status for the Mexican Duck is inconsistent with the species-status designation for the American Black Duck and the Mottled Duck, both of which hybridize with Mallards. The issue is clearly the interpretation of hybridization and what it means for designating species status. As the AOU (1998:xiv) noted, undisputed biological species "long retain the capacity for at least limited interbreeding with other species, even non-sister taxa." The AOU also notes that "essential (lack of free interbreeding) rather than complete reproductive isolation has been and continues to be the fundamental operating criterion for species status." Thus the critical question becomes how much interbreeding is "enough" to warrant lumping of the 2 species/subspecies involved. Ducks are a particularly perplexing case here, because, as Johnsgard (1960:25) noted, "without doubt, waterfowl of the family Anatidae have

provided the greatest number and variety of bird hybrids originating from both natural and captive conditions"; Grant and Grant (1992) also addressed this point. Gray (1958) listed about 400 hybrid combinations among the Anatidae, which was far more than for any other bird family. Where related species come together in the wild, hybridization occurs, yet species status has been maintained, such as for the Mottled Duck and the American Black Duck. More study is needed for Mexican Ducks, especially those from the northern parts of their range, but I believe the existing evidence and the above perspective on hybridization leans strongly toward species status.

Breeding. About 98% of the Mexican Duck population occurs in Mexico (Williams 1980), where their breeding range is in the interior highlands (with elevations of 900–2,500 m) from Chihuahua and Durango southward to the Trans-Mexican Volcanic Belt that extends from Jalisco and Nayarit east through México and Morelos; they also occur locally as far as western Tamaulipas (Aldrich and Baer 1970, Howell and Webb 1995). During a 6-year study (1973–78) in Mexico, Williams (1980) reported that Mexican Ducks occurred in 16 states on the Mexican plateau, but 85%–90% were found in the western central highlands, where the states of Jalisco, Guanajuato, and Michoacán meet. Their breeding range coincided with the range of pine-oak, mesquite-grassland, desert, and tropical deciduous zones. At the northern end of the interior highlands (Chihuahua and Durango), Howell and Webb (1995) listed Mexican Ducks as common residents in the western portion of the interior plateau, chiefly from Chihuahua to Durango and southward, and extending eastward during the winter. Populations are small and dispersed in the sparsely watered northern highland states of Chihuahua and Durango, but they also inhabit other dry areas within their breeding range in Mexico (Johnsgard 1961b, Aldrich and Baer 1970).

The northern edge of their breeding range extends into the United States, where Mexican

Ducks occur in Arizona, New Mexico, and Texas. Museum records farther north (in Nebraska and Colorado) appear to be hybrids (Hubbard 1977). In Arizona, these ducks are found throughout much of Cochise County (especially in Sulphur Springs Valley) in the extreme southeast, then northward along the San Simon River drainage into adjacent Graham County, and then west along the Gila River drainage at least to the central part of that county (Corman and Wise-Gervais 2005). They also occur fairly regularly to the west, along the lower San Pedro River to its intersection with the Gila River in eastern Pinal County, and at 1 location along the San Carlos River, which drains southward into the Gila River. In New Mexico, fair numbers of Mexican Ducks are found across the southern part of the state, from Hidalgo County in the extreme southwest to southcentral Otero County and then northward along the Rio Grande valley to the Bosque del Apache NWR, with scattered reports into central New Mexico, to about the Albuquerque area (Sandy Williams). More recently, they have been observed in the lower Pecos Valley of Eddy County in the southeast, and occasionally north to the vicinity of Bitter Lake NWR in adjacent Chaves County. Records from Rio Arriba County in the northern part of the state are old ones, as the Albuquerque area is roughly the northern limit for these ducks. In Texas, they regularly occur in the Trans-Pecos Region (Balmorhea Lake) of southwestern Texas and along the Rio Grande valley, and then east as far as Starr and Hidalgo Counties, with occasional reports from the coast (Rappole and Blacklock 1985, Mary Gustafson).

Winter. Mexican Ducks are resident throughout their range, although there is evidence that they make nomadic movements in response to water availability. In Texas, Mexican Ducks winter east of their breeding range, where they are regularly encountered below Falcon Dam and elsewhere in the lower Rio Grande valley (Lockwood et al. 2008).

In Mexico, an analysis of data from Midwinter Surveys conducted by the U.S. Fish and Wildlife Service in the interior highlands from 1960 to 2000 revealed that 84% of the Mexican Ducks were found in the central highlands and 16% in the northern highlands. More recent (1991–2000) percentages increased to 31% in the northern highlands and declined to 69% in the central highlands (Pérez-Arteaga et al. 2002). The increase in the northern highlands appears to be the result of large irrigation projects developed since the 1970s that created suitable wetland habitat for Mexican Ducks, and a greater abundance of agricultural foods, which the ducks readily consume (Scott and Reynolds 1984).

Pérez-Arteaga et al. (2002) used Midwinter Survey data from 1991 to 2000 (conducted at 3-year intervals, in 1991, 1994, 1997, and 2000) to identify 15 sites in the interior highlands (4 in the northern part and 11 in the central section) that collectively held 70%–75% of the Mexican Ducks recorded during those surveys, and contained, on average, about 18% of an estimated population of 55,500. Eight sites were situated in a relatively small area of the western central highlands along the Río Lerma drainage (East Atotonilco, Cabadas, West Yuriria, Lago de Cuitzeo, Irapuato, Presa Solís, and Presa Tepuxtepec). The greatest number (6) of sites were located in the state of Guanajuato, with 3 others in Michoacán and 3 in Chihuahua. The sites with the highest average number of Mexican Ducks from 1991 to 2000 were Laguna Bustillos in Chihuahua (1,579); the Cabadas wetland complex in Jalisco, Guanajuato, and Michoacán (1,236); the Languillo wetland complex in Jalisco and Aguascalientes (1,099); and Presa Solís in Chihuahua (868).

Migration. Although resident throughout most of their range, there is some evidence of migration among northern populations along the Rio Grande valley, where Aldo Leopold noted that Mexican Ducks "pass southward early in the [hunting] season" (quoted in Palmer 1976a). Williams (1980), however, noted that Mexican Ducks could be found throughout the year in all parts of their range; although local and seasonal movements occurred, regular north–south migration did not. In Arizona, Brown (1982) considered Mexican Ducks to be a "nomadic resident." After a review of the available literature, Pérez-Arteaga et al. (2002:36) also concluded that "there is no evidence for migratory movements, with breeding and wintering records occurring throughout its range." Howell and Webb (1995), however, indicated an expanded winter range for Mexican Ducks, eastward from the highlands through Coahuila to western Nuevo León and southwestern Tamaulipas.

MIGRATION BEHAVIOR

Nothing specific is known, but Mexican Ducks are probably nomadic during the nonbreeding season, congregating wherever suitable water conditions are found.

Molt Migration. In Mexico, postbreeding birds favored larger lakes and permanent marshes with extensive stands of emergent vegetation (Williams 1980). Presumably many birds molt at these sites, but the distances traveled from their breeding areas are not well known.

HABVITAT

Mexican Ducks inhabit arid and semiarid environments, where they use a wide array of water areas, ranging from large reservoirs and natural lakes to small irrigation canals, seasonal and permanent wetlands, and man-made water sites (Williams 1980). They occupy virtually all available habitats, with "few, if any, water areas within their range not visited by Mexican Ducks at some time during the year" (Williams 1980:192). They also appear to use habitat in proportion to its availability. During July and August 1976 in Chihuahua and Durango, Mexican Ducks ($n = 173$) were surveyed on 44 water areas, of which 84% were man-made and 16% were natural. They used 33 of the 44 areas (75%), of which 83% were man-made and 17% were natural: 49% small (<5 ha) presas (man-made areas created by erecting dams in places where water can

then be impounded) and ponds; 35% large presas and lakes; 12% ditches, canals, streams, and rivers; and 4% wet pastures. They are well adapted to take advantage of the agricultural habitats and associated man-made water bodies characteristic of their principal range, the interior highlands of Mexico (Pérez-Arteaga et al. 2002).

Habitat use varied seasonally, with permanent lakes, large reservoirs, and larger permanent streams being especially important during the long dry season, when Mexican Ducks congregated in these areas in flocks of several hundred to thousands, with 5,000 ducks in the largest flock observed. These larger water bodies are also important to molting birds. During the breeding season, some ducks use the margins of these large water areas, but their principal breeding sites are seasonal water areas such as small reservoirs, farm ponds, temporary marshes, and intermittent streams that acquire water during the summer rainy season. Mexican Ducks also readily adapt and use man-made water sites (e.g., farm ponds, stock ponds, and storage reservoirs) that are established in arid areas.

In Sulphur Springs Valley in Arizona, Mexican Ducks make widespread use of pump-back ponds, which are excavated at the end of irrigated fields to collect runoff water and recycle it back onto the crops (O'Brien 1975). Both nesting and brood-rearing Mexican Ducks selected ponds with permanent water levels, which usually had dense shoreline vegetation. Pump-back ponds increased in number during the 1960s and 1970s and facilitated the expansion of the ducks' range in the valley. There were 103 such ponds south of Willcox in 1973, but these habitats are being lost as irrigation sprinklers replace gravity-flow irrigation (Swarbrick 1975).

POPULATION STATUS

The Mexican Duck population in the United States is very small, but not well surveyed. Hubbard (1977) estimated a population of 650–900 *diazi*-like ducks in the United States—150–200 in Arizona, 200–300 in New Mexico, and 300–400 in Texas—but no systematic surveys for Mexican Ducks have been conducted throughout their range. Swarbrick (1975) estimated that there were about 150 resident Mexican Ducks in Sulphur Springs Valley in Arizona during studies conducted in 1973–74; O'Brien (1975) estimated that there were about 200 or so nesting pairs in Arizona during the early 1970s. In New Mexico, the current population is estimated to be 500–1,000 birds, based on various spring counts and other sighting reports (Sandy Williams). Christmas Bird Count data averaged 126 Mexican Ducks in Arizona from 2000/2001 to 2009/10, with a high of 315 in 2004/5. Over the same period, an average of 36 were recorded in New Mexico and 157 in Texas, with a high of 305 in 2006/7.

Such a small population and the associated destruction of their habitat prompted the U.S. Fish and Wildlife Service to list the Mexican Duck as an endangered species in 1967, when it was still considered to be a species separate from the Mallard. It was removed from the Endangered Species List in 1978, following its classification as a subspecies, primarily based on the work of Hubbard (1977), because the Endangered Species Act does not allow for the protection of a single population within a species, or of 1 phenotype (the general appearance of a species, as a product of both its genes and its environment) within a particular geographic segment.

About 98% of the Mexican Duck population occurs in Mexico. An aerial survey in the interior highlands from May to early June 1978 counted 38,890 adults, which (corrected for deficiencies in coverage) yielded an overall population estimate of 55,500: 86% in the western and 5% in the eastern central highlands, and 9% in the northern highlands (Williams 1980). A potential postbreeding population was estimated at 78,000. A different spring aerial survey in 1978 yielded a population estimate of 55,000 *diazi*-like birds in Mexico (Scott and Reynolds 1984).

During the winter, Pérez-Arteaga et al. (2002) determined population trends for Mexican Ducks in the interior highlands from Midwinter Surveys conducted by the U.S. Fish and Wildlife Service at 1-year intervals from 1960 to 2000 ($n = 21$ years of surveys) and then at 3-year intervals from 1991 to 2000 ($n = 4$ years of surveys). The resultant population indices averaged 13,762 in the yearly surveys (from 1960 to 2000), of which 84% were in the central highlands and 16% in the northern highlands. In contrast, populations surveyed at 3-year intervals averaged 12,249 from 1991 to 2000, of which 31% were in the northern highlands and 69% were in the central highlands. The long-term (1960–2000) trend showed a statistically significant annual increase of 2.5%/year, compared with a nonsignificant increase of 1.7% in the 3-year intervals (1991–2000). The long-term trend also indicated an average annual increase of 7.7% in the northern highlands, while there was no statistically significant change over time for the central highlands. The largest numbers recorded during the surveys were 49,510 Mexican Ducks in 1988 and 44,433 in 1982. More recent Midwinter Surveys in the interior highlands estimated 8,448 Mexican Ducks in 2003 and 18,585 in 2006 (Thorpe et al. 2006). Pérez-Arteaga et al. (2002) also found that at a local level, long-term trends at sites in the northern highlands showed statistically significant annual increases at Laguna Bustillos (25.9%), Laguna Mexicanos (20.4%), Laguna de Santiaguillo (16.9%), and Laguna de Babícora (13.9%). In the central highlands, statistically significant increases occurred at Languillo (15.3%), Zacapu (13.4%), and Presa Solís (8.9%) while 2 sites showed statistically significant declines: −5.2% at Lago de Chapala (1960–2000) and −11.8% at Lerma (1991–2000).

Pérez-Arteaga et al. (2002) concluded that Mexican Ducks were well adapted to the agricultural landscape that now characterizes the interior highlands, as agriculture has created new wetland habitats and a constantly available food source. It appeared likely that the population had never been much larger than at present, and that it was not under any serious threat.

HARVEST

Small numbers of Mexican ducks are harvested within the southwestern United States, primarily in southeastern Arizona, southern New Mexico, and western Texas. The size of the harvest in Mexico is unknown. Bent (1923) commented on the cautiousness of Mexican Ducks, noting that they would not swim in toward decoys, as did Mallards, but rather fed by themselves some distance away. Williams (1980:243) also remarked that flocks of Mexican Ducks "exhibited exceptional wariness." O'Brien (1975:19) noted that "at most times of the year Mexican Ducks are extremely wary and readily flush to approaching danger."

BREEDING BIOLOGY

Behavior

Mating System. Mexican Ducks are monogamous. In Mexico, Williams (1980) believed that 85% of the females were paired in May and probably all by June. Pairs were observed from April through September, as well as in December, and the bonds were suspected to last throughout the year. He thought that Mexican Ducks paired or re-paired soon after completing reproductive activities and the subsequent wing molt. Among individuals Williams could classify, 86% were paired in April ($n = 61$), 84% in May ($n = 384$), 77% in June ($n = 720$), 67% in July ($n = 671$), 28% in August ($n = 893$), and 38% in December ($n = 21$). In Sulphur Springs Valley in Arizona, Swarbrick (1975) only observed courtship displays on a single occasion, which also might indicate that adults remain paired or quickly re-pair.

Little is known about the duration of the pair bond, although it may be quite lengthy. In Arizona, O'Brien (1975) noted that pair bonds persisted until just before the appearance of broods. In New Mexico, Lindsey (1946) saw a male attending a female that had recently begun incubation, and another still attending his mate when the eggs were about to hatch. Williams (1980) reported an instance in Mexico of a male accompanying a female with a brood, as did Swarbrick (1975) in Arizona. There were also several sightings of a male and a female accompanying broods in New Mexico (Sandy Williams). Such observations suggest that the pair bond is rather strong, although studies with marked birds will be necessary to determine the extent of re-pairing or even year-round pair bonds.

Sex Ratio. The sex ratio (males:females) appears to be fairly even, but data are not extensive. Of museum specimens collected during all seasons in Mexico and the United States and examined by Hubbard (1977), the sex ratio of those determined to be Mexican Ducks or nearly / very nearly Mexican Ducks ($n = 137$) was 48.2% male and 51.8% female. Of other specimens collected in Mexico and categorized as Mexican Ducks or very nearly Mexican Ducks ($n = 79$), the sex ratio was 51.9% males and 48.1% females. Brown (1982) reported a sex ratio of 105 males to 100 females (51.2% males).

Site Fidelity and Territory. There is no information on site fidelity for Mexican Ducks. At San Simon Cienega in New Mexico, however, Bevill (1970) estimated the home range of Mexican Ducks ($n = 5$) to be 0.8–4.4 km long; he concluded that home-range size varied with the availability of habitat, the carrying capacity of the area, and the defense of their sites by paired males. In Mexico, Williams's (1980) observations led him to conclude that the general pattern of home ranges and territories in Mexican Ducks was similar to that of their close relatives (i.e., Mallards). He could not determine a specific home-range size but remarked on their "seemingly large size."

Courtship Displays. Courtship displays by Mexican Ducks have not been studied per se, but Williams (1980) considered them to be basically identical to those of Mallards. He observed displays ($n = 86$) between mated pairs, in trios made up of a pair and an intruding male, and in courting groups of 3–11 individuals (average = 5.4 birds) that contained more males than females (although sex could not always be determined). Social displays in groups of Mexican Ducks (*n* = 56) were characterized by much activity: short *rushes* across the water, *splashing, wing-flapping,* and *ducking*. Williams (1980:92) noted, "At times these activities took on almost a carnival atmosphere, with cartwheels and wing-flops across the surface, short jumps and hops into and out of the group, and birds diving below the surface or churning across the surface." These collective activities often attracted other ducks to the group, with loafing birds flying overhead quickly joining such assemblages.

During the months of his study (May–Sep and Dec), Williams also observed aerial *pursuit flights* in Mexican Ducks. Of 245 such flights involving 2–6 birds, 66% were with 3 birds, 18% with 2 birds, and 16% with 4–6 birds. Flights were characterized as *courtship-flights, exploratory-flights,* and *attempted-extra-pair-copulation-flights. Courtship-flights* primarily occurred during the prebreeding season in May and made up 91% of all *2-bird-flights* observed during that month. Such flights were typically high and graceful and appeared to serve as a way to maintain the pair bond. In contrast, flights of 3 or more birds (*n* = 200) were divided into *expulsion-flights* and *attempted-extra-pair-copulation-flights.* Most *3-bird-flights* appeared to be *expulsion-flights,* while flights of 4 or more birds were classified as *attempted-extra-pair-copulation-flights.* (Such behavior is also common in Mallards and summarized in that account.)

Nesting

Nest Sites. Mated pairs search for suitable nesting habitat by making *exploratory-flights,* with the female in the lead (Williams 1980). Like Mallards, Mexican Ducks are generalists in their selection of nest sites, although very few nest locations have been described, in part because they appear to be very widely dispersed; moreover, nests are usually well concealed. In the state of Jalisco in Mexico, nests (*n* = 4) were in a crop field, a pasture, a woodland, and a marsh (Williams 1980). The nests were in dense cover that was 55–180 cm tall, and they were woven into or placed at the base of some relatively firm support that, collectively, gave the nests a structurally solid appearance. On average, the nests were 197 m from water (range = 50–480 m). Williams provided detailed descriptions for each of these 4 nests, and he suspected that other nest sites occurred directly over water or as far as 0.5 km away.

Williams (1980) also gave descriptions for 16 nest sites in the United States, some of which were only probable nest sites. Lindsey (1946), however, presented descriptions of actual sites in New Mexico: (1) in a clump of willows (*Salix wrightii*) surrounded by water; (2) in a clump of cattails (*Typha latifolia*) surrounded by shallow water; (3) on dry ground, in grass 30 cm tall at the base of a small ash tree (*Fraxinus* sp.) about 6 m from the edge of a slough; (4) in a moist level meadow, characterized by common three-square bulrush (American bulrush; *Scirpus americanus*) and saltgrass (*Distichlis stricta*) arched over the nest for concealment, 41 m from a small stream; (5) in a moist meadow 70 m from open water, with the actual nest very well concealed among dead saltgrass that dropped over the nest from 36 cm above, and with a distinct runway also arched over by grasses and cocklebur (*Xanthium saccharatum*); (6) among tufts of rush (*Juncus balticus*), but otherwise poorly concealed from above and accessed by 2 short runways; and (7) 161 m from water, in a tall clump of sedges (*Carex praegracilis*) and grass (*Hordeum jubatum*) within a general area of dense spikerush (*Eleocharis rostellata*). One nest-site location had apparently been used for several years, as the well-packed foundation consisted of very worn nest material.

A nest in New Mexico measured 20.3 × 14.0 cm across the inside downy cup and 33.6 × 26.7 cm outside the downy rim. A shallow 30.5 × 22.9 cm lining of compacted grass fragments was below the down layer, with a shallow scoop in the ground beneath that.

Clutch Size and Eggs. In Mexico (1973–78), the average size of 18 clutches located by Williams (1980) was 6.7 eggs (range = 2–10 eggs), and he reported earlier records of 8 nests from the Lerma marshes that averaged 7.9 eggs. Measurements of 23 eggs from New Mexico averaged 56.8 × 41.2 mm, with a range of 53.3–59.6 × 40.0–43.2 mm (Lindsey 1946). Lindsey also noted that the average measurement for 71 eggs from the U.S. National Museum was 55.2 × 41.0 mm.

Mexican Duck eggs are white, faintly colored with bluish green, but Lindsey (1946) noted that none of 45 eggs from 9 clutches of then *A. diazi diazi* showed as strong a green tint as a set of *A. diazi novimexicana* eggs. There are no studies on the laying rate for Mexican Ducks, but it probably is similar to that of Mallards.

Incubation and Energetic Costs. Williams (1980) assumed 26 days for incubation. There are no studies on the reproductive energetics or incubation rhythms of Mexican Ducks.

Nesting Chronology. In the southwestern United States, Mexican Ducks breeding before June generally did so because of the presence of man-made water supplies (Aldrich and Baer 1970). In Mexico, nest initiation was linked to the annual wet/dry cycle, with June marking the beginning of the rainy season in their major breeding range in the Jalisco-Guanajuato-Michoacán region (western central highlands), compared with July farther north in Chihuahua. As the rainy season commenced, pairs rapidly dispersed from the flocks that mainly had gathered during the dry period to use seasonal wetlands (Williams 1980). Of 45 known nest and brood records from Mexico summarized by Williams, 78% were initiated after 1 June, versus only 22% before 1 June. Nests initiated before the rainy season were in areas where some water and cover were available: in reservoirs, large marshes, or irrigated landscapes. The entire breeding season lasted about 8 months, from the first nests in mid-April until the final young fledged in November.

Nest Success. Few Mexican Duck nests have ever been located, so there is little information on nest success. Success may be fairly high, however, because their nests are widely dispersed and well concealed in thick cover (Williams 1980). Potential avian nest predators in Mexico are White-necked (Chihuahuan) Ravens (*Corvus cryptoleucus*) and Common Ravens (*Corvus corax*), while potential mammalian predators are raccoons (*Procyon lotor*), ring-tailed cats (*Bassariscus astutus*), coatis (*Nasua narica*), 3 genera of skunks (*Mephitis, Spilogale,* and *Conepatus*), coyotes (*Canis latrans*), gray foxes (*Urocyon cinereoargenteus*), and various rodents.

Brood Parasitism. Brood parasitism has not been studied, but is probably very low to nonexistent, because of the wide spacing and concealment of Mexican Duck nests.

Renesting. Bevill (1970) documented renesting in New Mexico, but there are no general studies over the ducks' entire distribution area. The long breeding season that characterizes the range of Mexican Ducks, however, allows ample time for renesting attempts, which would be expected, given the high renesting propensity of Mallards.

REARING OF YOUNG

Brood Habitat and Care. In Mexico, females reared their broods on permanent and seasonal reservoirs, large permanent marshes, small seasonal wetlands, rivers, and man-made canals, all usually with some dense escape cover (Williams 1980). Feeding occurred on open water, in emergent vegetation, and on floating mats of vegetation. In Sulphur Springs Valley in Arizona, shoreline cover seemed especially important for broods (Swarbrick 1975). At the Bosque del Apache NWR in Arizona, Zahm (1973) noted that high levels of brood production occurred where dense escape cover was adjacent to open-water areas.

The female is very protective of her brood, as exhibited by the "constant alertness" Williams (1980) observed. Females were unconcerned with livestock, but quickly led their broods to emergent cover when humans approached. Females also performed 1 or more distraction displays, especially when their broods were surprised. In Sulphur Springs Valley in Arizona, broods usually remained on the pond where they hatched, but broods were known to move among ponds in response to changes in water levels or to disturbance (O'Brien 1975).

Brood Amalgamation (Crèches). Broods observed in Mexico by Williams (1980) were uniformly aged, which indicates that brood amalgamation was uncommon, if it occurred at all. Individual broods usually remained apart from other Mexican Duck broods and did not associate with them if they approached. Only 1 of the 16 broods Williams watched was not attended by a parent.

Development. No information is available, but presumably development in Mexican ducks is similar to that in Mallards and Mottled Ducks.

RECRUITMENT AND SURVIVAL

In Mexico, broods of various ages averaged 4.9 young ($n = 15$), with a range of 3–9 (Williams 1980). Probable Class I broods averaged 4.8 ducklings ($n = 10$), and Class II and III broods ($n = 5$ combined) averaged 5.0. There is no information on duckling survival to fledging in Mexico. At San Simon Cienega in southwestern New Mexico (1968–69), 6 broods observed by Bevill (1970) averaged 5.7 young at fledging (range = 3–12). In Sulphur Springs Valley in Arizona (1972–74), the size of broods at about 75% of fledging age ($n = 124$) averaged 5.6 ducklings (range = 5.3–5.9), with a range of 1–13 young per brood (O'Brien 1975). The minimum population on the study area was estimated at 75 pairs, which would yield an annual production of 420 young, based on a fledging brood size of 5.6

There is no information on survival after fledging or survival in adult birds. There is a longevity record, however, of 5 years and 6 months for a hatching-year male banded and later shot in New Mexico.

FOOD HABITS AND FEEDING ECOLOGY

Although there is very little specific information available, the feeding habits of Mexican Ducks appear to be like those of Mallards. Bent (1923) reported that Mexican Ducks ate alfalfa shoots, corn, wheat, the larger seeds of weeds and grasses, and mollusks, but he gives no other information. He also noted that foraging Mexican Ducks kept a short distance between themselves and other ducks.

Breeding. There are no data on the food habits of adults or ducklings, but an increased reliance on animal matter would be expected for egg-laying females and rapidly growing young ducklings.

Migration and Winter. During the dry (nonbreeding) season in Mexico (Oct–early Jun), Mexican Ducks are concentrated on the remaining available water, but these sites usually lack vegetative cover, because of overgrazing, burning, plowing, and the general desiccation accompanying the dry season. Hence their primary foods are acquired from nearby fields of cultivated grains and other crops (Williams 1980). The most heavily used crops in the Jalisco-Guanajuato-Michoacán region were garbanzo beans and small grains (wheat, barley, and oats). During late fall and winter, they also ate corn and grain sorghum (Leopold 1959). Most birds departed for these fields about 30 minutes before the sun was over the local horizon and were foraging in them by daybreak. The majority did not return to their roosting lakes until about 1 hour after daybreak. Evening field-feeding flights occurred over a longer time period that, at times, encompassed 3.5 hours. Departures from the water areas where they roosted usually began in the late afternoon, with most then leaving the crop fields at about dusk. The distances between roosting lakes and foraging fields were usually 3–5 km, but some birds traveled as far as 15–20 km. Feeding efforts during the rest of the day were directed at animal foods.

MOLTS AND PLUMAGES

Little detailed information is available for Mexican Ducks. Palmer (1976a) assumed that the sequence was the same as in Mallards, but with differences in timing. Since Mottled Ducks are the closest relative of Mexican Ducks, however, the pattern and timing of molting may be more similar between these 2 species than between Mexican Ducks and Mallards (see the Mottled Duck account).

Observations in Mexico indicated that the flightless period of the wing molt occurred as early as mid-August, but most individuals were probably not flightless until September or later (Williams 1980). In Arizona, molting seemed to occur from mid-June to mid-August (Swarbrick 1975). In New Mexico, Mexican Ducks appeared to begin to molt in late June and continued into part of September (Bevill 1970).

CONSERVATION AND MANAGEMENT

The situation for Mexican Ducks is most dire in the United States, where habitat loss has been extensive. In southeastern Arizona, Brown (1985) noted that an expanding livestock industry began to alter pristine wetland habitats by the late 1880s, as marshes were drained or reduced in size. Those that remained were quickly devoid of cover, due to the high stocking rates of cattle, which continued through World War I, when dry farmsteads began to appear, reaching a peak in the 1920s. These farmsteads subsequently declined until after World War II, when the development of deep-well pumps allowed access to the underlying aquifer and large-scale agriculture arrived in the region, which continues today. Agriculture, however, brought with it farm ponds, stock ponds, and pump-back ponds, the latter developed to reuse surface-irrigation water. These man-made water bodies provide significant habitat for Mexican Ducks, although pump-back ponds are now being replaced with sprinkler irrigation systems. In contrast, natural wetlands have been lost or dramatically reduced in size, which creates a significant need for habitat restoration. The wetland situation is similar in other areas within the Mexican Duck's range in the southwestern United States.

In Mexico, wetlands in the interior highlands have been lost or substantially altered, as this area is intensively populated, farmed, and grazed; wetlands here have been extensively drained, and their water diverted for human needs (Leopold 1959, Dickerman 1963). The pattern of wetland destruction began with the original Native American inhabitants, was continued by the colonial Spanish, and then greatly accelerated during the twentieth century. Many of the best Mexican Duck habitats were associated with large lake basins, which had no outlets, but these have all been altered by reduced runoff (due to deforestation of the surrounding mountain areas), upstream reservoir construction, and overgrazing (Saunders and Saunders 1981). The extensive marshes located on the Río Lerma delta, at the eastern end of Lake Chapala in Jalisco and Michoacán, once covered about 1,000 km², but these have now been mostly converted to croplands. The lake itself, once the largest in Mexico, has also been dramatically reduced in size. Goldman (1951) considered the Lerma marshes, located east of Toluca in the Federal District (México D.F.), to be an important breeding area for Mexican Ducks, but these marshes have been almost completely usurped for agriculture. Lake Cuitzeo, once Mexico's second-largest lake, has lost most of its wetland habitat since water sources to the lake were diverted. The majority of the other large wetlands in the interior highlands have suffered the same fate as they, too, were drained for agriculture and their upstream water sources were diverted for irrigation (Arellano and Rojas 1956). Mexican Ducks have adapted nicely, however, to the expansion of agriculture and its associated man-made habitats (e.g., ponds and presas), so their current population is not under threat.

Genetic research with larger samples sizes from

throughout the range of Mexican Ducks is needed to assess the level of introgressive hybridization with Mallards. Currently, the only data on molecular genetics are from 4 birds in San Luis Potosí (McCracken et al. 2001) and a single bird from Arizona (Johnson and Sorenson 1999). There is also a significant need for more detailed studies on their demographics, behavior, and habitat use, especially in the agriculturally modified areas that now characterize most of the Mexican Duck's range.

Mottled Duck

Anas fulvigula (Ridgway 1874)

Left, Western Gulf Coast Population. *Top*, hen; *bottom*, drake
Right, Florida Population. *Top*, hen; *bottom*, drake

The Mottled Duck is a nonmigratory species very closely related to the American Black Duck and the Mallard. Mottled Ducks occur in 2 populations: Florida and Western Gulf Coast. Some authorities recognize these populations as distinct subspecies, *A. f. fulvigula* in Florida (also called the Florida Duck), and *A. f. maculosa* (sometimes known as the Gulf Coast Mottled Duck) along the western Gulf Coast. Subspecies status seems warranted, based on the genetic differentiation and geographic isolation between the 2 groups, which is supported by band-recovery data. The Florida Population largely occurs on inland freshwater areas throughout the Florida peninsula, while the Western Gulf Coast Population inhabits the coastal marshes and inland coastal prairies of Texas and Louisiana. The sexes are similar in plumage, but the adults can be readily separated by their bill coloration. Mottled Ducks have an extended breeding season, which can last from January through July and even into August. Peak nesting, however, generally occurs in March and April, but it is influenced by precipitation, which can affect nest initiation and the percentage of females that ultimately attempt to nest. Mottled Duck numbers are not large, with breeding population estimates of about 53,000 in Florida, 25,500 in Texas, and 129,000 in Louisiana. Breeding populations are very small in Mississippi and Alabama. Little is known about breeding densities in Mexico, where Mottled Ducks occur along the east coast as far south as the Alvarado Lagoons in the state of Veracruz. The Florida Population is under threat from hybridization by feral Mallards, as well as from an introduced population of Mottled Ducks into South Carolina that has now expanded to Georgia and exhibits some genetic exchange with the Florida Population. The Western Gulf Coast Population is jeopardized by substantial and ongoing habitat loss and the alteration of coastal wetlands and adjacent inland prairies. Mottled Ducks along the Gulf Coast also exhibit a high incidence of lead-shot ingestion, because they are year-round residents in some of the most heavily hunted areas of North America, although this problem seems to be slowly abating in response to the requirement that steel shot must now be used for waterfowl hunting.

Mottled Duck, Western Gulf Coast Population. Sexes are similar

Mottled Duck, Florida Population. Sexes are similar

IDENTIFICATION

At a Glance. At first glance, the sexes appear similar, with a mottled dark brown plumage that appears black at a distance, but is distinctly browner than that of American Black Ducks. The throat and the face are usually lightly streaked; the speculum is bluish; there is no white in the tail. Adult males generally have a yellow bill, while the bill of the female is orange yellow, with black spots

or a saddle on the upper mandible. When together, the female can also be distinguished from the male by her smaller size.

Adult Males. The head and the neck are buffy brown, with variable light streaking. The body plumage and the upperwing coverts are dark brown, with buffy edges that give these ducks a mottled appearance. In flight, the white lower wing coverts contrast sharply with the brownish body plumage. The speculum is bordered in front by a single black bar, and in the rear by a thin white bar that can be obscure or lacking on Florida birds (Palmer 1976a). The legs and the feet are bright orange during the breeding season, but dull to bright orange during nonbreeding periods. The bill is yellow during breeding, but yellow orange to olive during nonbreeding. The irises are dark brown.

Adult Females. The overall body coloration of females is similar to that of males, but the females have wider buffy-brown feather edges than the males, which gives them a subtle but distinctly lighter appearance (Gray 1993). Females are also noticeably smaller than males. The legs are dull orange, although some are bright orange in autumn (Gray 1993). The bill is a dull orange yellow to olive and, unlike that of the male, has brown or black spots or a saddle on the upper mandible, concentrated near the middle of the bill. The irises are brown.

Juveniles. Young Mottled Ducks are distinguished from immature Mallards by dark tail feathers and a uniformly dark belly. When compared with adult birds, juveniles have thin, wispy, lighter brown feathers, with a duller coloration on the tips. The legs are dull orange to orange. The bill is a dull brownish gray.

Ducklings. Although similar to Mallard ducklings, those of Mottled Ducks are pale yellow below and blackish brown above, with a dark eye stripe, an orange-yellow face, and an inconspicuous ear spot (Nelson 1993). The feet are variably dark, instead of the lighter feet in Mallards. The bill is olive.

Above:
Mottled Ducks in flight.
Dave Irving

Left:
Mottled Duck pair; note
the yellow bill of the
male (*foreground*).
Ian Gereg

Mottled Duck pair; note the orange bill of the female (*foreground*). *GaryKramer.net*

Voice. Mottled Duck vocalizations generally resemble those of Mallards. The males produce a single low, guttural *raeb* alarm call, with a 2-note *raeb-raeb* given during courtship (Johnsgard 1965). Females have a loud decrescendo call that usually consists of 6 notes, with the second note tending to be the loudest and highest pitched. Females sound an alarm with 3–4 harsh quacks; they give a *tickety-tickety-tickety* call when feeding and, at times, in flight.

Similar Species. Mottled Ducks are similar in size, shape, and appearance to American Black Ducks; the 2 species can be difficult to distinguish in flight, but American Black Ducks have silvery-white underwing linings that contrast sharply with their dark bodies. The bluish speculum of Mottled Ducks is also much paler than the deep purplish speculum of American Black Ducks. Close up, Mottled Ducks usually lack any dark streaking (present on American Black Ducks) on an otherwise buffy-brown chin; the Latin name *fulvigula* means "tawny throat." American Black Ducks are also uncommon to rare within the range of Mottled Ducks. Female Mallards have white in the tail and a blue speculum bordered by white in front and in back, compared with white only on the trailing edge of the speculum in Mottled Ducks.

Weights and Measurements. Stutzenbaker (1988) reported average weights of Mottled Ducks during autumn in Texas of 1,081 g for 116 males (range = 830–1,330 g) and 967 g for 136 females (range = 590–1,380 g). Wing length averaged 262 mm for males ($n = 26$) and 249 mm for females ($n = 27$). Bill length averaged 65 mm for males ($n = 219$) and 60 mm for females ($n = 221$). In Florida, the ingesta-free body weight of 86 adult males averaged 1,043 g (range = 876–1,241 g); 71 females averaged 934 g (range = 699–1,151 g; Moorman and Gray 1994). Bill length was noticeably smaller than for the Texas birds, averaging 54 mm for males ($n = 21$) and 51 mm for females ($n = 51$). No measurements of wing length were provided.

DISTRIBUTION

The Mottled Duck is part of the "Mallard complex," which, in North America, also includes the American Black Duck, the Mallard, and the Mexican Duck. All 4 species are closely related;

439

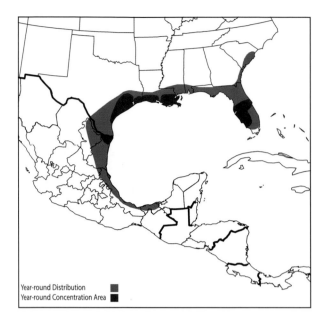

Year-round Distribution
Year-round Concentration Area

the DNA barcodes (short genetic markers used to identify different species) are virtually indistinguishable among the Mottled Duck, the American Black Duck, and the Mallard (Kerr et al. 2007). Mottled Ducks, American Black Ducks, and Mexican Ducks all have monochromatic plumage.

McCraken et al. (2001) suggest that Mottled Ducks were most closely related to Mexican Ducks, which (along with American Black Ducks) are descendants of the first monochromatic Mallard-like ancestor to invade North America from Asia. Johnsgard (1961b) considered Mottled Ducks to be a kind of Mallard and recognized 2 subspecies: the Florida Mallard (*Anas platyrhynchos fulvigula*) and the Mottled Mallard (*A. p. maculosa*). The American Ornithologists' Union (1998) recognizes the Mottled Duck as a distinct species (*Anas fulvigula*) with no subspecies, as did Livezey (1997), who included *A. f. maculosa* within *A. fulvigula*. Genetic studies conducted by McCracken et al. (2001), however, show distinct differences between the subspecies *A. f. fulvigula* and *A. f. maculosa*. They noted that an enduring geographic split has existed between the 2 subspecies for many years; gene flow between them is nonexistent, or at least undetectable. Williams et al. (2005) also found

limited gene flow between the Florida and Texas populations of Mottled Ducks. Each population possessed rare and unique alleles (DNA codings for specific genetic traits), and the loci (particular positions on a specific chromosome) for both allozymes (different forms of an enzyme coded by different alleles) and microsatellites (repeating sequences within a DNA fragment) exhibited differences (heterozygosities) in allele frequencies between populations. Over all loci, however, only 5%–6% of the variation was partitioned between populations. The Florida Population exhibited lower allozyme heterozygosity and allelic diversity than the Texas population, while microsatellite heterozygosities and allelic diversity were similar between the populations.

Extensive band-recovery data also have not revealed any evidence of mixing between the 2 subspecies. Of 2,075 recoveries from 19,294 Mottled Ducks banded in Florida during 1950–2010, almost all (99.4%) were in Florida, with none in Texas or Louisiana. The most distant recovery was a single bird in New Jersey. Similarly, none of the Mottled Ducks banded in Texas (*n* = 26,762) or Louisiana (*n* = 37,561) were recovered in Florida (1950–2010). Of the 4,564 recoveries from Mottled Ducks banded in Louisiana, most (90.9%) were in Louisiana, with a few (8.9%) in Texas. The most distant recovery was a single bird in South Dakota; none were recovered in Mexico. Of the 3,547 recoveries from Mottled Ducks banded in Texas, most (76.7%) were in Texas, but 22.2% were in Louisiana, indicating a stronger movement from Texas to Louisiana than Louisiana to Texas. In addition, there were 33 (0.9%) recoveries in Mexico. The farthest displacements were single birds recovered in Indiana and Tennessee.

Comparing subspecies, the western subspecies (*A. f. maculosa*), sometimes referred to as the Gulf Coast Mottled Duck, is slightly darker, the streaking on the cheeks and the neck is more pronounced, and the greenish-blue speculum has a more bluish cast when compared with *A. f. fulvigula*, also called the Florida Mottled Duck. Additionally, males of the Florida subspecies tend to have bright yellow bills, compared with olive yellow in the western subspecies. These 2 subspecies are managed and referred to as the Florida Mottled Duck Population and the Western Gulf Coast Mottled Duck Population.

The general range of Mottled Ducks extends from the Florida Peninsula west along the Gulf Coast, from Alabama, through the Pearl River marshes on the Mississippi and Louisiana border, to the Texas coast, and then south in Mexico to the Alvarado Lagoons near Veracruz. Of the 2 recognizable subspecies, the eastern subspecies (*fulvigula*) inhabits peninsular Florida, while the western subspecies (*maculosa*) is found along the Gulf Coast from Mobile Bay in Alabama south into Mexico.

Mottled Ducks from the Western Gulf Coast Population in Texas and Louisiana were introduced to the Santee River delta and the ACE basin (named after the Ashepoo, Combahee, and Edisto Rivers, which drain the area) of coastal South Carolina during 1975–82, where they established a population that has since spread to Georgia. This movement is significant, because some of these introduced birds have dispersed to Florida, which created a gene flow into the isolated Florida Population of Mottled Ducks (Weng 2006). Another significant threat to the genetic uniqueness of Mottled Ducks is hybridization with Mallards. In a genetic analysis of 225 Mottled Ducks and Mallards in Florida by Williams et al. (2005), 10.9% of the Mottled Ducks were hybrids, as were 3.4% of the Mallards. Hybridization rates of Mottled Ducks from different geographic areas of Florida varied from 0% to 24%. In contrast, genetics data could no longer differentiate between Mallards and Mottled Ducks in South Carolina; Weng (2006) had similar findings.

Breeding. In Florida, Mottled Ducks breed throughout most of the peninsula, from Alachua County south to Cape Sable, but they are most common in the prairie wetlands around Lake Okeechobee and in the Everglades Agricultural Area (EAA) south of the lake, where wetlands once occupied 33.6% of the EAA (Johnson et al. 1991). Relatively high densities also occur in the counties encompassing the St. Johns River. Densities in the prairies and EAA averaged 0.639 birds/km² in 1980–88, compared with only 0.163/km² outside those regions (1985). Ponds and ditches in suburban areas along both Florida coasts also contain high densities of Mottled Ducks. Over half of the Florida Population may occur in such urban/suburban areas (Bielefeld 2008).

Along the Gulf Coast, Mottled Ducks are uncommon permanent residents in the Mobile Bay area of Alabama (Imhof 1976), as well as in Mississippi, where the largest concentrations of breeding birds occur in the Pascagoula River Marsh (Toups and Jackson 1987). Farther west, moderate densities are found in the Rice Belt regions of southwestern Louisiana and southeastern Texas. In southwestern Louisiana, densities recorded via aerial surveys in 1984–85 were 0.23–1.11 birds/km² from April to July, highest (at 2.65/km²) in August, and 0.46–0.81/km² from October through March (Zwank et al. 1989).

Mottled Ducks are resident along the Gulf Coast and reach their highest densities in the fresh and intermediate marshes of the Louisiana coast and southeastern Texas (Bielefeld et. al. 2010). Along the eastern Texas coast, ground counts in April 1994–95 estimated about 105,000 breeding pairs (Ballard et al. 2001). Mottled Ducks are found at lower densities in coastal southern Texas and in Mexico, south to Laguna de Tamiahua on the coastal plain in the state of Veracruz (Stutzenbaker 1988); at times, small numbers occur as far south as the Alvarado Lagoons, also in Veracruz (Baldassarre et al. 1989).

Winter. Mottled Ducks are year-round residents throughout their range, so their winter and breeding ranges are the same.

Migration. Mottled Ducks are nonmigratory, although movements can occur in response to changes in habitat conditions, such as the drying of wetlands (Stutzenbaker 1988), with some individuals moving as far as 400 km from their banding sites. Although movements have been documented as far as 400 km from a banding site, Stutzenbaker remarked that few (if any) Mottled Ducks in southern Texas needed to move >161 km, even during extreme late-summer droughts. During the winter, Mottled Ducks breeding in the rice prairies of Texas shift to coastal marshes in southeastern Texas and Louisiana when rice fields are not available. In Florida, 105 recoveries of 668 Mottled Ducks from the Florida Population banded in 1969–70 demonstrated some mobility, but no general pattern indicative of migration (Fogarty and LaHart 1971). Of 690 recoveries of Mottled Ducks banded in Florida from 1977 to 1991, the average distance between the banding and the recovery sites was 38 km, again demonstrating that Mottled Ducks are not very mobile (Johnson et al. 1995). Overall, movements in Florida have been linked to surface water availability (Bielefeld and Cox 2006).

MIGRATION BEHAVIOR

Mottled Ducks do not undergo a molt migration, but they will move short distances to preferred molting marshes. Moorman et al. (1993) studied the molting ecology of Mottled Ducks in southwestern Louisiana (see Molts and Plumages).

HABITAT

Mottled Ducks inhabit fresh, brackish, and saline coastal marshes, as well as coastal and inland prairies and flooded rice fields. They generally select habitats with abundant emergent vegetation and water levels <15 cm deep, but they will use deeper areas during the wing molt, the winter dry season, and periodic droughts (Bielefeld et al. 2010).

During a 5-year study (1986–90) in Florida, Mottled Ducks seemed to prefer ditches (4 out of 5 years) and emergent wetlands (3 out of 5 years), but they avoided wet prairies and shrub and for-

ested wetlands (5 out of 5 years), open water (4 out of 5 years), and flooded uplands (4 out of 5 years; Johnson et al. 1991). Mottled Ducks in the upper St. Johns River basin in Florida used the thousands of ponds on ranches and farms, as well as irrigation reservoirs associated with citrus crops (Bielefeld and Cox 2006). During 2009–10, radiomarked female Mottled Ducks captured in urban/suburban areas near West Palm Beach, Florida, remained in those areas, where they used man-made pond and ditch habitats. In contrast, females captured in the EAA near Lake Okeechobee, a rural region, used storm-water-treatment areas, agricultural impoundments, and pasture ponds in the fall and early winter. As the dry season progressed, and ponds and storm-water-treatment areas dried up from drought conditions in midwinter through early spring, these ducks moved to more permanent marshes in the Everglades and around Lake Okeechobee, but they began to return to the EAA in late May (Bielefeld 2011). Mottled Ducks in the EAA also used flooded rice fields, ditches, and canals. Such data suggest that the freshwater marshes of Lake Okeechobee and the Everglades are important winter habitats for Mottled Ducks in southern Florida, both when dry conditions limit the availability of more ephemeral wetlands and, to some extent, during periods of normal water conditions.

In Texas and Louisiana, Mottled Ducks occurred in fresh, intermediate, and brackish ponds in the coastal marshes; emergent wetlands in the prairies; and agricultural lands, primarily rice fields (Grand 1988, Zwank et al. 1989). In the Rice Belt of southeastern Louisiana (1984–85), where 77% of the study area was in agricultural production, rice fields were preferred during April, May, and June, but fresh marshes were favored from August through March (Zwank et al. 1989). Their reduced use of flooded harvested rice fields in July, August, and October, however, may have reflected the researchers' difficulty in observing birds in this type of habitat. Mottled Ducks were never seen in harvested fields of soybeans, sorghum, or wheat, nor in woods or residential areas.

At the San Bernard National Wildlife Refuge (NWR) on the upper Texas coast, Mottled Ducks usually selected nontidal, fresh and brackish wetlands (Grand 1988). Within these general categories, Mottled Ducks chose intermediate–brackish marshes, impounded fresh marshes, and intermittent streams / sloughs. Solitary females and broods, however, were never observed in saline marshes or tidal riverine habitats. Nonbreeding birds used impounded fresh marshes, brackish marshes, and intermittent streams / sloughs. On freshwater sites at the Welder Wildlife Refuge in Texas (Oct–Dec 1973), Mottled Ducks favored shallow water with abundant emergent vegetation, a low pH, and low oxygen levels (White and James 1978).

On the Texas Chenier Plain NWR Complex (2004–5), Haukos et al. (2010) estimated that on 22,454 ha of coastal marshes, there were 18,830 ponds available to Mottled Duck pairs. Pond use was then assessed on 634 ponds that spanned the array of habitat types in the complex: 1.1% fresh ponds, 61.9% intermediate ponds, 35.3% brackish ponds, and 1.6% saline ponds. Mottled Ducks showed a distinct preference for ponds in fresh marshes (11.3% of the observations occurred on 1.1% of the available habitat), but they used other habitat types in proportion to their availability. The structure of the vegetation surrounding the ponds was also important. Mottled Ducks preferred fresh ponds with short surrounding vegetation that had been recently grazed by cattle, but avoided sites that were recently burned (<3 months). On average, ponds in fresh marshes measured 0.02 ha, with a water depth of 11.4 cm and 51.6% emergent plant cover. In contrast, ponds in intermediate marshes averaged 0.06 ha, with a water depth of 13.5 cm and 21.2% emergent cover. The averages for all ponds that were used were 0.12 ha in size and 9.9 cm in water depth, with 19.9% emergent plant coverage and a salinity of 2.81 ppt. Pond occupancy was estimated at 1.3% in 2004 and 2.5% in 2005.

Habitat and Body Condition

Moorman et al. (1992) studied the dynamics of body mass and composition of Mottled Ducks during fall (1 Oct–16 Nov) and winter (17 Nov–28 Feb) in 1987/88 and 1988/89 at the Rockefeller State Wildlife Refuge in southeastern Louisiana. During these time periods, both sexes replaced body mass and lipid reserves used during reproduction and molting. Adult males ($n = 69$) increased their body mass by 92–115 g and their lipids by 55–71 g from fall through late winter; the increases in adult females ($n = 48$) were 120 g of body mass and 85 g of lipids. Lipid reserves for both sexes would meet their existence energy requirements for 5–9 days during the winter. It is doubtful that such reserves are an adaption to winter conditions, however, because periods of extreme winter weather and any resultant potential food shortages are uncommon in coastal Louisiana, and they rarely persist for >2–3 days. Lipid reserves more likely are accumulated because the mild climate in Louisiana allows early breeding opportunities in the late winter and early spring. Mottled Ducks probably are able to take advantage of these breeding opportunities, because they do not incur the rigors associated with migration. The protein mass of males and females did not vary within winter periods, but differences of up to 39 g occurred between winters. These differences were associated with changes in gizzard mass and intestine length, which probably correlated with different habitat conditions and/ or diets between years. The patterns of immatures ($n = 62$) were similar to those of adults.

POPULATION STATUS

Mottled Duck numbers fluctuate with changing water conditions in Florida and along the Gulf Coast of Texas and Louisiana. Historical spring populations in Florida (1948–57) varied from a high of 30,000 in 1952 to a postdrought low of 7,000–10,000 in 1957 (Sincock 1957). Johnson et al. (1984) estimated an average fall population of 67,000 from 1977 to 1980 (range = 58,000–75,000).

In 1985–90, the spring population was estimated at 28,000 and the fall population at 56,000 (Johnson et al. 1991). The most recent estimate in Florida was 53,328 for the spring population, with survey data showing a weak increasing trend in spring densities from 1985 to 2006 (Bielefeld et al. 2010), but survey numbers also include Mallards and Mallard × Mottled Duck hybrids (Bielefeld 2008). Hence the Florida survey overestimates Mottled Duck numbers and needs correction. The Midwinter Survey (2000–2010) averaged 1,049 Mottled Ducks in Florida, 15 in South Carolina, and 19 in Georgia. Trend data from the annual Breeding Bird Survey show a statistically nonsignificant decline of 1.2%/year in Florida from 1966 to 2007, but a significant increase of 8.8%/year when examined from 1980 to 2007.

The Western Gulf Coast Population of Mottled Ducks is much larger than the Florida Population, but there have been no long-term surveys assessing the status of the former. Starting in 2008, however, aerial transect surveys of this population have been jointly conducted by the U.S. Fish and Wildlife Service, the Louisiana Department of Wildlife and Fisheries, and the Texas Parks and Wildlife Department. The survey methodology was modified during the first 3 years (2008–10) to achieve better precision in the visibility correction factor and, therefore, to improve population estimates; hence comparisons among years are not valid. In 2010 the survey covered 1,140 km² in Texas and Louisiana, was extrapolated to 59,280 km², and yielded a total breeding population estimate of 129,207: 103,707 in Louisiana and 25,500 in Texas. The survey is still experimental, however, and may see further changes in its design and estimation methods.

As for other estimates of the Western Gulf Coast Population, Esslinger and Wilson (2001) used a mark-recapture analysis of 1994–97 banding data to estimate an expected midwinter population of 646,067: 33.7% in the Mississippi River coastal wetlands, 26.2% in Louisiana's Chenier Plain, and 25.9% on the mid-Texas coast. Ballard et al. (2001)

estimated 105,000 breeding pairs along the Texas coast in 1994–95.

Mottled Ducks in Louisiana increased in number in the late 1960s and early 1970s, as hurricanes created many new ponds in the otherwise dense marsh vegetation, resulting in improvements in the habitat quality of coastal marshes and, therefore, an extension of their breeding range (Bellrose 1980). Extensive marsh loss in Louisiana more recently, however, is probably offsetting habitat gains made during the 1960s and 1970s. An analysis by Johnson (2009) indicated a serious decline in the Western Gulf Coast Population from 1994 to 2006 (see Recruitment and Survival). Mottled Ducks, however, increasingly are exploiting food and habitats associated with Louisiana's rice industry and are extending their breeding range northward into the coastal prairies.

Available data yield conflicting information on population trends, varying from steeply declining to stable, depending on which datasets are examined; the associated studies were summarized by Bielefeld et al. (2010). Midwinter inventories from 1971 to 2009 suggest a stable trend (0.7% increase/year) in Louisiana, a 2.8% decline in Texas, and an overall stable trend (0.5% annual decline) for the 2 states (Haukos 2009). Midwinter averages from 2000 to 2010 were 25,763 Mottled Ducks in Texas and 70,562 in Louisiana. In contrast, however, breeding-pair surveys on national wildlife refuges along the Texas coast indicate a steep decline of 12%/year from 1985 to 2009, but this trend may not reflect the broader Western Gulf Coast Population. Finite population growth rates (λ, estimated from life-stage projection matrices) for this population (1994–2005) averaged 0.85 for males (range = 0.65–1.15) and 0.79 for females (range = 0.49–1.16), which suggests a rapidly declining population (Johnson 2009). Collectively, these data indicate a steep, long-term (1971–2009) decline of Mottled Ducks in Texas, a stable trend in Louisiana, and a long-term, stable to slightly declining trend for the entire Western Gulf Coast Population. Trend data (1966–2007) from the annual Breeding Bird

Survey show nonsignificant declines of 1.1%/year in Louisiana and 3.8%/year in Texas. Trends from 1980 to 2007 were also not statistically significant in Louisiana (1.1%/year) and Texas (−2.1%/year).

Little is known about the population status of Mottled Ducks in Mexico, although the breeding population in northeastern Mexico is relatively low. Additional research is necessary to determine breeding densities farther south in Mexico.

Harvest

The average annual harvest of Mottled Ducks in the United States was 61,659 from 1999 to 2008. Most of the harvest was from the Western Gulf Coast Population (80.4%), with 19.6% from the Florida Population. Of the Western Gulf Coast Population harvest, 74.1% was from Louisiana, 22.7% from Texas, and the remainder (3.2%) from Arkansas, Mississippi, and Alabama. Nearly all (91.8%) of the Florida Population harvest was from Florida, with 5.5% in South Carolina, and 2.7% in Georgia and South Carolina. Haukos (2009) calculated annual harvest rates of Mottled Ducks in the Western Gulf Coast Population (1997–2008) as 9.5% for adult males, 7.7% for adult females, 17.8% for immature males, and 14.3% for immature females.

BREEDING BIOLOGY

Behavior

Mating System. Mottled Ducks are seasonally monogamous. It is suspected that they may renew the pair bond with the same mate, but this behavior is not confirmed. Mottled Ducks probably breed as yearlings, although the percentage of all females that breed varies with habitat conditions. In Florida, Mottled Duck pairs have been observed during every month of the year (Stieglitz and Wilson 1968). On the Gulf Coast, pairing can occur as early as August, with the percentage of females in pairs increasing through the fall and early winter. In southwestern Louisiana (1980–82), the chronology of pair formation, expressed as the percentage of females in pairs ($n = 1,059$), was 21.3% in August,

70.6% in September, 82.8% in October, 84.0% in November, 93.0% in December, 95.8% in January, and 100% in February (Paulus 1988). Pairs were long term, with some remaining together until July or early August, although the pair bond is generally broken some time during incubation. In Louisiana, pairs were dominant to unpaired birds, winning 94% of all interactions. Immatures were observed forming pairs at 5 months of age. Along the upper Texas coast, Stutzenbaker (1988) noted that 70%–80% of the Mottled Ducks were paired in December and February, compared with only about 10% in September.

Early pairing by Mottled Ducks in Louisiana may have minimized interspecific pairing with Mallards, which formed their pair bonds later (Paulus 1988). Although Mallards outnumbered Mottled Ducks by a 10:1 ratio in late winter, they did not arrive in large numbers until November or December, by which time most (93%–96%) of the Mottled Ducks were paired. Unpaired female Mottled Ducks associated with unpaired males 86% of the time, and unpaired males were more often observed with other unpaired Mottled Ducks, compared with being alone or with other species. Only 9 (0.4%) Mottled Ducks paired with another species: 4 males paired with female Mallards, 4 females with male Mallards, and 1 male with a female American Black Duck.

Courtship averaged only 1% ($n = 1,188$ hours of observation) of the time budget of Mottled Ducks in Louisiana, with similar results for both paired and unpaired birds (Paulus 1988). Courtship began in August, but the greatest amount of courtship activity occurred from October through December. Most courtship behavior took place during the early morning and late afternoon. Courting groups ($n = 21$) averaged 4.4 males (range = 3–7) and 2.2 females (range = 1–7). Of 63 copulations observed, 2 occurred in September ($n = 213$ hours), 23 in October ($n = 195$ hours), 11 in November ($n = 150$ hours), 9 in December ($n = 206$ hours), 13 in January ($n = 252$ hours), and 5 in February ($n = 155$ hours). Weeks (1969) reported courting groups in Louisiana averaging 6.4 individuals (range = 3–10), of which 60% were males. Courtship activity occurred along the edges of large marsh lakes, in open-water areas between stands of cordgrass (*Spartina patens*), and in small fresh to brackish ponds.

Sex Ratio. The sex ratio of Mottled Ducks is close to even, which may be related to their reduced dimorphism, long pair bonds, and low male–male competition (Brown 1982). In Louisiana, the sex ratio from August through February 1980–82 averaged 53% males (Paulus 1988). In Texas, Stutzenbaker (1988) reported a sex ratio of 56% males, which was based on observation of 346 Mottled Ducks in a rice field, and 58% males, which was based on 887 birds trapped and banded. The sex ratio of 4,613 Mottled Ducks harvested in Texas over a 12-year period (1964–75) was 54.6% males.

Site Fidelity and Territory. As a resident species, Mottled Ducks by definition exhibit strong fidelity to their breeding areas, usually not traveling far from their natal sites. In Florida, the average recovery distance (n = 105) from a banding site was 56.5 km, with most (71.4%) recoveries within 78.9 km (Fogarty and LaHart 1971). In Texas, 82% of the adults and 53% of the immatures were recovered within the county of banding (Stutzenbaker 1988).

In Florida (2009–10), the mean annual home-range size of 50 adult females captured in the rural EAA was 888 km², compared with only 54 km² for 45 females captured in urban areas near West Palm Beach (Bielefeld 2011). Home ranges become dramatically reduced during the breeding season, however, as documented during a 3-year study (2006–8) in coastal Texas (Rigby 2008). The mean home-range size of 25 females averaged 14.0 km² (100% minimum convex polygon method) during the breeding season (1 Feb–30 Jun), although the size varied, depending on the calculation method and the number of observations/female. In coastal Louisiana, breeding home ranges were between 42.5 and 132.0 ha (Weeks 1969).

Mottled Ducks do not defend their home range, but males will defend loafing and feeding areas used by the pair (Baker 1983). Loafing areas range from 10 to 130 ha in size but are usually <40 ha (Weeks 1969, Allen 1980).

Courtship Displays. Johnsgard (1965), Weeks (1969), and Paulus (1988) described courtship behaviors of Mottled Ducks, which appear to be identical to those of Mallards and American Black Ducks. Major male displays are the *head-shake* and *introductory-shake*, the latter followed by the *grunt-whistle* and, in turn, by *nod-swimming*. A *down-up* display is given by a male or a group of males and directed toward a given female. Other male displays are *head-up-tail-up* and *turn-the-back-of-the-head*, which is followed by *preen-behind-the-wing* and seems to be important in forming the pair bond. Females *incite* toward their chosen mate, a display indicative that a pair bond is formed. Displays associated with pair-bond maintenance are *inciting* by the female and concurrent *leading* by the male. Both males and females follow these displays with *mock-preening* and *ritualized drinking*.

Both sexes perform precopulatory *head-pumping*, which is quickly followed by copulation. Postcopulatory displays by the male involve swimming around the *bathing* female while *bridling* (rapidly pulling the head back and inward), *nod-swimming*, and/or *turning-the-back-of-the-head*.

Paulus (1988) observed the courtship behaviors of male Mottled Ducks in Louisiana (n = 566) and recorded the occurrence of displays: *introductory-shake* (26%), *grunt-whistle* (22%), *head-up-tail-up* (15%), *head-shake* (13%), *nod-swimming* (13%), and *preen-behind-the-wing* (8%); *down-up* (1%) and *jump-flights* (1%) were rarely observed. *Nod-swimming* immediately followed other displays 23% of the time, primarily (86%) after *head-up-tail-up*. The displays of males in courtship groups were often highly synchronized among 2–4 males.

Agonistic activities among Mottled Ducks in Louisiana were infrequent, with *chasing* as the predominant agonistic activity of pairs. In contrast, unpaired birds rarely *chased* other birds; unpaired males mostly gave *bill-threats* and unpaired females *incited*. Agonistic activities were most common in October and November, the peak courtship period. Mallards were observed courting near groups of courting Mottled Ducks, but the 2 species displayed to each other in only 2 out of 21 (9.5%) courtship groups.

In Texas, Stutzenbaker (1988) commonly observed *3-bird-flights* during the first 2 weeks of March, which were probably indicative of territorial defense and not extra-pair copulations. Such flights most commonly occurred when a mated pair flew low over the waiting area of a lone male, who then rose and pursued the paired female in "highly acrobatic flights" as the female attempted to evade the pursuing male. Flights generally only covered 0.75 km before the pursuing male returned to the waiting area. Paulus (1984c) reported that extra-pair copulations in Mottled Ducks were uncommon, most likely because of firm pair bonds that did not dissolve until well into incubation, and a strong attachment to the loafing area. Most extra-pair copulations probably were attempted by males whose females had nested successfully or were incubating.

Nesting

Nest Sites. Pairs search for suitable nest cover by flying low over nesting habitat, but the female searches on the ground for the exact nest site (Bielefeld et al. 2010). Nests in coastal marshes are usually situated on natural ridges or other areas of elevated topography. In contrast, nests in rice-production areas are on levees, road medians, fallow fields, and pastures (Rorabaugh and Zwank 1983).

Preferred nesting habitat in coastal marshes is characterized by tall, dense stands of grasses well above the high-tide mark, because nests at lower elevations are prone to flooding; areas with thick shrubs are avoided (Stutzenbaker 1988). Nests in dense stands of cordgrass (*Spartina* spp.) can be suspended in the vegetation, 75–90 cm above the ground (Rorabaugh and Zwank 1983, Stutzenbaker 1988). Females also add material to their nests in an effort to keep them above floodwaters.

In coastal Texas, Stutzenbaker (1988) found most nests ($n = 315$) in dense stands of cordgrass near permanent ponds within a marsh; 53 nests averaged 119 m from water (range = 15–219 m). The highest nest densities (a maximum of 1/1.4 ha) were on well-drained ridges, compared with 1/259 ha across all nesting habitats. All but 2 nests were located in supporting cordgrass 7.6–15.2 cm above the ground; they were completely concealed from above by overhanging cordgrass, and often by solitary shrubs of baccharis (*Baccharis halimifolia*) as well. Only 1 out of 315 nests in cordgrass habitat did not have any overhead cover. On the mid-Texas coast (2000–2002), nests ($n = 39$) were located in cordgrass prairie (54.2%), improved pasture (25.5%), and fallow rice fields (14.9%; Finger et al. 2003). Nests were typically found in robust vegetation >60 cm tall, with 75% overhead canopy coverage.

Completed nests on dredge-spoil islands on the eastern coast of Florida did not have a canopy (Stieglitz and Wilson 1968). Vegetative height at the nests averaged 86 cm (range = 15–244 cm). Plant types were recorded for the locations of 88 nests, with the majority (55.6%) found in seaside paspalum (seashore paspalum; *Paspalum vaginatum*) and 18.1% in broomsedge (*Andropogon* sp.). Paspalum stands normally were 38–61 cm tall, which provided excellent nest concealment. Nests averaged 8.5 m from water, with 78.9% within 3.0–12.2 m. The highest total nest density on a single island was 9.4/ha.

In contrast, Mottled Ducks nesting in east-central Florida used an array of nesting habitats: moist-soil plant communities, cattle pastures, sod fields, gravel-mining areas, pine-flatwood plant communities, citrus groves, and scrub-oak plant communities, with the choice of moist-soil areas apparently occurring only during a drought (Bielefeld 2002). One nest perched 1 m above the ground in dense paragrass (*Brachiaria mutica*). The mean distance to water for 25 nests averaged 170 m (range = 100–620 m), with most (84%) nests located closest to ditches and canals. The mean vegetative height at nest sites was 80.7 cm, and the mean canopy coverage was 90%. Both the mean height and the density of the vegetation at nest sites were greater than that observed at random locations around the nest.

Near Lake Okeechobee in Florida (1997–99), Mottled Duck nests (*n* = 25) were mostly (56%) located in hayfields, and 36% were in cattle pastures (Dugger et al. 2010). Vegetative height at nest sites (*n* = 24) averaged 68.8 cm, and the nests were well concealed from above, with 67% located in >76% overhead cover. Planted pasture grasses, such as bahiagrass (*Paspalum notatum*), were the most common vegetation at nest sites, occurring 72% of the time and dominant 62% of the time. Additional dominant plants (all at 25%) were sedges (*Carex* spp.), wiregrass (*Aristida stricta*), and saw palmetto (*Serenoa repens*). Distances from the nests (*n* = 25) to water averaged 188 m.

On 12 islands in the Atchafalaya River delta of Louisiana, nesting females preferred shrub-moderate habitats and avoided shrub-sparse and marsh habitats (Holbrook et al. 2000). Shrub-moderate habitats were characterized by herbaceous and woody vegetation <6 m tall, ground cover >30%, and a 10%–<50% cover at 1.5 m. Nest density (determined from plot surveys) averaged 3.9/ha, but it was 8.1/ha on the 4 islands that contained 91% of the nests; the highest density was 12.4/ha. In the Mississippi River delta of Louisiana (1998–99), nests (*n* = 279) were located in cattle pastures (47.7%), spoil islands (43.0%), and canal banks (5.7%; Walters et al. 2001).

On agricultural lands in southeastern Louisiana (1999–2000), female Mottled Ducks preferred to nest in permanent pastures with knolls (53%) and in idle fields (22%; Durham and Afton 2003). Of the 65 nests that were monitored, the average distance to water was 185 m; average vegetative height, 49 cm; average vegetative density, 7 (Robel pole index); and average plant species present, 10. In comparison, random locales averaged 32 cm in vegetative height and 5 (Robel pole index) in vegetative density. The number of plant species present (10) was the same at nest sites and random locations, but the composition of nest sites had a greater percentage of wax myrtle (*Morella cerifera*), basketgrass (*Dichanthelium acuminatum*), dewberries (*Rubus trivialis*), broomsedge, and Vasey's grass (*Paspalum urvillei*).

The female constructs the nest. Nests on dredge-spoil islands near Merritt Island on the eastern coast of Florida (1965–67) began with a bowl or scrape (*n* = 114), generally 5.1–7.6 cm deep and 15.2–20.3 cm wide (Stieglitz and Wilson 1968). The nearest vegetative material was then used to construct the nest, with seashore paspalum appearing in 71.6% of the completed nests located on spoil islands. Broomsedge was also commonly used as a nesting material. Females add material to the nest during laying and incubation, with down added after the fifth or sixth egg is laid. Completed nests in Florida averaged 26.2 cm (range = 20.3–33.0 cm) in diameter (Beckwith and Hosford 1955).

Clutch Size and Eggs. Clutch size in Florida averaged 9.4 eggs for 78 complete nests (range = 5–13), but before 1 May the average was 10.1 eggs, versus 8.9 after that date (Stieglitz and Wilson 1968). In coastal Texas, clutch size averaged 9.5 for 65 nests located in March and April (range = 8–13 eggs), and 8.0 (range = 5–10 eggs) for 26 nests located from late May through June (Stutzenbaker 1988). On islands in the Atchafalaya River delta of Louisiana (1994–96), clutch size for 307 nests averaged 9.2 eggs, but clutch size declined with later nest-initiation dates and differed among years (Johnson et al. 2002). The mean clutch size was smaller in 1994 (8.8) than in 1995 (9.1) or 1996 (9.3). The rate of clutch-size decline also varied among years: −0.03 eggs/day in 1994 to −0.06 eggs/day in 1995.

Mottled Duck eggs are oval to elliptical oval,

and they range in color from creamy white to lightly green-tinted white; a few eggs are tan colored (Stieglitz and Wilson 1968). In Louisiana, eggs from 98 nests averaged 54.9 × 40.8 mm and weighed 50.7 g (Johnson et al. 2002). Egg mass was not affected by the clutch size, the nest-initiation date, or interactions between the 2 factors. Individual females accounted for 51% of the variation in egg mass. The laying rate is 1 egg/day (Stieglitz and Wilson 1968).

Incubation and Energetic Costs. The incubation period was 25–26 days in Florida (Stieglitz and Wilson 1968) and 24–28 days in Texas (Stutzenbaker 1988). On average, females in Louisiana ($n = 12$) spent 83.0% of their time on their nests and took 1.8 recesses/day of 135 minutes each (Afton and Paulus 1992). During recesses (88.2 hours of observation), females averaged 61.7% of their time feeding, 15.4% preening, 10.4% resting, 0.1% alert, and 4.5% in forms of locomotion (Paulus 1984c). In Florida, 32% of the females feigned injury after being flushed off their nests by researchers (Stieglitz and Wilson 1968). Upon returning, the females usually landed "some distance" away before walking to the nest.

Early in the incubation period, the male waits for his mate at a particular place within their home range, but he will often fly to the incubating female when she leaves the nest (Stutzenbaker 1988). Males, however, have usually abandoned their females by mid- to late incubation. In Louisiana, males accompanying their females during incubation (58.6 hours of observation) spent 56.3% of their time feeding, 15.8% resting, 13.4% alert, 10.2% preening, and 3.3% in locomotion activities (Paulus 1984c).

Lipid reserves of males and females increased by as much as 51% during the postbreeding period (Sep–Jan), reaching a peak in January for males and during the period of rapid follicle growth for females (Moorman 1991, Moorman et al. 1992, Gray 1993). Lipids then declined dramatically during reproduction: 51% in males and 83% in females.

Lipid reserves for females were lowest at the end of incubation (23 g; $n = 5$) and remained at low levels during brood rearing.

Nesting Chronology. Mottled Ducks have an extremely long nesting season. Compared with duck species nesting farther north, Mottled Ducks initiated their nests very early, most likely because of the mild climatic conditions throughout their range during early winter and spring (Moorman et al. 1992). On the Mississippi River delta of Louisiana, the earliest nests were initiated in January, and the latest nests in July and even August; hence the nesting season can be as long as 7–8 months (Walters 2001). Most nests in Texas and Louisiana are initiated in March and April, although fall and winter rainfall affects nesting chronology.

During a 3-year study (1985–87) at the San Bernard NWR along the Texas coast, mean nest-initiation dates varied by as much as 68 days: 29 March in 1985, 7 May in 1986, and 28 February in 1987 (Grand 1992). The mean dates were latest during years with the lowest October–February rainfall, which affected habitat availability that correlated with water levels on refuge impoundments ($r = 0.58$–0.78). Reduced water levels may lower the availability of food for prenesting females, which would delay nesting. Along the mid-Texas coast (2000–2002), Mottled Ducks were also affected by precipitation, which varied from only 50% of the 30-year average in 2000 and 2002 to 18% above that average in 2001 (Finger et al. 2003). Subsequently, only 31%–33% of the females initiated nests during the 2 dry years, compared with 77% during the wet year. Female body weights were also lighter in the dry years (823–842 g; $n = 77$), compared with 885 g in the wet year ($n = 33$). On the Texas Chenier Plain NWR Complex on the upper Texas Coast (2006–8), the nesting propensity (inclination to nest) of 39 radiomarked females ranged from 15% to 63%; it, too, was reflective of wetland habitat conditions that were influenced by precipitation (Rigby 2008).

Other studies have also reported a highly variable nesting propensity among years, which was correlated with precipitation and wetland habitats. During a 3-year study (1994–96) in coastal Louisiana, the mean nest-initiation date, based on 324 nests, was 20 April, with an average span of 103 days between the first and the last nest (Johnson et al. 2002). Nesting chronology varied among years, however, with the mean dates of nest initiation being 30 April 1994, 14 April 1995, and 15 April 1996. The longest nesting season occurred during 1994, with the first nest initiated on 3 March and the last on 2 July (119 days); this nesting season was preceded by the most October–February rainfall. Although nesting chronology differed among years, the earliest and latest mean nest-initiation dates were separated by only 16 days. In the Rice Belt region of southwestern Louisiana, Durham and Afton (2006) concluded that nest initiation was stimulated when rice fields were flooded, which then provided loafing and feeding areas.

Near Lake Okeechobee in Florida (1997–99), the nesting propensity of 25 radiomarked females was 36%–56% during the wettest year (27 cm above the 30-year average) versus 22%–27% for 18 females during the driest year (79 cm below the average; Dugger et al. 2010). Mean nest-initiation dates here also varied with precipitation, averaging 31 March during the wettest year, compared with 20 May during the driest. Over all 3 years of the study, nests were initiated from 31 January to 4 July.

Nest Success. Mottled Duck nest success varies among years, locations, and habitat types within locations. In Florida (1965–67), the apparent nest success of dredge-spoil islands near Merritt Island was high, at 77% for 117 nests (Stieglitz and Wilson 1968), but spoil islands are not common nest sites for Mottled Ducks in Florida. In contrast, nest success was lower on inland sites: 32% in eastcentral Florida, 40.9% in southern Florida, and 9.4% in central Florida (Bielefeld et al. 2010). Near Lake Okeechobee (1997–99), nest success for 25 nests was only 9.5% (Dugger et al. 2010).

In Texas, apparent nest success was 24.7% for "closely observed nests" ($n = 146$) during the study by Stutzenbaker (1988), and 28% ($n = 108$) during an earlier study by Singleton (1953). In the Mississippi River delta, nest success (Mayfield Method) was 17.8% for 143 nests located in 1998, and 22.3% for 136 nests in 1999 (Walters et al. 2001). Nest success varied markedly among habitat types, however, with success highest on the Mississippi River levees (42.4%–56.7%) and spoil islands (21.3%–25.8%). Success was only 2.7%–7.0% in cattle pastures and 0.3%–4.3% on canal banks. On islands in the Atchafalaya River delta of Louisiana (1995–96), nest success for 228 nests on 6 islands averaged 40.4%, but ranged between 6.0% and 67.1% (Holbrook et al. 2000).

Nest success was low (only 6%) for 66 nests located on agricultural lands in southwestern Louisiana during 1999–2000 (Durham and Afton 2003). Coyotes (*Canis latrans*), striped skunks (*Mephitis mephitis*), and raccoons (*Procyon lotor*) were responsible for 77% of 52 unsuccessful nests, with the remainder caused by abandonment (15%), flooding (4%), cattle trampling (2%), and plowing (2%). On the mid-Texas coast (2000–2002), nest success for 100 female Mottled Ducks tagged with radiotransmitters was 9% and 38%, respectively, during the dry years of 2000 and 2002, but 62% during the wet year of 2001 (Finger et al. 2003). The authors stated that 39 unsuccessful nests were destroyed by predators (57%), "unknown causes" (18%), and flooding (10%). Depredation of the laying or incubating female caused 15% of all nest failures. Of the nests lost to predators, striped skunks and raccoons were suspected in most (73%) losses; another significant cause was the partial removal of clutches by rat snakes (*Elaphe obsoleta*), which led to nest failure.

Brood Parasitism. Brood parasitism appears to be rare in Mottled Ducks, probably because nests are well spaced in coastal marsh habitats and other upland sites. Away from the elevated cordgrass ridges in the Texas coastal marshes, which were frequently used, Stutzenbaker (1988) found only 1 nest/259 ha. On the upland prairie in Texas, Engeling (1950) noted that the closest 2 nests were 183 m apart.

Brood parasitism is reported in Mottled Ducks nesting on islands, however, because the distances between nests can be short. On a spoil island in Florida, 3 active nests occurred in a 4.6 m circle (Stieglitz and Wilson 1968). On a 22-ha island in the Atchafalaya River delta of Louisiana, 4 probable cases of intraspecific brood parasitism were noted among the 82 nests located (Johnson et al. 1996); 1–8 parasitic eggs were added to each parasitized nest, to yield final clutch sizes of 9–18. The parasitism rate on the island was minimally estimated at 5%, which is similar to rates reported for other puddle ducks in island-nesting situations (Rohwer and Freeman 1989). In the Mississippi River delta of Louisiana, where 43.0% of the Mottled Duck nests were located on islands, the parasitism rates were 9.0% for 178 nests in 1998 and 4.4% for 159 nests in 1999 (Walters et al. 2001).

There are no reports of interspecific brood parasitism, probably because no other ground-nesting puddle ducks commonly breed within the same range as Mottled Ducks, although numbers of breeding Mallards are increasing in Florida. On islands in the Mississippi River delta, however, Walters (2000) reported that 3 Laughing Gull (*Larus atricilla*) nests each contained 1 Mottled Duck egg.

Renesting. Stutzenbaker (1988) found that renesting was common among the Mottled Ducks he studied in Texas. Of 12 marked females, 9 renested at least once, with the new nest located an average of 91 m from the original. Using radiomarked females in Texas (2000–2002), Finger et al. (2003) reported an average clutch size of 8.6 eggs for first attempts ($n = 26$), 9.1 for second attempts ($n = 10$), and 7.5 for third attempts ($n = 2$). Females only renested more than once during a wet year. The renesting inter-val averaged 16 days, but ranged between 0 and 36 days. On an upland prairie site in Texas, 1 female renested 4 times, producing a total of 34 eggs in 5 nests (Engeling 1950).

REARING OF YOUNG

Brood Habitat and Care. In general, brood-rearing habitats for Mottled Ducks are coastal marshes, freshwater impoundments, or seasonal prairie wetlands characterized by an equal amount of open water and short emergent vegetation, with mudflats, exposed submerged vegetation, and islands that provide loafing sites for ducklings (Bielefeld et al. 2010). Water depths are usually <15 cm (Paulus 1984c, Stutzenbaker 1988, Gray 1993). In Florida, brackish marshes near Merritt Island on the eastern coast and Sanibel Island on the western coast were productive brood habitats, characterized by emergents such as needle rush (*Juncus roemerianus*) and cordgrass (*Spartina bakeri*), and submerged vegetation, such as widgeongrass (*Ruppia maritima*) and spiny naiad (*Najas marina*; LaHart and Cornwell 1970). On freshwater sites at the Kissimmee Chain of Lakes, ducklings pecked at the surface for invertebrates in dense stands of pennyworts (*Hydrocotyle* spp.) and widgeongrass.

On Lake Okeechobee in Florida (1989), radiomarked females with broods ($n = 21$) used areas of dense emergent cover during the day but more open areas at night, with 85% of the diurnal observations in vegetation ≥250 cm tall; only 20% of the nocturnal observations were in vegetation ≥20 cm tall (Gray 1993). Broods used emergent marshes (50%) and aquatic beds (50%), the latter especially at night. Aquatic beds were composed of mudflats or rooted plants (such as hydrilla), and had a water depth of 0–5 cm. At night, 45% of the brood locations were on mudflats or on prostrate beds of hydrilla, but they were within 50 m of cover 90% of the time. The majority (63%) of the brood observations were in water ≤5 cm deep, and no broods were observed in water >30 cm deep.

On the Texas Chenier Plain NWR Complex in Texas, Rigby (2008) used radiotelemetry to determine the movements and habitat use of 59 ducklings from 32 broods during 2006–8, although transmitters only remained on the ducklings for about 15 days. The mean brood home-range size averaged 40.5 ha ($n = 29$), with mean daily movements of 116–201 m. Movements were greatest in the year with the most rainfall. Broods used shallow (8–30 cm) water with a 50:50 water-to-emergent-vegetation ratio in coastal marshes, while moist-soil impoundments were more open.

Experiments with captive-reared ducklings exposed to different levels of salinity reflective of natural-habitat conditions in coastal Louisiana revealed 100% mortality at 18 ppt (slightly saline marsh), 90% at 15 ppt (strongly brackish marsh), and 10% at 12 ppt (brackish marsh); most mortality occurred in the first 5 days (Moorman et al. 1991). When ducklings were treated with 12 ppt and subsequently exercised, however, projected mortality rates increased to 70%. No ducklings died at salinities <12 ppt, but ducklings showed lethargy at 9 ppt (slightly brackish marsh). The authors concluded that the upper threshold of tolerable salinities for Mottled Duck ducklings was within the 9–12 ppt range, and marshes with <9 ppt provided the highest-quality brood habitat.

The young leave the nest in <24 hours and are led to brood-rearing habitat by the female (Paulus 1984, Stutzenbaker 1988). In Louisiana, brood mates marked with nasal saddles remained together until they were about 60–70 days old ($n = 120$; Paulus 1988). The daily time budget of females ($n = 259.3$ hours of observation) and their broods ($n = 1,226.5$ hours of observation) was quantified in southwestern Louisiana by Paulus (1984c). Ducklings devoted most of their time to feeding (57.7%) and resting (27.5%), with lesser amounts to preening (8.5%), being alert (3.3%), and walking (3.1%). Duckling fed 63% of the time during the day, compared with 46% at night. Females also devoted the most time to feeding (34.4%); the remainder was primarily spent resting (28.1%), remaining alert (19.8%), preening (10.5%), and locomoting (3.4%).

Brood Amalgamation (Crèches). The adoption of ducklings appeared to be common at Lake Okeechobee in Florida, with older broods containing more ducklings; 1 female accompanied 19 ducklings (Gray 1993). Adoption may not occur, however, in other areas of Florida, where brood densities are lower. In Louisiana, females with broods did not tolerate a close approach by other broods, pairs, or lone males during the day, but they were more tolerant of other broods at night (Paulus 1984c).

Development. Stutzenbaker (1988) provided elaborate drawings of and detailed text descriptions on the growth and plumage development of Mottled Ducks from hatching to 8 weeks of age, and body weight data to 15 weeks. Average duckling mass ($n = 15$) was 32.9 g at hatching, 129.8 g at 2 weeks, 345.5 g at 4 weeks, 637.1 g at 6 weeks, and 781.0 g at 9 weeks, which was about 75% of the average winter weight of hunter-shot birds. Weight at 16 weeks was 833.7 g. The first appearance of feathers is at 3 weeks, with the resulting juvenal plumage nearly complete at 5–6 weeks. Ducklings are difficult to distinguish from adults at 7 weeks, and the wings are completely covered with contour feathers by 8 weeks. Most birds can be sexed by mandible color at 10 weeks. The sheathed adult penis appears in 20–25 weeks.

In Florida, body mass of 7 ducklings from 1 brood reared in captivity averaged 30.9 g at day 1, 75.7 g at day 9, 150.1 g at day 16, and 235.9 g at day 27 (Beckwith and Hosford 1957). In Louisiana, captive Mottled Ducks ($n = 8$) reared in fresh water and fed ad libitum (an unlimited diet) averaged 712.1 g at 60 days (Moorman et al. 1991). Body composition was 6.4% lipids, 12.9% protein, 64.3% water, and 3.7% ash. Ducklings are capable of escape flights at 46–56 days, and sustained flights at 61–70 days (Stutzenbaker 1988, Moorman et al. 1991).

RECRUITMENT AND SURVIVAL

Rigby (2008) used data from studies in Texas and Louisiana, plus the methods and assumptions outlined in Cowardin and Johnson (1979), to calculate annual recruitment rates (the number of fledged females produced/adult female in the breeding population) of 29%–149%. Such wide variation probably reflects differing wetland habitat conditions, which influence nesting propensity and brood survival.

Duckling survival for Mottled Ducks also fluctuates with water conditions. During a wet year (2001) along the mid-Texas coast, duckling survival to 30 days was 41% and brood survival was 69% among 13 radiomarked females (Finger et al. 2003). In contrast, no broods survived during the dry year of 2000. Across all years of their study (2000–2002), total brood loss occurred in 31% of all broods, and it typically happened during the first 3 weeks after hatching ($n = 17$). Death of the adult female caused 40% of the total brood losses. On the upper Texas coast (2006–8), survival from hatching to 30 days old ($n = 59$ radiomarked ducklings from 32 broods) was 66.9% for ducklings and 34.5% for broods, but most duckling transmitters fell off before 15 days (Rigby 2008). On marshes at Lake Okeechobee in Florida (1989), duckling survival during the 56-day brood-rearing period ranged between 24% and 50% ($n = 19$–21), with a brood survivorship of 62%–82%; none of the 21 adult females died (Gray 1993). The high brood and female survivorship in this study may reflect the consistent water conditions that prevailed at Lake Okeechobee, compared with other wetlands used by breeding Mottled Ducks in Florida that are more susceptible to drought. Sample sizes were also small in this study. Ducklings are targeted by a wide array of predators in Florida, which prompted LaHart and Cornwell (1970:120) to remark that "a list of possible duckling predators would include most of the carnivorous, terrestrial, and aquatic vertebrates in Florida."

The survival of radiomarked females over a 15-week interval (26 Feb–10 Jun) near Lake Okeechobee was 90.2% in 1998 and 87.9% in 1999 (Dugger et al. 2010). During a 3-year study (1999–2002) of 101 radiomarked adult females in eastcen-

tral Florida, annual survival (30 Aug–29 Aug) was 21% in 1999–2000, 59% in 2000–2001, and 77% in 2001–2 (30 Aug–27 Jun; Bielefeld and Cox 2006). The overall survival rate was 52%, which was similar to the 50% estimated from banding data for Mottled Ducks in Florida (Johnson et al. 1995), as well as to the annual survival rates reported for the closely related American Black Ducks and Mallards. Survival rates varied seasonally: 85% during breeding, 68% postbreeding, 88% during the hunting season, and 100% in late winter. Mortality factors also varied during each year of the study; the primary causes were American alligators (*Alligator mississippiensis*; 17.8%), mammals (11.2%), raptors (10.2%), "unknown factors" (10.0%), and vehicles (2.2%).

Bielefeld and Cox (2006) found that Mottled Duck survival in Florida was lowest during the year when water levels were receding (1999–2000), in contrast to the other 2 years, when the wetlands were either already dry or water levels were increasing. The receding water levels concentrated wildlife in and around the remaining water areas, placing molting Mottled Ducks alongside alligators, which resulted in high alligator predation (30.9% vs. 0%–19.4% for the other 2 years). The lower water levels may also have contributed to the higher rates of mammalian and raptor predation observed, in comparison with the other 2 years of the study. During the extended drought year, when nearly all of the study-area wetlands were dry, Mottled Ducks responded by moving to suburban/urban wetlands, and many (56.2%) continued to use these sites even when conditions on the study area improved. The use of these wetlands resulted in lower hunting mortality as well as lower alligator mortality, because nuisance alligators were controlled in urban areas.

Mottled Ducks were also vulnerable to alligator predation in Louisiana during the molt, as Mottled Duck remains occurred in the stomachs of 20.9% of 43 alligators collected from shallow-water habitats in late summer (Elsey et al. 2004). In coastal Texas, Stutzenbaker (1988:149) also considered alligators to be "the single most efficient predator of adult Mottled Ducks and ducklings," especially when droughts reduce the availability of water and concentrate both Mottled Ducks and small alligators. During a 3-year study (2000–2002) on the mid-Texas coast, adult female survival averaged 75% ($n = 106$), with a range of 72%–87% (Finger et al. 2003). Of 21 female mortalities, mammalian predators accounted for at least 41%, and avian predators, 19% (Finger et al. 2003) A significant number (43%) of all mortalities occurred while the female was laying or incubating.

Johnson et al. (1995) estimated survival rates for the Florida Population of Mottled Ducks, based on 690 recoveries of 8,132 banded birds from 1971 to 1991. Mean annual survival rates were 54.8% for adult males, 50.3% for adult females, 90.9% for immature males, and 47.4% for immature females. Survival rates varied among years, but there was little evidence that survival or recovery rates varied between the sexes. The lack of differences in survival rates between males and females was related to the imprecision of the estimates, but it could also indicate that harvest risks are similar for both sexes, because their plumage similarity prevents hunter selectivity in the harvest.

Johnson (2009) determined survival and population growth rates for the Western Gulf Coast Population by analyzing 5,476 band recoveries from 44,127 Mottled Ducks banded in Texas and Louisiana from 1994 to 2006. Average annual survival rates were lower than those of most other dabbling ducks, including Mottled Ducks in Florida: 58% for adult males, 47% for adult females, 48% for juvenile males, and 37% for juvenile females. Survival rates varied widely among years and sex/age groups, from 19% to 80%. Average juvenile:adult ratios were higher for females (1.81) than for males (1.01). The geometric mean population growth (λ) was 0.85 for males and 0.79 for females. Male growth rates improved with better habitat conditions (in wet years), but those for females did not, primarily due to their lower survival during wet periods, which may have reflected the increased costs of nesting and renesting. Growth

rates were significantly <1 for both sexes, however, which indicated a rapidly declining population over the time period that was studied.

The longevity record for a Mottled Duck is 13 years and 5 months for an adult male banded and shot in Texas; the record in Florida is 9 years. Captive Mottled Ducks, however, can live >20 years (Bielefeld et al. 2010). Based on survival rates from banding data (Johnson et al. 2009) and Anderson's (1975) methodology, Bielefeld et al. (2010) estimated that average postbanding life expectancies for Mottled Ducks were 1.5 years for adult females, 1.4 years for immature females, 1.7 years for adult males, and 2.5 years for immature males.

FOOD HABITS AND FEEDING ECOLOGY

Mottled Ducks forage for seeds and aquatic invertebrates in shallow water that is generally <30 cm deep and is further characterized by emergent vegetation, aquatic plant beds, or temporary wetlands (Bielefeld et al. 2010). Greater amounts of invertebrates are consumed during the breeding season to meet the demands of egg production, while plant foods predominate during the fall and winter. Paulus (1984c) observed that Mottled Ducks feed by filtering the substrate and the surface as they sit on the water, with occasional feeding by tipping up or by standing in very shallow water. They rarely dive. Mottled Ducks also strip seeds from grasses and cultivated rice (Stutzenbaker 1988). They have been seen opportunistically feeding on small fish concentrated in drying wetlands (Tom Moorman).

The annual diet of Mottled Ducks, determined from the gizzard contents of 144 birds examined immediately northwest of Lake Okeechobee (1953–55), was 87.2% (aggregate volume) plant material and 12.8% animal material (Beckwith and Hosford 1957). Invertebrate use was highest during the breeding season (18.3% during the spring and 38.7% during the summer), but their consumption was probably underestimated, because invertebrates are underrepresented in gizzard samples (Swanson and Bartonek 1970). Nonetheless, invertebrate use during the fall and winter occurred at trace levels, except for a single winter (at 8.0%). Mottled Ducks consumed 77 species of plants, with their use varying in response to the extent and depth of surface water and the time of year. Some of the significant plant foods were dotted smartweed (*Persicaria punctata*), panicgrasses (*Panicum* spp.), and ragweed (*Ambrosia elatior*); important animal foods were various families of water beetles and snails.

Feeding frequently occurs in pairs or small flocks, but larger flocks of up to 3,000 have been observed in marshes and rice fields in late summer along the Texas coast and in southwestern Louisiana (Stutzenbaker 1988). Stutzenbaker also mentions the sighting of a feeding flight of about 40,000 Mottled Ducks in Cameron Parish, Louisiana, during the early fall in 1940.

Breeding. Aquatic invertebrates are the chief food source for breeding females. In Louisiana, esophageal contents revealed that the primary foods eaten by breeding adults were snails (*Physa* spp.), crayfish (*Procambarus* spp.), beetles (Corixidae, Dytiscidae), amphipods (Gammaridae), dragonfly (Libellulidae) nymphs, midge (Chironomidae) larvae, and mosquitofish (*Gambusia affinis*; Bielefeld et al. 2010). Invertebrates, chiefly midges and predaceous diving beetles (Dytiscidae), also made up 60.8% of the diet of 6 postbreeding flightless adults and juveniles in phosphate settling ponds in Florida (Montalbano 1980). According to Weeks (1969) and Stutzenbaker (1988), gizzard contents showed that molting birds in Texas and Louisiana consumed widgeongrass, spikerush (*Eleocharis* sp.), bulrush (*Scirpus* sp.), sea purslane (*Sesuvium portulacastrum*), and snails.

Mottled Duck ducklings, like other puddle ducks, primarily consume invertebrates early in life, which they obtain by dabbling when ≤21 days old and tipping up when ≥28 days old (Paulus 1984c). In Texas, Stutzenbaker (1988) used a small ($n = 13$) sample of gizzards and crops, along with feeding observations over a 10-year period (1970–79), to conclude that invertebrates formed 80% or more of ducklings' diets at <3 weeks old. In contrast, animal matter made up only 12.0% (by volume) of the foods eaten by ducklings aged 3–9 weeks, while 88.0% was plant material. The most important animal foods eaten during this time were small fish (7.0%) and insects (2.3%); significant plant foods were bulrush (*Scirpus californicus*) seeds (27.7%); longleaf pondweed (*Potamogeton americanus*; 12.1%); *Najas*, spikerush (*Eleocharis parvula*), and widgeongrass (15.7% combined); sago pondweed (*Potamogeton pectinatus*; 12.7%); and gulfcoast spikerush (*Eleocharis cellulosa*; 12.6%).

Migration and Winter. Stieglitz (1972) determined food habits from the gizzards of 85 Mottled Ducks collected at the Loxahatchee and the Merritt Island NWRs in Florida from November through December during 1960–66. Material from 36 species of plants made up 89.9% of the total food volume, and animal material made up just 10%, with plants appearing in 100% of the samples and animal foods in 51.2%. Of the plant foods, 55.8% were emergent species and 30.3% were submergent species. By volume, the most important plants were spiny naiad (21.7%), Tracey's beakrush (*Rhynchospora traceyi*; 18.3%), dotted smartweed (*Polygonum punctatum*; 16.4%), waterpepper (*Polygonum hydropiperoides*; 10.1%), and shoalgrass (*Diplanthera*; 4.7%). By frequency of occurrence, significant plant foods were sawgrass (*Cladium jamaicense*; 70.9%), wax myrtle (*Myrica cerifera*; 62.8%), spiny naiad (36.0%), Tracey's beakrush (24.4%), and widgeongrass (15.1%). By volume, the most important animal materials were snails (Gastropoda; 3.4%), small clams (Pelecypoda; 3%), and adult and larval insects (3.0%). There were differences in food habits, by volume, between the 2 national wildlife refuges, as plants made up 97.3% of the food volume in ducks collected in the freshwater system at Loxahatchee, compared with only 81.1% in ducks collected in the more brackish system at Merritt Island. Samples at Merritt Island contained a 7-times-greater volume of animal matter than those at Loxahatchee;

animal material appeared in 75.0% of all samples at Merritt Island, versus only 28.9% at Loxahatchee.

In Texas, Stutzenbaker (1988) examined 1,105 Mottled Duck gizzards collected during the hunting season (Nov–Jan) in 1963–72. Plant foods made up 99.1% of their diet by volume, but invertebrate remains occurred in 38% of the samples. Although an examination only of gizzards underestimates the consumption of invertebrates (Swanson and Bartonek 1970), Stutzenbaker's data are valuable in describing the plant foods eaten by Mottled Ducks during winter along this portion of the Texas coast. Of the 60 plant foods that were recognized, the most important (by volume) were the seeds of Gulfcoast spikerush (30.2%), cultivated rice (14.9%), bulrush (12.3%), and smartweed (*Polygonum hydropiperoides*; 12.1%). Of all foods eaten, 56% were derived from prairie marshes, 16% each from seasonally wet prairies and farmlands, and 12% from submerged and floating plant associations.

MOLTS AND PLUMAGES

The molts and plumages of Mottled Ducks have been described in detail by Johnson (1973), Palmer (1976a), Stutzenbaker (1988), Gray (1993), Moorman et al. (1993), and Bielefeld et al. (2010). From studies in Florida, Gray (1993) described 3 head, body, and tail molts/year—definitive basic, definitive alternate, and supplemental—based on recognizable patterns of plumages and significant changes in molting rates between spring, summer, and fall. Gray also noted that Mottled Ducks in Florida molted almost continuously throughout the year, although 10 out of 28 females with broods exhibited no molting. The differences among these plumages are subtle; for more information, consult Bielefeld et al. (2010).

Instead of first basic, Gray (1993) recognized definitive supplemental, which is acquired from August through December. In males, this plumage is very similar to definitive basic and definitive alternate. This plumage in females also resembles definitive alternate and definitive basic, but their supplemental plumage is slightly darker. Definitive alternate is similar to definitive supplemental and is acquired from December through March. Definitive basic is acquired from May through September and involves the wing molt. Some males begin the wing molt in late June, but most molt their remiges in mid-July. Females begin the wing molt later (Jul), due to the constraints of breeding. The duration of the flightless period is about 27 days, but escape flights are possible once the remiges are 75% regrown.

Juvenal plumage is completed about 56 days after hatching, and both sexes are similar in appearance. Juvenal plumage resembles definitive supplemental, but in the former the colors are lighter: light areas are a pale gray, and dark areas are dark gray to fuscous. In the closely related Mallard, juvenal plumage is reportedly followed by first basic, but there is no evidence of first basic in Mottled Ducks. Pyle (2005) noted that of the various puddle duck (*Anas*) specimens he examined, 22 contained no evidence of first basic, but instead retained their juvenal plumage; 19 other specimens only showed a prebasic molt involving 8%–19% of the head, the neck, the upper back, the breast, and the sides. Hence he argued that the traditional prebasic molt does not exist, probably because of confusion between what is an actual molt and what is only pigment deposition. Feather pigmentation arrives later in the juvenal molt process, which can be mistaken for first basic, yet feather formation and pigmentation in juvenal plumage can both be part of the same molt, as a given feather follicle is only activated once during this time period.

Moorman et al. (1993) studied body mass and composition of molting male (*n* = 15) and female (*n* = 16) Mottled Ducks at the Rockefeller Wildlife Refuge in coastal Louisiana to better understand the ecology associated with completion of the wing molt and the resultant flightless period. Both sexes entered the flightless period with equal lipid reserves (about 100 g), which had increased prior to entering the molt. Lipid reserves then declined to about 20 g near the end of the molt, with reductions in the reserves of both sexes negatively correlated with the length of the remiges ($r = -0.50-0.67$). Lipid reserves, however, could only meet 9 days (33%) of the existence energy required during the 27-day flightless period. Thus the lipid reserves accumulated before the wing molt probably allow Mottled Ducks to reduce certain activities, such as feeding, which would then lessen their exposure to predators. Male Mottled Ducks decreased their feeding time from 65% prior to molting to 9% during the flightless period (Paulus 1984). Another benefit of this ecological strategy is that as their lipid levels (and thus their overall body weights) decline, individuals are capable of short escape flights, because their wing loads are reduced, which again reduces the risk of predation.

CONSERVATION AND MANAGEMENT

Effective conservation measures for Mottled Ducks must address habitat loss and degradation problems at a range-wide scale. The majority of Mottled Duck habitats occur on private land, however, so educational programs should target private landowners and promote habitat conservation throughout the range of the species. One strategy is to work with landowners to implement land-use practices that favor Mottled Ducks (e.g., cattle grazing). There are also many landowner-incentive programs that can and are being brought to bear on wetland conservation issues across Mottled Duck range in both Florida and the western Gulf Coast: the Conservation Reserve Program, the Texas Prairie Wetlands Project, the Louisiana South Waterfowl Program, the U.S. Fish and Wildlife Service's Partners for Fish and Wildlife Program and their Coastal Program, and the U.S. Department of Agriculture's Wetland Reserve Program. There are also several private conservation groups actively protecting and/or managing habitat within the range of Mottled Ducks, such as Ducks Unlimited, the National Audubon Society, and the Nature Conservancy. The task at hand is large, however, and it will take a cooperative and sustained effort to preserve wetlands in Florida and to protect and enhance one of North America's premier wetland ecosystems:

the coastal wetlands, prairies, and estuaries along the western Gulf Coast.

Some of these habitat issues, though, as well as other concerns, are particular to the 2 geographically separate populations of Mottled Ducks.

Florida Population

Habitat Loss. Florida has experienced a tremendous increase in its human population, which, over the past 200 years, has at least partially fueled a reduction in the state's wetlands—from 54% of the surface area to 31%—although this area increased somewhat between 1998 and 2004 (Bielefeld et al. 2010). Most of the increase, however, was in the form of freshwater ponds in urban areas. Such ponds often lack the emergent cover preferred by Mottled Ducks, although large numbers of these ducks do use the ponds throughout the year. More positively, 6 Stormwater Treatment Areas were created in southern Florida, totaling 21,000 ha and containing diverse emergent and submergent plant communities. These newly formed wetlands are heavily used by Mottled Ducks (Bielefeld 2008).

Genetic Introgression. Genetic introgression from hybridization with feral Mallard populations, and with Mottled Ducks belonging to the Western Gulf Coast Population that were introduced into South Carolina, pose a very serious risk to the long-term genetic integrity of Mottled Ducks in Florida. This introgressive gene flow between populations is especially dire, because the hybrids can backcross with one or both of the parent populations. Such a situation is similar to the well-documented hybridization problems between Mallards and American Black Ducks in the northeastern United States (Ankney et al. 1987, Mank et al. 2004). A formal Environmental Assessment completed by the U.S. Fish and Wildlife Service in 2002 resulted in the issuance of depredation permits to the Florida Fish and Wildlife Conservation Commission, which allow the lethal control of feral Mallards in Florida during the summer months. Despite these actions, Mallard × Mottled Duck hybridization rates have reached 24% in some areas of Florida (eastcentral), although they are 0% in others (Williams et al. 2005). Nonetheless, because female Mottled Duck × Mallard hybrids are difficult to distinguish, introgressive hybridization seems unlikely to be well controlled.

Western Gulf Coast Population

Habitat Loss. Habitat loss is clearly the most serious factor affecting Mottled Duck populations in Texas and Louisiana. These 2 states are well known for the extent of their coastal wetland habitats—183,900 ha in Texas and 723,500 ha in Louisiana (Field et al. 1991)—but they are equally well known for the large-scale loss of wetland habitats that have occurred in their coastal ecosystems. Wetland loss is most severe in Louisiana, where shoreline erosion, saltwater intrusion, land subsidence, and reduced sediment loads caused by dams and/or levees on the Mississippi and Atchafalaya Rivers convert wetlands into unvegetated and more saline open-water areas. The estimated net loss of coastal marshes in Louisiana was 1,704 km^2 between 1978 and 2000, or 77.4 km^2/year (Barras et al. 2003). Projections from 2000 to 2050 are less severe, but they nonetheless indicate a net marshland loss of 1,329 km^2, or 26.6 km^2/year. In some coastal marsh areas, the shoreline has retreated 6–12 m/year (Esslinger and Wilson 2001).

Wetland loss in coastal Texas is not as severe as in Louisiana, but large areas of coastal wetlands have either been converted to rice culture or degraded by other means: the Intracoastal Waterway, the diversion of inflows, real-estate development, and urban and industrial encroachment (Stutzenbaker and Weller 1989). The cumulative result of these activities was the loss of >850 km^2 of coastal wetlands between 1955 and 1992 (Moulton et al. 1997). Climate change and the associated rise in sea levels will also increasingly affect coastal wetlands in both Texas and Louisiana through such factors as altered precipitation patterns, droughts, reduced flushing times, and increased salinity (Zing et al. 2003).

In addition to the loss and degradation of coastal wetlands, the amount of land conversion in the adjacent coastal prairies has been staggering: once covering nearly 3.4 million ha across both Texas and Louisiana, <1% of this ecosystem now remains (Smeins et al. 1992). The original coastal prairies were mostly tallgrass prairies, with some post oak (*Quercus stellata*) savanna on upland sites. The region's fertile soils, high rainfall, and 270-day growing season made the coastal prairies extremely attractive to agriculture, especially for rice (Esslinger and Wilson 2001). Rice agriculture can provide habitat for Mottled Ducks, because fields are flooded during the growing season and idled on 2–3 year rotations, but the number of hectares farmed has declined dramatically in both Texas and Louisiana: from an average of 253,052 ha/year in the 1970s (75% in Louisiana) to 186,588 ha/year in the 1980s, a decline of 25% (Esslinger and Wilson 2001). Rice agriculture increased somewhat in Louisiana from 1990 to 1998 (to 160,130 ha/year), but it continued to decline in Texas (to only 32,487 ha/year). The total in both states was only 184,000 ha planted in rice in 2008.

The loss of rice agriculture seriously affects waterfowl habitat, because abandoned fields are quickly colonized by Chinese tallow (*Triadica sebifera*) and other woody vegetation (such as baccharis), which are common on disturbed soils in the region. These invasive plants severely degrade the value of abandoned fields as nesting or feeding habitats. Unless abandoned areas are restored to their historical prairie conditions, the landscape will offer substantially reduced support for Mottled Ducks (Bielefeld et al. 2010).

Extensive and sustained projects designed to restore and/or maintain the hydrology and integrity of coastal marshes and prairies in Texas and Louisiana will be needed to provide Mottled Duck habitat, as well as habitat for the many species of waterfowl that rely on the Gulf Coast, especially during the winter. The Gulf Coast Joint Venture has established population targets for Mottled Ducks based on the 1971–2004 average from the

Midwinter Survey: 70,132 for Louisiana, 35,322 for Texas (coastal zone only), and 105,816 for the entire Western Gulf Coast Population (Wilson 2007). The Western Gulf Coast, as defined in the Joint Venture, stretches from the Mexican border at Brownsville, Texas, east to Mobile Bay in Alabama. Within the region, there are 6 initiative areas, each with plans that detail goals and approaches for habitat conservation: the Laguna Madre (Texas), the mid-Texas coast, Chenier Plain (southwestern Louisiana), the Mississippi River coastal wetlands (southeastern Louisiana), the coastal Mississippi wetlands, and Mobile Bay. The plans in Texas and Louisiana also extend inland to include the coastal prairies.

Lead Shot. The ingestion of lead shot has historically affected Mottled Ducks, but lead ingestion rates have declined somewhat since the implementation of regulations requiring the use of steel shot. Nonetheless, Mottled Ducks are a resident species in one of the most heavily hunted areas of North America: the coastal wetlands of Texas and Louisiana. Stutzenbaker (1988) estimated that about 15,500 kg of lead shot (72 million pellets) had been deposited on 1,618 ha of marsh at the J. D. Murphree Wildlife Management Area, and that >635 metric tons had been deposited annually in coastal waterfowl-use areas in Texas and Louisiana. Mottled Ducks may be differentially susceptible to lead-shot ingestion, because their nonmigratory status exposes them to deposited lead for a longer period of time than migratory waterfowl. Anderson et al. (1987) noted that the incidence of ingested shot (lead and steel) was highest (25.6%) for Mottled Ducks among the 21 species they examined in the Mississippi Flyway (1977–79), as did Moulton et al. (1988) in Texas. Mottled Ducks probably have the highest lead-shot-ingestion rates of any species of North American waterfowl (Stendell et al. 1979).

In coastal Louisiana, the incidence of lead shot was 26% in 611 Mottled Duck gizzards (Smith 1981). Stutzenbaker (1988) detected lead shot in 18% of 1,115 gizzards he examined from various locations in Texas. Ingestion rates ranged from 10% to 39% for 3,482 gizzards examined at the J. D. Murphree Wildlife Management Area in Jefferson County from 1963 to 1982 (Moulton et al. 1988). These authors also examined 1,974 gizzards from waterfowl shot in public hunting areas in Chambers and Jefferson Counties (on the lower Texas coast) over 6 hunting seasons between 1973 and 1984. They found an average ingestion rate of 28% for Mottled Ducks, which was the highest incidence among the 16 waterfowl species studied. The incidence of lead-shot ingestion, however, declined from 37.5% in 1973–75 to 24.7% in 1982–84, the period after the implementation of steel-shot regulations in 1978/79.

Merendino et al. (2005) examined trends in the incidence of lead shot in the gizzards of Mottled Ducks examined over a 16-year period (1987–2002) on the central Texas coast ($n = 1,292$) and the upper Texas coast ($n = 5,558$). Average ingestion rates were 7.3% on the central coast and 18.1% on the upper coast, although these averages were not statistically different. Ingestion rates declined, however, over the 16-year period. An analysis of lead levels in wing bones ($n = 68$) from Mottled Ducks collected in parts of Texas and Louisiana revealed an average lead level of 16.62 ppm for all regions and sex/age classes combined (Paine 1996). Lead levels of 10–20 ppm in bones indicate exposure to lead, and levels >20 ppm are clinical. Thus 22% of the adult birds had levels indicative of exposure and 28% had levels that were clinical; 24% of the hatching-year birds had exposure levels and 22% were clinical. Nonetheless, these lead levels in Mottled Duck bones were lower than those reported in 1972–73 for the Texas coast (54.4 ppm; $n = 55$) and the Louisiana coast (26.7 ppm; $n = 44$; Stendell et al. 1979). While there is an apparent decline in their exposure to lead, the levels in Mottled Ducks are still high and warrant continued monitoring, despite the fact that steel shot has not been used across most of the Mottled Ducks' range in Texas since 1983 (Moulton et al. 1988), and in Louisiana since 1987 (Merendino et al. 2005).

Blue-winged Teal

Anas discors (Linnaeus 1766)

Left, drake; *right*, hen

The Blue-winged Teal breeds only in North America, but its winter range is primarily in the Neotropics: from Mexico and the Caribbean south through Central America into northern South America, particularly Columbia and Venezuela. Their breeding range extends across a very broad swath of the northern United States and southern Canada, but the greatest numbers occur in eastern North Dakota and South Dakota and southern Saskatchewan. Blue-winged Teal are the third-most-abundant breeding duck in North America, after Mallards and scaup, averaging 5.7 million from 2001 to 2010. They are among the last ducks to return north in the spring, where their preferred breeding habitat contains seasonal and temporary wetlands surrounded by grassland nesting cover. The Blue-winged Teal is a relatively small puddle duck. The males aggressively defend territories during the breeding season, but most nesting does not occur until mid- to late May. The females rely heavily on macroinvertebrates during the egg-laying period, but Blue-winged Teal are otherwise omnivorous. In the fall, they are among the first ducks to migrate south, to wintering grounds outside the United States; hence the North American harvest of Blue-winged Teal is small. Their population status appears secure, but it is heavily linked to the fate of habitats in the Prairie Pothole Region (PPR).

Wings of Blue-winged Teal and
Cinnamon Teal hens are alike

Wings of Blue-winged Teal and
Cinnamon Teal drakes are alike

IDENTIFICATION

At a Glance. The male has a violet-gray head, with a bold white crescent on each side, between the eye and bill. In flight, Blue-winged Teal show a bright, powder-blue wing patch. Females are grayish overall, similar to female Cinnamon Teal.

Adult Males. The grayish head and the bold white facial crescent make the males of this species unlike any other duck. The chest and the sides are tan colored and sprinkled with dark brown dots. The brown feathers may hide the extensive blue on the shoulder when the wings are folded, but the blue patches are very visible when the birds are in flight. The speculum is green. There is a distinctive white patch in front of the black tail. The legs and the feet are yellowish, with grayish webbing. The bill is blackish. The irises are dark grayish brown.

Adult Females. The overall plumage of adult females is a mottled grayish brown, similar to the females of other teal species. The blue shoulder patches are less bright than those of the males, and they are the only really distinctive color marks on females. The speculum is green. The legs and

the feet are a dull yellowish brown. The bill is dark gray. The irises are brown.

Juveniles. Juveniles are very similar to adult females.

Ducklings. Ducklings show good contrast between their yellowish base color and brownish pattern color (Nelson 1993). The belly, the breast, and the head are generally yellow, with a brownish cap. The color pattern, a thin dark eye stripe, and a wide nail on the bill distinguish Blue-winged Teal ducklings from those of the closely related Cinnamon Teal.

Voice. Males have a characteristic single call that is a high-pitched whistled *peew* or a low-pitched nasal *paay*. A decrescendo begins with this single call, followed by a short series of evenly spaced *pew* notes (McKinney 1970). Both calls are frequently heard during the fall and early winter, but are only rarely voiced once pairs are formed. Females give a variety of *quacks*, similar to those of Mallards, but softer and more rapid (Bellrose 1980). Blue-winged Teal also frequently utter a *keck keck keck*.

Similar Species. In the field, female Blue-winged Teal are virtually indistinguishable from female Cinnamon Teal. Female Blue-winged Teal, however, have shorter necks, a dark eye line extending from the base of the bill toward the back of the head, a white spot on the throat, and another white spot at the base of the bill. The lack of buff markings on the tail can also help to distinguish female Blue-winged Teal from the females of other species of teal. Female American Green-winged Teal are distinctly smaller and have noticeably smaller bills. In low-light conditions, Northern Shovelers in flight, because of their rapid wing beat and low, darting flight pattern, can be mistaken for teal.

Weights and Measurements. Bellrose (1980) reported average body weights of 463 g for adult males (*n* = 35), 458 g for immature males (*n* = 146), 377 g for adult females (*n* = 129), and 390 g for immature females (*n* = 315). Wing length averaged

Male Blue-winged Teal
in flight.
RyanAskren.com

18.7 cm for adult males (*n* = 50), 18.5 cm for immature males (*n* = 49), 18.0 cm for adult females (*n* = 31), and 17.8 cm for immature females (*n* = 59). During the breeding season in North Dakota, body weight averaged 397 g for 313 males and 348 g for 255 females (Lokemoen et al. 1990b). During winter in the state of Yucatán in Mexico, body weight averaged 380 g for males (*n* = 110) and 340 g for females (*n* = 82; Thompson and Baldassarre 1990). Rohwer (1986) reported data on the weights of breeding females in Manitoba: 384 g on arrival (*n* = 26), and 380 g at the end of laying (*n* = 26). In North Dakota, female body weight averaged 329 g (*n* = 64) by late incubation (Loos 1999).

DISTRIBUTION

Blue-winged Teal breed only in North America, where they can be found over a broad area: from British Columbia east to southern Ontario and Québec, and then south along the Atlantic coast from New Brunswick to North Carolina (Johnsgard 1978). They also breed as far south as Cali-

fornia and the Gulf Coast. Some Blue-winged Teal winter along the Gulf Coast, but most winter much farther south, through Mexico and Central America into northern South America. Two subspecies of Blue-winged Teal were once suggested, a nominate subspecies (*Anas discors discors*) consisting of those breeding west of the Appalachian Mountains, and a second subspecies (*A. d. orphna*) consisting of those nesting along the Atlantic Seaboard from New Brunswick to Pea Island in North Carolina (Stewart and Aldrich 1956). These proposed subspecies, however, were not recognized (Livezey 1997).

Breeding. Although breeding over a large portion of North America, Blue-winged Teal are primarily birds of the northern prairies and parklands of the PPR in the United States and Canada, particularly the eastern Dakotas and the Prairie Provinces, where they are often the most abundant breeding duck. Over 6 years (1976–81) on 2 study areas (22.6 km² each) in the PPR in North Dakota, pair

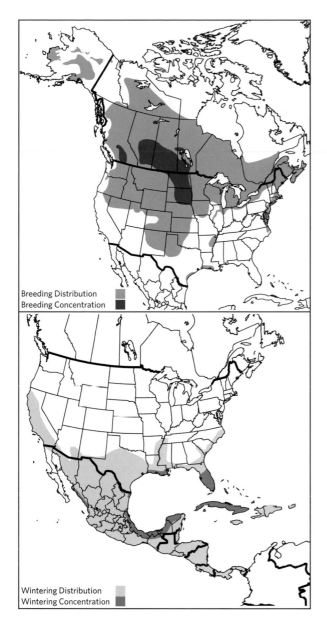

Breeding Distribution
Breeding Concentration

Wintering Distribution
Wintering Concentration

mented by the Waterfowl Breeding Population and Habitat Survey. Results from the 2010 Traditional Survey reported 6.3 million Blue-winged Teal: 61.1% in the eastern Dakotas; 21.5% in southern Saskatchewan; 7.9% in southern Alberta and Manitoba; 4.4% to the north, in the Northwest Territories, northeastern British Columbia, and central and northern Alberta; 0.06% in Alaska/Yukon; and 0.07% in northern Saskatchewan, northern Manitoba, and western Ontario (U.S. Fish and Wildlife Service 2010).

From 1955 to 2010, the distribution of Blue-winged Teal among regions in the Traditional Survey was 27.7% in southern Saskatchewan; 19.9% in North Dakota; 19.2% in South Dakota; 12.6% in southern Alberta; 7.9% in southern Manitoba; 5.9% in the Northwest Territories, British Columbia, and northern Alberta; 2.7% in northern Manitoba; 2.1% in Montana, and 0.03% each in Alaska/Yukon and western Ontario (Collins and Trost 2010).

Outside the PPR in the United States, Blue-winged Teal breed west of the Cascades, in Washington, Oregon, and south to northeastern California. From the Cascade Mountains east to the Rocky Mountains, Blue-winged Teal are largely (and sometimes exclusively) replaced by Cinnamon Teal. Blue-winged Teal do breed, however, in northern Nevada, northern Utah, and northern and eastern Colorado, as well as in western and central Kansas, northern Oklahoma, southern New Mexico, and the Texas Panhandle, as well as locally throughout Texas (Rohwer et al. 2002). In general, Cinnamon Teal breed in the drier regions of the West, while Blue-winged Teal prefer the more mesic northeasterly areas. Surveys in the western states, which are outside the Traditional Survey area, do not separate Blue-winged Teal from Cinnamon Teal, but averages from 2001 to 2010 were 38,350 teal in California, 31,105 in Oregon, 13,093 in Washington, 8,247 in Nevada, and 7,682 in Utah (Collins and Trost 2010).

Blue-winged Teal also breed throughout the midwestern states, eastward through parts of Ohio, Pennsylvania, and New York. Along the

density averaged 13.8/km² on the Koenig Study Area and 20.3/km² on the Woodworth Study Area (Lokemoen et al. 1990a). Pair density ranged from a low of 1.9–8.5/km² during a drought year, when only 31 basins on the study areas held water, to 13.4–23.2/km² during the wettest year, when 335 basins contained water.

The number of breeding Blue-winged Teal rapidly diminishes north of the PPR; very few breed in the Canadian Arctic or Alaska. Within the PPR, the population size of Blue-winged Teal is well docu-

Male Blue-winged Teal.
RyanAskren.com

Atlantic coast, breeding occurs from the Maritime Provinces south through New England and the Delmarva Peninsula (most of Delaware and parts of Maryland and Virginia) to northern coastal North Carolina. They also breed in Missouri, western Tennessee, and eastern Arkansas, south to the Gulf Coast of Louisiana (Barry Wilson), where they occur at low densities. Small numbers are found as far south as southern Texas, especially during wet years (McAdams 1987).

In the East, an average of 27,243 Blue-winged Teal were recorded during the Atlantic Flyway Breeding Waterfowl Plot Survey from 2003 to 2010: 34.5% in Delaware, 28.0% in New York, 18.6% in Pennsylvania, and 10.1% as far south as Virginia. Small numbers breed farther north along the Atlantic coast, from New England to the Maritime Provinces.

Winter. Blue-winged Teal winter farther south in greater numbers than any other duck in North America. Most overwinter in Mexico, the Caribbean, and Venezuela, but their wintering range extends throughout Central America, with some birds occurring as far south as Argentina and Chile

(Johnsgard 1978). In an extensive analysis of 44,186 band recoveries of Blue-winged Teal through 31 August 1980, 9,468 (21.4%) were reported in the Neotropics; of those, 37.4% were in South America, 28.0% in Mexico, 24.5% in the Caribbean, and 10.1% in Central America (Botero and Rusch 1988). The majority of recoveries in South America were in northeastern Columbia (52.4%) and Venezuela (30.8%), but recoveries also occurred near the Atlantic coasts of Guyana, Suriname, French Guiana, and northeastern Brazil, as well as in Ecuador and Peru, primarily in wetlands along the Pacific coast south to Lima. Most recoveries in Central America were in wetlands on the Pacific coast: the Gulf of Fonseca in Honduras, the Gulf of Nicoya in Costa Rica, and the Gulf of Panama. Of all recoveries in the Neotropics, 38.1% were of Blue-winged Teal banded in the Prairie Provinces of Alberta, Saskatchewan, and Manitoba; others were from states in the Central Flyway (23.4%) and the Mississippi Flyway (21.7%). Only 4.0% came from the Atlantic Flyway, and 1.1% from the Pacific Flyway.

Very few (< 25,000) Blue-winged Teal wintered in the United States (mostly Florida) until 1957,

Blue-winged Teal flock. *Thomas Marriage*

when Hurricane Audrey altered the extensive coastal marshes of Louisiana to create favorable winter habitat for Blue-winged Teal. Although the Midwinter Survey reports Blue-winged Teal numbers for the Atlantic Flyway, it reports combined numbers of Blue-winged Teal and Cinnamon Teal in the Mississippi, Central, and Pacific Flyways. The Midwinter Survey averaged 250,567 teal from 2000 to 2010: 52.9% in the Mississippi Flyway, 36.1% in the Central Flyway, 8.5% in the Atlantic Flyway, and 2.5% in the Pacific Flyway. In the Mississippi Flyway, 99.3% are recorded in Louisiana, and nearly all are Blue-winged Teal, because Cinnamon Teal are rare there in winter. Yearly numbers vary widely, however, in response to habitat conditions. Small numbers of Blue-winged Teal

can also be found along the coasts of Mississippi and Alabama. In the Central Flyway, most (99.8%) occur along the Texas Gulf Coast, but numbers fluctuate greatly among years (e.g., 27 in 1998, and 220,056 in 2005). As in Louisiana, most teal on the Texas coast are Blue-winged Teal. Small numbers can appear in the Midwinter Surveys in Missouri, Tennessee, Oklahoma, and New Mexico. Along the Atlantic coast, Blue-winged Teal winter in North Carolina, South Carolina, Georgia, and Florida, with most recorded in Florida. Overall, however, only a miniscule percentage of the Blue-winged Teal population winters in the United States.

Outside the United States, Midwinter Surveys have been conducted in Mexico by the U.S. Fish and Wildlife Service, beginning in 1947 (Saunders

465

and Saunders 1981), although Blue-winged Teal are not separated from the less abundant Cinnamon Teal. The survey is conducted in 3 regions of the country: the east coast, the interior highlands (northern and central Mexico), and the Pacific coast. Surveys were conducted at 1- to 3-year intervals until 1984, but since 1985 they usually have occurred at 3-year intervals. Summary data from 1961 to 2000 averaged 287,805 Blue-winged Teal and 63,330 Blue-winged/Cinnamon Teal in Mexico (Pérez-Arteaga and Gaston 2004). Almost all birds on the east coast are Blue-winged Teal (Cinnamon Teal are very uncommon there), while most on the Pacific coast are probably Cinnamon Teal. Of 2,671 Blue-winged/Cinnamon Teal tallied during a waterfowl-harvest survey in the state of Sinaloa in Mexico, 93% were Cinnamon Teal and 7% were Blue-winged Teal (Migoya 1989). Summary data from 1981 to 1994 averaged around 416,000 Blue-winged Teal and Blue-winged/Cinnamon Teal in Mexico, of which 233,000 (56.0%) occurred on the east coast, 126,000 (30.3%) on the Pacific coast, and 57,000 (13.7%) in the interior highlands.

During surveys on the east coast from 1978 to 2006, the number of Blue-winged Teal averaged 323,794. These birds were observed in the Tabasco Lagoons (54.8%), the Campeche-Yucatán Lagoons (28.0%), and the Alvarado Lagoons (11.4%). The latter surveys during this period recorded 380,446 Blue-winged Teal on the east coast in 2000 and 404,392 in 2006, of which 51.1% occurred in the Tabasco Lagoons and 35.0% were in the Campeche-Yucatán Lagoons. On the Pacific coast (1981–94), the largest concentrations of Blue-winged/Cinnamon Teal were found along the coast of Sinaloa, in the Pabellón (41.3%) and Topolobampo (20.6%) Lagoons, and to the south in Nayarit, at the Marismas Nacionales (20.6%). Most of the Blue-winged/Cinnamon Teal in the interior highlands of Mexico wintered in lakes and marshes across the central highlands, with the primary areas being Laguna Cuitzeo (31.6%), Cabadas (14.0%), East Chapala (12.3%), Zacapu (10.5%), and Lago Yuriria (8.8%).

Blue-winged Teal winter throughout the Caribbean islands, wherever suitable habitat is available. Most winter in Cuba, because it is the largest island by far and contains the most natural wetland habitats, as well as extensive rice fields. The Republic of Cuba, moreover, consists of the 2 principal islands (Cuba and Isla de la Juventud) and 4,195 smaller islands, with a total area of 110,926 km² (Denis 2006). In the Republic, estuarine wetland complexes alone total 9,500 km². The largest wetland ecosystem in the Caribbean is the 452,000 ha Ciénaga de Zapata (Zapata Swamp) on the southwestern coast of Cuba, which contains about 250 bird species, 21 of which are endemic. Other major wetlands in the Republic are Ciénaga de Lanier on Isla de la Juventud (126,200 ha), Ciénaga de Birama (57,048 ha), and the Archipélago de Sabana-Camagüey, which is a 3,400 km² complex along 465 km of Cuba's northern coast.

Migration. Blue-winged Teal have one of the most complex migration patterns among all of the ducks in North America (Bellrose 1980). In late summer, many immatures and some adults move north from southern breeding areas in Montana and the Dakotas to the Prairie Provinces of Canada. Blue-winged Teal gather for their fall migration in relatively small groups at countless staging areas that are widely dispersed over the northern Great Plains, which may explain why concentrations of migrants are not found north of their wintering grounds. A sizable number of Blue-winged Teal, more than other prairie-nesting ducks, migrate directly eastward. Of 3,789 recoveries of Blue-winged Teal banded in the PPR, 4.8% were recovered in New England, Ontario, Québec, and the Maritime Provinces (Sharp 1972). From these eastern locations, Blue-winged Teal fly south and eventually out over the Atlantic Ocean to wintering areas in the Caribbean, the West Indies, and northern South America. Bellrose (1980) noted 2 major migration corridors for Blue-winged Teal in the West. The first extends from Saskatchewan south-southeast across the eastern Great Plains to eastern Texas and Louisiana, and the second stretches from Manitoba and Minnesota to Florida. Blue-winged Teal are uncommon to rare migrants in the Pacific Flyway, perhaps because of the presence of Cinnamon Teal, an ecological equivalent in that region.

During fall migration, large concentrations of Blue-winged Teal occur in Kansas from September to early October, the Texas Panhandle from mid-August to mid-September, and coastal Texas and Louisiana from mid-August through late October (Rohwer et al. 2002). By late October, most have left for their wintering grounds in Central and South America. Blue-winged Teal migrating in eastern North America begin to depart their prairie breeding grounds in early August, moving eastward through Minnesota and south to Florida before reaching their wintering grounds in the Caribbean and South America (Botero and Rusch 1988). In the West, Blue-winged Teal leave Oregon from early August through mid-October (Gilligan et al. 1994), with the first individuals arriving on their wintering grounds, extending from Mexico to Panama, in September (Migoya 1989, Ridgely and Gwynne 1989, Howell and Webb 1995).

During spring migration, Blue-winged Teal have departed Panama by April (Ridgley and Gwynne 1989), and Mexico by May (Howell and Webb 1995). They begin to arrive along the coasts of Texas and Louisiana by late February, with their numbers increasing greatly during March (Bellrose 1980). Spring migration through the southern tier of states largely occurs during the last 2 weeks of March; the middle tier, during the first 2 weeks of April; and the northern tier and the southern Canadian prairie breeding grounds, by the last week of April. By mid-May, most migrants have reached their breeding areas.

In North Dakota, an examination of long-term (1936–68 and 1965–77) datasets from 2 study areas revealed median arrival dates of 11–12 April and 15–16 April, with arrival delayed by 1.07 days for each Celsius-degree difference in mean temperature during the arrival period (Hammond and Johnson 1984). Arrival dates ranged from 1 to 25 April. Dane (1966) noted arrivals at Delta Marsh in Manitoba on 24 April in 1962 and 13 April in 1963. In British Columbia, Blue-winged Teal reach their breeding grounds from late April to early May (Butler and Campbell 1987).

MIGRATION BEHAVIOR

Blue-winged Teal are among the very first ducks to fly south in the fall, and the last species to return north in the spring. Based on observations by Bellrose (1980), migration is nocturnal; most departures occur between sunset and dark, although migrants can be observed during early daylight hours if they are not yet at their final destination. Migratory flights are thought to be long, but this supposition has not been confirmed by radiotelemetry studies. A hatching-year Blue-winged Teal banded on the Athabasca River delta in Alberta, however, was recovered 1 month later at Lake Maracaibo in Venezuela, a distance of about 6,000 km, traveling at a rate of 200 km/day. Two individuals banded in central Illinois were recovered 2 days later on the Alabama coast, having traveled an average distance of 543 km/day (Bellrose 1958). The maximum known travel distance (>6,400 km) was for an individual banded in Manitoba and recovered near Lima, Peru (Aldrich 1949).

Blue-winged Teal migrate in small flocks of 10–40 birds (Palmer 1976a), with most of the fall migration associated with the passage of a cold front (Owen 1968). On departure days, Blue-winged Teal in Illinois fed more during midday and exhibited preflight swimming maneuvers, as well as increased wing flapping (Owen 1968). During fall, adult males migrate earlier than females or immatures, with most flocks after mid-September composed largely of adult hens and immatures (Bellrose 1980).

Blue-winged Teal probably make a trans–Gulf of Mexico flight to their primary wintering grounds on the east coast of Mexico, and Thompson and Baldassarre (1990) estimated a caloric expenditure of 4,530 kJ for an 1,800 km round-trip migration from the coast of Louisiana to the Yucatán Peninsula. The daily caloric expenditure there during winter was estimated at 342 kJ/day. During winter 1987/88 in brackish coastal lagoons on the Yucatán Peninsula, the body composition of females remained fairly constant, but the lipid content of males increased before their spring departure. In contrast, both sexes exhibited declines in lipids from early to late winter 1986/87 (57.2% for males and 39.8% for females). The differences in lipid cycles appeared to be influenced by tidal patterns between years, which made foraging difficult. Strong winds associated with intense winter storm activity in 1986/87 caused tides to be above the 80% mean maximum high-tide level 23.4% of the time, versus only 2.8% in 1987/88. The energy savings for a Blue-winged Teal to migrate and winter in the Yucatán Peninsula, compared with coastal Louisiana, was equivalent to about a 22.6-day energy supply in the Yucatán, which may partially explain some of the benefits of wintering in the Neotropics, even after considering the migration costs.

Molt Migration. By midsummer, some males and nonbreeding females will concentrate on large marshes, which indicates that they move to these areas for the wing molt (Sterling 1966, McHenry 1971). Only small numbers of Blue-winged Teal breed at Freezeout Lake in Montana, but 5,000–10,000 have been observed there during the wing molt (Bellrose 1980). Many molting Blue-winged Teal moved onto the Camas National Wildlife Refuge (NWR) in Idaho, where Oring (1964) reported the first males in basic plumage on 6 July, and the first flightless birds on 12 July. Other concentrations of molting Blue-winged Teal are documented at Pel and Kutawagan Marshes in Saskatchewan (Sterling 1966) and Delta Marsh in Manitoba (DuBowy 1985a). In southwestern Manitoba, McHenry (1971) noted that most Blue-winged Teal were flightless between 10 July and 10 August.

The percentage of the Blue-winged Teal population that participates in molt migration is not known. McHenry (1971) believed that most males

at Delta Marsh molted in the vicinity of their breeding grounds, although many molting teal were concentrated on Whitewater Lake, about 160 km to the south. Successfully breeding females usually will molt on their breeding areas (Oring 1964).

HABITAT

On arrival to breeding areas in North Dakota, Blue-winged Teal preferred temporary and seasonal wetlands with an abundance of invertebrate foods, but they shifted to more permanent waters later in the summer, as the seasonal wetlands dried up (Swanson et al. 1974). In North Dakota (1967–69), breeding pairs mainly used seasonal wetlands (61.2%) and semipermanent wetlands (31.2%), with pair density highest on seasonal wetlands (122/km²) and temporary wetlands (113/km²; Kantrud and Stewart 1977). Stewart and Kantrud (1973) reported that 76% of 2,322 breeding pairs observed during another North Dakota study (1967–69) occurred on seasonal and semipermanent wetlands. Population density drops dramatically, however, when temporary and seasonal wetlands are dry. High pair densities are also found in hemi-marsh conditions, which occur when cover-to-water ratios are 50:50 (Kaminski and Prince 1984). During a 2-year study (1981 and 1987) outside the PPR, breeding Blue-winged Teal in southern Ontario mostly used habitats created by beavers (*Castor canadensis*; 21.6%–29.7%), as well as emergent marshes (9.8%) and ponds (8.8%–21.6%), but they avoided open-water lakes (Merendino et al. 1995).

On migration, Blue-winged Teal also prefer shallow-water habitats, such as playa lakes, which are shallow wetlands that dot the southern High Plains of the Texas Panhandle and adjacent New Mexico (Bolen et al. 1989). Fall migrants in the rice prairie region of Texas and Louisiana are frequently found in flooded rice fields, which are also extensively used during the winter (Hobaugh et al. 1989). Wintering Blue-winged Teal in coastal Texas inhabited 36 wetland types, which represented 86.9% of the available wetland habitat (Anderson et al. 2000). Density was highest in wetlands characterized by open water interspersed with shrubs or emergent vegetation, especially palustrine scrub-shrub wetlands that were man-made, had low salinity, and were ≥5 ha in size.

On the Yucatán Peninsula in Mexico, wintering Blue-winged Teal occurred on the Celestun estuary, a shallow (<2 m) brackish lagoon characterized by a salinity of 8–24 ppt; submerged vegetation dominated by widgeongrass (*Ruppia maritima*), muskgrass (*Chara* spp.), and nitella (*Nitella* spp.); and a surrounding mangrove forest (Thompson and Baldassarre 1991). From 1 October to 30 March, Blue-winged Teal in the estuary on average spent 43.3% of their time feeding, 28.7% locomoting, 20.8% resting, and 5.6% preening. Resting and preening were the dominant activities in red mangroves (*Rhizophora mangle*), where branches and prop roots provided the only places where teal and other waterfowl could leave the water. Unlike ducks wintering farther north, however, Blue-winged Teal and other species of dabbling ducks in the Yucatán Peninsula exhibited few daily or seasonal changes in their activities throughout the fall and winter, most likely due to the consistently mild temperatures on the peninsula. Blue-winged Teal and other dabbling ducks wintering there experienced temperatures below their lower critical temperature only 0%–4.2% of the time, in contrast to 14%–96% in Texas and 41%–99% in Alabama (Thompson and Baldassarre 1991).

POPULATION STATUS

Next to Mallards and scaup (Greater and Lesser combined), Blue-winged Teal are the third-most-abundant breeding duck in North America. Populations fluctuate from year to year, however, in response to water and habitat conditions. The 2010 Traditional Survey reported 6.3 million Blue-winged Teal, which was 36% above the long-term average of 4.7 million. Low periods include 1961–68, when their population averaged only 3.7 million, and 1981–93, when it averaged 3.6 million. The

lowest recorded population was 2.8 million (1990), and the highest, 7.4 million (2000 and 2009). The population averaged 5.4 million between 1991 and 2000, and 5.7 million from 2001 to 2010. Because of these large-scale fluctuations, however, there is no discernable long-term trend. In their analysis of Midwinter Survey data from Mexico, Pérez-Arteaga and Gaston (2004) reported a significant increase of 3.4% annually from 1961 to 2000, but a decline of 3.5% annually from 1981 to 2000.

The recovery of the breeding population after the drought years of the late 1980s and 1990s was quite dramatic, particularly on the mixed-grass prairie landscape of the Dakotas. The population surged from 384,000 in the 1990s, the lowest point during the drought, to 3.8 million in 2000, a 900% increase. During the same period, the population on the mixed prairies in Canada increased by 154%; on the parklands, by 24%; and the shortgrass prairies, by 3%.

Harvest

Among the major species of game ducks in North America, Blue-winged Teal are the least exposed to hunting, because they are very early fall migrants, and most of the population winters outside North America. Although these ducks are sought by hunters, most of the harvest occurs during a special September teal season, designed to offer some sport harvest before almost all Blue-winged Teal have left North America for their wintering grounds in the Neotropics. The direct recovery rate (1951–99) averaged only 2.0% for Blue-winged Teal, compared with 7.3% for Mallards, 6.0% for Gadwalls, 5.9% for Redheads, and 3.4% for Lesser Scaup (Frank Bellrose). Bellrose further calculated that harvests (retrieved and unretrieved) accounted for only 11% of all annual mortality for Blue-winged Teal, which is very low among the game ducks.

Harvest estimates also do not separate Cinnamon Teal and Blue-winged Teal, although most of the estimates reflect the more abundant Blue-winged Teal. The combined harvest in the United

States averaged 955,146 teal from 1999 to 2008: 57.1% in the Mississippi Flyway, 29.3% in the Central Flyway, and distinctly smaller amounts in the Atlantic (7.6%) and the Pacific (6.0%) Flyways. In the Mississippi Flyway harvest, 50.3% were harvested in Louisiana and 15.6% in Minnesota. In the Central Flyway, most were harvested in Texas (60.2%), with <10% in any other state. From 1999 to 2008, the harvest in Canada averaged 36,572 teal, with a high of 44,292 in 2008. Over this time period, 25.7% were harvested in Saskatchewan, 21.4% in Ontario, 19.7% in Manitoba, 19.0% in Alberta, 6.8% in New Brunswick, and 7.4% in the remaining provinces.

Little published data are available for harvests outside the United States. A hunting survey conducted from 1987 to 1993 throughout Mexico in areas where waterfowl hunting was a traditional activity estimated 57,422 waterfowl killed, of which only 8.9% (5,103) were Blue-winged Teal (Kramer et al. 1995). At 2 major duck-hunting clubs in the state of Sinaloa on the west coast of Mexico, Blue-winged Teal made up only 1.1% of the duck harvest during the 1987/88 hunting season (Migoya 1989, Migoya and Baldassarre 1993).

BREEDING BIOLOGY

Behavior

Mating System. Blue-winged Teal are seasonally monogamous, and almost all nest as yearlings (Dane 1965). General observations indicate that a few pairs form in late December / early January, but most pairing does not occur until April/May (Bluhm 1988). Of 4,417 Blue-winged Teal studied on the Yucatán Peninsula from 1 October to 31 March 1986–88, none were paired (Thompson and Baldassarre 1992), but Glover (1965) noted that about 60% were paired when they arrived in northwestern Iowa in late March. McHenry (1971) reported that all Blue-winged Teal were paired on arrival farther north in southwestern Manitoba. Moreover, the pair bond is strong. Most pairs did not dissolve until the third week of incubation in southeastern Idaho (Oring 1964). In northwestern

Iowa, pair bonds were terminated between days 11 and 19 of incubation, but renesting females retained their original mates for the same length of time as those nesting only once (Strohmeyer 1967). In Manitoba, McHenry (1971) believed that the duration of the pair bond in Blue-wing Teal correlated more with the lateness of the nesting season than with the period of incubation, because most males deserted their females during the last 2 weeks in June, regardless of the stage of incubation.

Sex Ratio. Sex ratio (males:females) data from field observations (mostly in the spring) was 55.5% males for 5,554 Blue-winged Teal in North Dakota, 59.6% for 3,782 in Iowa, and averaged 59.7% during April and May for 491 birds observed in Manitoba (Bellrose et al. 1961). During winter in the Yucatán Peninsula of Mexico, the sex ratio was 3.1:1, or 75.6% males, which clearly indicates some differences in the distribution of the sexes during the nonbreeding season (Thompson and Baldassarre 1992).

Site Fidelity and Territory. Blue-winged Teal exhibit very little fidelity to their breeding areas, and instead readily pioneer into new areas when habitat conditions are suitable. In North Dakota, the homing rate of 159 marked females (1977–81) averaged only 7.2%, compared with 49.4% for Mallards and 61.8% for Gadwalls (Lokemoen et al. 1990a), and it did not vary with a female's age or prior nesting success. In Manitoba, Sowls (1955) found that Blue-winged Teal had the lowest return rate of the 5 prairie-breeding ducks he studied: only 14% of 58 banded adult females came back to the study areas, and none of the 30 juvenile females returned.

Like the related Northern Shovelers, Blue-winged Teal are territorial during the breeding season, which provides the female with a place to feed that is free from harassment by conspecifics and protects paternity by the territorial male. Paired males are not known to participate in extra-pair courtships by leaving their territories to court other females, and attempted extra-pair copulations are uncommon, compared with other dabbling ducks (Rohwer et al. 2002). Of 130 *pursuit flights* observed by Stewart and Titman (1980), <3% involved attempts at extra-pair copulation; most were territorial in nature.

In southwestern Manitoba, the territories of 11 pairs averaged 0.69 ha and included 1.5 water areas (Stewart and Titman 1980). Paired males were intolerant toward all other Blue-winged Teal; they established and maintained territories by an array of *threat* and *rush* behaviors, as well as *pursuit flights* (Titman and Seymour 1981). Of 456 aggressive interactions observed on territories in Manitoba, 28% featured *threats*, 43% involved *rushes*, and 29% resulted in *pursuit flights. Pursuit flights*, the most successful means of expelling territorial intruders, increased dramatically during territory delineation and the nest-establishment period. Of 130 *pursuit flights* by the territorial male, 37% were targeted toward another pair or a female, 9% were toward a paired male, and 54% were toward an unpaired male or a male of unknown status (Stewart and Titman 1980). *Pursuit flights* were usually short: 96% lasting <30 seconds, 89% rising <20 m, and 65% traveling <300 m. The main component of the *threat* was *hostile pumping*, which was usually accompanied by a peeping call. Established territories were well defined and remained stable from the time of nest-site selection until the third week of incubation. In areas where Blue-winged Teal overlapped with Cinnamon Teal (e.g., central Washington), aggressive interactions between the 2 species were frequent (Connelly and Ball 1984). Male Blue-winged Teal were more aggressive, however, which provided this species with access to more preferred habitats.

Home-range size is larger than territory size. In southwestern Manitoba, home-range size averaged 6.9 ha (*n* = 41), varying between 0.6 and 31.8 ha (McHenry 1971). In northeastern South Dakota, Evans and Black (1956) reported 11 home ranges averaging 36 ha, with the largest at 104 ha.

Courtship Displays. Most of the information on displays was described by McKinney (1970). Males utter *repeated-calls*, which are a series of 2–20 or more quiet, clear whistles (*pew . . . pew . . . pew*), or a deeper, more nasal *pay . . . pay . . . pay* in some individuals. *Repeated-calls* are usually given as single or double notes from a stationary posture, with the head fairly erect, and are probably homologous to the *burp* given by other dabbling ducks (Johnsgard 1965). *Jump-flights* are seen in active courting parties of males and seem to attract a female's attention. *Turn-the-back-of-the-head* is also a common male display. Other male displays include *lateral-dabbling* and *head-dipping*, along with various *shakes* and *preening* movements (McKinney 1970). A female will *incite*, a display typically given while next to her mate or preferred male and directed toward a rejected or harassing male (Rohwer et al. 2002). Females utter a multisyllabic *chattering-call* when *inciting*. The most noticeable precopulatory display is *mutual-head-pumping*, performed by both members of the pair. The female *bathes* after copulation, while the male adopts a posture lateral to the female and displays with the bill angled downward, the neck feathers erected, the wings raised slightly, the tail wagged, and the feet paddled rapidly, while uttering a loud, high-pitched whistle (*beeew*) and a nasal *baaay*.

In Iowa, Glover (1956) noted nuptial flights soon after Blue-winged Teal arrived in late March. These early flights involved 4 or 5 males and a single female, but included only 2 or 3 males and a single female as migration progressed into the first 2 weeks of April. Most birds were paired by mid-April to May, at which time only occasional flights were observed.

Nesting
Nest Sites. The female selects the final nest site, but she is accompanied by the male during an inspection of the general nesting habitat, which usually occurs during the early morning or late evening (Glover 1956). Nests are almost always located in upland grass cover and are concealed from the top and sides. Of 1,972 Blue-winged Teal nests monitored on the Woodworth Study Area in North Dakota, 56% were concealed from all sides and 39%

from 3 sides (Higgins et al. 1992). Total concealment from the top occurred at 32% of the nests, with 54% half concealed from that direction.

Blue-winged Teal nests (>5,000) were located in a variety of habitats across a broad range of the PPR in the United States: 39.1% in planted cover, 11.8% in grasslands, 8.4% in haylands, 15.9% in odd areas (patches of cover <2 ha in size with an array of features), and 12.9% in wetlands (Klett et al. 1988). In a large sample (>870) from the PPR in Canada (1982−85), Blue-winged Teal nests were located in grasslands (37%), other grassy habitats in odd areas (33%), and rights-of-way (21%; Greenwood et al. 1995). Reynolds et al. (2001) noted that their greatest nesting preference was for planted grassland cover. At the Woodworth Study Area in North Dakota (1966−81), 67% of 1,973 Blue-winged Teal nests were found in grass cover and 11% in forb cover (Higgins et al. 1992).

In 1949, the 173 nests located on the Ruthven Area in northwestern Iowa primarily (76.3%) were in stands of bluegrasses (*Poa* spp.), with 23.7% in sedge meadows (Glover 1956). Of 276 nests discovered in mixed-grass prairie habitat in North Dakota, most were found in grass cover: 37% in western wheatgrass (*Agropyron smithii*), 14% in blue grama (*Bouteloua gracilis*), 10% in green needlegrass (*Stipa viridula*), 8% in Kentucky bluegrass (*Poa pratensis*), and 4% in smooth bromegrass (*Bromus inermis*; Duebbert et al. 1986). Nesting-habitat use was maximized where the height / density reading of residual vegetation was 0.5 dm or higher. At Lostwood NWR in North Dakota (1980−88), nesting Blue-winged Teal selected vegetation between 1.5 and 2.5 dm but avoided vegetation <0.5 dm (Kruse and Bowen (1996). Nest densities were lower in areas where grazing and spring burns decreased the amount of favorable nesting habitat by reducing vegetation height. During a 1976−81 study at 2 sites in North Dakota, the nest density of Blue-winged Teal reached 140.7/km² in dry wetlands, 117.1/km² in sites located along canals, and 67.3/km² in areas seeded for grasses (Lokemoen et al. 1990a).

Blue-winged Teal nests have been located as far as 1.6 km from the water's edge. In Iowa, nests (*n* = 186) were found between 8.3 and 192.2 m from open water, with an average distance of 72.6 m; 80% were <91.4 m away (Glover 1956). In southeastern Alberta, 162 Blue-winged Teal nests averaged 22.6 m from water, and they had an average light-penetration measure of 40% (Keith 1961).

Egg laying often starts when the nest is barely a scrape in the ground. Plant material—usually grass, but always vegetation within reach—is used to line the nest bowl, with down feathers generally (82% of 134 nests) added after 4 or more eggs are deposited (Glover 1956). Measurements from 186 nests in Iowa averaged 19.6 cm in outside diameter, 13.5 cm in inside diameter, 5.6 cm in bowl depth, and 2.0 cm under the bowl (Glover 1956). Out of 173 nests, 68.8% had an overhead canopy and 9.2% showed evidence of an entrance ramp. All nests were >0.3 m above water level, with a mean of 0.9 m. Observations of 63 females indicated that nest building occurred in the morning, from 07:00 to 10:00, with the time to final construction varying from a few days to more than a week.

Clutch Size and Eggs. Summary data from 45 nesting studies throughout the breeding range of Blue-winged Teal yielded a clutch-size range of 6−14 eggs and an average of 10.1 for 5,634 nests (Rohwer et al. 2002). During Bennett's (1938) classic study of nesting Blue-winged Teal in Iowa, clutch size averaged 9.3 in 341 normal nests. Glover (1956) noted that the mean clutch size in 87 active nests in Iowa was 8.0, with the number incubated ranging from 7 to 13. During a 6-year study (1976−81) in North Dakota, the clutch size of Blue-winged Teal averaged 10.4 eggs and did not vary either with the year or a female's age (Lokemoen et al. 1990a), although clutch size declined an average of 0.06/day as the nesting season progressed. During a 3-year study (1961−63) at Delta Marsh in Manitoba, clutch size averaged 10.5−11.0 eggs for 100 nests initiated before 4 June versus 7.9−8.7 in 55 clutches completed after 4 June (Dane 1965). Clutch size also varied with the age of the laying female; yearlings averaged 10.5 eggs, and older birds, 11.4 eggs (Dane 1965). Clutch size in the PPR in the Dakotas (1993–95) averaged 10.8 for 1,805 nests (Krapu et al. 2004c).

Blue-winged Teal eggs are subelliptical, smooth, and creamy tan in color, with little or no variation in shade (Glover 1956). Bent (1923) reported average measurements for 93 eggs from various collections as 33.4 × 46.4 mm, and Glover's (1956) findings were 33.9 × 47.1 mm for 142 eggs in Iowa. The average for 1,124 eggs in North Dakota was 46.4 × 33.3 mm (Rohwer et al. 2002). Eggs are almost always laid at a rate of 1/day (Strohmeyer 1967).

Incubation and Energetic Costs. Dane (1966) reported an average incubation period of 24.3 days (range = 23–27 days) for 15 nests placed in an incubator at 0–4 days of development. In North Dakota, the incubation period averaged 23.1–23.2 days during a 2-year study (1995–96), with a range of 19–28 days (Feldheim 1997). The length of incubation declined as the nesting season progressed, with females instead increasing their incubation constancy, but the incubation period was unaffected by clutch size. Bennett (1938) reported an incubation period of 21–23 days in Iowa.

Incubation constancy averaged 81.1% (range = 66.1%–88.4%) for 100 females monitored in North Dakota (Loos 1999). Females averaged 3.0 recesses/day for a total of 97.6 minutes, with recesses becoming more frequent later in the incubation period. Females on nest recesses in Iowa averaged 60.0% of their time feeding (Miller 1976). Incubating females are often difficult to flush from their nests, and late in incubation they will give a *distraction* display, flapping their wings over the ground or water in an attempt to divert predators (McKinney 1970). Glover (1956) noted that 75% of the females defecated on the eggs when disturbed on their nests.

In Iowa, females departing their nests usually flew directly toward their mates, who were on nearby waiting areas (Glover 1956). Of 59 waiting areas described by Glover, 54 (92%) were located in places where >75% of the shoreline was covered with vegetation, and 59% were situated near a prominent object, such as a rock, a log, or an old muskrat (*Ondatra zibethicus*) house. Waiting stations were also often in close proximity to the nest site.

Blue-winged Teal had a higher nest attentiveness during the laying period than did later-nesting species, such as Gadwalls and Lesser Scaup (Loos and Rohwer 2004). Females incubated their eggs an average of 6.8 hr/day between the laying of eggs 2 and 10, and females with smaller clutches increased their nest attentiveness more rapidly than those with larger clutches. By initiating incubation prior to clutch completion, females reduce the eggs' exposure to cold and thus maintain egg viability while minimizing the time when they are susceptible to nest predators.

There is no specific study of reproductive energetics in Blue-winged Teal, but Arnold and Rohwer (1991) provided significant insights from their discussion of clutch-size limitations in dabbling ducks. The caloric requirement for females during egg production was estimated at 390 kJ/day, and the cost of producing 1 egg was 275 kJ. Formation of a 9-egg clutch thus required 2,475 kJ, which was 31.2% of a female's total caloric expenditure during the egg-formation period. Stored lipids contributed <1% of the cost of egg formation. Gloutney and Clark (1991) found that nearly all the nutrients required for egg production and incubation are acquired by females while on the breeding grounds, probably because their small body size does not permit storage of as much in absolute reserves as do larger species, such as Mallards.

In northwestern Iowa, female Blue-winged Teal lost 6.2 g of their body mass/day from the time the seventh egg was laid until the fifth day of incubation (70 g total), after which the rate of weight loss declined (Harris 1970). Of this 70-g weight loss, only 47% reflected the use of energy reserves; 53% resulted from atrophy of the ovary and oviduct.

The mobilization of lipid reserves was indicated by an increase in the plasma level of free fatty acids.

Nesting Chronology. Because of their late arrival on breeding areas, Blue-winged Teal are among the last dabbling ducks to nest. In northwestern Iowa, the nesting season began during the first week of May and lasted until the first week of August, a period of about 100 days ($n = 186$ nests). Peak nest initiation occurred during the last week of May (Glover 1956). At Delta Marsh in Manitoba (1962–63), nesting peaked during the week of 21–27 May in 1962, compared with 2 peaks during May 1963, probably because cold weather delayed nesting by some females (Dane 1966). Sowls (1955) reported peak nesting at Delta Marsh around 13–19 May in 1949 and 1950, but nest initiation was delayed about a week during the colder spring of 1950 ($n = 112$ nests). At the Lostwood NWR in North Dakota (1980–88), the average nest-initiation date was 24 May for 435 nests, with 90% initiated between 11 May and 14 June (Kruse and Bowen 1996).

Examination of long-term (1936–68 and 1965–77) datasets from 2 study areas (Salyer and Woodworth) in North Dakota revealed that nest initiation by Blue-winged Teal began about 24.5 days after arrival (7–13 May), but variations of around a week were not uncommon. The date of nest initiation was correlated ($r = -0.563$) with average temperatures between 16 April and 27 May, with a 2-day delay in nest initiation for each Celsius-degree change in average temperatures. Median hatching dates were 4–5 July at the Salyer Study Area and 29 June at the Woodworth Study Area. The mean date of nest initiation was 28 May in the PPR in Canada during a 1982–85 study (Greenwood et al. 1995).

Early dates of nest initiation were 24 April in Iowa (Strohmeyer 1967); 25 April in South Dakota (Evans and Black 1956); 5 May at Redvers, Saskatchewan (Stoudt 1971); 9 May at Delta Marsh in Manitoba (Sowls 1955); 10 May in North Dakota (Greenwood et al. 1995); and 22 May at Brooks, Alberta (Keith 1961). Few nests are started after mid-July. At Delta Marsh, Dane (1965) found that yearlings began nesting 5 days later than older females.

Nest Success. Bellrose (1980) reported an overall nest success of 21% for 9,523 nests monitored during an array of studies across the breeding range of Blue-winged Teal. Among the earliest nesting studies of Blue-winged Teal were those on the Ruthven area in northwestern Iowa, including the classic life-history work by Logan Bennett. The apparent nest success of 223 normal nests located at Ruthven from 1932 to 1936 was 59.6% (Bennett 1938), compared with 21.4% for 173 nests located in 1949 (Glover 1956). In southeastern Alberta (1953–57), Keith (1961) reported a nest-success rate of 42% for 202 Blue-winged Teal nests: 41.8% for 153 first nests and 42.9% for 49 first renests.

Nest success for 7,166 Blue-winged Teal nests monitored during 3 time periods from 1966 to 1984 at various study areas in the PPR in the Dakotas and Minnesota averaged 17.6%, but ranged between 10% and 29% (Klett et al. 1988). Averages for success (determined from 5,393 nests) were 15.3% in grasslands, 15.0% in idle grasslands, and only 8.3% in croplands. Mammalian predation was the major (54%–85%) cause of nest loss among all 5 species of puddle ducks studied.

In the PPR in Canada (1982–85), the success of 872 Blue-winged Teal nests averaged 15% (Greenwood et al. 1995). Among the unsuccessful nests, 72% were destroyed by predators, 14% were abandoned (due to several causes), and 2% were destroyed by farm equipment. The remains of 64 Blue-winged Teal, 65.0% of which were females, were found on the study areas. Across the Dakotas, Sargeant et al. (1984) reported evidence of 385 Blue-winged Teal (30%) among the 1,293 duck remains found at the dens of red foxes (*Vulpes vulpes*). Greenwood et al. (1995) stated that habitat conditions in the PPR had changed dramatically over the past 100 years; much of the region was less suitable for duck production than in the past.

Reynolds et al. (2001), however, noted that the daily survival rate of 3,165 Blue-winged Teal nests monitored in the PPR in the Dakotas and Montana (1992–95) averaged 95.4% in planted cover associated with the Conservation Reserve Program and 95.3% in planted cover on Waterfowl Production Areas. Daily survival was most strongly correlated with the percentage of perennial cover ($r = 0.43$). Recruitment of Blue-winged Teal was also 32% higher with Conservation Reserve Program cover on the landscape, compared with croplands. Hence major government programs are capable of at least partially offsetting the long-term effects of habitat alteration in the PPR.

Although nest-success rates are low for Blue-winged Teal, and the nest-success rates for prairie-nesting ducks in general has declined about 0.5%/year from 1935 to 1992, late-nesting species, such as Blue-winged Teal and Gadwalls, experienced higher nest success than early-nesting species, like Mallards and Northern Pintails (Beauchamp et al. 1996). More vegetative cover, which is available later in the nesting season, is thought to conceal nests better from predators. The overall decline in the nesting success of ducks in the PPR, however, was similar in both parkland and grassland regions, indicating a broad-scale cause, despite predator and habitat differences between the regions.

As with other ducks nesting in the PPR, there are numerous nest predators of Blue-winged Teal. Sargeant et al. (1993) identified 20 avian and mammalian predators in the PPR, of which 9 species of mammals and 7 species of birds commonly preyed on waterfowl nests and/or adults and ducklings. Of 1,409 unsuccessful Blue-winged Teal nests on the Woodworth Study Area in North Dakota (1966–81), 87.6% were destroyed by mammals, 2.3% by birds, 2.6% by humans/machinery, 2.0% by fire, and 1.6% by unknown causes (Higgins et al. 1992). Researchers determined that the probable causes of predation for 956 nests were mainly red foxes (63.8%) and striped skunks (*Mephitis mephitis*; 27.2%); other predators were ground squirrels (*Spermophilus* spp.; 3.5%), raccoons (*Procyon lotor*; 2.2%), gulls (*Larus* spp.; 2.2%), and badgers (*Taxidea taxus*; 0.6%).

On the Ruthven area in northwestern Iowa (1948–49), predators were responsible for the loss of 100 out of 186 (53.7%) Blue-winged Teal nests under observation by Glover (1956). The principal predators of these 100 nests were striped skunks (47%), mink (*Mustela vison*; 19%), red foxes (18%), and raccoons (10%). Skunks were the primary nest predators in Alberta (Keith 1961). Skunks are especially significant nest predators, because they forage in upland grassland habitats managed for nesting waterfowl (Larivière and Messier 2000), which is the principal nesting habitat of Blue-winged Teal. In contrast, a study of 30 radiomarked raccoons in North Dakota found that they seldom used upland habitats (e.g., pasturelands, haylands, or idle upland cover); hence an upland nest-site location significantly decreases the probability of depredation by these mammals (Fritzell 1978).

Brood Parasitism. Brood parasitism is rare in Blue-winged Teal, most likely because their nests are dispersed and well concealed. On the Ruthven area in northwestern Iowa, Glover (1956) reported 7 out of 186 (3.8%) Blue-winged Teal nests parasitized by Ring-necked Pheasants (*Phasianus colchicus*), and 1 nest was parasitized by a Northern Shoveler. None of the parasitized nests were successful. Earlier on the Ruthven area, Bennett (1938) noted that 4.7% of 289 nests were parasitized by Ring-necked Pheasants, with parasitism being more prevalent during years of high pheasant populations. Hamilton (1957) stated that a Brown-headed Cowbird (*Molothrus ater*) parasitized a Blue-winged Teal nest.

Renesting. The most thorough study of the renesting ecology of Blue-winged Teal is the work of Strohmeyer (1967) on the Ruthven area in northwestern Iowa. At least one-third of the yearlings and half or more of the older females renested if their first clutches were destroyed during egg laying. In contrast, only 6.7% of the females renested if their clutches were lost during incubation. Almost 36% of all females renested following the loss of their first nest, 31% after losing their second, and 43% after losing their third (i.e., a fourth nest initiation). The result of all this renesting activity was that 16% of all successful nests were renests, and ducklings from renests survived as well as early-hatched ducklings.

In an earlier study on the Ruthven area, Glover (1956) reported that 52 out of 186 (28%) Blue-winged Teal nests were renests, of which 26% were successful. Bennett (1938) noted 14.8% nest success for renesting Blue-winged Teal at Ruthven. In southeastern Alberta (1953–57), renesting occurred 55% of the time for Blue-winged Teal, compared with 100% for Mallards, 82% for Gadwalls, 75% for Northern Shovelers, and 39% for Lesser Scaup (Keith 1961). Sowls (1955) considered Blue-winged Teal to be the poorest renesting species among the 5 species of dabbling ducks he studied at Delta Marsh in Manitoba.

Clutch size in renests can be difficult to assess without marked females, because renests can be as well constructed as initial nests, although there is often less down in renests. Nonetheless, as with other renesting waterfowl, clutch size for renests tends to be smaller than that for initial nests. At Delta Marsh in Manitoba, Sowls (1955) noted that clutch size for 54 Blue-winged Teal nests completed before 15 June averaged 10.6 eggs, compared with 8.8 for 42 nests completed after 15 June, with the latter group probably including many renests. In another Delta Marsh study (1961–63), clutch size averaged 10.9 for 100 nests initiated before 4 June versus 8.3 for 55 nests begun after 4 June (Dane 1966). In Iowa, Bennett (1938) reported an average clutch size of 9.3 for 223 normal nests and 4.3 for 27 renests. Strohmeyer (1967) noted an average clutch size of 10.7 for initial nests, 9.2 for first renests, and 8.3 for second renests.

Females in the 5 dabbling duck species studied in Manitoba by Sowls (1955), including Blue-winged Teal, waited at least 3 days before renesting, and they waited an average of 0.6 days before renesting for each additional day of incubation at the time their nest was destroyed. In Iowa, Blue-winged Teal continued to lay additional eggs without interruption, up to a loss of the first 6 eggs (Strohmeyer 1967). After the destruction of nests with 8 or more eggs, however, some time elapsed before new nests were initiated. For nests lost after 1 day of incubation, the renesting interval averaged 6 days, compared with 7.5 days for nests lost after 10 days of incubation. Females that lost nests during their last week of incubation did not renest.

In Iowa, 44 renesting Blue-winged Teal located their new nest sites an average of 182 m from their initial nests, with a maximum distance of 678 m (Strohmeyer 1967). Five renests observed by Sowls (1955) averaged 247 m from earlier nest sites.

REARING OF YOUNG

Brood Habitat and Care. Brood habitat for Blue-winged Teal generally consists of flooded vegetation. Observations of 1,115 Blue-winged Teal broods over 18 years in the PPR in the Dakotas recorded most (55%) occurrences on semipermanent wetlands, and 32% on seasonal wetlands (Duebbert and Frank 1984). At the Woodworth Study Area in North Dakota (1965–75), 82% of the broods were on wetlands <5.0 ha in size (Higgins et al. 1992). Brood densities were highest in tilled ephemeral wetlands (73.0/40.5 ha), and lower on seasonal wetlands (14.2) and semipermanent wetlands (11.0). In eastern Washington, most (58%) observations of Blue-winged / Cinnamon Teal broods (*n* = 53) occurred in areas with water depths of 0.4–0.9 m, and aquatic plant cover indexed at 41%–110% (average = 62%; Monda and Ratti 1988). In eastern North Dakota (1993–95), out of 27 Class I–II broods, 66.7% were on seasonal wetlands, 25.9% on permanent wetlands, and 7.4% on temporary wetlands (Krapu et al. 2001).

In South Dakota, Class I Blue-winged Teal broods used emergent cover for >50% of the daylight period, with the greatest amount of swimming activity in the afternoon and evening (Rin-gelman and Flake 1980). Feeding activity for both Class I and Class II broods peaked in the morning and evening, at about 40%–70% of their activity budgets. The mean duration of activity bouts was 42 minutes for Class I, 54 minutes for Class II, and 103 minutes for Class III. Class I broods were most visible in the evening, while Class II and III broods were more visible in the morning. Brood visibility decreased when wind speeds were >24 km/hr and the air temperature was >23°C.

In Iowa, Bennett (1938) observed between 5,000 and 6,000 ducklings during his 1932–36 study: 30% occurred in stands of various bulrush (*Scir-pus*) species in water 0.3–0.6 m deep; about 25% in stands of bulrush/bur-reed (*Sparganium eurycar-pum*) in water 0.3–0.5 m deep, and 20% in shallow-water stands of river bulrush (*Scirpus fluviatilis*) and cattails (*Typha latifolia*) that retained water in late summer.

Bennett also reported that all young had hatched within 4 hours of the first egg being pipped (except for dead or weak ducklings). The ducklings were dry in 3–4 hours, after which they began to travel to water, arriving within 12 hours after hatching. The female provides all parental care for the brood, but some males may briefly accompany broods that hatch early in the breeding season (Rohwer et al. 2002). Late-hatched broods are often aban-doned by the female.

Brood Amalgamation (Crèches). The extent of brood amalgamation is not well known in Blue-winged Teal, but apparently it does occur. Bennett (1938:72–73) noted the "willingness of ducklings to abandon their mother for some other brooding female."

Development. Dane (1965) provided growth and development data on Blue-winged Teal, based on 9 males and 9 females he reared in captivity and fed ad libitum (an unlimited diet). Body mass of the females averaged 87 g at 7 days, 111 g at 14 days, 205 g at 21 days, 258 g at 28 days, 301 g at 35 days,

and 298 g at 42 days. The males averaged 89 g at 7 days, 126 g at 14 days, 221 g at 21 days, 277 g at 28 days, 319 g at 35 days, and 315 g at 42 days. Primary growth in both sexes was complete in 23–25 days.

In North Dakota, weights of wild Blue-winged Teal averaged 27.2 g in pipped eggs ($n = 22$) and 18.1 g at 3 days ($n = 21$; Lokemoen et al. 1990b). By 17.5 days old, females ($n = 5$) averaged 162.0 g and males ($n = 8$), 166.2 g. At 26 days, the females ($n = 13$) averaged 203.5 g, and the males ($n = 9$), 206.7 g. Near fledging (38.5 days), the females ($n = 93$) averaged 281.9 g, and the males ($n = 97$), 301.3 g. Duckling development was best described by a logistic growth curve, with an asymptote of 365 g for females and 384 g for males. The growth rate for Blue-winged Teal was 0.0750, compared with 0.0700 for Gadwalls and 0.0692 for Mallards. Fledging occurred at about 42 days (Bennett 1938).

RECRUITMENT AND SURVIVAL

During his 1932–36 studies at the Ruthven area in northwestern Iowa, Bennett (1938) estimated that out of 100 nests, 55% of the ducklings reached flight stage, or 3.1 young/breeding pair. This estimate would vary annually, especially in response to the highly changeable water conditions (e.g., drought) that characterize much of the principal breeding range of Blue-winged Teal. From 1965 to 1981 on the 4 km² Woodworth Study Area in southcentral North Dakota, the minimum number of broods observed averaged 74/year, but varied from 12 to 165 (Higgins et al. 1992).

In an analysis of breeding-population data from 1963 to 2001, Frank Bellrose found that breeding numbers of Blue-winged Teal during 16 dry years declined 34% on the mixed-grass prairies and 39% in the shortgrass prairies. In contrast, populations in the parklands north of the prairies declined only 7% during those dry years. Blue-winged Teal populations in the vast boreal forest of westcen-tral Canada, however, increased an average of 76% during the dry years, and populations in the subarctic deltas rose by an average of 47%, probably as a result of birds displaced from the prairies. The production of breeding Blue-winged Teal in these suboptimal northern habitats, however, was probably significantly lower than in the prairies, as the former regions contained a substantial number of nonbreeding and unsuccessfully breeding birds.

Bellrose also estimated a 52% overall survival rate for Blue-winged Teal, based on an analysis of 409,006 individuals banded in North America from July to September (1978–96). Survival was 63% for adult males, 53% for adult females, 50% for immature males, and 43% for immature females. Geis et al. (1963) determined annual survival rates for Blue-winged Teal banded on prairie breeding areas at 58.3% for adult males, 47.3% for adult females, and 28.3% for immatures.

Blue-winged Teal generally have higher annual mortality rates than other dabbling ducks, despite very low harvest rates. These higher mortality rates may be the result of their long, often overwater migrations to Neotropical wintering areas, or perhaps of increased predation during the winter at those sites.

Longevity records for Blue-winged Teal are 23 years and 3 months for an after-hatching-year male banded in 1983 in Saskatchewan and shot in Cuba in 2005, and 22 years and 4 months for an hatching-year female banded in 1960 in Vermont and shot in Québec in 1982.

FOOD HABITS AND FEEDING ECOLOGY

Blue-winged Teal are omnivores, usually feeding in shallow (<20 cm deep) wetlands (Gammonley and Fredrickson 1995). Bennett (1938) summarized food-habits data from his work in Iowa (*n* = 45, including 26 ducklings), as well as from 21 Blue-winged Teal collected in Mexico, and 319 collected by Mabbott (1920) in 29 states and 4 provinces during every month except January. The diet of these teal (by frequency of occurrence) was 75.0% plant matter and 25.0% animal matter. The most important plant foods were Cyperaceae (21.1%), Najadaceae (14.3%), and Gramineae (14.1%); the most

significant animal foods were mollusks (14.1%) and insects (9.2%). The study by Mabbott is exhaustive in its identification of the many food items found in the stomachs of Blue-winged Teal.

Foraging methods are dabbling with only the bill submersed (bill-dipping), direct pecking of individual food items on the water's surface (hawking), dabbling with the head submersed, and tipping up; diving is rare. In eastern Washington, the foraging modes of breeding Blue-winged Teal in open-water habitats were bill-submersed dabbling (41%), head-submersed dabbling (21%), hawking (36%), and tipping up (3%; Connelly and Ball 1984). In emergent vegetation, the major foraging methods were bill-submersed dabbling (84%) and head-submersed dabbling (14%). Blue-winged Teal also spent more time foraging in open water than did the closely related Cinnamon Teal, which were more likely to feed within emergent vegetation. The bill lamellae of Blue-winged Teal only permit the consumption of foods ≥1.5 mm in diameter; hence they are limited in their ability to eat small crustaceans, such as cladocerans, copepods, and ostracods (Swanson et al. 1974). In contrast, such foods are available to Northern Shovelers, which have much finer-spaced lamellae. Postflightless male Blue-winged Teal in Manitoba spent 85.9% of their time feeding, with dabbling (40.6%) and picking (34.2%) as the dominant foraging methods (DuBowy 1985a). In South Dakota, Class I ducklings fed primarily at the water's surface (61%) and by bill-dipping (37%), while Class II ducklings almost exclusively fed by ducking their heads underwater (48%) and bill-dipping (43%). Class III ducklings fed by bill-dipping (61%) and ducking their heads underwater (39%; Ringelman and Flake 1980).

Flightless Blue-winged Teal in southwestern Québec preferred habitats dominated by cattails and interspersed with open water containing submerged vegetation (Courcelles and Bédard 1979). Preflightless male Blue-winged Teal spent 68.6% of their time feeding, compared with 85.9% during the postflightless period (DuBowy 1985a). Males increased their fat index during the preflightless period, which then declined sharply during the early part of the flightless period (0%–33% regrowth of the remiges), before largely leveling off, and then increasing after the remiges were fully grown (DuBowy 1985c).

Breeding. Dirschl (1969) reported food habits at monthly intervals for 126 adult Blue-winged Teal (sexes combined) collected on the Saskatchewan River delta from May through September in 1964 and 1965. Their overall diet (identified food items) averaged 47.8% plant matter and 52.2% animal matter. The use of plant material was much greater in August (69.8%) and September (83.3%), however, than in May through July (18.8%–34.4%), while the use of animal material was 65.8%–81.2% from May through July. Gastropods (snails) were important animal foods (42.8%–47.9%) in May and June, compared with only 15.6%–26.7% during the other 3 months. The seeds of shoreline and emergent plants—such as sedges (*Carex* spp.), spikerushes (*Eleocharis* spp.), bur-reeds (*Sparganium* spp.), and bulrushes—were used throughout the 5-month period (18%–35%/month). Of foods consumed just during 1 month, chironomid (midge) larvae (primarily *Chironomus plumosus*) were only eaten in May (14.9%), and leeches (Hirudinea) just in July (61.5%).

Swanson et al. (1974) separated food-habits data by males and females during their 5-year study (1967–71) of the feeding ecology of breeding Blue-winged Teal (*n* = 46 males, 61 females) in southcentral North Dakota from mid-April through June. Animal foods made up 89% of their diet during the entire breeding season, of which 35% were insects; 32% were gastropods; and 19% were crustaceans, such as amphipods and cladocera. Females consumed more animal food than did males (91% vs. 85%), and they appeared to feed more consistently. Changes in the diets occurred in both sexes, with the percentage of animal matter increasing from 45% in April to 81% in May, before leveling off at 95% for the remainder of the breeding season.

The foods consumed varied among the habitat types from which birds were collected. Thus mollusks and insects were the dominant foods on temporary and seasonal wetlands, while crustaceans were the principal items on more saline wetlands in the outwash plains. Blue-winged Teal foraged in and around plant parts on or near the water's surface, and fed on the bottom by tipping up. They also fed over deep water, where they could glean invertebrates from submerged vegetation or consume emerging insects from the water's surface. Blue-winged Teal and other puddle ducks are known to forage at night over open water to take advantage of emerging mayflies (Ephemeroptera) and chironomids (Swanson and Sargeant 1972).

Swanson and Meyer (1977) demonstrated the significance of aquatic macroinvertebrates in the diet of female Blue-winged Teal (*n* = 20), which averaged 99% during the egg-laying period in North Dakota. Important animal foods were insects (44%), gastropods (38%), and crustaceans (14%). Dipterans, especially chironomid larvae, were the most significant (32%) insect food. Food selection was affected by fluctuating water levels, with teal shifting from a diet high in gastropods on seasonal wetlands to a diet dominated by chironomids on semipermanent wetlands when water levels were receding.

Bennett (1938) provided data on the foods consumed by 26 ducklings (from 3 days to 6 weeks old) that he collected in northwestern Iowa during July and August 1936. Plant material was found in all 26 stomachs, while animal matter was found in 12 (46%). The seeds of bulrushes, a sedge (*Carex riparia*), and sago pondweed (*Potamogeton pectinatus*) were the most frequently occurring (73%–100%) plant foods, while the only animal materials were insects (35%) and mollusks (19%).

Migration and Winter. The diets of prebreeding Blue-winged Teal began to shift from a plant-dominated diet in the winter to a greater use of animal matter during the breeding season. At the Mingo NWR in Missouri, the diet of Blue-winged Teal (*n* = 10 males, 10 females) using seasonally flooded impoundments averaged 65% animal matter and 35% plant matter (Taylor 1978). The most important animal foods were gastropods (24%) and insects (22%). Eleven plant foods were identified, but none formed >8% of the diet. In southwestern Louisiana, paired and unpaired male Blue-winged Teal ($n = 79$) collected during spring migration in 1990–91 primarily (56.4%–86.1%) consumed animal matter, of which midges were most important (34.1%–65.2%), although gastropods (7.5%–18.1%) were also eaten (Manley et al. 1992). The remainder (13.5%–43.5%) of their diet consisted of seeds from a variety of emergent wetland plants. There was no difference in the animal foods eaten by paired and unpaired males, and the consumption of animal matter was not correlated with molting intensity. Sampling the available food resources revealed that animal foods were simply preferred to plant foods.

The diet (aggregate percentage) of postbreeding adult male Blue-winged Teal in Manitoba ($n = 30$) was 57.0% animal matter and 43.0% plant matter (DuBowy 1985a). Gastropod (snail) larvae (19.5%) and midge (Chironomidae) larvae (15.4%) were the most important animal foods. Seeds (32.1%) and vegetation (10.9%) made up the plant foods. During fall migration in Illinois (1977–81), smartweeds (*Polygonum* spp.) and wild millet (*Echinochloa crus-galli*) were the dominant (88% combined total) foods, although the study also included about 20% American Green-winged Teal (Havera 1999). During the early fall on playa lakes in Texas (1966), the diet (aggregate percentage) of 11 Blue-winged Teal was 91.9% plant material and 8.1% animal matter (Rollo and Bolen 1969). Seeds from sorghum (*Sorghum vulgare*; 52.0%) and wild millet (28.0%) were the most important plant foods. For 9 Blue-winged Teal collected in the same area during fall 1976, wild millet was the most important (75.2%) food item, but waste corn was also used (23.2%), as the birds made field-feeding flights to nearby harvested fields (Sell 1979). In an analysis of the food habits for Blue-winged Teal and American

Green-winged Teal (combined) collected in Louisiana (Nov–Jan 1960/61), their diet (on average) was 54.9% carbohydrates, 18.2% protein, and 17.8% fiber, with small amounts of calcium and phosphorus (Bardwell et al. 1962).

Blue-winged Teal ($n = 56$) wintering in coastal wetlands on the Yucatán Peninsula in Mexico from mid-December to mid-March consumed 96.9% plant matter (Thompson et al. 1992). Nearly all (96.6%) of the plant material contained high-carbohydrate tubercles from muskgrass, a macroalgae that dominated the submerged vegetation in the area. In contrast, 4 birds collected during early winter (1 Oct–15 Dec) primarily (98.0%) consumed gastropods. The food habits of 84 Blue-winged Teal collected at 2 sites over 3 time periods in Costa Rica, however, reflected the omnivorous diet of Blue-winged Teal (Botero and Rusch 1994). The seeds of waterlilies (*Nymphaea* spp.) were the most important food at 1 site (62%), while gastropods (73%) and water boatmen (Corixidae; 16%) were the most significant foods at the other site. There were no differences between the sexes in the percentages of plant and animal foods consumed.

MOLTS AND PLUMAGES

Bennett (1938), Palmer (1976a), and Rohwer et al. (2002) have all described the molts and plumages of Blue-winged Teal.

Acquisition of juvenal plumage begins when the ducklings are about 2 weeks old and is completed about 4 weeks later, when the birds reach flight stage. At this stage, both sexes are brownish, with a buffy head and neck; the sides of the head and the front of the neck are finely streaked. In Iowa, this plumage was completed by mid-August, at which time most birds were flying. Juvenal plumage immediately grades into first basic by late August and early September, but it is usually interrupted by migration.

First basic involves the body feathers, the retrices, and the innermost wing feathers, but the juvenal wing is retained until late winter or early spring. Males in first basic have a gray head, with a white or pale gray color on the sides of the head and the neck. The upperparts are dark; the underparts are white; the sides are brownish gray. Females in first basic are similar to the males, but they have a white area at the base of the bill.

First alternate is acquired from late fall through early winter and involves all feathering, although the juvenal wing is still retained. Some individuals will acquire much of this plumage in early fall, but the molt then seems to be interrupted by migration. Males in this plumage are usually not distinguishable from males in definitive alternate. Females exhibit a similar timing in their acquisition of first alternate, which resembles definitive alternate.

Definitive basic in males is acquired during the summer, beginning by early June, and involves the feathers on the wings. Hence there is a flightless period, which averaged 21 days in captive males (Oring 1964) and 26–36 day in the wild (McHenry 1971). Males in definitive basic resemble the females, but they then acquire definitive alternate beginning in the fall, with this plumage complete by January. Females acquire definitive basic over an extended period that begins in the spring but is interrupted by the breeding season. The head, the body, and the tail feathers are replaced before nesting, while the wing feathers (remiges) are usually molted after the young fledge. Definitive alternate is then acquired from late summer through fall.

CONSERVATION AND MANAGEMENT

Overharvest is not much of a conservation concern for Blue-winged Teal, because their early-fall migration sharply limits hunter kills in North America, and harvests are not large in Mexico, Central America, or South America. Habitat loss and degradation, however, are continual problems throughout the principal breeding range of Blue-winged Teal, the PPR (see the Mallard account). Within the PPR, the small size of the seasonal and temporary wetlands that are so important to breeding Blue-winged Teal renders these habitats more vulnerable to drainage than larger, more permanent wetlands.

Wintering habitats along the Texas and Louisiana Gulf Coast have been and continue to be subjected to loss and degradation from an array of sources (Chabreck et al. 1989, Stutzenbaker et al. 1989), although much of the extensive wintering waterfowl habitat along the East Coast is in good condition, especially in areas used by Blue-winged Teal (Baldassarre et al. 1989). Except for studies in the Yucatán Peninsula in Mexico, virtually nothing is known about the wintering ecology of Blue-winged teal in the Neotropics. Additionally, although the few surveys available indicate a small harvest in the Neotropics, more extensive harvest surveys are warranted over the broad winter range used by Blue-winged Teal, especially for their major wintering areas in Venezuela and Columbia.

Cinnamon Teal

Anas cyanoptera septentrionalium (Vieillot 1816)

Left, drake; *right,* hen

The Cinnamon Teal is endemic to the Western Hemisphere, where 5 subspecies are recognized: 1 (*A. c. septentrionalium*) in North America, and 4 in South American (*A. c. borreroi, A. c. cyanoptera, A. c. orinomus,* and *A. c. tropica*). The North American subspecies breeds on freshwater wetlands in the Intermountain Region (which includes the Great Basin) and the Central Valley of California, but also uses the alkaline wetlands so characteristic of the Great Basin. They rarely are observed east of the Mississippi River. Cinnamon Teal are 1 of 3 North American species in the blue-winged duck group, along with Blue-winged Teal and Northern Shovelers (there are 7 species worldwide). The cinnamon plumage of the male is unmistakable, while the female is almost indistinguishable from the closely related Blue-winged Teal. Except for early work on marshes near the Great Salt Lake (Spencer 1953), Cinnamon Teal were not well studied until the 1990s, on the Colorado Plateau on Arizona (Gammonley 1996a). Population estimates are not accurate, in part because Cinnamon Teal are not separated from Blue-winged Teal during surveys; the breeding population may be not much more than 260,000, which would render Cinnamon Teal among the least abundant of the dabbling ducks. Most are early migrants to wintering areas in Mexico; hence the North American harvest is not extensive.

Wings of Cinnamon Teal and
Blue-winged Teal hens are alike

Wings of Cinnamon Teal and
Blue-winged Teal drakes are alike

IDENTIFICATION

At a Glance. Adult males are unmistakable, with a deep cinnamon head, neck, and underparts. The female closely resembles the female Blue-winged Teal. Hybridization occurs between Cinnamon Teal and Blue-winged Teal where their breeding ranges overlap on the Great Plains. While hybrids can occur, there are few actual records from wild birds (Bolen 1978). Spencer (1953) thought that hybrid Cinnamon × Blue-winged Teal were "extremely uncommon."

Adult Males. The brilliant cinnamon-red breeding plumage on the head and the body make the male Cinnamon Teal unique among all waterfowl. The back, the rump, the uppertail coverts, and the tail are a dull brown. In flight, the cinnamon-red head and body, the black tail, and the powder-blue wing shoulders create striking contrasts in color. The wing coloration is almost identical to that of a Blue-winged Teal, with black primaries, powder-blue upperwing coverts, and a bright green speculum. The legs are black; the feet are yellow to orange. The bill is black. The irises become a deep reddish orange when males are about 2 months old.

Adult Females. Female Cinnamon Teal are nearly identical to female Blue-winged Teal, and they cannot be reliably separated in the field, except at close range. The plumage of Cinnamon Teal, however, is a richer brown; the white lore spot and the eye line are less distinct; and the bill is larger and more spatulate. The legs and the feet of female Cinnamon Teal are yellowish; the bill is a dark slate gray. The irises are hazel.

Juveniles. Young birds are female-like, but they often have a bleached straw-like coloration during the first fall, which is lighter and less mottled than that of females. The legs and the feet are grayish.

Ducklings. Cinnamon Teal ducklings are yellowish, patterned with brownish, and with a brownish eye stripe (Nelson 1993). The long bill is grayish.

Voice. Cinnamon Teal are less vocal than other dabbling ducks, with calls that are similar to those of Blue-winged Teal and Northern Shovelers, which are the other blue-winged ducks in North America (there are 7 species worldwide). McKinney (1970) described the *repeated call* of males, given during courtship, as a rolling rattled *rrar . . . rrar . . . rrar . . . rrar.* The decrescendo call of the females is usually several or many syllables that have a hoarse rattling quality, although not as long or as deep as that of Northern Shovelers.

Similar Species. Males are readily distinguished from the rusty-feathered male Ruddy Duck by their dark bill and lack of a large white cheek patch. Females are very similar to female Blue-winged Teal. Female American Green-winged Teal at rest are separable from both female Cinnamon Teal and males in basic plumage by their smaller size, noticeably smaller bill, and grayish feet. In flight, American Green-winged Teal do not have blue shoulder patches on the coverts.

Weights and Measurements. Gammonley (1996) reported average body weights for Cinnamon Teal of 383 g for males ($n = 44$) and 372 g for females ($n = 69$). Culmen length averaged 44.2 mm for

Above:
Cinnamon Teal pair.
GaryKramer.net

Left:
Cinnamon Teal pair
in flight.
John C. Avise

Male Cinnamon Teal.
GaryKramer.net

males and 42.9 mm for females. Wing length averaged 191 mm for males and 182 mm for females.

DISTRIBUTION

The Cinnamon Teal is a New World species for which most authorities recognize 5 subspecies (Johnsgard 1978, Gammonley 1996), although Livezey (1997a) only recognized it as a single species with no subspecies. The subspecies vary considerably in their plumage characteristics (Wilson et al. 2008, Wilson et al. 2010). The descriptions here are from information in Johnsgard (1978) and Gammonley (1996); the common names are from Johnsgard (1978).

The Cinnamon Teal (*A. c. septentrionalium*), also referred to as the Northern Cinnamon Teal, occurs in western North America, from southern Canada to Mexico, and it is the subspecies described in this account. The Southern Cinnamon Teal (*A. c. cyanoptera*) is found throughout western South America (from Brazil to Patagonia), from sea level to midelevations; it is a rare resident on the Falkland Islands. Males have black spots on the breast. The Andean Cinnamon Teal (*A. c. orinomus*) is the largest subspecies, inhabiting the

Andes Mountains of Peru and Bolivia (Lake Titicaca) south to Argentina and Chile, where they are residents at elevations of 3,800 m in the páramo zone. Males are comparatively paler than the other subspecies, and the bill is relatively long. The Borrero Cinnamon Teal (*A. c. borreroi*) is known only from the eastern slope of the Andes in Columbia, at elevations of 1,000–3,600 m. The Tropical Cinnamon Teal (*A. c. tropica*) is the smallest subspecies, occurring in the lowlands of Columbia. The sexes are similar in size. Males have black spots on the upper breast, the sides, and the flanks. Little else is known about the various South American species.

Breeding. Cinnamon Teal are abundant to locally common breeding ducks in suitable habitat in the Great Basin and other arid portions of the Intermountain Region in the United States, as well as north to southern British Columbia and Alberta. Important breeding areas in the Great Basin / U.S. Intermountain Region are the marshes surrounding the Great Salt Lake (Bellrose 1980), the Malheur Lake / Summer Lake region of eastern Oregon (Gilligan et al. 1994), Ruby Lake and Carson Sink

485

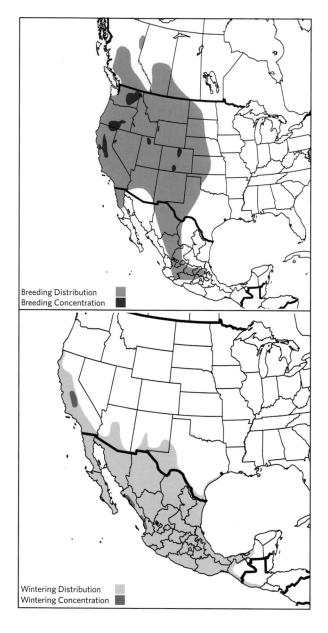

Breeding Distribution
Breeding Concentration

Wintering Distribution
Wintering Concentration

Cinnamon Teal are common breeding ducks in the Central Valley of California (Small 1994) and the Playa Lakes Region on the southern High Plains of Texas and New Mexico (Smith 2003). Cinnamon Teal also breed in suitable areas of Idaho (western and southern), Montana (mostly west of the Rocky Mountains), and Wyoming, mainly west of the Bighorn Mountains (Gammonley 1996b). In Mexico, Cinnamon Teal are uncommon to fairly common summer residents, breeding locally on the northern Baja Peninsula and throughout the interior highlands south to the central Trans-Mexican Volcanic Belt (Howell and Webb 1995).

Good estimates of their distribution and abundance are lacking for most areas, because Cinnamon Teal are found outside the region covered by the Traditional Survey; in other surveys they are combined with Blue-winged Teal. The Pacific Flyway Data Book (Collins and Trost 2010) lists average breeding populations for Blue-winged / Cinnamon Teal from 2001 to 2010: 13,093 in Washington, 31,105 in Oregon, 38,350 in California, 8,247 in Nevada, and 7,682 in Utah. Virtually all the birds recorded in California are Cinnamon Teal, with 1996–2010 averages of 55% in the Central Valley and 45% in the northwestern part of the state (Dan Yparraguirre). The breeding population of Cinnamon Teal in Colorado is estimated at around 24,000, with about half occurring in the San Luis Valley (Kingery 1998).

Bellrose (1980:287) had reported 150,000 breeding Cinnamon Teal in Utah, with "many of them favoring the marshes bordering east and north shores of Great Salt Lake," but that situation has clearly changed. Aerial surveys of breeding waterfowl in the Great Salt Lake area from 1990 to 2010 averaged only 22,411 Cinnamon Teal (range = 9,778–47,616). Managers in this region have noted that Cinnamon Teal are much less numerous since the floods of 1983–90 (Stephens 1990), with the consequent loss of historical saltmarshes, due to the invasive plant phragmites (common reed; *Phragmites australis*) and a different predator commu-

in Nevada (Alcorn 1988), the Central Valley of California (Small 1994), the scablands of eastern Washington (Connelly and Ball 1984), and the San Luis Valley of Colorado (Szymczak 1986). In British Columbia, about 30,000 breed on the interior plateau, mostly (22,000) in the Northern Rockies Bird Conservation Region (BCR), but also the Great Basin BCR to the south (Canadian Intermountain Joint Venture Technical Committee 2010).

nity that is now dominated by raccoons (*Procyon lotor*) and red foxes (*Vulpes vulpes*; Tom Aldrich), which are significant predators of ground-nesting puddle ducks (Sargeant et al. 1993).

Winter. The winter range of Cinnamon Teal extends from northwestern California and the southern Sacramento Valley south into Mexico, along both Baja California and the mainland west coast, but also sporadically in almost all of Mexico except the Yucatán Peninsula (Howell and Webb 1995). Cinnamon Teal also winter in central and southern Arizona, the Playa Lakes Region of Texas and New Mexico, and south into central Mexico, reaching all the way to southern Guatemala. Most (>90%), however, winter in the large wetland complexes along the mainland west coast of Mexico, especially Pabellón and Topolobampo Lagoons (Gammonley 1996b). An average of 63,330 Blue-winged / Cinnamon Teal were recorded during winter surveys in Mexico from 1961 to 2000 (Pérez-Arteaga and Gaston 2004).

The Midwinter Survey in the United States combines Cinnamon Teal and Blue-winged Teal in the Mississippi, Central, and Pacific Flyways, where the 2000–2010 average was 229,160. Only a small portion (2.7%) of this total occurred in the Pacific Flyway, where most were found in the Central Valley of California (Heitmeyer et al. 1989). Nearly all those recorded in the Central and Mississippi Flyways were seen along the Gulf Coast, where most teal (but not all) are Blue-winged Teal. Fedynich et al. (1989) noted a 9.7% homing rate of Cinnamon Teal to the Playa Lakes Region on the southern High Plains of Texas.

Of 502 band recoveries from 1990 to 2010 (57% direct recoveries), 86.1% were in California and 6.8% were in Mexico; Nicaragua was the southernmost recovery location (1 bird). During this time period, 7,133 Cinnamon Teal were banded, mostly (58%) in California. Of 327 band recoveries in Mexico, 54% were in states on the mainland west coast, 21% were in the interior highlands, 16% were

in the Baja Peninsula, and 9% were in states on the Gulf Coast. Almost 44% of the Mexican recoveries came from 3 states on the west coast: Sinaloa, Jalisco, and Michoacán. The 16% recovered on the Baja Peninsula contrasts with the near-absence of Blue-winged / Cinnamon Teal recorded in midwinter aerial surveys, which suggests that Cinnamon Teal migrated over the Baja Peninsula to winter along the mainland west coast and interior highlands.

Migration. Most Cinnamon Teal undergo a short- to moderate-distance migration, although birds may be resident within the southern portion of their breeding range in California's San Joaquin Valley and in Mexico (Gammonley 1996b). Like Blue-winged Teal, Cinnamon Teal are early migrants, with males and unsuccessful females beginning their migration in late summer or early fall (Spencer 1953, Gammonley 1996b). Nearly all Cinnamon Teal have departed their northern breeding grounds by late October, with most arriving in Mexico by November (Bellrose 1980, Brown 1985). In Utah, adult males were the first to depart the Great Salt Lake marshes, with few remaining there after mid-September (Spencer 1953).

The closeness of the breeding and wintering areas for much of the Cinnamon Teal population limits any definition of migration corridors, but the marshes around the Great Salt Lake in Utah are important staging areas for those Cinnamon Teal breeding in the valleys of the northcentral Rocky Mountains, before then migrating directly to Mexico. Some birds from Utah and Arizona may move to the Central Valley of California before traveling south into Mexico (Bellrose 1980, Brown 1985), as 2 out of 22 (9.1%) recoveries of Cinnamon Teal banded in Utah (1990–2010) were in California.

Cinnamon Teal are early spring migrants, with most birds on their breeding areas by early May (Bellrose 1980). Spencer (1953) noted a peak arrival in late April at the Ogden Bay marshes in Utah, while Hohman and Ankney (1994) observed peak migration in early March in the Tulare basin in the San Joaquin Valley of California. Cinnamon Teal do not commonly stray from their normal range, but they are occasionally found to the east; nonetheless, they are rare in the Atlantic Flyway.

MIGRATION BEHAVIOR

Not much is known about the migration behavior of Cinnamon Teal, although they tend to migrate in smaller flocks (usually 10–25 birds) than most species of dabbling ducks (Bellrose 1980). They have been observed in diurnal migration (Spencer 1953), but most Cinnamon Teal usually migrate at night (Bellrose 1980). They often mix with other species of dabbling ducks during migration, including Gadwalls, Blue-winged Teal, and Northern Shovelers (Evarts 2005).

Molt Migration. There is no pronounced molt migration in Cinnamon Teal. Postbreeding males congregate in small flocks during the molt. In Arizona, the majority of males molt on the breeding grounds (Gammonley 1996a), while those breeding in the Great Salt Lake area generally fly to marshes in Idaho (Cox 1993). Females molt on the breeding grounds during brood rearing (Palmer 1976a). According to Oring (1964), males at the Camas National Wildlife Refuge (NWR) in southeastern Idaho left the breeding grounds to molt by the third week of incubation. Spencer (1953) reported that molting Cinnamon Teal in northern Utah preferred small wetlands with dense bulrushes (*Scirpus* spp.) or other emergent plants.

HABITAT

Cinnamon Teal breed on wetlands of various types and sizes, ranging from freshwater wetlands to the highly alkaline systems characteristic of the Great Basin. In eastern Washington, breeding birds exhibited a preference for wetlands with dense stands of emergent vegetation, and they used the emergent zone more than open water (Connelly 1977). Wintering birds occupied similar habitats. On the lower Colorado River, wintering Cinnamon Teal were associated with areas of submerged vegetation, as were other species of dabbling ducks (Anderson and Ohmart 1988). At the Turnbull NWR in eastern Washington, Blue-winged/Cinnamon Teal adults with broods used seasonal and semipermanent wetlands, where they often found food on top of the submerged vegetation (Monda and Ratti 1988). Much (58%) of their habitat use was in water <0.4–0.9 m deep, but their niche overlap was high with other puddle ducks and most diving ducks. Concentrations of wintering Cinnamon Teal in the northern San Joaquin Valley of California were highest on shallow wetlands (Colwell and Taft 2000).

POPULATION STATUS

Cinnamon Teal are one of the least abundant dabbling ducks found in North America, although their population numbers are difficult to assess, because most surveys do not separate Cinnamon Teal from Blue-winged Teal. Bellrose (1980) estimated a breeding population of 260,000–300,000, and a fall population of 500,000–600,000. The Pacific Flyway Data Book (Collins and Trost 2010) reports a breeding population of 100,414 Blue-winged/Cinnamon Teal for the Pacific Flyway states. Wetlands International (2006) lists a breeding population of 260,000, which probably was taken from the lower figure reported by Bellrose (1980).

In the United States, the Breeding Bird Survey indicated a slight decline of 2.06% from 1990 to 2010. Christmas Bird Counts averaged 4,655 Cinnamon Teal from 2000 to 2008, of which 95% were found in California. In their analysis of wintering waterfowl populations in Mexico, Pérez-Arteaga and Gaston (2004) estimated a 50% decline in the numbers of Cinnamon Teal from 1991 to 2000. Outside North America, population estimates are 25,000–100,000 for *A. c. cyanoptera*, 10,000–100,000 for *A. c. orinomus*, and <10,000 for *A. c. tropica*. The *borreroi* subspecies is critically endangered, with <250 individuals (Johnsgard 2010a).

Harvest

Unfortunately, harvest estimates in North America do not differentiate between Blue-winged and Cinnamon Teal, but the combined harvests of these teal in the United States averaged 955,146 birds from 1999 to 2008, with 57.1% occurring in the Mississippi Flyway and 29.3% in the Central Flyway, in comparison to only 7.6% in the Atlantic Flyway and 6.0% in the Pacific Flyway. In Canada, the harvest from 1999 to 2008 averaged 36,572 birds. Over this time period, 25.7% were harvested in Saskatchewan, 21.4% in Ontario, 19.7% in Manitoba, and 19.0% in Alberta. U.S. and Canadian harvest estimates are very misleading, because they include the far more abundant Blue-winged Teal. The harvest in Mexico is not well known, but it is assumed to be small, based on hunter surveys of the overall waterfowl harvest in Mexico (Migoya and Baldassarre 1993, Kramer et al. 1995).

BREEDING BIOLOGY
Behavior

Mating System. Cinnamon Teal breed as yearlings and are seasonally monogamous. In Utah, captive birds observed by Spencer (1953) commenced courtship activities leading to pairing in late February, as their alternate (breeding) plumage neared completion. Wild birds wintering in Utah in January were not paired and were not observed courting. Many spring migrants arrived as paired birds in March, but pairing continued until early May. Hohman and Ankney (1994) noted that males attaining breeding plumage early have better success in attracting a mate. The peak period of pair formation occurs from late February to March, but pairing continues into May (Gammonley 1996b). Oring (1964) reported that 25% of male Cinnamon Teal deserted their females in the second week of incubation, and 63% during the third week; 12% remained in attendance until the eggs were pipped.

Sex Ratio. The sex ratios (males:females) were 1.19:1 during the breeding season at Ogden Bay on the Great Salt Lake in Utah (Spencer 1953), and 1.27:1 in an Arizona breeding population (Gammonley 1996b).

Site Fidelity and Territory. Although little information is available, some Cinnamon Teal females are known to exhibit site fidelity when conditions are favorable on their breeding sites. In Arizona, Gammonley (1996a) noted that 2 out of 5 marked adult females returned to the same breeding area the next year.

Paired males were aggressive in their defense of both their females and their waiting areas, beginning from the start of nest initiation through the third week of incubation (Gammonley 1996b). Defense was by the male only, and other species were tolerated, except for the closely related Blue-winged Teal. In Utah, territories were not established until after the nest site was selected, with the defended area usually being small (<30 m²) and either including the nest site or ≤100 m from the nest (Spencer 1953). In Arizona, the home range of 16 marked pairs averaged 8.3 ha, with home ranges often overlapping and pairs feeding and loafing in close proximity (Gammonley 1996a).

Courtship Displays. Courtship among Cinnamon Teal occurs in small groups that include at least 1 female (Spencer 1953, Gammonley 1996b), with most displays similar to those of other blue-winged ducks, such as Northern Shovelers and Blue-winged Teal (McKinney 1970). McKinney also noted that *turn-the-back-of-the-head* was one of the commonest displays associated with pair formation, as well as *preen-dorsally, lateral-dabbling (mock-feeding), repeated-calls, body-shake, head-dip,* and *belly-preen*. Spencer (1953) stated that *bobbing-and-bowing-of-the-head* was the most frequent and persistent male courtship display, at times lasting for 1 hour or longer. Females *incite* by swimming in front of a preferred male while either *head-pumping* or opening their bill toward the male. McKinney (1970) only observed 1 short *jump-flight* in a late-spring courting party, but *jump-flights* are apparently not uncommon within groups of unpaired males competing for females (Connelly 1977, Gammonley 1996b). Spencer (1953:62) noted that aerial flights where 2 or more males pursued 1 female were associated with courtship, commenting that "one cannot help but be impressed with the whistling wings and tortuous, agile flight during these chases." Copulation occurs on the water and is preceded by precopulatory *head-pumping*. *Head-pumping* is also the most common aggressive display by males, accounting for >90% of all agonistic interactions observed by Gammonley (1996a).

Cinnamon Teal, however, appear to be less aggressive in social interactions with Blue-winged teal, as Connelly and Ball (1984) observed in eastern Washington; 71% of 121 interspecific social interactions were initiated by Blue-winged Teal, despite the fact that they were outnumbered 3 to 1 by Cinnamon Teal. The initiator also won 90% of all interactions where the outcome could be determined (Connelly 1977). Further, Connelly and Ball (1984) found that Blue-winged Teal used actively hostile displays in 56% of the interspecific interactions, compared with only 22% by Cinnamon Teal. Cinnamon Teal used *hostile-pumping* (66%–67%) and *chasing* (22%–26%) during both intraspecific interactions with other Cinnamon Teal and interspecific interactions with Blue-winged Teal.

Nesting

Nest Sites. Pairs make reconnaissance flights over marsh areas to choose the general locale of the nest site, but the female alone selects the exact location (Spencer 1953). Females nest near water in dense vegetation, with cover density apparently a more important prerequisite for the nest site than the type of vegetation. At Ogden Bay in Utah, Cinnamon Teal nests ($n = 286$) were primarily (53.9%–80.6%) located in saltgrass (*Distichlis stricta*) at 2 sites, with nest density always greater where stands of saltgrass were broken up by clumps of cattails (*Typha* spp.), bulrushes, or tall weeds, which perhaps provided escape cover during the brood-rearing period (Spencer 1953). At a third site, however, 46.3% of 145 nests were situated in bulrush

(*Scirpus olneyi*) stands; no nests were found in saltgrass, which was only present in small amounts. Nests in spikerushes (*Eleocharis* spp.) and bulrushes (*Scirpus* spp.) were often located under the mat of dead stems from the previous year, with females accessing their nests via tunnels under the matted vegetation.

During an extensive study (1964–90) of duck nesting at the Monte Vista NWR in Colorado, most (73.9%) teal nests (Cinnamon Teal, Blue-winged Teal, and American Green-winged Teal) were located in stands of Baltic rush (*Juncus balticus*), which was the dominant vegetative type on the refuge (Gilbert et al. 1996). At the Tule Lake and the Lower Klamath NWRs on the Oregon/California border, vegetation height was <60 cm for 75% of 40 nests, with 40% <1 m from water and 100% <50 m (Miller and Collins 1954). Nesting close to water may be an important nest-site selection criterion for Cinnamon Teal, as 60% of 15 nests located in the Humboldt Bay area of California were situated <1 m from water (Wheeler and Harris 1970), and 44% of 57 nests in Arizona were <10 m from water (Myers 1982).

Nest bowls measured by Spencer (1953) averaged 18.3 cm in diameter and 6.9 cm deep. Unless located in dense matted vegetation, the nest bowl was excavated in the earth and usually lined with dead grasses and other plant stems available at the site. Down is primarily added during early incubation, but the amount varies among females. Most nests eventually had a 5 cm rim of down.

Clutch Size and Eggs. Bellrose (1980) summarized data from multiple studies and reported that the clutch size from normal nests averaged 8.9 eggs ($n = 1,368$), with a range of 4–16, although some studies may have included Blue-winged Teal clutches. A few specific studies also reported mean clutch sizes: 9.7 for 52 nests in Utah (Spencer 1953), 9.3 for 76 nests in California (Miller and Collins 1954), and 8.7 for 281 teal nests (mostly Blue-winged Teal and Cinnamon Teal) in Colorado (Gilbert et al. 1996).

Cinnamon Teal eggs are usually subelliptical and smooth, with a slight gloss. Their color varies from a pale pinkish buff to nearly white (Bent 1923). In Utah, Spencer (1953) reported egg measurements of 46. 4 × 34. 6 mm for 46 eggs in 6 clutches. Bent (1923) reported 46.9 × 34.1 mm for 90 nests measured from various collections. Fresh egg mass averaged 27.0 g ($n = 11$) in Utah (Spencer 1953), and the mass of 28 eggs in Arizona averaged 30.9 g (Gammonley 1995b). Spencer (1953) noted that egg laying was most frequent in the morning (between 08:00 and 10:00), at an interval of 1 egg/ day, and that midseason clutches were sometimes larger than early- or late-season clutches. Early in the season, the first 3–4 eggs could be deposited at 3-day intervals.

Incubation and Energetic Costs. Incubation begins within 24 hours of laying the final egg and lasts 21–25 days (Spencer 1953). During incubation, females lose about 13% of their body weight (Afton and Paulus 1992). Gammonley (1995b) found that after egg laying, females retained 10.6 g of stored lipids, which was equivalent to about 7% of the energy needed for incubation. The incubation constancy of 2 females monitored in the San Joaquin Valley of California averaged 77.0%–79.9% and was positively correlated with the day of incubation (Hohman 1991). Females took 3.7–4.7 recesses/day, lasting 1.1–1.4 hours. Feeding was probably the dominant activity of females during nest recesses; in eastern Washington, female Cinnamon Teal and Blue-winged Teal foraged 63.1%– 66.9% of the time while off the nest (Connelly and Ball 1984).

On the Colorado Plateau (elevation 1,937–2,819 m) in Arizona, females arrived in relatively lean condition, with 24.5 g of stored lipids, but they gained an average of 44 g of body mass between arrival and the initiation of rapid follicle growth, which included 12 g of protein and 9 g of lipids (Gammonley 1995b). Overall, females gained 3.07 g of body lipids/g of reproductive lipids produced during rapid follicle growth, while body lip-

ids declined by 0.63 g/g of reproductive lipids used during laying. A female producing a 10-egg clutch containing 40 g of lipids would thus use 23.7 g of body lipids during clutch production. About 30% of the energy needed to produce a typical clutch came from stored lipids, or about 10% of the total energy required (if existence energy is also included). Females did not reduce the protein levels in their body during egg laying. Hence breeding Cinnamon Teal in this study depended largely on dietary nutrient sources to meet the energetic costs of egg production.

Stored lipids may also contribute less to the reproductive costs for Cinnamon Teal and other small-bodied dabbling ducks, compared with larger species like Mallards, because the lipid reserves of the latter species are both absolutely and relatively larger than in their smaller conspecifics. The reserves of the smaller species also account for a smaller percentage of clutch-formation costs. Regression slopes of lipid reserves used during clutch production were −0.59 for Cinnamon Teal and −0.42 for Blue-winged Teal, compared with −0.72 in the larger Northern Shovelers and −1.04 in Mallards (Gammonley 1995b).

Nesting Chronology. The nesting chronology for Cinnamon Teal varies with elevation and latitude. In Arizona, nests are initiated 2–4 weeks later at elevations above 2,500 m, compared with those at 1,900–2,400 m (Piest 1982, Gammonley 1996b). Hence peak nest initiation in Arizona ranged from 24 May to 26 June, with the last nests initiated as early as 5 June and as late as early August, again influenced by elevation (Myers 1982, Gammonley 1996a). At Ogden Bay in Utah, the nesting season began in late April (first initiations) and ended in late July (last hatch; Spencer 1953). In the Humboldt Bay area of California, the nesting season extended from late March to late July (Wheeler and Harris 1970). Nest-initiation dates for Cinnamon Teal in the San Luis Valley of Colorado were 16 May in 1995 and 10 May in 1996 (Laubhan and Gammonley 2000).

Nest Success. In most studies of nest success for Cinnamon Teal, some nests were no doubt those of Blue-winged Teal. Nonetheless, Bellrose (1980) noted an average nest success of 32% for 1,707 Cinnamon Teal nests monitored during 19 studies in California, Idaho, Montana, and Utah. During their landmark research (1937) at the Bear River marshes along the Great Salt Lake in Utah, Williams and Marshall (1938) observed a nest success of 62.2% for 524 nests. Spencer (1953) reported a nest-success rate of 45.4% for 229 nests from the Ogden Bay marshes, also along the Great Salt Lake. During 2 years (1951, 1953) at Honey Lake Valley in California, nest success averaged 55.3% for 147 nests (Hunt and Naylor 1955). More recently, Piest (1982) noted a 65.2% nest-success rate for 119 nests in Arizona, and Gilbert et al. (1996) reported 54.5% for 241 nests at the Monte Vista NWR in Colorado (that total also included 13 American Green-winged Teal nests).

The percentage of Cinnamon Teal eggs hatching from successful nonparasitized nests (hatching success) is normally high, with reports of 84.1% by Williams and Marshall (1938), 86.7% by Hunt and Naylor (1955), and 93.6% by Gammonley (1996b). Hatching success at Ogden Bay in Utah, however, was only 43% (Spencer 1953), probably because California Gulls (*Larus californicus*) took many eggs from the nests without destroying entire clutches. California Gulls removed 29.7% of the eggs laid, and striped skunks (*Mephitis mephitis*), 8.1% (Spencer 1953). Probable and confirmed predators of nesting females were mink (*Mustela vison*), weasels (*Mustela* spp.), coyotes (*Canis latrans*), red foxes, Great-horned Owls (*Bubo virginianus*), and Peregrine Falcons (*Falco peregrinus*).

Brood Parasitism. Interspecific brood parasitism occurs in Cinnamon Teal, at least in locations where habitat conditions and species composition facilitate such behavior. At Ogden Bay, along the Great Salt Lake in Utah, Redheads parasitized 24% of 129 Cinnamon Teal nests in 1949 and 22% of 170 nests in 1950; 1 nest was parasitized by a Northern

Shoveler (Spencer 1953). A later study (1972–74) at Farmington Bay, also along the Great Salt Lake, reported that Redheads parasitized 41% of 474 Cinnamon Teal nests, Ruddy Ducks parasitized 2%, and both Redheads and Ruddy Ducks parasitized 7% (Joyner 1976). Mean clutch size was 7.7 eggs in nonparasitized nests, compared with 6.9 in parasitized nests, with 59.0% of the eggs hatching in nonparasitized nests versus 45.3% in parasitized nests. Joyner proposed that the high rates of brood parasitism observed at Farmington Bay may have occurred because of several species crowding into a preferred habitat. Most egg loss was due to displacement from the nest by the parasitizing female, with displacement averaging 16.7% in parasitized nests, compared with only 4.1% in nonparasitized nests. Increasing the number of parasitic eggs also decreased the number of Cinnamon Teal eggs that hatched. Joyner (1973) noted that Cinnamon Teal parasitized 1 Northern Pintail nest, 7 Redhead nests, and 1 Ruddy Duck nest. Conspecific brood parasitism is reported, but such behavior is not well studied (Sayler 1992).

Renesting. Like other puddle ducks, Cinnamon Teal will renest if a first nest is destroyed. In Lassen County in California, 6 out of 48 Cinnamon Teal renested after their nests were destroyed, with 1 female renesting twice (Hunt and Anderson 1966). About 5–6 days elapsed between the loss of nonincubated eggs and the start of new nests. First clutches averaged 10 eggs ($n = 16$), while second clutches averaged 8.3 ($n = 6$).

REARING OF YOUNG

Brood Habitat and Care. During a study of brood-habitat use and niche overlap among 9 duck species at the Turnbull NWR in eastern Washington, Cinnamon Teal, Blue-winged Teal, and Gadwalls formed a tight group in the dabbling duck category, with high (94%) overlap in each of the 12 niche dimensions that were measured (Monda and Ratti 1988). These ducks fed over dense beds of submerged vegetation, usually at depths <0.9 m. Cinnamon Teal ducklings are highly mobile, leaving the nest within 24 hours of hatching and feeding from the first day. Gammonley (1996b) noted that females usually remained with their young until fledging. At Ogden Bay in Utah, Spencer (1953) observed larger brood sizes where there was a juxtaposition between good nesting cover and good brood-rearing cover. Spencer also thought that female Cinnamon Teal were the best mothers of the waterfowl he observed. They kept their broods near escape cover, where the ducklings took refuge at the first sign of danger while the female diverted the intruder's attention by feigning injury.

Brood Amalgamation (Crèches). Brood amalgamation has not been reported in Cinnamon Teal.

Development. There is little information on growth in Cinnamon Teal, although their development is probably similar to that of the closely related Blue-winged Teal and Northern Shoveler. Spencer (1953), however, reported that among several hand-reared Cinnamon Teal, juvenile feathering was nearly complete at 6 weeks. The young were capable of flight at 7 weeks, at which time the irises of the males were changing from brown to red.

RECRUITMENT AND SURVIVAL

There are only a few studies on brood survival in Cinnamon Teal, and none that have used radiotelemetry to obtain estimates from marked females or ducklings. As with virtually all ducks, most brood mortality for Cinnamon Teal occurs during the first few weeks after hatching, although the incidence of total brood loss is not known. In Utah, Spencer (1953) reported average brood sizes as 10 for Class I ducklings ($n = 9$ broods), compared with only 4.5 in Class III (near fledging). In Arizona, Gammonley (1996a) reported average brood sizes of 9.7 in Class I ($n = 65$), 7.3 in Class II ($n = 68$), and 6.9 in Class III ($n = 55$), and he noted that brood size was not correlated with hatching date. The overall number of young fledged/pair was 2.0. The survival rates of immatures and adults are also not well known, because few Cinnamon Teal

are banded. Kozlik (1972), however, used data on Cinnamon Teal banded in California to calculate survival rates of 29% for immatures and 46% for adults. The longevity record for a Cinnamon Teal is 10 years and 6 months for an after-hatching-year female banded and shot in California.

FOOD HABITS AND FEEDING ECOLOGY

Cinnamon Teal feed on plant and animal foods in shallow-water areas along the margins of wetlands, as well as over beds of submerged aquatic vegetation surrounded by emergent marsh plants. They spend more time foraging in emergent vegetation beds than do Blue-winged Teal, which prefer open water (Connelly and Ball 1984). Dabbling at the surface of the water is the most common foraging method, but they also tip up and feed with the head below the water (Gammonley 1996b). Pairs feed in close association. Groups will engage in social feeding, which also occurs in Northern Shovelers but not in Blue-winged Teal (Johnsgard 1965). Females have a higher density (12.32/cm) of lamellae than males (11.92/cm), which may facilitate the partitioning of food resources (Nudds and Kaminski 1984).

Breeding. From their spring arrival through the egg-laying period, 52 paired female Cinnamon Teal in Arizona consumed 31.7% plant matter (seeds) and 68.3% animal matter, primarily dipterans (35.8%, including 21.1% midge larvae), and gastropods (15.5%; Gammonley 1995b). In contrast, the foods eaten by 34 paired males were 52.5% animal matter and 47.5% plant matter (mostly seeds). Dipterans (32.6%, including 20.8% midge larvae) and gastropods (8.4%) were also the most important animal food for males. The types of seeds consumed varied by location, but preferred species were sedges (*Carex* spp.), spikerushes, smartweeds (*Polygonum* spp.), and bulrushes. Paired females also spent more time feeding than did their mates (56%–65% vs. 48%–52%).

Animal foods formed >71% of the diet (aggregate volume) and were found in >92% of 46 prenesting male Cinnamon Teal (paired and unpaired) collected in the Tulare basin in the San Joaquin Valley of California (Hohman and Ankney 1994). There were no differences in the percentages of animal matter in the diets of paired and unpaired males: 21.2%–32.2% midge larvae, 8.0%–20.25% water boatmen (Corixidae), and 15.0%–15.8% water fleas (Cladocera).

Migration and Winter. The diets of 28 spring-migrant Cinnamon Teal collected from 4 habitat types at the Bosque del Apache NWR in New Mexico averaged 45.0% animal matter and 55.0% plant matter (Thorn and Zwank 1993). In contrast, the diets of 27 Cinnamon Teal collected during the fall averaged 90.8% plant material and only 9.2% animal material. During the spring, midges (Chironomidae) were the most important (17.4%–54.2%) animal food, depending on the habitat type. During the fall, the most significant plant foods were wild millets (*Echinochloa* spp.; 27.9%–46.7%) and sprangletop (*Leptochloa fascicularis*; 18.2%). Individuals collected during the spring in habitats dominated by alkali bulrush (*Scirpus maritimus*) / three-square bulrush (American bulrush; *Scirpus americanus*) or by saltgrass consumed the largest percentage (80.1%–81.6%) of animal foods. Feeding sites in those 2 habitats were characterized by shallow water (<15 cm) and short (<29 cm), moderately dense vegetation. In contrast, individuals collected during the fall from habitats with annual plants and cattails (*Typha* spp.) / hardstem bulrush (*Scirpus acutus*) consumed the highest percentage (86.1%–86.6%) of plant foods. Feeding sites in these habitats were deeper (>23 cm) and the vegetation taller (>86 cm) than in the alkali bulrush / three-square (American) bulrush and the saltgrass habitats.

Important foods for 25 male Cinnamon Teal collected during the postbreeding season in habitats surrounding the Great Salt Lake in Utah were midge larvae (17%), the seeds of widgeongrass (*Ruppia maritima*; 12%), and bulrushes (12%; Cox 1993). At Ogden Bay in Utah, the food habits of 24 Cinnamon Teal collected during the fall were dominated by seeds: 37.8% by volume from salt-marsh bulrush (*Scirpus paludosus*), and 19.2% from saltgrass (Spencer 1953). Important invertebrate foods were gastropods (41.2%) and beetles (23.5%). Cinnamon Teal / Blue-winged Teal using playa lake habitats on the southern High Plains of Texas ate wild millet (*Echinochloa crus-galli*), as well as corn obtained by field-feeding in dry fields (Sell 1979). During the winter in habitats on the coastal plain in the state of Sinaloa in Mexico, the stomach contents of 12 Cinnamon Teal shot by hunters were 66.6% plant material and 33.4% animal material (Migoya and Baldassarre 1993). Seeds from alkali bulrush dominated (58.3%) the plant foods, while dipteran larvae (Anthomyiidae and Muscidae; 16.7%) and midge larvae (8.3%) were the most important animal foods.

MOLTS AND PLUMAGES

Adult Cinnamon Teal have 2 plumages/year: basic and alternate. Palmer (1976a) and Gammonley (1996b) should be consulted for further details.

Juvenal plumage is obtained in about 7 weeks, at which time the appearance of the sexes is similar, with a buffy-colored face and neck, dark brown upperparts, and a brown or tan breast. The juvenal wing feathers are retained until the following summer. First basic is an incomplete molt that immediately succeeds juvenal plumage and is completed in early fall, although the entire tail may not be replaced until late fall or early winter. Males in this plumage are similar to those in definitive basic, although short dark streaks on the head and the neck are conspicuous on close inspection. Aside from the streaks, females in first basic are like the males. First alternate is acquired by late fall or early winter and is generally indistinguishable from definitive alternate, except for the juvenal wing.

Definitive basic is acquired from late May through late September and involves all of the remiges (the wing molt); hence there is a flightless period of about 21 days. Males in definitive basic are similar to the females, but with a more rufous color overall and a reddish or yellowish eye color. Males begin to acquire definitive alternate immediately after the wing molt, but the basic wing is retained. This plumage is usually completed between late September and early November. In Utah, however, Spencer (1953) reported some males with only partial cinnamon plumage as late as mid-January, and some captive males did not assume complete alternate plumage until mid-March.

Females acquire definitive basic over a long period that starts in the spring, usually before departure from their wintering grounds, but the associated molt is usually interrupted by breeding, although it may continue at low intensity into the laying period (Gammonley 1996a). The retrices, and sometimes the remiges, are not molted until after brood rearing, occurring concurrently with initiation of the prealternate molt. The subsequent definitive alternate plumage is acquired from late summer through early fall and retained until the prebasic molt in late winter / spring.

CONSERVATION AND MANAGEMENT

Much of the breeding range of Cinnamon Teal is in arid areas of the Intermountain Region (the Great Basin) or the Central Valley of California, where problems associated with water scarcity are increasing in severity as wetlands compete for water with municipalities, agriculture, and industry (Gilmer et al. 1982, Kadlec and Smith 1989). Brown et al. (2001:36) noted the "enormous human-driven competition for water in [the] Great Basin."

Boron, arsenic, and selenium were detected in Cinnamon Teal eggs collected in the San Joaquin valley of California, but not at levels sufficient to affect reproduction (Hothem and Welsh 1994). High levels of selenium, however, are known to impair reproductive success and cause deformities in the embryos of aquatic birds nesting in contaminated areas, such as the Kesterson Reservoir area of the San Joaquin valley (Ohlendorf et al. 1989).

During the second half of the twentieth century, the breeding range of Blue-winged Teal expanded westward into breeding areas dominated by Cinnamon Teal. Both species have a similar breeding chronology, but Blue-winged Teal males are more aggressive and dominate Cinnamon Teal males. Connelly and Ball (1984) suggested that Blue-winged Teal have a competitive advantage in social interactions and foraging efficiency. Thus, although there appears to be some partitioning of food resources, there may be reason for concern, because these 2 taxa hybridize (Bolen 1978).

More research is clearly needed on Cinnamon Teal, in particular on adult and immature survival rates, especially because this species is among the least abundant of the dabbling ducks. Additional research on their breeding ecology is also warranted, given the broad breeding distribution of Cinnamon Teal and their adaptations for breeding in an arid environment.

Northern Shoveler

Anas clypeata (Linnaeus 1758)

Left, hen; *right*, drake

The Northern Shoveler is the most widely distributed of 4 species of Shovelers worldwide, with a breeding range that extends across most of the Northern Hemisphere. Commonly called "Spoonbills," their primary breeding range in North America is the Prairie Pothole Region (PPR) in Canada and the United States, although substantial numbers breed in the Central Valley of California. Wintering birds mainly occur along the Gulf Coast, the Central Valley of California, and in wetlands along both coasts of Mexico, as well as sites in the interior highlands. Their specialized bill is adapted to strain small zooplankton from the water, especially water fleas (Cladocera), which are a favorite food during the breeding season. Northern Shovelers are also among the most territorial of all dabbling ducks, with males aggressively defending a specified territory that appears to ensure the availability of a high-quality food supply for their mates. Northern Shovelers are a common dabbling duck in suitable habitat, with an average population in the Traditional Survey area of 3.6 million from 2000 to 2010. The 2010 population was 4.1 million. Northern Shovelers are considered to be poor tasting and thus are not highly prized by hunters; the 1999–2008 harvest averaged only 542,485 in the United States and 22,221 in Canada.

Northern Shoveler. *Left*, hen; *right*, drake

IDENTIFICATION

At a Glance. Adult males are immediately identified by the green head and neck, the large bill, the white breast, and the chestnut-brown sides. The female is a mottled brownish gray overall. Both sexes have an extensive blue patch on the wing coverts. In flight the massive bill (usually angled downward 30°–45° in flight and at rest), coupled with a short tail and a long neck, give Northern Shovelers a humpbacked, top-heavy appearance that is totally unlike that of other dabbling ducks.

Adult Males. Males in alternate (breeding) plumage are strikingly colored, with a green head and neck that are separated from the chestnut belly and sides by the white chest and breast. The green head is darker than that of Mallards, and in poor light it often appears black. Males in early fall have a partial white crescent on each side of the face. A white flank spot is noticeable in front of the black undertail coverts. The wings show a bright gray-blue patch on the coverts that is separated from a brilliant green speculum by a tapered white stripe. The tertials are long, pointed, and black, with a white midrib. White stripes, formed by the white scapulars, extend from the upper breast along the margin of the gray-brown back. The blue wing patches are visible in flight, but they are usually hidden by the scapulars when these ducks are on the water. When males spring into the air to take flight, they show a prominent black stripe (peninsula) extending about midway up the white back. The legs and the feet are bright orange; the bill is

black. The irises of males are yellow, contrasting sharply with the dark head.

Adult Females. The brownish-gray females are similar to, but smaller than, female Mallards. They are also distinguished by their large spatulate bill and blue wing coverts. The wings are similar to those of adult males, but both the speculum and the white bar are duller, and the coverts are dull bluish gray instead of blue. The legs and the feet are orange. The bill is dusky in the center, with yellow-orange margins and small black dots. The irises are brown or yellowish brown.

Juveniles. Young Northern Shovelers are nearly indistinguishable from females, but they can be identified by a dark crown and back of the neck that contrast with their paler and more speckled underparts. The legs and the feet of juveniles are dull orange; the bill is dusky. The irises are brown or yellowish brown.

Ducklings. A yellowish belly and undertail region contrasts with the grayish brown upperparts of Northern Shoveler ducklings (Nelson 1993). The head is a dusky yellow, with a brownish cape; the bill is noticeably long.

Voice. Northern Shovelers are relatively quiet. On the water or in flight, however, males produce a nasal *chugh chugh* and a guttural *took took* that are loud and distinctive; the latter vocalization is also given during courtship displays (Palmer 1976a). The female has a low-pitched *quack* that is not commonly heard, except in alarm. When pairs are

Northern Shovelers in a typical feeding behavior. *John C. Avise*

flushed during the breeding season, females give 3–7 loud, evenly spaced quacks (McKinney 1970). The wings produce a clattering noise as the birds take off in flight.

Similar Species. Males can be confused with male Mallards because both species share a green head, but Mallards have a distinct white neck band. Mallards also have a brown breast, compared with the white breast of Northern Shovelers. Male Northern Shovelers have black bills, as opposed to the yellow bills of male Mallards. Females are similar to female Mallards, but the former are grayish, have blue wing patches, and a large bill. In low-light conditions, Northern Shovelers in flight can be mistaken for teal, because of the former's rapid wing beat and low, darting flight.

Northern Shovelers are closely related to the 2 other species of blue-winged ducks that occur in North America: Blue-winged Teal and Cinnamon Teal. A number of hybrids have been reported between Northern Shovelers and Blue-winged Teal (Palmer 1976a).

Weights and Measurements. Bellrose (1980) reported average body weights of 680 g for adult males ($n = 21$), 635 g for adult females ($n = 20$), 635 for immature males ($n = 89$), and 590 g for immature females ($n = 118$). Wing length averaged 244 mm in adult males ($n = 114$) and 229 mm in adult females ($n = 58$).

DISTRIBUTION

Of the 4 species of Shovelers throughout the world, Northern Shovelers have the broadest distribution, stretching across the Northern Hemisphere.

498

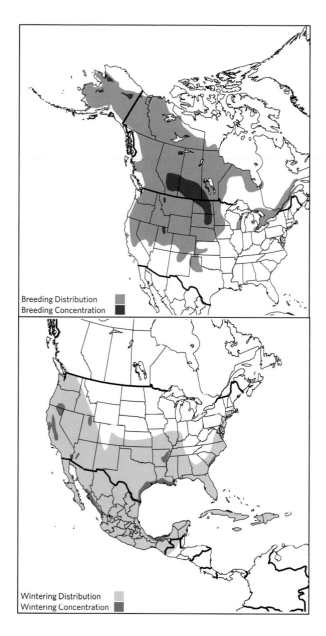

Breeding Distribution
Breeding Concentration

Wintering Distribution
Wintering Concentration

In contrast, the other Shoveler species are each restricted to a single continent: the Red Shoveler (*Anas platalea*) to southern South America, the Cape Shoveler (*A. smithii*) to southern Africa, and the Australasian Shoveler (*A. rhynchotis*) to Australia. Outside North America, Northern Shovelers breed from the United Kingdom eastward through Europe (except northern Scandinavia) and most of Asia (Johnsgard 1978). Their wintering range extends from the southern parts of their breeding range to northern and eastern Africa, the

Persian (Arabian) Gulf, India, Myanmar (Burma), southern China, Japan, and the Philippines.

Breeding. In North America, Northern Shovelers are widespread over the western part of the continent, but they only occur at discontinuous breeding locations in the east (DuBowy 1996). They breed farther south and west than most other species of dabbling ducks, except the closely related Cinnamon Teal. Their core breeding range is the PPR in Alberta, Saskatchewan, Manitoba, Montana, North Dakota, and South Dakota, but they are also found northward, up through the Canadian parklands (Johnson and Grier 1988). They reach their highest breeding densities in the prairies and parklands of the Prairie Provinces and central North Dakota (Johnson and Grier 1988). To the northwest, they breed throughout much of Alaska and the Yukon. Elsewhere in the West, they breed in the Central Valley of California (Small 1994) and, except in the PPR, sporadically in most other states west of the Mississippi River (DuBowy 1996).

The western populations are well tracked by the Waterfowl Breeding Population and Habitat Survey. Results from 2010 Traditional Survey reported 4.1 million Northern Shovelers: 38.3% in the eastern Dakotas; 33.4% in the southern Prairie Provinces; 15.4% in Alaska and the Yukon (including Old Crow Flats); 7.3% in the Northwest Territories, northeastern British Columbia, and central and northern Alberta; and 5.4% in Montana and the western Dakotas (U.S. Fish and Wildlife Service 2010). The distribution of Northern Shovelers by strata in the Traditional Survey from 1955 to 2010 was 30.0% in southern Saskatchewan; 16.2% in southern Alberta; 13.0% in North Dakota; 12.3% in Alaska/Yukon; 9.4% in the Northwest Territories, British Columbia, and northern Alberta; 8.7% in South Dakota; 4.7% in southern Manitoba; 3.9% in Montana; 1.0% in northern Manitoba; 0.7% in northern Saskatchewan; and 0.04% in western Ontario (Collins and Trost 2010). Outside the Traditional Survey area, breeding Northern Shoveler

Male Northern Shoveler; note the distinctive spatulate bill.
Ian Gereg

populations in the western states from 2001 to 2010 averaged 24,960 in California, 22,980 in Oregon, 5,917 in Washington, 1,492 in Utah, and 439 in Nevada (Collins and Trost 2010).

East of the Mississippi River, Northern Shovelers are rare, breeding locally in low numbers. Robbins (1991) observed them in southeastern Wisconsin. They also occur in the Lake Erie marshes (Peterjohn 1989) and much of southeastern Ontario, northeast along the St. Lawrence River Plain to Québec, with a few breeding records in northern Ontario along the southern coasts of Hudson and James Bays (Ross and North 1983). They also breed in eastern and northern New Brunswick, Prince Edward Island, and possibly Nova Scotia (DuBowy 1996). Farther south, they are very local in New York (McGowan and Corwin 2008) and Pennsylvania (Brauning 1992).

Winter. The winter range of Northern Shovelers extends over a broad area of the United States, the Caribbean, Mexico, and Central America. A few Northern Shovelers winter as far north as the Fraser River delta and southeastern Vancouver Island in British Columbia. Major wintering areas are the Central Valley of California, the Texas and Louisiana Gulf Coast, Mississippi, and parts of Mexico.

The Midwinter Survey averaged 996,713 Northern Shovelers from 2000 to 2010: 52.6% in the Pacific Flyway, 32.4% in the Mississippi Flyway, 13.9% in the Central Flyway, and 1.0% in the Atlantic Flyway. In the Pacific Flyway, most (93.9%) were in California, primarily the Central Valley: 48% in the Sacramento Valley and 38% the San Joaquin Valley in 2010. In the Mississippi Flyway, Northern Shovelers were found in Louisiana (66.9%), Mississippi (18.2%), and Arkansas (9.7%). Northern Shovelers in the Central Flyway were primarily in coastal Texas (94.5%), but a few were as far north as Colorado (1.1%) and Kansas (0.9%).

South of the United States, the west coast of Mexico wintered an average of 213,000 Northern Shovelers from 1981 to 2005. Almost 80% were evenly distributed among 3 areas: Pabellón Lagoon, the Marismas Nacionales, and Topolobampo Lagoon. Along the east coast of Mexico, an average of 35,000 birds wintered between the Rio Grande delta and the Yucatán lagoons, with most found on the Tabasco Lagoons (40%) and the Rio Grande delta (37%). The interior highlands of

Right:
Northern Shoveler female with her brood; note that the bill shape of the ducklings is not yet spatulate.
C. G. Massie-Taylor

Below:
Northern Shovelers in flight.
Clair de Beauvoir

Mexico wintered around 94,000 Northern Shovelers, with about 20% in the northern highlands and 80% in the central highlands, largely between Guadalajara and Mexico City (1981–2005). The Midwinter Survey in 2006 recorded 36,645 Northern Shovelers from throughout the interior highlands (Thorpe et al. 2006).

South of Mexico, the Northern Shoveler is considered to be a fairly common duck in northwestern Costa Rica (Stiles and Skutch 1989), but there are only rare reports from South America. Vilella and Baldassarre (2010) did not record any Northern Shovelers during their extensive waterbird surveys in the Llanos (grassland plains) of Venezuela. Northern Shovelers regularly winter in the western Caribbean, but they are rare in the eastern Caribbean (Evans 1990). There are, however, no reliable estimates of wintering numbers from these areas. In Puerto Rico, they occur frequently, but in small numbers, during the winter (Francisco Vilella).

Migration. Among the dabbling ducks, Northern Shovelers are early migrants in the fall and late migrants in the spring. Most of the population travels through the central and western parts of North America; <3% migrate east of the Mississippi River (DuBowy 1996). Migration corridors exist along the Mississippi and Missouri River valleys, as well as through the central Great Plains, to either the Gulf Coast or Mexico.

Fall migration begins in early September, but arrival to their main wintering areas often does not occur until December. In British Columbia, fall migration started in early August and extended to late October, with a peak from September through early October (Campbell et al. 1990). Northern Shovelers apparently migrate later along the coast, because of the moderating effect of the marine climate, compared with an earlier migration through the Great Plains states. Robbins and Easterla (1992) observed migration through Missouri from September until late November, with a peak in early November. James and Neal (1986) reported arrival to Arkansas from late August to mid-December, with a peak in early November.

Spring departure from wintering areas in Arkansas occurred between mid-February and late April, but peaked during March (James and Neal 1986). Departure from Missouri took place from early March to late May, but peaked in early April (Robbins and Easterla 1992). Arrival on their prairie breeding grounds (North Dakota, Manitoba, and Alberta) went from mid-April to early May (Sowls 1955, Poston 1974, DuBowy 1980). At Delta Marsh in Manitoba, the mean arrival date over a 12-year period (1939–50) was 15 April, ranging between 9 and 27 April (Sowls 1955). In southeastern Saskatchewan, Northern Shovelers arrived from 1 to 15 May (Stoudt 1971). Keith (1961) reported that 7 April was the average for first arrivals near Brooks, Alberta, during 1953–56, but the timing varied by 8 days. In British Columbia, peak arrival occurred from early April to late May (Campbell et al. 1990). Arrival at Yellowknife, Northwest Territories (about 1,300 km north of Brooks), averaged 8 May (1962–64), with major influxes occurring between 12 and 14 May (Murdy 1964). Gabrielson and Lincoln (1959) reported arrival on the Yukon delta in Alaska as 11–16 May.

MIGRATION BEHAVIOR

Little is known about the migration behavior of Northern Shovelers, but migration occurs during both day and night. They usually migrate singly or in small flocks of up to about 25 individuals (McKinney 1970, Poston 1974).

Molt Migration. Northern Shovelers typically do not undergo a distant molt migration; instead, they molt on the breeding grounds or travel to larger lakes nearby. In Manitoba, DuBowy (1985a) observed that the males had deserted their mates by late June and formed all-male flocks of 10–40 birds; nonbreeding females occasionally joined these flocks. In Alberta, Poston (1974) noted the first molting males in early June, about the time hatching began, with molting apparent on all individuals by mid-June. The population began to leave the study area between 18 and 25 June, with all males having departed by mid-July. Within a 24 km radius of the study area, 4 lakes of about 0.8–1.5 km² in size contained as many as 2,500 molting Northern Shovelers during late July and August, although some may have been young birds. Unlike other dabbling ducks, Northern Shovelers are secretive and solitary during the flightless period of the wing molt, often seeking shelter in the dense vegetation along lakeshores, where they remained hidden even at night (DuBowy 1980, 1985a).

HABITAT

Northern Shovelers breed on small wetlands throughout the PPR and the Canadian parklands, but they also use wetlands in the closed boreal forest and the Intermountain Region states (Bellrose 1980). They usually prefer the edges of shallow-water wetlands. In North Dakota (1967–69), breeding pairs of Northern Shovelers ($n = 608$) mostly used seasonal (59.2%) and semipermanent wetlands (33.1%; Kantrud and Stewart 1977). Pair densities were 36.2/km² on temporary wetlands, 34.4/km² on seasonal wetlands, and 22.7/km² on semipermanent wetlands. At Delta Marsh in Manitoba (1976–78), Kaminski and Prince (1984) found that breeding-pair densities were positively correlated with the percentage of forest cover (except in 1977), the vegetation:water ratio, emergent habitat richness, shoreline development, the percentage of common reed (*Phragmites communis*), the percentage of whitetop (*Scolochloa festucacea*) / sedge (*Carex atherodes*), and the percentage of cattails (*Typha* spp.) / bulrushes (*Scirpus* spp.). Over a broad area of the prairies and parklands (1955–81), the numbers of breeding Northern Shovelers were correlated with pond densities ($r = 0.388$), but this relationship tended to be positive in the southern portion of the breeding range, and negative in the northern portion (Johnson and Grier 1988).

Poston (1974) noted Northern Shovelers use of large shallow ponds during migration, while wintering birds have been reported from an array of habitats. In coastal Texas (1991–93), they occupied 37 different wetland types, representing 81.2% of the available habitat (Anderson et al. 2000). Density was highest in lacustrine littoral wetlands that were seasonally flooded, and in estuarine subtidal wetlands. On freshwater sites in coastal Texas, they were found on ponds averaging 25 cm deep in 1982–83 and 50 cm deep during 1983–84 (Tietje and Teer 1988). In contrast, the water depth of estuarine ponds and salt flats used by Northern Shovelers averaged 14 cm in 1982–83 and 15 cm in 1983–84. Most of these sites were also characterized by very high salinity and dense beds of widgeongrass (*Ruppia maritima*) until late December. Thereafter the amount of widgeongrass was markedly reduced, because of lower water levels and cold temperatures. On the Yucatán Peninsula in Mexico, Northern Shovelers used a shallow (<2 m) brackish coastal estuary (Celestun Lagoon) that featured submerged widgeongrass, muskgrass (*Chara* sp.), and nitella (*Nitella* sp.), and was surrounded by mangrove swamps (Thompson et al. 1991). During fall and winter (1 Oct–30 Mar) in the Celestun estuary, Northern Shovelers averaged 43.6% of their time feeding, 25.2% resting, 22.0% in locomotion, and 7.8% preening.

POPULATION STATUS

In 2010, the breeding population of Northern Shovelers in the Traditional Survey area was 4.1 million, which was 76% above the long-term average of 2.3 million (U.S. Fish and Wildlife Service 2010). The breeding population has expanded substantially (61%), from 1.7 million in the 1960s to 2.8 million in the 1990s, with most of the increase in the 1990s occurring during a 5-year period (1993–97), when the population doubled from 2.0 to 4.1 million. The rise in population numbers since the mid-1990s is undoubtedly due to favorable breeding conditions, with nearly all of the increase noted within the PPR. From 2000 through 2010, the population averaged 3.6 million, with a high of 4.6 million in 2007 and a low of 2.3 million in 2002.

Harvest

Northern Shovelers are generally not considered to be very palatable, perhaps due to their invertebrate-dominated diet, and thus are not highly prized by hunters. The Northern Shoveler harvest in the United States averaged 542,485 from 1999 to 2008: 38.7% in the Pacific Flyway, 38.5% in the Mississippi Flyway, 20.0% in the Central Flyway; and just 2.8% in the Atlantic Flyway. California was the leading harvest state in the Pacific Flyway (70.4%), while Louisiana (30.2%) and Arkansas (28.8%) were the major states in the Mississippi Flyway. In Canada, the harvest from 1999 to 2008 averaged only 22,221 Northern Shovelers. Over this time period, 35.5% were harvested in Saskatchewan, 33.3% in Alberta, 17.2% in Manitoba, and 15% in the remaining provinces. The high was 30,793 birds in 2005.

There are few harvest surveys in Mexico, but a survey from 1987 to 1993 in the states and specific locations where waterfowl hunting was a traditional activity estimated 57,422 total waterfowl harvested, of which only 5.9% (3,360) were Northern Shovelers (Kramer et al. 1995). At 2 major duck-hunting clubs in the state of Sinaloa on the west coast of Mexico, Northern Shovelers made up 19.8% of a harvest of 10,447 ducks at 1 location, but only 4.2% at the other (Migoya 1989, Migoya and Baldassarre 1993).

BREEDING BIOLOGY

Behavior

Mating System. Northern Shovelers are seasonally monogamous, with most individuals breeding during the first spring after hatching. During a long-term study (1976–92) in Latvia, however, the percentage of yearlings that were breeding averaged 70% (Blums et al. 1996). Northern Shovelers remain paired longer than most species of North American dabbling ducks, often well into the pre-

basic molt (DuBowy 1985a). As with other species that are late in acquiring breeding plumage, they do not begin to establish pair bonds until the winter season. Courtship commonly takes place during spring migration. McKinney (1967) noted that captive Northern Shovelers formed pair bonds from November to May. Northern Shovelers observed in coastal North Carolina were involved in courtship activities beginning in late December and continuing into January, with 66% paired in January and 96% paired by February (Hepp and Hair 1983). Poston (1974), however, considered pair bonds to be weak when Northern Shovelers arrived on their breeding grounds in southwestern Alberta. Males vary in the length of time they remain with their mates after incubation begins. Of 9 pairs observed by Oring (1964) at the Camas National Wildlife Refuge (NWR) in southeastern Oregon, 2 deserted their females on the first day of incubation, 6 remained until the third week, and 1 remained until hatching.

Sex Ratio. The sex ratio of 9,068 Northern Shovelers observed at various locations during spring in North America averaged 60.1% males (Bellrose et al. 1961). During winter in North Carolina, the percentage of males was 64.2%–70.1% for 2,122 Northern Shovelers observed from November through March in 1978/79 and 1979/80 (Hepp and Hair 1984). During the winter in the Yucatán Peninsula of Mexico, the sex ratio of 2,830 Northern Shovelers was 1.8:1, or 64.6% males (Thompson and Baldassarre 1992).

Site Fidelity and Territory. At Delta Marsh in Manitoba, 42% of 19 adult females marked at their nests returned in a subsequent year, and 1 female returned to the same area for 3 years (Sowls 1955). Only 8% (*n* = 12) of the yearlings returned. During a 4-year study (1965–68) on grasslands near Calgary, Alberta, 15% of 20 marked adult females returned the following year, but only 4 out of 116 (3.5%) marked juvenile females did; 2 out of 19 (10.5%) marked adult males, but only 1 out of 134 (0.7%) marked juvenile males, came back the next year (Poston 1974). Return rates, based on 118 females marked with nasal saddles at the St. Dennis NWR in southcentral Saskatchewan (1982–93), averaged 29.1% for adults and 17.6% for juveniles (Arnold and Clark 1996).

Northern Shovelers return to their breeding grounds in small flocks, which break up as paired birds disperse over the available nearby ponds. In Alberta, arriving pairs spread out over the study area, but the males did not become intolerant of other males until the pairs settled on their home ranges, which generally occurred 1–2 weeks after arrival (Poston 1974). Near Delta Marsh in Manitoba, Seymour (1974a) observed that arriving pairs were not aggressive, often sitting within 1.5 m of each other, but threat displays increased by the second week, when males defended mobile females, but not a territory per se. By the third week, however, aggression increased further and pairs became isolated from each other. In Alberta, the home range of 8 Northern Shovelers each contained a core area, the nest site, and 3–13 peripheral ponds, with the core area occupied by the pair 60%–90% of the time (Poston 1974). The apparent function of the core area was for feeding, loafing, and isolation of the pair. Home-range size averaged 31 ha, but ranged between 8 and 52 ha. Parts of most home ranges overlapped, and some were completely within the boundaries of others. Nest cover, peripheral ponds, and waiting areas were all shared, but only 1 pair at a time occupied the core area. In Manitoba, Seymour (1974a) noted that all territorial defenses during laying and incubation occurred within 90 m of the major loafing spot within the home range; hence the maximum effective size of territories was only about 0.9 ha. Although the pair bond might dissolve, most males remained on their home ranges until the eggs hatched.

The male aggressively defends the core area of the territory, especially during the laying period, which appears to ensure that a high-quality food supply is available for the female (McKinney 1973, Seymour 1974a). Food resources on the territory appear critical in providing the resources needed for successful incubation by females (Ankney and Afton 1988). The male clearly defends the female, but he also defends the territory in her absence, as Seymour (1974b) demonstrated in an experiment using a female decoy. In an analysis of 266 *pursuit flights* of territorial Northern Shovelers at Delta Marsh in Manitoba, the defending male returned to the core loafing area 96.2% of the time, and in most (94.1%) of the encounters, the pursued birds typically did not return to the territory of the defender within 15 minutes (Seymour 1974c). In highly territorial Northern Shovelers, such flights serve to reserve space (territory), and are not reflective of territorial males pursuing other females for potential extra-pair copulations. In 39% of the *pursuit flights* observed by Seymour (1974), only males were involved, and 91% were with just 2 birds. Males also defend breeding areas for a longer time period than other dabbling ducks, often until the final week of incubation (Seymour 1974, Afton 1979a, DuBowy 1980). In additional to *pursuit flights*, defensive males on the water will occasionally fight, which can involve lunging past each other with wings flapping (McKinney 1967, 1970).

Courtship Displays. Courtship behaviors of Northern Shovelers have been fairly well described by Hori (1962), Johnsgard (1965), McKinney (1967, 1970), and Seymour (1974a). The various calls uniquely associated with courtship are also well documented in Northern Shovelers. Male courtship displays are often accompanied by *repeated calls* and *fast calls*, which are a series of *took-took* notes often associated with various displays. Male displays include *turn-the-back-of-the-head*, *lateral-dabbling*, *head-dip*, *up-end*, *wing-flaps*, and *jump-flights*. The *jump-flight* is quite noticeable, with the calls being given several times and the male then taking off toward the female with a noisy rattle of wings (Johnsgard 1965). The male lands near the female and often immediately tips up. *Head-pumping*, accompanied by *took* calls, precedes copulation. McKinney (1970) noted a wheezy *whe* or *thic* or muffled *took* notes when males were copulating, and a loud, nasal *paaay* call after copulation. Females *incite* and have a slightly descending decrescendo call, *quack-gack-gack-ga-ga*, with the last 2 notes muffled (Johnsgard 1965). The male decrescendo call is heard most frequently during the fall and winter (rarely thereafter) and is a *paaaay . . . took . . . took-took*, with the latter notes being quieter and lower pitched (DuBowy 1996).

Northern Shovelers can exhibit aggressive behavior in winter, as Thompson and Baldassarre (1992) documented on the Yucatán Peninsula of Mexico during the winters of 1986–88. Of 347 aggressive interactions, 71% were directed at other Northern Shovelers, 14% at Blue-winged Teal, and 13% at Northern Pintails. Aggressive behaviors were *threat* displays, characterized by *open-bill-threats* and *jabs* (71.4%); *chases* on the water (14.6%); the *supplanting* of 1 individual by another without physical confrontation (11.3%); and actual fighting, which involved *biting* and *striking* with the wings (2.8%). In contrast, there was little aggression in all-male flocks of 10–40 Northern Shovelers that were observed on the breeding grounds in Manitoba (DuBowy 1980).

Nesting

Nest Sites. The female selects the nest site, but the male can be in attendance. In Alberta, females searched for potential nests sites by pecking and probing the vegetation, while the male wandered in a wide area about his female or remained inactive (Poston 1974). In the PPR in Canada (1982–85), 82% of 616 Northern Shoveler nests were located in upland habitat, 16% in dry wetland habitat, and 2% in wet wetland habitat (Greenwood et al. 1995). In central North Dakota, Northern Shovelers nested in a variety of habitats: 45.2% in planted cover, 14.3% in odd areas (patches of cover <2 ha in size with an array of features), 12.1% in wetlands, 10.0% in grasslands, and 6.8% in haylands (Klett et al. 1988).

Grasses of many kinds form the preferred vegetative cover for nests. Of 20 Northern Shoveler nests in Alberta, pastures were the most frequent site, but nests were also located in fields of alfalfa (*Medicago sativa*), spikerush (*Eleocharis macrostachya*), rush (*Juncus balticus*), and grasses (*Agropyron* spp.; Poston 1974). Nests were rarely <91 m from the core area of the home range. In grazed mixed-grass prairies in North Dakota, Northern Shoveler nests were most abundant in cool-season grasses, especially western wheatgrass (*Agropyron smithii*) and green needlegrass (*Stipa viridula*; Duebbert et al. 1986). At the Lower Klamath and the Tule Lake NWRs in California (1952), Northern Shoveler nests (*n* = 39) were located in upland sites with a cover of low grasses (Miller and Collins 1954). Of 67 nests located during a follow-up study in this area during 1957, 76.1% were in fields and 22.4% were on islands (Rienecker and Anderson 1960). No nests were found in marshes during either study. Elsewhere in California, Miller and Collins (1954) stated that nests (*n* = 39) were located in grass cover (56%), mustards (*Brassica* spp.; 18%), and nettle (*Urtica* sp.; 10%). Most (64%) nests were situated 2.7–45 m from water, with 28% >45 m from water. Almost all (98%) nests were in vegetation 0–0.6 m tall (with 82% in vegetation 0–0.3 m tall), and the nests were usually surrounded on 3 sides by vegetation.

Clutch Size and Eggs. Mean clutch size was 10.7 eggs for 35 nests in California (Miller and Collins 1954), 10.0 for 79 nests at Delta Marsh in Manitoba (Afton 1979b), and 10.3 for 669 nests in the PPR in North Dakota and South Dakota (Krapu et al. 2004c). Clutch size for 24 nests in South Dakota averaged 10.2 (Duebbert and Lokemoen 1976). In southeastern Alberta (1953–57), clutch size averaged 10.1 for 45 nests (Keith 1961).

Northern Shoveler eggs are elliptical ovate. In Manitoba, Afton (1979b) recorded measurements of 51.8 × 36.5 mm for 486 eggs. Fresh egg mass averaged 38.4 g (*n* = 150). Bent (1923) noted egg measurements of 52.3 × 37.1 mm, with the color varying from pale olive buff to pale green gray. According to Bent, Northern Shoveler eggs were almost identical to those of Mallards and Northern Pintails in color and in their smooth texture, but not in size or shape. Eggs are laid at a rate of 1/day, although occasionally a day may be skipped.

Incubation and Energetic Costs. The incubation period was calculated at 25 days in Saskatchewan (Clark et al. 1988) and 23 days in Manitoba (Afton 1979a). The incubation constancy of 12 females observed in Manitoba averaged 84.6%, with females averaging 2.4 recesses/day of 93.8 minutes each (Afton 1979b). During recesses, on average, females spent 68.3% of their time feeding, compared with 57.1% during laying, 58.4% during prelaying, and 68.9% during spring arrival (Afton 1979a). Incubation constancy increased when it rained and when the ambient temperature declined, which may reduce the potential of embryo injury from the eggs being chilled (Afton 1980). Females also rarely left their nests between the hours of 10:00 and 13:00, and thus were in attendance to protect eggs from the direct effects of solar radiation and high midday temperatures. Loos and Rohwer (2004) found that incubating Northern Shovelers and other prairie-nesting ducks generally increased their incubation constancy as laying progressed, but that Northern Shovelers did so more rapidly than conspecifics that nested later, perhaps because early-nesting species need to spend more time keeping eggs viable in the face of the cold temperatures that are more likely to be encountered at the beginning of the nesting period. Northern Shovelers maintain a lower overall incubation constancy than other ducks, however, which is probably an adaptation to obtain the food necessary to meet the energy demands of incubation (Afton 1980). Females used warm afternoons to spend long periods of time feeding, when there would be little risk of mortality for embryos. The small body size of Northern Shovelers requires them to leave the nest more frequently to feed.

The energetics of reproduction in Northern Shovelers is fairly well known, based on 2 geographically different studies. In the PPR in Mani-

toba, the lipid reserves of females declined by 0.72 g for every gram deposited in their eggs, while protein increased by 0.1 g/g deposited in the eggs (Ankney and Afton 1988). Because the protein-to-lipid ratio is about 14:1 in the diet of Northern Shovelers and protein is easily obtained during the spring, this study concluded that lipids, and not proteins, limited egg production, and that the amount of lipids stored by the time a female begins rapid follicle growth (the precursor to egg laying) is the key factor affecting her clutch size. Incubating females lost 4.46 g of body weight/day, or about 18% of their initial body weight, while their lipid levels declined by 1.68 g/day, which was about 38% of their total body-weight loss. Females clearly used the food resources obtained during nest recesses to meet most of the demand of incubation, as their stored reserves only satisfied about 26% of the energy needed for incubation. Therefore food resources available on individual territories are crucial for successful incubation.

In contrast to the situation in the PPR, Northern Shovelers breeding at Minto Flats in Alaska used an insignificant (2%) amount of stored lipids for egg production, and neither their protein level nor their lipids declined as laying progressed (Mac-Cluskie and Sedinger 2000). One explanation for this difference was that the high rates of invertebrate productivity in Alaska, and 22 hours of daylight, enable Northern Shovelers there to ingest greater quantities of nutrients, and thus maintain their energy and nutrient balances during egg production. Nutrients did not limit clutch size for this population, although females initiating clutches early had acquired more lipids than later-nesting females. Incubating females used lipids at a rate similar to those in Manitoba, despite starting with 8% more lipids and spending 11.8% less time on their nests than females in the latter locale (Mac-Cluskie and Sedinger 1999). Incubating females at Minto Flats averaged 4.0 recesses/day, nearly twice that of dabbling ducks in Manitoba, and had an incubation constancy of 72.8%, 18% lower than that in Manitoba, although their recess durations were similar. Such data suggested that females in Alaska had more difficulty obtaining food during incubation than did females on the prairies. Thus having enough energy to successfully complete incubation and brood rearing may be the ultimate factor that controls clutch size for this population, not the initial acquisition of energy for egg production (Mac-Cluskie and Sedinger 2000).

Nesting Chronology. Northern Shovelers are intermediate nesters, initiating their nests between early-nesting species (e.g., Mallards and Northern Pintails) and late-nesting species (e.g., the closely related Blue-winged Teal). In North Dakota (1993–95), the median date for nest initiation was 21 April for Northern Shovelers, compared with 8 April for Mallards, 10 April for Northern Pintails, 27 April for Blue-winged Teal, and 3 May for Gadwalls (Krapu et al. 2004c). During a 4-year study (1982–85) in both prairie and parkland habitats in Canada, the median nest-initiation date was 25 May (Greenwood et al. 1995). The average length of the nest-initiation period was 16 days, with the availability of wetlands in May, April–June temperatures and precipitation, and nest success all correlated with the length of the nest initiation period ($r^2 = 0.44$). This period was extended by 0.13 days for each 1.0 cm increase in April rainfall.

Rienecker and Anderson (1960) presented hatching dates for 63 Northern Shoveler nests in the Klamath basin in northern California (1957). A 25-day incubation period was used to backdate to nest-initiation times, revealing that the earliest nests (3.2%) were begun on 17 April, with peak initiation (72.3%) from 8 to 18 May, and the last nests (1.6%) started on 17 June. During a 3-year study (1991–93) at Minto Flats in Alaska, peak nest initiation occurred from 27 to 29 May (MacCluskie and Sedinger 2000). Near Calgary in Alberta, 20 nests located in 1965–68 were begun between 9 May and 19 June, with most nests initiated from mid-May to mid-June (Poston 1974). Slightly to the southwest (near Brooks, Alberta), the mean date of nest initiation for Northern Shovelers ranged between 24 April and 18 May from 1953 to 1957 (Keith 1961). Early nesting of all waterfowl on the study area was highly ($r = -0.99$) correlated with a heat-sum index, obtained by summing the degrees above freezing for daily maximum temperatures.

Nest Success. As with other dabbling ducks, nest success of Northern Shovelers varies both spatially and temporally. Over a broad area in the PPR in Canada (1982–85), nest success was only 12%, with most (72%) nests destroyed by predators (Greenwood et al. 1995). Over a wide swath in the PPR in the United States, nest success averaged 10% in 1966–74 ($n = 94$), 16% in 1975–79 ($n = 503$), and 13% in 1980–84 ($n = 274$; Klett et al. 1988). Predators were again the main cause of nest losses in all regions and habitats studied, although farming operations were responsible for appreciable nest losses in croplands and haylands. The most significant nest predators common to all regions studied were red foxes (*Vulpes vulpes*), striped skunks (*Mephitis mephitis*), mink (*Mustela vison*), raccoons (*Procyon lotor*), badgers (*Taxidea taxis*), and Franklin's ground squirrels (*Spermophilus franklinii*; Sargeant et al. 1993).

In contrast to these low levels of nest success from primarily unmanaged habitats, within fields of undisturbed grass/legume cover planted under the auspices of the Cropland Adjustment Program in South Dakota, the hatching success of duck nests, including those of Northern Shovelers, was 69% in 1971, 58% in 1972, and 32% in 1973 (Duebbert and Lokemoen 1976). Nest losses were attributed to red foxes (27%), raccoons (15%), badgers (8%), striped skunks (5%), avian predators (21%), and unknown predators (24%). Reynolds et al. (2001) evaluated the effect of the Conservation Reserve Program (CRP)—a major national program that provides financial incentives to farmers for planting undisturbed grass cover in place of agricultural crops—on duck recruitment (1992–95) in the PPR in the United States by comparing the nest success of 5 species of dabbling ducks in CRP fields and in nearby Waterfowl Production Areas (WPA).

The daily survival rate for Northern Shoveler nests was 94.4% for 662 nests on CRP lands, similar to the 95.0% for 368 nests on WPA lands. The estimated nest success for all 5 species, however, was 46% higher on CRP landscapes, compared with estimates from model simulations that replaced CRP lands with croplands. Their findings were very significant, demonstrating that because of the CRP, an estimated 12.4 *million* additional waterfowl were recruited from the study area over the 1992–97 period, with perennial grass cover being the prime factor in improving nesting success.

In other studies, nest success was 89.7% for 39 nests on the Lower Klamath and the Tule Lake NWRs in California in 1952, and 94.0% for 67 nests in 1957; 3.0%–7.7% were deserted and 2.6%–3.0% were destroyed (Reinecker and Anderson 1960). Of the 389 eggs laid in 1952, 91.5% hatched, 4.4% contained dead embryos, 1.3% were infertile, and 2.8% were missing or unaccounted for (Miller and Collins 1954). Of 80 destroyed nests in the Woodworth Study Area in North Dakota (1966–81), red foxes were the principal predator, responsible for >71% of the nest losses (Higgins et al. 1992). Sergeant et al. (1984) reported that Northern Shovelers made up 9% of 1,783 duck carcasses found at red fox dens in the Dakotas; in the eastern Dakotas, where predation was the greatest, an average of 90% of all Northern Shoveler carcasses were females. Where Northern Shovelers nested near California Gull (*Larus californicus*) colonies on the Farmington Bay Waterfowl Management Area in Utah, the gulls took 13% of the Shoveler eggs, especially from nests that were not well concealed (Odin 1957). In Alberta, American Crows (*Corvus brachyrhynchos*) and Black-billed Magpies (*Pica hudsonia*) were the main nest predators, along with striped skunks and long-tailed weasels (*Mustela frenata*; Poston 1974).

Brood Parasitism. Both intraspecific and interspecific brood parasitism has been reported in Northern Shovelers, but neither form generally appears to be extensive, except perhaps at high densities in habitats such as islands. Of 389 eggs deposited in 35 Northern Shoveler nests in California, only 14 (3.6%) were laid by other species: either Ring-necked Pheasants (*Phasianus colchicus*) or other ducks (Miller and Collins 1954). On islands in North Dakota, however, 5 out of 9 Northern Shoveler nests were parasitized by other nesting ducks, compared with no parasitism of 13 nests located on peninsulas (Lokemoen 1991). Clutch size averaged 9.2 eggs in parasitized nests versus 10.4 for nonparasitized nests, but the difference was not statistically significant. In contrast, brood parasitism for 4 other dabbling duck species reduced both their clutch size and their nesting success.

Renesting. At Delta Marsh in Manitoba (1946–50), 7 out of 33 (21%) marked Northern Shoveler females renested, and 1 female renested twice (Sowls 1955). The average distance between the locations of the first and the second nests averaged 325 m (range = 41–233 m). Renesting females (n = 24 renests for all species on the study area) waited at least 3 days before initiating their second nest, but the average delay was 0.62 days for each day of incubation at the time of nest loss. Clutch size was 12 eggs in 2 initial nests, compared with 8 and 9 for 2 renests, and 8 for a third nest. During his 5-year study (1953–57) near Brooks, Alberta, Keith (1961) calculated a renesting rate of 75% for Northern Shovelers. He estimated that clutch size averaged 10.7 eggs for first nests and 9.4 for renests (n = 28).

REARING OF YOUNG

Brood Habitat and Care. Observations of 343 Northern Shoveler broods over 18 years in the PPR in the Dakotas recorded that they occurred on semipermanent wetlands (53%), seasonal wetlands (31%), and sewage lagoons (8%; Duebbert and Frank 1984). At the Woodworth Study Area in North Dakota (1965–75), the density of Northern Shoveler broods was highest on seasonal wetlands, although others were found on impoundments and semipermanent wetlands. At Strathmore in

Alberta, Northern Shoveler broods were observed on virtually all available wetland types, including narrow roadside ditches, irrigation canals, and semipermanent marshes, but use was greater on permanent wetlands >0.7 ha in size, especially by older broods (Poston 1969). Three marked females with Class Ia broods were first seen on the core ponds used by the breeding pair, but observations of these and other broods on the study area indicated that broods seldom remained on 1 wetland for >7–10 days (Poston 1974).

Development. Southwick (1953) provided data on the growth and development of captive-reared Northern Shovelers hatched from wild eggs. Ducklings averaged 23 g at hatching, 53 g at 1 week, 135 g at 2 weeks, 242 g at 3 weeks, 303 g at 4 weeks, and 370 g at 5 weeks. The sample size was 6 from hatching through week 3, 4 in week 4, and 2 in week 5; the ducklings were almost fully feathered in 5 weeks. Payn (1941) reported that captive ducklings were fully feathered at 6 weeks, but the flight feathers had not emerged from their quills. Hochbaum (1944) reported that captive-reared Northern Shovelers fledged in 52–60 days. In the wild, fledging was reported at 50 days in Saskatchewan (Clark et al. 1988) and 47–54 days in South Dakota (Gollop and Marshall 1954).

RECRUITMENT AND SURVIVAL

There is little specific information on recruitment in Northern Shovelers, but most brood mortality occurs in the first 2 weeks, as is the case for other duck species. Based on 913 brood observations in northern California, Miller and Collins (1954) reported a decline in brood size from 10.2 at hatching to 7.3 (−28%) in week 1, with brood size around the time of fledging (week 7) averaging 7.3 ducklings.

The limited number of Northern Shovelers that are banded restricts any ability to determine accurate recruitment and survival rates. Age ratios in the U.S. harvests, however, indicate fluctuating recruitment, with variations from 0.8 young/adult to 1.99 young/adult for 2005–9 (Raftovich et al.

2010). Annual survival, based on a small sample of 136 local/hatching-year birds banded during 1948–61 in Minnesota, was 39% for adults and 29% for immatures (Lee et al. 1964). A marked-resighting study of 118 female Northern Shovelers and other dabbling ducks on the St. Dennis National Wildlife Area in southcentral Saskatchewan (1982–93) yielded annual survival rates of 51% for adults and 32% for immatures (Arnold and Clark 1996). In Europe, a 27-year study (1964–93) of female Northern Shovelers noted survival rates of 58% for adults and 38% for immatures (Blums et al. 1996). In the United Kingdom, Boyd (1962) reported an adult survival rate of 56%.

The longevity record is 16 years and 7 months for a male Northern Shoveler banded in Nevada in 1952 as a hatching-year bird and shot in California in 1969.

FOOD HABITS AND FEEDING ECOLOGY

As in all Shovelers, the Northern Shoveler has a unique bill structure characterized by densely packed lamellae that are highly adapted for straining small invertebrate foods, which are sought-after items, from the water. Dabbling at the water's surface is the preferred feeding method, and then foraging with the head underwater, although they will tip up for food and, on rare occasions, dive (Kear and Johnsgard 1968, McKinney 1970). In Finland, Pöysä (1985) recorded Northern Shovelers feeding in July in water depths of 13.5–24.2 cm. Breeding Northern Shovelers spent 65.7% of their feeding time dabbling and 25.2% with their heads underwater, compared with 97%–99% dabbling during the postbreeding season and 76.9% dabbling during the winter (DuBowy 1996). At 4 sites in western France, straining (dabbling) was the dominant feeding method during early winter (66%–87%) and midwinter (40%–87%), but not late winter (20%–69%), when dipping (18%–46%) and upending (8%–55%) were the more common foraging behaviors (Guillemain et al. 2000). Straining predominated (>85%) at all sites during nighttime feeding.

Straining for food may be preferred because it allows these ducks to use their specially adapted bills while remaining on the surface and staying alert for nonaquatic predators, in comparison with deep foraging, where the head would be immersed (Pöysä 1986). Moreover, straining allows Northern Shovelers to filter water continuously, while birds foraging with their heads below the water interrupt their total feeding time to breathe and to be vigilant (Nudds and Bowlby 1984). The ducklings are also capable of straining small food items. In comparison with other species, captive Northern Shoveler ducklings were especially adept at feeding on zooplankton, such as water fleas (*Daphnia*). Collias and Collias (1963:8) stated that when "hundreds" of young *Daphnia* (about 20% the size of adults) were placed in a pan with Northern Shoveler, American Wigeon, and Gadwall ducklings, the Shoveler ducklings "soon cleaned up almost all of them alone," while the other species scarcely attempted to feed.

Northern Shovelers often are observed feeding singly or in groups, either by dabbling or by partially to fully immersing their heads as they slowly move forward, their bills sweeping from side to side, as they strain small swimming invertebrates (e.g., cladocerans) and seeds from the water (DuBowy 1985a). Using standard metabolic equations, a 700 g Northern Shoveler needs to consume 204 kcal of energy to meet its daily energetic needs, which could be satisfied by consuming about 233 g (dry weight) of cladocerans, 60–80 g of insects, or 910 g of vascular plants (DuBowy 1996).

Breeding. At Delta Marsh in Manitoba, the diets of 25 male Northern Shovelers were 99% animal matter: 85.5% water fleas (Cladocera), and 12.9% midge (Chironomidae) larvae (DuBowy 1985a). The cladocerans were largely *Daphnia* and related genera, which occurred in dense concentrations. DuBowy argued that the specialized feeding adaptations of Northern Shovelers allowed them to efficiently exploit such a ubiquitous and dense food resource, and thus to evolve their highly territorial breeding strategy. In southern Manitoba, the diets of prelaying males ($n = 3$) and females ($n = 14$) contained about 90% animal matter, primarily gastropods and crustaceans (zooplankton). The females, however, consumed more gastropods during the prelaying period than did the males (59% vs. 41%), while the males consumed more crustaceans (13% vs. 24%). During the laying period, the females ($n = 23$) again consumed more gastropods (55% vs. 32%) than the males ($n = 10$). The males increased their consumption of plant material from 10% during the prelaying period to 34% during laying, while this percentage did not change for the females (7%–9%). In North Dakota, the diet of 15 female Northern Shovelers was 99% animal matter: gastropods (40%), cladocerans (33%), other small crustaceans (21%), and aquatic insects (5%; Swanson et al. 1979).

Diurnal fasting has also been reported in molting males (DuBowy 1985c, 1997), which may occur because of a summer decline in the abundance of cladocerans, a preferred food. Hence, after abandoning their mates in early summer, males forage almost constantly (84.2%) during the preflightless period to gain weight when cladocerans are abundant (DuBowy 1885a). Male Northern Shovelers ($n = 81$) collected during the breeding season at Delta Marsh in Manitoba (1977–78) increased their body weights by an average of 21.8% during the prebasic molt (DuBowy 1985c). Body weights then decreased significantly in the flightless period, to the low levels observed when the birds arrived in the spring; during the postflightless period they again increased, to 49.4% above spring weights. Skin weights, which reflect stored lipid reserves, increased by 56.3% from spring to the preflightless period (0%–33% regrowth of the primaries), decreased significantly during the flightless period, and then increased by 97% above spring weights prior to fall migration. These stored lipid reserves allowed males to prepare for the midsummer decline of cladocerans, with males using their lipid reserves during the wing molt rather than switching to feeding on less profitable foods (DuBowy

1985c). With these reserves, males could fast for up to 2 weeks (DuBowy 1997).

Migration and Winter. The winter diets of Northern Shovelers appear to be more variable than those during the breeding season, and they can differ by habitat. In coastal Texas (1982–84), in saltwater habitats their diets contained more invertebrates, compared with more seeds and vegetation in freshwater sites (Tietje 1986, Tietje and Teer 1988). The 122 specimens collected from freshwater habitats contained some animal matter, such as cladocerans (21%), gastropods (13%), and copepods (8%), as well as plant matter, such as vegetation from coontail (*Ceratophyllum demersum*; 14%), the seeds of pondweeds (*Potamogeton* spp.; 7%), and coontail seeds (5%). In contrast, the 134 Northern Shovelers collected from saline habitats primarily ate animal foods—ostracods (24%), Foraminifera (a saltwater protozoan; 16%), gastropods (15%), copepods (7%), and fish (7%)—along with the seeds and vegetation of widgeongrass (10%). Body condition measurements generally were higher for all sex and age classes of Northern Shovelers in freshwater habitats versus saltwater habitats, except during the record cold winter of 1983/84. That winter the birds on saltwater ate small, cold-stunned fish, while the condition of the birds on freshwater declined, because of the reduced availability of plankton and macroinvertebrates and the unavailability of fish.

The food habits of Northern Shovelers during 2 winters on the Great Salt Lake in Utah (2005–6) were reported by Vest and Conover (2011) and provide a contrast with other areas, due to the unique high salinity of the lake: around 13% salinity in areas used by waterfowl, compared with 3.5% salinity in the open ocean. Animal foods (88.6% aggregate biomass) dominated the diets of 241 birds: mostly brine shrimp (*Artemia franciscana*) cysts (51.8%) and adults (20.2%), as well as the larvae of brine flies (Ephydridae; 7.8%), and both adults (4.8%) and eggs (2.5%) of water boatmen (Corixidae). Brinefly larvae and pupae can be found at densities of >5,000/m² in the substrate; brine shrimp occur in the water, where densities can reach 2,500/m³ for adults and >20,000/m³ for cysts. Plant foods made up only 11.4% of their diets, mostly (7.2%) the seeds of alkali bulrush (*Scirpus maritimus*).

On the Yucatán Peninsula in Mexico, wintering Northern Shovelers (*n* = 7) primarily (98.6%) consumed gastropods in early winter, but their late winter diets (*n* = 4) contained corixids (49.9%), gastropods (23.2%), and tubercles from the macroalga *Chara* (24.7%; Thompson et al. 1992). On a major wintering area in western France, sites with the most Northern Shovelers were characterized by birds straining the water's surface for zooplankton (Guillemain et al. 2000). At other sites, this straining behavior was replaced by deeper foraging methods, perhaps because the ducks were at least partially feeding on benthic macroinvertebrates. They initially concentrated at sewage works, where they showed a diel pattern of foraging in response to the vertical movements of zooplankton in the water, but they abandoned these sites later in the winter, once the zooplankton populations had declined.

MOLTS AND PLUMAGES

Molts and plumages of Northern Shovelers are described in detail by Southwick (1953), Palmer (1976a), and DuBowy (1985b, 1999), which should be consulted for further information.

The first feathers of the juvenal plumage appear at about 3 weeks, with ducklings fully feathered in approximately 6 weeks, except for the emergence of the flight feathers. Northern Shovelers in juvenal plumage appear brownish overall. Juvenal plumage is quickly replaced by first basic during the fall, from about mid-September to November. First basic replaces much of the body feathering, but not the juvenal remiges. Birds in first basic have a buffy to white head and neck, and a variably brownish belly and flanks. The back is blackish brown. First basic is immediately replaced by first alternate, which is acquired from late fall to January. In males, the first signs of this plumage are dull green "freckles" on the cheeks, the throat, and the chin. First alternate is similar to definitive alternate, although males in first alternate can be distinguished by the black vermiculations on the feathers posterior to the vent.

Males acquire definitive basic during the summer. The associated molt involves all feathering; hence, there is a 3- to 4-week flightless period. Males in definitive basic appear similar to females. The prealternate molt begins before the flightless period is completed and can continue through fall migration. Adult females acquire most of their basic plumage during the spring, but delay their wing molt until after nesting, sometimes even until their arrival on the wintering grounds. Definitive basic in females is similar to definitive alternate, but more brownish.

CONSERVATION AND MANAGEMENT

No management activities are conducted specifically for Northern Shovelers, but broad-scale habitat protection and management policies and their associated programs, particularly in the PPR, will benefit breeding Northern Shovelers. Significant amounts of wetland habitats are protected as Waterfowl Production Areas, and by 1992 the Conservation Reserve Program (initiated in 1985) had converted about 1.9 million ha of croplands to upland nesting cover in the PPR (Reynolds et al. 2001). The Prairie Pothole Joint Venture grassland programs also protect critical upland habitats in the region. In Canada, habitat-protection programs include Ducks Unlimited Canada's Prairie Care, the National Soil Conservation Program, permanent cover programs, Greencover Canada, the Saskatchewan Agriculture Conservation Cover Program, the Rural Tax Assessment Program (in Manitoba and Saskatchewan), and drainage and flood-control programs to reduce wetland drainage.

At a broad scale, the expiration of contracts for lands in the Conservation Reserve Program will reduce upland nesting cover for waterfowl. Climate change also may significantly affect waterfowl in the PPR, with predictions that drier climates could reduce the number and distribution of seasonal wetlands, as well as the open phase of semipermanent wetlands (Johnson et al. 2005). There have not been many studies on contaminants in Northern Shovelers, although DuBowy (1989) demonstrated that Northern Shovelers at the Kesterson NWR in California could hypothetically ingest large amount of selenium. Bioaccumulation of selenium increased 3-fold with progression up the food chain: from algae and vegetation, to invertebrates, and then to fish.

Northern Pintail

Anas acuta (Linnaeus 1758)

Left, hen; *right,* drake

A medium-sized dabbling duck, the male Northern Pintail is striking and is easily identified by the long neck and tail feathers, the long pointed wings, and an overall streamlined appearance; the female is brownish gray. Northern Pintails are circumpolar in their distribution; in North America they are most abundant in the Prairie Pothole Region (PPR) and in Alaska, especially in coastal tundra habitats such as the Yukon-Kuskokwim (Y-K) delta. Their primary wintering range is in central California and along the Gulf Coast south through Mexico and into Central America, with major wintering concentrations in the Central Valley of California, the mainland west coast of Mexico, the east coast of Mexico, and the Gulf Coast of Texas and Louisiana. As early-fall and early-spring migrants, they are among the first nesting ducks, preferring open grasslands with an abundance of seasonal wetlands, as well as coastal tundra wetlands. Pair bonds are formed during the winter, but males desert their mates early in the incubation period to seek extra-pair copulations with other females. The continental Northern Pintail population has declined dramatically since the 1970s, however, more so than for any other species of dabbling duck. The 2010 breeding population of 3.5 million was 13% below the long-term average of 4.0 million, and 29% below the objective of 5.6 million called for in the North American Waterfowl Management Plan. More alarming, perhaps, is the fact that their breeding populations in the PPR are not recovering, despite years of increased numbers of May ponds. Poor recruitment appears to be the major factor suppressing the Northern Pintail population, due to the long-term loss of prairie nesting habitat. In addition, Northern Pintails have a greater propensity than other puddle ducks to nest in stubble fields of grain, where early cultivation destroys virtually all their nests. They are not vigorous renesters, which, combined with their nesting-habitat preferences, adds to the recruitment problems affecting Northern Pintails. Management recommendations and efforts have focused on improving their nesting habitat.

Northern Pintail hen

Northern Pintail drake

IDENTIFICATION

At a Glance. In male Northern Pintails, the key identifying features are the long central tail feathers and the brilliant white breast that extends in a thin line onto the chocolate-brown head. The female is a mottled brown, with a bluish-gray bill, as in the male. Their overall body appearance is trim and streamlined, with their long narrow wings being more gull-like in shape than those of other ducks. These features, along with their swift flight, led Bellrose (1980:262) to dub them the "greyhound of the air." Northern Pintails also float higher on the water than do other dabbling ducks.

Adult Males. The head and the throat are a rich chocolate brown, while the back of the long neck is vermiculated with black. The breast, the belly, and the ventral region are bright white or, occasionally, a light cream. The white on the breast extends in a narrow streak along the side of the neck. The sides and the back are also gray; the black upper- and undertail coverts contrast with the white flanks and belly. In flight, the iridescent green or greenish-black speculum, bordered by white below and buff above, is readily visible. The pair of long black central tail feathers extends well beyond the tail and gives the Northern Pintail one of its most popular common names, "Sprig." Only male Long-tailed Ducks have comparably long tail feathers. The legs and the feet are bluish gray, with darker webs. The bill is bluish gray, with a black stripe along the central ridge that includes the nail. The irises are brown or yellowish brown.

Adult Females. Female Northern Pintails are a mottled brown overall, similar to many other female dabbling ducks. Closer examination reveals that the upper-body feathers are dark brown, while the head and the lower-body feathers are a noticeably lighter buff or gray, spotted with tan or fuscous, which gives the female a streaked appearance. The dull brown or bronze speculum, sometimes tinged with green, is not iridescent; it is bordered in front with buff, and behind with white. The speculum is usually not visible when females are at rest, but it is helpful in identifying them when in flight. The legs and the feet are slate gray. The bill is bluish gray, blotched with black. The irises are brown.

Left:
Northern Pintail male in a comfort movement. *Declan Troy, Troy Ecological Research Associates*

Below:
Female Northern Pintail defending her brood. *GaryKramer.net*

Male Northern Pintails
fighting.
RyanAskren.com

Juveniles. Young Northern Pintails are similar in plumage to adult females, but they have a plainer head and body, and darker upperparts. They gradually achieve adult plumage during their first fall and into early winter. The legs and the feet are bluish gray or greenish gray. The irises are brown.

Ducklings. Ducklings are characterized by a faintly yellowish tinge to an otherwise white base color, patterned with a darker brownish gray that is more reddish than in other species of *Anas* (Nelson 1993). The feet and the bill are bluish gray or olive gray.

Voice. Male Northern Pintails give a wheezy whistled *whee* note, heard throughout the year, to signal danger as well as to secure a pair bond with a female. During courtship, males also produce high pitched *ee hee* or *geeegee* notes, preceded by a loud whistle as part of the *burp* call. Females produce a gravelly *kuk-kuk* or a squealing noise, given singly.

Similar Species. The brownish color of female Northern Pintails is similar to that of female Mallards, Gadwalls, Northern Shovelers, Blue-winged Teal, and American Green-winged Teal, but Pintail females have a slenderer body, a longer neck, long slender wings, and a bluish-gray bill.

Weights and Measurements. Austin and Miller (1995) presented extensive body-measurement data from Northern Pintails collected in the Sacramento Valley of California. Body weights averaged 1,006 g for adult males ($n = 188$), 887 g for adult females ($n = 151$), 961 g for juvenile males ($n = 26$), and 843 g for juvenile females ($n = 41$). Bill length averaged 51.2 mm in adult males ($n = 191$), 46.8 mm in adult females ($n = 152$), 51.5 mm in juvenile males ($n = 26$), and 46.8 mm in juvenile females ($n = 41$). Wing length averaged 274.5 mm in adult males ($n = 189$), 255.3 mm in adult females ($n = 150$), 265.8 mm in juvenile males ($n = 25$), and 252.4 mm in juvenile females ($n = 40$).

DISTRIBUTION

Northern Pintails inhabit the Northern Hemisphere, where they range farther over the earth's surface than any other species of waterfowl. Their circumpolar breeding range extends from a few areas in Greenland to Iceland, across northern Europe, and eastward across northern Asia to

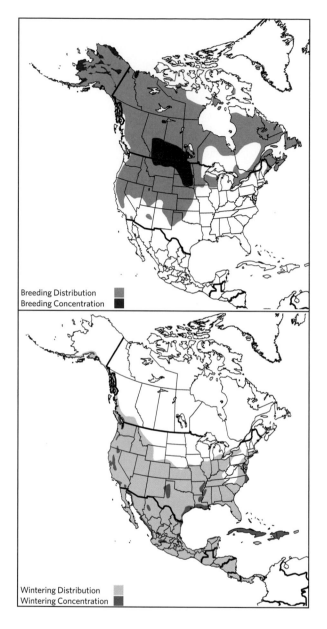

Breeding Distribution
Breeding Concentration

Wintering Distribution
Wintering Concentration

the Great Lakes and the St. Lawrence River valley. Eurasian birds winter in central Africa, the Persian (Arabian) Gulf, India, southeastern Asia, southern China, the Philippines, the island of Borneo, Malaysia, and islands in the mid- and southern Pacific Ocean, such as the Marshall Islands, Palmyra Island, and the Hawaiian Islands. North American birds winter in the southern United States, especially along the Gulf Coast, as well as in the Playa Lakes region of Texas and the Central Valley of California. They are abundant in Mexico and south through parts of Central America, but they are rare in northern South America (Johnsgard 1978, Made and Burn 1988).

Johnsgard (1978) recognized 3 subspecies of pintails: the Northern Pintail; the Kerguelen or Eaton's Pintail (*A. a. eatoni*), and the Crozet Pintail (*A. a. drygalskii*). Eaton's Pintails are found only on the Kerguelen Islands in the southern Indian Ocean and resemble small female Northern Pintails. These pintails are a threatened taxon, because of introduced feral cats. Crozet Pintails occur nearby, on the Crozet Islands. Livezey (1997) recognized Eaton's Pintail as a completely separate species (*Anas eatoni*) that included the Crozet Pintail subspecies.

Breeding. Northern Pintails have a vast breeding range in North America; it covers virtually all of Alaska, including the Aleutian Islands west to Amchitka Island, as well as nearly all of western and central Canada. The Arctic is an important breeding area for this species, where they are by far the most abundant dabbling duck. In the conterminous United States, they occur throughout the PPR, as well as east of the Cascade Mountains in Washington, Oregon, and northern California, south to the Central Valley. An isolated breeding population occurs in central Arizona. Their range in eastern North America is also extensive, stretching along the lowlands of Hudson and James Bays, and from northwestern Québec eastward to Labrador and Newfoundland and southward through the Maritime Provinces and along the St. Lawrence

Kamchatka and the Commander Islands. They also nest as far south as southern Europe, central Siberia in the west, and southern Siberia in the east. In North America, their widespread breeding range extends from the Aleutian Islands across Alaska and mainland Canada to the Maritime Provinces, as well as to Victoria and Banks Islands, and southward to southern California, east to southern Colorado, northwestern New Mexico, and the Texas Panhandle, across the PPR in the United States south to central Kansas, and east to

Courtship flight of Northern Pintails; note the female (*bottom*). *GaryKramer.net*

River valley, with small local populations in southern Ontario, central Wisconsin, eastern Michigan, northwestern Ohio, northeastern New York, and the northeastern coast of Maine (Austin and Miller 1995).

In 2010, the Traditional Survey recorded 3.5 million Northern Pintails and noted good water conditions at 6.7 million ponds, which was 34% above the long-term (1974–2009) average (U.S. Fish and Wildlife Service). The greatest (35.1%) numbers of Northern Pintails were in the eastern Dakotas, although a significant (33.2%) segment was farther north—in Alaska and the Yukon (including Old Crow Flats)—which indicates the importance of northern breeding areas for the continental population. The breeding population in Alaska averaged 1.1 million from 2000 to 2010: 37.0% on the Y-K delta, 15.3% on Yukon Flats, 13.8% on the Seward

Peninsula, and 9.8% at Kotzebue Sound. There was also a 1986–2006 average of 220,494 Northern Pintails estimated from separate aerial surveys of the Arctic Coastal Plain of Alaska (Larned et al. 2010); the 2009 estimate was 171,023. Traditional Survey data in 2010 indicated that 33.2% of all Northern Pintails were in Alaska/Yukon; 35.1% in the eastern Dakotas; 16.9% in the southern prairies of Alberta, Saskatchewan, and Manitoba; and 9.6% in the Northwest Territories, northeastern British Columbia, and northern and central Alberta. The distribution of Northern Pintails by region in the Traditional Survey from 1955 to 2010 was 29.2% in southern Saskatchewan; 22.9% in Alaska/Yukon ; 14.0% in the Dakotas, 9.2% in the Northwest Territories, British Columbia, and northern Alberta; and <3.0% each in the 8 other surveyed areas (Collins and Trost 2010).

Northern Pintails opportunistically pioneer into suitable habitat when it becomes available (Hochbaum and Bossenmaier 1972). Some will initially use breeding habitats on the southern portions of their range but, if conditions there are not suitable, will rapidly shift northward; others will bypass prairie and parkland habitats entirely and move directly northward (Johnson and Grier 1988). Northern Pintails are also very sensitive to drought conditions, to which they respond by moving farther north. Smith (1970) was first to notice this behavior, stemming from an analysis that revealed a strong negative correlation ($r = -0.91$) between the number of prairie wetlands in the prairies and parklands of Alberta and Saskatchewan and the portion of the Northern Pintail population moving farther north during 1959–68. Smith also demonstrated that an index of annual Northern Pintail reproduction had a statistically significantly negative correlation ($r = -0.92$) with the proportion of the population moving north. Other studies have corroborated that drought-displaced Northern Pintails experience lower reproductive rates (Calverley and Boag 1977, Derksen and Eldridge 1980, Hestbeck 1995).

Breeding population estimates for Northern Pintails can be misleading when based simply on averages from long-term datasets, because of their movements in response to habitat conditions. An analysis (1963–2001) of yearly estimates of pond numbers in the mixed-prairie landscape of Canada and the United States revealed that the Northern Pintail breeding population there averaged 62% lower during 16 drought years, compared with the other 23 years with better wetland conditions. Parkland habitats had 45% fewer Northern Pintails during drought, and shortgrass-prairie areas had 49% fewer birds. During the dry years their numbers occasionally went up in the boreal forest, but they always increased in the taiga (50%) and the arctic deltas (94%) of Canada (Frank Bellrose).

The dramatic extent of drought displacement into Alaska was reported by Derksen and Eldridge (1980), who noted that during years of severe drought on the prairies (1976–77), the 1977 survey estimate of Northern Pintails in Alaska was 123% higher than in 1976 and 87% above the then long-term average. The 1977 breeding population in Alaska made up 48% of the continental population that year, compared with 20% during the previous 10 years. The authors also found that Northern Pintail sex ratios recorded on the Arctic Coastal Plain were dominated by males during years with normal water conditions, when they gathered there following molt migration. In contrast, the sex ratios were more balanced during droughts, indicating that females either did not attempt breeding or were unsuccessful. In other western states, the breeding populations from 2001 to 2010 averaged 2,867 in Washington, 5,679 in Oregon, 572 in California (where the average was 3,986 from 1994 to 2010), and 238 in Nevada (Collins and Trost 2010). The Eastern Survey average was only 919 Northern Pintails from 1996 to 2010: 38.0% in Ontario; 37.2% in Labrador and Newfoundland, and 11.5% in Québec

Winter. Although small numbers of Northern Pintails winter as far north and west as the Alaska Peninsula and the Aleutian Islands (Gabrielson and Lincoln 1959), their major wintering areas are the Central Valley of California; the Gulf Coast marshes of Texas and Louisiana; the rice-growing areas of Texas, Arkansas, and Louisiana; and both mainland coasts of Mexico, as well as the interior highlands. They occur throughout Central America, where they are rare and local, but they are sometimes common during the winter in Panama. They are rare in northern South America (Hilty 2003). These ducks are common in Cuba, uncommon in Puerto Rico, and rare elsewhere in the West Indies (Raffaele et al. 2003).

The Midwinter Survey in the United States averaged 2.5 million Northern Pintails from 2000 to 2010: 51.7% in the Pacific Flyway, 24.2% in the Central Flyway, 21.8% in the Mississippi Flyway, and only 2.3% in the Atlantic Flyway. In the Pacific Flyway, 86.0% were in California (the Central

Valley), 6.2% in Oregon, and 5.5% in Washington. In the Central Flyway, nearly all (96.2%) were in Texas (the Playa Lakes Region and the Gulf Coast); in the Mississippi Flyway, 73.5% were in Louisiana (the Rice Belt and coastal marshes), with 12.8% in Arkansas. Northern Pintails in the Atlantic Flyway primarily were in the coastal region of the Carolinas (71.4%). Their principal wintering areas in the United States were in California (44.5%), Texas (23.2%), and Louisiana (16.1%). The Christmas Bird Count averaged 375,536 Northern Pintails in the United States from 2000/2001 to 2009/10.

In Mexico, substantial numbers of Northern Pintails have always wintered in the coastal lagoons and adjacent rice fields of the mainland west coast. An average of 198,311 was recorded during Midwinter Surveys conducted at 3-year intervals from 1982 to 2006, but the average was only 81,100 for the latter 3 surveys (2000, 2003, and 2006). Especially important areas are in the state of Sinaloa: the Topolobampo and Pabellón Lagoons, and the Dimas wetland complex to the south. All 3 sites are large wetland complexes, with total areas of 400 km² at Topolobampo, 800 km² at Pabellón, and 120 km² at Dimas (Kramer and Migoya 1989). The number of Northern Pintails along the east coast of Mexico averaged 110,570 during Midwinter Surveys conducted annually from 1978 to 1982 and then at 3-year intervals from 1982 to 2006. Most were in the lower Laguna Madre (62.3%), with 17.1% in the Rio Grande delta, and lesser numbers (2.8%–5.8%) in each of the 5 survey units south to the Yucatán Peninsula. In the interior highlands, the number of Northern Pintails recorded on 8 surveys conducted at 3-year intervals between 1985 and 2006 averaged 76,187, divided about equally between the northern and central highlands (Thorpe et al. 2006). Some of the more important Northern Pintail wintering areas in the interior highlands are Laguna San Rafael and Laguna Mexicanos in the state of Chihuahua, Laguna de Chapala in Jalisco, and the Cabadas wetlands complex along the border of Jalisco and Michoacán.

Northern Pintails exhibit strong fidelity to their wintering sites, determined by an analysis of 12,562 recoveries of 119,954 birds banded during winter in 1950–87 at 14 reference areas (Hestbeck 1993a). An index of site fidelity (the percentage of band recoveries, after the year of banding, that occurred in the same region where the birds were banded) was 81.3% in the Central Valley of California, 70.3% in northwestern Utah, 67.0% in Chesapeake Bay, 66.0% in western Washington, 64.9% on the Gulf Coast, and 5.3%–38.6% at the remaining 9 sites. Fidelity was generally greater to areas near the coast and to large bodies of water. The overall results, however, strongly suggested that Northern Pintails wintered as fairly distinct populations, and that winter fidelity to some sites was more stable than breeding-ground affiliations. Reinecker (1987) demonstrated that wintering fidelity was greater for females (46%) than for males (39%). Fedynich et al. (1989) noted a very low homing rate to the Playa Lakes Region of Texas (<1%), but many Northern Pintails use this region as a stopover in flights to wintering areas on the Gulf Coast or farther south into the highlands of Mexico.

Migration. Northern Pintails are the most paradoxical of ducks in their seasonal migration: they are among the first ducks to migrate south in the fall, and some of the first to migrate north in the spring. The males undertake fall migration earlier than the females, and those migrating longer distances to wintering grounds in Mexico and Central America begin their migration earlier than those traveling a shorter distance (Rienecker 1987a).

In general, Northern Pintails across the breadth of North America initiate fall migration by late August–September, with some departing from arctic breeding grounds in Alaska at the same time others are arriving on wintering grounds in California, Texas, and Louisiana. Passage through British Columbia starts in mid-August and extends through October (Butler and Campbell 1987). In California, the most frequented wintering area for Northern Pintails in the world, Miller (1987) noted that these ducks arrived in the Sacramento Valley by the first week of August, with about 315,000 present by mid-September; peak numbers were >1.5 million. Farther south, they begin to arrive in October on both the west coast of Mexico (in Sinaloa; Migoya and Baldassarre 1995) and the east coast (in Yucatán; Thompson and Baldassarre 1990). In Texas, they are found along the Gulf Coast from September through mid-March (Smith 1968).

An analysis of 4,204 recoveries for 54,519 Northern Pintails banded in Alaska from 1980 to 2010 demonstrates that the majority winter in the Pacific Coast states, especially in the Central Valley of California, and travel along the coastal migration corridor. Excluding the 645 recoveries in Alaska (15.3%), most of the recoveries were in California (49.6%), Oregon (11.4%), and Washington (10.4%). Recoveries were made in 32 states and 6 provinces, but only 1 was in the Atlantic Flyway (Delaware); 80 (2.2%) were in Mexico, and 10 (0.3%) were in Asia.

In the PPR, band-recovery data show a migration along the Mississippi River to wintering areas on the Gulf Coast. There were 4,915 recoveries of 64,809 Northern Pintails banded in the Dakotas from 1980 to 2010. Most (78.4%) of the recoveries occurred outside the Dakotas: 26.4% were in Louisiana, 18.7% in Texas; 5.2% in California and 2.8% in Mexico. Recoveries were made in 38 states and 8 provinces, and 6 were in Russia. Unlike the results for Northern Pintails banded in Alaska, 32 recoveries (0.8%) of Dakota-banded birds were in the Atlantic Flyway, mostly (40.6%) in North Carolina.

In the East, summer tracking (Jun–Jul) of female Northern Pintails ($n = 46$) fitted with satellite transmitters found most (89%) of them above 50° N latitude, which is a line that runs just south of James Bay (Malecki et al. 2006). Primary areas used during the summer were the coasts of Hudson and James Bays in Ontario and Québec, and southern Ungava Bay in Québec. By August, females began moving southward, arriving in southeastern Ontario, southwestern Québec, and the northern part of New York by mid-September, although 19 out of 23 (68%) females remained above 50° N latitude in mid-September, with movement into the St. Lawrence region continuing through late October. The first birds ($n = 2$) did not arrive on their winter range (below 40° N latitude) until late October, with the number of arrivals increasing through October and early November; fewer arrived through December. Bellrose (1980) reported that a small number of Northern Pintails appeared in the New Jersey / Chesapeake Bay area in September, but significant numbers were not present until mid-October, and peak numbers did not reach there until late November. They arrived 2–3 weeks later at points farther south along the Atlantic coast.

In the spring, Northern Pintails depart their wintering areas in late January and early February and move northward through March. Although Austin and Miller (1995) stated that spring migration routes were not well known, this situation has changed dramatically from new information made available by satellite and traditional radiotelemetry studies across the winter range of Northern Pintails. In California, Fleskes et al. (2002a) reported data from a major study (1991–93) involving the movements of 416 females that were marked with radiotransmitters from late August to early October in the San Joaquin Valley. Of the 385 females remaining in the Central Valley during the winter, 83% moved from the San Joaquin Valley northward to areas such as the Sacramento Valley, with most departing in December. Movements also started at the beginning of the hunting season and reflected a complex pattern that related to a bird's age, body mass, and capture location, as well as to the year and the weather. The probability of individuals leaving the San Joaquin Valley was 57% greater on days with rainfall, and more frequent movements occurred out of the San Joaquin Valley during years when there was more rain and fog and fewer southerly winds.

In a more recent study (2000–2003) of spring migration from the Central Valley of California, Miller et al. (2005a) used satellite-tracking for

136 adult females to document annual departure dates and key stopover areas along the spring migration corridors. Northern Pintail females started leaving during January or early February, with peak departures occurring from late February to mid-March. Mean departure dates were in early March each year, and virtually all had left by early April, although 77%–87% had subsequently stopped in northeastern California (the Klamath basin), southcentral Oregon (the Malheur basin), or northwestern Nevada. From these locations, 7%–23% exhibited a long stay (60 days) before migrating directly to Alaska (over the Pacific Ocean) between late April and early May, while 0%–28% also migrated to Alaska, but flew along the coast. Another 17%–39% had a moderate stay (>1 month) before departing between mid-March and mid-April to prairie Canada, southern Alberta, and interior British Columbia, with many moving on to northern Canada and Alaska. Most (32%–50%), however, exhibited a short stay (<2–3 weeks) before migrating in March to prairie Canada after stops in southern Idaho and western Montana, and some then went on to northern Canada and Alaska. Migration strategies were generally related to the geographic destination for nesting, but were modified in response to wetland abundance in the midcontinental prairie region and record cold temperatures during 2002. The annual result of all movements was that 50%–70% of all marked Northern Pintails with active transmitters resided in Alaska by mid-June.

Data from 31 Northern Pintails tracked via satellite from the middle Rio Grande valley of New Mexico, the Playa Lakes Region of Texas, and the Texas Gulf Coast indicated departures that were up to a month later (late Feb–late Mar) than for birds wintering in California (Haukos et al. 2006). Those wintering in New Mexico and the central highlands of Mexico used 2 main routes: (1) the Rio Grande valley to the San Luis Valley in southcentral Colorado and on to the Dakotas and Canadian prairies; and (2) northeast to the Playa Lakes Region and then on to southwestern Kansas, where

they joined with Northern Pintails that wintered in the Playa Lakes Region, before continuing north through the Dakotas and into southern Saskatchewan. Northern Pintails wintering in the Playa Lakes Region used 3 principal spring migration routes: (1) north-northeast to western and central Kansas, with many stopping at the Cedar Bluffs Wildlife Area, and then on to the Nebraska Sandhills and western Rainwater Basin; (2) southeastern and eastern Colorado; and (3) Kansas and Nebraska directly to the Dakotas and southern Saskatchewan, in the Saskatoon/Regina area. A few continued farther north to breeding areas in Alberta, British Columbia, the Yukon, and Alaska. Those wintering along the Texas Gulf Coast traveled north following the eastern edge of the Central Flyway through eastern Nebraska and the Dakotas and then on to southern Manitoba and Saskatchewan. Wetland availability in the PPR influenced nesting destinations, but individuals settled across a wide swath of northern North America.

In Nebraska's Rainwater Basin, Pearse et al. (2011b) monitored 71 female Northern Pintails: 31 radiomarked there, and the rest radiomarked by collaborators in the Playa Lakes Region and the Gulf Coast and rice-prairie areas of Texas. Northern Pintails first arrived in Nebraska on 7 March in 2003 and 18 February in 2004, with average arrival dates of 18 March in 2003 and 12 March in 2004. The residency time for individuals varied markedly (1–40 days), although yearly means were similar, at 9.5 days within the region. Diurnal locations were most often in palustrine wetlands (72%), although Northern Pintails also were found in riverine wetlands (7%), lacustrine wetlands (3%), municipal sewage lagoons and irrigation reuse pits (6%), and croplands (10.5%). The majority (57%) of their palustrine wetland use was in marshes containing 25%–75% emergent vegetation. Evening field-feeding flights averaged 4.3 km, of which 72% were to cornfields.

In the East, satellite tracking of 55 female Northern Pintails marked on their coastal winter range between New Jersey and northern Florida in

2003–5 revealed that spring migration occurred from March to May (Malecki et al. 2006). Migrating females traveled along 2 primary corridors, located east and west of the Great Lakes, with those females migrating to the west exhibiting a greater affinity for winter locations south of North Carolina. Birds to the east staged near Delaware Bay (the northern portion of their winter range) and in the southern Lake Ontario / St. Lawrence River plain. Females using the eastern corridor departed in early May for more northern locations in eastern Canada; females using the western corridor, toward the PPR, were more dispersed, although important stopover activity occurred from north-central Ohio northward to western Lake Erie. Females arrived on midcontinental breeding areas in mid- to late April.

MIGRATION BEHAVIOR

Satellite transmitters placed on female Northern Pintails during midwinter (2000–2003) in the Central Valley of California, fall and winter (2002 and 2003) in the Playa Lakes Region and the Gulf Coast of Texas, and early fall (2002 and 2003) in the middle Rio Grande valley of New Mexico (2001/2 and 2002/3) have provided significant new information on spring migration behavior (Miller et al. 2005b). Data from 17 females along 21 flight paths revealed an average ground speed of 77 km/hr (range = 40–122 km/hr). At a typical but hypothetical altitude of 1,460 m, 17 out of 21 flight paths occurred in tailwinds averaging 55 km/hr, which indicated that Northern Pintails migrated at airspeeds that maximized their range while minimizing energy expenditure. They increased their ground speed 61% with a tailwind, compared with a 27% decrease in headwinds. Most (19 out of 21) flights occurred partly or entirely at night. The longest distance flown was 2,926 km, from Goose Lake in southern Oregon to the Kenai Peninsula in Alaska, which would have required 38 hours to complete nonstop, assuming an average ground speed of 77 km/hr; the maximum in-flight time per day, however, was 5.5 hours.

Molt Migration. Northern Pintails undergo the longest molt migration among dabbling ducks, usually moving south of their breeding areas (Salomonsen 1968). Most males and unsuccessful females migrate for the molt; successful females molt on or near their brood-rearing marshes. Preferred molting areas selected by males are large shallow marshes with extensive emergent and submerged aquatic vegetation (Salomonsen 1968, Anderson and Sterling 1974). Early-molting males migrate before later-molting ones (Sterling 1966).

Bellrose (1980:273) reviewed molt migration by Northern Pintails, noting their major molting areas and the timing of their arrival to those sites. In the marshes surrounding the Great Salt Lake, "tens of thousands" of adult males in full breeding plumage arrived between 1 and 10 June. Thousands arrived at lakes on the Camas National Wildlife Refuge (NWR) in northeastern Idaho during late July and early August, just prior to entering the wing molt (Oring 1964). In the Pel and Kutawagan Marshes (north of Regina, Saskatchewan), a major Northern Pintail molting area, the first postbreeding males appeared as early as mid-May, and females began to gather by late July (Sterling 1966, Anderson and Sterling 1974). Other important Northern Pintail molting areas are the Mackenzie River delta in the Northwest Territories and the Y-K delta in Alaska.

HABITAT

Breeding Northern Pintails nest in open habitats characterized by low vegetation and seasonal to semipermanent wetlands (Austin and Miller 1995). In North Dakota (1967–69), most pairs occupied seasonal wetlands (40%), cultivated wetland basins (24%), or semipermanent wetlands (19%); the use of more permanent wetlands and water bodies was low. Pair density averaged 2.8/km² (Stewart and Kantrud 1973). Breeding-pair densities were correlated with the number and area of total wetlands/km², as well as the number and area of seasonal and semipermanent wetlands ($r > 0.90$; Stewart and Kantrud 1974).

Wintering birds use a wide array of freshwater and intertidal habitats; they also readily forage in agricultural fields, consuming waste grain, such as corn, rice, and soybeans. Significant information on winter habitat use stems from a radiotelemetry study over 3 winters (1989–92) in the Ensenada del Pabellón wetland complex on the coast of Sinaloa, where 1.5 million Northern Pintails were recorded in February 1990 (Migoya et al. 1994). Each winter 47–59 marked females were monitored to determine activity budgets and habitat-use patterns.

Northern Pintails used 4 major habitats: fresh-brackish marshes, mangrove mudflats, ephemeral ponds, and a 5 km² reservoir. They also fed in nearby rice fields at night. The study revealed that the use of these wetlands by Northern Pintails varied markedly within and among winters. Over the 3 winters, the time spent feeding ranged from 3.6% on the reservoir to 28.7% in fresh-brackish marshes, while resting times were greatest in mangrove mudflats (68.4%) and least in fresh-brackish marshes (40.5%). Reservoirs were used most often for locomotion (23.7%) and preening (18.2%). Activity levels differed markedly, however, among winters within habitats. Feeding in ephemeral ponds was nonexistent in 1989/90, when rainfall was only 3.4 cm; occupied 7.0% of their time in 1990/91, when rainfall was 4.0 cm; but rose to 43.2% in 1991/92, when 9.2 cm of rainfall created thousands of hectares of flooded ponds and fields in the area. The activity budget across the 3 winters and 4 habitats was dominated by resting (46.5%), with lesser amounts of time spent feeding (19.5%), preening (17.1%), and locomoting (12.9%).

Habitat-use patterns also varied among winters. During 1989/90, the birds were primarily found in fresh-brackish marshes (73%–90%) and mangrove mudflats (10%–27%), while use of ephemeral ponds was rare (1%) and the reservoirs were not used. In contrast, during 1990/91, 19%–53% of their time was spent in fresh-brackish marshes, 5%–9% on mangrove mudflats, and 33%–48% on reservoirs. Within a winter period, the use of ephemeral ponds was only 1%–2% during arrival and midwinter, but

39% during departure. The study also found that the number of wintering Northern Pintails in the Pabellón area was correlated with the availability of rice, as they selected resting habitats in close proximity to planted rice.

In contrast, Northern Pintails wintering on the east coast of Mexico, in Yucatán (1986–88), exclusively chose an open, brackish estuary surrounded by a mangrove forest (Thompson and Baldassarre 1991). Unlike behaviors in Sinaloa, time budgets here did not vary among the arrival period, early winter, and late winter, nor during the various times of day. Because neither rice nor any other agricultural foods were available, however, feeding was the dominant activity (42.1%–44.9% vs. 19.5% in Sinaloa). Other activities were locomotion (20.7%–23.6%), resting (19.0%–22.8%), and preening (7.5%–9.6%). In Sinaloa, habitat changes caused the birds to vary their time budgets accordingly, but those in Yucatán had consistent time budgets during the winter, because the coastal estuary was a stable habitat throughout that entire period.

In the Sacramento Valley of California (1979–82), wintering Northern Pintails used national wildlife refuges within the rice-growing region of the valley (Miller 1987). Habitat features on the refuges consisted of managed marshes, with permanent ponds making up <5% of the refuges' marsh habitat. The marshes were flooded in the summer, so they contained a mix of moist-soil plants. The ducks' diurnal time budgets differed from those during an unusually dry winter in 1980/81, and an unusually wet winter in 1981/82 (Miller 1985). Time spent feeding and loafing was higher during midwinter in the dry winter, suggesting that food availability could become a limiting factor during dry years. Unlike findings in other studies, here the females fed and loafed more than the males, while the males swam and courted more. Feeding was more frequent in flooded marsh habitats during August and September, but their preference was for flooded rice fields during February and March; also, less time was needed to obtain food in these fields, compared with marshes. Elsewhere in the Central Valley, Northern Pintails on seasonal marshes occupied open-water habitat almost entirely during the day, but switched to a near-exclusive use of densely vegetated marshes at night (Euliss and Harris 1987).

In southwestern Louisiana, habitat use by 272 radiomarked females was determined over 2 wintering periods, 1991/92 and 1992/93 (Cox and Afton 1997). Diurnal use of refuges was higher during the hunting season than in the periods immediately before or after it; nocturnal use was low (<14%) and declined late in the winter. The ducks extensively (68%–93%) foraged in rice fields and fallow agricultural fields at night, and differential habitat use between the 2 winters was correlated with the abundance of food in these areas.

Collectively, these winter studies demonstrated that Northern Pintails are flexible in their use of habitats during the winter, rely on both natural and agricultural foods, and primarily engage in resting and feeding, although courtship and pair formation also occur at this time (see Mating System).

Habitat and Body Condition

Major studies of the relationship between the body condition in Northern Pintails and habitat conditions are especially insightful, because they provide comparative data among most of the primary wintering sites. During 3 winters in the Sacramento Valley of California (Aug–Mar 1979–82), Northern Pintails were lightweight when they arrived (Aug–Sep), with small lipid reserves and reduced muscle mass (Miller 1986). They then steadily gained weight, with lipids and protein mass peaking in October and November. Protein mass then remained high (but variable) for the remainder of winter, but body mass and lipid levels declined to lows in December and January before increasing again in February and March. Such midwinter declines were probably related to the increased time spent in courtship behavior and the subsequent decrease in diurnal feeding time. Midwinter body weight, lipid levels, and protein levels were significantly lower during the dry winter of 1980/81,

compared with 1979/80 and 1981/82. Body weights in January were 60–70 g less in the dry winter: a 16.4% reduction in January versus November for males, and a 15.7% decrease for females. In comparison, these losses were only 3.2%–3.6% and 8.2%–9.2% in the 2 wet winters. Northern Pintails wintering in the Sacramento Valley have a winter survival strategy that responds to a mild climate and predictable food supplies. They can recover body condition after arrival, but then body mass and lipids are reduced during courtship in midwinter. In late winter they again increase their feeding time, regain their reserves, and (for females) complete the prebasic molt before spring migration. Poor habitat conditions associated with dry winters, however, can adversely affect their body condition.

Along the lower Texas Gulf Coast, Ballard et al. (2006) reported the winter condition of 260 adult Northern Pintails collected on the Laguna Madre from October through February 1997/98 (a dry winter) and 1998/99 (a normal wet winter). During both winters, the birds were heaviest on arrival in October. Their average body mass in February of the wet winter declined by 105 g (12%) in males and 138 g (15%) in females. During the dry winter, however, body mass in February had decreased by 217 g in males and 165 g in females, and the ducks departed the coast about 20% lighter than when they arrived. Lipid levels remained relatively stable during the wet winter, but declined 65% during the dry winter. Furthermore, during the dry winter the average lipid levels of females were 18% lower than during the wet winter, and 28% lower for males. The authors argued that Northern Pintails may choose to winter at southerly latitudes because the mild climate does not require them to store large lipid reserves, and food resources are reliable, albeit of poor quality. Departure from the wintering grounds at reduced body weights, however, increases their reliance on staging areas to accumulate the reserves needed for migration and reproduction. Smith and Sheeley (1993a) reported a similar annual variation in Northern Pintail body weight and condition between a wet and a dry winter in the Playa Lakes Region of Texas.

In Yucatán, the southern terminus of their wintering range, body mass and composition were examined for 18 males and 21 females collected from October through March 1986–88 (Thompson and Baldassarre 1990). Males increased their mass by 8.5% from arrival (Oct–Nov) to early winter (Dec), but it then declined by 7.0% through late winter (Jan–mid-Feb); females, however, did not increase their body mass until late winter (18.2% between early and late winter). Lipids followed a similar pattern, but protein levels hardly changed during the winter. Mean body mass over the entire winter period in Yucatán was 9.9%–21.9% lighter for males and 4.2%–22.0% lighter for females than the mean body mass of Northern Pintails wintering in the Sacramento Valley of California (Miller 1986). In addition, lipid levels were 16%–17% of body mass in Mexico (Thompson and Baldassarre 1990), versus 22%–39% in California. The generally mild temperatures in Yucatán probably negated any need for large lipid reserves to survive periods of cold weather, but the reserves were more than adequate (6 days) to meet food shortages caused by storm tides, which made food unavailable for up to 2 days. Northern Pintails wintering in Yucatán, as opposed to those farther north in coastal Louisiana, incurred a hypothetical round-trip flight of 1,800 km, with a migrational caloric cost of 6,632 kJ. Temperatures in Yucatán were very mild throughout the winter, however, so Northern Pintails there spent <4.2% of their time below their lower critical temperature. Even with migration costs, wintering in Yucatán saved Northern Pintails 14.5 days of daily energy acquisition as opposed to wintering in coastal Louisiana.

POPULATION STATUS

Among all species of North American ducks, the Northern Pintail population has declined most dramatically over the past 40 years, which is of grave concern to managers. Peaks and lows measured during the Traditional Survey have been progressively lower since 1955–56. The 2010 breeding population of 3.5 million was 13% below the long-term average of 4.0 million, and 29% below the objective of 5.6 million called for in the North American Waterfowl Management Plan (NAWMP). More alarmingly, breeding Northern Pintail populations in the PPR are not recovering in response to increased numbers of May ponds, as had historically occurred. Following the extended drought of the 1980s and early 1990s, May ponds attained record-high levels in 1996 and 1997, but Northern Pintail numbers only increased by 30%, 19% below the long-term average and 36% below the goal of the NAWMP (Miller and Duncan 1999). In contrast, breeding populations of all other dabbling ducks rebounded to levels above those set in the NAWMP. The Northern Pintail population decline has led to restrictive hunting seasons for this species since 1987.

The extremely large breeding populations of 9.8 and 10.4 million that occurred in the Traditional Survey area in 1955 and 1956, respectively, may not be accurate. Nonetheless, the population declined to a low of 3.2–4.2 million during the dry years of 1961–65, but then partially recovered to average 5.6 million (range = 3.9–7.0 million) in the 1970s. Following the drought years of the late 1980s and early 1990s, however, the population crashed to a low of 1.8 million birds in 1991. During the wet years of the mid- to late 1990s, the population averaged 3.0 million, with a high of 3.6 million in 1997. The drought in 2002 again caused their numbers to plunge to 1.8 million. The population has averaged 2.9 million (range = 2.2–3.5 million) from 2003 through 2010. In Mexico, Midwinter Survey data also showed significant declines. The long-term (1961–2000) average of 563,000 decreased by 2.1%/year, but the short-term (1981–2000) average of 399,015 fell more precipitously, by 7.9%/year (Pérez-Arteaga and Gaston 2004).

Outside North America, the population of Northern Pintails is estimated at about 2.0 million: 750,000 from their breeding range in northeastern Europe and northern Siberia, 700,000 from central Siberia and Asia, 200,000–300,000 from eastern Siberia, and 60,000 from northern Europe and western Siberia (Wetlands International 2006).

Harvest

The Northern Pintail is a major game duck in North America, although the harvest has declined with the implementation of restrictive hunting regulations. In the 1960s and 1970s, the Northern Pintail ranked third in the continental waterfowl harvests, after Mallards and American Green-winged Teal, but declined to fifth in the 1980s and seventh in the 1990s. They have ranked about sixth in the harvest since 2000.

The average harvest in the United States was 435,618 from 1999 to 2008: 46.1% in the Pacific Flyway, 29.2% in the Mississippi Flyway, 20.4% in the Central Flyway, and 4.3% in the Atlantic Flyway. Of the Pacific Flyway harvest, 54.1% were taken in California, 15.8% in Oregon, and 10.9% in Washington. In Canada, the harvest from 1999 to 2008 averaged 51,477, with a high of 57,038 in 2002. Over this time period, 26.8% were harvested in Saskatchewan, 20.0% in Alberta, 19.0% in Manitoba, 11.6% in Ontario, and 22.6% in the remaining provinces. The harvest in Mexico is insignificant from a continental perspective (Kramer et al.1995).

BREEDING BIOLOGY

Behavior

Mating System. Northern Pintails are seasonally monogamous, with most probably breeding in their first year if water conditions are favorable. When prairie water areas are scarce, however, and the birds displace northward, nearly all of the nonbreeding birds, especially the males, are undoubtedly yearlings. Northern Pintails begin to form pair bonds in the fall and early winter. Females appeared to choose males based on their attentiveness and the intensity of their courtship, as well as on the brightness of their plumage, especially those with whiter breasts and longer scapulars (Sorensen and Derrickson 1994). Females also preferred 2-year-old males, compared with yearlings, as the

adult males were more colorful overall, and they courted more aggressively. In addition, this species has a well-developed system for extra-pair copulations (McKinney et al. 1983). Northern Pintails might well be considered the most promiscuous of dabbling ducks, readily deserting their mates and being tolerant of other males. At both Delta Marsh in Manitoba and the Camas NWR in Idaho, males abandoned their mates a few days after incubation began (Sowls 1955, Oring 1964). Seemingly paired males will leave their mates to join courting groups, only to return to their mates later (Michael Miller).

In the Sacramento Valley of California (1980–82), courtship began in October, but aerial courtship flights were not observed until November (Miller 1985). Courtship was most intense in December and January (up to 2 hr/day for males), usually early in the day (predawn to midmorning); copulations were most frequent (49%) in January. Courtship flights averaged 4.2 males/female. Few aerial displays occurred during December in a dry winter, but one-third of all courtship displays were aerial during December in a wet winter. Courtship activity in December in the dry winter was one-third to one-half that in January, when normal rainfall resumed, and less than half that observed during December in the wet winter. In North Carolina, courtship activity was highest during January (Hepp and Hair 1983). Courtship continued during spring migration and on the breeding grounds.

About 55% of the Northern Pintails in the Sacramento Valley (1982–83) were paired by mid-November, 83% by mid-December, 93% by late December through January, and 96% by mid-February. In contrast, only 11% were paired by December in the coastal wetlands of North Carolina, but 84% were paired by January, and 100% by February (Hepp and Hair 1984). In sharp contrast, only 2.5% of the female Northern Pintails observed in Yucatán in Mexico (1986–88) were paired by February (Thompson and Baldassarre 1992), despite a sex ratio that was 84.4% males; females there were also more likely to initiate aggressive encounters with males, and they won more of those encounters. The authors concluded that better mates (dominant males) perhaps wintered farther north, and there was no benefit for them in pairing with low-quality local males.

Sex Ratio. The sex ratio of Northern Pintails is male biased, but both spatial and temporal variation in sex ratios during the winter indicate a differential migration and distribution of the sexes. Wintering populations in the Sacramento Valley in California contained 94% males in August, 76% in November, and 53% in January (Miller 1985). Farther south, in the state of Sinaloa in Mexico (1989–92), the sex ratio was 43% males on arrival (Nov–mid-Dec), 48% in midwinter (mid-Dec–late Jan), and 70% in late winter (late Jan–mid-Mar; Migoya et al. 1994). In contrast to winter, the percentage of males observed during the spring on 9 areas in North America averaged 59.4% (Bellrose et al. 1961). Sex ratios in Alaska were male dominated (67%–79%) in early spring, but near parity (53%–57%) at the start of nesting (Austin and Miller 1995).

Site Fidelity and Territory. Where habitats are fairly stable, Northern Pintails exhibit strong site fidelity. On the Y-K delta in Alaska, data from 13,645 Northern Pintails banded between 1990 and 2001 revealed that females returned to the banding site more frequently than males, and the return rates were correlated with the number of May ponds on the Canadian prairies (Nicholai et al. 2005). The return rates of females were 89.9% (77.4% for males) during years with low May-pond numbers, compared with 94.3% (87.2% for males) when May-pond numbers were high. On Delta Marsh in Manitoba, 39% of the adults and 13% of the immatures returned in subsequent years (Sowls 1955). Because of the unstable water conditions characteristic of the PPR and the sensitive response of Northern Pintails to those conditions, they may completely abandon one area when habitats deteriorate and move to another. Newly flooded lands north of Winnipeg, Manitoba, were immediately used by breeding Northern Pintails (Hochbaum

and Bossenmaier 1972), the key here being a large area of new nesting habitat and the birds' elastic response to favorable water conditions.

The home range of breeding Northern Pintail pairs is the largest reported for any species of dabbling duck, which reflects their social behavior and preference for seasonal wetlands. In North Dakota, the average home-range sizes of radio-marked individuals were large, varying from 579 ha for unpaired males ($n = 5$), to 896 ha for paired males ($n = 8$), and 480 ha for paired females ($n = 15$; Derrickson 1978). Maximum home-range sizes ranged between 1,067 and 1,477 ha, with maximum lengths ranging from 1.5 km for paired females to 5.8 km for paired males. The home ranges of pairs that included both the prenesting and nesting periods ($n = 7$) averaged 508 ha, compared with a 167 ha home range of pairs that covered only the nesting period ($n = 4$).

A home-range stability index demonstrated the mobility of Northern Pintails during the breeding season: 32% exhibited stability; 39%, temporary stability; and 29%, no stability (Derrickson 1977). This mobility was reflected in the number of wetlands used, which averaged 17 for paired males, 12 for unpaired males, and 11 for females, but ranged between 3 and 27. Among paired males, temporary stability, if achieved, tended to occur during the laying period. Female home-range sizes decreased from 605 ha during prenesting to 124 ha while nesting. Males do not defend a specific territorial area within their home ranges, but they have been observed defending the immediate area around their mate. Smith (1968) noted that paired males usually chase away other Northern Pintails that approach within 3.0–4.5 m.

Courtship Displays. Courtship displays and the pair formation of Northern Pintails were described by Lorenz (1951–53), Johnsgard (1965), and Smith (1968). On the water, the common sequence of displays given by groups of males is *chin-lift*, *burp*, *grunt-whistle*, and *head-up-tail-up* or *turn-the-back-of-the-head*. Northern Pintails also have a very characteristic aerial courtship that involves a female and several males. Northern Pintails have a unique form of *burp* display, characterized by raising the head vertically, with the neck extended and the bill tilted slightly downward, while giving a *whee* call that rises in inflection. Females *incite*, which serves to show the selection of a given male and the rejection of others in the group. *Inciting* is accompanied by a *kuk-kuk-kuk-kuk* or *rrr-rrr-rrr* call. *Turn-the-back-of-the-head*, often performed by the male in association with *inciting*, involves holding the head up while moving away from the female and erecting a streak of brown feathers along the neck, which contrasts with the white sides. Some displays are given during aerial courtship flights. Males, but not females, exhibit *head-pumping* as a precopulatory display. *Bridling* is a common postcopulatory display by males, while *nod-swimming* is relatively rare. On the breeding grounds, *pursuit-flights* become prevalent when egg laying commences; they involve paired males chasing the females of other mated pairs, a behavior that tends to space females apart during the egg-laying period (Smith 1968).

Nesting

Nest Sites. Females lead their mates on morning and evening scouting trips to seek a suitable nesting site 3–6 days before egg laying (Derrickson 1977). Northern Pintails, more than any other ducks, select very open areas, where vegetation is either low or sparse, for their nests. The nest can sometimes be located on bare earth or in stubble fields. Northern Pintails also commonly nest far from water, again in contrast to other ducks.

Klett et al. (1988) observed that of 1,219 Northern Pintail nests located over 3 time periods (1966–74, 1975–79, 1980–84) across a broad area of the PPR in the United States, 53.8% were in planted cover, 15.2% in idle grasslands, and 12.7% in grasslands. Northern Pintails averaged far more (55.0%) nest initiations in croplands than 4 other species of dabbling ducks (2.3%–8.3%), and they had the highest (5.4%) preference for cropland

nesting cover, compared with other dabbling ducks (0.2%–0.3%). Reynolds et al. (2001) also found that Northern Pintails had the highest (2.8%) nest preference for croplands among the 5 species of dabbling ducks they studied, but their most frequently chosen nesting sites were in planted cover (37.6%) and haylands (19.2%).

During a 4-year study (1982–85) across a wide expanse of prairie Canada that located 841 Northern Pintail nests, of those in parkland habitat, 34% were found in croplands, 27% in rights-of-way, 18% in odd areas (patches of cover <2 ha in size with an array of features), and 11% in brush (Greenwood et al. 1995). In prairie habitat, Northern Pintails used croplands (45%), brush (31%), and rights-of-way (15%).

On the Y-K delta in Alaska (1991–93), 69% of 787 Northern Pintail nests were located on slough banks in highly saline tidal areas, where there were few mammalian predators (Grand et al. 1997). The minimum nest density was 7.2–13.4/km² (Flint and Grand 1996a), and 94% of all nests were in low-, intermediate-, or high-sedge meadows. Females probably preferred these nest sites because they were more elevated and were better drained early in the nesting season. Northern Pintails using levees typically selected nest sites on the top (35%) or on the side nearest the channel (48%). Vegetation at the nest site varied greatly, but was low in height.

At Tule Lake and Lower Klamath NWRs in California (1957), most Northern Pintail nests ($n = 216$) were found in saltbush (*Atriplex* sp.; 32.4%) and grasses (29.2%; Rienecker and Anderson 1960), but 29.5% of 44 nests located in 1952 occurred in mustards (*Sisymbrium* sp. and *Brassica* sp.; Miller and Collins 1954). Of the 216 nests, 49.5% were in grasslands, 36.1% were on islands, and 14.4% were on dikes.

Near Brooks in Alberta (1981–84), nests ($n = 154$) were situated around shallow bodies of water in grazed mixed-prairie habitat (Duncan 1987a). Northern Pintails also nested on islands, at a density of 10.4 nests/ha, compared with 0.16–0.17/

ha on the mainland. The mean distance of mainland nests from water was 781 m in 1983 ($n = 71$) and 1,126 m in 1984 ($n = 27$). Some females nested as far as 3 km from water, but in general, they exhibited little tendency to nest >1,600 m from a shoreline. In the same general area, Keith (1961) found that nest distances to water averaged 50 m. In southern Alberta (1994–96), 72% of 57 nests were located ≤100 m from water (Guyn and Clark 1999). Derrickson (1977) reported an average distance of 190 m in North Dakota. Smith (1968) and McKinney (1973) suggested that females nested far from water to avoid harassment by males. Duncan (1987a), however, argued that because the distance from the nest to water did not decline with later nesting dates, when most males had left the study area, nesting far from water instead was a female choice that reduced potential nest predation.

The nest is usually a simple bowl of grasses or other vegetation: 19–26 cm in outside diameter, 13–19 cm in inside diameter, 0–18 cm in height, and 6–10 cm deep (Fuller 1953). In response to storm tides on the Copper River delta in Alaska, however, these birds built up their nests to 10.2–15.2 cm above the initial nest height (Hansen 1961). The first egg is often laid while the nest is just a scrape (Alisauskas and Ankney 1992a).

Clutch Size and Eggs. Northern Pintails lay small clutches in comparison with other dabbling ducks. Bellrose (1980) reported 3–14 eggs in completed clutches, with an average of 7.8 ($n = 1,276$ nests). In the PPR in the Dakotas (1993–95), the average clutch size for 681 nests was 7.7 eggs, which was 1–3 eggs less than for other dabbling ducks (Krapu et al. 2004c). On the Woodworth Study Area in North Dakota (1966–81), clutch size for 99 nests averaged 7.8. During a 4-year study (1981–84) near Brooks, Alberta, the average clutch size was 6.9 for 369 nests (Duncan 1987b). Clutch size in the parklands of Alberta (1952–65) averaged 7.2 for 73 nests (Smith 1978). On the Y-K delta in Alaska (1991–93), clutch size averaged 7.6 for 795 nests and did not differ among years (Flint and Grand 1996a). Clutch

size declined with nest-initiation dates, though, at a rate of 0.09 eggs/day. Similarly, in southern Alberta (1994–96), clutch size for 217 nests averaged 7.2, but ranged between 7.1 and 7.3, and declined in a curvilinear fashion with the nest-initiation date (Guyn and Clark 2000).

In their review of the effect of day length on clutch size, Krapu et al. (2002) noted that clutch size varied little among 4 different sites: California, North Dakota, Saskatchewan, and Alaska. The predicted decline in the numbers of eggs/clutch over the nesting season was similar for all 4 locations, despite nesting seasons that ranged from 42 days in Alaska to 70 in California. The authors argued that a reduction in nutrient availability during the nesting season contributed to a higher rate of decline in clutch size in Alaska than in temperate regions. Northern Pintails in Alaska that nested early laid large first clutches, but clutch size shrank rapidly thereafter, and breeding terminated early. This reproductive strategy was thought to be adaptive, because young that hatch the earliest have the highest survival rates (Grand and Flint 1996b).

Northern Pintail eggs are smooth and vary between elliptical and subelliptical in shape (Palmer 1976a). Their color is greenish yellow or tending toward grayish, although white eggs can occur. On the Y-K delta in Alaska, egg dimensions ($n = 448$) averaged 54.0 × 37.5 mm, and egg mass ($n = 60$) averaged 42.8 g (Craig Ely). On Minto Flats in interior Alaska, egg dimensions were 53.7 × 37.6 mm for 806 eggs (Petrula 1994). In Alberta, the dimensions of 166 eggs were 52.7 × 37.4 mm (Duncan 1987c). Egg size was strongly correlated with both fresh egg weight ($r^2 = 0.89$) and the weight of 1-day-old ducklings ($r^2 = 0.89$), but body weight of the female was only weakly correlated with egg size ($r^2 = 0.11$). Egg size was not correlated with laying date or clutch size and did not differ between adults and yearlings. Individuals, however, tended to lay consistently sized eggs, with no significant heritability for egg size. Captive females fed a diet of 29% protein laid larger eggs than those fed a 14% protein diet. In Alberta, the mean egg volume was 39.85 cm³ (Guyn and Clark 2000), slightly larger than the 38.6 cm³ reported in Alaska (Flint and Grand 1996a), but there was no relationship between clutch size and egg size. Eggs are laid early in the morning, at a rate of 1/day (Alisauskas and Ankney 1992a).

Incubation and Energetic Costs. The incubation period is 22–24 days (Austin and Miller 1995). Contrary to the common assumption that incubation begins within 24 hours of laying the last egg, Loos and Rohwer (2004) discovered that Northern Pintails breeding within the PPR in southwestern Manitoba and North Dakota began incubating their eggs prior to the completion of the full clutch. Females averaged 41.8 hours on their nests between the laying of eggs 2 through 9, with attendance increasing as clutch size increased.

Hoover (2002) noted that female Northern Pintails in North Dakota averaged 81% of each day attending to their nests during incubation, with 2 to 3 recesses/day. Afton (1978) reported that the incubation attentiveness for a renesting Northern Pintail was 86.3%. Burris (1991) stated that females in Alaska ($n = 22$) averaged 22.2% of each day off their nests. Derrickson (1977) noted that males remain on nearby wetlands during the laying period, but depart to join small groups of other males as soon as incubation commences.

On the Y-K delta in Alaska (1990–91), the lipid reserves of breeding females declined by 2.58 g/g allocated to reproduction, and protein declined by 0.20 g (Esler and Grand 1994b). The commitment of lipids varied, however, from a high of 3.35 g early in the nesting season ($n = 85$) to 0 g for renests occurring about 40 days later. Northern Pintails used more lipid reserves to form their first clutches than any other duck species studied to date. The levels of lipid reserves were also correlated with the date of nest initiation. Moreover, females depleted their lipid reserves in excess of the costs needed for egg production, because some lipids were used for maintenance over much of the nesting season. This strategy of lipid use is intermediate in comparison with other temperate-nesting puddle ducks, which deplete their lipid reserves in amounts equal to or less than the requirements for their clutches. Mann and Sedinger (1993) observed a similar pattern for Northern Pintails nesting in interior Alaska (Minto Flats), noting that stored proteins provided 21%–62% of the protein costs of egg production and suggesting that the amount of available protein limited clutch size.

In North Dakota (1969–71), Krapu (1974a) reported that breeding females arrived with large subcutaneous and visceral fat reserves, but such reserves were largely depleted in early nesting efforts. Visceral fat stores were used by mid-May, and renesting females after mid-May had body weights averaging 25% less than prenesting females in April. This decline in lipid reserves increased the dependency of renesting females on food resources in wetlands to meet the nutritional demands of producing a second clutch.

Nesting Chronology. Along with Mallards, Northern Pintails are among the earliest-nesting ducks. In a summary of nesting chronology, Krapu et al. (2002) reported that Northern Pintails initiate nests in mid-March in California, mid-April in North Dakota and Saskatchewan, and mid-May in Alaska. The duration of the nesting period was 70 days in California, 60 days in North Dakota, 66 days in Saskatchewan, and 42 days in Alaska. Spring weather conditions can delay nest initiation, as was documented in the parklands of Alberta by Smith (1971). During a spring with 35.5 cm of snowfall, blizzards, and below-freezing temperatures between late April and the first week of May, Northern Pintails initiated nests on 10 May, compared with 10–30 April in other years (1952–65).

During a 4-year study (1981–84) near Brooks, Alberta, Northern Pintails initiated nests over a 9-week span in 1982 and 1983, but only over 5 weeks in 1984 (Duncan 1987b). The mean date of nest initiation was 18–25 April, but experienced females begin nesting an average of 5 days earlier than yearlings (Duncan 1987c). Two peaks of nest initiation were observed: the first resulted from adults nesting in mid-April, and a second (minor) peak from yearlings nesting in May. Body weight at capture was not correlated with laying date ($r^2 = 0.02$; $n = 112$). There was also no relationship between laying date and the mean daily temperature, or the mean daily minimum temperature, or cumulative daily mean temperatures >0°C. Smith (1968), however, reported that freezing temperatures delayed nesting by Northern Pintails in southern Alberta. In another study (1994–96), Northern Pintails in southern Alberta typically initiated nests over a 9-week period, with the first nests ($n = 292$) appearing in mid-April, and a median nest-initiation date between 13 and 24 May (Guyn and Clark 2000). Across a broad area in prairie Canada (1982–85), the median date of nest initiation was 13 May, with a nesting interval that varied among years but ranged from 26 April to 10 June (Greenwood et al. 1995). The average length of the nest-initiation period was 26 days, and it increased by 0.12 days for each 1.0 cm of precipitation in May. On the Y-K delta in Alaska (1991–93), the first nests were initiated around 15 May each year, with a nesting interval of 44–47 days (Flint and Grand 1996a). The mean date of nest initiation was 2–6 June.

Nest Success. Nest success of Northern Pintails during 3 time periods (1966–74, 1975–79, 1980–84) across 5 regions of the PPR in the United States was low, reaching only 7%–10% (Klett et al. 1988). Success was highest in idle grasslands (18%–27%), and then in grasslands (10%–19%); in croplands it was 5%–11%. Only Mallards had lower (6%–8%) overall rates of nest success. In the Dakotas and Montana (1992–95), daily survival rates were 92.6% for 799 nests on Conservation Reserve Program lands and 93.1% on Waterfowl Production Areas (Reynolds et al. 2001). These translate to nest-success rates of 11.8% and 9.8%, respectively, assuming a 23-day incubation period and 7 days to produce the clutch. The amount of perennial cover had the highest correlation with nesting success ($r = 0.36$), but Northern Pintails nonetheless had

the lowest survival rates of the 5 species of dabbling ducks that were observed. During a 3-year study (1994–96) in southern Alberta, nest success ranged between 19.5% and 29.5%, but success was considerably higher on islands (53.3%–67.5%) than in uplands (6.3%–18.0%; Guyn and Clark 2000). In the parklands of Alberta (1952–65), nest success averaged 43% for 73 nests (Smith 1971).

In interior Alaska (1989–91), the nest success of Northern Pintails averaged only 3.8% (Petrula 1994). A 3-year nesting study (1991–93) on the Y-K delta in Alaska is revealing, however, because habitat there is pristine in comparison with the heavily impacted PPR. Nonetheless, nest success ($n = 795$ nests) differed greatly among years, ranging from lows of 11% in 1993 and 18% in 1992 to a high of 43% in 1991 (Flint and Grand 1996a). Nest success was highest for nests initiated early in the season, and declined to near zero for nests begun at the end of the season. Although nest success in some years was similar to that on the PPR, the mean nest-success rate of 24% across the 3 years, and the very high nest-success rate of 43% in 1991, demonstrated that Northern Pintail nest success on the Y-K delta is generally higher than that on the PPR. Predation caused most (36.0%–69.6%) of the nest losses on the delta; another factor was flooding, which destroyed 15.2% of the nests in 1991. Arctic foxes (*Alopex lagopus*), mink (*Mustela vison*), Glaucous Gulls (*Larus hyperboreus*), Mew Gulls (*Larus canus*), Long-tailed Jaegers (*Stercorarius longicaudus*), and Parasitic Jaegers (*Stercorarius parasiticus*) were known nest predators. Annual variation in the degree of predation was correlated with annual and seasonal variations in nest success, as avian predators caused a greater loss of ducklings when hatching was late (mid-Jul) and alternative prey sources were less available.

Several studies have revealed that nest losses due to predators are correlated with the degree of nest concealment. In research on California Gull (*Larus californicus*) predation on waterfowl nests in Utah, Odin (1957) calculated the loss rate of well-concealed Northern Pintail nests at 7.7%, compared with 26.7% for partially concealed nests and 46.2% for poorly hidden nests. In analyzing the nest loss at Ogden Bay in Utah, Fuller (1953) reported a 21% loss of well-concealed nests, versus 79% for poorly concealed ones.

Birds and mammals are about equally responsible for nest destruction. Other important avian predators besides gulls are American Crows (*Corvus brachyrhynchos*) and Black-billed Magpies (*Pica hudsonia*). Among mammals, striped skunks (*Mephitis mephitis*) are significant nest predators, because they commonly forage in the upland grassland habitats Northern Pintails use for nesting. Nest losses are also caused by other mammalian predators, including coyotes (*Canis latrans*), red foxes (*Vulpes vulpes*), mink, raccoons (*Procyon lotor*), badgers (*Taxidea taxus*), and ground squirrels (*Citellus* sp.; Austin and Miller 1995). Red foxes, the principal nest predator at the Woodworth Study Area in North Dakota, were responsible for 69% of 74 nest predations (Higgins et al. 1992). Among 1,293 duck carcasses found at red fox dens in the Dakotas, 26% were Northern Pintails, which had the highest vulnerability index among all prairie ducks (Sargeant et al. 1984). In southern Saskatchewan, the adult female mortality rate from raptors (14.1%) was greater than that due to red foxes (1.1%), collisions with power lines (1.1%), or unknown factors (3.9%; Richkus et al. 2005); these results differed from the prior research findings of Sargeant et al. (1984). Raptors may have been the major cause of deaths in southern Saskatchewan because of the otherwise limited exposure of females during incubation, due to their use of open habitats.

Northern Pintails are especially vulnerable to nest losses from farming operations, because they use croplands as nesting habitat. During a 4-year study (1982–85) across a broad area of the PPR in Canada, 17% of all Northern Pintail nest losses were from farm equipment, compared with only 2%–3% for the 4 other species of dabbling ducks that were observed (Greenwood et al. 1995). Over-

all nest success for Northern Pintails was only 7%. In an earlier study on the Portage Plains in Manitoba, Milonski (1958) reported that farming operations destroyed 57% of all Northern Pintail nests in 1956 and 41% in 1957. The largest losses were caused by cultivation, as well as by disking, mowing, plowing, and harrowing. In Alberta, farming operations were responsible for 18.8% of 32 destroyed Northern Pintail nests (Smith 1971).

Brood Parasitism. I found no reports of intraspecific brood parasitism among Northern Pintails in the literature, perhaps because Northern Pintail nests are often spaced far apart. In California, Rienecker and Anderson (1960) noted that only 6 out of 216 (2.8%) Northern Pintail nests were parasitized, 4 of them by Ring-necked Pheasants (*Phasianus colchicus*). Duck species parasitizing Northern Pintail nests include Redheads, Ruddy Ducks, Mallards, Common Goldeneyes, and Blue-winged Teal (Weller 1959). Of 24 parasitized Northern Pintail nests studied by Joyner (1976) in Utah, 20 were parasitized by Redheads, 3 by Ruddy Ducks, and 1 by both species. Redheads added an average of 2.8 eggs to each parasitized nest, and Ruddy Ducks added 4.0. Parasitism did not significantly reduce the clutch size of Northern Pintails, but egg hatchability was 43.0% in parasitized nests, versus 66.4% in nonparasitized nests. In North Dakota, where Redheads were the principal nest parasite, parasitism did reduce clutch size, which averaged 5.4 eggs in parasitized nests, compared with 8.0 in nonparasitized nests (Lokemoen 1991). Also, more Northern Pintail nests were interspecifically parasitized on islands (12 out of 17) than on peninsulas (0 out of 14).

Renesting. Although renesting was fairly extensive in North Dakota (Derrickson 1977), the renesting rate for Northern Pintail females is generally lower than that for other dabbling ducks. At Delta Marsh in Manitoba, 30% of 62 marked females renested, 1 even after losing her brood; 2 renested a third time after the destruction of their second nests (Sowls

1955). Clutch size for 13 renesting females averaged 8.6 eggs for initial nests and 6.6 for first renests; 2 females renested twice, with an average clutch size of 7.5. Renests were an average of 258 m from first nests. In southern Alberta, 55% of 20 marked females renested once, and 1 individual renested a second time (Guyn and Clark 2000). The nest-initiation dates of renesting females varied from 20 April to 21 May, with the interval between the first and the second nests ranging from 2 to 29 days. The stage at which the initial nest was destroyed varied from the laying period to day 11 of incubation. Renesting was deemed to be costly, however, because late-nesting females produced fewer offspring; hence females must balance the tradeoff between maximizing their reproductive effort in initial clutches with the risk of having smaller broods if they renest.

In another Alberta study, Duncan (1987b) reported a very low renesting rate (4%) for 127 color-marked females induced to renest by removing their first clutches, and no radiomarked females (n = 17) renested. In contrast, the renesting rate was 50% for captive females fed a diet of either 14% or 29% protein, which implies that food availability somehow limited renesting in wild birds. Water conditions in late summer during this study were average, and Duncan suggested that the renesting rate might have been higher during very wet years. From his research in parkland habitat near Redvers, Saskatchewan, Stoudt (1971) believed that Northern Pintails seldom renested, and Smith (1971) considered Northern Pintails at Lousana, Alberta, to be less prone to renest than Mallards and Canvasbacks. Collectively, these studies indicate regional disparities in renesting by Northern Pintails that are perhaps influenced by water availability, as Bellrose (1980) suggested.

In contrast to the generally low rates of renesting in the PPR, renesting on the Y-K delta in Alaska (1994–95) was 56% for 39 radiomarked female Northern Pintails; 3 females renested twice (Grand and Flint 1996a). The mean interval between the

first and the second nests was 11.4 days, but it ranged between 7 and 26 days (n = 22 females). The median distance between the first and the second nests was 276 m, but 3 females moved >5 km. Mean clutch size was 8.2 eggs for 51 first nests and 6.3 for 15 renests. For the same females, clutch size declined by 2.3 eggs between the first and the second nests (n = 15). Late-nesting females renested less frequently than early nesters, and both nest success and fledgling survival declined with renests. This study contrasted sharply with that of Calverly and Boag (1977), who concluded that arctic-nesting Northern Pintails had a lower reproductive output than their prairie counterparts, due to smaller clutch sizes and a lower renesting effort.

REARING OF YOUNG

Brood Habitat and Care. Preferred brood habitat for Northern Pintails includes shallow wetlands interspersed with open-water areas and emergent vegetative cover. Observations of 393 broods in the Dakotas (1958–63, 1967–78) revealed that semipermanent wetlands were used most frequently (50%); other habitats were seasonal wetlands (31%) and sewage lagoons (10%; Duebbert and Frank 1984). On stock ponds in South Dakota, broods (n = 22) chose ponds with a long shoreline and associated small bays, as well as an emergent vegetation coverage of 5%–95% (Mack and Flake 1980).

On the Y-K delta in Alaska (1991–93), 80% of 60 radiomarked females nesting in saline habitats moved their broods to less saline areas, while females nesting in preferred brood-rearing habitat did not move their broods (Grand et al. 1997). Preferred habitats were productive and moderately saline, and broods moved an average of 1,951 m to reach those sites. Broods avoided the freshest-water ponds, where invertebrate productivity was lowest. Brood use was highest in high-sedge (33.1%–34.5%) and high-graminoid (19.0%–28.5%) meadow habitats, except during 1993, when intermediate-sedge meadows were used most (36.4%).

Northern Pintails in the PPR often nest far from water and thus lead their broods farther overland to brood habitats than do other prairie ducks. The ducklings are well equipped for long overland journeys, however, as 40 captive ducklings given no food survived an average of 5 days (Krapu 1974a). At Oak Hammock Marsh in Manitoba, Guinn and Batt (1985) reported that females are very attentive to their young and seldom leave them, often not until their ducklings have attained flight. A minimum of 37 females averaged 60.8% of their time in self-maintenance activities (such as feeding and comfort movements), and 34.6% in parental-care activities (such as keeping alert, leading, and following). Feeding was the predominant (51.9%) self-maintenance activity, and following the brood was the most frequent (18.2%) parental-care activity. Time spent on parental care did not vary with brood size, but the total amount of parental care was higher for Class I broods than for Class II. Females devoted more time to feeding and less to parental care later in the season, most likely because they needed to recover the weight they lost during incubation, acquire enough energy to complete the molt, and prepare for fall migration.

Brood Amalgamation (Crèches). Brood amalgamation has not been reported for Northern Pintails.

Development. Day-old ducklings in Manitoba ($n = 19$) weighed an average of 25.5 g (Smart 1965a), and 26.8 g in Utah ($n = 25$; Fuller 1953). In a study of the growth and development of semicaptive Northern Pintails (4 males and 4 females), the ducklings averaged 28 g at hatching and grew most rapidly between weeks 1 and 4; after 7 weeks, the ducklings were 24 times heavier than at hatching (Blais et al. 2001). Adult size was reached in 117 days. Bill length grew very rapidly during the first 3 weeks and was nearly to adult size after 40 days. The first feather sheaths appeared at about 12 days, with juvenal plumage fully developed around 110 days. Young birds were capable of short flights in 52 days. In South Dakota, Gollop and Marshall (1954) found that Northern Pintail broods reached flight stage in 46–57 days.

RECRUITMENT AND SURVIVAL

In southern Alberta (1994–96), Guyn and Clark (1999) determined brood and duckling survival for 57 radiomarked females. Brood survival ranged between 72.3% and 88.2% and did not vary among the 3 years. A 30-day duckling survival rate of 65.2% in 1994 was among the highest ever reported for ducks, and it was probably influenced by greater water availability for broods that year. Survival was 42.4%–43.8% during the other 2 years of the study, with brood sizes at the hatch averaging 6.7. Duckling mortality in all years was greatest (60%–76%) during the first 7 days after hatching, and duckling survival did not vary with a female's age or the distance of the nest from water. Both brood and duckling survival declined with later hatching dates; successful broods (broods with at least one duckling surviving to 30 days) hatched about 10 days earlier than unsuccessful broods (5 vs. 15 Jun).

On the Y-K delta in Alaska (1991–93), the 30-day survival rates of 751 ducklings from 111 radiomarked Northern Pintails ranged from 3.9% to 14.5%; brood survival was 18%–45% (Grand and Flint 1996b). Most (72%–89%) duckling mortality occurred during the first 10 days after hatching. Brood size averaged 6.9 at hatch, with an annual range of 5.9–7.2. Duckling survival was lowest for late-hatched broods, declining at a rate of 0.6% for each day's delay in hatching; hence early-nesting females produced a large part of each year's annual recruitment. The Alberta and Alaska studies both demonstrate that there are wide annual and spatial variations in Northern Pintail recruitment.

Flint et al. (1998a) used these brood-survival data and data from other Northern Pintail studies on the Y-K delta (Flint and Grand 1996a; Grand and Flint 1996a, 1996b) to develop an insightful model of productivity. Their model assumed that nest success was 25%; duckling survival, 11%; and the probability of renesting, 56%. The results revealed that hen success (the percentage of females fledging ≥1 duckling) was 39%, but only 13% of the breeding females produced all of the fledged young. Each Northern Pintail female only produced 0.16 young/year, with early nesters hatching many more young than late nesters or renesters. Combining this estimate with first year (51.5%) and adult (61.5%) survival rates determined for Northern Pintails banded in California (1948–79), Rienecker (1987b) calculated a population growth rate (λ) of only 0.697, which indicates a rapid decline for this population. Changes in adult survival rates had the greatest effect on population growth, compared with relative changes in first-year survival and reproductive success. Nest success and duckling survival would need to increase to 40% to achieve population stability.

There have been numerous survival studies of adult and immature Northern Pintails, with researchers attempting to determine the cause of the population decline. Band-recovery data now exist from many breeding and wintering sites, and they have been used to provide a continental perspective on survival. Hestbeck (1993b) estimated annual survival rates using recovery data from Northern Pintails banded on their wintering grounds throughout North America (1950–80): 63.2%–80.6% for males and 42.1%–76.9% for females. In a more recent analysis of 352,252 banding records and 24,370 recovery records from 1970 to 2003, Rice et al. (2010) found that survival varied with age, sex, banding region (western, central, and eastern), and time, with significant interactions between time and age, and time and region. The average annual survival rate was 75.9% for adult males, 65.3% for immature males, 65.0% for adult females, and 53.6% for immature females. Trends suggested that annual variation in survival rates was not the cause of the initial decline in the continental Northern Pintail population, nor was it the dominant factor preventing the population from increasing. Problems with recruitment were the more likely underlying factors depressing Northern Pintail numbers.

Relative to individual studies from single lo-

cations, an analysis of band-recovery data from 13,645 Northern Pintails banded on the Y-K delta (1990–2001) estimated annual survival rates of 77.6% for males and 60.2% for females from this important breeding site (Nicolai et al. 2005). Rickus et al. (2005) studied adult female survival during the nesting season in southern Saskatchewan (1998–2000) and provided additional data from another significant Northern Pintail breeding area that contains typical prairie breeding habitat. Survival of 140 radiomarked females over a 75-day nesting period (30 Apr–14 Jul) was 80.6%, did not vary among years or between age classes (adults and immatures), and was not correlated with a female's body condition.

The numerous winter-survival studies of Northern Pintails also provide valuable data, because Northern Pintails show strong fidelity to their wintering areas. In the Central Valley of California, the most significant wintering area for Northern Pintails in North America, Fleskes et al. (2002b) examined the seasonal survival of 191 radiomarked immature females and 228 adult females from September to March in 1991–94. Survival rates were 75.6% for adults and 65.4% for immatures, and were much lower than the rate of 97.4% for 190 radiomarked adult females to the north, in the Sacramento Valley of California (Miller et al. 1995). The disjunct in survival rates between the 2 areas was attributed to differences in the availability of refuges and other sanctuaries, types of feeding habitats, waterfowl populations, and hunting pressure. While about 25% of the wetland habitat on wildlife areas and national wildlife refuges in the Sacramento Valley was closed to hunting, only 6% was closed in the San Joaquin Valley. Northern Pintails in the Sacramento Valley also had access to rice fields, a habitat that contained high-nutrient food, which fulfilled their energetic requirements more quickly and provided a certain degree of refuge from hunters.

In a later study, Fleskes et al. (2007) determined the August–March survival of 163 adult and 128 hatching-year females marked with radiotransmitters in the Sacramento Valley, and another 885 adult females radiomarked throughout the Central Valley. Survival estimates were examined in relation to flooded habitat, January population abundance, hunter days, and a combination of these variables that indexed hunting pressure. Survival was also estimated for two periods, early (1987–94) and late (1998–2000), when habitat increased by 39%, January duck abundance by 45%, and hunter days by 21%; the length of the duck-hunting season went from 59 to 100 days; and Northern Pintail bag limits were unchanged (1 bird/day). Survival was higher during the late period versus the early period for Northern Pintails radiomarked in each region: the Sacramento Valley (93.2% vs. 87.6%), Suisun Marsh (86.6% vs. 77.0%), and the San Joaquin Valley (86.6% vs.76.9%). Most deaths (72%) were from hunting; a lower hunting-pressure index, a higher January population, and a greater amount of habitat were all associated with reduced winter mortality. The authors recommended that regulations and habitat management continue in a manner that minimizes natural mortality but allows harvest at a level that encourages the management of waterfowl habitat in the Central Valley.

Farther south in the state of Sinaloa in Mexico, the winter survival rate of 163 radiomarked females did not differ among 3 winters (2 Nov–16 Feb 1989–92), and averaged 91% for combined first-year and adult birds (Migoya and Baldassarre 1995). Hunting was the primary cause of mortality (6.1%–10.7%); there was only 1 natural death. The high survival rate observed on this important wintering area was probably the result of abundant habitat and low hunting pressure.

In the Playa Lakes Region of Texas, Moon and Haukos (2006) studied fall/winter survival and the causes of mortality for 159 radiomarked females in 2002/3 and 168 females in 2003/4. Survival was 92.5% in 2002/3, but only 69.4% in 2003/4. Most mortality occurred during the hunting season: 88% in 2002/3 and 34% in 2003/4. Age class and capture time did not affect survival during either winter period, but there was a positive correlation

between body mass at the time of capture and survival during 2003/4. The lower rate of survival in 2003/4 probably occurred because fewer wetlands were available.

In southwestern Louisiana, Cox et al. (1998b) reported survival rates for 320 radiomarked females during 3 wintering periods: 1990/91, 1991/92, and 1992/93. Adult survival (71.4%) was higher than that of immatures (55.0%), primarily because immatures had a higher hunting mortality rate than adults (28.7% vs. 13.0%). Of 70 deaths, 61% were confirmed hunting mortalities. Survival did not differ in relation to the year or to body condition, and hunting mortality did not differ in relation to the year, to body condition, or to the region. Hunting mortality was high, despite conservative Northern Pintail harvest regulations during the study (a 30-day season, with a limit of 1 bird/bag). High hunting mortality was attributed in part to the fact that Northern Pintails left the few refuges in southwestern Louisiana to feed on privately owned agricultural fields, and in part because experienced hunters had a long tradition of hunting Northern Pintails in Louisiana. In comparison with the posthunting period, females were 20.9 times more likely to be killed during their first hunting season, and 17.6 times more likely during their second.

In contrast to the situation in southwestern Louisiana, Lee et al. (2007) radiomarked adult Northern Pintail males and females wintering in the middle Rio Grande valley in New Mexico (2001/2 and 2002/3) and found that natural mortality contributed more to their low survival rate than did hunting, although the sample size ($n = 69$) was smaller than in other studies. Winter survival (27 Oct–2 Mar) did not vary by year, sex, time, or body condition, and the overall estimate was 59.7%. Survival also did not differ between the hunting and the nonhunting periods. The survival estimate for adult females was 63.9%, which was 5.5%–28.6% lower than published estimates for adult females in 5 other geographic regions within their wintering range. The small sample size may have biased these results, and abundant food and water resources in a refuge setting probably reduced hunting mortality. Nonetheless, these low overall survival rates indicated that natural causes of mortality had a greater influence on survival than what was reported from other wintering areas.

Raveling and Heitmeyer (1985) correlated Northern Pintail population size and recruitment with habitat conditions (May ponds and winter precipitation in the Central Valley of California) and harvest scenarios, using data from 1955 to 1985. Their results revealed the multidimensional complexity associated with the population dynamics of such a highly mobile species. Annual changes in the size of the Northern Pintail breeding population were most strongly related to spring habitat conditions, especially when the population size was small and spring conditions were wet. Population size was less dependent on harvest, although, among the harvest variables (season length, bag limits, and total harvest), total harvest had the greatest relationship to population size. Annual recruitment (as indexed by age ratios in the harvest) was most related to breeding habitat conditions when spring seasons were dry and the population size was small, but it was most related to winter habitat conditions the previous year when spring seasons were wet and population size was large. Harvest variables were not related to recruitment, except in years following dry winters in the Central Valley; harvest rates of adults increased in dry winters, which may have reduced overall recruitment the following spring. Recruitment following dry winters, however, was even more strongly related to spring habitat conditions than to harvest variables the previous winter. Hence, in addition to providing breeding habitat, sustaining and enhancing the quality of late-winter habitat is an important component of Northern Pintail management.

Longevity records for Northern Pintails are 22 years and 3 months for a hatching-year male banded and shot in Saskatchewan, and 21 years and 4 months for an adult male banded in California and shot in Idaho.

FOOD HABITS AND FEEDING ECOLOGY

Northern Pintails forage by picking, dabbling, and tipping up; they rarely dive. They are commonly seen foraging in very shallow water, where they walk along and pick or filter food items from the substrate. Their long necks allow them to reach the bottom in shallow water, where they are adept at separating small seeds from the sediments (Krapu 1974a). Breeding Northern Pintails often are found feeding in shallow temporary wetlands and flooded fields. They will also feed on the surface during late spring and summer, when midge (Chironomidae) pupae move to the surface in large numbers. Wintering birds commonly are found in brackish coastal wetlands and mudflats. They also readily use agricultural fields, especially flooded rice fields, as well as flooded and dry waste cornfields. In shallow-water areas, they often feed with American Green-winged Teal and shorebirds.

Breeding. Food use during the breeding season, especially among female Northern Pintails, switches dramatically from a diet dominated by plant material to one dominated by animal material. During the spring and summer in North Dakota (1969–71), female Northern Pintails ($n = 39$) feeding in shallow nontilled wetlands on average consumed 20.8% plant matter and 79.2% animal matter (Krapu 1974b). Animal foods included aquatic dipterans (primarily larvae; 27.9%), larval and adult beetles (Coleoptera; 22.5%), fairy shrimp (Anostraca; 10.9%), and earthworms (Oligochaeta; 10.1%). In contrast, males ($n = 14$) ate an average of 70.0% plant material and 30.0% animal material. The high protein content in the female diet was needed to meet the demands of egg production. Breeding female Northern Pintails ($n = 61$) increased the percentage of animal foods in their diets from 56.0% during egg-follicle development to a high of 77.1% during egg laying, but animal matter declined to only 28.9% during the postlaying period (Krapu 1974a). Earthworms were the first invertebrates eaten by Northern Pintails breeding near shallow temporary wetlands, as the flooding of these previously dry areas moved considerable numbers of earthworms to the surface, where they were readily available. The availability of these and other invertebrate taxa was seen as a key factor governing the breeding response of Northern Pintails to the flooding of temporary wetlands: reduced breeding following drought "is understandable as the major foods of egg-producing hens are not available" (Krapu 1974a:286).

The diets of 8 flightless young Northern Pintails in North Dakota (1969–70) averaged 81.1% animal matter and 18.9% plant matter (Krapu and Swanson 1978). The most important animal foods were snails (48.9%), dipterans (flies and midges; 11.4%), and beetles (6.1%). In contrast, the diets of 15 flying young contained 57.1% animal matter and 42.9% plant matter. Dipterans again were the most significant animal food (39.0%); leeches (Hirudinea) and water fleas (Cladocera), made up 6.7% each in their diets. Preferred plant matter included widgeongrass seeds (26.2%) and pondweeds (*Potamogeton* spp.; 6.5%).

Sugden (1973) described the food habits and feeding ecology of Northern Pintail ducklings (*n* = 144) in Alberta (1963–67). During the first 5 days after hatching, the ducklings consumed surface invertebrates (73%), which they obtained by surface pecking. Ducklings 0–5 days old also ate aquatic invertebrates below the water's surface and on the substrate bottom (25%), but consumed only 2% plant material. The consumption of surface invertebrates declined rapidly for ducklings 6–15 days old (14%–19%), and was only 1%–4% from day 16 to fledging. The food composition in the diets of prefledging ducklings (0–41 or more days) was 4% surface invertebrates, 63% aquatic invertebrates, and 33% plant matter. Plant matter only made up 2%–9% between days 0 and 15, but it averaged 34% from day 16 to fledging. Significant invertebrate foods were snails (36% dry weight) and aquatic insects (26%), of which the most important group was midges (larvae and pupae; 16%). The feeding sites of Class I ducklings (1–18 days old) were open water (43%), emergent vegetation (32%), submerged plants (21%), and mudflats (5%), while Class II ducklings (19–43 days old) used open water (31%), emergent vegetation (32%), and submerged plants (37%). Class III ducklings (44–51 days old) used open water (48%), emergent vegetation (19%), and submerged plants (33%).

Migration and Winter. During spring migration in the Rainwater Basin in Nebraska (1998–99), the diets of 130 female Northern Pintails differed, depending on if the birds were collected while returning from field-feeding flights or using local wetlands (Pearse et al. 2011b). Corn dominated (84.1% dry mass) the diets of 93 birds collected on roosting wetlands after returning from field-feeding. Wetland feeders (*n* = 37) also consumed corn (54.1%), but in addition they ate the seeds of smartweed (*Polygonum* spp.; 22.5%) and millet (*Echinochloa*; 12.5%). Smartweed seeds were the most frequently (78%) chosen food, and corn less so (57%). Midge larvae were the most common (24%) animal food.

During the fall and winter, Northern Pintail diets are dominated by plant foods. On the Stikine River delta in southeastern Alaska, fall-migrating Northern Pintails (*n* = 31) collected from September through November 1976–78 fed on 71.8% plant material and 28.1% animal material. Preferred foods (by percentage of dry weight) were the seeds of sedges (*Carex* spp.; 57.6%), bivalve clams (Pelecypoda; 20.1%), and the seeds of marestail (*Hippuris* spp.; 5.5%) (Hughes and Young 1982). Stomach contents of birds (*n* = 37) collected during fall migration in Illinois (1978–81) along the Mississippi and Illinois Rivers contained 99.3% plant matter (Havera 1999). The most important foods were 5 species of smartweeds (32.3%), corn (19.3%), and brittle naid (*Najas minor*; 14.1%). On managed tidal impoundments in South Carolina, plant foods dominated the diet of Northern Pintails (*n* = 81) collected during the hunting seasons of 1972/73 and 1973/74 (Landers et al. 1976). The most important plant foods (by percentage of volume) were redroot (*Lachnanthes caroliniana*;

18.8%), dotted smartweed (*Polygonum punctatum*; 17.7%), fall panicgrass (*Panicum dichotomiflorum*; 13.3%), warty panicgrass (*Panicum verrucosum*; 11.6%), and redroot sedge (*Cyperus erythrorhizos*; 11/1%). In southern Louisiana, fall foods in the crops of Northern Pintails collected in 1961 consisted of 98.6% plant material and 1.4% animal material (Glasgow and Bardwell 1962). Seeds from freshwater grasses were the most important food items by volume (95.2%), particularly fall panicum (fall panicgrass; 25.2%), brownseed paspalum (*Paspalum plicatulum*; 21.9%), Walter's millet (*Echinochloa walteri*; 13.5%), and bagscale grass (American cupscale; *Sacciolepis striata*; 8.1%). In a tidal estuary on the Yucatán Peninsula of Mexico, their primary winter food was the tubercles of muskgrass (*Chara* sp.), which composed >99% of their diet (Thompson et al. 1988).

Wintering Northern Pintails commonly exploit agricultural foods such as rice and corn. During fall and winter 1991/92 and 1992/93 at the Lacassine NWR in southwestern Louisiana, radiomarked female Northern Pintails made regular evening flights to forage in rice fields; 96% of 205 flights occurred between a half hour before to an hour after sunset, with an average one-way distance of 8.7–24.4 km that lasted 16.4–32.9 minutes (Cox and Afton 1996). The associated transport cost for these feeding flights was 8%–20% of their energy intake from rice. The distance and duration of the flights increased as winter progressed, indicating that food resources were preferentially depleted at distances nearest the refuge.

In contrast, field-feeding flights to cornfields by wintering Northern Pintails in the Playa Lakes Region of Texas (Sep–Mar 1980–82) seldom exceeded 5 km, with 15 km being the longest recorded flight (Baldassarre and Bolen 1984). Northern Pintails underwent 2 field-feeding flights/day, in mixed flocks with American Green-winged Teal, Mallards, and American Wigeon. On average the morning flight was initiated 52 minutes before sunrise, although most (75%–90%) departed 44 minutes before sunrise. The last birds returned

from the morning flight at 16.5 minutes before sunrise, with the average duration of the flight being 23 minutes. The evening flight for most birds was initiated 25 minutes after sunset and, on average, was 37 minutes long. As in Louisiana, however, the flight duration increased as winter progressed.

Feeding-ecology studies of Northern Pintails from 3 major wintering areas—the Central Valley of California, the Texas Gulf Coast, and the Playa Lakes Region of Texas—are collectively significant, because they provide a description of their winter feeding ecology across a broad geographic area, as well as a comparison between inland freshwater habitats and coastal habitats. In the Sacramento Valley of California (Aug–Mar 1979–82), Northern Pintail diets on refuge habitats were ≥97% plant material from August through January, but invertebrate use increased to 28.9%–65.6% during February and March (*n* = 187; Miller 1987). Rice was the dominant food used while feeding off refuges from October through January (99.7%) and again in February and March (63%), but common barnyardgrass (*Echinochloa crus-galli*) also formed 31% of the diet in the latter period. Besides rice, the most significant plant foods were the seeds of common barnyardgrass, southern naiad (*Najas guadalupensis*), swamp timothy (*Heleochloa schoenoides*), and smartweeds, while important animal foods were snails (Gastropoda), and the larvae of midges (Diptera) and beetles (Coleoptera).

Miller and Newton (1999) used these data to estimate the daily caloric requirements needed by Northern Pintails in the Sacramento Valley between mid-August and mid-March, which was 794–1,180 kJ/day for males and 700–1,044 kJ/day for females. These were met by consuming 49–82 g of food/day from rice fields and wetlands, an amount equal to 5.9%–8.3% of male body mass and 6.0%–8.1% of female body mass. Extrapolating from these numbers, a large population of wintering Northern Pintails in the Sacramento Valley could consume 11.4 million kg of food from rice fields and 2.9 million kg from wetlands, the equivalent of 18.6% (41,500 ha) of the harvested rice fields and 9.0% of the wetlands in the Sacramento Valley. In the absence of commercially grown rice, management would need to provide about 10,000 ha of additional wetlands to satisfy the winter food demands of Northern Pintails.

Elsewhere in the Central Valley of California, Northern Pintail diets ($n = 262$ birds) on 4 seasonal marshes included 72.3% plant material and 27.7% animal material during an October through February study in 1979–82, but their food use distinctly shifted toward animal matter as winter progressed (Euliss and Harris 1987). The most important plant foods were swamp timothy (37%) and common barnyardgrass (12%), while midge (Chironomidae) larvae were the favored animal food (22%). The number of Northern Pintails was greatest on swamp timothy and alkali bulrush (*Scirpus paludosus*) marshes during the day, while croplands were usually preferred at night.

Along the Texas Gulf Coast, Northern Pintails winter in the Laguna Madre, a saline habitat that, in comparison with freshwater habitats, is characterized by a limited amount of aquatic plants. An early study there reported that wintering Northern Pintails ($n = 47$) subsisted almost exclusively on shoalgrass (*Diplanthera wrightii*), which formed 88.1% of their diet, with snails making up only 2.0% (McMahan 1970); inland, however, rice was a major food item, as well as other seeds. Ballard et al. (2004) conducted a later study along the lower Texas Gulf Coast, in response to declining freshwater habitats and rice production. Their study collected 253 Northern Pintails over 2 winters (1997/98 and 1998/99) and compared the composition and nutritional quality of those diets to the diets of birds wintering on inland freshwater habitats. While 11 plant taxa and 23 animal taxa were consumed, 5 food items made up >70% of what they ate: the rhizomes and foliage of shoalgrass (*Halodule wrightii*); the seeds of widgeongrass (*Ruppia maritima*); amphipods (*Gammarus* spp.); and dwarf surf clams (*Mulinia lateralis*). The nutritional composition of this diet was 7.0%–17.8% protein, 11.1%–42.9% carbohydrates,

6.2%–23.8% fiber, 0.6%–2.2% fat, and 21.6%–74.8% ash. The true metabolizable energy content of this diet varied between and within winters, ranging from 1.86 kJ/g for males during late winter in a dry year to 5.02 kJ/g for females during early winter in a wet year. Hence, to maintain their body mass, Northern Pintails needed to ingest 226–527 g (dry mass) of food/day during a dry winter, compared with 156–288 g/day during a wet winter—a nearly 3-fold difference. In contrast, wintering Northern Pintails in freshwater habitats required only 83–148 g/day when feeding on rice and <94 g/day when feeding on waste corn. Thus reductions in rice acreage along the Texas Gulf Coast may adversely affect Northern Pintails in this important wintering area by making them rely more on saline habitats for food.

Wintering Northern Pintails in the Playa Lakes Region of Texas extensively ate waste corn during the winter (Baldassarre and Bolen 1984). Food-habit studies (1984–86) revealed that although corn made up 90.1% of the diet of Northern Pintails killed by hunters, since the birds were returning from field-feeding flights when they were shot, the diets of Northern Pintails observed feeding on playa lakes included just 27.1% corn, 49.6% nonagricultural seeds, and 20.5% animal foods (Sheeley and Smith 1989). The seeds of smartweeds (11.0%), barnyardgrass (10.4%), and dock (*Rumex* spp.; 9.8%) were the primary plant foods, while gastropods (12.3%) and dipterans (6.2%) were significant animal foods.

MOLTS AND PLUMAGES

Palmer (1976a), Bellrose (1980), and Austin and Miller (1995) have described the molts and plumages of Northern Pintails; they should be consulted for further details. Additional information on molts and the associated ecology in fall and winter is available for Northern Pintails in the Sacramento Valley of California (Miller 1986a) and in the Playa Lakes Region of Texas (Smith and Sheeley 1993b).

Acquisition of juvenal plumage begins in week 3, with the primaries beginning to emerge by week

4, and the wing speculum visible in week 5. Fledging of wild birds in South Dakota occurred in 46–57 days (Gollop and Marshall 1954); in Alaska it reportedly was 36 days at Yukon Flats and 43 days at Minto Flats (Bellrose 1980). Northern Pintails in juvenal plumage are brown to buffy overall, but the upperparts of the males are darker and less well marked than in the females, and their speculums are different. Juvenal plumage is quickly followed by the acquisition of first basic, which is usually completed by late August or September, but the juvenal wing is retained. The sexes are similar in first basic. The head is buffy and streaked; the mantle, the rump, and the sides grayish, flecked with white or buff. The underparts are whitish, flecked with dark brown.

First basic is rapidly succeeded by first alternate, which, in both sexes, includes all the feathering except the juvenal wing. First alternate is acquired from September until late fall or early winter (Jan). In California, the number of feather groups molting in males peaked in October and November and declined in December. There was no molt from January through March. For females this molt peaked in September and October, declined during December and January, and then peaked once more with the prebasic molt in February. Both sexes acquire first alternate earlier when fall rainfall is above average. First alternate in both sexes resembles definitive alternate, although the scapulars and the central tail feathers of males are not as elongated as in definitive alternate.

For males, the prealternate molt leading to definitive alternate occurred from September through December in California, peaking in October with 30%–40% new feathers. The molt declined rapidly thereafter, with an average of <1 feather group showing pin feathers in December and January; only the scapulars and the tail feathers were molted in December. In the Playa Lakes Region of Texas, the prealternate molt of males, as well as adult females, also peaked in October. Not much molting occurred in December, and molting was negligible from January through March.

During years of above-average rainfall in both California and Texas, when more playas become flooded, the availability of noncultivated foods in playa lakes probably facilitate an earlier molt in Northern Pintails.

Females acquire definitive basic in the winter (Jan–Mar); the remiges are not replaced until after breeding. The associated prebasic molt peaked during February in California, but not until March for birds on the southern High Plains in Texas. In California, 9 radiomarked adult females were sedentary for about 36 days, with a flightless period of around 25 days. Daily survival during the 36-day period was 99.3% (Miller et al. 1992). Females molted near brood-rearing marshes, but some migrated to staging areas for the wing molt. Males acquire definitive basic during the summer (Jun–Jul) and undergo a flightless period of 27–30 days (Oring 1964, Cox 1993). Females are brownish gray in definitive basic; males resemble the females but are more brownish.

Males acquire definitive alternate immediately after definitive basic, but the accompanying molt does not include the remiges. This molt is usually of low intensity on the breeding grounds but accelerates during fall migration and into early winter. The intensity of the prealternate molt peaked in October in both Texas and California and was complete by January. The central retrices can reach 22 cm in length. Females acquire definitive alternate in late summer to early fall, but this plumage is not completed until early winter. Females in definitive alternate are grayer than those in definitive basic.

CONSERVATION AND MANAGEMENT

In their comprehensive review of the decline in the Northern Pintail population, Miller and Duncan (1999) identified recruitment as the major factor precipitating the decline. Northern Pintail age ratios in the harvest, calculated at 10-year intervals from 1961 to 1997, ranged from a high of 1.26 young/adult in the 1960s to a low of 0.93 in the 1980s, the lowest among 6 species of dabbling ducks (except for Mallards in the 1990s). The 2005–9 average was only 1.08. The nest success of all upland-nesting ducks decreases as the percentage of croplands increases on the landscape (Greenwood et al. 1995), and Miller and Duncan (1999:796) implicated the loss of nesting cover since the 1950s and 1970s as another significant factor in "reducing Pintail productivity to levels below the threshold needed to maintain populations." In addition, Northern Pintails nest in grain stubble, more so than any other species of duck, and early cultivation destroys virtually all of these nests. Miller and Duncan (1999:796) noted that "the stubble-nesting characteristic probably is most responsible for the failure of Pintails to recover." Lastly, because Northern Pintails are not persistent renesters, they are unlikely to recover from large-scale losses of their first nest attempts. Krapu et al. (2002) also noted that the conversion of grasslands to crops on primary prairie breeding grounds has reduced the hatching rates of clutches laid early in the nesting season, which, when added to their limited capacity to renest in late spring, probably has contributed to reduced Northern Pintail populations.

Northern Pintail populations continued to decline, despite years with record numbers of May ponds on the prairies in the 1990s, and they have not yet recovered. Key management recommendations by Miller and Duncan (1999) were (1) protecting existing mixed-grass prairie habitat; (2) restoring vast areas of grasslands in the Canadian prairies, similar to what was achieved by the Conservation Reserve Program in the United States; (3) encouraging cattle ranching in mixed-grass prairie habitats, but in a manner that provides productive nesting habitat; and (4) encouraging cultivation practices that minimize spring tillage (e.g., no-till and fall-seeded crops). Predator management could also play a role, if habitat approaches are not successful in reversing the population decline.

In 1997, 3 outbreaks of avian botulism probably killed more than 350,000 Northern Pintails in North America. Miller and Duncan (1999) acknowledged the potential for disease to adversely affect Northern Pintail populations, and they called for estimates of the numbers that are annually exposed to lakes harboring perennial botulism. Friend et al. (2001) hypothesized that the continued suppression of Northern Pintail populations was not surprising, because these ducks are often the dominant species affected in major epizootics, not only of avian botulism, but also of avian cholera. Austin and Miller (1995) also noted the importance of botulism and avian cholera as mortality factors affecting Northern Pintail populations. Mora et al. (1987) assessed levels of organochlorines in 86 Northern Pintails collected in California and Mexico, but they found that the levels were not high enough then to adversely affect reproduction or survival.

In sum, the decline of the Northern Pintail population in North America since the 1970s is unquestionable, and it is the most severe among all species of dabbling ducks. This decline—which is steep, long term, and shows little sign of improvement—has occurred despite the fact that Northern Pintails occupy a vast breeding and wintering range, have a diverse and adaptable diet, and have comparatively high survival rates. Adult mortality did not increase during the years of population decline; in fact, hunting losses actually decreased. Therefore, the evidence implies that over the long term, diminishing annual recruitment, which has resulted in a reduction in the fall population size, has lowered the North American breeding population of Northern Pintails.

American Green-winged Teal
Anas carolinensis (Linnaeus 1758)

Left, drake; *right*, hen

The American Green-winged Teal is the smallest dabbling duck in North America. Males are readily identified by their small size, chestnut head, and grayish sides, while females are a grayish brown overall. Although the American Green-winged Teal is closely related to the Eurasian Green-winged Teal (*Anas crecca*) that occurs across Eurasia, the 2 species are separated by significant genetic variation and subtle plumage differences. The Aleutian Green-winged Teal (*A. c. nimia*), resident on the Aleutian Islands, is an insular subspecies recognized by some authorities. American Green-winged Teal have one of the widest breeding distributions of all North American waterfowl. Unlike other puddle ducks, however, their principal breeding range is not in the grasslands of the Prairie Pothole Region (PPR), but farther north in the parklands, boreal forest, and tundra river deltas across most of Alaska and Canada. Their wintering range is also extensive, from coastal southeastern Alaska and British Columbia south through Baja California, east across a major portion of the United States, then south through all of Mexico to the Guatemalan border, and across to the Caribbean islands. The small bill of American Green-winged Teal has very densely packed lamellae, which allows them

American Green-winged Teal. *Left*, hen; *right*, drake

to select small seeds and invertebrates that are often obtained in shallow (0–10 cm) water or on mudflats. Groups of males usually display to 1 female, in a social courtship that is vigorous and thus highly visible to observers during the winter and early spring. While the nonbreeding ecology of American Green-winged Teal is well known, there is no major study of their breeding ecology. Their nests are widely dispersed and, among all the puddle ducks, are some of the best concealed, often being placed in woodlands and shrublands. They are an abundant duck in North America; the 2010 estimate of 3.5 million breeding birds was 78% above the long-term average of 1.9 million. They are also the second-most-harvested duck in the United States, after the Mallard.

IDENTIFICATION

At a Glance. The smallest dabbling duck in North America, American Green-winged Teal are about the size of a Rock (Domestic) Pigeon (*Columba livia*). The chestnut head, with a dark green ear patch, and the black bill readily identify the male. The female is similar to other female teal, but has a smaller bill and white undertail coverts. The belly is white in both sexes. Both sexes also lack blue on the upperwing, a feature that is striking in both Cinnamon Teal and Blue-winged Teal.

Adult Males. Males are among the most colorful of the dabbling ducks. At rest they are identified by the chestnut-red head, with a dark green crescent sweeping backward from the eye. A conspicuous vertical white stripe extends from the breast to the shoulder and, at a distance, is the most noticeable field mark. The chest is pinkish brown, with black spots. The back, the sides, and the flanks are a vermiculated gray, separated from the chest by the vertical white bar. The black rump contrasts with the undertail coverts, which are buff on the outside and black in the center. The uppertail coverts are black. The green speculum is usually visible, but at times it is covered by the flank feathers. The legs and the feet are gray. The bill is black. The irises are dark brown.

Adult Females. Females are grayish brown overall, which is a similar coloration to that of other species of female teal. The sides of the head and the throat are paler than the crown. The back and the rump are slightly darker and less mottled than the breast, the sides, and the flanks. Females have a light brown eye ring, a dark eye line extending to the back of the head, and (often) a white spot behind the bill. The legs and the feet are a light gray. Black speckles on the bluish-black bill distinguish females from males in all plumages. The irises are dark brown.

Juveniles. Juveniles resemble adult females.

Ducklings. Ducklings have a very dark olive-brown color on the back and on parts of the sides, and yellow on the belly and the head (Nelson 1993). The brown eye stripe and cheek stripe are usually prominent, with the cheek patch sometimes large enough to form an ear patch. The legs and the feet are blackish.

Voice. Males give a shrill clear whistle, not unlike that of a spring peeper (*Pseudacris crucifer*; Palmer

Male American
Green-winged Teal.
GaryKramer.net

1976a). Females typically *quack*, with a decrescendo call that is 4–7 repeated *quacks* of decreasing volume. Other calls are given in association with courtship displays.

Similar Species. Female American Green-winged Teal are distinguished from female Cinnamon Teal and female Blue-winged Teal by their smaller overall size, smaller bill, and darker and rounded head. American Green-winged Teal also do not have the blue upperwing patches characteristic of other teal.

Weights and Measurements. Bellrose (1980) reported average body weights of 322 g for adult males ($n = 113$), 308 g for adult females ($n = 79$), 327 g for immature males ($n = 332$), and 290 g for immature females ($n = 265$). Average wing length was 18.5 cm for adult males ($n = 86$), 17.8 cm for adult females ($n = 51$), 18.3 cm for immature males ($n = 66$), and 17.5 cm for immature females ($n = 71$). Palmer (1976a) reported an average bill length of 38.2 mm for males ($n = 12$) and 36.5 mm for 11 females. During the fall and winter on the southern High Plains of Texas (Sep–Mar 1980–82), body weights averaged 350 g (range = 250–450 g) for adult males ($n = 1,538$), 326 g (range = 235–405 g) for adult females ($n = 305$), 332 g (range = 215–430 g) for juvenile males ($n = 1,455$), and 305 g (range = 210–400 g) for juvenile females ($n = 857$; Baldassarre and Bolen 1986). Birds attained their maximum body weight in December, just prior to weather-induced periods of caloric shortages. Measurements of adult males ($n = 108$) collected during this study averaged 361 g in body weight, 188 mm in wing length, 42 mm in bill length, and 53 mm in tarsus length (Guy Baldassarre). Adult females ($n = 50$) averaged 326 g in body weight, 178 mm in wing length, 36 mm in bill length, and 51 mm in tarsus length. Hatching-year males ($n = 30$) averaged 339 g in weight, 184 mm in wing length, 42 mm in bill length, and 53 mm in tarsus length. Hatching-year females ($n = 32$) averaged 320 g in weight, 177 mm in wing length, 37 mm in bill length, and 48 mm in tarsus length. Second-year males ($n = 48$) averaged 309 g in weight, 183 mm in wing length, 41 mm in bill length, and 53 mm in tarsus length, while second-year females ($n = 55$) averaged 310 g in weight, 175 mm in wing length, 34 mm in bill length, and 48 mm in tarsus length.

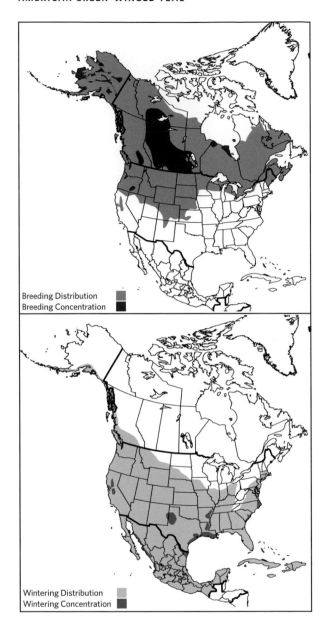

Breeding Distribution
Breeding Concentration

Wintering Distribution
Wintering Concentration

DISTRIBUTION

The American Ornithologists' Union (1998) uses *Anas crecca* as the scientific name for the Green-winged Teal, within which they recognize 2 subspecies: the American Green-winged Teal (*A. c. carolinensis*) and the Eurasian Green-winged Teal (*A. c. crecca*). Johnsgard (1978) recognized a third subspecies, the Aleutian Green-winged Teal (*A. c. nimia*). Livezey (1997), however, recognized 2 distinct species: the American Green-winged

Teal (*Anas carolinensis*) and the Eurasian Green-winged Teal (*Anas crecca*), with the latter including *nimia*; he considered *A. crecca* to be a superspecies (a biogeographical term used when the forms involved are basically distinct and have adjacent ranges that do not overlap significantly).

A molecular analysis by Johnson and Sorenson (1999) revealed an extremely high (5.8%) genetic divergence between Eurasian Green-winged Teal and American Green-winged Teal, despite their nearly identical plumages, a divergence similar to the 5.7% genetic distance between Mallards and Northern Pintails. Hence these authors also recognized 2 separate species of Green-winged Teal (*A. crecca* and *A. carolinensis*), which agreed with the morphologically based conclusion of Livezey (1991). In North America, the genetic diversity of American Green-winged Teal is apparently high, with birds examined from the southern High Plains of Texas having an average heterozygosity (different alleles [DNA codings for specific genetic traits] at 1 or more loci [particular positions on a specific chromosome]) of 0.11 and an average number of alleles/locus of 2.48 (Rhodes et al. 1991).

The Eurasian Green-winged Teal is a rare visitor to eastern North America. Their native breeding range is across Eurasia, from Iceland and the United Kingdom through Europe and Asia north to 70° N latitude, and from the tundra through the steppes and south to the Mediterranean Sea (Johnsgard 1978, Fox 2005). Their wintering range extends from the southern portion of their breeding range to northern and central Africa and across Asia and India to southern Japan. The Aleutian Green-winged Teal is resident on the Aleutian Islands, from their western tip east to Akutan Island, where it is the most abundant dabbling duck (Gabrielson and Lincoln 1959, Murie 1959). Gabrielson and Lincoln also stated that Frank Beals found these teal to be common to plentiful at all seasons of the year, from Amchitka Island to Unimak Island. Support for Aleutian Green-winged Teal being true residents came from exten-

American Green-winged Teal pair. The female is stretching—a typical comfort movement. *GaryKramer.net*

sive surveys on the northcentral Alaska Peninsula (1976–80), which did not find an Aleutian Green-winged Teal among the American Green-winged Teal (Gill et al. 1981).

Male Eurasian Green-winged Teal lack the vertical white line seen on the flank of the American Green-winged Teal; instead they have a horizontal white stripe along the scapulars, between the gray flanks and the back. They also have a cream line extending in a half-circle from the bill, above the green eye swatch, to the back of the head. Aleutian Green-winged Teal resemble Eurasian Green-winged Teal, but are slightly larger.

Breeding. American Green-winged Teal have one of the most extensive breeding ranges of all North American waterfowl, with a distribution generally similar to that of American Wigeon. Unlike most puddle ducks, however, their principal breeding range is not the PPR, but rather the boreal forest and the deciduous parklands farther north. This breeding range extends across Alaska and almost all of Canada, south through Washington, Oregon, and northern California, east through the Inter-

mountain Region (including part of northcentral Arizona) and the PPR to Minnesota, northern Wisconsin, and northern Michigan, then along the St. Lawrence River valley, and south through much of New England to Maryland and Virginia (Johnson 1995).

The principal breeding range of American Green-winged Teal encompasses the boreal forest, taiga, and river deltas in northwestern Canada and Alaska. Although American Green-winged Teal are small, they are quite cold tolerant and (except for Northern Pintails and American Wigeon) are the most abundant dabbling duck on arctic breeding grounds. The Traditional Survey recorded 3.5 million American Green-winged Teal in 2010, with 42.1% in the Northwest Territories, British Columbia, and northern Alberta, and 27.4% in Alaska/ Yukon. In Alaska, large numbers inhabit the survey strata associated with the coastal deltas of western Alaska (Bristol Bay, the Yukon-Kuskokwim delta, the Seward Peninsula, and Kotzebue). In 2009, 55.3% of 649,600 American Green-winged Teal in Alaska occurred in these areas, 59% of which were on the Yukon-Kuskokwim (Y-K) delta. Yukon

Male American
Green-winged Teal
performing a *down-up*
courtship display.
Ian Gereg

Flats is a significant breeding area in the interior of Alaska, as is the Copper River delta on the southeastern coast. Important areas in Canada are the Mackenzie delta in the Northwest Territories, Old Crow Flats in the Yukon, the Athabasca River delta in Alberta, and the Saskatchewan River delta in Manitoba.

In contrast to these more northern areas, a much smaller percentage of the breeding population occurs to the south. Only 16.6% were in the PPR in southern Alberta, Saskatchewan, and Manitoba; 9.7% in the eastern Dakotas; and just 3.0% in northern Saskatchewan, Manitoba, and Ontario (U.S. Fish and Wildlife Service 2010). From 1955 to 2010, the distribution of American Green-winged Teal by region in the Traditional Survey was 39.3% in the Northwest Territories, British Columbia, and northern Alberta; 20.3% in Alaska/Yukon ; 12.9% in southern Saskatchewan; 9.6% in southern Alberta; 5.1% in northern Saskatchewan; 4.0% in

northern Manitoba; 2.6% in southern Manitoba 2.1% in North Dakota; 1.6% in South Dakota 1.2% in Montana; and 1.3% in western Ontario (Collins and Trost 2010). In other western states, breeding populations estimated from 1990 to 2010 averaged 3,440 in Washington, 5,876 in Oregon, 3,986 in California, and 238 in Nevada (Collins and Trost 2010).

The Atlantic Flyway Breeding Waterfowl Plot survey averaged 56,599 American Green-winged Teal from 2003 to 2010, and 11,878 breeding pairs. Breeding-pair numbers were highest in New York (42.9%), and 17.8% were in Pennsylvania (Klimstra and Padding 2010). The Eastern Waterfowl Survey, which covers parts of Ontario and all of Québec, Labrador, Newfoundland, the Maritime Provinces, Maine, and a small part of New York, averaged 241,000 American Green-winged Teal from 1990 to 2009. The 2010 total was 256,000, 6% above the long-term average (U.S. Fish and Wildlife Service

Male American
Green-winged Teal
in flight.
RyanAskren.com

2010). Most (>50%) of these were tallied in Québec, where the 2003–7 average was 151,000 (Lepage and Bordage 2010).

Winter. The wintering range of American Green-winged Teal is very widespread, extending from coastal southeastern Alaska, British Columbia, and the Pacific coastal states south through Baja California, and moving east through much of the United States. The northern limits of their winter range go through Montana, eastern South Dakota and eastern Iowa, southern Illinois to Ohio and Virginia, and then north along the Atlantic coast to southern New England (Johnson 1995). They winter in all states to the south of this demarcation, as well as south through nearly all of Mexico to the Guatemalan border (Howell and Webb 1995). They are common in Cuba, fairly common in Hispaniola, less so on other islands in the Greater Antilles, and vagrant in the Lesser Antilles (Evans 1990, Raffaele et al. 2003). They occupy areas where open

water is available, and move south as these water areas freeze over, but otherwise they are very cold tolerant.

The Midwinter Survey in the United States averaged 1.9 million American Green-winged Teal from 2000 to 2010: 43.9% in the Mississippi Flyway, 26.6% in the Central Flyway, 24.6% in the Pacific Flyway, and only 4.8% in the Atlantic Flyway. In the Mississippi Flyway, 90.1% were in Louisiana, mostly along the coast and in the Rice Belt. In the Pacific Flyway, 84.1% were in California (the Central Valley) and 11.9% in Oregon, while nearly all (96.4%) Central Flyway birds were in Texas, largely along the coast and on the playa lakes in the Texas Panhandle. In the Atlantic Flyway they primarily (54.2%) were found in coastal North Carolina, with 12.3% in Florida. Overall, 39.6% of all American Green-winged Teal occurred in Louisiana (coastal area and the Rice Belt), and 25.7% in Texas (coastal area and the Playa Lakes Region). The Christmas Bird Count averaged 67,009 American Green-winged Teal in the United States from 2000/2001 to 2009/10.

Mexico is also an important wintering area for American Green-winged Teal. Most occupy wetlands along the mainland west coast, particularly in the extensive coastal lagoons in the state of Sinaloa (Pabellón, Topolobampo, and Bahía Santa María). Surveys conducted at 3-year intervals from 1982 to 2006 on this coast averaged 119,000 birds (Conant and King 2006). The most recent totals have been smaller, however, at <56,000 during the 1997, 2000, 2003, and 2006 surveys. In the interior highlands, the number of American Green-winged Teal recorded on 8 surveys conducted at 3-year intervals between 1985 and 2006 averaged 82,480, with most in the central highlands: 45,105 in 2006, and 39,842 in 2003 (Thorpe et al. 2006). Lago de Chapala in Jalisco is a major wintering locale in the central highlands, but American Green-winged Teal have also occurred in most of the 23 areas surveyed. In contrast, only 18,018 were observed in the northern highlands in 2006, and 6,614 in 2003, with Laguna de Babícora in Chihuahua being an especially important site. On the east coast of Mexico, an average of only 21,387 American Green-winged Teal were counted during surveys conducted from 1978 to 2006, with most (85.9%) found in the state of Tamaulipas, from the Tamesí and Pánuco River deltas (Tampico Lagoons) north to the Rio Grande delta.

Migration. An analysis of band-recovery data through March 2001 revealed that about 75% of the American Green-winged Teal from Alaskan breeding grounds migrated along the coastal corridor to wintering areas in British Columbia and the Pacific coastal states (Frank Bellrose). The remaining 25% migrated through the plains of Alberta and the United States, where most were divided equally between Texas and Louisiana. In contrast, birds from the Northwest Territories migrated more to the Central Flyway (42%) and the Mississippi Flyway (28%) than the Pacific Flyway (22%). American Green-winged Teal from Alberta migrated to the Central Flyway (35%), the Pacific coastal states (31%), and the Mississippi Flyway (24%), with 10% in the Rocky Mountain states. Saskatchewan birds largely migrated down the Central Flyway (42%) to Texas, and the Mississippi Flyway to Louisiana (34%), but 18% migrated to California via Utah. From Manitoba, about 60% migrated to the Mississippi Flyway (primarily Louisiana) via the Mississippi River corridor, and 35% to Texas in the Central Flyway. To the east, American Green-winged Teal banded in Ontario migrated to the Mississippi Flyway (66%) and Atlantic Flyway (33%). Slightly >90% of those banded in Québec and the Maritime Provinces remained in the Atlantic Flyway.

The fall migration in most parts of the continent is a protracted affair, beginning by early September and continuing until freezeup, with peak numbers in the northern states occurring from early to mid-October (Bellrose 1980). Birds arrive in the central regions by late September; some birds reach southern areas as early as late September, but the larger influxes begin during the first 2 weeks of November, with a peak in the first 2 weeks of December. Gabrielson and Lincoln (1959) reported that the birds departed in September from breeding grounds in northern Alaska, but Quinlan and Baldassarre (1984) noted that they had arrived on the southern High Plains of Texas by mid-September. They reach coastal southwestern Louisiana in early October (Rave and Baldassarre 1989, Johnson and Rohwer 1998). In British Columbia, migration is initiated in September, but peak staging numbers do not occur until late October into November (Butler and Campbell 1987).

American Green-winged Teal begin to depart their wintering areas by early February for regions immediately to the north (Bellrose 1980). They generally leave the southern High Plains in the Texas Panhandle between 20 February and 8 March (Baldassarre et al. 1988), although Rave and Baldassarre (1991) collected American Green-winged Teal in southwestern Louisiana between 25 February and 31 March. On the Atchafalaya River delta in Louisiana, few American Green-winged Teal remained by late March (Johnson and Rohwer 1998). As in the fall, spring migration is protracted, with departures continuing well into April. Arrival to the middle parts of the United States occurs in early March, with arrival to the southernmost breeding areas beginning in early April and increasing steadily until mid-May (Bellrose 1980).

American Green-winged Teal rapidly migrate through British Columbia during March and April (Butler and Campbell 1987). The earliest arrivals at Delta Marsh in Manitoba over a 15-year period occurred between 5 and 12 April (5 years), 13 and 20 April (7 years), and 21 and 28 April (3 years; Hochbaum 1955). They were observed near Yellowknife, Northwest Territories, between 25 April and 4 May, with a major influx between 5 and 15 May (Murdy 1964). In Alaska, arrival on the Y-K delta occurred between 12 and 23 May 1969–72 (Mickelson 1973).

MIGRATION BEHAVIOR

American Green-winged Teal migrate in large flocks, mostly at night (Bellrose 1980). Unlike the comparatively few but large staging areas used by diving ducks, however, American Green-winged Teal have literally hundreds of areas that provide suitable habitat during migration. Both banding and marking data from American Green-winged Teal captured during the fall and winter on playa lakes on the southern High Plains of Texas (1980–82) indicate that this area is an important staging area as well as a major wintering site. Observations of 1,343 American Green-winged Teal marked with individually numbered patagial tags revealed that the birds were mobile at this time; resightings averaged only 12% from mid-November to mid-March. Resighting data also revealed a distinct passage of females through the area before December: just 10% of those marked before 1 December were resighted, compared with 27% of the females marked after that date. Birds that remained on the study area did not exhibit any differential mobility between the sexes in response to cold weather,

even during a period when temperatures averaged 13°C below the 40-year mean. These data showed that birds accumulating sufficient lipid reserves remained on the southern High Plains for the winter.

Recovery data from 4,200 banded birds indicated that other American Green-winged Teal probably passed through the southern High Plains on their way to wintering areas along the Gulf Coast of Texas and Louisiana or interior Mexico (Baldassarre et al. 1988). Indirect recoveries ($n = 104$), however, indicated that juveniles, especially juvenile males, were more likely to be found away from the study area in subsequent years, almost always west of 110° w longitude and at an incidence 4.6 times that of adult birds. Omitting recaptures at the banding site, 12 (32.4%) of the indirect recoveries of juvenile males occurred west of about 100° w longitude (11 in California and 1 in Nevada), which strongly indicated young males used a different wintering area in subsequent years after their initial banding in Texas. Young males are most likely to be unsuccessful at attaining mates (see Mating System), and thus may disperse westward from the breeding areas and follow a different migration corridor that leads to California.

Molt Migration. After deserting their females, male American Green-winged Teal move to large marshes or lakes prior to the wing molt. At the Camas National Wildlife Refuge (NWR) in southeastern Idaho, the first males deserted their females on 12 June, the molt began on 3 July, and the first flightless males were observed on 11 July (Oring 1964). At Delta Marsh in Manitoba, Hochbaum (1944) noted that molting males from considerable distances away arrived between mid-June and early July. Some females arrived later, because more flightless females were observed in August than during the breeding season. Other large marshes or lakes used by molting American Green-winged Teal are the Pel-Kutawagan Marshes in south-central Saskatchewan; Niska Lake in westcentral Saskatchewan; and Macallum, Gypsy, and Canoe Lakes, which are located north of the parklands between the Athabasca River and the Saskatchewan delta (Bellrose 1980). Some American Green-winged Teal migrate south to molt; Peterjohn (1989) observed flocks of 20–100 flightless birds during July in Ohio, a considerable distance south of their major breeding grounds.

HABITAT

A general habitat description for breeding American Green-winged Teal is shallow lakes, marshes, and ponds with emergent vegetation (American Ornithologists' Union 1998). Breeding birds reach high densities in wooded wetlands within the parklands, the boreal forest, and northern river deltas. In general, they prefer the wooded wetlands of the parklands over prairie potholes (Moisan 1967). In British Columbia, Paquette and Ankney (1996) noted that wetland use by breeding birds was positively correlated with water-chemistry measurements, wetland size, the percentage of emergent cover, and areas measuring 0–1 m deep, which collectively indicated that wetland fertility was a significant factor affecting breeding distribution at both a landscape and a geographic level. At Delta Marsh in Manitoba, the use of experimentally constructed wetlands over a 10-year period (1980–89) revealed spring and summer use of densely vegetated places with midrange water depths, compared with much shallower areas in the fall (Murkin et al. 1997). The shift in habitat use by American Green-winged Teal during the fall was unique among the duck species studied and probably occurred because their small size, feeding methods, and preference for seeds limit them to shallow-water habitats (see Food Habits and Feeding Ecology).

In South Dakota, pairs ($n = 35$) made heavy use of dugouts (38.2%) and stock ponds (27.3%); in natural wetlands, they often chose semipermanent potholes (35.1%; Ruwaldt et al. 1979). In southcentral North Dakota, 64.0% of the pairs ($n = 214$) occurred on seasonal wetlands, and 22.0% on semipermanent wetlands (Kantrud and Stewart 1977). The density of breeding pairs was 13.1/km²

on seasonal wetlands, 10.5/km² on seasonal-type wetlands tilled for agriculture, and 6.0/km² on temporary wetlands. In comparisons with 6 other puddle duck species, summer habitat overlap in the PPR in North Dakota was greatest (74.2%–80.3%) between American Green-winged Teal and Blue-winged Teal, and least (36.9%–44.5%) with Gadwalls and American Wigeon (DuBowy 1988). Winter habitat overlap in California was highest (44.1%–52.7%) with Mallards, but much lower (16.1%–33.0%) with the 5 other species.

Wintering American Green-winged Teal use habitat complexes, preferring areas of both fresh and brackish wetlands (marshes, shallow lakes, estuaries, and flooded fields), where they are commonly seen foraging in shallow water. During the winter in the San Joaquin Valley of California, the densities of American Green-winged Teal and Northern Pintails were greatest on shallow wetlands (Colwell and Taft 2000). On managed impoundments at the Kern NWR in the Central Valley of California, daytime habitat use by wintering American Green-winged Teal primarily occurred on impoundments dominated by swamp timothy (*Heleochloa schoenoides*) and alkali bulrush (*Scirpus paludosus*), except during February, October, and November 1980, when the birds were proportionately distributed among cropland impoundments (Euliss and Harris 1987). At night, ponds dominated by swamp timothy received >60% use on a monthly basis, and ponds dominated by barnyardgrass (*Echinochloa crus-galli*) received 33%. No teal were ever flushed from the open-water areas of the impoundments at night.

During the fall and winter in southwestern Louisiana (1985–86), wintering American Green-winged Teal were found in 6 major habitat types, including tidal mudflat areas where they foraged for macroinvertebrates (Rave and Baldassarre 1989). They did not consume rice, which was not in the vicinity of the study area, or other agricultural foods. A time-budget analysis revealed that the most important activities across all habitats were resting (45.4%), feeding (33.3%), preening (11.4%), and locomotion (8.6%), but activity patterns varied among habitat types in response to water levels and the habitat structure. Feeding time was lowest (17%) in brackish impounded marshes, moderate (35%) in fresh unimpounded marshes, and highest (41%) in intermediate impounded marshes, where food was more abundant and readily available. Resting averaged 35%–40% of their time budgets in intermediate impounded marshes, brackish tidal flats, and fresh unimpounded marshes, but 52%–62% in other habitats, where structural features, such as clumps of vegetation and small islands, provided them with sites for loafing, preening, and sleeping. The functional role of each habitat remained relatively constant over the winter period, underlining the importance of wetland complexes to wintering American Green-winged Teal. An exception occurred in late October 1985, however, when 40 cm of rain fell on the study area. The resultant high water levels meant food was unavailable in the managed impoundments; the teal moved 25 km away, where they found food in the newly flooded pastures.

Habitat and Body Condition

American Green-winged Teal ($n = 516$) wintering in the Playa Lakes Region in the southern High Plains of Texas (1980–82) arrived in September and steadily increased their body mass through December: on average, by 5% for adult males, 12% for adult females, 10% for juvenile males, and 6% for juvenile females (Baldassarre et al. 1986). Body mass then decreased, with the lowest levels occurring in February: on average, a loss of 14% for adult males, 16% for adult females, 11% for juvenile males, and 14% for juvenile females. Body mass increased again, by 3%–10%, from February to March. Most of these changes reflected a greater lipid mass, which increased most from September to December: 67% for adult males, 150%–219% for adult females, 161% for juvenile males, and 73% for juvenile females. Midwinter lipid levels were 21.7%–23.4% of body mass. Caloric requirements for American Green-winged Teal in January were about 414–439 kJ/day, which the birds met by field-feeding twice daily in nearby cornfields, where they obtained about 15 g of corn/flight (30 g/day), yielding a true metabolizable intake of about 502 kJ/day.

The pattern of lipid and body-mass cycles in American Green-winged Teal reflected habitat conditions encountered on the southern High Plains from their arrival in September, when the weather was still warm, to December through February, when 85%–96% of their time was spent below their lower critical temperature; they departed in late February and early March. The authors reasoned that storing the maximum caloric reserves just before the period when nutrient shortages were most likely to be encountered is an adaptive strategy, because these reserves could then be progressively used throughout the winter rather than maintained at high levels, which would have incurred many foraging and metabolic costs. It is also adaptive to use reserves in this fashion, because the probability of encountering a nutrient shortage decreases as winter advances and spring approaches. Hence the decline in lipid reserves (and body mass) and the associated adjustments to foraging strategy for American Green-winged Teal during the winter probably reflected an adaptation to, rather than a consequence of, winter conditions in Texas.

In contrast to the situation in Texas, the body mass and lipid levels of American Green-winged Teal wintering in Louisiana (1985–85; $n = 227$) did not change over the winter (Rave and Baldassarre 1991). Moreover, the winter lipid levels of birds in Louisiana, which fed solely on noncultivated foods, were 26%–50% lower than those in Texas, which relied heavily on waste corn. American Green-winged Teal wintering in Louisiana were exposed to temperatures averaging 5.6°C–6.3°C warmer than those experienced by birds in Texas; the latter also encountered more severe weather conditions, such as snowfall, ice, and high winds, that demanded extra caloric reserves. Thus winter severity probably influences the different patterns of lipid cycles between these 2 major wintering locations for American Green-winged Teal. In addition, dietary differences between the sites also affect lipid patterns. Teal feeding on carbohydrate-rich waste corn in Texas can store more lipids than birds using a higher-protein diet of noncultivated foods in Louisiana, which may be a key proximate factor in their being able to winter in more northerly and harsher environments, such as the southern High Plains of Texas.

POPULATION STATUS

Within the Traditional Survey area, there was a general long-term upward trend in the average size of the American Green-winged Teal breeding population from 1955 to 2010: 1.41 million in the 1960s, 1.86 million in the 1970s, 1.78 million in the 1980s, 2.1 million in the 1990s, and 2.79 million from 2000 to 2010. The 2010 estimate of 3.5 million was 78% above the long-term average of 1.9 million. The American Green-winged Teal population within this area has probably prospered, while other dabbling duck populations have not, because so many of the former breed in the boreal forest and tundra wetlands, which have not experienced the well-known habitat loss occurring in the PPR to the south. The Eastern Survey averaged 241,000 birds from 1990 to 2009, and the 2010 tally was 256,000, which was 6% above the long-term average. There were no long-term or short-term trends in wintering numbers of American Green-winged Teal in Mexico; the Midwinter Surveys there averaged 203,565 birds from 1961 to 2000 (long term), and 207,633 from 1981 to 2000 (short term; Pérez-Arteaga and Gaston 2004). In the United States, the Midwinter Survey numbers show an upward movement in numbers, but no apparent trends, due to large year-to-year fluctuations.

Outside North America, the population estimate for Eurasian Green-winged Teal is 2.25–3.27 million: 500,000 in northern and northwestern Europe, 750,000–1,375,000 in northeastern Europe and Siberia, 400,000 in western and central Siberia, and 600,000–1,000,000 in eastern Siberia and China (Wetlands International 2006). The population of Aleutian Green-winged Teal is estimated at 10,000.

Harvest

American Green-winged Teal are a major game duck species in North America, ranking second or third in the U.S. harvest, after Mallards and (occasionally) Wood Ducks and Northern Pintails. The average harvest in the United States was 1,640,028 birds from 1999 to 2008: 41.1% in the Mississippi Flyway, 29.7% in the Pacific Flyway, 19.7% in the Central Flyway, and 9.6% in the Atlantic Flyway. The majority of the harvest in the Mississippi Flyway was taken in Louisiana (35.6%), Arkansas (18.9%), and Minnesota (8.7%). In the Pacific Flyway, 60.7% of the harvest was California, 12.1% in Oregon, 8.9% in Utah, and 8.1% in Washington.

From 1999 to 2008, the harvest of American Green-winged Teal in Canada averaged 95,672, with a high of 154,757 in 1999. Over this time period, 33.4% of the harvest was in Québec, 23.5% in Ontario, 8.3% in Manitoba, 8.0% in Nova Scotia, 7.1% in New Brunswick, and 19.7% in the remaining provinces.

The harvest in Mexico is not well known, but it is assumed to be small. Kramer et al. (1995) estimated an average annual harvest of 14,408 from 1987 to 1993, which was 25.1% of all the waterfowl harvested in that country. Harvest surveys conducted at 2 duck-hunting clubs at Bahía Santa María and Pabellón Bay in the state of Sinaloa during the 1987/88 hunting season reported 12,437 ducks (55% American Green-winged Teal) harvested at 1 club and 33,361 ducks (15% American Green-winged Teal) at the other (Migoya and Baldassarre 1993).

BREEDING BIOLOGY

Behavior

Mating System. Most American Green-winged Teal breed as yearlings, but many first-year males probably do not find mates. Pairing occurs on the wintering grounds, but this happens later than with most dabbling ducks, perhaps because the small body size of American Green-winged Teal does not allow them to accumulate sufficient lipid reserves before initiating courtship activities and yet have enough remaining for use during periods of cold weather (Baldassarre and Bolen 1986). Their small size would also make it difficult to sustain the costs of pair-bond maintenance throughout the winter, in addition to coping with colder temperatures. Hence American Green-winged Teal delay pairing until well into winter and early spring.

At the Rockefeller Wildlife Refuge in southwestern Louisiana (1985–86), courtship activity was not observed until December, with only 7% of the birds paired in January, 35% in February, and 59% in March (Rave and Baldassarre 1989). Paired and unpaired teal spent similar amounts of time in courtship behavior from October through February (0.0%–0.4%), but unpaired birds devoted more time to courtship (1.6% vs. 0.2%) and agonistic behaviors (0.3% vs. 0.2%) during March. In a study on the Atchafalaya River delta of Louisiana (1994–95), American Green-winged Teal arrived in early October, but courtship and pairing behaviors did not occur until late December (Johnson and Rohwer 1998). Paired females were first observed during the second week of January, with pairing then increasing steadily; 81% of the females ($n = 62$) observed in March were paired. Intraspecific aggressive interactions with clear outcomes ($n = 84$) were nearly always (99%) won by the initiator, and paired birds won most (83%) of their aggressive encounters with unpaired birds ($n = 12$).

On the southern High Plains of Texas (1981–82), from mid-September through October, courtship activity occurred 0.2% of the time for males and females, but from then until mid-March it was 0.4%–4.1% for males and 0.1%–1.7% for females. For both sexes it was highest from February through mid-March: 4.1% for males, and 1.7% for females (Quinlan and Baldassarre 1984). Courtship activity tended to occur only in the early morning from September to February, but took place throughout the day from February through mid-

March. In coastal North Carolina, 0%–2% were paired by November–December, 34% by January, and 80% by February (Hepp and Hair 1984). In the Central Valley of California, 80% were paired by January (Miller et al. 1988). Palmer (1976a) noted that males nearly always left their mates by the time incubation began.

Sex Ratio. Moisan et al. (1967) calculated an average adult sex ratio of 1.2 males/female (54.6% male) between 1961 and 1963. The sex ratio for immatures averaged 0.9 males/female (47.4% male). Bellrose et al. (1961) concluded that the sex ratio for immatures was close to 50:50, with any variations caused by differences in seasonal movements. Of 4,265 American Green-winged Teal banded in British Columbia, 52% were males (Munro 1949). Of 250 observed from late April through early May in Manitoba, the sex ratio averaged 53.7% males (Bellrose et al. 1961). Observations of wintering American Green-winged Teal in coastal Louisiana revealed a preponderance (74.3%) of males, but the ratio varied over the winter period (Tamisier 1976).

Site Fidelity and Territory. American Green-winged Teal have been studied extensively during the winter, but comparatively little research has been conducted during the breeding season, probably because they breed at low densities throughout their range. Hence there is very little information on philopatry to the breeding grounds. During the winter on the southern High Plains of Texas, the homing rate was only 1.4% in the second winter after banding (Fedynich et al. 1989). Similarly, there are limited data on the home-range sizes of breeding birds, although radiotracking of 3 marked pairs in southern Alberta revealed that there was extensive overlap, with no territorial defense behavior. The home range for males was 6–70 ha when the female was laying eggs or incubating, and 50–100 ha when the male was with the female or after a nest loss. The home range of females was 7–20 ha during laying and incubation, and 50–100 ha when with the male or after a nest loss (Johnson 1995).

Courtship Displays. In general, American Green-winged Teal are very active during courtship, with plenty of movement and vocalizations. Much of the display is given in the context of social courtship, where multiple males display toward 1 female (and sometimes more). McKinney (1965a:118) noted that a social-courtship group was easily detected "by the loud and often continuous whistling calls of the males, audible for at least a half mile in calm weather." He also noted that "the activity of the group is incessant, the males circling around the female, performing displays, chasing, and avoiding each other in a bewilderingly complex pattern of interactions." In Manitoba, social courtship groups were commonly observed during the spring and usually consisted of <10 males, but 1 group contained 25 males about a single female. McKinney (1965a, 1975) provided detailed descriptions of the courtship displays of American Green-winged Teal.

The *burp* display is a loud whistled *tliu*, given while raising the head, during which the feathers on the back and the wings are momentarily vibrated in a shuddering motion. The *burp* often occurs before other displays and is usually repeated 3–10 times. The *grunt-whistle* is preceded by 1 or 2 *head-shakes* and accompanied by a single loud and whistled *tliu* and a quiet grunting sound. During this readily visible display, the male arches the neck and the back out of the water, with the bill pointing downward, but then the bill is flicked rapidly toward the female to direct a string of water droplets in her direction. The display lasts 0.8–1.0 seconds and is given broadside to the female, usually from 0.6–1.8 m away. The *grunt-whistle* is usually immediately followed by a *head-flick* and a simultaneous *tail-wag*. Other common male displays are *head-up-tail-up, turn-toward-female, turn-the-back-of-the-head, head-shake, bill-up, bill-down, down-up, jump-flight, nod-swimming,* and *bridling. Bridling* is performed on land and is characterized by moving the head backward and protruding the chest while giving a single whistle. Females *incite*, a repeated sideways movement of

the bill in the direction of the chosen male, while giving a harsh rattling call. Other female displays are *nod-swimming* and a decrescendo call of 4–7 repeated *quacks*. Both sexes perform a mutual pre-copulatory *head-pumping*, after which the female assumes a *prone* position and the male mounts. Both sexes also engage in postcopulatory *bathing*.

Nesting

Nest Sites. American Green-wing Teal nest in low densities, and their nests are well concealed; hence comparatively few nests have been located for such an abundant and widespread species. During a 1993–2000 study across a broad expanse of aspen parklands in prairie Canada, 17,361 nests of 8 species of ducks were located, but only 383 (2.2%) were those of American Green-winged Teal (Emery et al. 2005). Nonetheless, that sample size is the largest reported from a single study and revealed that 19.8% of all American Green-winged Teal nests were in unmanaged grasslands, 19.1% in managed but idle parklands, 17.0% in planted cover, 15.7% in woodlands, and 13.3% in shrublands.

Near Brooks, Alberta, 86% of the 22 nests Keith (1961) found were constructed in beds of rushes (*Juncus* spp.), at an average distance of 19.8 m from water. American Green-winged Teal also had the best-concealed nests of the 12 duck species he studied, with an average of only 32% light penetration at the nest site. Of 54 nests located on the Woodworth Study Area in southeastern North Dakota (1966–81), 39% were in grass, 33% in brush, and 26% in forbs; 49% were classified as occurring in good to excellent cover, and 44% were completely concealed from above (Higgins et al. 1992). At the Lower Souris NWR in North Dakota, 30 nests were situated in mixtures of grasses and herbs, as well as in grasses alone and in sweetclovers (*Melilotus* spp.; Palmer 1976a). Farther north in British Columbia, Munro (1949) described the birds' preferred nesting habitats as grasslands, sedge meadows, and dry hillsides with aspen (*Populus tremuloides*) or brush thickets or open woods that were adjacent to ponds or sloughs. Nest sites on the Alaska Peninsula were on ponds and tidal flats (Gill et al. 1981).

Nest construction is an ongoing process, with additional materials added by the female throughout egg laying. Down is not added until the female begins incubation (Johnson 1995).

Clutch Size and Eggs. There are limited data on clutch size for American Green-winged Teal, because they nest at such low densities that their nests often make up <1% of the total waterfowl nests located during field studies. Nevertheless, data from 341 completed clutches were obtained during 1993–2000 in parkland Canada (Emery et al. 2005) and 2002–10 in prairie and parkland Canada (Ducks Unlimited Canada Institute for Wetland and Waterfowl Research). Clutch size averaged 9.1 eggs (range – 4–14 eggs), with 91.5% ranging between 7 and 11 eggs. In southeastern Alberta (1953–57), Keith (1961) reported an average clutch size of 8.7 for 18 nests. At the Lower Souris NWR in North Dakota, clutch size in 28 nests averaged 8.4 eggs and ranged between 5 and 11; 20 nests (71%) contained 7–10 eggs (Palmer 1976a). At the Woodworth Study Area in southeastern North Dakota (1966–81), the average size of 21 successful clutches was 9.1 eggs (Higgins et al. 1992).

American Green-winged Teal eggs are dull white, cream, or pale olive buff and are ovate in shape (Bent 1923). Measurements for 1 egg each from 20 clutches averaged 45.87 × 33.76 mm (Palmer 1976a). Alisauskas and Ankney (1992a) observed a laying rate of 1/day.

Incubation and Energetic Costs. Palmer (1976a) noted that incubation lasted about 21 days, with reports of 20–23 days in the wild, and 23–24 days in captivity. Bent (1923) and Delacour (1956) gave the incubation period as 21–23 days. The female covers the nest with down or other nesting material before departing on nest recesses (Munro 1949). The incubation constancy of a single female monitored in Manitoba was 79.4%, with an average of 3.6 recesses/day of 82.5 minutes each.

Feeding was the dominant (65.1%) activity of a different female observed by Munro (1949) during 3 nest recesses, which indicates that such a small-bodied duck species would rely more on food acquired in the breeding area to meet the caloric needs of incubation, rather than use stored reserves. There are no studies, however, of the reproductive energetics in American Green-winged Teal.

Nesting Chronology. American Green-winged Teal initiate their nests about midway through the nesting cycle of waterfowl. During a 1993–2000 waterfowl-nesting study in parkland Canada (Emery et al. 2005) and a 2002–10 study in prairie and parkland Canada (Ducks Unlimited Canada Institute for Wetland and Waterfowl Research), American Green-winged Teal clutches (n = 507) were initiated between 28 April and 14 July. The majority (87.9%) were begun between 5 May and 22 June, with weekly initiations ranging between 10.3% and 14.8% over that time period; of these, 42.0% were initiated between 19 May and 1 June. At the Woodworth Study Area in southeastern North Dakota (1966–81), 99% of 54 nests were initiated between 6–12 May and 1–7 July (Higgins et al. 1992). At Yellowknife, Northwest Territories, the period of nest initiation varied from 28 to 34 days, between 29 May and 1 July (Murdy 1964). On Yukon Flats in Alaska (1989–91), the earliest nest initiation dates were between 10 and 24 May, and the latest nest initiation dates were between 15 and 28 June (n = 12 nests; Grand 1995).

In British Columbia, the earliest date for the appearance of downy young was 20 June, and the latest was 10 August (Munro 1949). Keith (1951) noted that cold weather may delay nest initiation. On subarctic taiga near Yellowknife, Northwest Territories (1962–65), the mean hatching date was 21–30 June (n = 132; Toft et al. 1984).

Nest Success. Very few American Green-winged Teal nests have been located, not only because they are well concealed, but also in large part because most studies of nesting puddle ducks occur in the PPR, which is outside of this species' main breeding range. During a 3-year nesting study in the PPR in South Dakota, only 3 American Green-winged Teal nests occurred among the 620 waterfowl nests that were found (Duebbert and Lokemoen 1976).

Nonetheless, in studies with reasonable sample sizes, the lowest estimate of nest success for American Green-winged Teal was 7.9% for 24 nests located near Lousana, Alberta, between 1952 and 1965 (Smith 1971), and the highest was 77% for 30 nests on the Lower Souris NWR in North Dakota (Palmer 1976a). On Yukon Flats in Alaska, the estimated nest success of 9 duck species collectively, including American Green-winged Teal, was only 12.5%. Higgins et al. (1992) studied 56 nests on the Woodworth Study Area in North Dakota (1966–81) and estimated a nest-success rate of 39.3% by the apparent success method and 15.6% by the Mayfield method. The probable cause of losses at 20 of these nests was predation by red foxes (*Vulpes vulpes*) and striped skunks (*Mephitis mephitis*), at 8 nests (40%) each, with other predators being badgers (*Taxidea taxus*), ground squirrels (*Citellus* spp.), gulls (*Larus* spp.), and raccoons (*Procyon lotor*). Of 1,293 duck carcasses found at red fox dens in the Dakotas, 2.6% were American Green-winged Teal (Sargeant et al. 1984).

At Lake Mývatn in Iceland (1961–70), Bengtson (1972b) reported a hatching success of 79% for 71 nests of Eurasian American Green-winged Teal, but 10.9% of the eggs from successful nests did not hatch. Of those unhatched eggs, 75% were infertile, 20% disappeared, and 5% were dead embryos.

Brood Parasitism. Brood parasitism has not been documented for American Green-winged Teal in North America, but is undoubtedly low, due to the low density and high degree of concealment of their nests. At Lake Mývatn in Iceland, 0.9% of 227 Eurasian Green-winged Teal nests were parasitized by other duck species, but there was no evidence of intraspecific brood parasitism (Bengtson 1972b).

Renesting. Little is known about the renesting potential of American Green-winged Teal. None-

theless, during waterfowl-nesting studies across parts of prairie Canada in 1993–2000 (Emery et al. 2005) and prairie and parkland Canada in 2002–10 (Ducks Unlimited Canada Institute for Wetland and Waterfowl Research), 32.9% of 507 American Green-winged Teal clutches were initiated late in the spring (9–29 Jun), which implies that there were many second nesting attempts.

Keith (1961) determined that 15 pairs on his study area in southeastern Alberta produced 21 nests, which again indicated that some renesting undoubtedly occurs. Among the species of ducks he studied, the 1.4 nests produced/pair of American Green-winged Teal was about midway between the figures for Mallards (2.15/pair) and Redheads (0.45/pair). Females are known to renest as late as early July if their first nest or brood is destroyed (Toft et al. 1984). At Lake Mývatn in Iceland (1961–70), clutch size of Eurasian Green-winged Teal averaged 10.3 eggs for 44 early-initiated normal nests and 9.6 for 40 late-initiated ones, but 8.5 for 16 known renests (Bengtson 1972b). An estimated 75% of the failed nesters attempted to renest.

REARING OF YOUNG

Brood Habitat and Care. At the Woodworth Study Area in southeastern North Dakota (1965–75), American Green-winged Teal broods occurred on seasonal and semipermanent wetlands (Higgins et al. 1992). The highest densities were recorded on wetlands 2.01–5.00 ha in size, followed by those measuring 0.76–1.00 ha, and then 0.10–0.25 ha. Otherwise there are very few data on habitat use during the brood-rearing period.

Females provide all the brood care, as males desert their mates at some point during incubation. Munro (1949) noted the vigorous defense of broods by attending females, with defensive displays involving short distraction flights and the female swimming away from intruders while beating her wings on the water; both actions were accompanied by a "continuous, excited quacking." Koskimies and Lahti (1964) stated that newly hatched to 1-day-old American Green-winged Teal ducklings were the most cold sensitive of the 10 duck species the authors studied; these ducklings were unable to maintain their core body temperature when the air temperature fell below 10°C.

Brood Amalgamation (Crèches). Munro (1949) observed potential brood amalgamation in the Cariboo parklands of British Columbia, where 3 females accompanied a brood of 19 young, and another instance where 3 females accompanied 14 young. Brood amalgamation has otherwise not been reported but is probably rare, due to the low nesting density of American Green-winged Teal.

Development. There are no studies on growth and development in American Green-winged Teal. Nelson (1993) reported an average weight of 15.1 g for 5 ducklings within 1 day of hatching, and Koskimies and Lahti (1964) reported an average of 16.8 g for 1-day-old Eurasian Green-winged Teal ($n = 6$). Bellrose (1980) cites several studies reporting fledging in 34–35 days.

RECRUITMENT AND SURVIVAL

As with other ducks, brood size for American Green-winged Teal decreases with the age of the ducklings; most mortality occurs during the first week. At the Woodworth Study Area in southeastern North Dakota (1966–81), brood size averaged 7.0 for Class I ducklings in July, but 4.8 by August; Class II averaged 6.3; and Class III averaged 4.0 (Higgins et al. 1992). The overall average during the 17-year study was 7.0 in July and 5.1 in August. In British Columbia, Munro (1949) observed an average brood size of 6.2 for 48 broods seen in July and 17 broods observed in August. He attributed this relatively high survival rate to the vigorous brood defenses by the females. For Eurasian Green-winged Teal, a study from a coastal peatland complex in westcentral Wales reported that 151 out of 240 (63%) ducklings from 53 broods <5 days old became fully fledged or attained adult size. The mean brood size was 6.2 for broods <5 days old ($n = 53$), and 3.8 at fledging ($n = 45$).

The mean mortality rate for 40 broods was 41% between hatching and fledging (Fox 1986).

Toft et al. (1984) analyzed data for 154 American Green-winged Teal broods observed on subarctic taiga near Yellowknife, Northwest Territories (1962–65), to examine patterns and possible reasons for the seasonal decline in brood size. The mean size for Class I broods (1–15 days old) was 5.5 ($n = 154$), but broods observed before 6 July ($n = 56$) were larger (6.3) than broods observed from 6 to 19 July (5.3; $n = 40$) or after 19 July (4.3; $n = 26$). The seasonal decline averaged 0.07 ducklings/brood/day, with most of it due to the smaller size of replacement clutches, compared with first clutches.

A major analysis of survival based on 3,633 recoveries from 63,380 American Green-winged Teal banded between 1950 and 1987 revealed an average survival rate of 55% for males and 51% for females, which were within the ranges reported for other puddle ducks (Chu et al. 1995). Moisan et al. (1967) calculated an average annual mortality rate of 63% for the continental American Green-winged Teal population: 70% for immatures and 50% for adults. Data for 82,314 banded American Green-winged Teal and associated recoveries throughout North America (1978–96) yielded an annual survival rate of 50.5% across the 4 sex and age classes (Frank Bellrose): an average of 55% for adult males, 54% for adult females, 51% for immature males, and 42% for immature females. These similar survival rates between adult males and adult females are unlike those for other puddle ducks, where mortality for females is greater, due to a higher degree of predation when they are nesting, especially in the PPR (Sargeant et al. 1984, Cowardin et al. 1985). Perhaps the more equal mortality rates between the sexes in adult American Green-winged Teal result from a large part of the population nesting outside the PPR, with nests that are well dispersed and well concealed.

In Europe, an analysis of wing recoveries from hunter-killed Eurasian Green-winged Teal revealed a high (89%) percentage of juveniles in the northern Finland harvest, compared with only 58%

farther south in western France, which indicates a high degree of juvenile mortality as the birds are migrating southward (Guillemain et al. 2010). Based on the declining numbers of juveniles harvested from north to south and given an annual survival rate of 48.5% for adult teal in France, the authors calculated that only 14.7% of the juveniles survived the 3-month autumn migration period (a 52.8% monthly survival rate). In contrast, monthly survival rates during August–November for adults were estimated at 94.2%. The authors also cited studies of Eurasian Green-winged Teal that reported annual adult survival rates of 50%–59% in the United Kingdom and 36%–45% in other European countries.

Longevity records of wild birds are 20 years, 3 months for a female American Green-winged Teal banded as an adult in Oklahoma and shot in Missouri; and 16 years, 10 months for a Eurasian Green-winged Teal (Cramp and Simmons 1977).

FOOD HABITS AND FEEDING ECOLOGY

A generalized characterization of the food habits of American Green-winged Teal comes from the early work of Mabbott (1920), who examined the contents of 653 stomachs from specimens collected throughout the year over a broad area in North America and reported that plant material, particularly small seeds, was the dominant (90.7%) component of their diets. The most important plant foods were seeds from several species of sedges (Cyperaceae; 38.8%), especially bulrushes (*Scirpus* spp.). Sedges occurred in 530 out of 653 (81%) stomachs and formed the sole component in 51 of them (8%). Other important plant materials were pondweeds (*Potamogeton*; 11.5%); the seeds of grasses (11.0%), such as panicums (*Panicum*) and millets (*Echinochloa*); and the seeds of smartweeds (*Polygonum*; 5.3%). Insects, mainly dipterans (fly larvae), were the most common (4.6%) animal food.

The food habits of American Green-winged Teal during 2 winters (2005–6) on the Great Salt Lake in Utah were reported by Vest and Conover (2011) and provide a contrast with studies from other areas, due to the unique, highly saline nature of the lake (about 13% in areas used by waterfowl); the salinity of the open ocean is about 3.5%. Animal foods dominated (94.1% aggregate biomass) the diets of 157 birds: mostly brine shrimp (*Artemia franciscana*) cysts (79.5%) and the larvae of brine flies (Ephydridae; 10.8%). Brine fly larvae and pupae can occur at densities of >5,000/m^2 in the substrate, while brine shrimp occur in the water, where the densities of adults can reach 2,500/m^3, and those of cysts can be >20,000/m^3.

Due to their small size and bill structure, American Green-winged Teal often forage on mud flats and adjacent shallow water, more so than any of the other puddle duck species. Their small bills, with densely packed lamellae, allow them to select small seeds and aquatic invertebrates that they glean from these shallow-water environments. Almost all food-habit studies note their consumption of small food items readily procured from shallow water. During the summer in North Dakota (1984–85), 43.9%–44.4% of 331 American Green-winged Teal foraged in open water, away from emergent vegetation and with water depths of 0–10 cm (DuBowy 1988). Wintering birds in California (1984–85) foraged in dense emergent vegetation (46.5%–52.6%), but also at water depths of 0–10 cm. During the winter on tidal mudflats in the Atchafalaya River delta in Louisiana, American Green-winged Teal fed by dabbling (70% of the time) and head-dipping (28%); tipping up was rare (Johnson and Rohwer 2000). The percentage of diurnal time spent foraging on the mudflats averaged 68%.

At Lake Simpele in southeastern Finland, about 75% of 7,572 Eurasian Green-winged Teal observed from May to August fed at depths of only 0–5 cm, where they used foraging methods suitable in shallow water (Pöysä 1983): bill-submerged (54.1%), neck-submerged (16.8%), straining from the surface (20.3%), and upending (8.8%). In a guild (a group using the same ecological resource in the same way) with 5 other species of dabbling ducks,

Pöysä (1983) reported that bill morphology accounted for 78.5% of the total variation in niche distribution among the species he studied, but species with similar bill morphologies had varying neck lengths and foraged differently. Northern Pintails also fed in shallow water (80.5%), but they upended much more than teal (61.4% vs. 8.8%). During the summer in North Dakota, dipping the head below the surface was the most common (79%–81%) foraging method for American Green-winged Teal, while dabbling occurred most (77%–89%) during the winter in California (DuBowy 1988).

Relative to bill structure, Nudds and Bowlby (1984) noted that the lamellae of American Green-winged Teal are more finely spaced (13.28/cm) than those in most other dabbling ducks except Northern Shovelers (21.48/cm). Such dense lamellae no doubt allowed wintering American Green-winged Teal foraging on a mudflat in coastal southwestern Louisiana to feed heavily on meiofauna, which are the small invertebrates at the mud/water interface often not visible without magnification (Gaston 1992). In their comparison of lamellae density and food use among 7 dabbling duck species, Nudds and Bowlby (1984) concluded that American Green-winged Teal were specialists with respect to both prey-size selection and foraging methodology; generalist foragers, like Mallards, had a lamellae density of only 7.96/cm. The authors' analysis of prey size / caloric distribution in the diets of the dabbling ducks they studied found that American Green-winged Teal differed from all other species except American Wigeon. DuBowy (1988) also noted that during the summer in North Dakota, the dietary overlap of American Green-winged Teal was greatest (56.6%) with American Wigeon, compared with an overlap range from 9.7% (Northern Shovelers) to 39.9% (Northern Pintails) for the 5 other species of dabbling ducks. Dietary overlap during the winter was highest (45.9%) with Cinnamon Teal, but low (1.7%–17.1%) with other species. Overlap in foraging behavior during the summer was greatest (63.5%–92.2%) with Northern Pintails and Mallards, while this overlap during the winter

in California was greatest (81.9%–94.4%) with Northern Shovelers.

Breeding. There are almost no data on the food habits of American Green-winged Teal during the breeding season. During the spring in Maine, the stomachs of 12 American Green-winged Teal (and 1 Blue-winged Teal) contained 93.9% plant matter and 6.1% animal matter (Coulter 1955). Sedges (Cyperaceae) were the predominant (63.4%) plant food; others were smartweed seeds (8.2%) and the seeds of bur-reed (*Sparganium* spp.; 6.4%). Earthworms were the most important (5.0%) animal food. DuBowy (1988) collected 2 American Green-winged Teal during the summer in North Dakota (1985) and found that plant material made up 82.8% of their diet: mainly pondweed seeds (50.3%) and smartweed seeds (32.1%). Munro (1949) reported that insect matter formed 60%–100% of the food in the stomachs of 2 downy young ducklings collected in British Columbia.

Migration and Winter. In contrast to the breeding season, the food habits and feeding ecology of American Green-winged Teal are well known during migration and winter and reflect their use of small seeds and invertebrates. During the fall in southeastern Alaska (1976–77), their diet ($n = 55$) was 83.0% plant material and 16.9% animal material (Hughes and Young 1982). The seeds of sedges (*Carex* spp.) were the most important (55.0%) plant food; and marestail (*Hippuris* spp.) nutlets made up 7.9%. Ostracods (Ostracoda) were the most significant (11.3%) animal matter. During the fall at Swan Lake in British Columbia, the stomachs of 46 adult American Green-winged Teal contained 80.4% plant material and 19.6% animal material (Munro 1949). Bulrush seeds were by far the most important (50.7%) plant food, while insects were the most significant (11.8%) animal food. During the fall in Illinois (1978–79), 218 gizzards contained 96.9% plant material, generally small seeds (Havera 1999). Redroot flatsedge (*Cyperus erythrorhizos*) was the most important (46.8%) plant food, as well as several other sedge species (20.7%).

During the winter (Oct–Feb) in the Central Valley of California (1979–81), the esophagi of 173 American Green-winged Teal contained 62.3% plant seeds and 37.6% animal matter (Euliss and Harris 1987). The small seeds of plants such as swamp timothy (24.7%) and barnyardgrass (11.3%) were the most important plant foods, while midges (Chironomidae) were the primary (27.1%) animal food. These birds were highly opportunistic, generally shifting their food preferences in response to food availability. They also commonly foraged at night, as 93% of 101 birds collected after dark contained fresh food in their esophagi. During the winter on natural wetlands in coastal Louisiana, 55%–83% fed at night (Rave and Baldassarre 1989), as did birds in rice fields farther north in Louisiana (Tamisier 1976). On managed impoundments in coastal South Carolina during the winter (1972/73, 1973/74), American Green-winged Teal (*n* = 130) primarily (96.9%) consumed plant material, which was dominated by small seeds (Landers et al. 1976). Their preferred plants were 2 species of panicgrasses (*Panicum*; 34.0%), 3 species of sedges (*Cyperus*; 33.6%), and dotted smartweed (*Polygonum punctatum*; 18.7%).

Wintering American Green-winged Teal in coastal southwestern Louisiana used an array of wetland types, where they foraged entirely on noncultivated foods (see Habitat). In contrast, however, nonbreeding American Green-winged Teal on the playa lakes in Texas (1981–82) had access to agricultural grain (waste corn remaining after the harvest) and exploited that food resource heavily by flying twice daily to nearby cornfields (Baldassarre and Bolen 1984). Time-budget data revealed that the highest (23.2%) feeding levels in playa lakes occurred when the birds arrived in mid-September and October, and then declined to about 10% for the remainder of the wintering period, through mid-March (Quinlan and Baldassarre 1984). The greater foraging times in early fall may have occurred because heavy rainfall flooded emergent vegetation, with the birds preferring the readily available and nutritionally complete diet offered by noncultivated foods over corn. Corn is low (11%) in protein content; it also lacks calcium, as well as several important amino acids necessary in waterfowl diets, especially during the molt (Baldassarre et al. 1983).

The easily obtained calories from waste corn probably reduced the overall feeding time for the Texas birds, in comparison with American Green-winged Teal foraging on noncultivated foods in coastal Louisiana (about 10% vs. 33%), which allowed more time for energy-conserving activities, such as resting. Activity patterns did not vary between males and females in either Texas or Louisiana, although teal in Louisiana spent more time feeding in the morning (42.1%) than in late afternoon (29.0%). In addition to feeding, the fall/winter time budget of American Green-winged Teal in Louisiana was dominated by resting (45.4%). Resting was also the primary (54.4%) fall/winter activity in the much colder Playa Lakes Region of Texas, reaching 64%–67% during the coldest parts of winter, presumably because the birds were conserving energy. In the Rice Belt region of Louisiana, wintering American Green-winged Teal fed in rice fields during the night and returned to loafing areas during the day, where they spent 8–9 hours sleeping, often on beds of floating vegetation while facing into the sun (Tamisier 1976).

MOLTS AND PLUMAGES

Palmer (1976a) provides a detailed description of the molts and plumages in American Green-winged Teal, and Jackson (1991) has information on the plumages of female teal species.

Juvenal plumage is completed shortly after flight is attained, at about 35 days. Birds in juvenal plumage are buffy tan and resemble adult females, although females in juvenal plumage have more pronounced spotting and streaking on the belly. First basic is a partial body molt acquired in August and September, but the juvenal remiges and retrices are retained. This plumage is quite female-like and almost indistinguishable from juvenal plumage. The birds immediately molt first basic and acquire first alternate, with most feathering replaced from late September through October, although the last feathers may not appear until November (or even Dec and Jan for late-hatched birds). This plumage retains the juvenal wing but otherwise resembles definitive alternate. Quinlan and Baldassarre (1984) noted that American Green-winged Teal were completing this molt on arrival at their migration/wintering grounds on the southern High Plains of Texas. These ducks subsequently spent the most time preening their feathers (21.1% in mid-Sep through Oct vs. 5.2% through to mid-March). Similarly, American Green-winged Teal wintering in coastal southwestern Louisiana spent the most time (16.1%) preening in October and November (Rave and Baldassarre 1989).

The acquisition of definitive basic in males usually begins in June and is completed in July, with the remiges replaced last. In northern Sweden, the flightless period associated with the wing molt of male Eurasian Green-winged Teal ($n = 126$) was about 21 days, and males can lose 10%–19% of their body mass then (Sjöberg 1988). During the wing molt, the growth rate of the ninth primary averaged 4.8 mm/day ($n = 14$), and growth of the first secondary averaged 3.5 mm/day ($n = 11$). The birds could fly when the ninth primary reached 76% of its final length. Males in definitive basic resemble females in definitive alternate. Definitive alternate in males is then acquired during the fall and largely completed by late October. This plumage will be worn by them until the following summer. Females acquire definitive basic during the spring, primarily in March, and retain that plumage into the summer breeding season. The remiges are not replaced until late summer or even early fall, when definitive alternate is acquired.

CONSERVATION AND MANAGEMENT

American Green-winged Teal populations have remained remarkably stable over more than 5 decades (1955–2010), which included 4 droughts and annual harvests ranging from 1.6 to 2.4 million birds. These comparatively stable numbers probably reflect a very broad breeding distribution that is largely outside of the heavily impacted grasslands habitats in the PPR. Their well-concealed nests may also contribute to their high level of productivity. The population trend from 1970 to 2003 was increasing, and the 2004 updated North American Waterfowl Management Plan established a breeding population goal of 1.9 million in the midcontinent (the Traditional Survey region). Hence it is significant that the 2010 estimate from the Traditional Survey area was 3.5 million, which was 78% above the long-term average of 1.9 million.

American Green-winged Teal are usually the second-most-heavily harvested ducks in the United States, after Mallards. The wintering ecology of American Green-winged Teal is well known. In their monograph on American Green-winged Teal distribution and population dynamics, however, Gaston et al. (1967:44) remarked, "A thorough study of the breeding biology of Green-winged Teal in North America has not yet been conducted." Over 40 years later, there is still no detailed study of their breeding ecology conducted with a large number of marked individuals; such research would be desirable.